food and culture

a reader

second edition

edited by

carole counihan and penny van esterik

Routledge
Taylor & Francis Group

NEW YORK AND LONDON

Please visit the companion website at
www.routledge.com/textbooks/9780415977777

First published 1997
by Routledge

Second edition published 2008
by Routledge
270 Madison Ave, New York, NY 10016

Simultaneously published in the UK
by Routledge
2 Park Square, Milton Park, Abingdon, Oxon OX14 4RN

Routledge is an imprint of the Taylor & Francis Group, an informa business

Typeset in Minion by
RefineCatch Limited, Bungay, Suffolk
Printed and bound in the United States of America on acid-free paper by
Sheridan Books, Inc.

Library of Congress Cataloging in Publication Data
CIP data has been applied for

ISBN10: 0–415–97776–2 (hbk)
ISBN10: 0–415–97777–0 (pbk)

ISBN13: 978–0–415–97776–0 (hbk)
ISBN13: 978–0–415–97777–7 (pbk)

food and culture

a reader

second edition

Food touches everything important to people: it marks social difference and strengthens social bonds. Common to all peoples, yet it can signify very different things from table to table.

Food and Culture takes a global look at the social, symbolic, and political-economic role of food. The stellar contributors to this reader examine some of the meanings of food and eating across cultures, with particular attention to how men and women define themselves differently through their foodways. Crossing many subjects, this innovative, first-of-its-kind in the field includes perspectives of anthropology, history, psychology, philosophy, politics, and sociology.

Carole Counihan is Professor of Anthropology at Millersville University in Pennsylvania, and co-editor-in-chief of *Food and Foodways*. Her earlier books include *Around the Tuscan Table: Food, Family, and Gender in Twentieth-century Florence, Food in the USA*, and *The Anthropology of Food and Body: Gender, Meaning and Power*.

Penny Van Esterik is Professor of Anthropology at York University in Toronto, Canada, where she teaches nutritional anthropology, in addition to doing research on food and globalization in Southeast Asia. She is a founding member of WABA (World Alliance for Breastfeeding Action), and writes on infant and young child feeding, including *Beyond the Breast-Bottle Controversy*.

ROUTLEDGE TITLES OF RELATED INTEREST

American Commodities in an Age of Empire by Mona Domosh

Branding New York by Miriam Greenberg

Chewing Gum: The Fortunes of Taste by Michael Redclift

Food in the USA: A Reader, edited by Carole Counihan

Georg Simmel: REMBRANDT: An Essay in the Philosophy of Art translated and edited by Alan Scott and Helmut Staubmann

Point of Purchase: How Shopping Changed American Culture by Sharon Zukin

The American Suburb: The Basics by Jon Teaford

The Anthropology of Food and Body by Carole Counihan

The Suburb Reader edited by Becky Nicolaides and Andrew Wiese

Where Stuff Comes From: How Toasters, Toilets, Cars, Computers and Many Other Things Come To Be As They Are by Harvey Molotch

Contents

Foreword from *The Gastronomical Me*, M. F. K. Fisher,   xi
Preface to the Second Edition   xii
Acknowledgments   xiv

Introduction to the Second Edition
   Carole Counihan and Penny Van Esterik   1

Foundations

1. The Problem of Changing Food Habits   17
    Margaret Mead

Mead's early government work explores recommendations for changing
American food habits and establishes the importance of food studies.

2. Toward a Psychosociology of Contemporary Food Consumption   28
    Roland Barthes

French structuralists explain how food acts as a system of communication
and provides a body of images that mark eating situations.

3. The Culinary Triangle   36
    Claude Lévi-Strauss

This classic structuralist statement, often critiqued, shows how food
preparation can be analyzed as a triangular semantic field, much like
language.

4. Deciphering a Meal   44
    Mary Douglas

Hebrew dietary laws are not irrational but rather precoded messages about
purity, defilement, and holiness as wholeness.

5. The Abominable Pig   54
    Marvin Harris

Materialists like Harris reject symbolic and structuralist explanations and
explain food prohibitions based on economic and ecological utility.

6. The Nourishing Arts   67
    Michel de Certeau and Luce Giard

The "practice of everyday life" includes how French women constitute
tradition as they carry out daily meal preparation.

7. The Recipe, the Prescription, and the Experiment 78
   Jack Goody

   Shopping lists, menus and recipes are among the earliest and most
   enduring evidence of written instructions for food use, reflecting significant
   advances in human knowledge.

8. Time, Sugar, and Sweetness 91
   Sidney W. Mintz

   Colonialism made high-status sugar produced in the Caribbean into a
   working class staple.

9. Anorexia Nervosa and its Differential Diagnosis 104
   Hilde Bruch

   Renowned eating disorder psychiatrist Bruch defines true anorexia nervosa
   as involving distorted body image, inaccurate perception of hunger,
   hyperactivity, and an overwhelming sense of ineffectiveness.

Gender and Consumption

10. Fast, Feast, and Flesh: The Religious Significance of Food to
    Medieval Women 121
    Caroline Walker Bynum

    Medieval women used food for personal religious expression, including
    giving food away, exuding foods from their bodies, and undertaking fasts to
    gain religious and cultural power.

11. The Appetite as Voice 141
    Joan Jacobs Brumberg

    The origins of anorexia nervosa can be found in the nineteenth century
    fasting of Victorian girls, who used control of appetite as an important form
    of self-expression.

12. Anorexia Nervosa: Psychopathology as the Crystallization
    of Culture 162
    Susan Bordo

    Anorexia nervosa can be viewed as a culturally over-determined
    psychological disorder resulting from longstanding cultural ideologies
    related to mind-body dualism, control, and gender power.

13. Feeding Hard Bodies: Food and Masculinities in Men's Fitness
    Magazines 187
    Fabio Parasecoli

    Men's fitness magazines define masculinity through discussions of food
    and body, increasingly involving men in concerns about constructing
    corporeal perfection and regulating consumption to build muscle and
    strength.

14. The Overcooked and Underdone: Masculinities in Japanese
    Food Programming   202
      T. J. M. Holden

    Cooking shows featuring male chefs predominate on Japanese television
    and propagate one-dimensional definitions of masculinity based on power,
    authority, and ownership of consumer commodities.

15. Japanese Mothers and *Obentōs*: The Lunch-Box as Ideological
    State Apparatus   221
      Anne Allison

    Japanese mothers, in preparing elaborate lunch-boxes for their preschool
    children, reproduce state ideologies of power.

16. Conflict and Deference   240
      Marjorie DeVault

    In feeding others, women sometimes reproduce their own subordination by
    deferring to men's preferences and thus reinforce the "naturalness" of
    women's service and undermine progress toward reciprocal nurturance.

17. Feeding Lesbigay Families   259
      Christopher Carrington

    Because feeding work is complex, laborious, and highly gendered, it is
    problematic in lesbigay families because a full accounting of it would
    destroy illusions of equality and call into question masculinity of gay men
    who do it and femininity of lesbians who do not.

Food and Identity Politics

18. How to Make a National Cuisine: Cookbooks in Contemporary
    India   289
      Arjun Appadurai

    Cookbooks written for an Anglophone audience tell unusual tales about
    the development of a national cuisine, the boundaries of edibility, and the
    logic of meals in post-colonial India.

19. "Real Belizean Food": Building Local Identity in the Transnational
    Caribbean   308
      Richard Wilk

    Transformations in Belizean food from colonial times to the present
    demonstrate transnational political, economic and culinary influences that
    have affected the ways Belizean people define themselves and their nation.

20. Let's Cook Thai: Recipes for Colonialism   327
      Lisa Heldke

    Cultural food colonialism is reproduced by food adventurers who seek out
    ethnic foods to satisfy their taste for the exotic other.

21. More than Just the "Big Piece of Chicken": The Power of Race,
     Class, and Food in American Consciousness   342
         Psyche Williams-Forson

Ethnographic, historical, and literary research reveals not only controlling
and damaging stereotypes about African Americans and chicken but also
the ways Black women have used chicken as a form of resistance and
community survival.

22. *Mexicanas'* Food Voice and Differential Consciousness in the San
     Luis Valley of Colorado   354
         Carole Counihan

Food-centered life histories portray the voices and perspectives of
traditionally muted Hispanic women of rural southern Colorado whose
food stories reveal differential behaviors and consciousness which promote
empowerment.

23. Rooting Out the Causes of Disease: Why Diabetes is So Common
     Among Desert Dwellers   369
         Gary Paul Nabhan

Skyrocketing adult-onset diabetes among desert dwelling Seri Indians of
Northern Mexico suggests that changes in diet have caused this major
health problem and that traditional desert foods—especially legumes, cacti
and acorns—are protective.

24. Slow Food and the Politics of Pork Fat: Italian Food and European
     Identity   381
         Alison Leitch

The Slow Food Movement has emerged as an important political force to
preserve traditional, artisan foods such as lard from Colonnata, which
serves as a case study revealing the politics of Slow Food in the context of
European market economies.

25. Taco Bell, Maseca, and Slow Food: A Postmodern Apocalypse for
     Mexico's Peasant Cuisine?   400
         Jeffrey M. Pilcher

Italy's Slowfood Movement offers strategies for the maintenance of
traditional, local, and sustainable Mexican food, which is threatened by
the economies of scale and market dominance of multinational giants like
Taco Bell.

26. The Raw and the Rotten: Punk Cuisine   411
         Dylan Clark

Punk cuisine—based on scavenged, rotten, and/or stolen food—challenges
the hierarchy, commodification, toxicity, and environmental destruction of
the capitalist food system.

27. Salad Days: A Visual Study of Children's Food Culture   423
  Melissa Salazar, Gail Feenstra, and Jeri Ohmart

This chapter is available with color illustrations online at www.routledge.com/textbooks/9780415977777. Photographic documentation of children's self-serve salads at Northern California elementary schools was designed to assess the nutritional content of children's meals, but also conveyed rich information about children's tastes, food aesthetics, and definitions of appropriate meals.

Political Economy of Food: Transformation and Marginalization

28. The Chain Never Stops   441
  Eric Schlosser

The mistreatment of meatpacking workers in the United States is linked to the high rates of trauma in this dangerous industry and reveals general problems of corporate food production.

29. Whose "Choice"? "Flexible" Women Workers in the Tomato Food Chain   452
  Deborah Barndt

"Flexible," part-time, low-wage female labor is an increasingly important component of the global food economy that insures profits for agribusinesses, fast food corporations, and supermarkets, but threatens the livelihood and food security of women and families.

30. The Politics of Breastfeeding: An Advocacy Update   467
  Penny Van Esterik

The commodification of baby food has had severe consequences, but advocacy groups actively resist the promotional tactics of transnational food and pharmaceutical companies.

31. The Political Economy of Obesity: The Fat Pay All   482
  Alice Julier

The culture-wide denigration of the "obesity epidemic" is not only due to its health consequences, but also to the political and economic benefits to the food corporations, the diet industry, and the health professions.

32. Of Hamburger and Social Space: Consuming McDonald's in Beijing   500
  Yunxiang Yan

In Beijing Chinese consumers associate fast food with being American and being modern. They enjoy the standardization of meals, the hospitable service, the democratic environment, and the cleanliness, which create a desirable space to socialize and linger.

33. "Plastic-bag Housewives" and Postmodern Restaurants?: Public and
     Private in Bangkok's Foodscape   523
         Gisèle Yasmeen

     Bangkok women can pick up small plastic bags of excellent quality
     traditional dishes that go with rice at local vendors near their home or
     workplace.

34. The Political Economy of Food Aid in an Era of Agricultural
     Biotechnology   539
         Jennifer Clapp

     The advent of genetically modified organisms (GMOs) has seriously
     affected food aid, even in the context of famine and extreme hunger.

35. Street Credit: The Cultural Politics of African Street Children's
     Hunger   554
         Karen Coen Flynn

     In Mwanza, Tanzania, homeless children acquire food by working at
     market stands and restaurants, scavenging garbage, stealing, trading sex
     for food, and begging. The importance of charity suggests rethinking Sen's
     entitlement theory explanation of hunger.

36. Want Amid Plenty: From Hunger to Inequality   572
         Janet Poppendieck

     Because of great need, many U.S. volunteers feed the hungry, but charity
     not only fails to solve the underlying causes of hunger—poverty and
     inequality—but contributes to it by offering token rather than structural
     solutions and taking the government off the hook.

Contributors   582
Credit Lines   589
Index   593

Foreword

People ask me: Why do you write about food, and eating and drinking? Why don't you write about the struggle for power and security, and about love, the way others do?

They ask it accusingly, as if I were somehow gross, unfaithful to the honor of my craft.

The easiest answer is to say that, like most other humans, I am hungry. But there is more than that. It seems to me that our three basic needs, for food and security and love, are so mixed and mingled and entwined that we cannot straightly think of one without the others. So it happens that when I write of hunger, I am really writing about love and the hunger for it . . . and warmth and richness and fine reality of hunger satisfied . . . and it is all one.

I tell about myself, and how I ate bread on a lasting hillside, or drank red wine in a room now blown to bits, and it happens without my willing it that I am telling too about the people with me then, and their other deeper needs for love and happiness.

There is food in the bowl, and more often than not, because of what honesty I have, there is nourishment in the heart, to feed the wilder, more insistent hungers. We must eat. If, in the face of that dread fact, we can find other nourishment, and tolerance and compassion for it, we'll be no less full of human dignity.

There is a communion of more than our bodies when bread is broken and wine drunk. And that is my answer, when people ask me: Why do you write about hunger, and not wars or love?

M. F. K. Fisher

Preface to the Second Edition

When *Food and Culture: a Reader* was first published in 1997, it helped define and legitimize the field of food and culture studies. Since that time, the field has experienced explosive growth across myriad disciplines and topics. The field thus needs more than ever a book that defines foundational work and introduces innovative and exciting research. This new edition aims to serve as a continuing guide to the field by identifying classic articles that provide the field's intellectual roots and by introducing key issues and themes that demonstrate its ongoing vitality.

This book offers a comprehensive overview of some of the best work on food and culture to general readers as well as to scholars and students in the humanities and social sciences. The reader is organized into four sections: foundations of food studies, gender and consumption, food and identity politics, and the political economy of food. Among the foundational articles, we have retained work by Margaret Mead, Roland Barthes, Claude Lévi-Strauss, Mary Douglas, Hilde Bruch, Jack Goody, and Sidney Mintz, and added a seminal piece by Michel De Certeau and Luce Giard. These authors are some of the most influential early thinkers who established the significance and impact of food for studying the human condition.

In the first edition we had important articles by Joan Jacobs Brumberg, Caroline Bynum, Susan Bordo, Marjorie DeVault and Anne Allison on women's relationship to eating, feeding, and fasting across time and space, which we have retained in the new edition. In addition, we have expanded our consideration of gender by adding two new articles on masculinity—Fabio Parasecoli's on American men's eating and body-building and T. J. M. Holden's on male identity in Japanese food television. We have also added Christopher Carrington's groundbreaking article on food work and gender power in lesbigay households.

All the articles in the section "Food and Identity Politics" are new to this edition, a fact that reflects the exciting growth in this area. Papers explore fascinating questions about the relation of food to gender, race, class, nation, and personhood through a variety of case studies. Arjun Appadurai's influential piece on cookbooks and the construction of a national Indian cuisine makes a nice contrast to Lisa Heldke's work on ethnic cookbooks as vehicles of culinary colonization. Richard Wilk explores the ongoing construction of Belizean food in the increasingly transnational Caribbean. Three articles look at how minority groups construct identity and cope with socio-economic subordination through manipulating food: Psyche Williams-Forson's on African Americans' chicken consumption, Carole Counihan's on Hispanic women's cooking and food voice, and Gary Paul Nabhan's on desert-dwelling Indians' efforts to combat diabetes through reclaiming traditional diet.

Alison Leitch examines the consequences of the politicization of food through the Slow Food movement in Italy, and Jeffrey Pilcher examines the potential impact of Slow Food on Mexican culinary traditions. Dylan Clark examines how punk cuisine refutes the capitalist food system by espousing vegetarianism, dumpster diving, and locally produced organic foods. Salazar, Feenstra and Ohmert use photographic methodology to examine the distinctive cuisine of U.S. middle school students and to illustrate how visual methods contribute to food studies.

The last section on "The Political Economy of Food" contains an updated version of Penny Van Esterik's work on the complex politics of breast-feeding advocacy and adds eight new articles. These include two on the dehumanization, deskilling, and danger of food labor: Eric Schlosser's on meat-packing and Deborah Barndt's on the tomato food chain. Two articles look at fast-food in Asia—Yungxiang Yan's on McDonald's in Beijing and Gisèle Yasmeen's on street foods in Bangkok. Jennifer Clapp introduces the important subjects of GMOs (genetically modified organisms) and their impact on food aid. This section concludes with the political economy of over- and under-consumption in Alice Julier's sweeping article on obesity and Karen Coen Flynn's and Janet Poppendieck's papers on hunger in Africa and the United States.

This updated volume not only introduces readers to significant issues in food studies but also to important theories, including symbolic, structuralist, materialist, feminist, and political–economic approaches. It presents a variety of foodways research methods from ethnography, history, photography, psychology, geography, and mass media studies. The volume is outstanding for its detailed case studies covering significant theoretical and methodological innovations, its inclusion of influential classics and pioneering new work, its breadth of coverage, and its readability.

Carole Counihan and Penny Van Esterik
July 20, 2007

Acknowledgments

We wish to thank all the colleagues and friends who used the first edition of this book and gave us formal and informal feedback over the years which contributed in many ways to the shape of this second edition. We thank the following people for responding to our survey about the book and for their insights on food studies: Pauline Adema, Gary Allen, Denise Amon, Linda Murray Berzok, Janet Chrzan, Alfredo Manual Coelho, Jon Deutsch, Karen Coen Flynn, Carla Guerron Montero, Annie Hauck-Lawson, Jon Holtzman, Lynn Houston, Samantha Kwan, Christopher O'Leary, and Kyla Wazana Tompkins. Many other colleagues have inspired us over the years through their work in food studies, not only all the contributors in this book, but also Meredith Abarca, Ken Albala, Arlene Avakian, Warren Belasco, Anni Bellows, Rachel Black, Amy Bentley, Jennifer Berg, Anne Bower, Martin Bruegel, Antonella Campanini, Alberto Capatti, Piero Garofalo, Allen Grieco, Barbara Haber, Stacy Jameson, Steven Kaplan, Tom Kelly, Ronald LeBlanc, Laura Lindenfeld, Mario Montaño, Marion Nestle, Ramona Lee Perez, Carlo Petrini, Krishnendu Ray, Cinzia Scaffidi, Bruno Scaltriti, Jeffery Sobal, and David Sutton. We extend special thanks to the anonymous outside reviewers who offered excellent feedback on our proposal for revising this book.

We thank Bill Germano, formerly of Routledge, who was our editor for the first edition of this reader and who enthusiastically launched us on the second edition. For their help in moving the book into print, we are grateful for the assistance of our editor Steve Rutter, editorial assistant Anne Horowitz, cover designer Mark Lerner, production editor Alf Symons, and copy-editor Rosalind Wall.

At Millersville University, we thank the sociology-anthropology department faculty and especially department secretary Barbara Dills; Dean of Humanities and Social Sciences John Short and his staff; Provost Vilas Prabhu; President Francine McNairy; the Faculty Grants Committee; the lively and creative anthropology majors; and the three student assistants who helped with this project, Emily Robinson, Amanda Altice, and especially Shannon Dempsey. Shannon was the editorial manager of this project and without her patience, sense of humor, organizational skills, and attention to detail, it would never have been completed.

From Carole: special thanks to the "Catholic Girls Writing Group"—Kathy Brown, Theresa Russell-Loretz, Barb Stengel and Tracey Weis—friends and spiritual guides who have provided support for writing and so much else. For being a great family, many thanks to my husband and colleague, Jim Taggart, my sons Ben and Will Taggart, my step-daughter Marisela Taggart, and her husband Kraig Singleton and children Julian, William and Kristina. Thanks to my siblings Ted, Chris, Steve, and Sue, a sister for the ages.

From Penny: special thanks to Jillian Olivierre and Mandeep Kohli for research assistance at York University. My students in nutritional anthropology courses here at York are a constant source of stimulation and new insights. In addition, I want to acknowledge an intellectual debt to my co-author on forthcoming books on food related topics, Richard O'Connor, whose theoretical insights into the human condition spill over into all my work, including this reader. For support with this, and all my other projects, thanks to my husband and colleague, John Van Esterik.

Finally, we would like to dedicate this second edition to Mary Douglas (1921–2007), whose recent passing is an occasion to remember her contributions to food studies. The questions she raised about religious food prohibitions, culture and risk, consumption practices and food as a system of communication remain as significant today as when she first raised them.

Introduction to the Second Edition

Carole Counihan and Penny Van Esterik

As we set out to do a new edition of *Food and Culture: A Reader* we did not fully comprehend the magnitude of the task. We knew the field of food studies had mushroomed since we compiled the first edition in 1997, but we had not fully taken stock of the massive expansion until we began to select articles and take a systematic look at the state of food and culture research. We were astounded at how food has permeated almost every scholarly field and become a widely accepted research topic. What does this enormous expansion mean? Why has food become such a mainstream topic of research today when thirty years ago it was marginalized as a scholarly focus?

Just since the turn of the millennium the growth in scholarly books is amazing. A quick and by no means exhaustive bibliographic search turns up scores of recent food books in fields as diverse as film studies (Ferry 2003, Bower 2004, Keller 2006)[1], architecture (Franck 2003, Horwitz & Singley 2006), philosophy (Korsmeyer 2002, Heldke 2003, Singer 2006), psychology (Conner & Armitage 2002), and geography (Carney 2001, Yasmeen 2006), not to mention the vast literature in food's traditional fields of nutrition, home economics, and agriculture. Countless new texts abound on food in literature, from the study of eating and being eaten in children's literature (Daniel 2006), to food symbols in early modern American fiction (Appelbaum 2006) and classical Arab literature (Van Gelder 2000), to post-Freudian analysis of literary orality (Skubal 2002). In its more longstanding disciplinary homes, food continues to fascinate, so we find texts exploring the history of food from the Renaissance banquet (Albala 2007a) to the future of food (Belasco 2006), from the U.S. (Williams-Forson 2006) to Italy (Capatti & Montanari 2003) and all of Europe (Flandrin & Montanari 2000), to the history of many specific foods including beans (Albala 2007b), turkey (Smith 2006), chocolate (Coe & Coe 2000), salt (Kurlansky 2003), and spices (Turner 2004). Sociologists have not hesitated to stir the food studies pot (Ray 2004) and anthropologists have continued to produce work on topics as varied as hunger in Africa (Flynn 2005), children's eating in China (Jing 2000), food and memory in Greece (Sutton 2001), and Japan's largest fish market (Bestor 2004).

These examples provide some measure of the many texts that have been published in the last decade. Why has the field exploded so? We would like to suggest several reasons for this explosion. Without a doubt feminism and women's studies have contributed to the growth of food studies by legitimizing a domain of human behavior so heavily associated with women over time and across cultures. The recognition of food as a feminist issue has expanded beyond the subject of eating disorders (cf. Gremillion 2003) to cover the wide diversity of feminist approaches to food and women's food stories, a topic explored in section two of the reader. A second reason is the politicization of food and the expansion of social movements linked to food. This has created an increased awareness of the links between consumption and production beginning with new books on food and agriculture (e.g. Magdoff et al. 2000, Guthman 2004), as well as more interdisciplinary work on food politics (Nestle 2003). A third reason is that once food became a legitimate topic of scholarly research, its novelty, richness, and scope provided limitless grist for the scholarly mill—as food links body and soul, self and other, the personal and the political, the material and the symbolic. Moreover as food shifts from being local and known to being global and unknown it has been transformed into a potential symbol of fear and anxiety (Ferrieres 2005), as well as of morality (Telfer 2005). For many North Americans, concerns about food security have become concerns about food safety.

Scholars have found food a powerful lens of analysis and written insightful books about a range of compelling contemporary issues: diaspora and immigration (Gabaccia 1998, Ray 2004); nationalism, globalization, and local manifestations (Barndt 1999, Wilk 2006a, 2006b); culinary tourism (Long 2003); gender and ethnic identity (Inness 2001a, 2001b, 2001c, Abarca 2006, Williams-Forson 2006); class, work, and economic power (Schlosser 2002, Thompson and Wiggins 2002); social justice and human rights (Wenche Barth and Kracht 2005, Kent 2007)[2]; modernization and dietary change (Watson 1997, Counihan 2004); food safety and contamination (Friedberg 2004, Nestle 2004, Schwartz 2004); and taste perception (Korsmeyer 2002, 2005). Many of these subjects have important material dimensions, which have also been studied by archaeologists, folklorists and even designers, as food leaves its mark on the human environment.

The explosion of the field of food studies is also reflected in new and continuing interdisciplinary journals such as *Agriculture and Human Values; Appetite; Culture and Agriculture; The Digest; Food and Foodways; Food, Culture and Society; Gastronomic Sciences/Scienze Gastronomiche; Gastronomica; Nutritional Anthropology;* and *Slow.* Hundreds of websites inform food professionals, researchers and the general public. Ground-breaking documentary films such as *Fast Food Nation, Supersize Me, The Future of Food,* and *The Real Dirt on Farmer John* have called attention to problems in our food system and efforts to redress them. Food advocacy is reflected in food movements that promote organic, local, fairly traded, and slow food, revitalizing vegetarianism and freeganism (Van Esterik 2005), and decrying what fast, processed food has done to our bodies and communities. Of particular interest is how food-focused social movements interact with one another, and with academic research (Belasco 2007).

Given the vastness of the food studies repast, there is no way this book can offer up a complete meal. Rather, we envision it as an appetizer to introduce the field to the reader, a taste of the diverse array of scrumptious intellectual dishes that await further pursuit. We have chosen articles that are high quality and that explore issues of enduring importance written by some of the leading food studies scholars. "Write a book with legs," our editor urged us in 1997—and we did. But those legs have taken food studies in exciting new directions in the last decade, and this revised reader reflects these changes.

The second edition retains classic papers reflecting the foundations of food studies, and provides an interdisciplinary collection of cutting-edge articles in the social sciences that combine theory with ethnographic and historical data. We hope our readers will find this second edition engages even more deeply with both past and present scholarship on food and culture. From the first reader, we retain the wise words of M. F. K. Fisher, and reaffirm that food touches everything and is the foundation of every economy, marking social differences, boundaries, bonds, and contradictions—an endlessly evolving enactment of gender, family and community relationships.[3]

Foundations

Many of the papers we have retained are considered classics and are valuable for demonstrating the history of the field. The selection demonstrates the centrality of cultural anthropology to the development of food studies.[4] We begin this section with Margaret Mead's report on "The Problem of Changing Food Habits." This piece illustrates the originality of Mead's work on food policy and her inspiration to others to conduct applied nutritional research. Mead, following Audrey Richards' pioneering work, was one of the earliest anthropologists to articulate the centrality of foodways to human culture and thus to social science.[5]

We have included the classic articles of Barthes and Lévi-Strauss on food's ability to convey meaning. While many have critiqued the specifics of Lévi-Strauss' "culinary triangle," and he himself later revised his formulations in his 1978 book, *The Origin of Table Manners*, this piece remains a classic structuralist statement. Barthes' ruminations on "The Psychosociology of Contemporary Food Consumption" also constitute a seminal account of the semiotic and symbolic power of foodways. Mary Douglas builds on the work of Barthes and Lévi-Strauss to explain Jewish dietary law, a much debated topic in the field of food and culture and an excellent case study for examining food taboos and symbolism. While Douglas explains Jewish food prohibitions on the basis of the religious conception of holiness based on wholeness, Marvin Harris rejects semiotic interpretations of the abomination of pigs and offers a cultural materialist explanation based on economic and ecological utility. The mystery of food taboos, even when the prohibited foods are available, nutritious, and "edible", is a test case for exploring the gustatory selectivity of all human groups, and a wonderful example of how the same cultural phenomenon can be explained from different theoretical viewpoints.

We have added to the foundations section Mintz's memorable paper on time, sugar and sweetness in the Caribbean, an appropriate tribute to his influence on the field. Mintz shows how the rich controlled access to desirable high status sugar until it was produced in sufficient quantity to become a staple rather than a luxury consumed only by the elite; this transformation—and the processes of slavery, global trade, and worker exploitation on which it depended—changed the course of human

history. In this section we have also included several additional papers that merit
the status of classics. These include Jack Goody's wonderful "The Recipe, the Prescription, and the Experiment" (rather than his article on Industrial Food, which
we had in the first edition). Also new to this edition is De Certeau and Giard's paper
on "The Nourishing Arts", which offers an example of the detailed ethnographic
work on food being produced in the European tradition of food ethnology. Last,
is an important early paper by Hilde Bruch in which she developed the definitive
diagnosis of anorexia nervosa as characterized by distorted body image, inaccurate
hunger perception, hyperactivity, and an overwhelming sense of ineffectiveness.
In rethinking and updating the "Foundations of Food Studies", we recognize the
significant contributions these authors have made to food studies by introducing
basic definitions and conceptual tools used by later scholars.

Gender and Consumption

Our second section, "Gender and Consumption," recognizes the productive cross-
fertilization between food studies and gender studies. Since the publication of our
first edition, a plethora of books and articles have demonstrated this cross-
fertilization, including Sherrie Inness' three edited volumes, Abarca's (2006) study of
culinary chats among Mexican and Mexican-American working class women, Witt
(1999) and Williams-Forson's (2006) books on African-American food and gender,
and Avakian and Haber's (2005) interdisciplinary edited collection on feminist
approaches to food studies. While many works focus on women and food, there has
been new interest in how food and masculinity construct each other, represented
especially in Julier and Lindenfeld's (2005a, 2005b) edited special issue of *Food and
Foodways* on "Masculinities and Food". We have included papers by Parasecoli and
Holden from that issue here. To avoid limiting understandings of eating to heterosexual populations, we have incorporated Carrington's paper on food, gender
identity, and power in gay and lesbian households.

A key issue in Western women's relationship to food for hundreds of years has
been their unremitting fasting. We have retained Caroline Bynum's striking discussion of how medieval women used food to gain religious and cultural power. By
giving food to the poor, exuding milk from their bodies, and relentlessly fasting, they
were able to subvert the economic control of husbands and the religious authority of
male priests to commune directly with God.[6] Following Bynum's article, Brumberg's
important piece on the "Appetite as Voice" demonstrates the historical endurance
of women's fasting to attain influence, and it sets the stage for Bordo's paper on
contemporary anorexia nervosa. Bordo shows how pathological eating patterns
reflect cultural ideologies surrounding mind-body dualism, the moral value on self-
control, and the association of women with the body that needs to be controlled.[7]
These papers on food refusal as a form of social alienation highlight the importance
of food sharing as a form of communion and alliance, as Mauss (1925) so clearly
demonstrated in his ground breaking study of *The Gift*.

Men do not escape cultural manipulation through ideologies of food and body, as
Parasecoli's article on "Feeding Hard Bodies" demonstrates. He offers a challenge to

contemporary researchers to pay more attention to the subversion of male power through heightened insecurity about food consumption. Holden's article reveals how television can provide exciting new opportunities for foodways research. His demonstration of Japanese food television's broadcasting of narrow definitions of masculinity based on power, authority, and consumerism makes a nice contrast to Allison's article on how women reproduce Japanese definitions of subservient femininity through their construction of children's lunch-boxes, or *obentōs*. DeVault and Carrington look at the gendered nature of feeding work in heterosexual and lesbigay families respectively. Interestingly, both find that food work is associated with femininity and subservience. While the cooks in both sets of families often enact deference by catering to partners' preferences, lesbigay couples are more likely to implicitly acknowledge the subordinating dimensions of that practice by denying the extent of the feeders' work and the inequality it implies.[8]

Food and Identity Politics

In our third section, "Food and Identity Politics", we include articles that consider the expression of race, class, nation, and personhood through food production and consumption. Several articles deal with the construction of national identity through the contested transformation of cuisine. Appadurai looks at how cookbooks written for an Anglophone audience in post-colonial India define national cuisine, the boundaries of edibility, and the logic of meals. In a similar vein but through different kinds of data, Wilk examines historical transformations in Belizean food resulting from colonialism and globalization, which demonstrate the ways Belizean people define themselves and their nation. Reversing the lens, Heldke examines how food adventurers at home reproduce "cultural food colonialism" by seeking and cooking ethnic foods to satisfy their taste for the exotic other without actually encountering real others on their own terms. She raises important questions about the meaning of "authentic" food and about recipe authorship and ownership, challenging scholars and cookbook writers to think about their responsibility to the native cooks whose recipes they appropriate.

Williams-Forson, Nabhan, and Counihan examine how race and ethnicity play into diet and cuisine to affect identity, health, and power. Using ethnographic, historical, and literary research on chicken consumption and stereotypes associated with it, Williams-Forson examines the harmful effects of controlling images on African Americans but also uncovers ways that Black women have used chicken as a source of income, a delicious meal, and a bonding force to resist oppression and foster community and cultural survival. Counihan uses food-centered life histories to document the voices of traditionally muted Hispanic women of rural southern Colorado. Their stories challenge notions that Mexican-American women are compliant housewives complacently accepting subservient feeding roles. They reveal differential behaviors and attitudes towards food work that promote empowerment. Nabhan uses his ethnobotanical research among the desert dwelling Seri Indians of Northern Mexico to investigate the causes of rapidly increasing adult-onset diabetes. Listening to elders who asserted that diabetes was non-existent two generations earlier and thus rejecting simple genetic explanations for the disease, Nabhan suggests that dietary changes towards high-sugar, low fiber, rapidly digested foods have caused this major health problem and that traditional, slow release, desert foods are protective. This research provides support from a nutritional and health perspective for global efforts to promote local foods and traditional agriculture.

Both Leitch and Pilcher examine the ideology and practices of the Slow Food movement to foster local, sustainable, and just food production. Leitch looks at Slow Food's work in its country of origin, Italy, through a fascinating case study of pork fat—*lardo di Colonnata*—in the Carrara region, famous for its marble on which the lard is cured in humid underground cellars. Achievement of protected status for this traditional product raised questions of national autonomy and identity in the context of the European Union's efforts to impose universal hygienic standards. Pilcher investigates the relevance of the Slow Food Movement to Mexico's culinary traditions and sees similar issues to those confronted in Italy as Mexican peasant producers strive for living wages to produce traditional varieties of maize and tortillas against competition from global giants like Taco Bell.

Back in the U.S., in a grubby restaurant called the Black Cat Café, Dylan Clark employs perspectives from Lévi-Strauss and Marx to examine punk cuisine, a form of what today some call freeganism (Edwards 2006), a cuisine based on rotten, raw, scavenged, or stolen food. Punks consciously choose these counter-cultural foods to challenge the oppressive dimensions of the capitalist food system with its unequal food access, its environmentally dangerous production methods, and its wastefulness. Like vegetarians, punks offer an alternative ethic and practice of consumption to demonstrate how eating is an ideological as well as physical act. Salazar, Feenstra and Ohmart's article offers a fascinating methodological approach in visual food studies to investigate children's expression of their culinary identity. Photographic research on children's construction of salads at several Northern California elementary schools that had recently instituted salad bars, was designed to assess the nutritional content of children's meals. But as the authors analyzed the photographs they made unexpected discoveries about children's portrayal of identity, taste, aesthetics, and appropriate meals. This chapter is available online, with photography in color, at www.routledge.com/textbooks/9780415977777.

Political Economy of Food: Transformation and Marginalization

The last section of this reader, as in the first edition, remains grounded in the political economy of food, but reflects the ever more sophisticated research done in the last decade on how contemporary food systems are changing. The articles demonstrate that food commodification is deeply implicated in perpetuating and concealing gender, race and class inequalities, while transforming cultures. Case studies on tomatoes (Barndt), meats (Schlosser) and baby foods (Van Esterik) exemplify some of the transformations based on industrial processing of basic foods. Schlosser carries forward the work he did in his renowned *Fast Food Nation* (2002) to examine the many health dangers suffered by meat-packing workers—including broken bones, muscle strain, burns, and severed limbs—resulting from exhausting and monotonous labor for low wages, few benefits, and high turnover. Worker exploitation results from the concentration of the meat-packing industry, its reliance on immigrant labor, its concerted resistance to unionizing efforts, and its political power.

Barndt's article reflects her long-term project on the tomato food chain (Barndt 2002). It shows how agribusinesses, fast food giants, and supermarket chains increasingly rely on flexible, part-time, low-wage female labor, which enables them to generate huge profits at the expense of women workers who lack health and other benefits, cannot earn a living wage, and must constantly juggle their lives to accommodate their ever-changing work schedules. Van Esterik describes how the commodification of infant food through the international marketing of infant formula has had severe economic, cultural, and health consequences. But, as this paper and many others in this section show, resistance is continuous and effective; women have actively protested the actions of transnational pharmaceutical and food companies promoting industrially processed foods.

The lens of political economy provides fascinating insights into the current obsession with obesity, as Julier shows. Taking a critical functionalist approach, she shows who benefits from blaming the obese for their weight: the government, the diet food and supplement industries, bariatric medical practitioners, and exercise businesses. Blaming the obese draws attention away from the broader social and economic causes, such as the inadequacy of current food distribution systems and the excess of unhealthy food available to the poor.[9] Even food aid has been affected by the concentration of power in the hands of global food industries. Clapp shows how several African societies exercise their rights to limit the import of genetically modified foods even in the face of famine and extreme hunger.

In Asia, the original home of fast food, Yan and Yasmeen offer contrasting approaches to convenience food in Beijing and Bangkok. Yan stresses that the advantages of consuming modern foods like hamburgers have more to do with where they are consumed and what they signify than what they taste like. In Bangkok, working women can pick up plastic bags of traditional Thai dishes, such as curries and other specialties, from street vendors to serve at home on rice prepared in electric rice cookers timed to the family's arrival.

The street provides food for children in Tanzania, as well, but not enough food, as Flynn's ethnographic research on homeless boys and girls in the city of Mwanza shows. These children do whatever they can to eat, including working sporadically at food businesses, scavenging on the streets or in the dump, stealing, exchanging sex for food, and begging. The role of charity in their food acquisition suggests ways to rethink Sen's entitlement theory of food insecurity. Hunger and handouts are not unknown in North American communities. In the final article, Poppendieck looks at the role of charity in combating food insecurity in the U.S. While it plays a critical role in temporarily abating hunger, it fails to address the poverty and structural inequality that are its real underlying causes (Lapp and Collins 1986, Fitchen 1988, Poppendieck 1998).

Food is a particularly powerful lens on capital, labor, health and the environment. Taken together, these papers force us to re-examine the interconnections between the availability of cheap food in North America and the conditions of its production in other parts of the world. Food advocacy is a growing arena for political activism, as the success of Italy's Slow Food Movement shows. Food unites all humans; its lack strikes a painful chord among haves and have-nots alike. Progress towards social justice can only come through a concerted effort on the part of social activists everywhere to end world hunger and bring about universal access to nutritious and adequate food.

Cross-cutting Themes

Throughout the four sections of the reader, several themes emerge that can structure how readers approach the book. Theory and method constitute one important theme. While all of the articles are embedded in theory, some explicitly identify theoretical positions: semiotic (Barthes), structuralist (Lévi-Strauss), symbolic (Douglas), materialist (Harris), Marxist (Clark), critical functionalist (Julier), psychological (Bruch), and feminist (Bordo, Counihan, Van Esterik, Williams-Forson).

Articles also employ different methodologies, providing readers with a wealth of information about the different means of investigating the role of food in history and culture. A number of articles use ethnographic approaches, for example Allison, Carrington, DeVault and Yan use interviews and participant-observation, and Counihan uses food-centered life histories. Analysis of cultural symbols and meanings is employed by Douglas and Bordo, the former from an anthropological and the latter from a philosophical perspective. Parasecoli and Holden analyze the mass media, while Appadurai and Goody mine recipes for their rich data about cultures of food.[10] Bruch uses methods of psychotherapy, Nabhan of ethnobotany, and Yasmeen of geography, while Brumberg, Bynum, and Williams-Forson employ fine-grained historical research. Salazar, Feenstra and Ohmart introduce photography as a potential method for studying contemporary foodways. Together these articles provide readers with a rich sampling of diverse theories and methods to help them think about and conduct their own research.

Another cross-cutting theme is food as a means of communication. Because of food's multi-sensorial properties of taste, touch, sight, sound, and smell, it has the ability to communicate in a variety of registers and constitutes a form of language (Barthes). Definitions of acceptable and prohibited foods (Lévi-Strauss, Douglas, Harris), stereotypes associating certain groups with certain foods (Williams-Forson), consumption of foods to express belonging (Clark, Heldke) or attain desired states (Parasecoli, Yan), and use of food narratives to speak about the self (Counihan) are all ways that food communicates.

Food as an index of power relations is another significant theme in this book. Gender power is one important axis of hierarchy which is addressed not only in the section on gender and consumption, but also in other articles, many of which display the intertwining of gender with race, ethnicity, class, and national power. Gender, race, and/or class are central in Williams-Forson's paper on African American women's contested relationship with chicken, Counihan's article on rural *Mexicanas*' diverse approaches towards food preparation, and in Heldke's examination of first world consumers' "adventure cooking and eating" through ethnic cookbooks.

Barndt's and Schlosser's examination of the exploitation of workers, Van Esterik's analysis of the complex power dynamics involved in breast vs. bottle feeding of infants, Julier's analysis of obesity, and Flynn's study of the different tactics to acquire food employed by homeless boys and girls in Mwanza, Tanzania all demonstrate how food can be used to make unequal power relations visible.

Access to food is the most basic human need and its denial is a terrible measure of human powerlessness, an issue addressed in different ways by Nabhan's examination of Native Americans, and Flynn and Poppendieck's insightful studies of food insecurity. With the increasing commodification and globalization of food, power issues are revealed not only in access to food but also in the production of local, culturally meaningful foods whose endurance is key to cultural survival, as Wilk, Nabhan, Leitch, Pilcher, and Clapp demonstrate.

Concluding Thoughts

We would like to close with some questions for readers to consider as they read the book. What is it about food that makes it an especially intriguing and insightful lens of analysis? What questions about foodways still need to be addressed? How have food regimes changed through time? How does the universal need for food bind individuals and groups together? What are the most serious problems in the global food system and what causes them? Which political, economic, social, and ideological constructions of food contribute to inequality and which to social justice? What sorts of changes on the personal, community, national, or international level could contribute to a more equitable food system?

As you embark on this journey through food studies, we leave you with the words of M. F. K. Fisher who affirmed that "with our gastronomical growth will come, inevitably, knowledge and perception of a hundred other things, but mainly of ourselves."[11]

Notes

1. Some recent articles on food and film are Baron 2003, Johnson 2002, Van Esterik 2006.
2. Two insightful articles on food and human rights are Bellows 2003, Van Esterik 1999b.
3. M.F.K. Fisher (1954, 1961, 1983) is one of the most lyrical food writers who has inspired countless others.
4. The development of research interests in food in anthropology is as old as the discipline. Early anthropologists recognized the central role of food in different cultures, most notably Audrey Richards (1932, 1939), but also Raymond Firth (1934), Bronislaw Malinowski (1935), M. and S. L. Fortes (1936), and Cora DuBois (1941). Anthropology continues to make important contributions—both ethnographic and theoretical—to the field today. Some influential books on the anthropology of food are Anderson 1988, 2005, Counihan 1999, 2004, Dettwyler 1994, Fink 1998, Goody 1982, Kahn 1986, Kulick and Meneley 2005, Meigs 1984, Mintz 1985, 1996, Nichter 2000, Ohnuki-Tierney 1993, Pollock 1992, Watson 1997, Weismantel 1988, Wilk 2006a, 2006b.
5. See Spang (1988) on anthropologists' work on food during World War II.
6. Some books that examine the religious and ideological dimensions of fasting and dieting are Adams 1990, Bell 1985, Bynum 1987, Griffith 2004, Sack 2005, and Vandereycken and Van Deth 1994.
7. The following are influential studies of women's food restriction: Bruch 1973, 1978, Brumberg 1988, Nichter 2000, Thompson 1994.
8. On the complex relationship between gender, cooking, and power, see Avakian 1997, Avakian and Haber 2005, Charles and Kerr 1988, Counihan 2004, DeVault 1991, Inness 2001a, 2001b, 2001c, Williams-Forson 2006, and Witt 1999.

9. On the construction of obesity in the U.S. and across cultures see De Garine and Pollock 1995, Kulick and Meneley 2005, Massara 1989, Millman 1980, Sobal and Maurer 1999a, 1999b, Sobal and Stunkard 1989, and Styles 1980.
10. See Bower 1997 for a range of approaches to recipes and community cookbooks.
11. Fisher (1954: 350), from "How to Cook a Wolf."

References

Abarca, Meredith (2006) *Voices in the Kitchen: Views of Food and the World from Working-Class Mexican and Mexican American Women*. College Station: Texas A & M University Press.

Adams, Carol (1990) *The Sexual Politics of Meat: A Feminist-Vegetarian Critical Theory*. New York: Continuum.

Albala, Ken (2002) *Eating Right in the Renaissance*. Berkeley: University of California Press.

Albala, Ken (2007a) *The Banquet: Dining in the Great Courts of Late Renaissance Europe*. Champagne-Urbana: University of Illinois Press.

Albala, Ken (2007b) *Beans: A History*. Oxford: Berg.

Anderson, E. N. (1988) *The Food of China*. New Haven: Yale University Press.

Anderson, E. N. (2005) *Everyone Eats: Understanding Food and Culture*. New York: NYU Press.

Appelbaum, Robert (2006) *Aguecheek's Beef, Belch's Hiccup, and Other Gastronomic Interjections: Literature, Culture, and Food Among the Early Moderns*. Chicago: University of Chicago Press.

Avakian, Arlene Voski (ed.) (1997) *Through the Kitchen Window: Women Explore the Intimate Meanings of Food and Cooking*. Boston: Beacon.

Avakian, Arlene Voski and Barbara Haber (eds) (2005) *From Betty Crocker to Feminist Food Studies: Critical Perspectives on Women and Food*. Amherst: University of Massachusetts Press.

Barndt, Deborah (ed.) (1999) *Women Working the NAFTA Food Chain: Women, Food and Globalization*. Toronto: Second Story Press.

Barndt, Deborah (2002) *Tangled Routes: Women, Work, and Globalization on the Tomato Trail*. Lanham, Md: Rowman and Littlefield.

Baron, Cynthia (2003) "Food and Gender in *Bagdad Café*." *Food and Foodways* 11, 1: 49–74.

Belasco, Warren (2006) *Meals to Come: A History of the Future of Food*. Berkeley: University of California Press.

Belasco, Warren (2007) *Appetite for Change*. Ithaca: Cornell University Press. 2nd ed.

Bell, David and Gill Valentine (1997) *Consuming Geographies: We Are Where We Eat*. New York: Routledge.

Bell, Rudolph M. (1985) *Holy Anorexia*. Chicago: University of Chicago.

Bellows, Anne (2003) "Exposing Violences: Using Women's Human Rights Theory to Reconceptualize Food Rights." *Journal of Agricultural and Environmental Ethics*, 16: 249–279.

Bestor, Theodore C. (2004) *Tsukiji: The Fish Market at the Center of the World*. Berkeley: University of California Press.

Bower, Anne (ed.) (1997) *Recipes for Reading: Community Cookbooks, Stories, Histories*. Amherst: University of Massachusetts Press.

Bower, Anne (ed.) (2004) *Reel Food: Essays on Food and Film*. New York: Routledge.

Bruch, Hilde (1973) *Eating Disorders: Obesity, Anorexia Nervosa, and the Person Within*. New York: Basic Books.

Bruch, Hilde (1978) *The Golden Cage: The Enigma of Anorexia Nervosa*. New York: Vintage.

Brumberg, Joan Jacobs (1988) *Fasting Girls: the Emergence of Anorexia Nervosa as a Modern Disease*. Cambridge: Harvard University Press.

Bynum, Caroline Walker (1987) *Holy Feast and Holy Fast: The Religious Significance of Food to Medieval Women*. Berkeley: University of California Press.

Capatti, Alberto and Massimo Montanari (2003) *Italian Cuisine: A Cultural History*. New York: Columbia University Press.

Carney, Judith A. (2001) *Black Rice: The African Origins of Rice Cultivation in the Americas*. Cambridge, MA: Harvard University Press.

Charles, Nicki and Marion Kerr (1988) *Women, Food and Families*. Manchester: Manchester University Press.

Coe, Sophie D. and Michael D. Coe (2000) *The True History of Chocolate*. London: Thames and Hudson.

Conner, Mark and Christopher Armitage (2002) *The Social Psychology of Food*. London: Open University.

Counihan, Carole (1999) *The Anthropology of Food and Body: Gender, Meaning and Power*. New York: Routledge.

Counihan, Carole (2004) *Around the Tuscan Table: Food, Family and Gender in Twentieth Century Florence*. New York: Routledge.

Daniel, Carolyn (2006) *Voracious Children: Who Eats Whom in Children's Literature*. New York: Routledge.

De Garine, Igor and Nancy Pollock (eds) (1995) *The Social Aspects of Obesity*. New York: Gordon and Breach.

Dettwyler, Katherine (1994) *Dancing Skeletons: Life and Death in West Africa.* Prospect Heights, IL: Waveland.

DeVault, Marjorie L. (1991) *Feeding the Family: The Social Organization of Caring as Gendered Work.* Chicago: University of Chicago Press.

Du Bois, Cora (1941) "Attitudes toward Food and Hunger in Alor," in *Language, Culture, and Personality,* edited by Leslie Spier, et al. Menasha, WI: Sapir Memorial Publication Fund, pp. 272–281.

Edwards, Ferne (2006) Dumpster Dining. *Alternatives Journal,* 32, 3: 16–17.

Ferrieres, Madeleine (2005) *Sacred Cow, Mad Cow: A History of Food Fears.* New York: Columbia University Press.

Ferry, Jane (2003) *Food in Film: A Culinary Performance of Communication.* New York: Routledge.

Fink, Deborah (1998) *Cutting into the Meatpacking Line: Workers and Change in the Rural Midwest.* Chapel Hill: University of North Carolina Press.

Firth, Raymond (1934) "The Sociological Study of Native Diet." *Africa,* 7, 4: 401–414.

Fisher, M. F. K. (1954) *The Art of Eating.* Cleveland: World Publishing.

Fisher, M. F. K. (1961) *A Cordiall Water: a Garland of Odd and Old Receipts to Assuage the Ills of Man and Beast.* New York: North Point Press.

Fisher, M. F. K. (1983) *As They Were.* New York: Vintage.

Fitchen, Janet M. (1988) "Hunger, Malnutrition and Poverty in the Contemporary United States: Some Observations on Their Social and Cultural Context." *Food and Foodways,* 2, 3: 309–333.

Flandrin, Jean Louis and Massimo Montanari (eds) (1999) *Food: a Culinary History,* English edition edited by Albert Sonnenfeld. New York: Penguin.

Flynn, Karen Coen (2005) *Food, Culture, and Survival in an African City.* New York: Palgrave Macmillan.

Fortes, Myer and S. L. Fortes (1936) "Food in the Domestic Economy of the Tallensi." *Africa,* 9, 2: 237–276.

Franck, Karen A. (ed.) (2003) *Food and Architecture.* Chichester, U.K.: John Wiley & Sons.

Friedberg, Susanne (2004) *French Beans and Food Scares: Culture and Commerce in an Anxious Age.* New York: Oxford University Press.

Gabaccia, Donna (1998) *We Are What We Eat: Ethnic Foods and the Making of Americans.* Cambridge: Harvard University Press.

Goody, Jack (1982) *Cooking, Cuisine and Class: A Study in Comparative Sociology.* New York: Cambridge.

Gremillion, Helen (2003) *Feeding Anorexia.* Durham: Duke University Press.

Griffith, R. Marie (2004) *Born Again Bodies: Flesh and Spirit in American Christianity.* Berkeleley: University of California Press.

Guthman, Julie (2004) *Agrarian Dreams: The Paradox of Organic Farming in California.* Berkeley: University of California Press.

Heldke, Lisa (2003) *Exotic Appetites: Ruminations of a Food Adventurer.* New York: Routledge.

Horwitz, Jamie and Paulette Singley (2006) *Eating Architecture.* Cambridge, MA: MIT Press.

Inness, Sherrie A. (ed.) (2001a) *Cooking Lessons: The Politics of Gender and Food.* New York: Rowman and Littlefield

Inness, Sherrie A. (ed.) (2001b) *Pilaf, Pozole, and Pad Thai: American Women and Ethnic Food.* Amherst, MA: University of Massachusetts Press.

Inness, Sherrie A. (ed.) (2001c) *Kitchen Culture in America: Popular Representations of Food, Gender, and Race.* Philadelphia: University of Pennsylvania Press.

Jing, Jun (ed.) (2000) *Feeding China's Little Emperors: Food, Children, and Social Change.* Stanford: Stanford University Press.

Johnson, Ruth D. (2002) "The Staging of the Bourgeois Imaginary," in *The Cook, the Thief, His Wife, and Her Lover* (1990). *Cinema Journal,* 41, 2: 19–40.

Julier, Alice and Laura Lindenfeld (2005a) Mapping Men onto the Menu: Masculinities and Food. *Food and Foodways,* 13 (1–2): 1–16.

Julier, Alice and Laura Lindenfeld (eds) (2005b) Masculinities and Food. Special double issue of *Food and Foodways,* 13 (1–2).

Kahn, Miriam (1986) *Always Hungry, Never Greedy: Food and the Expression of Gender in a Melanesian Society.* Cambridge: Cambridge University Press.

Keller, James R. (2006) *Food, Film and Culture: A Genre Study.* Jefferson, North Carolina: McFarland.

Kent, George (2007) *Freedom from Want: The Human Right to Adequate Food.* Washington, DC: Georgetown University Press.

Korsmeyer, Carolyn (2002) *Making Sense of Taste: Food and Philosophy.* Ithaca, NY: Cornell 2002.

Korsmeyer, Carolyn (ed.) (2005) *The Taste Culture Reader: Experiencing Food and Drink.* Oxford: Berg.

Kulick, Don and Anne Meneley (eds) (2005) *Fat: The Anthropology of an Obsession.* New York: Penguin.

Kurlansky, Mark (2003) *Salt: A World History.* New York: Penguin.

Lappé, Frances Moore and Joseph Collins (1986) *World Hunger: Twelve Myths.* New York: Grove Press.

Lévi-Strauss, Claude (1978) *The Origin of Table Manners.* London: Jonathan Cape.

Long, Lucy (ed.) (2003) *Culinary Tourism.* Lexington: University of Kentucky Press.

Magdoff, Fred, John Bellamy Foster, and Frederick H. Buttel (eds) (2000) *Hungry for Profit: The Agribusiness Threat to Farmers, Food, and the Environment.* New York: Monthly Review Press.

Malinowski, Bronislaw (1935) *Coral Gardens and Their Magic: A Study of the Methods of Tilling the Soil and of Agricultural Rites in the Trobriand Islands.* New York: American Book Co. (2 volumes).

Massara, Emily Bradley (1989) *Que Gordita. A Study of Weight Among Women in a Puerto Rican Community.* Brooklyn, NY: AMS Press.

Mauss, Marcel (1967 [orig. 1925]) *The Gift: Forms and Functions of Exchange in Archaic Societies.* New York: Norton.

Meigs, Anna S. (1984) *Food, Sex, and Pollution: A New Guinea Religion.* New Brunswick, NJ: Rutgers University Press.

Millman, Marcia (1980) *Such a Pretty Face: Being Fat in America.* New York: Norton.

Mintz, Sidney W. (1985) *Sweetness and Power: The Place of Sugar in Modern History.* New York: Penguin.

Mintz, Sidney (1996) *Tasting Food, Tasting Freedom: Excursions into Eating, Culture, and the Past.* Boston: Beacon.

Nestle, Marion (2003) *Food Politics: How the Food Industry Influences Nutrition and Health.* Berkeley: University of California Press.

Nestle, Marion (2004) *Safe Food: Bacteria, Biotechnology, and Bioterrorism.* Berkeley: University of California Press.

Nestle, Marion (2007) *What to Eat.* San Francisco: North Point Press.

Nichter, Mimi (2000) *Fat Talk: What Girls and their Parents Say about Dieting.* Cambridge: Harvard University Press.

Ohnuki-Tierney, Emiko (1993) *Rice as Self: Japanese Identities through Time.* New Brunswick, NJ: Princeton University Press.

Pollock, Nancy (1992) *These Roots Remain: Food Habits in Islands of the Central and Eastern Pacific since Western Contact.* Honolulu: University of Hawaii Press.

Poppendieck, Janet (1998) *Sweet Charity? Emergency Food and the End of Entitlement.* New York: Penguin.

Ray, Krishnendu (2004) *The Migrant's Table: Meals and Memories in Bengali-American Households.* Philadelphia: Temple University Press.

Richards, Audrey I. (1932) *Hunger and Work in a Savage Tribe.* London: Routledge.

—— (1939) *Land, Labour and Diet in Northern Rhodesia: An Economic Study of the Bemba Tribe.* Oxford: Oxford University Press.

Sack, Daniel (2005) *Whitebread Protestants: Food and Religion in American Culture.* New York: Palgrave.

Schlosser, Eric (2002) *Fast Food Nation: The Dark Side of the All-American Meal.* New York: Harper Collins.

Schwartz, Maxime (2004) *How the Cows Turned Mad: Unlocking the Mysteries of Mad Cow Disease.* Berkeley: University of California Press.

Skubal, Susan (2002) *Word of Mouth: Food and Fiction After Freud.* New York: Routledge.

Singer, Peter and Jim Mason (2006) *The Way We Eat: Why Our Food Choices Matter.* Emmaus, PA: Rodale Press.

Smith, Andrew (2006) *The Turkey: An American Story.* Champagne-Urbana: University of Illinois Press.

Sobal, Jeffery and Donna Maurer (eds) (1999a) *Interpreting Weight: The Social Management of Fatness and Thinness.* Hawthorne, NY: Aldine Transaction.

Sobal, Jeffery and Donna Maurer (eds) (1999b) *Weighty Issues: Fatness and Thinness As Social Problems.* Hawthorne, NY: Aldine de Gruyter.

Sobal, Jeffery and Albert J. Stunkard (1989) "Socioeconomic Status and Obesity: A Review of the Literature." *Psychological Bulletin,* 105, 2: 260–275.

Spang, Rebecca L. (1988) "The Cultural Habits of a Food Committee." *Food and Foodways,* 2, 4: 359–391.

Styles, Marvalene H. (1980) "Soul, Black Women and Food," in *A Woman's Conflict, the Special Relationship between Women and Food,* edited by Jane Rachel Kaplan. Englewood Cliffs, NJ: Prentice-Hall, pp. 161–176.

Sutton, David E (2001) *Remembrance of Repasts: An Anthropology of Food and Memory.* Oxford: Berg.

Symons, Michael (2003) *A History of Cooks and Cooking.* Champagne-Urbana: University of Illinois Press.

Telfer, Elizabeth (2005) *Food for Thought: Philosophy and Food.* New York: Routledge.

Thompson, Becky (1994) *A Hunger So Wide and So Deep: American Women Speak Out on Eating Problems.* Minneapolis: University of Minnesota Press.

Thompson, Charles D. and Melinda F. Wiggins (2002) *The Human Cost of Food: Farmworkers' Lives, Labor and Advocacy.* Austin: University of Texas Press.

Turner, Jack (2004) *Spice: the History of a Temptation.* New York: Vintage Books.

Van Esterik, Penny (1996) "The Cultural Context of Breastfeeding and Breastfeeding Policy." *Food and Nutrition Bulletin,* 17(4): 422–431.

Van Esterik, Penny (1997) "Women and Nurture in Industrial Societies." *Proceedings of the Nutrition Society,* 56 (1B): 335–343.

Van Esterik, Penny (1999a) "Gender and Sustainable Food Systems: a Feminist Critique," in *For Hunger-Proof Cities: Sustainable Urban Food Systems,* edited by M. Koc, R. MacRae, L. Mougeot and J. Welsh. Ottawa: IDRC, pp. 157–161.

Van Esterik, Penny (1999b) "Right to Food; Right to Feed; Right to be Fed: the Intersection of Women's Rights and the Right to Food." *Agriculture and Human Values*, 16: 225–232.

Van Esterik, Penny (2005) "No Free Lunch." *Agriculture and Human Values*, 22: 207–208

Van Esterik, Penny (2006) "Anna and the King: Digesting Difference." *Southeast Asian Research*, 14 (2): 289–307.

Van Esterik, Penny (2006) "From Hunger Foods to Heritage Foods: Challenges to Food Localization in Lao PDR." In *Fast Food/Slow Food: The Cultural Economy of the Global Food System*, edited by R. Wilk. Lanham, Md.: Altimira Press, pp. 83–96.

Vandereycken, Walter and Ron Van Deth (1994) *From Fasting Saints to Anorexic Girls: The History of Self-Starvation*. New York: New York University Press.

Van Gelder, Geert Jan (2000) *God's Banquet: Food in Classical Arabic Literature*. New York: Columbia University Press.

Watson, James L. (ed.) (1997) *Golden Arches East: McDonald's in East Asia*. Stanford: Stanford University Press.

Weismantel, M. J. (1988) *Food, Gender and Poverty in the Ecuadorian Andes*. Philadelphia: University of Pennsylvania Press.

Wenche Barth, Eide and Uwe Kracht (eds) (2005) *Human Rights and Development*. Antwerp: Intersentia.

Wilk, Richard (2006a) *Home Cooking in the Global Village: Caribbean Food from Buccaneers to Ecotourists*. Oxford: Berg.

Wilk, Richard (ed.) (2006b) *Fast Food/Slow Food: The Cultural Economy of the Global Food System*. Lanham, Md.: Altamira.

Williams-Forson, Psyche A. (2006) *Building Houses out of Chicken Legs: Black Women, Food, and Power*. Chapel Hill: University of North Carolina Press.

Witt, Doris (1999) *Black Hunger: Food and the Politics of U.S. Identity*. New York: Oxford University Press.

Yasmeen, Gisèle (2006) *Bangkok's Foodscape*. Bangkok: White Lotus Press.

Foundations

1 The Problem of Changing Food Habits
Margaret Mead

2 Toward a Psychosociology of Contemporary Food Consumption
Roland Barthes

3 The Culinary Triangle
Claude Lévi-Strauss

4 Deciphering a Meal
Mary Douglas

5 The Abominable Pig
Marvin Harris

6 The Nourishing Arts
Michel de Certeau and Luce Giard

7 The Recipe, the Prescription, and the Experiment
Jack Goody

8 Time, Sugar, and Sweetness
Sidney W. Mintz

9 Anorexia Nervosa and its Differential Diagnosis
Hilde Bruch

1

The Problem of Changing Food Habits

Margaret Mead

Problems of changing food habits cut across ordinary discipline lines, in addition to involving contributions from both pure and applied sciences. There is a mass of literature and recorded experimentation on many aspects of the problem, ranging from studies of soil agronomy which illuminate the question of whether the habit of eating locally grown food is or is not the most nutritionally valuable behavior, through data on the content of diets, data on the relationship between purchasing power and diet, studies of historically changing diets, animal experiments in individual taste and preference and their relation to nutrition, and records of the cultural integration of food through case histories of individuals whose gastrointestinal disorders can be shown to be systematically related to the way in which learning to eat was combined with other types of learning.

As the Committee's task was to integrate existing materials and devise new ways of tapping existing knowledge on the problem of cultural change, a primary requirement was to develop a point of view, an approach which could make systematic use of additions to knowledge in all of the fields from which research results could be expected. New information is continually coming out on the perishable quality of vitamins in vegetables on a steam table; on a new experiment in a rat's preference for the food which he ate less often regardless of whether it was a more or less nutritious food; on the relative palatability of different varieties of the same food; or, on the other hand, on detailed studies of purchasing habits of subcultural groups, or of striking shifts in the consumption of different foods under various sorts of pressure (advertising, wartime shortages, etc.); on new experiments in the relationship between anxiety and acidity in the stomach. All of these findings have to be fitted together to provide a systematic and coherent scientific background for recommendations directed toward changing the dietary pattern of American culture.

Furthermore these findings have to be examined also to provide orientation for the Committee's responsible task of estimating clearly the relationship between any particular change which may be introduced in the dietary patterns, or the culturally standardized methods of inculcating food habits, and the impact of such a change on other parts of our culture or the culture of other peoples of the world with whom we come into contact through lend-lease, relief and rehabilitation procedures. For example, an increased knowledge of nutrition which is acquired by mothers without any concordant alterations in their methods of child training may produce feeding problems which have a far more serious effect upon the child's development than the less perfect diet which the child ate without pressure. Gifts of relief white flour, while temporarily stemming famine, may fasten upon a population which traditionally ate whole grains a habit with disastrous repercussions for their future health. The way in which substitutions necessitated by war are presented to the public may have later effects on the acceptance of those foods. Authoritarian methods used in enforcing nutritional standards may endanger democratic participation in other community activities. Alterations in the source of the food, as when individual food tickets are given to children in emergency mass feeding situations, may result in a breakdown of parental authority which was primarily sanctioned by the food-giving functions of the family. Only by putting each recommended innovation and the methods suggested for bringing it about against the total cultural picture, is it possible to guard against initiating changes which, while nutritionally desirable in the narrow sense, may be socially undesirable in a wider sense.

The principal discipline represented among the Committee members is cultural anthropology, and the conceptions of cultural anthropology have been used in developing the approach. Food habits are seen as the culturally standardized set of behaviors in regard to food manifested by individuals who have been reared within a given cultural tradition. These behaviors are seen as systematically interrelated with other standardized behaviors in the same culture. In attempting to estimate the strength of any given item of behavior, for example, preference for meat, aversion to milk, etc., this item is not treated as isolated, but is referred to the total complex of behaviors which constitute the food habits. Similarly, in considering methods of change, of innovation or alteration in existing patterns, recourse is had not merely to traditional psychological pronouncements on learning, but also to the habits of learning of Americans in 1943, and furthermore to learnings or resistances which involve the body as directly as do habits connected with food. Moreover, the interaction between the culturated individual and his environment has two aspects in any consideration of food habits, interaction with the food producing and food distributing systems, that is, adjustments to the physical environments, and interaction between the individual organism and the actual food. While cultural factors are expected to account in very large degree for the food habits of mankind, there is also the possibility that combinations of foods may exert a certain degree of coercion upon physiological responses, so that the constitution of foods themselves must also be taken into account.

A dynamic description of the total food habits pattern of a culture or sub-culture can be approached in a number of different ways. A minute survey of the food eaten at any given time by adults may be combined with careful observations and experimentally instituted attempts to change that pattern. For example, if we had a complete picture of the current eating habits of the members of an upstate New York community, adequate verbal accounts of the grounds upon which they rationalized their current procedures, and records of attempts to introduce new foods and alter both their meal pattern and the combinations in which foods were habitually eaten, together with verbatim records of acceptances and refusals and reasons given, it would be possible to construct an adequate description of the contemporary pattern, not only its overt content but also its deeper emotional content, of the terms in which the people of that community seek and accept food, the fears and repugnance which deter them from eating other food, the situations in which they will share food, or accept food differently prepared or served, and of the states of mind which might result from any drastic alteration in their traditional food habits.

Another approach to the dynamics of the food complex is to study the ways in which, within any given culture, good habits are inculcated in the growing child. Such an approach involves descriptions of the post partum procedures, breast feeding, supplementary feeding, weaning techniques, sanctions invoked to narrow the child's acceptance within socially approved limits, sanctions invoked to widen the child's acceptance to include all socially prescribed foods, and ways in which gifts of food, threats of deprivation of food, and situations involving food are integrated in the system of character formation. On the basis of such analysis, it would be possible to prescribe the lines which would have to be followed if effective change were to occur, and the implications of changed food habits for the rest of the personality.

Neither type of study is yet available in any complete form. During the last two years two field expeditions, one consisting of seven field workers under the auspices of the University of Chicago,[1] and one of two field workers under the auspices of the Nutrition Division[2] of the Office of Defense Health and Welfare Services, have made preliminary studies of this kind which provide us with the most extensive body of data yet collected on intensively observed small communities in a modern culture. Neither study approximates the sort of completeness which can be obtained on a primitive society. Preliminary memoranda have been prepared, using living primary sources, on the culturally differentiated food habits of certain subcultural groups in the United States (Harris 1942; Nizzardini & Joffe 1942; Benet & Joffe 1943; Joffe 1943b; Pirkova-Jacobson & Joffe 1943).

As there was no immediate prospect of financing intensive cultural studies of American food habits, it was decided to devise a flexible instrument by which a quick appraisal of some of the more significant motivations could be arrived at. Professor Kurt Lewin (Joffe 1943b),[3] of the State University of Iowa, developed a test based upon intensive interviewing, which he tested out upon some 2300 school children in Cedar Rapids, Iowa. This test can be used rapidly with groups of school children in any area and brings into relief such dynamic relationships as the degree of association between food which is disliked and food which is healthy; which types of food are presided over by the mother, the father, contemporaries, teachers, and health authorities; how food behavior is expected to change with increasing maturity and independence, and so on.

A related instrument which can be used in defining the content and pattern of regional food practices has been devised by W. Franklin Dove (Dove 1935; 1941; 1943). Dr. Dove's many other papers cover an integrated attack upon the whole problem of regional food patterns, nutritive value of foods, and the possibility of altering natural foods so that they make a more ideal contribution to a given food problem.

From these various sources of data, combined with other partial sources of material on diet, resistance to change, rationalizations of resistance, and so on, it is possible to identify various important social psychological characteristics of the American food pattern such as: the role of European peasant conceptions of status which have given an importance to white bread, much sugar, meat every day, and so on; the Puritan tradition of a connection between food which is healthful and food which is disliked, and the tendency in communities with a Puritan tradition to use food for purposes of reward and punishment and to handle delicious food as the reward for eating healthful, but disliked food; the equally definite Southeastern food pattern in which the emphasis is not upon health and duty but upon personal taste and a personal relationship between the eater and his food. Other tendencies which may be identified include the emphasis upon appearance of food rather than taste, an increasing preference for refined, purified, highly processed foods in which there is a minimum of waste material, and a parallel emphasis upon purity, packaging, etc. Culturally standardized objections to complex food dishes in which the constituents cannot be identified may also be referred, as may the characteristics listed above, to the situation in which people with very widely different food habits have found themselves in close association with each other, dependent upon alien cooking, alien serving, alien ownership and management of food distributing agencies. An investigation conducted by the Committee into ways in which emergency feeding could make maximum allowance for cultural differences in food habits (Committee on Food Habits 1942) showed that the most practical way of avoiding giving offense to anyone in a mixed group is to cook single foods with a minimum of seasoning and serve all condiments separately. Contemporary cafeteria procedures in American and the large development of self-selected types of meals are an example of a social institution which is adapted to a variety of mutually incompatible food habits. It is probable that many other characteristic American attitudes toward foods, including taboos on all subjects which may arouse disgust during eating, may be referred to the experience of different mutually unacceptable food patterns. Up until the last year, with the exception of some types of company-owned towns, very few Americans under forty have ever had the experience of having money to buy food which, however, could not be obtained. The attitude of many European peasants who treat bread as sacred and guard against a single crumb falling on the floor has vanished in a country in which food was the certainty and money the uncertainty.

The patterning of food habits in other cultures is not only a significant source of materials from which useful abstractions may be drawn (Richards 1932; Malinowski 1935; Mead 1938; Dubois 1941) and a necessary background for understanding subcultural groups in the United States (Harris 1942; Nizzardini & Joffe 1942; Benet & Joffe 1943; Joffe 1943b; Pirkova-Jacobson & Joffe 1943); the food habits of other peoples are also significant for us because of the task of feeding liberated and war torn countries. Not only must the rations which we send be adapted to existing supplies of food, to carefully ascertained nutritional requirements and to the exigencies of transportation, but also, if they are to be of maximum use, the rations must be adapted to the food habits of the inhabitants of the various countries. The Committee has set up a series of cultural food experiments in which samples of a variety of concentrated emergency rations[4] are tested and worked over by groups of nationals of the various European countries to whom we may expect eventually to send food. On the basis of recommendations developed out of these test situations it will be possible to send with supplies of emergency rations detailed instructions in the native language as to the most acceptable method of cooking them.

As world food plans develop, the more data which we have upon ways in which existing supply can be adapted to traditional patterns so as to provide good nutrition, the more effective world-wide policies for improving nutrition will be.

In any study of food habits, it is important to define the patterns into which the available foods are arranged, such as number and form of meals, and the cultural—as opposed to the nutritional—equivalences which can be invoked within these patterns. So while a nutritional substitute for meat may be milk, the cultural substitute may be a "casserole" in which the container has been substituted for the thing contained, and food that is in no sense a substitute for a protein will be accepted when presented on the table in a suitably shaped container.[5]

Meal patterns are equally arbitrary and important, and alterations in the time or designation of a meal may mean severe nutritional dislocations, as when some Eastern Europeans, upon immigrating to America, dropped the second breakfast, or when odd-shift workers eat three meals, none of which is breakfast;[6] and so the foods which customarily appear only at a breakfast table, fruit juice, cereal with milk, and eggs, tend to disappear from the diet. Shopping habits may have equally far reaching effect upon family food habits; for instance, since rationing, increased shopping by men and children of high school age has been reported. In the North American food pattern, the father presides over meat and fish, the mother over milk, vegetables, fruit juices and liver, while adolescents tend to demonstrate their independence by refusing to eat what is good for them. A shift in shopping habits resulting from the difficulties of using rationing coupons, increasing or irregular employment of women, and so on, may therefore also affect the actual content of the meal, for, as Professor Lewin's study[7] has demonstrated, the person who controls the channel controls the diet of the family.

Recent studies of the effect of methods of food preparation on vitamin loss (Joffe 1943a; Nagel & Harris 1943) have pointed up the close relationship between habits of food preparation, such as the sharecropper's habit of leaving greens cooking on the stove while going to work in the fields, (Reder undated) or the forehanded housewife's habit of preparing vegetables before breakfast, and the nourishment which families will finally receive from the food prepared in this way. Similarly, the growth of institutional feeding, with increased dependence on the steam table, may have important nutritional implications so that a carefully cultivated habit of selecting a balanced meal which would be efficacious under one condition of food preparation may be wholly inadequate under other circumstances. Where meals have been set in a pattern of family life, either with the southern emphasis upon catering to the tastes of each member of the family or the northern emphasis upon moral overhauling of family members' behavior, conditions which isolate the individual and force him to eat alone may have important nutritional implications: the food associated with family living may be rejected entirely or overconsumed.

Studies like those of Dr. Curt Richter (Richter 1936; Richter & Eckert 1937) on the ability of individual animals to select their own optimum diets also suggest that the nutrition of a people may be affected by habits of rigid food service in which each individual is served an exactly balanced meal without regard to individual preference. Such habits of food service as the individual plate and later types of blue plate service which are even invading the home all serve arbitrarily to standardize the food eaten by one person. Although it may be argued that such an arbitrarily balanced meal would be superior to the meals habitually eaten by the worst fed third of our population, there is a danger that the conventions of a balanced meal—plate service—may become established in the higher income levels and sift down as a style to the lower income levels, without the necessary knowledge to see that the meal is really balanced. The sort of individual adjustment which human beings, as well as rats, may conceivably make to an inadequate diet, such as reduction in caloric intake when the protective foods fell below a certain minimum, would then be ruled out by a food habit which had been nutritionally meaningful at a different income level. The whole tendency to train children in terms of "eating what's on your plate" introduces the same type of rigid category in the individual's relationship to his food.

When the culturally standardized dietary behavior has been described and analyzed, it is possible to consider the implications of experiments such as that of Festinger (Festinger 1943) on the preference manifested for the food less often eaten, or the relevancy of experiments in transfer of learning, etc., and to integrate them so that they will help in determining how to change American food habits and the results of such changes. Thus in every consideration, we do not think of an abstract human being eating an abstract food, but of particular human beings, members of an identifiable subculture of the United States, eating particular foods with definite qualities in addition to the socio-psychological values which have been assigned to that food by the culture. We do not ask, "How can we change food habits?" but rather, "How can we change the food habits of a community of second generation Americans of Polish, or Italian, or Hungarian extraction, where both men and women work in the mills and the average grade completed is the fifth?" or "How can we change the food habits of southern sharecroppers whose food habits are tied into a one crop method of production, type of credit allowed by the stores, rolling stores, habits of catering to individual preference and assigning ill health to the effects of particular foods, who live in a caste society where there is a characteristic rejection of any food identified as 'animal,' i.e., likely to lower the eater, etc.?" For such groups, definite methods can be discussed. Upon what ages of children, through what agencies, through what media should nutritionally valuable habits be urged? What sanctions should be used? How should new foods be presented or old, nutritionally undesirable foods be disparaged? Who should become the surrogate of the new nutrition knowledge, the mother, the teacher, the physician, the baseball hero, or moving picture actress, a puppet character like "Little Jackie" of the Dental Hygiene Division, South Carolina Public Health Service, the "government"? Should the emphasis be upon converting each individual to purposeful, careful eating or upon altering the style of American meals so that individuals will be well fed without having to exercise conscious and unremitting nutritional vigilance? (Mead 1943b; 1943c).

Asking questions like these shifts the deliberations from such questions as, "Isn't the radio a good medium, shouldn't it be used more?" or "Wouldn't prestige and status be the easiest way to influence people?" to precise problems which can be stated in such a way that they can be answered. These precise questions become, "Under what conditions will individuals, with a known character structure, with known attitudes toward and ideas about food, eating a known diet, and with other known behaviors which will be affected by and will affect their food habits, tend to resist or accept an alteration in food preparation, food content, food proportions?" The task of applied science then is to set up a program for controlling a social process in such a way that the desired changes will occur, instead of a program aimed at the reform of identified individuals.

While it is possible to predict upon the basis of our available knowledge the general lines which acceptance of or resistance to change will follow, the Committee has found it necessary, in making specific recommendations, to make current studies of American attitudes on such matters as reduction in meat supply, relationship between food and morale, and so on. The underlying attitudes which appear in these materials remain constant, and the studies are designed to reveal the situational impact of news, world events, and strictly contemporary conditions. A method of analysis of verbatim materials collected by the voluntary work of individuals and institutions throughout the country has been developed by Rhoda Métraux,[8] so that it is possible for the Committee to bring knowledge of basic cultural processes and American social psychological behavior to bear upon concrete problems. This method is not related to a Gallup poll. There is no intention of counting or getting an estimate of the number of individuals who would say they would be willing to drink more or less milk or eat soy bean products; the aim of the analysis is to show how existing situations are being interpreted by people of culturally standardized attitudes. So that by asking a question such as "How do you think people would feel if they were asked to eat more bread?" it is possible to work out whether more bread, as at present valued by Americans to whom it is not a principal food but to which an aura still attaches, will be regarded as a deprivation or an indulgence, a request to "eat something that is good for you," or a "reduction in the standard of living."

Recent developments in psychosomatic medicine have provided another approach to the problem. By studying the characteristic personality structure of individuals who manifest different marked gastrointestinal disorders or who come to clinics for treatment of obesity, food allergies and asthmas (Deutsch 1939), and anorexias (Rose 1943), it is possible to gain insight into the dynamics of the normal character development within our society and the role which food plays in that development. These extreme cases can serve as barometers of tendencies which appear, in less severe forms, in the general population. The delicate interrelationship between the way in which eating is culturally stylized and the expression of anxiety, fear, dependency, etc., can be shown from such detailed studies as Dr. Bruch's on obesity[9] and Drs. Wolff and Mittelman's on the way in which such conditions shift over time.

In addition to the long-time contributions to our understanding of the part which learning to eat plays in character development, clinical studies can be utilized as indicators of the immediate reactions of selected parts of the population to threats of food shortage, or drastic alterations in diet. Both Dr. Bruch's[10] recent researches and a cooperative project now under way with the Emotions and Food Therapy Section of the American Dietetic Association (Mead 1943d), in which food clinicians are making systematic observations of striking alterations in the behavior of clinic patients, are designed to tap this source of information. From psychiatric investigations we may also hope to get material on the more inarticulate bases of acceptances and rejections of food (Mead 1943e).

While it is clear that any final adjustment of the food habits of a nation to the current findings of nutrition can best be established by a basic alteration in the culturally defined style of what is a meal and what is food, exigencies of wartime conditions have made it necessary to resort to special measures to accomplish immediate changes and adjustments to shortages and substitutes. Here the Committee was confronted with an already established program of directed social change. The officials of the change were trained home economists, some 15,000 members of a profession which developed about 35 years ago, in which many of the pioneers are still alive and, despite a membership from all over the country, in which the ethos and occupational style is quite homogeneous. In all directed efforts to alter existing food habits, the home economist is in a key position, whether to give food demonstrations, calculate new menus to fit shortages, set up new methods of food preservation, direct the professional propagandist of newspaper or radio, or train the neighborhood leader who is to carry word of mouth messages into homes not reached by other media. Preparation of materials in a form which could be used by home economists and experiments in procedures which would facilitate their tasks have therefore been an essential part of the Committee's work. A series of experiments by Professor Lewin,[11] have been designed to demonstrate methods of group decision in which the home economist could function most perfectly as an expert, with a minimum call upon her for the performance of miscellaneous activities not as directly dependent upon her special training.[12] The interdependence of the method of instruction at the top of a chain of communication and the reception which the information will receive at the bottom is also of importance in implementing the nutritional program (Mead 1942a; Committee on Food Habits 1943).

Types of community organization most appropriate for the dissemination of information about food have also an important bearing on the wartime adjustment to food conditions and so the problem has been raised as to the relative efficiency of different forms. The study made by Mr. Koos at Cornell University Medical College in 1942[13] suggested that lines of friendship were not the best lines for the diffusion of nutrition information, and this finding was confirmed by the studies of Cussler and de Give (Cussler 1943). This resistance may be systematically related to the phrasing of changes of food habits in moral terms and the objection to the exploitation of pleasant friendship relationships to pass on morally sanctioned information. The study of Passin et al. (Bennett et al. 1942) also showed that exchange of recipes for festive dishes was the only food content of personal relationships in the Southern Illinois community which they studied. While the block plan[14] (Mittleman & Wolff 1942), which invests a neighbor with a government sanction of patriotism and patriotic license for intrusion into domestic affairs, is probably more suitable than friendship lines, here also nutrition information can be discussed more efficiently if the emphasis is upon adjustment of meals to wartime conditions rather than upon eating correctly, upon helping a woman to adjust a process rather than urging her to be good. The neighborhood and block leader may show almost as much resistance to passing on information to her neighbors as she does to passing on information to her friends if that information is cast in terms which suggest that she is trying to reform her neighbor. If, on the other hand, nutrition information comes as news of ways in which food shortages or food shifts can be dealt with efficiently, such information will be passed on much more willingly and effectively.

During the entire war and immediate post-war period it will be necessary to implement two tasks: to devise ways in which the health of the people may be maintained by the most skillful use of the existing food supplies; to present an increased utilization of knowledge of nutrition in such a way that it does not become associated with the wartime period of deprivation, to be rejected later (Joffe 1943b). The long-time task is to alter American food habits so that they are based upon tradition which embodies science and to do so in such a way that food habits at any period are sufficiently flexible to yield readily to new scientific findings. In order to accomplish this goal, the food habits of the future will have to be sanctioned not by authoritarian statements which breed rigid conformity rather than intelligent flexibility, but by a sense of responsibility on the part of those who plan meals for others to eat. At the same time it will be necessary to invent channels through which new findings can be readily translated into the meal planning of the woman on the farm, in the village, and in the city. To devise such a system of education, communication, and change which will link the daily habits of the people to the insight of the laboratory, and at the same time contribute to the development of a culture which produces individuals who are generally better adjusted as well as specifically better fed, is a task which requires a recognition of the total cultural equilibrium. Further, the application of science to improve eating habits may become empty and meaningless if it is not paralleled by efforts to apply science so as to increase the supply and adequate distribution of food. Efforts to better the nutrition of the world simply by altered production and distribution will also fall short of their goals unless corresponding and congruent changes are made in the patterns of consumption. The science of applied nutrition stands at the point (Mead 1943c) where a variety of techniques has been developed to deal with different aspects of the problem, and the most pressing problem is the integration of all of these techniques.

Notes

1. See "Social process and dietary change," by Herbert Passin and John W. Bennett on page 113 of this report.
2. See "Outline of studies on food habits in the rural southeast," by Margaret T. Cussler and Mary L. de Give on page 109 of this report.
3. See "Forces behind food habits and methods of change," by Kurt Lewin on page 35 of this report.
4. See "Tests of acceptability of emergency rations," by Natalie F. Joffe, on page 104 of this report.
5. For the interpretation of data demonstrating the importance of size and shape of the meal item, I am indebted to Mr. John B. Lansing.
6. See "A study of the effect of odd-shifts upon the food habits of war workers," by Gladys Engel-Frisch on page 82 of this report.
7. See "Forces behind food habits and methods of change," by Kurt Lewin on page 35 of this report.
8. See "Qualitative attitude analysis. A technique for the study of verbal behavior," by Rhoda Métraux on page 86 of this report.
9. See bibliography on page 72 of this report.
10. See "Adjustment to dietary changes in various somatic disorders," by Hilde Bruch and Marjorie Janis on page 66 of this report.
11. See "Forces behind food habits and methods of change," by Kurt Lewin on page 35 of this report.
12. See "Summary of the Liaison Sessions on 'The wartime roles of the nutritionist' and 'Supplementing the role of the nutritionist at the household level,' " on page 158 of this report.

13. See "A study of the use of the friendship pattern in nutrition education," by Earl L. Koos on page 74 of this report.
14. See "A summary of a study of some personality factors in block leaders in low income groups" on page 105 of this report.

References

Benet, Sula M., and Joffe, Natalie F. *Some Central European food patterns and their relationship to wartime problems of food and nutrition. Polish food patterns.* Washington, D. C., Committee on Food Habits, National Research Council (February 1943). 14 p. Mimeographed.

Bennett, John W., Smith, Harvey L., and Passin, Herbert. Food and culture in Southern Illinois. A preliminary report. *Am. Soc. Rev.,* 7: 645–660 (1942).

Committee on Food Habits, National Research Council. *The relationship between food habits and problems of wartime emergency feeding.* The Committee (May 1942). 9 p. Mimeographed.

Committee on Food Habits, National Research Council. *A study of some personality factors in block leaders in low income groups.* The Committee (June 1943). 20 p. Mimeographed.

Cussler, Margaret T. (a) Cultural sanctions of the food pattern in the rural south-east (1943). 354 p. (Unpublished thesis, Radcliffe College). (b) Chapter VIII.

Deutsch, Felix. The choice of organ in organ neurosis. *Internat. J. Psychoanalysis,* 20: nos. 3 and 4 (1939).

Dove, W. Franklin. *A study of the relation of man and animals to the environment.* Reprinted from the Annual Report of the Maine Agricultural Experiment Station for 1935. 18 p.

Dove, W. Franklin. *A study of the relation of man and animals to the environment* VI. Reprinted from the Annual Report of the Maine Agricultural Experiment Station for 1941, 441–458.

Dove, W. Franklin. On the linear arrangement of palatability of natural foods with an example of varietal preference in Leguminosae and Cruciferae by a new, rapid laboratory method. *J. Nutrition,* 25: 447–462 (1943).

Dubois, Cora. Attitudes toward food and hunger in Alor. *In* Spier L., and others, ed. *Language, culture, and personality: Essays in memory of Edward Sapir,* Menasha, Wisconsin, Sapir Memorial Publication Fund (1941), 272–281.

Festinger, Leon. Development of differential appetite in the rat. *J. Exper. Psychol.,* 32: 226–234 (1943).

Harris, R. H. *Effect of restaurant cooking and dispensation on the vitamin content of foods.* Washington, D. C., Committee on Nutrition in Industry, Food and Nutrition Board, National Research Council (September 1942). 8 p. Mimeographed. (A report of the Subcommittee on vitamin losses during the preparation of foods in industrial restaurants.)

Joffe, Natalie F. *Some Central European food patterns and their relationship to wartime programs of food and nutrition.* Washington, D. C., Committee on Food Habits, National Research Council (January 1943a), 3 p. Mimeographed.

Joffe, Natalie F. *Some Central European food patterns and their relationship to wartime problems of food and nutrition. Hungarian food patterns.* Washington, D. C., Committee on Food Habits, National Research Council (February 1943b) 10 p. Mimeographed.

Lewin, Kurt. A group test for determining the anchorage points of food habits. Washington, D. C., Committee on Food Habits, National Research Council (June 1942a) 21 p. Mimeographed.

Lewin, Kurt. *The relative effectiveness of a lecture method and a method of group decision for changing food habits.* Washington, D. C., Committee on Food Habits, National Research Council (June 1942b) 9 p. Mimeographed.

Malinowski, Bronislaw. *Coral gardens and their magic.* London, Allen & Unwin, (1935), 2 v.

Mead, Margaret. *The mountain Arapesh, an importing culture.* Anthropological Papers of the American Museum of Natural History, 36: 139–149 (1938).

Mead, Margaret. The problem of training the volunteer in community war work. *Sch. and Soc.,* 56: 520–522 (1942a).

Mead, Margaret. Reaching the last woman down the road. *J. Home Econ.,* 34: 710–713 (1942b).

Mead, Margaret. Changing food habits. *In The nutrition front, report of the New York State Joint Legislative Committee on Nutrition,* 37–43 (1943a).

Mead, Margaret. Dietary patterns and food habits. *J. Am. Diet. Assoc.,* 19: 1–5 (1943b).

Mead, Margaret. The factor of food habits. *Ann. Am. Acad. Polit. Soc. Science,* 225; 136–141 (1943c).

Mead, Margaret. Food therapy and wartime food problems. *J. Am. Diet. Assoc.,* 19: 201–202 (1943d).

Mead, Margaret. The problem of changing food habits: with suggestions for psychoanalytic contributions. *Bull. Menninger Clinic,* 7: 57–61 (1943e).

Mittelman, Bela, and Wolff, Harold G. Experimental studies on patients with gastritis, duodenitis and peptic ulcer. *Psychosom. Med.,* 4: 5–61 (1942).

Nagel, A. H., and Harris, R. S. Effect of restaurant cooking and service on vitamin content of foods. *J. Am. Diet. Assoc.,* 19: 23–25 (1943).

Nizzardini, G., and Joffe, Natalie F. *Italian food patterns and their relationship to wartime problems of food and nutrition.* Washington, D. C., Committee on Food Habits, National Research Council (August 1942). 22 p. Mimeographed.

Pirkova-Jakobson, Svatava, and Joffe, Natalie F. *Some Central European food patterns and their relationship to wartime problems of food and nutrition. Czech and Slovak food patterns.* Washington, D. C., Committee on Food Habits, National Research Council (February 1943). 14 p. Mimeographed.

Reder, Ruth, Chairman, Southern Cooperative Project, Dept. Agric. Chem. Res., Stillwater, Okla. *Ascorbic acid content of turnip greens II. Effect of length of time of cooking.* (National Cooperative Project Progress Notes.) 2 p. Mimeographed.

Richards, Audrey I. *Hunger and work in a savage tribe; a functional study of nutrition among Southern Bantu.* London, Routledge (1932). 238 p.

Richter, C. P. Increased salt appetite in adrenalectomized rats. *Am. J. Physiol.,* 115: 155–161 (1936).

Richter, C. P., and Eckert, John F., Increased calcium appetite of parathyroidectomized rats. *Endocrin.,* 21: 50–54 (1937).

Rose, John A. Eating inhibitions in children in relation to anorexia nervosa. *Psychosom. Med.,* 5: 117–124 (1943).

Sweeny, Mary. Changing food habits. *J. Home Econ.,* 34: 457–462 (1942).

2

Toward a Psychosociology of Contemporary Food Consumption

Roland Barthes

The inhabitants of the United States consume almost twice as much sugar as the French.[1] Such a fact is usually a concern of economics and politics. But this is by no means all. One needs only to take the step from sugar as merchandise, an abstract item in accounts, to sugar as food, a concrete item that is "eaten" rather than "consumed," to get an inkling of the (probably unexplored) depth of the phenomenon. For the Americans must do something with all that sugar. And as a matter of fact, anyone who has spent time in the United States knows that sugar permeates a considerable part of American cooking; that it saturates ordinarily sweet foods, such as pastries; makes for a great variety of sweets served, such as ice creams, jellies, syrups; and is used in many dishes that French people do not sweeten, such as meats, fish, salads, and relishes. This is something that would be of interest to scholars in fields other than economics, to the psychosociologist, for example, who will have something to say about the presumably invariable relation between standard of living and sugar consumption. (But is this relation really invariable today? And if so, why?)[2] It could be of interest to the historian also, who might find it worthwhile to study the ways in which the use of sugar evolved as part of American culture (the influence of Dutch and German immigrants who were used to "sweet-salty" cooking?). Nor is this all. Sugar is not just a foodstuff, even when it is used in conjunction with other foods; it is, if you will, an "attitude," bound to certain usages, certain "protocols," that have to do with more than food. Serving a sweet relish or drinking a Coca-Cola with a meal are things that are confined to eating habits proper; but to go regularly to a dairy bar, where the absence of alcohol coincides with a great abundance of sweet beverages, means more than to consume sugar; through the sugar, it also means to experience the day, periods of rest, traveling, and leisure in a specific fashion that is certain to have its impact on the American. For who would claim that in France wine is only wine? Sugar or wine, these two superabundant substances are also institutions. And these institutions necessarily imply a set of images, dreams, tastes, choices, and values. I remember an American hit song: *Sugar Time*. Sugar is a time, a category of the world.[3]

I have started out with the example of the American use of sugar because it permits us to get outside of what we, as Frenchmen, consider "obvious." For we do not see our own food or, worse, we assume that it is insignificant. Even—or perhaps especially—to the scholar, the subject of food connotes triviality or guilt.[4] This may explain in part why the psychosociology of French eating habits is still approached only indirectly and in passing when more weighty subjects, such as life-styles, budgets, and advertising, are under discussion. But at least the sociologists, the historians of the present—since we are talking only about contemporary eating habits here—and the economists are already aware that there is such a thing.

Thus P.H. Chombart de Lauve has made an excellent study of the behavior of French working-class families with respect to food. He was able to define areas of frustration and to outline some of the mechanisms by which needs are transformed into values, necessities into alibis.[5] In her book *Le mode de vie des familles bourgeoises de 1873 à 1953*, M. Perrot came to the conclusion that economic factors played a less important role in the changes that have taken place in middle-class food habits in the last hundred years than changing tastes; and this really means ideas, especially about nutrition.[6] Finally, the development of advertising has enabled the economists to become quite conscious of the ideal nature of consumer goods; by now everyone knows that the product as bought—that is, experienced—by the consumer is by no means the real product; between the former and the latter there is a considerable production of false perceptions and values. By being faithful to a certain brand and by justifying this loyalty with a set of "natural" reasons, the consumer gives diversity to products that are technically so identical that frequently even the manufacturer cannot find any differences. This is notably the case with most cooking oils.[7]

It is obvious that such deformations or reconstructions are not only the manifestation of individual, anomic prejudices, but also elements of a veritable collective imagination showing the outlines of a certain mental framework. All of this, we might say, points to the (necessary) widening of the very notion of food. For what is food? It is not only a collection of products that can be used for statistical or nutritional studies. It is also, and at the same time, a system of communication, a body of images, a protocol of usages, situations, and behavior. Information about food must be gathered wherever it can be found: by direct observation in the economy, in techniques, usages, and advertising; and by indirect observation in the mental life of a given society.[8] And once these data are assembled, they should no doubt be subjected to an internal analysis that should try to establish what is significant about the way in which they have been assembled before any economic or even ideological determinism is brought into play. I should like to give a brief outline of what such an analysis might be.

When he buys an item of food, consumes it, or serves it, modern man does not manipulate a simple object in a purely transitive fashion; this item of food sums up and transmits a situation; it constitutes an information; it signifies. That is to say that it is not just an indicator of a set of more or less conscious motivations, but that it is real sign, perhaps the functional unit of a system of communication. By this I mean not only the elements of *display* in food, such as foods involved in rites of hospitality,[9] for all food serves as a sign among the members of a given society. As soon as a need is satisfied by standardized production and consumption, in short, as soon as it takes on the characteristics of an institution, its function can no longer be dissociated from the sign of that function. This is true for clothing;[10] it is also true for food. No doubt, food is, anthropologically speaking (though very much in the abstract), the first need; but ever since man has ceased living off wild berries, this need has been highly structured. Substances, techniques of preparation, habits, all become part of a system of differences in signification; and as soon as this happens, we have communication by way of food. For the fact that there is communication is proven, not by the more or less vague consciousness that its users may have of it, but by the ease with which all the facts concerning food form a structure analogous to other systems of communication.[11] People may very well continue to believe that food is an immediate reality (necessity or pleasure), but this does not prevent it from carrying a system of communication; it would not be the first thing that people continue to experience as a simple function at the very moment when they constitute it into a sign.

If food is a system, what might be its constituent units? In order to find out, it would obviously be necessary to start out with a complete inventory of all we know of the food in a given society (products, techniques, habits), and then to subject these facts to what the linguists call transformational analysis, that is, to observe whether the passage from one fact to another produces a difference in signification. Here is an example: the changeover from ordinary bread to *pain de mie* involves a difference in what is signified: the former signifies day-to-day life, the latter a party. Similarly, in contemporary terms, the changeover from white to brown bread corresponds to a change in what is signified in social terms, because, paradoxically, brown bread has become a sign of refinement. We are therefore justified in considering the varieties of bread as units of signification—at least these varieties—for the same test can also show that there are insignificant varieties as well, whose use has nothing to do with a collective institution, but simply with individual taste. In this manner, one could, proceeding step by step, make a compendium of the differences in signification regulating the system of our food. In other words, it would be a matter of separating the significant from the insignificant and then of reconstructing the differential system of signification by constructing, if I may be permitted to use such a metaphor, a veritable grammar of foods.

It must be added that the units of our system would probably coincide only rarely with the products in current use in the economy. Within French society, for example, bread as such does not constitute a signifying unit: in order to find these we must go further and look for certain of its varieties. In other words, these signifying units are more subtle than the commercial units and, above all, they have to do with subdivisions with which production is not concerned, so that the sense of the sub-division can differentiate a single product. Thus it is not at the level of its cost that the sense of a food item is elaborated, but at the level of its preparation and use. There is perhaps no natural item of food that signifies anything in itself, except for a few deluxe items such as salmon, caviar, truffles, and so on, whose preparation is less important than their absolute cost.

If the units of our system of food are not the *products* of our economy, can we at least have some preliminary idea of what they might be? In the absence of a systematic inventory, we may risk a few hypotheses. A study by P.F. Lazarsfeld[12] (it is old, concerned with particulars, and I cite it only as an example) has shown that certain sensorial "tastes" can vary according to the income level of the social groups interviewed: lower-income persons like sweet chocolates, smooth materials, strong perfumes; the upper classes, on the other hand, prefer bitter substances, irregular materials, and light perfumes. To remain within the area of food, we can see that signification (which, itself, refers to a twofold social phenomenon: upper classes/ lower classes) does not involve kinds of products, but flavors: *sweet* and *bitter* make up the opposition in signification, so that we must place certain units of the system of food on that level. We can imagine other classes of units, for example, opposite substances such as dry, creamy, watery ones, which immediately show their great psychoanalytical potential (and it is obvious that if the subject of food had not been so trivialized and invested with guilt, it could easily be subjected to the kind of "poetic" analysis that G. Bachelard applied to language). As for what is considered tasty, C. Lévi-Strauss has already shown that this might very well constitute a class of oppositions that refers to national characters (French versus English cuisine, French versus Chinese or German cuisine, and so on).[13]

Finally, one can imagine opposites that are even more encompassing, but also more subtle. Why not speak, if the facts are sufficiently numerous and sufficiently clear, of a certain "spirit" of food, if I may be permitted to use this romantic term? By this I mean that a coherent set of food traits and habits can constitute a complex but homogeneous dominant feature useful for defining a general system of tastes and habits. This "spirit" brings together different units (such as flavor or substance), forming a composite unit with a single signification, somewhat analogous to the suprasegmental prosodic units of language. I should like to suggest here two very different examples. The ancient Greeks unified in a single (euphoric) notion the ideas of succulence, brightness, and moistness, and they called it *yávos*. Honey had *yávos*, and wine was the *yávos* of the vineyard.[14] Now this would certainly be a signifying unit if we were to establish the system of food of the Greeks, even though it does not refer to any particular item. And here is another example, modern this time. In the United States, the Americans seem to oppose the category of sweet (and we have already seen to how many different varieties of foods this applies) with an equally general category that is not, however, that of salty—understandably so, since their food is salty and sweet to begin with—but that of *crisp* or *crispy*. *Crisp* designates everything that crunches, crackles, grates, sparkles, from potato chips to certain brands of beer; *crisp*—and this shows that the unit of food can overthrow logical categories—*crisp* may be applied to a product just because it is ice cold, to another because it is sour, to a third because it is brittle. Quite obviously, such a notion goes beyond the purely physical nature of the product; *crispness* in a food designates an almost magical quality, a certain briskness and sharpness, as opposed to the soft, soothing character of sweet foods.

Now then, how will we use the units established in this manner? We will use them to reconstruct systems, syntaxes ("menus"), and styles ("diets")[15] no longer in an empirical but in a semantic way—in a way, that is, that will enable us to compare them to each other. We now must show, not that which is, but that which signifies. Why? Because we are interested in human communication and because communication always implies a system of signification, that is, a body of discrete signs standing out from a mass of indifferent materials. For this reason, sociology must, as soon as it deals with cultural "objects" such as clothing, food, and—not quite as clearly— housing, structure these objects before trying to find out what society does with them. For what society does with them is precisely to structure them in order to make use of them.

To what, then, can these significations of food refer? As I have already pointed out, they refer not only to display,[16] but to a much larger set of themes and situations. One could say that an entire "world" (social environment) is present in and signified by food. Today we have a tool with which to isolate these themes and situations, namely, advertising. There is no question that advertising provides only a projected image of reality; but the sociology of mass communication has become increasingly inclined to think that large-scale advertising, even though technically the work of a particular group, reflects the collective psychology much more than it shapes it. Furthermore, studies of motivation are now so advanced that it is possible to analyze cases in which the response of the public is negative. (I already mentioned the feelings of guilt fostered by an advertising for sugar which emphasized pure enjoyment. It was bad advertising, but the response of the public was nonetheless psychologically most interesting.)

A rapid glance at food advertising permits us rather easily, I think, to identify three groups of themes. The first of these assigns to food a function that is, in some sense, commemorative: food permits a person (and I am here speaking of French themes) to partake each day of the national past. In this case, this historical quality is obviously linked to food techniques (preparation and cooking). These have long roots, reaching back to the depth of the French past. They are, we are told, the repository of a whole experience, of the accumulated wisdom of our ancestors. French food is never supposed to be innovative, except when it rediscovers long-forgotten secrets. The historical theme, which was so often sounded in our advertising, mobilizes two different values: on the one hand, it implies an aristocratic tradition (dynasties of manufacturers, *moutarde du Roy*, the Brandy of Napoleon); on the other hand, food frequently carries notions of representing the flavorful survival of an old, rural society that is itself highly idealized.[17] In this manner, food brings the memory of the soil into our very contemporary life; hence the paradoxical association of gastronomy and industrialization in the form of canned "gourmet dishes." No doubt the myth of French cooking abroad (or as expressed to foreigners) strengthens this "nostalgic" value of food considerably; but since the French themselves actively participate in this myth (especially when traveling), it is fair to say that through his food the Frenchman experiences a certain national continuity. By way of a thousand detours, food permits him to insert himself daily into his own past and to believe in a certain culinary "being" of France.[18]

A second group of values concerns what we might call the anthropological situation of the French consumer. Motivation studies have shown that feelings of inferiority were attached to certain foods and that people therefore abstained from them.[19] For example, there are supposed to be masculine and feminine kinds of food. Furthermore, visual advertising makes it possible to associate certain kinds of foods with images connoting a sublimated sexuality. In a certain sense, advertising eroticizes food and thereby transforms our consciousness of it, bringing it into a new sphere of situations by means of a pseudocausal relationship.

Finally, a third area of consciousness is constituted by a whole set of ambiguous values of a somatic as well as psychic nature, clustering around the concept of *health*.

In a mythical way, health is indeed a simple relay midway between the body and the mind; it is the alibi food gives to itself in order to signify materially a pattern of immaterial realities. Health is thus experienced through food only in the form of "conditioning," which implies that the body is able to cope with a certain number of day-to-day situations. Conditioning originates with the body but goes beyond it. It produces *energy* (sugar, the "powerhouse of foods," at least in France, maintains an "uninterrupted flow of energy"; margarine "builds solid muscles"; coffee "dissolves fatigue"); *alertness* ("Be alert with Lustucru"), and *relaxation* (coffee, mineral water, fruit juices, Coca-Cola, and so on). In this manner, food does indeed retain its physiological function by giving strength to the organism, but this strength is immediately sublimated and placed into a specific situation (I shall come back to this in a moment). This situation may be one of conquest (alertness, aggressiveness) or a response to the stress of modern life (relaxation). No doubt, the existence of such themes is related to the spectacular development of the science of nutrition, to which, as we have seen, one historian unequivocally attributes the evolution of food budgets over the last fifty years. It seems, then, that the acceptance of this new value by the masses has brought about a new phenomenon, which must be the first item of study in any psychosociology of food: it is what might be called nutritional consciousness. In the developed countries, food is henceforth *thought out*, not by specialists, but by the entire public, even if this thinking is done within a framework of highly mythical notions. Nor is this all. This nutritional rationalizing is aimed in a specific direction. Modern nutritional science (at least according to what can be observed in France) is not bound to any moral values, such as asceticism, wisdom, or purity,[20] but on the contrary, to values of *power*. The energy furnished by a consciously worked out diet is mythically directed, it seems, toward an adaptation of man to the modern world. In the final analysis, therefore, a representation of contemporary existence is implied in the consciousness we have of the function of our food.[21]

For, as we said before, food serves as a sign not only for themes, but also for situations; and this, all told, means for a way of life that is emphasized, much more than expressed, by it. To eat is a behavior that develops beyond its own ends, replacing, summing up, and signalizing other behaviors, and it is precisely for these reasons that it is a sign. What are these other behaviors? Today, we might say all of them: activity, work, sports, effort, leisure, celebration—every one of these situations is expressed through food. We might almost say that this "polysemia" of food charac-terizes modernity; in the past, only festive occasions were signalized by food in any positive and organized manner. But today, work also has its own kind of food (on the level of a sign, that is): energy-giving and light food is experienced as the very sign of, rather than only a help toward, participation in modern life. The snack bar not only responds to a new need, it also gives a certain dramatic expression to this need and shows those who frequent it to be modern men, managers who exercise power and control over the extreme rapidity of modern life. Let us say that there is an element of "Napoleonism" in this ritually condensed, light, and rapid kind of eating. On the level of institutions, there is also the business lunch, a very different kind of thing, which has become commercialized in the form of special menus: here, on the con-trary, the emphasis is placed on comfort and long discussions; there even remains a trace of the mythical conciliatory power of conviviality. Hence, the business lunch emphasizes the gastronomic, and under certain circumstances traditional, value of the dishes served and uses this value to stimulate the euphoria needed to facilitate the transaction of business. Snack bar and business lunch are two very closely related work situations, yet the food connected with them signalizes their differences in a perfectly readable manner. We can imagine many others that should be catalogued.

This much can be said already: today, at least in France, we are witnessing an extraordinary expansion of the areas associated with food: food is becoming incorporated into an ever-lengthening list of situations. This adaptation is usually made in the name of hygiene and better living, but in reality, to stress this fact once more, food is also charged with signifying the situation in which it is used. It has a twofold value, being nutrition as well as protocol, and its value as protocol becomes increasingly more important as soon as the basic needs are satisfied, as they are in France. In other words, we might say that in contemporary French society *food has a constant tendency to transform itself into situation.*

There is no better illustration for this trend than the advertising mythology about coffee. For centuries, coffee was considered a stimulant to the nervous system (recall that Michelet claimed that it led to the Revolution), but contemporary advertising, while not expressly denying this traditional function, paradoxically associates it more and more with images of "breaks," rest, and even relaxation. What is the reason for this shift? It is that coffee is felt to be not so much a substance[22] as a circumstance. It is the recognized occasion for interrupting work and using this respite in a precise protocol of taking sustenance. It stands to reason that if this transferral of the food substance to its use becomes really all-encompassing, the power signification of food will be vastly increased. Food, in short, will lose in substance and gain in function; this function will be general and point to activity (such as the business lunch) or to times of rest (such as coffee); but since there is a very marked opposition between work and relaxation, the traditionally festive function of food is apt to disappear gradually, and society will arrange the signifying system of its food around two major focal points: on the one hand, activity (and no longer work), and on the other hand, leisure (no longer celebration). All of this goes to show, if indeed it needs to be shown, to what extent food is an organic system, organically integrated into its specific type of civilization.

Notes

This article originally appeared in "Vers une psycho-sociologie de l'alimentation moderne" by Roland Barthes, in *Annales: Économies, Sociétés, Civilisations* no. 5 (September–October 1961), pp. 977–986. Reprinted by permission of *Annales.*

1. Annual sugar consumption in the United States is 43 kg. per person; in France 25 kg. per person.
2. F. Charny, *Le sucre*, Collection "Que sais-je?" (Paris: P. U. F., 1950), p. 8.
3. I do not wish to deal here with the problem of sugar "metaphors" or paradoxes, such as the "sweet" rock singers or the sweet milk beverages of certain "toughs."
4. Motivation studies have shown that food advertisements openly based on enjoyment are apt to fail, since they make the reader feel guilty (J. Marcus-Steiff, *Les études de motivation* [Paris: Hermann, 1961], pp. 44–45).
5. P. H. Chombart de Lauwe, *La vie quotidienne des familles ouvrières* (Paris: C.N.R.S., 1956).
6. Marguerite Perrot, *Le mode de vie des familles bourgeoises, 1873–1953* (Paris: Colin, 1961). "Since the end of the nineteenth century, there has been a very marked evolution in the dietary habits of the middle-class families we

have investigated in this study. This evolution seems related, not to a change in the standard of living, but rather to a transformation of individual tastes under the influence of a greater awareness of the rules of nutrition" (p. 292).

7. J. Marcus-Steiff, *Les études de motivation*, p. 28.

8. On the latest techniques of investigation, see again J. Marcus-Steiff, *Les études de motivation*.

9. Yet on this point alone, there are many known facts that should be assembled and systematized: cocktail parties, formal dinners, degrees and kinds of display by way of food according to the different social groups.

10. R. Barthes, "Le bleu est à la mode cette année: Note sur la recherche des unités signifiantes dans le vêtement de mode," *Revue française de sociologie* 1 (1960): 147–162.

11. I am using the word *structure* in the sense that it has in linguistics: "an autonomous entity of internal dependencies" (L. Hjelnislev, *Essais linguistiques* [Copenhagen, 1959], p. 1).

12. P. F. Lazarsfeld, "The Psychological Aspect of Market Research," *Harvard Business Review* 13 (1934): 54–71.

13. C. Lévi-Strauss, *Anthropologie structurale* (Paris: Plon, 1958), p. 99.

14. H. Jeanmaire, *Dionysos* (Paris: Payot), p. 510.

15. In a semantic analysis, vegetarianism, for example (at least at the level of specialized restaurants), would appear as an attempt to copy the appearance of meat dishes by means of a series of artifices that are somewhat similar to "costume jewelry" in clothing, at least the jewelry that is meant to be seen as such.

16. The idea of social *display* must not be associated purely and simply with vanity; the analysis of motivation, when conducted by indirect questioning, reveals that worry about appearances is part of an extremely subtle reaction and that social strictures are very strong, even with respect to food.

17. The expression *cuisine bourgeoise*, used at first in a literal, then in a metaphoric way, seems to be gradually disappearing while the "peasant stew" is periodically featured in the photographic pages of the major ladies' magazines.

18. The exotic nature of food can, of course, be a value, but in the French public at large, it seems limited to coffee (tropical) and pasta (Italian).

19. This would be the place to ask just what is meant by "strong" food. Obviously, there is no psychic quality inherent in the thing itself. A food becomes "masculine" as soon as women, children, and old people, for nutritional (and thus fairly historical) reasons, do not consume it.

20. We need only to compare the development of vegetarianism in England and France.

21. Right now, in France, there is a conflict between traditional (gastronomic) and modern (nutritional) values.

22. It seems that this stimulating, re-energizing power is now assigned to sugar, at least in France.

3

The Culinary Triangle

Claude Lévi-Strauss

Linguistics has familiarized us with concepts like "minimum vocalism" and "minimum consonantism" which refer to systems of oppositions between phonemes of so elementary a nature that every known or unknown language supposes them; they are in fact also the first oppositions to appear in the child's language, and the last to disappear in the speech of people affected by certain forms of aphasia.

The two concepts are moreover not really distinct, since, according to linguists, for every language the fundamental opposition is that between consonant and vowel. The subsequent distinctions among vowels and among consonants result from the application to these derived areas of such contrasts as compact and diffuse, open and closed, acute and grave.

Hence, in all the languages of the world, complex systems of oppositions among phonemes do nothing but elaborate in multiple directions a simpler system common to them all: the contrast between consonant and vowel which, by the workings of a double opposition between compact and diffuse, acute and grave, produces on the one hand what has been called the "vowel triangle":[1]

$$a$$
$$u \qquad i$$

and on the other hand the "consonant triangle":

$$k$$
$$p \qquad t$$

It would seem that the methodological principle which inspires such distinctions is transposable to other domains, notably that of cooking which, it has never been sufficiently emphasized, is with language a truly universal form of human activity: if there is no society without a language, nor is there any which does not cook in some manner at least some of its food.

We will start from the hypothesis that this activity supposes a system which is located—according to very difficult modalities in function of the particular cultures one wants to consider—within a triangular semantic field whose three points correspond respectively to the categories of the raw, the cooked and the rotted. It is clear that in respect to cooking the raw constitutes the unmarked pole, while the other two poles are strongly marked, but in different directions: indeed, the cooked is a cultural transformation of the raw, whereas the rotted is a natural transformation. Underlying our original triangle, there is hence a double opposition between *elaborated/unelaborated* on the one hand, and *culture/nature* on the other.

No doubt these notions constitute empty forms: they teach us nothing about the cooking of any specific society, since only observation can tell us what each one means by "raw," "cooked" and "rotted," and we can suppose that it will not be the same for all. Italian cuisine has recently taught us to eat *crudités* rawer than any in traditional French cooking, thereby determining an enlargement of the category of the raw. And we know from some incidents that followed the Allied landings in 1944 that American soldiers conceived the category of the rotted in more extended fashion than we, since the odor given off by Norman cheese dairies seemed to them the smell of corpses, and occasionally prompted them to destroy the dairies.

Consequently, the culinary triangle delimits a semantic field, but from the outside. This is moreover true of the linguistic triangles as well, since there are no phonemes *a, i, u* (or *k, p, t*) in general, and these ideal positions must be occupied, in each language, by the particular phonemes whose distinctive natures are closest to those for which we first gave a symbolic representation: thus we have a sort of concrete triangle inscribed within the abstract triangle. In any cuisine, nothing is simply cooked, but must be cooked in one fashion or another. Nor is there any condition of pure rawness: only certain foods can really be eaten raw, and then only if they have been selected, washed, pared or cut, or even seasoned. Rotting, too, is only allowed to take place in certain specific ways, either spontaneous or controlled.

Let us now consider, for those cuisines whose categories are relatively well-known, the different modes of cooking. There are certainly two principal modes, attested in innumerable societies by myths and rites which emphasize their contrast: the roasted and the boiled. In what does their difference consist? Roasted food is directly exposed to the fire; with the fire it realizes an unmediated conjunction, whereas boiled food is doubly mediated, by the water in which it is immersed, and by the receptacle that holds both water and food.

On two grounds, then, one can say that the roasted is on the side of nature, the boiled on the side of culture: literally, because boiling requires the use of a receptacle, a cultural object; symbolically, in as much as culture is a mediation of the relations between man and the world, and boiling demands a mediation (by water) of the relation between food and fire which is absent in roasting.

The natives of New Caledonia feel this contrast with particular vividness: "Formerly," relates M. J. Barrau, "they only grilled and roasted, they only 'burned' as the natives now say . . . The use of a pot and the consumption of boiled tubers are looked upon with pride . . . as a proof of . . . civilization."

A text of Aristotle, cited by Salomon Reinach (*Cultes, Mythes, Religions*, V, p. 63), indicates that the Greeks also thought that "in ancient times, men roasted everything."

Behind the opposition between roasted and boiled, then, we do in fact find, as we postulated at the outset, the opposition between nature and culture. It remains to discover the other fundamental opposition which we put forth: that between elaborated and unelaborated.

In this respect, observation establishes a double affinity: the roasted with the raw, that is to say the unelaborated, and the boiled with the rotted, which is one of the two modes of the elaborated. The affinity of the roasted with the raw comes from the fact that it is never uniformly cooked, whether this be on all sides, or on the outside and the inside. A myth of the Wyandot Indians well evokes what might be called the paradox of the roasted: the Creator struck fire, and ordered the first man to skewer a piece of meat on a stick and roast it. But man was so ignorant that he left the meat on the fire until it was black on one side, and still raw on the other ... Similarly, the Poconachi of Mexico interpret the roasted as a compromise between the raw and the burned. After the universal fire, they relate, that which had not been burned became white, that which had been burned turned black, and what had only been singed turned red. This explanation accounts for the various colors of corn and beans. In British Guiana, the Waiwai sorcerer must respect two taboos, one directed at roast meat, the other red paint, and this again puts the roasted on the side of blood and the raw.

If boiling is superior to roasting, notes Aristotle, it is because it takes away the rawness of meat, "roast meats being rawer and drier than boiled meats" (quoted by Reinach, *loc. cit.*).

As for the boiled, its affinity with the rotted is attested in numerous European languages by such locutions as *pot pourri, olla podrida*, denoting different sorts of meat seasoned and cooked together with vegetables; and in German, *zu Brei zerkochetes Fleisch*, "meat rotted from cooking." American Indian languages emphasize the same affinity, and it is significant that this should be so especially in those tribes that show a strong taste for gamey meat, to the point of preferring, for example, the flesh of a dead animal whose carcass has been washed down by the stream to that of a freshly-killed buffalo. In the Dakota language, the same stem connotes putrefaction and the fact of boiling pieces of meat together with some additive.

These distinctions are far from exhausting the richness and complexity of the contrast between roasted and boiled. The boiled is cooked within a receptacle, while the roasted is cooked from without: the former thus evokes the concave, the latter the convex. Also the boiled can most often be ascribed to what might be called an "endo-cuisine," prepared for domestic use, destined to a small closed group, while the roasted belongs to "exo-cuisine," that which one offers to guests. Formerly in France, boiled chicken was for the family meal, while roasted meat was for the banquet (and marked its culminating point, served as it was after the boiled meats and vegetables of the first course, and accompanied by "extraordinary fruits" such as melons, oranges, olives and capers).

The same opposition is found, differently formulated, in exotic societies. The extremely primitive Guayaki of Paraguay roast all their game, except when they prepare the meat destined for the rites which determine the name of a new child: this meat must be boiled. The Caingang of Brazil prohibit boiled meat for the widow and widower, and also for anyone who has murdered an enemy. In all these cases, prescription of the boiled accompanies a tightening, prescription of the roasted a loosening of familial or social ties.

Following this line of argument, one could infer that cannibalism (which by definition is an endo-cuisine in respect to the human race) ordinarily employs boiling rather than roasting, and that the cases where bodies are roasted—cases vouched for by ethnographic literature—must be more frequent in exo-cannibalism (eating the body of an enemy) than in endo-cannibalism (eating a relative). It would be interesting to carry out statistical research on this point.

Sometimes, too, as is often the case in America, and doubtless elsewhere, the roasted and the boiled will have respective affinities with life in the bush (outside the village community) and sedentary life (inside the village). From this comes a subsidiary association of the roasted with men, the boiled with women. This is notably the case with the Trumai, the Yagua and the Jivaro of South America, and with the Ingalik of Alaska. Or else the relation is reversed: the Assiniboin, on the northern plains of North America, reserve the preparation of boiled food for men engaged in a war expedition, while the women in the villages never use receptacles, and only roast their meat. There are some indications that in certain Eastern European countries one can find the same inversion of affinities between roasted and boiled and feminine and masculine.

The existence of these inverted systems naturally poses a problem, and leads one to think that the axes of opposition are still more numerous than one suspected, and that the peoples where these inversions exist refer to axes different from those we at first singled out. For example, boiling conserves entirely the meat and its juices, whereas roasting is accompanied by destruction and loss. One connotes economy, the other prodigality; the former is plebeian, the latter aristocratic. This aspect takes on primary importance in societies which prescribe differences of status among individuals or groups. In the ancient Maori, says Prytz-Johansen, a noble could himself roast his food, but he avoided all contact with the steaming oven, which was left to the slaves and women of low birth. Thus, when pots and pans were introduced by the whites, they seemed infected utensils; a striking inversion of the attitude which we remarked in the New Caledonians.

These differences in appraisal of the boiled and the roasted, dependent on the democratic or aristocratic perspective of the group, can also be found in the Western tradition. The democratic Encyclopedia of Diderot and d'Alembert goes in for a veritable apology of the boiled: "Boiled meat is one of the most succulent and nourishing foods known to man. . . . One could say that boiled meat is to other dishes as bread is to other kinds of nourishment" (Article "Bouilli"). A half-century later, the dandy Brillat-Savarin will take precisely the opposite view: "We professors never eat boiled meat out of respect for principle, and because we have pronounced *ex cathedra* this incontestable truth: boiled meat is flesh without its juice. . . . This truth is beginning to become accepted, and boiled meat has disappeared in truly elegant dinners; it has been replaced by a roast filet, a turbot, or a matelote" (*Physiologie du goût*, VI, ¶2).

Therefore if the Czechs see in boiled meat a man's nourishment, it is perhaps because their traditional society was of a much more democratic character than that of their Slavonic and Polish neighbors. One could interpret in the same manner distinctions made—respectively by the Greeks, and the Romans and the Hebrews— on the basis of attitudes toward roasted and boiled, distinctions which have been noted by M. Piganiol in a recent article ("Le rôti et le bouilli," *A Pedro Bosch-Gimpera*, Mexico City, 1963).

Other societies make use of the same opposition in a completely different direction. Because boiling takes place without loss of substance, and within a complete enclosure, it is eminently apt to symbolize cosmic totality. In Guiana as well as in the Great Lakes region, it is thought that if the pot where game is boiling were to overflow even a little bit, all the animals of the species being cooked would migrate, and the hunter would catch nothing more. The boiled is life, the roasted death. Does not world folklore offer innumerable examples of the cauldron of immortality? But there has never been a spit of immortality. A Cree Indian rite admirably expresses this character of cosmic totality ascribed to boiled food. According to them, the first man was commanded by the Creator to boil the first berries gathered each season. The cup containing the berries was first presented to the sun, that it might fulfill its office and ripen the berries; then the cup was lifted to the thunder, whence rain is expected; finally the cup was lowered toward the earth, in prayer that it bring forth its fruits.

Hence we rejoin the symbolism of the most distant Indo-European past, as it has been reconstructed by Georges Dumézil: "To Mitra belongs that which breaks of itself, that which is cooked in steam, that which is well sacrificed, milk . . . and to Varuna that which is cut with the axe, that which is snatched from the fire, that which is ill-sacrificed, the intoxicating soma" (*Les dieux des Germains*, p. 60). It is not a little surprising—but highly significant—to find intact in genial mid-nineteenth-century philosophers of cuisine a consciousness of the same contrast between knowledge and inspiration, serenity and violence, measure and lack of measure, still symbolized by the opposition of the boiled and the roasted: "One becomes a cook but one is born a roaster" (Brillat-Savarin); "Roasting is at the same time nothing, and an immensity" (Marquis de Cussy).

Within the basic culinary triangle formed by the categories of raw, cooked and rotted, we have, then, inscribed two terms which are situated: one, the roasted, in the vicinity of the raw; the other, the boiled, near the rotted. We are lacking a third term, illustrating the concrete form of cooking showing the greatest affinity to the abstract category of the cooked. This form seems to us to be smoking, which like roasting implies an unmediated operation (without receptacle and without water) but differs from roasting in that it is, like boiling, a slow form of cooking, both uniform and penetrating in depth.

Let us try to determine the place of this new term in our system of opposition. In the technique of smoking, as in that of roasting, nothing is interposed between meat and fire except air. But the difference between the two techniques comes from the fact that in one the layer of air is reduced to a minimum, whereas in the other it is brought to a maximum. To smoke game, the American Indians (in whose culinary system smoking occupies a particularly important place) construct a wooden frame (a buccan) about five feet high, on top of which they place the meat, while underneath they light a very small fire which is kept burning for forty-eight hours or more. Hence for one constant—the presence of a layer of air—we note two differentials which are expressed by the opposition *close/distant* and *rapid/slow*. A third differential is created by the absence of a utensil in the case of roasting (any stick doing the work of a spit), since the buccan is a constructed framework, that is, a cultural object.

In this last respect, smoking is related to boiling, which also requires a cultural means, the receptacle. But between these two utensils a remarkable difference appears, or more accurately, is instituted by the culture precisely in order, it seems, to create the opposition, which without such a difference might have remained too ill-defined to take on meaning. Pots and pans are carefully cared for and preserved utensils, which one cleans and puts away after use in order to make them serve their purpose as many times as possible; but the buccan *must be destroyed immediately after use*, otherwise the animal will avenge itself, and come in turn to smoke the huntsman. Such, at least, is the belief of those same natives of Guiana whose other symmetrical belief we have already noted: that a poorly conducted boiling, during which the cauldron overflowed, would bring the inverse punishment, flight of the quarry, which the huntsman would no longer succeed in overtaking. On the other hand, as we have already indicated, it is clear that the boiled is opposed both to the smoked and the roasted in respect to the presence or absence of water.

But let us come back for a moment to the opposition between a perishable and a durable utensil which we found in Guiana in connection with smoking and boiling. It will allow us to resolve an apparent difficulty in our system, one which doubtless has not escaped the reader. At the start we characterized one of the oppositions between the roasted and the boiled as reflecting that between nature and culture. Later, however, we proposed an affinity between the boiled and the rotted, the latter defined as the elaboration of the raw by natural means. Is it not contradictory that a cultural method should lead to a natural result? To put it in other terms, what, philosophically, will be the value of the invention of pottery (and hence of culture) if the native's system associates boiling and putrefaction, which is the condition that raw food cannot help but reach spontaneously in the state of nature?

The same type of paradox is implied by the problematics of smoking as formulated by the natives of Guiana. On the one hand, smoking, of all the modes of cooking, comes closest to the abstract category of the cooked; and—since the opposition between raw and cooked is homologous to that between nature and culture—it represents the most "cultural" form of cooking (and also that most esteemed among the natives). And yet, on the other hand, its cultural means, the buccan, is to be immediately destroyed. There is striking parallel to boiling, a method whose cultural means (the receptacles) are preserved, but which is itself assimilated to a sort of process of auto-annihilation, since its definitive result is at least verbally equivalent to that putrefaction which cooking should prevent or retard.

What is the profound sense of this parallelism? In so-called primitive societies, cooking by water and smoking have this in common: one as to its means, the other as to its results, is marked by duration. Cooking by water operates by means of receptacles made of pottery (or of wood with peoples who do not know about pottery, but boil water by immersing hot stones in it): in all cases these receptacles are cared for and repaired, sometimes passed on from generation to generation, and they number among the most durable cultural objects. As for smoking, it gives food that resists spoiling incomparably longer than that cooked by any other method. Everything transpires as if the lasting possession of a cultural acquisition entailed, sometimes in the ritual realm, sometimes in the mythic, a concession made in return to nature: when the result is durable, the means must be precarious, and vice-versa.

This ambiguity, which marks similarly, but in different directions, both the smoked and the boiled, is that same ambiguity which we already know to be inherent to the roasted. Burned on one side and raw on the other, or grilled outside, raw within, the roasted incarnates the ambiguity of the raw and the cooked, of nature and culture, which the smoked and the boiled must illustrate in their turn for the structure to be coherent. But what forces them into this pattern is not purely a reason of form: hence the system demonstrates that the art of cooking is not located entirely on the side of culture. Adapting itself to the exigencies of the body, and determined in its modes by the way man's insertion in nature operates in different parts of the world, placed then between nature and culture, cooking rather represents their necessary articulation. It partakes of both domains, and projects this duality on each of its manifestations.

But it cannot always do so in the same manner. The ambiguity of the roasted is intrinsic, that of the smoked and the boiled extrinsic, since it does not derive from things themselves, but from the way one speaks about them or behaves toward them. For here again a distinction becomes necessary: the quality of naturalness which language confers upon boiled food is purely metaphorical: the "boiled" is not the "spoiled"; it simply resembles it. Inversely, the transfiguration of the smoked into a natural entity does not result from the nonexistence of the buccan, the cultural instrument, but from its voluntary destruction. This transfiguration is thus on the order of metonymy, since it consists in acting as if the effect were really the cause. Consequently, even when the structure is added to or transformed to overcome a disequilibrium, it is only at the price of a new disequilibrium which manifests itself in another domain. To this ineluctable dissymmetry the structure owes its ability to engender myth, which is nothing other than an effort to correct or hide its inherent dissymmetry.

To conclude, let us return to our culinary triangle. Within it we traced another triangle representing recipes, at least the most elementary ones: roasting, boiling and smoking. The smoked and the boiled are opposed as to the nature of the intermediate element between fire and food, which is either air or water. The smoked and the roasted are opposed by the smaller or larger place given to the element air; and the roasted and the boiled by the presence or absence of water. The boundary between nature and culture, which one can imagine as parallel to either the axis of air or the axis of water, puts the roasted and the smoked on the side of nature, the boiled on the side of culture as to means; or, as to results, the smoked on the side of culture, the roasted and the boiled on the side of nature:

$$
\begin{array}{cc}
 & \text{RAW} \\
 & \text{roasted} \\
(-) & \quad (-) \\
\textit{Air} & \quad \textit{Water} \\
(+) & \quad (+) \\
\text{smoked} & \quad \text{boiled} \\
\text{COOKED} & \quad \text{ROTTED}
\end{array}
$$

The operational value of our diagram would be very restricted did it not lend itself to all the transformations necessary to admit other categories of cooking. In a culinary system where the category of the roasted is divided into roasted and grilled, it is the latter term (connoting the lesser distance of meat from fire) which will be situated at the apex of the recipe triangle, the roasted then being placed, still on the air-axis, halfway between the grilled and the smoked. We may proceed in similar fashion if the culinary system in question makes a distinction between cooking with water and cooking with steam; the latter, where the water is at a distance from the food, will be located halfway between the boiled and the smoked.

A more complex transformation will be necessary to introduce the category of the fried. A tetrahedron will replace the recipe triangle, making it possible to raise a third axis, that of oil, in addition to those of air and water. The grilled will remain at the apex, but in the middle of the edge joining smoked and fried one can place roasted-in-the-oven (with the addition of fat), which is opposed to roasted-on-the-spit (without this addition). Similarly, on the edge running from fried to boiled will be braising (in a base of water and fat), opposed to steaming (without fat, and at a distance from the water). The plan can be still further developed, if necessary, by addition of the opposition between animal and vegetable foodstuffs (if they entail differentiating methods of cooking), and by the distinction of vegetable foods into cereals and legumes, since unlike the former (which one can simply grill), the latter cannot be cooked without water or fat, or both (unless one were to let the cereals ferment, which requires water but excluded fire during the process of trans-formation). Finally, seasonings will take their place in the system according to the combinations permitted or excluded with a given type of food.

After elaborating our diagram so as to integrate all the characteristics of a given culinary system (and no doubt there are other factors of a diachronic rather than a synchronic nature; those concerning the order, the presentation and the gestures of the meal), it will be necessary to seek the most economical manner of orienting it as a grille, so that it can be superposed on other contrasts of a sociological, economic, esthetic or religious nature: men and women, family and society, village and bush, economy and prodigality, nobility and commonality, sacred and profane, etc. Thus we can hope to discover for each specific case how the cooking of a society is a language in which it unconsciously translates its structure—or else resigns itself, still unconsciously, to revealing its contradictions.

Note

Translated from the French by Peter Brooks.

1.   On these concepts, see Roman Jakobson, *Essais de linguistique générale.* Editions de Minuit, Paris (1963).

4

Deciphering a Meal

Mary Douglas

In *Purity and Danger*[1] I suggested a rational pattern for the Mosaic rejection of certain animal kinds. Ralph Bulmer has very justly reproached me for offering an animal taxonomy for the explanation of the Hebrew dietary laws. The principles I claimed to discern must remain, he argued, at a subjective and arbitrary level, unless they could take account of the multiple dimensions of thought and activity of the Hebrews concerned.[2] S. J. Tambiah has made similarly effective criticisms of the same shortcoming in my approach.[3] Both have provided from their own fieldwork distinguished examples of how the task should be conducted. In another publication I hope to pay tribute to the importance of their research. But for the present purpose, I am happy to admit the force of their reproach. It was even against the whole spirit of my book to offer an account of an ordered system of thought which did not show the context of social relations in which the categories had meaning. Ralph Bulmer let me down gently by supposing that the ethnographic evidence concerning the ancient Hebrews was too meager. However, reflection on this new research and methodology has led me to reject that suggestion out of hand. We know plenty about the ancient Hebrews. The problem is how to recognize and relate what we know.

New Guinea and Thailand are far apart, in geography, in history, and in civilization. Their local fauna are entirely different. Surprisingly, these two analyses of animal classification have one thing in common. Each society projects on to the animal kingdom categories and values which correspond to their categories of marriageable persons. The social categories of descent and affinity dominate their natural categories. The good Thailand son-in-law knows his place and keeps to it: disordered, displaced sex is reprobated and the odium transferred to the domestic dog, symbol of dirt and promiscuity. From the dog to the otter, the transfer of odium is doubled in strength. This amphibian they class as wild, counterpart-dog. But instead of keeping to the wild domain it is apt to leave its sphere at flood time and to paddle about in their watery fields. The ideas they attach to incest are carried forward from the dog to the otter, the image of the utterly wrong son-in-law. For the Karam the social focus is upon the strained relations between affines and cousins. A wide range of manmade rules sustains the categories of a natural world which mirrors those anxieties. In the Thailand and Karam studies, a strong analogy between bed and board lies unmistakably beneath the system of classifying animals. The patterns of rules which categorize animals correspond in form to the patterns of rules governing human relations. Sexual and gastronomic consummation are made equivalents of one another by reasons of analogous restrictions applied to each. Looking back from these examples to the classifications of Leviticus we seek in vain a statement, however oblique, of a similar association between eating and sex. Only a very strong analogy between table and altar stares us in the face. On reflection, why should the Israelites have had a similar concern to associate sex with food? Unlike the other two examples, they had no rule requiring them to exchange their womenfolk. On the contrary, they were allowed to marry their parallel first cousins. E. R. Leach has reminded us how strongly exogamy was disapproved at the top political level,[4] and within each tribe of Israel endogamy was even enjoined (Numbers 36). We must seek elsewhere for their dominant preoccupations. At this point I turn to the rules governing the common meal as prescribed in the Jewish religion. It is particularly interesting that these rules have remained the same over centuries. Therefore, if these categories express a relevance to social concerns we must expect those concerns to have remained in some form alive. The three rules about meat are: (1) the rejection of certain animal kinds as unfit for the table (Leviticus 11; Deuteronomy 14), (2) of those admitted as edible, the separation of the meat from blood before cooking (Leviticus 17: 10; Deuteronomy 12: 23–7), (3) the total separation of milk from meat, which involved the minute specialization of utensils (Exodus 23: 19; 34: 26; Deuteronomy 14: 21).

I start with the classification of animals whose rationality I claim to have discerned. Diagrams will help to summarize the argument (first outlined in *Purity and Danger*, 1966). First, animals are classified according to degrees of holiness (see Figure 4.1). At the bottom end of the scale some animals are abominable, not to be touched or eaten. Others are fit for the table, but not for the altar. None that are fit for the altar are not edible and vice versa, none that are not edible are sacrificeable. The criteria for this grading are coordinated for the three spheres of land, air, and water. Starting with the simplest, we find the sets as in Figure 4.2.

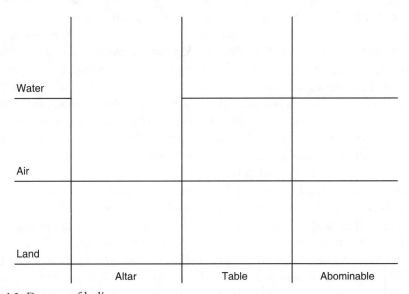

Figure 4.1 Degrees of holiness.

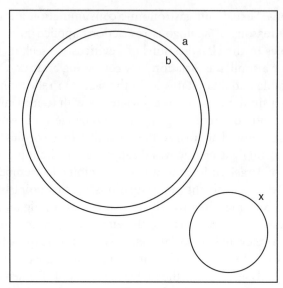

Figure 4.2 Denizens of the water (a) insufficient criteria for (b); (b) fit for table; (x) abominable: swarming.

Water creatures, to be fit for the table, must have fins and scales (Leviticus 13: 9–12; Deuteronomy 14: 19). Creeping swarming worms and snakes, if they go in the water or on the land, are not fit for the table (Deuteronomy 14: 19; Leviticus 11: 41–3). "The term swarming creatures (*shéreç*) denotes living things which appear in swarms and is applied both to those which teem in the waters (Genesis 1: 20; Leviticus 11: 10) and to those which swarm on the ground, including the smaller land animals, reptiles and creeping insects."[5] Nothing from this sphere is fit for the altar. The Hebrews only sanctified domesticated animals and these did not include fish. "When any one of you brings an offering to Jehovah, it shall be a domestic animal, taken either from the herd or from the flock" (Leviticus 1: 2). But, Assyrians and others sacrificed wild beasts, as S. R. Driver and H. A. White point out.

Air creatures (see Figure 4.3) are divided into more complex sets: set (a), those which fly and hop on the earth (Leviticus 11: 12), having wings and two legs, contains two subsets, one of which contains the named birds, abominable and not fit for the table, and the rest of the birds (b), fit for the table. From this latter subset a sub-subset (c) is drawn, which is suitable for the altar—turtledove and pigeon (Leviticus 14; 5: 7–8) and the sparrow (Leviticus 14: 49–53). Two separate sets of denizens of the air are abominable, untouchable creatures; (f), which have the wrong number of limbs for their habitat, four legs instead of two (Leviticus 9: 20), and (x), the swarming insects we have already noted in the water (Deuteronomy 14: 19).

The largest class of land creatures (a) (see Figure 4.4) walk or hop on the land with four legs. From this set of quadrupeds, those with parted hoofs and which chew the cud (b) are distinguished as fit for the table (Leviticus 11: 3; Deuteronomy 14: 4–6) and of this set a subset consists of the domesticated herds and flocks (c). Of these the first born (d) are to be offered to the priests (Deuteronomy 24: 33). Outside the set (b) which part the hoof and chew the cud are three sets of abominable beasts: (g) those which have either the one or the other but not both of the required physical

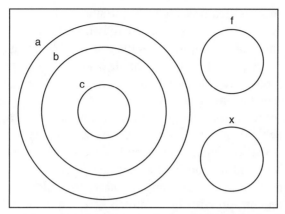

Figure 4.3 Denizens of the air (a) fly and hop: wings and two legs; (b) fit for table; (c) fit for altar; (f) abominable: insufficient criteria for (a); (x) abominable: swarming.

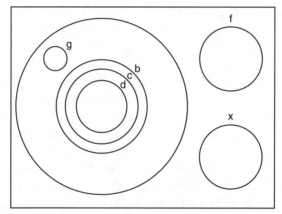

Figure 4.4 Denizens of the land (a) walk or hop with four legs; (b) fit for table; (c) domestic herds and flocks; (d) fit for altar; (f) abominable: insufficient criteria for (a); (g) abominable: insufficient criteria for (b); (x) abominable: swarming.

features; (f) those with the wrong number of limbs, two hands instead of four legs (Leviticus 11: 27 and 29: 31; and see Proverbs 30: 28); (x) those which crawl upon their bellies (Leviticus 11: 41–4).

The isomorphism which thus appears between the different categories of animal classed as abominable helps us to interpret the meaning of abomination. Those creatures which inhabit a given range, water, air, or land, but do not show all the criteria for (a) or (b) in that range are abominable. The creeping, crawling, teeming creatures do not show criteria for allocation to any class, but cut across them all.

Here we have a very rigid classification. It assigns living creatures to one of three spheres, on a behavioral basis, and selects certain morphological criteria that are found most commonly in the animals inhabiting each sphere. It rejects creatures which are anomalous, whether in living between two spheres, or having defining features of members of another sphere, or lacking defining features. Any living being which falls outside this classification is not to be touched or eaten. To touch it is to be defiled and defilement forbids entry to the temple. Thus it can be summed up fairly by saying that anomalous creatures are unfit for altar and table. This is a peculiarity of the Mosaic code. In other societies anomaly is not always so treated. Indeed, in some, the anomalous creature is treated as the source of blessing and is specially fit for the altar (as the Lele pangolin), or as a noble beast, to be treated as an honorable adversary, as the Karam treat the cassowary. Since in the Mosaic code every degree of holiness in animals has implications one way or the other for edibility, we must follow further the other rules classifying humans and animals. Again I summarize a long argument with diagrams. First, note that a category which divides some humans from others also divides their animals from others. Israelites descended from Abraham and bound to God by the Covenant between God and Abraham are distinguished from all other peoples and similarly the rules which Israelites obey as part of the Covenant apply to their animals (see Figure 4.5). The rule that the womb opener or first born is consecrated to divine service applies to firstlings of the flocks and herds (Exodus 22: 29–30; Deuteronomy 24: 23) and the rule of Sabbath observance is extended to work animals (Exodus 20: 10). As human and animal firstlings are to God, so a man's own first born is unalterably his heir (Deuteronomy 21: 15–17). The analogy by which Israelites are to other humans as their livestock are to other quadrupeds develops by indefinite stages the analogy between altar and table.

Since Levites who are consecrated to the temple service represent the first born of all Israel (Numbers 3: 12 and 40) there is an analogy between the animal and human firstlings. Among the Israelites, all of whom prosper through the Covenant and observance of the Law, some are necessarily unclean at any given time. No man or woman with issue of seed or blood, or with forbidden contact with an animal classed as unclean, or who has shed blood or been involved in the unsacralized killing of an animal (Leviticus 18), or who has sinned morally (Leviticus 20), can enter the temple. Nor can one with a blemish (Deuteronomy 23) enter the temple or eat the flesh of sacrifice or peace offerings (Leviticus 8: 20). The Levites are selected by pure descent from all the Israelites. They represent the first born of Israel. They judge the cleanness and purify the uncleanness of Israelites (Leviticus 13, 14, 10: 10; Deuteronomy 21: 5). Only Levites who are without blemish (Leviticus 21: 17–23) and without contact with death can enter the Holy of Holies. Thus we can present these rules as sets in Figures 4.6 and 4.7. The analogy between humans and animals is very clear. So is the analogy created by these rules between the temple and the living body. Further analogies appear between the classification of animals according to holiness (Figure 4.1) and the rules which set up the analogy of the holy temple with its holier and holier inner sanctuaries, and on the other hand between the temple's holiness and the body's purity and the capability of each to be defiled by the self-same forms

Under the Covenant		
Human	Israelites	others
Nonhuman	their livestock	others

Figure 4.5 Analogy between humans and nonhumans.

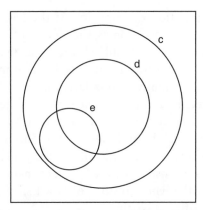

Figure 4.6 The Israelites (c) under the Covenant; (d) fit for temple sacrifice: no blemish; (e) consecrated to temple service, first born.

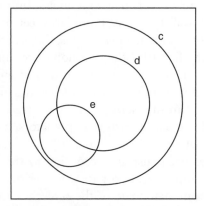

Figure 4.7 Their livestock (c) under the Covenant; (d) fit for temple sacrifice: no blemish; (e) consecrated to temple service, first born.

of impurity. This analogy is a living part of the Judeo-Christian tradition which has been unfaltering in its interpretation of New Testament allusions. The words of the Last Supper have their meaning from looking backward over the centuries in which the analogy had held good and forward to the future celebrations of that meal. "This is my body . . . this is my blood" (Luke 22: 19–20; Mark 14: 22–4; Matthew 26: 26–8). Here the meal and the sacrificial victim, the table and the altar are made explicitly to stand for one another.

Lay these rules and their patternings in a straight perspective, each one looking forward and backward to all the others, and we get the same repetition of metonyms that we found to be the key to the full meaning of the categories of food in the home. By itself the body and its rules can carry the whole load of meanings that the temple can carry by itself with its rules. The overlap and repetitions are entirely consistent. What then are these meanings? Between the temple and the body we are in a maze of religious thought. What is its social counterpart? Turning back to my original analysis (in 1966) of the forbidden meats we are now in a much better position to assess intensity and social relevance. For the metonymical patternings are too obvious to ignore. At every moment they are in chorus with a message about the value of purity and the rejection of impurity. At the level of a general taxonomy of living beings the purity in question is the purity of the categories. Creeping, swarming, teeming creatures abominably destroy the taxonomic boundaries. At the level of the individual living being impurity is the imperfect, broken, bleeding specimen. The sanctity of cognitive boundaries is made known by valuing the integrity of the physical forms. The perfect physical specimens point to the perfectly bounded temple, altar, and sanctuary. And these in their turn point to the hard-won and hard-to-defend territorial boundaries of the Promised Land. This is not reductionism. We are not here reducing the dietary rules to any political concern. But we are showing how they are consistently celebrating a theme that has been celebrated in the temple cult and in the whole history of Israel since the first Covenant with Abraham and the first sacrifice of Noah.

Edmund Leach, in his analysis of the genealogy of Solomon, has reminded us of the political problems besetting a people who claim by pure descent and pure religion to own a territory that others held and others continually encroached upon.[6] Israel is the boundary that all the other boundaries celebrate and that gives them their historic load of meaning. Remembering this, the orthodox meal is not difficult to interpret as a poem. The first rule, the rejection of certain animal kinds, we have mostly dealt with. But the identity of the list of named abominable birds is still a question. In the Mishnah it is written: "The characteristics of birds are not stated, but the Sages have said, every bird that seizes its prey (to tread or attack with claws) is unclean."[7] The idea that the unclean birds were predators, unclean because they were an image of human predation and homicide, so easily fits the later Hellenicizing interpretations that it has been suspect. According to the late Professor S. Hooke (in a personal communication), Professor R. S. Driver once tried out the idea that the Hebrew names were onomatopoeic of the screeches and calls of the birds. He diverted an assembly of learned divines with ingenious vocal exercises combining ornithology and Hebrew scholarship. I have not traced the record of this meeting. But following the method of analysis I have been using, it seems very likely that the traditional predatory idea is sufficient, considering its compatibility with the second rule governing the common meal.

According to the second rule, meat for the table must be drained of its blood. No man eats flesh with blood in it. Blood belongs to God alone, for life is in the blood.[8] This rule relates the meal systematically to all the rules which exclude from the temple on grounds of contact with or responsibility for bloodshed. Since the animal kinds which defy the perfect classification of nature are defiling both as food and for entry to the temple, it is a structural repetition of the general analogy between body and temple to rule that the eating of blood defiles. Thus the birds and beasts which eat carrion (undrained of blood) are likely by the same reasoning to be defiling. In my analysis, the Mishnah's identifying the unclean birds as predators is convincing.

Here we come to a watershed between two kinds of defilement. When the classfica-tions of any metaphysical scheme are imposed on nature, there are several points where it does not fit. So long as the classifications remain in pure metaphysics and are not expected to bite into daily life in the form of rules of behavior, no problem arises. But if the unity of Godhead is to be related to the unity of Israel and made into a rule of life, the difficulties start. First, there are the creatures whose behavior defies the rigid classification. It is relatively easy to deal with them by rejection and avoidance. Second, there are the difficulties that arise from our biological condition. It is all very well to worship the holiness of God in the perfection of his creation. But the Israelites must be nourished and must reproduce. It is impossible for a pastoral people to eat their flocks and herds without damaging the bodily completeness they respect. It is impossible to renew Israel without emission of blood and sexual fluids. These problems are met sometimes by avoidance and sometimes by consecration to the temple. The draining of blood from meat is a ritual act which figures the bloody sacrifice at the altar. Meat is thus transformed from a living creature into a food item.

As to the third rule, the separation of meat and milk, it honors the procreative functions. The analogy between human and animal parturition is always implied, as the Mishnah shows in its comment on the edibility of the afterbirth found in the slaughtered dam: if the afterbirth had emerged in part, it is forbidden as food; "it is a token of young in a woman and a token of young in a beast."[9] Likewise this third rule honors the Hebrew mother and her initial unity with her offspring.

In conclusion I return to the researches of Tambiah and Bulmer. In each case a concern with sexual relations, approved or disapproved, is reflected on to the Thailand and Karam animal classifications. In the case of Israel the dominant concern would seem to be with the integrity of territorial boundaries. But Edmund Leach has pointed out how over and over again they were concerned with the threat to Israel's holy calling from marriages with outsiders. Foreign husbands and foreign wives led to false gods and political defections. So sex is not omitted from the meanings in the common meal. But the question is different. In the other cases the problems arose from rules about exchanging women. In this case the concern is to insist on not exchanging women.

Perhaps I can now suggest an answer to Ralph Bulmer's question about the abhorrence of the pig.

> Dr Douglas tells us that the pig was an unclean beast to the Hebrew quite simply because it was a taxonomic anomaly, literally as the Old Testament says, because like the normal domestic animals it has a cloven hoof, whereas *un*like other cloven-footed beasts, it does not chew the cud. And she pours a certain amount of scorn on the commentators of the last 2,000 years who have taken alternative views and drawn attention to the creature's feeding habits, etc.

Dr Bulmer would be tempted to reverse the argument and to say that the other animals are prohibited as part of an elaborate exercise for rationalizing:[10]

> the prohibition of a beast for which there were probably multiple reasons for avoiding. It would seem equally fair, on the limited evidence available, to argue that the pig was accorded anomalous taxonomic status because it was unclean as to argue that it was unclean because of its anomalous taxonomic status.

On more mature reflection, and with the help of his own research, I can now see that the pig to the Israelites could have had a special taxonomic status equivalent to that of the otter in Thailand. It carries the odium of multiple pollution. First, it pollutes because it defies the classification of ungulates. Second, it pollutes because it eats carrion. Third, it pollutes because it is reared as food (and presumably as prime pork) by non-Israelites. An Israelite who betrothed a foreigner might have been liable to be offered a feast of pork. By these stages it comes plausibly to represent the utterly disapproved form of sexual mating and to carry all the odium that this implies. We now can trace a general analogy between the food rules and the other rules against mixtures: "Thou shalt not make the cattle to gender with beasts of any other kind" (Leviticus 19: 19). "Thou shalt not copulate with any beast" (Leviticus 18: 23). The common meal, decoded, as much as any poem, summarizes a stern, tragic religion.

We are left the question of why, when so much else had been forgotten[11] about the rules of purification and their meaning, the three rules governing the Jewish meal have persisted. What meanings do they still encode, unmoored as they partly are from their original social context? It would seem that whenever a people are aware of encroachment and danger, dietary rules controlling what goes into the body would serve as a vivid analogy of the corpus of their cultural categories at risk. But here I am, contrary to my own strictures, suggesting a universal meaning, free of particular social context, one which is likely to make sense whenever the same situation is perceived. We have come full-circle to Figure 1, with its two concentric circles. The outside boundary is weak, the inner one strong. Right through the diagrams summarizing the Mosaic dietary rules the focus was upon the integrity of the boundary at (b). Abominations of the water are those finless and scaleless creatures which lie outside that boundary. Abominations of the air appear less clearly in this light because the unidentified forbidden birds had to be shown as the widest circle from which the edible selection is drawn. If it be granted that they are predators, then they can be shown as a small subset in the unlisted set, that is as denizens of the air not fit for table because they eat blood. They would then be seen to threaten the boundary at (b) in the same explicit way as among the denizens of the land the circle (g) threatens it. We should therefore not conclude this essay without saying something more positive about what this boundary encloses. In the one case it divides edible from inedible. But it is more than a negative barrier of exclusion. In all the cases we have seen, it bounds the area of structured relations. Within that area rules apply. Outside it, anything goes. Following the argument we have established by which each level of meaning realizes the others which share a common structure, we can fairly say that the ordered system which is a meal represents all the ordered systems associated with it. Hence the strong arousal power of a threat to weaken or confuse that category. To take our analysis of the culinary medium further we should study what the poets say about the disciplines that they adopt. A passage from Roy Fuller's lectures helps to explain the flash of recognition and confidence which welcomes an ordered pattern. He is quoting Allen Tate, who said: "Formal versification is the primary structure of poetic order, the assurance to the reader and to the poet himself that the poet is in control of the disorder both outside him and within his own mind."[12]

The rules of the menu are not in themselves more or less trivial than the rules of verse to which a poet submits.

Notes

I am grateful to Professor Basil Bernstein and to Professor M. A. K. Halliday for valuable suggestions and for criticisms, some of which I have not been able to meet. My thanks are due to my son James for working out the Venn diagrams used in this article.
[. . .]

1. Mary Douglas (1966), *Purity and Danger: An Analysis of Concepts of Pollution and Taboo*, London, Routledge.
2. Ralph Bulmer (1967), "Why Is the Cassowary not a Bird? A Problem of Zoological Taxonomy among the Karam of the New Guinea Highlands," *Man, new sex*, 2, 5–25.
3. S. J. Tambiah (1969), "Animals Are Good to Think and Good to Prohibit," *Ethnology*, 7: 423–59.
4. E. R. Leach (1969), "The Legitimacy of Solomon," *Genesis as Myth and Other Essays*, London, Jonathan Cape.
5. S. R. Driver and H. A. White, *The Polychrome Bible, Leviticus*, v.1. fn. 13.
6. Leach, "Legitimacy of Solomon."
7. H. Danby, trans. (1933), *The Mishnah*, London, Oxford University Press, p. 324.
8. See Jacob Milgrom (1971), "A Prolegomena to Leviticus 17: 17," *Journal of Biblical Literature*, 90, II: 149–56. This contains a textual analysis of the rules forbidding eating flesh with the blood in it which is compatible with the position herein advocated.
9. *Ibid*, p. 520.
10. Bulmer, "Why Is the Cassowary not a Bird?" p. 21.
11. Moses Maimonides (1904), *Guide for the Perplexed*, trans. M. Friedlander, London, Routledge, first ed., 1881.
12. Roy Fuller (1971), *Owls and Artificers: Oxford Lectures on Poetry*, London, Deutsch, p. 64.

5

The Abominable Pig

Marvin Harris

An aversion to pork seems at the outset even more irrational than an aversion to beef. Of all domesticated mammals, pigs possess the greatest potential for swiftly and efficiently changing plants into flesh. Over its lifetime a pig can convert 35 percent of the energy in its feed to meat compared with 13 percent for sheep and a mere 6.5 percent for cattle. A piglet can gain a pound for every three to five pounds it eats while a calf needs to eat ten pounds to gain one. A cow needs nine months to drop a single calf, and under modern conditions the calf needs another four months to reach four hundred pounds. But less than four months after insemination, a single sow can give birth to eight or more piglets, each of which after another six months can weigh over four hundred pounds. Clearly, the whole essence of pig is the production of meat for human nourishment and delectation. Why then did the Lord of the ancient Israelites forbid his people to savor pork or even to touch a pig alive or dead?

> Of their flesh you shall not eat, and their carcasses you shall not touch; they are unclean to you (Lev. 11: 1) . . . everyone who touches them shall be unclean.
>
> (Lev. 11:24)

Unlike the Old Testament, which is a treasure trove of forbidden flesh, the Koran is virtually free of meat taboos. Why is it the pig alone who suffers Allah's disapproval?

> These things only has He forbidden you: carrion, blood, and the flesh of swine.
>
> (Holy Koran 2, 168)

For many observant Jews, the Old Testament's characterization of swine as "unclean" renders the explanation of the taboo self-evident: "Anyone who has seen the filthy habits of the swine will not ask why it is prohibited," says a modern rabbinical authority. The grounding of the fear and loathing of pigs in self-evident piggishness goes back at least to the time of Rabbi Moses Maimonides, court physician to the Islamic emperor Saladin during the twelfth century in Egypt. Maimonides shared with his Islamic hosts a lively disgust for pigs and pig eaters, especially Christian pigs and pig eaters: "The principal reason why the law forbids swine-flesh is to be found in the circumstance that its habits and food are very filthy and loathsome." If the law allowed Egyptians and Jews to raise pigs, Cairo's streets and houses would become as filthy as those of Europe, for "the mouth of a swine is as dirty as dung itself." Maimonides could only tell one side of the story. He had never seen a clean pig. The pig's penchant for excrement is not a defect of its nature but of the husbandry of its human masters. Pigs prefer and thrive best on roots, nuts, and grains; they eat excrement only when nothing better presents itself. In fact, let them get hungry enough, and they'll even eat each other, a trait which they share with other omnivores, but most notably with their own masters. Nor is wallowing in filth a natural characteristic of swine. Pigs wallow to keep themselves cool; and they much prefer a fresh, clean mudhole to one that has been soiled by urine and feces.

In condemning the pig as the dirtiest of animals, Jews and Moslems left unexplained their more tolerant attitude toward other dung-eating domesticated species. Chickens and goats, for example, given motivation and opportunity, also readily dine on dung. The dog is another domesticated creature which easily develops an appetite for human feces. And this was especially true in the Middle East, where dung-eating dogs filled the scavenging niche left vacant by the ban on pigs. Jahweh prohibited their flesh, yet dogs were not abominated, bad to touch, or even bad to look at, as were pigs.

Maimonides could not be entirely consistent in his efforts to attribute the abstention from pork to the pig's penchant for feces. The Book of Leviticus prohibits the flesh of many other creatures, such as cats and camels, which are not notably inclined to eat excrement. And with the exception of the pig, had not Allah said all the others were good to eat? The fact that Maimonides's Moslem emperor could eat every kind of meat except pork would have made it impolitic if not dangerous to identify the biblical sense of cleanliness exclusively with freedom from the taint of feces. So instead of adopting a cleaner-than-thou attitude, Maimonides offered a proper court physician's theory of the entire set of biblical aversions: the prohibited items were not good to eat because not only was one of them—the pig—filthy from eating excrement but all of them were not good for you. "I maintain," he said, "that food forbidden by the Law is unwholesome." But in what ways were the forbidden foods unwholesome? The great rabbi was quite specific in the case of pork: it "contained more moisture than necessary and too much superfluous matter." As for the other forbidden foods, their "injurious character" was too self-evident to merit further discussion.

Maimonides's public health theory of pork avoidance had to wait seven hundred years before it acquired what seemed to be a scientific justification. In 1859 the first clinical association between trichinosis and undercooked pork was established, and from then on it became the most popular explanation of the Jewish and Islamic pork taboo. Just as Maimonides said, pork was unwholesome. Eager to reconcile the Bible with the findings of medical science, theologians began to embroider a whole series of additional public health explanations for the other biblical food taboos: wild animals and beasts of burden were prohibited because the flesh gets too tough to be digested properly; shellfish were to be avoided because they serve as vectors of typhoid fever; blood is not good to eat because the bloodstream is a perfect medium for microbes. In the case of pork this line of rationalization had a paradoxical outcome. Reformist Jews began to argue that since they now understood the scientific and medical basis of the taboos, pork avoidance was no longer necessary; all they had to do was to see to it that the meat was thoroughly cooked. Predictably, this provoked a reaction among Orthodox Jews, who were appalled at the idea that the book of God's law was being relegated to the "class of a minor medical text." They insisted that God's purpose in Leviticus could never be fully comprehended; nonetheless the dietary laws had to be obeyed as a sign of submission to divine will.

Eventually the trichinosis theory of pork avoidance fell out of favor largely on the grounds that a medical discovery made in the nineteenth century could not have been known thousands of years ago. But that is not the part of the theory that bothers me. People do not have to possess a scientific understanding of the ill effects of certain foods in order to put such foods on their bad-to-eat list. If the consequences of eating pork had been exceptionally bad for their health, it would not have been necessary for the Israelites to know about trichinosis in order to ban its consumption. Does one have to understand the molecular chemistry of toxins in order to know that some mushrooms are dangerous? It is essential for my own explanation of the pig taboo that the trichinosis theory be laid to rest on entirely different grounds. My contention is that there is absolutely nothing exceptional about pork as a source of human disease. All domestic animals are potentially hazardous to human health. Undercooked beef, for example, is a prolific source of tapeworms, which can grow to a length of sixteen to twenty feet inside the human gut, induce a severe case of anemia, and lower the body's resistance to other diseases. Cattle, goat, and sheep transmit the bacterial disease known as brucellosis, whose symptoms include fever, aches, pains, and lassitude. The most dangerous disease transmitted by cattle, sheep, and goats is anthrax, a fairly common disease of both animals and humans in Europe and Asia until the introduction of Louis Pasteur's anthrax vaccine in 1881. Unlike trichinosis, which does not produce symptoms in the majority of infected individuals and rarely has a fatal outcome, anthrax runs a swift course that begins with an outbreak of boils and ends in death.

If the taboo on pork was a divinely inspired health ordinance, it is the oldest recorded case of medical malpractice. The way to safeguard against trichinosis was not to taboo pork but to taboo undercooked pork. A simple advisory against undercooking pork would have sufficed: "Flesh of swine thou shalt not eat until the pink has been cooked from it." And come to think of it, the same advisory should have been issued for cattle, sheep, and goats. But the charge of medical malpractice against Jahweh will not stick.

The Old Testament contains a rather precise formula for distinguishing good-to-eat flesh from forbidden flesh. This formula says nothing about dirty habits or unhealthy meat. Instead it directs attention to certain anatomical and physiological features of animals that are good to eat. Here is what Leviticus 11: 1 says:

> Whatever parts the hoof and is cloven footed and chews the cud among animals, you may eat.

Any serious attempt to explain why the pig was not good to eat must begin with this formula and not with excrement or wholesomeness, about which not a word is said. Leviticus goes on to state explicitly of the pig that it only satisfies one part of the formula. "It divideth the hoof." But the pig does not satisfy the other part of the formula: "It cheweth not the cud."

To their credit, champions of the good-to-eat school have stressed the importance of the cud-chewing, split-hoof formula as the key to understanding Jahweh's abomination of the pig. But they do not view the formula as an outcome of the way the Israelites used domestic animals. Instead they view the way the Israelites used domestic animals as an outcome of the formula. According to anthropologist Mary Douglas, for example, the cud-chewing, split-hoof formula makes the split-hoof but non-cud-chewing pig a thing that's "out of place." Things that are "out of place" are dirty, she argues, for the essence of dirt is "matter out of place." The pig, however, is more than out of place; it is neither here nor there. Such things are both dirty and dangerous. Therefore the pig is abominated as well as not good to eat. But doesn't the force of this argument lie entirely in its circularity? To observe that the pig is out of place taxonomically is merely to observe that Leviticus classifies good-to-eat animals in such a way as to make the pig bad to eat. This avoids the question of why the taxonomy is what it is.

Let me attend first to the reason why Jahweh wanted edible animals to be cud-chewers. Among animals raised by the ancient Israelites, there were three cud-chewers: cattle, sheep, and goats. These three animals were the most important food-producing species in the ancient Middle East not because the ancients happened capriciously to think that cud-chewing animals were good to eat (and good to milk) but because cattle, sheep, and goats are ruminants, the kind of herbivores which thrive best on diets consisting of plants that have a high cellulose content. Of all domesticated animals, those which are ruminants possess the most efficient system for digesting tough fibrous materials such as grasses and straw. Their stomachs have four compartments which are like big fermentation "vats" in which bacteria break down and soften these materials. While cropping their food, ruminants do little chewing. The food passes directly to the rumen, the first of the compartments, where it soon begins to ferment. From time to time the contents of the rumen are regurgitated into the mouth as a softened bolus—the "cud"—which is then chewed thoroughly and sent on to the other "vats" to undergo further fermentation.

The ruminant's extraordinary ability to digest cellulose was crucial to the relationship between humans and domesticated animals in the Middle East. By raising animals that could "chew the cud," the Israelites and their neighbors were able to obtain meat and milk without having to share with their livestock the crops destined for human consumption. Cattle, sheep, and goats thrive on items like grass, straw, hay, stubble, bushes, and leaves—feeds whose high cellulose content renders them unfit for human consumption even after vigorous boiling. Rather than compete with humans for food, the ruminants further enhanced agricultural productivity by providing dung for fertilizer and traction for pulling plows. And they were also a source of fiber and felt for clothing, and of leather for shoes and harnesses.

I began this puzzle by saying that pigs are the most efficient mammalian converters of plant foods into animal flesh, but I neglected to say what kinds of plant foods. Feed them on wheat, maize, potatoes, soybeans, or anything low in cellulose, and pigs will perform veritable miracles of transubstantiation; feed them on grass, stubble, leaves, or anything high in cellulose, and they will lose weight.

Pigs are omnivores, but they are not ruminants. In fact, in digestive apparatus and nutrient requirements pigs resemble humans in more ways than any mammal except monkeys and apes, which is why pigs are much in demand for medical research concerned with atherosclerosis, calorie-protein malnutrition, nutrient absorption, and metabolism. But there was more to the ban on pork than the pig's inability to thrive on grass and other high-cellulose plants. Pigs carry the additional onus of not being well adapted to the climate and ecology of the Middle East. Unlike the ancestors of cattle, sheep, or goats, which lived in hot, semiarid, sunny grasslands, the pig's ancestors were denizens of well-watered, shady forest glens and riverbanks. Everything about the pig's body heat-regulating system is ill suited for life in the hot, sun-parched habitats which were the homelands of the children of Abraham. Tropical breeds of cattle, sheep, and goats can go for long periods without water, and can either rid their bodies of excess heat through perspiration or are protected from the sun's rays by light-colored, short fleecy coats (heat-trapping heavy wool is a characteristic of cold-climate breeds). Although a perspiring human is said to "sweat like a pig," the expression lacks an anatomical basis. Pigs can't sweat—they have no functional sweat glands. (Humans are actually the sweatiest of all animals.) And the pig's sparse coat offers little protection against the sun's rays. Just how does the pig keep cool? It does a lot of panting, but mostly it depends on wetting itself down with moisture derived from external sources. Here, then, is the explanation for the pig's love of wallowing in mud. By wallowing, it dissipates heat both by evaporation from its skin and by conduction through the cool ground. Experiments show that the cooling effect of mud is superior to that of water. Pigs whose flanks are thoroughly smeared with mud continue to show peak heat-dissipating evaporation for more than twice as long as pigs whose flanks are merely soaked with water, and here also is the explanation for some of the pig's dirty habits. As temperatures rise above thirty degrees celsius (eighty-six degrees Fahrenheit), a pig deprived of clean mudholes will become desperate and begin to wallow in its own feces and urine in order to avoid heat stroke. Incidentally, the larger a pig gets, the more intolerant it becomes of high ambient temperatures.

Raising pigs in the Middle East therefore was and still is a lot costlier than raising ruminants, because pigs must be provided with artificial shade and extra water for wallowing, and their diet must be supplemented with grains and other plant foods that humans themselves can eat.

To offset all these liabilities pigs have less to offer by way of benefits than ruminants. They can't pull plows, their hair is unsuited for fiber and cloth, and they are not suited for milking. Uniquely among large domesticated animals, meat is their most important produce (guinea pigs and rabbits are smaller equivalents; but fowl produce eggs as well as meat).

For a pastoral nomadic people like the Israelites during their years of wandering in search of lands suitable for agriculture, swineherding was out of the question. No arid-land pastoralists herd pigs for the simple reason that it is hard to protect them from exposure to heat, sun, and lack of water while moving from camp to camp over long distances. During their formative years as a nation, therefore, the ancient Israelites could not have consumed significant quantities of pork even had they desired it. This historical experience undoubtedly contributed to the development of a traditional aversion to pig meat as an unknown and alien food. But why was this tradition preserved and strengthened by being written down as God's law long after the Israelites had become settled farmers? The answer as I see it is not that the tradition born of pastoralism continued to prevail by mere inertia and ingrown habit, but that it was preserved because pig raising remained too costly.

Critics have opposed the theory that the ancient Israelite pork taboo was essentially a cost/benefit choice by pointing to evidence of pigs being raised quite successfully in many parts of the Middle East including the Israelite's promised land. The facts are not in dispute. Pigs have indeed been raised for ten thousand years in various parts of the Middle East—as long as sheep and goats, and even longer than cattle. Some of the oldest Neolithic villages excavated by archaeologists—Jericho in Jordan, Jarmo in Iraq, and Argissa-Magulla in Greece—contain pig bones with features indicative of the transition from wild to domesticated varieties. Several Middle Eastern pre-Bronze Age villages (4000 B.C. to 2000 B.C.) contain concentrated masses of pig remains in association with what archaeologists interpret as altars and cultic centers, suggestive of ritual pig slaughter and pig feasting. We know that some pigs were still being raised in the lands of the Bible at the beginning of the Christian era. The New Testament (Luke) tells us that in the country of the Gadarenes near Lake Galilee Jesus cast out devils from a man named Legion into a herd of swine feeding on the mountain. The swine rushed down into the lake and drowned themselves, and Legion was cured. Even modern-day Israelites continue to raise thousands of swine in parts of northern Galilee. But from the very beginning, fewer pigs were raised than cattle, sheep, or goats. And more importantly, as time went on, pig husbandry declined throughout the region.

Carlton Coon, an anthropologist with many years of experience in North America and the Levant, was the first scholar to offer a cogent explanation of why this general decline in pig husbandry had occurred. Coon attributed the fall of the Middle Eastern pig to deforestation and human population increase. At the beginning of the Neolithic period, pigs were able to root in oak and beech forests which provided ample shade and wallows as well as acorns, beechnuts, truffles, and other forest floor products. With an increase in human population density, farm acreage increased and the oak and beech forests were destroyed to make room for planted crops, especially for olive trees, thereby eliminating the pig's ecological niche.

To update Coon's ecological scenario, I would add that as forests were being destroyed, so were marginal farmlands and grazing lands, the general succession being from forest to cropland to grazing land to desert, with each step along the way yielding a greater premium for raising ruminants and a greater penalty for raising swine. Robert Orr Whyte, former director general of the United Nations Food and Agricultural Organization, estimated that in Anatolia the forests shrank from 70 percent to 13 percent of the total land area between 5000 B.C. and the recent past. Only a fourth of the Caspian shore-front forest survived the process of population increase and agricultural intensification; half of the Caspian mountainous humid forest; a fifth to a sixth of the oak and juniper forests of the Zagros Mountains; and only a twentieth of the juniper forests of the Elburz and Khorassan ranges.

If I am right about the subversion of the practical basis of pig production through ecological succession, one does not need to invoke Mary Douglas's "taxonomic anomaly" to understand the peculiarly low status of the pig in the Middle East. The danger it posed to husbandry was very tangible and accounts quite well for its low status. The pig had been domesticated for one purpose only, namely to supply meat. As ecological conditions became unfavorable for pig raising, there was no alternative function which could redeem its existence. The creature became not only useless, but worse than useless—harmful, a curse to touch or merely to see—a pariah animal. This transformation contrasts understandably with that of cattle in India. Subject to a similar series of ecological depletions—deforestation, erosion, and desertification—cattle also became bad to eat. But in other respects, especially for traction power and milk, they became more useful than ever—a blessing to look at or to touch—animal godheads.

In this perspective, the fact that pig raising remained possible for the Israelites at low cost in certain remnant hillside forests or swampy habitats, or at extra expense where shade and water were scarce, does not contradict the ecological basis of the taboo. If there had not been some minimum possibility of raising pigs, there would have been no reason to taboo the practice. As the history of Hindu cow protection shows, religions gain strength when they help people make decisions which are in accord with preexisting useful practices, but which are not so completely self-evident as to preclude doubts and temptations. To judge from the Eight-fold Way or the Ten Commandments, God does not usually waste time prohibiting the impossible or condemning the unthinkable.

Leviticus consistently bans all vertebrate land animals that do not chew the cud. It bans, for example, in addition to swine, equines, felines, canines, rodents, and reptiles, none of which are cud-chewers. But Leviticus contains a maddening complication. It prohibits the consumption of three land-dwelling vertebrates which it specifically identifies as cud-chewers: the camel, the hare, and a third creature whose name in Hebrew is *shāphān*. The reason given for why these three alleged cud-chewers are not good to eat is that they do not "part the hoof":

> Nevertheless, these shall ye not eat of them that chew the cud . . . the camel because he . . . divideth not the hoof. And the *shāphān* because he . . . divideth not the hoof. . . . And the hare, because he . . . divideth not the hoof.
>
> (Lev. 11: 4–6)

Although strictly speaking camels are not ruminants, because their cellulose-digesting chambers are anatomically distinct from those of the ruminants, they do ferment, regurgitate, and chew the cud much like cattle, sheep, and goats. But the classification of the hare as a cud-chewer immediately casts a pall over the zoological expertise of the Levite priests. Hares can digest grass but only by eating their own feces—which is a very uncud-like solution to the problem of how to send undigested cellulose through the gut for repeated processing (the technical term for this practice is coprophagy). Now as to the identity of the *shāphān*. As the following stack of Bibles shows, *shāphān* is either the "rock badger," "cherogrillus," or "cony":

Bibles Translating Shāphān as "Rock Badger"

The Holy Bible. Berkeley: University of California Press.
The Bible. Chicago: University of Chicago Press, 1931.

The New Schofield Reference Library Holy Bible (Authorized King James Version). New York: Oxford University Press, 1967.

The Holy Bible. London: Catholic Truth Society, 1966.

The Holy Bible. (Revised Standard Version). New York: Thomas Nelson and Sons, 1952.

The American Standard Bible. (Reference Edition). La Habra, CA: Collins World, 1973.

The New World Translation of the Holy Scriptures. Brooklyn, NY: Watchtower Bible and Tract Society of Pennsylvania, 1961.

Bibles Translating Shāphān as "Cony"

The Pentateuch: The Five Books of Moses. Edited by William Tyndale. Carbondale: Southern Illinois University Press, 1967.

The Interpreter's Bible: The Holy Scriptures. 12 vols. New York: Abingdon Press, 1953.

The Holy Bible. King James Version (Revised Standard Version). Nashville: Thomas Nelson and Sons, 1971.

Holy Bible. Authorized version. New York: Harpers.

Holy Bible. Revised. New York: American Bible Society, 1873.

Modern Readers Bible. Edited by Richard Moulton. New York: Macmillan, 1935.

Bibles Translating Shāphān as "Cherogrillus"

Holy Bible. (Duay, translated from Vulgate.) Boston: John Murphy and Co., 1914.

The Holy Bible. (Translated from the Vulgate by John Wycliffe and his followers.) Edited by Rev. Josiah Forshall and Sir Frederick Madden. Oxford: Oxford University Press, 1850.

All three terms refer to a similar kind of small, furtive, hoofed herbivore about the size of a squirrel that lives in colonies on rocky cliffs or among boulders on hilltops. It has two other popular aliases: "dassie" and "damon." It could have been any of these closely related species: *Hyrax capensia*, *Hyrax syriacus*, or *Procavia capensis*. Whichever it was, it had no rumen and it did not chew the cud.

This leaves the camel as the only bona fide cud-chewer that the Israelites couldn't eat. Every vertebrate land animal that is not a ruminant was forbidden flesh. And only one vertebrate land animal that is a ruminant, the camel, was forbidden. Let me see if I can explain this exception as well as the peculiar mixup about hares and shāphān.

My point of departure is that the food laws in Leviticus were mostly codifications of preexisting traditional food prejudices and avoidances. (The Book of Leviticus was not written until 450 b.c.—very late in Israelite history.) I envision the Levite authorities as undertaking the task of finding some simple feature which good-to-eat vertebrate land species shared in common. Had the Levites possessed a better knowledge of zoology, they could have used the criterion of cud-chewing alone and simply added the proviso, "except for the camel." For, as I have just said, with the

exception of the camel, all land animals implicitly or explicitly forbidden in
Leviticus—all the equines, felines, canines, rodents, rabbits, reptiles, and so forth—
are nonruminants. But given their shaky knowledge of zoology, the codifiers could
not be sure that the camel was the only undesirable species which was a cud-chewer.
So they added the criterion of split hooves—a feature which camels lacked but which
the other familiar cud-chewers possessed (the camel has two large flexible toes on
each foot instead of hooves).

But why was the camel not a desirable species? Why spurn camel meat? I think the
separation of the camel from the other cud-chewers reflects its highly specialized
adaptation to desert habitats. With their remarkable capacity to store water, with-
stand heat, and carry heavy burdens over great distances, and with their long eye-
lashes and nostrils that shut tight for protection against sandstorms, camels were the
most important possession of the Middle Eastern desert nomads. (The camel's hump
concentrates fat—not water. It acts as an energy reserve. By concentrating the fat in
the hump, the rest of the skin needs only a thin layer of fat, and this facilitates
removal of body heat.) But as village farmers, the Israelites had little use for camels.
Except under desert conditions, sheep and goats and cattle are more efficient converters of cellulose into meat and milk. In addition, camels reproduce very slowly.
The females are not ready to bear offspring and the males are not ready to copulate
until six years of age. To slow things down further, the males have a once-a-year
rutting season (during which they emit an offensive odor), and gestation takes twelve
months. Neither camel meat nor camel milk could ever have constituted a significant
portion of the ancient Israelites' food supply. Those few Israelites such as Abraham
and Joseph who owned camels would have used them strictly as a means of transport
for crossing the desert.

This interpretation gains strength from the Moslem acceptance of camel meat. In
the Koran, pork is specifically prohibited while camel flesh is specifically allowed. The
whole way of life of Mohammed's desert-dwelling, pastoral Bedouin followers was
based on the camel. The camel was their main source of transport and their main
source of animal food, primarily in the form of camel milk. While camel meat was
not daily fare, the Bedouin were often forced to slaughter pack animals during their
desert journeys as emergency rations when their regular supplies of food were
depleted. An Islam that banned camel flesh would never have become a great world
religion. It would have been unable to conquer the Arabian heartlands, to launch its
attack against the Byzantine and Persian empires, and to cross the Sahara to the Sahel
and West Africa.

If the Levite priests were trying to rationalize and codify dietary laws, most of
which had a basis in preexisting popular belief and practice, they needed a taxonomic
principle which connected the existing patterns of preference and avoidance into a
comprehensive cognitive and theological system. The preexisting ban on camel meat
made it impossible to use cud-chewing as the sole taxonomic principle for identifying land vertebrates that were good to eat. They needed another criterion to exclude
camels. And this was how "split hooves" got into the picture. Camels have conspicuously different feet from cattle, sheep, or goats. They have split toes instead of split
hooves. So the priests of Leviticus added "parts the hoof" to "chews the cud" to make
camels bad to eat. The misclassification of the hare and *shāphān* suggest that these

animals were not well known to the codifiers. The authors of Leviticus were right about the feet—hares have paws and *Hyrax* (and *Procavia*) have tiny hooves, three on the front leg and five on the rear leg. But they were wrong about the cud-chewing—perhaps because hares and shāphān have their mouths in constant motion.

Once the principle of using feet to distinguish between edible and inedible flesh was established, the pig could not be banned simply by pointing to its nonruminant nature. Both its cud-chewing status and the anatomy of its feet had to be considered, even though the pig's failure to chew the cud was its decisive defect.

This, then, is my theory of why the formula for forbidden vertebrate land animals was elaborated beyond the mere absence of cud-chewing. It is a difficult theory to prove because no one knows who the authors of Leviticus were or what was really going on inside their heads. But regardless of whether or not the good-to-eat formula originated in the way I have described, the fact remains that the application of the expanded formula to hare and *shāphān* (as well as to pig and camel) did not result in any dietary restrictions that adversely affected the balance of nutritional or ecological costs and benefits. Hare and *shāphān* are wild species; it would have been a waste of time to hunt them instead of concentrating on raising far more productive ruminants.

To recall momentarily the case of the Brahman protectors of the cow, I do not doubt the ability of a literate priesthood to codify, build onto, and reshape popular foodways. But I doubt whether such "top-down" codifications generally result in adverse nutritional or ecological consequences or are made with blithe disregard of such consequences. More important than all the zoological errors and flights of taxonomic fancy is that Leviticus correctly identifies the classic domesticated ruminants as the most efficient source of milk and meats for the ancient Israelites. To the extent that abstract theological principles result in flamboyant lists of interdicted species, the results are trivial if not beneficial from a nutritional and ecological viewpoint. Among birds, for example, Leviticus bans the flesh of the eagle, ossifrage, osprey, ostrich, kite, falcon, raven, nighthawk, sea gull, hawk, cormorant, ibis, water-hen, pelican, vulture, stork, hoopoe, and bat (not a bird of course). I suspect but again cannot prove that this list was primarily the result of a priestly attempt to enlarge on a smaller set of prohibited flying creatures. Many of the "birds," especially the sea birds like pelicans and cormorants, would rarely be seen inland. Also, the list seems to be based on a taxonomic principle that has been somewhat overextended: most of the creatures on it are carnivores and "birds of prey." Perhaps the list was generated from this principle applied first to common local "birds" and then extended to the exotic sea birds as a validation of the codifiers' claim to special knowledge of the natural and supernatural worlds. But in any event, the list renders no disservice. Unless they were close to starvation and nothing else was available, the Israelites were well advised not to waste their time trying to catch eagles, ospreys, sea gulls, and the like, supposing they were inclined to dine on creatures that consist of little more than skin, feathers, and well-nigh indestructible gizzards in the first place. Similar remarks are appropriate vis-à-vis the prohibition of such unlikely sources of food for the inland-dwelling Israelites as clams and oysters. And if Jonah is an example of what happened when they took to the sea, the Israelites were well advised not to try to satisfy their meat hunger by hunting whales.

But let me return to the pig. If the Israelites had been alone in their interdictions of pork, I would find it more difficult to choose among alternative explanations of the pig taboo. The recurrence of pig aversions in several different Middle Eastern cultures strongly supports the view that the Israelite ban was a response to recurrent practical conditions rather than to a set of beliefs peculiar to one religion's notions about clean and unclean animals. At least three other important Middle Eastern civilizations—the Phoenicians, Egyptians, and Babylonians—were as disturbed by pigs as were the Israelites. Incidentally, this disposes of the notion that the Israelites banned the pig to "set themselves off from their neighbors," especially their unfriendly neighbors. (Of course, after the Jews dispersed throughout pork-eating Christendom, their abomination of the pig became an ethnic "marker." There was no compelling reason for them to give up their ancient contempt for pork. Prevented from owning land, the basis for their livelihood in Europe had to be crafts and commerce rather than agriculture. Hence there were no ecological or economic penalties associated with their rejection of pork while there were plenty of other sources of animal foods.)

In each of the additional cases, pork had been freely consumed during an earlier epoch. In Egypt, for example, tomb paintings and inscriptions indicate that pigs were the object of increasingly severe opprobrium and religious interdiction during the New Kingdom (1567–1085 B.C.). Toward the end of late dynastic times (1088–332 B.C.) Herodotus visited Egypt and reported that "the pig is regarded among them as an unclean animal so much so that if a man in passing accidentally touches a pig, he instantly hurries to the river and plunges in with all his clothes on." As in Roman Palestine when Jesus drove the Gadarene swine into Lake Galilee, some Egyptians continued to raise pigs. Herodotus described these swineherds as an in-marrying pariah caste who were forbidden to set foot in any of the temples.

One interpretation of the Egyptian pig taboo is that it reflects the conquest of the northern pork-eating followers of the god Seth by the southern pork-abstaining followers of the god Osiris and the imposition of southern Egyptian food preferences on the northerners. The trouble with this explanation is that if such a conquest occurred at all, it took place at the very beginning of the dynastic era and therefore does not account for the evidence that the pig taboo got stronger in late dynastic times.

My own interpretation of the Egyptian pig taboo is that it reflected a basic conflict between the dense human population crowded into the treeless Nile Valley and the demands made by the pig for the plant foods that humans could consume. A text from the Old Kingdom clearly shows how during hard times humans and swine competed for subsistence: "food is robbed from the mouth of the swine, without it being said, as before 'this is better for thee than for me,' for men are so hungry." What kinds of foods were robbed from the swine's mouth? Another text from the Second Intermediate period, boasting of a king's power over the lands, suggests it was grains fit for human consumption: "The finest of their fields are ploughed for us, our oxen are in the Delta, wheat is sent for our swine." And the Roman historian, Pliny, mentions the use of dates as a food used to fatten Egyptian pigs. The kind of preferential treatment needed to raise pigs in Egypt must have engendered strong feelings of antagonism between poor peasants who could not afford pork and the swineherds who catered to the tastes of rich and powerful nobles.

In Mesopotamia, as in Egypt, the pig fell from grace after a long period of popularity. Archaeologists have found clay models of domesticated pigs in the earliest settlements along the lower Tigris and Euphrates rivers. About 30 percent of the animal bones excavated from Tell Asmar (2800–2700 B.C.) came from pigs. Pork was eaten at Ur in predynastic times, and in the earliest Sumerian dynasties there were swineherds and butchers who specialized in pig slaughter. The pig seems to have fallen from favor when the Sumerians' irrigated fields became contaminated with salt, and barley, a salt-tolerant but relatively low-yielding plant, had to be substituted for wheat. These agricultural problems are implicated in the collapse of the Sumerian Empire and the shift after 2000 B.C. of the center of power upstream to Babylon. While pigs continued to be raised during Hammurabi's reign (about 1900 B.C.), they virtually disappear from Mesopotamia's archaeological and historical record thereafter.

The most important recurrence of the pig taboo is that of Islam. To repeat, pork is Allah's only explicitly forbidden flesh. Mohammed's Bedouin followers shared an aversion to pig found everywhere among arid-land nomadic pastoralists. As Islam spread westward from the Arabian Peninsula to the Atlantic, it found its greatest strength among North African peoples for whom pig raising was also a minor or entirely absent component of agriculture and for whom the Koranic ban on pork did not represent a significant dietary or economic deprivation. To the east, Islam again found its greatest strength in the belt of the semiarid lands that stretch from the Mediterranean Sea through Iran, Afghanistan, and Pakistan to India. I don't mean to say that none of the people who adopted Islam had previously relished pork. But for the great mass of early converts, becoming a Moslem did not involve any great upending of dietary or subsistence practices because from Morocco to India people had come to depend primarily on cattle, sheep, and goats for their animal products long before the Koran was written. Where local ecological conditions here and there strongly favored pig raising within the Islamic heartland, pork continued to be produced. Carlton Coon described one such pork-tolerant enclave—a village of Berbers in the oak forests of the Atlas Mountains in Morocco. Although nominally Moslems, the villagers kept pigs which they let loose in the forest during the day and brought home at night. The villagers denied that they raised pigs, never took them to market, and hid them from visitors. These and other examples of pig-tolerant Moslems suggest that one should not overestimate the ability of Islam to stamp out pig eating by religious precept alone if conditions are favorable for pig husbandry.

Wherever Islam has penetrated to regions in which pig raising was a mainstay of the traditional farming systems, it has failed to win over substantial portions of the population. Regions such as Malaysia, Indonesia, the Philippines, and Africa south of the Sahara, parts of which are ecologically well suited for pig raising, constitute the outer limits of the active spread of Islam. All along this frontier the resistance of pig-eating Christians has prevented Islam from becoming the dominant religion. In China, one of the world centers of pig production, Islam has made small inroads and is confined largely to the arid and semiarid western provinces. Islam, in other words, to this very day has a geographical limit which coincides with the ecological zones of transition between forested regions well suited for pig husbandry and regions where too much sun and dry heat make pig husbandry a risky and expensive practice.

While I contend that ecological factors underlie religious definitions of clean and unclean foods, I also hold that the effects do not all flow in a single direction. Religiously sanctioned foodways that have become established as the mark of conversion and as a measure of piety can also exert a force of their own back upon the ecological and economic conditions which gave rise to them. In the case of the Islamic pork taboos, the feedback between religious belief and the practical exigencies of animal husbandry has led to a kind of undeclared ecological war between Christians and Moslems in several parts of the Mediterranean shores of southern Europe. In rejecting the pig, Moslem farmers automatically downgrade the importance of preserving woodlands suitable for pig production. Their secret weapon is the goat, a great devourer of forests, which readily climbs trees to get at a meal of leaves and twigs. By giving the goat free reign, Islam to some degree spread the conditions of its own success. It enlarged the ecological zones ill suited to pig husbandry and removed one of the chief obstacles to the acceptance of the words of the Prophet. Deforestation is particularly noticeable in the Islamic regions of the Mediterranean. Albania, for example, is divided between distinct Christian pig-keeping and Moslem pig-abominating zones, and as one passes from the Moslem to the Christian sectors, the amount of woodland immediately increases.

It would be wrong to conclude that the Islamic taboo on the pig caused the deforestation wrought by the goat. After all, a preference for cattle, sheep, and goats and the rejection of pigs in the Middle East long antedated the birth of Islam. This preference was based on the cost/benefit advantages of ruminants over other domestic animals as sources of milk, meat, traction, and other services and products in hot, arid climates. It represents an unassailably "correct" ecological and economic decision embodying thousands of years of collective wisdom and practical experience. But as I have already pointed out in relation to the sacred cow, no system is perfect. Just as the combination of population growth and political exploitation led to a deterioration of agriculture in India, so too population growth and political exploitation took their toll in Islamic lands. If the response to demographic and political pressures had been to raise more pigs rather than goats, the adverse effects on living standards would have been even more severe and would have occurred at a much lower level of population density.

All of this is not to say that a proseletyzing religion such as Islam is incapable of getting people to change their foodways purely out of obedience to divine commandments. Priests, monks, and saints do often refuse delectable and nutritious foods out of piety rather than practical necessity. But I have yet to encounter a flourishing religion whose food taboos make it more difficult for ordinary people to be well nourished. On the contrary, in solving the riddle of the sacred cow and abominable pig, I have already shown that the most important food aversions and preferences of four major religions—Hinduism, Buddhism, Judaism, and Islam—are on balance favorable to the nutritional and ecological welfare of their followers.

References

Coon, Carleton (1951) *Caravan*. New York: Henry Holt.
Douglas, Mary (1966) *Purity and Danger: An Analysis of Concepts of Pollution and Taboo*. New York: Praeger.

6

The Nourishing Arts

Michel de Certeau and Luce Giard

What follows very much involves the (privileged?) role of women in the preparation of meals eaten at home. But this is not to say that I believe in an immanent and stable feminine nature that dooms women to house-work and gives them a monopoly over both the kitchen and the tasks of interior organization.[1] Since the time when Europe left its geographic borders in the sixteenth century and discovered the difference of other cultures, history and anthropology have taught us that the sharing of work between the sexes, initiation rites, and diets, or what Mauss calls "body techniques,"[2] are reliant on the local cultural order and, like it, are changeable. Within a certain culture, a change of material conditions or of political organization can be enough to modify the way of conceiving or dividing a particular kind of everyday task, just as the hierarchy of different kinds of housework can be transformed.

The fact that there are still *women* in France who in general carry out the everyday work of doing-cooking stems from a social and cultural condition and from the history of mentalities; I do not see the manifestation of a feminine essence here. If, in this study, we judged it necessary to become interested in this example of cultural practice rather than another, it is because of the central role it plays in the everyday life of the majority of people, independent of their social situation and their relationship to "high culture" or to mass industrial culture. Moreover, alimentary habits constitute a domain where tradition and innovation matter equally, where past and present are mixed to serve the needs of the hour, to furnish the joy of the moment, and to suit the circumstance. With their high degree of ritualization and their strong affective investment, culinary activities are for many women of all ages a place of happiness, pleasure, and discovery. Such life activities demand as much intelligence, imagination, and memory as those traditionally held as superior, such as music and weaving. In this sense, they rightly make up one of the strong aspects of ordinary culture.

Entrée

As a child, I refused to surrender to my mother's suggestions to come and learn how to cook by her side. I refused this women's work because no one ever offered it to my brother. I had already chosen, determined my fate: one day, I would have a "real profession"; I would do math or I would write. These two paths seemed closely linked to me, as if they called out to one another. Age, travel, and books seemed to guarantee that one day, by dint of work and practice, it would be possible for me to attain these writings of words and numbers that were destined to fill my life.

Having left home early on, I did, as did many others, an apprenticeship in communal meals, in institutional food lacking both taste and identity, and in noisy, depressing cafeterias. My only memory is of the omnipresence of potatoes, sticky rice, and meats that could not be named, which, to my mind, perpetuated the survival of ancient animal species: only their great genetic age seemed to justify their degree of toughness. I thus discovered, *a contrario*, that up to that point, I had been very well fed, that no one had ever measured out the amount of fruits and cheeses I received, and that the prosperity of a family was expressed first in its daily diet. But for a long time, I still regarded as elementary, conventional, and pedestrian (and therefore a bit stupid) the feminine savoir faire that presided over buying food, preparing it, and organizing meals.

One day finally, when I was twenty, I got my own small apartment, apart from school barracks, that included a rudimentary but sufficient facility in which to prepare my meals. I discovered myself invested with the care of preparing my own food, delighted with being able to escape from the noise and crowds of college cafeterias and from the shuttling back and forth to face preordained menus. But how was I to proceed? I did not know how to do anything. It was not a question of waiting for or asking advice from the women in the family because that would have implied returning to the maternal hearth and agreeing to slip back into that discarded feminine model. The solution seemed obvious: just like everything else, these sorts of things could be learned in books. All I had to do was find in a bookstore a source of information that was "simple," "quick," "modern," and "inexpensive," according to my then naive vocabulary. And in order to secure the means to do so (at least, so I thought), I undertook the close study of a paperback cookbook devoid of both illustrations and "feminine" flourishes. To my mind, this absence endowed the book with eminent practical value and sure efficiency.

From the groping experience of my initial gestures, my trials and errors, there remains this one surprise: I thought that I had never learned or observed anything, having obstinately wanted to escape from the contagion of a young girl's education and because I had always preferred my room, my books, and my silent games to the kitchen where my mother busied herself. Yet, my childhood gaze had seen and memorized certain gestures, and my sense memory had kept track of certain tastes, smells, and colors. I already knew all the sounds: the gentle hiss of simmering water, the sputtering of melting meat drippings; and the dull thud of the kneading hand. A recipe or an inductive word sufficed to arouse a strange anamnesis whereby ancient knowledge and primitive experiences were reactivated in fragments of which I was the heiress and guardian without wanting to be. I had to admit that I too had been provided with a woman's knowledge and that it had crept into me, slipping past my mind's surveillance. It was something that came to me from my body and that integrated me into the great corps of women of my lineage, incorporating me into their anonymous ranks.

I discovered bit by bit not the pleasure of eating good meals (I am seldom drawn to solitary delights), but that of manipulating raw material, of organizing, combining, modifying, and inventing. I learned the tranquil joy of anticipated hospitality, when one prepares a meal to share with friends in the same way in which one composes a party tune or draws: with moving hands, careful fingers, the whole body inhabited with the rhythm of working, and the mind awakening, freed from its own ponderousness, flitting from idea to memory, finally seizing on a certain chain of thought, and then modulating this tattered writing once again. Thus, surreptitiously and without suspecting it, I had been invested with the secret, tenacious pleasure of *doing-cooking*.

When this became clear in my mind, it was already too late; the enemy was on the inside. It then became necessary to try to explain its nature, meaning, and manner to myself in the hopes of understanding why that particular pleasure seems so close to the "pleasure of the text," why I twine such tight kinship ties between the writing of gestures and that of words, and if one is free to establish, as I do, a kind of reciprocity between their respective productions. Why seek to satisfy, with one as with the other, the same central need to *spend* [*dépenser*], to dedicate a part of one's lifetime to that of which the trace must be erased? Why be so avid and concerned about inscribing in gestures and words the same fidelity to the women of my lineage?

There have been women ceaselessly doomed to both housework and the creation of life, women excluded from public life and the communication of knowledge, and women educated at the time of my grandmothers' generation, of whom I would like to retain a living and true memory. Following in their footsteps, I have dreamed of practicing an impoverished writing, that of a *public writer* who has no claim to words, whose name is erased. Such writing targets its own destruction and repeats, in its own way, that humble service to others for whom these non-illustrious women (no one knows their names, strength, or courage anymore) represented for generations basic gestures always strung together and necessitated by the interminable repetition of household tasks performed in the succession of meals and days, with attention given to the body of others.

Perhaps that is exactly what I am seeking in my culinary joys: the reconstruction, through gestures, tastes, and combinations, of a *silent legend* as if, by dint of merely living in it with my hands and body, I would succeed in restoring the alchemy of such a history, in meriting its secret of language, as if, from this stubborn stomping around on Mother Earth, the truth of the word would come back to me one day. Or rather, a writing of words, reborn, that would finally achieve the expression of its wonderful debt and the impossible task of being able to return its favor. Women bereft of writing who came before me, you who passed on to me the shape of your hands or the color of your eyes, you whose wish anticipated my birth, you who carried me, and fed me like my great-grandmother blinded with age who would await my birth before succumbing to death, you whose names I mumbled in my childhood dreams, you whose beliefs and servitudes I have not preserved, I would like the slow remembrance of your gestures in the kitchen to prompt me with words that will remain faithful to you; I would like the poetry of words to translate that of gestures; I would like a writing of words and letters to correspond to your writing of recipes and tastes. As long as one of us preserves your nourishing knowledge, as long as the recipes of your tender patience are transmitted from hand to hand and from generation to generation, a fragmentary yet tenacious memory of your life itself will live on. The sophisticated ritualization of basic gestures has thus become more dear to me than the persistence of words and texts, because body techniques seem better protected from the superficiality of fashion, and also, a more profound and heavier material faithfulness is at play there, a way of being-in-the-world and making it one's home.

Innumerable Anonymous Women

> I began with a few very precise images of my childhood: seeing my mother next to the kitchen sink, my mother carrying packages. I did not want to do a kind of naturalism, but rather, with a very stylized image, to attain the very essence of reality.[3]
>
> When all is said and done, *Jeanne Dielman* is a hyperrealist film about the use of time in the life of a woman bound to her home, subjected to the imposed conformity of everyday gestures . . . I thus revalued all these gestures by giving back to them their actual duration, by filming in sequence and static shots, with the camera always facing the character, whatever the character's position. I wanted to show the right value of women's everyday life. I find it more fascinating to see a woman—who could represent all women—making her bed for three minutes than a car chase that lasts twenty.[4]
>
> But with regard to my cinema, it seems to me that the most appropriate word for it is *phenomenological*: it is always a sequence of events, of tiny actions described in a precise way. And what interests me precisely is this relationship with the immediate glance, the way one looks at those tiny actions that are going on. It is also a relationship with strangeness. Everything is strange to me; everything that does not surface is strange. It is a strangeness linked to a knowledge, linked to something that you have always seen, which is always around you. This is what produces a certain meaning.[5]

These sharp-edged sentences and effective images of Chantal Akerman translate almost too well the intention of this study on the Kitchen Women Nation [*le peuple féminin des cuisines*]. In this voice, in its gentleness and its violence, I recognized the same necessity of returning to triviality in order to break through the entrapment. It represents the same will to learn how to detach one's view from that of "high culture," this inherited background, much praised among the residents of good neighborhoods. It also represents the same distance with regard to the "popular culture" whose naive praises one sings all the better while one is burying or despising those people who gave birth to it. It represents the same refusal to denigrate a "mass culture" of which one deplores the mediocrity produced on the industrial scale, all the while sharing in the advantages that this industry provides. Thus, a will to turn one's eyes toward contemporary people and things, toward ordinary life and its indeterminate differentiation. A wish to rediscover the "taste for the anonymous and innumerable germination"[6] and everything that constitutes the *heart* of it. A will to see the fragile frost of habits, the shifting soil of biases in which social and user circulations are inserted and where shortcuts are to be guessed at. A will to accept as worthy of interest, analysis, and recording the ordinary practices so often regarded as insignificant. A will to learn to consider the fleeting and unpretentious ways of operating that are often the only place of inventiveness available to the subject: they represent precarious inventions without anything to consolidate them, without a language to articulate them, without the acknowledgment to raise them up; they are bricolages subject to the weight of economic constraints, inscribed in the network of concrete determinations.

At this level of social invisibility, at this degree of cultural nonrecognition, a place for women has been granted, and continues to be, as if by birthright, because no one generally pays any attention to their everyday work: "these things" must be done, someone has to take care of them; this someone will preferably be a woman, whereas in the past it was an "all-purpose maid," whose title alone best describes her status and function. These jobs, deprived as they are of visible completion, never seem likely to get done: the upkeep of household goods and the maintenance of family bodies seem to fall outside the bounds of a valuable production; only their absence garners attention, but then it is a matter of reprobation. As the healthy verve of the Quebec women sings it, "Mom don't work 'cause she got too much to do!"[7]

Like all human action, these female tasks are a product of a cultural order: from one society to another, their internal hierarchy and processes differ; from one generation to the next in the same society, and from one social class to another, the techniques that preside over these tasks, like rules of action and models for behavior that touch on them, are transformed. In a sense, each operator can create her *own style* according to how she accents a certain element of a practice, how she applies herself to one or another, how she creates her personal way of navigating through accepted, allowed, and ready-made techniques. From thus drawing on common savoir faire, each perfect homemaker ends up giving herself a manner suitable for playing one chronological sequence on top of another and for composing, on given themes, *ne varietur*, music of variations that are never determined in a stable form.

Culinary practices situate themselves at the most rudimentary level, at the most necessary and the most unrespected level. Traditionally in France, the responsibility for them falls almost exclusively on women and these tasks are the object of ambivalent feelings: the value of French cuisine is enhanced when compared to that of neighboring countries; the importance of diet in raising children and care for the family is emphasized in the media; the responsibility and role of the housewife as primary buyer and supplier for the household are stressed. At the same time, people judge this work to be repetitive and monotonous, devoid of intelligence and imagination; people exclude it from the field of knowledge by neglecting dietary education in school programs. Yet, except for residents from certain communities (convents, hospitals, prisons), almost all women are responsible for cooking, either for their own needs or in order to feed family members or their occasional guests.

In each case, *doing-cooking* is the medium for a basic, humble, and persistent practice that is repeated in time and space, rooted in the fabric of relationships to others and to one's self, marked by the "family saga" and the history of each, bound to childhood memory just like rhythms and seasons. This women's work has them proliferate into "gesture trees" (Rilke), into Shiva goddesses with a hundred arms who are both clever and thrifty: the rapid and jerky back and forth movement of the whisk whipping egg whites, hands that slowly knead pastry dough with a symmetrical movement, a sort of restrained tenderness. A woman's worry: "Will the cake be moist enough?"; a woman's observation: "These tomatoes are not very juicy, I'll have to add some water while they cook." A transmission of knowledge: "My mother (or aunt or grandmother) always told me to add a drop of vinegar to grilled pork ribs." A series of techniques [*tours de main*] that one must observe before being able to imitate them: "To loosen a crêpe, you give the pan a sharp rap, like this." These are multifaceted activities that people consider very simple or even a little stupid, except in the rare cases where they are carried out with a certain degree of excellence, with extreme refinement—but then it becomes the business of *great chefs*, who, of course, are men.

Yet, from the moment one becomes interested in the process of culinary production, one notices that it requires a multiple memory: a memory of apprenticeship, of witnessed gestures, and of consistencies, in order, for example, to identify the exact moment when the custard has begun to coat the back of a spoon and thus must be taken off the stove to prevent it from separating. It also calls for a programming mind: one must astutely calculate both preparation and cooking time, insert the various sequences of actions among one another, and set up the order of dishes in order to attain the desired temperature at the right moment; there is, after all, no point in the apple fritters being just right when the guests have barely started on the hors d'oeuvres. Sensory perception intervenes as well: more so than the theoretical cooking time indicated in the recipe, it is the smell coming from the oven that lets one know if the cooking is coming along and whether it might help to turn up the temperature. The creative ingenuity of cleverness also finds its place in culinary production: how can one make the most out of left-overs in a way that makes everyone believe that it is a completely new dish? Each meal demands the invention of an alternative ministrategy when one ingredient or the appropriate utensil is lacking. And when friends make a sudden, unexpected appearance right at dinnertime, one must improvise without a score and exercise one's combinatory capacities. Thus, entering into the vocation of cooking and manipulating ordinary things make one use intelligence, a subtle intelligence full of nuances and strokes of genius, a light and lively intelligence that can be perceived without exhibiting itself, in short, *a very ordinary intelligence.*

These days, when the job one has or seeks in vain is often no longer what provides social identity, when for so many people nothing remains at the end of the day except for the bitter wear and tear of so many dull hours, the preparation of a meal furnishes that rare joy of producing something oneself, of fashioning a fragment of reality, of knowing the joys of a demiurgic miniaturization, all the while securing the gratitude of those who will consume it by way of pleasant and innocent seductions. This culinary work is alleged to be devoid of mystery and grandeur, but it unfurls in a complex montage of things to be done according to a predetermined chronological sequence: planning, organizing, and shopping; preparing and serving; clearing, putting away, and tidying up. It haunts the memories of novelists, from the fabulous excesses of Rabelais's heroes, all busy eating, digesting, and relieving themselves,[8] to the "long lists of mounds of food" of Jules Verne,[9] passing through the "bourgeois cuisine" of Balzac's creatures,[10] the recipes of Zola, and the tasty simmering dishes of Simenon's concierges.[11]

Listen to these men's voices describing women's cooking, like Pierre Bonte's simple people, whose hearty accents [on an early morning radio talk show] used to populate city mornings with good savages:

You see, this soup, made with beans, is what we call, of course, a bean soup, but you shouldn't think there are only beans in it. My wife made it this morning. Well, she got up at seven o'clock, her pot of water was on the wood-burning stove—she put her beans on to soak last night—then she added two leeks chopped very fine and some nice potatoes; she put all that together, and when it started boiling, she put her salt pork in. An hour before serving it to us, after three and a half to four hours of cooking, she made a fricassee for it. A fricassee is made in a pan with bacon drippings. She browns an onion in it, and when the onion is nice and golden, she makes up a nice flour roux and then puts it all in the soup.[12]

I will admit it myself: I still dream about the rice croquettes and the fritters that nice children in the Comtesse de Ségur's books used to eat for dinner as a reward for good behavior; I was less well behaved than them and these unknown dishes, which seemed to me adorned with exotic flavors, were never served at our family table. But, taken out of its literary dressing and stripped of its fleeting ennoblement, culinary work finds itself once again in dreary reality. This women's work, without schedule or salary (except to be paid off through service to others), work without added value or productivity (men have more important things to calculate), work whose success is always experienced for a limited duration (the way a soufflé just out of the oven, balancing in a subtle equilibrium, in this glorious peak, is already wavering well before it finally collapses). Yes, women's work is slow and interminable. Women are extremely patient and repeat the same gestures indefinitely:

Women, they peel potatoes, carrots, turnips, pears, cabbages, and oranges. Women know how to peel anything that can be peeled. It's not hard to do. You learn when you're very young, from mother to daughter: "Come and help peel some potatoes for dinner, dear." . . . Women peel potatoes every day, noon and night; carrots and leeks too. They do it without complaining to themselves or to their husbands. Potatoes, they're a woman's problem . . . Women's domain is that of the table, food, and the potato. It's a basic vegetable, the least expensive; it's the one about which you say little but that you peel and prepare in a thousand ways. How am I going to serve the potatoes tonight? That's what you call a domestic problem. Supply. How much importance should we give supply . . .? "You do everything so well, honey. I love your potatoes," says the man. "Will you make me French fries tomorrow?" And the woman makes fries. The stakes are high, higher than the discussion itself: not to make him unhappy so that he'll still want me, as much as he wants my fries. And the next day, she peels again, vegetable after vegetable; she chops and slices them into small, patient, meticulous, and identical pieces. She does this so that everything is good and also pretty, well presented. Something that is well presented makes you hungry. Then you'll want to feed yourself, to feed off of me. "I'm hungry for you," says the man. "You're pretty enough to eat. I want to munch on you—one day," says the man. "I'm hungry for the food you give me."[13] [. . .]

Earthly Foods

Why do we eat? The primary reason: to satisfy the energy needs of a living organism. Like other animal species, humans must submit to this necessity throughout life; but they distinguish themselves from other animals by their practice of elected periods of abstinence (voluntarily, or to save money, or in periods of shortage) that can go so far as the observance of a rigorous and prolonged fast (Ramadan in Islam or African rites of initiatory purification) or as far as a steadfast refusal of all food (anorexic behavior or hunger strikes driven by a political will that consists of opposing the symbolic and tangible counterviolence of torture of one's own body as a response to the violence of power or of the established order).

This daily intake of food is not undifferentiated. In quality as well as quantity, it must satisfy certain imperatives (the makeup of intake and the relative proportions of various nutrients) at the risk of not being able to ensure the maintenance of good health for the individual, protection against cold or infectious agents, and the capacity to sustain steady physical activity.[14] Below a certain quantitative threshold, a paltry *subsistence intake* ensures the temporary survival of a weakened organism, which reabsorbs in part its own tissues to feed itself, loses strength and resistance and, if this situation persists, enters a state of undernourishment.[15]

Similarly, a diet unbalanced in both quality and diversity leads to malnutrition. Daily intake must provide a sufficient supply of proteins, a quantity proportionate to the weight, age, and physical activity of the individual; the makeup of this supply (proteins from both animal and plant sources) also enters into the equation. In addition to the three classes of nutrients (proteins, fats, and carbohydrates), alimentary intake must also provide certain indispensable elements (vitamins, minerals, and both amino and fatty acids), but according to a system of subtle proportions and interrelations: thus, the same nutrient "can represent the gain of one vitamin at the loss of another";[16] likewise, mineral needs cannot be isolated from each other because their metabolism is interdependent; certain elements must be combined in order to allow assimilation, and so on.

The history of medicine itemizes an entire list of illnesses caused by deficiency or resulting from the poor quality of absorbed foods, such as the *mal des ardents*, common during the Middle Ages and the Renaissance, whose descriptive name concerns ergotic poisoning from ergoticized rye flour, that is, rye flour that contains a parasitic fungus; the *new plague* of the Crusades, which was the vitamin deficiency associated with scurvy; *Saint Quentin's grand mal* during the Hundred Years' War, an edema of famine caused by the lack of proteins; pellagra caused by a lack of vitamin B_3, common in rural areas where corn provides the food base (though certain local practices, such as the liming of corn in Mexico, have avoided it in Latin America), and so on.[17]

There does not exist one precise description of alimentary intake that is always appropriate for human beings; the needs for proteins, minerals, and vitamins vary according to the size, weight, sex, and climatic life conditions of individuals (living conditions, clothing, protection from inclement weather), as well as the intensity of their activity and the stage of life (growth, pregnancy, breast-feeding, adulthood, old age). We now know that a state of malnutrition coming at certain periods of life has profound and lasting consequences: thus, the undernourishment of a nursing infant slows brain growth and leads to irreversible disorders of the structure and function of the brain, and hence to effects on mental ability.[18] We know about the temporary sterility in women of childbearing age caused by severe undernourishment (for example, the amenorrhea associated with famines, observed during prolonged sieges in cities such as Leningrad during World War II).[19] The lack of animal proteins considerably increases the risk of toxemia in pregnant women;[20] the nursing period requires a surplus in both quantity and quality of food intake, to protect the mother and the infant.

To all these traditional illnesses must be added, in the overfed West, the illnesses of abundance or excess that are at the heart of the new *maladies of civilization*. Epidemiology has documented some troubling facts.

1. The growing frequency of cancers and disorders of the intestines seems to be in direct correlation with the impoverishment of diets rich in cellulose and vegetable fiber (resulting from a drop in the consumption of cereals and a taste for white bread and refined flour);[21] this is why bakers have been encouraged to offer "bran bread," which is less refined, and "specialty breads" made from different grains.

2. The extent of cardiovascular diseases appears linked to, among other things, diets too rich in glucides and certain types of fats.[22]

3. The increase in cases of death by cancer in France, from 1950 to 1967, differs depending on the region (the highest rate exists in the north and the lowest in the Mediterranean region), and seems to correlate to diet (the lowest consumption of fruit is in the north as well as the high use of butter for cooking instead of vegetable oil).[23]

That should suffice to underline what is essentially at stake for health in the composition of an alimentary diet and how many subtle and multiple requirements there are to reconcile with the pleasures of the table—something that at first glance seems simple and natural.[24] If things are much more complicated than they first appear, it is because there is no standard popular wisdom in this field: beyond all economic limitations, certain culinary traditions encourage by choice diets that are deficient or dangerous for certain of their members. Thus, we know about the strange story of the kuru that selectively affects the women and children in the Fore tribe in New Guinea: this always fatal illness is a slow viral attack of the central nervous system, transmitted by the consumption, reserved as a choice gift to warriors' wives, of the brains of killed enemies. A genetic factor is most likely responsible for the persistence of the virus and for the absence of an adapted immune response.[25]

Moreover, there is no innate wisdom in the individual: to be convinced, it suffices to perform a short inquiry among one's family circle; this would allow us to quickly assemble an amazing assortment of stupidity. Even well-educated people, attentive as they are to precise and exact information in other fields, shamelessly talk about foods that are "easily digestible," "light," "fortifying," or "good for kids." Such talk becomes a sundry mix of hearsay, old wives' tales, baseless prejudices, and vague pieces of information gleaned anywhere. Here, what is obvious loses its primary clarity and collides with the absence of man's internal regulation of alimentary behaviors, because human practices are more flexible and adaptable than those of animals, but they also make humans more vulnerable.[26] Through ignorance, lack of concern, cultural habit, material shortage, or personal attitude (if one considers, for example, the complementary behaviors of bulimia and anorexia), people can ruin their health by imposing on themselves deficient or excessive diets, and go as far as to die from what they eat, "digging their graves with their teeth," as it were—not to mention accidental or intentional poisonings (in these cases, success stories often remain anonymous).

No more so than any other elements of material life, food is not presented to humans in a natural state. Even raw or picked from a tree, fruit is already a *cultured foodstuff*, prior to any preparation and by the simple fact that it is regarded as being edible. Nothing is more variable from one human group to another than the notion of what is edible: one has only to think about the dog, spurned in Europe but appreciated in Hong Kong; or grasshoppers, considered disgusting here yet highly prized in the Maghreb; or the worms savored in New Guinea.[27] Closer to home, there is the offal lovingly simmered in Latin countries but despised in the United States, with, moreover, differences of national tradition inside Europe itself: certain cuisines prefer brains and others tripe, but would not eat lamb spleen or *amourettes*, fried spinal marrow.[28] Sometimes the necessity or the contagion of exoticism pushes us to eat elsewhere what we would never consider eating at home, but we have also seen people reduced to famine who allow themselves to die rather than eat unusual foods, such as the African villages in a rural area suffering from a long drought that gave to their animals the powdered milk that international relief organizations distributed among them.[29]

There exist a complex geography and a subtle economy of choices and habits, of likes and dislikes. Food involves a primary need and pleasure, it constitutes an "immediate reality," but "substances, techniques, and customs all enter into a system of significant differences,"[30] a system that is coherent and illogical. Humans do not nourish themselves from natural nutrients, nor from pure dietary principles, but from *cultured* food-stuffs, chosen and prepared according to laws of compatibility and rules of propriety unique to each cultural area (in the Maghreb, for example, poultry is stuffed with dried fruits and in England currant jam is served with a roast, whereas French cuisine practices a strict separation between sweet and savory). Foodstuffs and dishes are arranged in each region according to a detailed code of values, rules, and symbols,[31] around which is organized the alimentary model characteristic of a cultural area in a given period. In this detailed code, more or less well known and followed, the organizer of the family meal will draw on inspiration, on her purchase and preparation possibilities, on her whim and the desires of her "guests." But sometimes lassitude overtakes her in the face of the ephemeral, perishable character of her task. Even her "successes," in the words of her "customers," will not seem to justify the trouble they cost her: "It's such a mess, and then in almost no time, everything disappears. I find that hopeless" (from Marie Ferrier's interview of Irène).

This wavering, a fugitive moment of discouragement, the Kitchen Women Nation knows it all too well, but does its best not to give in to it. Tomorrow will be the day for another meal, another success. Each invention is ephemeral, but the succession of meals and days has a durable value. In the kitchen, *one battles against time*, the time of this life that is always heading toward death. The nourishing art has something to do with the art of loving, thus also with the art of dying. In the past, in the village, a burial was the chance for an extended family reunion around a solid meal, serious and joyful, after the interment. People thus began the work of mourning by sharing earthly foods. In the past, death was a part of life; it seems to me that it was less alarming that way.

Notes

1. Luce Giard, "La fabrication des filles," *Esprit* (June 1976): 1108–23; and in collaboration, "Note conjointe sur l'éminente relativité du concept de femme," *Esprit* (June 1976): 1079–85.

2. Marcel Mauss, *Sociology and Psychology*, trans. Ben Brewster (London: Routledge and Kegan Paul, 1979), 95–123.

3. Chantal Akerman, interview in *Télérama*, cited in *Études* (April 1976): 564.

4. Chantal Akerman, a conversation with Jacques Siclier, *Le Monde*, January 22, 1976; this interview was preceded by an article by Louis Marcorelles, "Comment dire chef-d'œuvre au féminin?"

5. Interview with Chantal Akerman, *Cahiers du cinéma*, no. 278 (July 1977): 41.

6. Freddy Laurent, *La Revue nouvelle* (March 1974): 296.

7. Le Théâtre des cuisines, *Môman travaille pas, a trop d'ouvrage!* (Montreal: Éditions du Remue-ménage, 1976).

8. On Rabelais, Noëlle Châtelet, *Le Corps à corps culinaire* (Paris: Seuil, 1977), 55–92.

9. "This narrative, but also all the others, and this one with no exception, are interlarded with long lists of mounds of food, as in Dickens, Rabelais, Cervantes . . . For Verne, as for those writers, there is a naive and simple fantasy of feeling full, the horror of emptiness . . . nature is the mother and she provides food. She is full everywhere, as Leibniz said, and she cannot be hungry. Man is the hole in Nature, he is the hunger of the world" (Michel Serres, *Jouvences sur Jules Verne* [Paris: Minuit, 1974], 176).

10. Robert Courtine, *Balzac à table* (Paris: Laffont, 1976).

11. Robert Courtine, *Le Cahier de recettes de Mme Maigret* (Paris: Laffont, 1974), with a preface by Georges Simenon; and *Zola à table: trois cents recettes* (Paris: Laffont, 1978).

12. Pierre Bonte, *Le Bonheur est dans le pré* (Paris: Stock, 1976), 232. See the book review by Catherine B. Clément, "Pierre Bonte et ses philosophes du matin: le Christophe Colomb de Chavignol," *Le Monde*, February 15–16, 1976.

13. Nicole, "Les pommes de terre," *Les Temps modernes*, issue titled "Les femmes s'entêtent" (April–May 1974): 1732–34.

14. Arlette Jacob, *La Nutrition* (Paris: PUF, "Que sais-je?" 1975); contains a clear and precise account of these questions.

15. Jean Trémolières, "Dénutrition," in *Encyclopaedia universalis* (Paris: Encyclopaedia Universalis France, 1968).

16. Jacob, *La Nutrition*, 108–16.

17. Jean Claudian, "L'alimentation," in Michel François, ed., *La France et les Français* (Paris: Gallimard, Pléiade, 1972), 152–53; and Maurice Aymard, "Pour l'histoire de l'alimentation: quelques remarques de méthode," *Annales ESC* 30 (1975): 435, 439–42.

18. John Dobbing, "Malnutrition et développement du cerveau," *La Recherche*, no. 64 (February 1976): 139–45; and Ciba Foundation Symposium, *Lipids, Malnutrition and the Developing Brain* (Amsterdam: North Holland, 1972).

19. Emmanuel Le Roy Ladurie, "L'aménorrhée de famine (XVIIᵉ–XXᵉ siècle)," *Annales ESC* 24 (1969), special issue, *Histoire biologique et société*, 1589–1601.

20. Tom Brewer, *Metabolic Toxemia of Late Pregnancy* (Springfield, Ill.: C. C. Thomas, 1966); Tom Brewer, "Consequences of Malnutrition in Human Pregnancy," *Ciba Review: Perinatal Medicine* (1975): 5–6.

21. Dr. Escoffier-Lambiotte, "Vers une prévention des affections et des cancers intestinaux?" *Le Monde*, September 17, 1975. On the great number of stomach cancers in Japan linked, it would seem, to a certain type of food diet, see J.-D. Flaysakier, "Au Japon, le cancer de l'estomac: un exemple réussi de prévention," *Le Monde*, October 3, 1979.

22. M. D., "L'alimentation et la préservation de la santé," *Le Monde*, September 29, 1976.

23. Dr. Escoffier-Lambiotte, "Graisses alimentaires et fibres végétales," *Le Monde*, February 16, 1977, according to the dissertation by F. Meyer (Lyons).

24. Henri Dupin, *l'Alimentation des Français: Évolution et problèmes nutritionnels* (Paris: Éditions Sociales Françaises, 1978).

25. This discovery earned its author, Daniel Carleton Gajdusek, a Nobel Prize in 1976; see Edmond Schuller, "Virologie tous azimuts: du cerveau de l'anthropophage . . ." *La Recherche*, no. 73 (December 1976): 1061–63.

26. Jacob, *La Nutrition*, 117–19.

27. Yvonne Rebeyrol, "Y a-t-il encore des explorateurs?" *Le Monde*, September 14, 1977. This article details François Lupu's trip into the Sepik Valley [New Guinea], where people eat the parasitic worms of sago palm trees.

28. Léo Moulin, *L'Europe à table: Introduction à une psychosociologie des pratiques alimentaires* (Paris and Brussels: Elsevier Sequoia, 1975), 20–26, 129–30, 136–38.

29. In this specific case, it seems that biological factors come into play: European ethnic groups may be the only ones to retain until adulthood the capacity to produce an enzyme necessary for the complete digestion of raw milk.

30. Roland Barthes, "Pour une psychosociologie de l'alimentation contemporaine," in Hémardinquer, *Pour une histoire*, 309–10.

31. Aymard, "Pour l'histoire de l'alimentation," 431ff.

7

The Recipe, the Prescription, and the Experiment

Jack Goody

3 *Witch.* Scale of dragon, tooth of wolf;
Witches' mummy; maw, and gulf,
Of the ravin'd salt-sea shark;
Root of hemlock, digg'd i' th' dark;
Liver of blaspheming Jew;
Gall of goat, and slips of yew,
Sliver'd in the moon's eclipse;
Nose of Turk, and Tartar's lips;
Finger of birth-strangled babe,
Ditch-deliver'd by a drab,
Make the gruel thick and slab:
Add thereto a tiger's chaudron,
For th' ingredience of our cauldron.

> *Note.* "Egyptian mummy, or what passed for it, was formerly a regular part of the *Materia Medica*"; "Turks and Tartars were not only regarded as types of cruelty . . . but . . . they were *unchristened*, and hence valued by the witches."

Macbeth, Act IV, Scene I

The organization of the kind of non-familial household required by a king's court, a standing army or an educational college tends to generate lists and tables of a variety of kinds, both of the food to be purchased, prepared and consumed (the rations), and of the individuals who are entitled to these benefits (the personnel). These lists not only "reflect" certain aspects of the social organization; they also determine other aspects, in that they have certain implicit features (such as hierarchy and lateral placement) that influence behavior, as well as other explicit features that prescribe it. Moreover through these formulations they make possible what would present enormous difficulties in an oral society, both in the field of social organization as well as in the growth of knowledge. The development of major branches of medicine, slow as it has been, depends upon the ability to examine critically a list of ingredients, their mode of employment, and reports of the results, then to share and diffuse this knowledge for the use of practitioners and their patients.

There are three kinds of nominal list which it is convenient to distinguish at this point. Firstly there is the nominal roll, which lists the members of a group (e.g. a school class or army platoon) who are or should be present, or who are entitled to certain privileges (e.g. the list of the members of a guild or the Fellows of a college). In this type falls the ration list of any army, since entrance on the roll and participation in the activities of the organization are two sides of the same coin. Where they are not, we get a selective list, sometimes self-inscribed, which indicates those members who have put their names down or who have been chosen for a certain activity, such as dining in Hall or going out on patrol. Thirdly the retrospective, as distinct from the prospective, list gives the names of those individuals who were in fact present, who did attend, a type of list which may obviously be used outside the kinds of "corporate" situations mentioned before.

When the Fellows of St John's College, Cambridge, dine together each evening, two pieces of paper lie upon the table, the menu and the dining-list.

The list of diners is an example of the second type of nominal list mentioned; it is self-inscribed and indicates, both to the kitchens and to other Fellows, who has decided to dine that evening. It is worth examining since it displays a number of the characteristics of lists in general. Firstly it insists upon an explicit hierarchy; the list has to be arranged so that the items, names of persons in this case, are one above another, from top to bottom. It is true that horizontal "strings" of names, found in the plurality of authors of some academic papers, place one item *before* another, thus implying priority, or seniority, running from left to right. But left to right is a more "conventional" movement than that from top to bottom. Indeed, systems of writing that begin at the right and move in the opposite direction, such as Arabic and Hebrew, clearly have a different order of priority. No-one, I think, has employed a system of writing that starts at the foot of the page and moves to the top, then jumps to the bottom again. The vertical hierarchy is more compelling, more insistent, than horizontal differentiation; it is the difference between head and foot in one context and right and left on the other.

Indeed, so compelling is the idea of hierarchy in a written list that extraordinary steps have to be taken to avoid the implications of higher and lower, first and last, with their associations of differential power or responsibility. To get over the problem, the signatories of a letter have to place their names in a circle, producing a Round Robin of the kind sent to Dr Johnson by his friends when they wanted to make some slight protest about the use of Latin for his obituary of the dramatist, Oliver Goldsmith (page 80), to which the Doctor replied that he would "never consent to disgrace the walls of Westminster Abbey, with an English inscription." In this way the Round Robin has similar functions to the Round Table, namely, to avoid the linear hierarchy implicit in the list as part of a figurative table and the seating plan as part of a physical one.

The hierarchy of the dining-list is distinct, with the Master at the top followed by the President (known by their titles, rather than their names), then other Fellows appearing in strict order of seniority. Because it is written down in the form of a list and placed upon the table at which the meal is eaten, the hierarchy is always visibly present in front of every diner. Indeed during the meal, Fellows can be seen consulting this list, not merely to ascertain the names of guests or younger Fellows, but to confirm their own position in the daily ranking. The hierarchy is therefore more linear than in most oral societies. It imposes itself in an immediate way, making the ordering more "meaningful" to the actors than it has any right to be, seniority for this purpose depending simply upon date of election. But the list in the form of a single column also gives a peculiar shape to the hierarchy; unlike the positions in

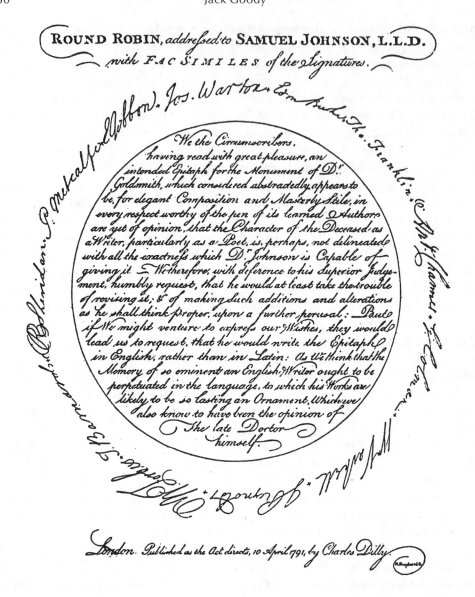

a table of organization, no one can be equal to any one else. There is no horizontal placement; every item must be vertically ordered, even if this means forcing hierarchy on what is in practice a looser form of social organization.

I do not want to imply that unilineal ranking is only to be found in written cultures. Clearly many forms of spatial and temporal ordering may place individuals in distinctive ranked positions of this kind e.g. the file of Gonja chiefs that greet the paramount at the annual Damba celebration, or the recipients of the kava cup at a Tongan festival; a central phase of mass ceremonies is often devoted to such ordering. My point is that the list provides an ordering whether you want it or not.

The list of those dining has particular as well as general characteristics. Towards the lower end, individuals begin to be given initials as well as titles and surnames. In earlier days, lists of county cricket teams, and even international ones, used to indicate who was paid as a professional and who played as an amateur by giving the names of the latter with their initials, and the former without; so that an entry read H. D. G. Leveson Gower in one case, and Hobbs on the other. Each year, an important fixture in the cricketing calendar was the match of Gentlemen against Players, those with initials against those without. In the case of the list of diners, it is the Fellows who have no initials, and those "with dining rights" who are defined in a more specific manner. It would perhaps be a waste of time to seek out the reasons for this distinction, the use or non-use of initials being simply an arbitrary "diacritical" feature to set apart the two categories. But it might be suggested that the Fellows, who are known to each other, require minimal marking, the others maximum, for both in cricket and in Hall, Smiths (and others with similar names) are excluded from the rule when there is more than one of them present: in case of ambiguity initials are provided.

One further discriminating feature in the dining-list is the lateral displacement of certain names, that of guests. They appear partly in a different column and partly under the name of their host, indicating their protected but interstitial status.

Finally, the most distinctive feature of the usual list is its exclusively male character. Some lists contain the names of women. These would appear not as Fellows but (and only recently) as guests and very occasionally in the category of those having dining rights. Otherwise women in the College, as the list indicates by implication, occupy the non-academic roles, serving at table, making beds, typing letters. The differentiation, indeed discrimination, between the roles of the sexes is radical, and is clearly illustrated in the dining-list.

The second piece of paper lying on the table is the menu, the list of rations as distinct from the list of those entitled to them. Conventional in form, it presents the shape of the meal to come. The shape is a regular one; variation falls within a restricted range, although deliberate changes do occur over the longer run. The first course consists of the soup, or an equivalent preparatory dish such as antipasto or *hors d'oeuvres*; it is followed by the main course, a hot dish of meat and vegetables, occasionally fish (especially on Fridays); cold meat, vegetarian dishes and alternatives to pork are available on request. For the third course three alternatives are announced, namely, a fruit pie, cheese and a sweet. In addition, the menu specifies the wine to be served, but not the beer which may be taken instead. This omission indicates two aspects of the menu. Firstly, there is no need to specify constants; the menu deals, and enables the individuals concerned to deal, with variables. Secondly, though the first reason probably covers the omission of beer, the inclusion of the foreign at the expense of the local is typical of the language of this and most high-status English menus, with their frequent recourse to restaurant French.

The role of the written menu in a restaurant is somewhat different since it specifies not the variations offered diachronically (from day to day) to a fixed Fellowship, but the alternatives offered synchronically (on one day) to a variable clientele. In the first place it acts as an advertisement, when placed in the window to attract custom. Secondly, it helps the proprietor to offer and the client to choose among a wider range of dishes. Some years ago there used to be a small Italian restaurant in Frith Street, Soho, where there was no menu; the waiter came up to the table and asked:

"Whadyawan?"
To which the standard answer was,
"Whadyagot?"

He then reeled off the lists of *plats du jour* and shouted your answer down the stair-well to the kitchen below. Such a procedure is possible in small "family" restaurants where the range of dishes is limited. But the offer of a wider choice *necessitates* a recourse to writing, because even if the waiter could recall a much longer (and frequently-changing) list, his client would be unable to make up his mind among these dishes because he cannot scan the choices available to him. Alternatively, he would be likely to plump for the first or the last dish named, or possibly to seize upon a name that he had heard before. An arbitrary or conservative choice would be made because of the difficulties of scanning.

The use of the written menu for the sale of meals in restaurants and (to a lesser extent) in hotels is easy to understand. But its adoption for high-status dinners is equally interesting because it implies not only the complete separation between consumers and preparers, between table and kitchen, but also a marked separation between the servers (standing, mobile) and the consumers (sitting, immobile), since information about the content of the meal is mediated impersonally by a piece of paper. Of course, matters of prestige enter directly into the whole situation. Written communication is valued above oral; the menu in itself has a high status, the symbol of luxury and of choice, of the great rather than the little cuisine. The menu is appropriate to the special occasion, the non-domestic meal, the repast prepared by men rather than women.

The two pieces of paper placed on the High Table of St John's College represent only the surface structure; behind it lie two other schedules of differing kinds that serve to organize and frame the meal. The first is the board that carries the graces said before and after dinner. However well he knows the grace, the designated speaker holds this board in front of him, and half reads, half recites the prayer. The rendering itself provides a continual subject for comment at the dining tables, a conversational opener used almost as frequently as the weather or the passage of the term. Any alteration is seen either as an unconscious mistake or else as a deliberate attempt to alter a highly ritualized pattern of speech.

The grace itself is a variant of one recited in many Cambridge Colleges. The existence of these variations shows that some "creative" innovation has taken place in the past. However, what is presently required is strict conformity to a written original, which together with the comments of the listeners (whose model is the identical printed sheet), serves to correct any deviation.

The consecration not only asks for the blessings of the universal God upon this particular meal; it also intercedes on behalf of those benefactors of the College who have provided for the repast. In this way the communal meal is marked off from ordinary life and the consumption of food and wine made into something more than either hunger or gluttony demands; it provides a "sacred frame."

The fourth schedule that organizes the meal, in a secular rather than a sacred way, is the recipe, one of a number of lists that adorn the kitchens, including the nominal roll of available staff (the "strength" rather than the "establishment"), and the duty roster that allocates them particular times of work. Behind the recipe lies the shopping list, that is, the list of objects required to implement the recipe, which at once precedes and succeeds the in-gathering of the unprepared (not necessarily raw) food. Since the immediate source of food may be neither the market nor the field, but rather the storehouse where it has found its way as the result of either primary production or the collection of tribute, lists of tribute and stores are in themselves closely connected with the shopping list, though they consist of statements of what has been done rather than prescriptions (the suffix "pre" combined with the verb "write" indicating its nature) for what should be done. The shopping list is a way of constructing the future schedule of an individual or group, a schedule of movements in space and in time, relative to the requirements of the household economy; it is not simply a statement of what one wants, for writing permits a person to re-order the items in relation to the source of supply; articles to be bought at the greengrocer's can be grouped together, and this set put next to that to be bought at the baker's, the shop next door; thus the shopping list can be made to represent future movements of a fairly precise kind.

Basically, we have three types of list concerned with the meal itself, one connected with the "production" of the food for the kitchen, i.e. the shopping list, one connected with the preparation of the food, i.e. the recipe, and one with the consumption of food, i.e. the menu. Each of these is a PLAN in the sense given to that word by Miller et al. (1960) and they are plans in which the externalization of the organizing mechanism makes it more determinative, more enduring, more inclusive and more formal. The first and the third I have briefly considered. Let me now turn to the recipe, which has implications over a wider field than the meal, since it lists not only what ingredients are required but what actions have to be carried out in order to produce the menu. Etymologically the word derives from the Latin verb, to take (as does the cognate, receipt), and was originally used by physicians in the abbreviated form of R or Rx "to head prescriptions, and hence applied to these and similar formulae." But the medical meaning was soon supplemented by the culinary usage, and by 1500 we find "recipe brede gratyd, and eggis" (*Babees Book*, Harl. MS. 5401) which, in the eighteenth century became generalized to "a statement of the ingredients and procedure necessary for the making or compounding of some preparation, esp. a dish in cookery; a receipt". Indeed the word "receipt" is essentially an early version of the same, being used in its medical sense in Chaucerian times, and defined by the Oxford English Dictionary in virtually identical terms, the context once again being medical as well as culinary. A usage of 1703 refers to "Medicinal and Cookery receipts collected from the best authors".

The point about a recipe or a receipt that emerges from these dictionary definitions and literary usages is their essentially written character. The recipes are collected in one place, classified, then serve as a reference book for the doctor or the cook, for the sick or the hungry, as in Dryden's line, "The Patients, who have open before them a Book of admirable Receipts for their Diseases" (Dryden, *Juvenal* Ded., 1697). For recipes, once collected, have then to be tried: "There is not a receipt in the whole extent of chemistry which I have not tried" (Malkin, *Gil Blas* XII. iv para 5, 1809). And once recipes are set down and tried in this manner, then some turn out better, or more approved, than others: Steele writes of "The most approved Receipt now extant for the Fever of the Spirits" (*Spectator* No. 52 para 3). In these effective remedies, common elements can easily be identified and their effects assessed: "In all good receites and medicynes Amomium is ofte ido" (Trevisa, *Barth*. De P. R. XVII, viii. Bodl. M.S.). Experiment, assessment, the isolation of common elements, all are encouraged by the written recipes, whose very existence changes the course and nature of teaching: "A Booke . . . teaching the waye of making diverse good and excellent Receiptez" (Sloane M.S. 2491 If. 73, c. 1500). For the recipe now exists independently of the teacher; it has become depersonalized and acquired a more general, universal quality.

The written character of the recipe is brought out in the second, and nowadays more usual, meaning of receipt. As early as 1442 we find the Rolls of Parliament referring to the "bookes of receyte," containing the "*written* acknowledgement of money or goods received into possession or custody." The goods of course might be tribute, taxes, gifts or alms, but the common characteristic is that their acceptance should be confirmed by means of a written document (or at the very least a signature at the foot of a printed one), which can then serve as evidence that the transaction has taken place, either to the employer of an intermediary or to a court of law, both situations assuming a certain degree of social or physical distance between the transacting parties, a distance that writing itself promotes. For this purpose a receipt-book may be used and in the seventeenth century, significantly, this same term was used for a book of medical or cooking recipes.

The recipe or receipt, then, is a written formula for mixing ingredients for culinary, medical or magical purposes; it lists the items required for making preparations destined for human consumption. Clearly oral cultures also follow relatively standard procedures for such purposes. What then is gained or lost by committing such instructions to writing?

Let us look at some of the very earliest examples. Towards the end of the third millennium B.C.E., a Sumerian physician decided to collect and record his most prized medical prescriptions. Kramer describes this tablet as the first pharmacopoeia. In one way the claim is nonsense; earlier man certainly had his *materia medica* related to the classifications of plants, animals and minerals. What we have here is the first *written* pharmacopoeia, the importance of which was that it provided the starting point for a series of incremental developments associated with the storage, categorization and development of medical information.

The text presented by Kramer indicates not only the materials employed but also the mode of employment, e.g. salves applied externally (often mixed with wine), filtrates also applied externally (often prepared by decoction), liquids to be taken internally (often mixed with beer); indeed, some of the chemical operations were similar to those used for producing alkali ash and fat. Here we shift from the simple list to the recipe itself, which consists of two parts, the list plus the instructions, the ingredients plus the action. The instructions are of two kinds, one for its preparation and one for its consumption, corresponding to the recipe and the mode of employment, the former characteristic of cooking, the latter of medicine. Indeed, in the latter case (and occasionally in the former), the mode of preparation may be deliberately hidden from the outside world, either by a form of secret writing or by sketchy instructions whose meaning will be understood only by members of the craft. In the present case, the recipes lack proper indications of quantities, an omission which may have been a deliberate way of concealing the secrets of the trade. For these recipes were often connected with crafts that required special preparations for their particular tasks. The dyer's craft, for example, probably developed in the Neolithic but became much more complex in the urban civilizations of the Near East. Early Egyptian pictures show coats of varied hue, including striped cloths of many colours. In the temple workshops of Mesopotamia, the sacred vestments of the gods and priests had to be given the correct colours and patterns. Egyptian dyers had books of recipes (though we know them only from the third century A.D.). "They contain the notes of an alchemist from ancient recipe books describing the dyeing of cloth with alkanet, safflower, saffron, kermes, madder, and woad" (Forbes 1954: 249), the specification of which would clearly encourage the elaboration of colour terminology. We also find recipes for the manufacture of chemicals such as alum needed in dyeing. But while it was useful to write down such recipes so that a more elaborate technology could be developed and communicated, the aim was also to restrict communication to the in-group, the initiated. Indeed it was to prevent identification by the uninitiated that in Akkadian the substance was deliberately referred to by a pseudonym, "lichen from the tamarisk" (Forbes 1954: 263). The secrecy of such recipes was doubtless associated with an early structure of guilds in which trade secrets were passed down only between members. The written recipe-book is more open to the world, so that deliberate steps have to be taken to preserve its secrets.

In so far as they constituted lists, these books of prescriptions led to the kind of explicit arranging and rearranging of data that we earlier observed with proverbs and more generally and more generatively with concepts and classes, rules and procedures. But being also prescriptions for taking action, they formulated a programme as well, leading to an extension of the repertoires of specialist and layman alike, as well as to "trying out" recipes (i.e. "testing," "experimenting") and to a comparative assessment of the results.

The first approximation to recipes in the culinary sense comes from the northern Semitic city of Mari, on the upper Euphrates. In the documents found in Room 5 of the royal palace, dating from the earlier part of the second millennium, there were 300 tablets representing lists of produce (cereals, oil, sesame, honey, wine, dates); of livestock, clothing and precious metals, as well as lists of personnel and inventories (Birot 1960: 245). In type the texts are similar to those found in Room 100 except that the number dealing with comestibles is much higher, amounting to about two-thirds (Bottéro 1956). For Room 5 was in the part of the palace, given the name of *quartier de l'intendance*, which held the archives relating to food for the royal table (Bottéro 1958).

The texts from Room 5 consisted largely of lists of comestibles issued for the king's table, *le repas du roi*, provided for his family, his domestic staff and many of the palace officials, who were also officials of the realm. This meal consisted of solid food (*akalum*) and fermented drink (*šikarum*). In addition, other lists made provision for the *kispum*, or meal for the dead kings; food for the living and the dead was recorded in a similar manner. Some of the lists end with the phrase "for the table of the king", and others "pour les esprits infernaux." An example of a short list of the former kind is the following:

1 *kur* 4 *qa* of bread KUM;
1 *kur* 10 *qa* of leavened bread(?);
70 *qa* of cake;
28 *qa* of *šipku*;
2 *qa* of *arsânu*;
7 *qa* of oil
2 *qa* of honey;
8 *qa* of sesame (Birot 1964: 35)

But in addition to these lists of food issued, we also find some dealing with the elaboration of the different constituents which initiate us into the activities of the kitchens themselves (Birot 1964: 1), though in a very summary way.[1] These proto-recipes, indicating actions as well as objects, verbs (or adverbs) as well as nouns (or pronouns), took a very simple form:

1   pot of honey
for the cake (Birot 1964: 97)

followed by a date and a seal. In other words they have shifted from a mere list of comestibles to specifying what the constituents are to be used for, i.e. to prescribing a set of actions. So the two facets of a culinary recipe take the same kind of shape as the medical prescription; the list of constituents is followed by a summary description of the processes to which they are to be subjected (often using shorthand technical terms relevant only to this specific *written* context). While writing down recipes does not have quite the same relevance for the growth of knowledge as recording the preparation of other concoctions (with which it is at first closely associated), nevertheless the recipe has important implications for the extension and differentiation of the repertoires of cooking that accompanied the differentiation of culture and society associated with the technological developments of the Bronze Age, when cooking took on a specific "class" aspect.[2] Equally, the elaborateness of the modern cuisine depends essentially upon the literacy of its practitioners. There the recipe (as in purchase, the receipt) reigns supreme, essentially a literate product. The production of collections of recipes (i.e. cookbooks) is a significant part of the book trade, forming a steady sale for any publisher so engaged. In the daily and weekly press, individual recipes form a standard ingredient of the "women's page." Even the rural areas of our own culture are often dominated by the book, though this might take the form of an exercise book filled with recipes gathered from home, neighbours and newspapers; I recently found half a dozen volumes of this kind of recipe-book when going through the belongings of a maternal aunt in north-east Scotland. Nevertheless there were and are certain urban–rural differences in Europe which, while not abso-lute in any sense, embody the reliance on the shop and market as distinct from the garden and the farm and learning by word of book rather than by word of mouth.

Country cuisine of course often depends upon an implicit recipe in the sense that each meal requires the assemblage of the ingredients followed by a phase of pre-paration based upon "tradition" rather than cookbook. The methods learnt in the context of family living are acquired by direct rather than indirect means, by observation rather than by reading, by watching rather than by the kind of testing that might have constituted a possible model for the genesis of the experimental method; indeed the dissemination of recipes and books of household management among the English middle classes was a fact not unconnected with the dissemination of knowledge by the Royal Society. Knowledge of cooking acquired by participation rather than by instruction is necessarily conservative (in one sense of the word) and tied to the ingredients readily available, placing less emphasis on the fulfilment of a set of written "orders" and more upon the utilisation of the contents of the cupboard in an improvisation upon certain *recettes de base* (base recipes). It is the constraints and freedom of the *bricoleur* as opposed to those of the scientist, where exact measurement may be a prelude to discovery and invention.

The attraction of regional cooking is precisely that it is tied to what grandmother did (*"les gaufres de mémé"*) and to the "natural" ingredients she used rather than to the recipes that are diffused by writing (or rather, by print, since the standardization of typography also spelt out the standardization of food) and whose wide range of ingredients was made possible first by the size of the steamship and later by the speed of the aeroplane, which in turn transformed the esoteric into the quotidian.

Until the advent of printing, the European literature on cooking was essentially a literature for the Court or for noble households. As such, it was eclectic in a regional sense, taking recipes from here and there and elaborating them into dishes "fit for a king," to use the words of an English nursery rhyme, though the same idea is found in early Arab literature (see Rodinson 1949, Roden 1968). For the upper classes, literature (as for example, the Roman treatise entitled the *Deipnosophists*) helped to bring together the recipes (or more usually the outlines of recipes) gathered from different places and hence requiring the import of any exotic ingredients native to those localities. At this level the existence of a specific written recipe for, say, mullet in hummus, exercises a constraining influence on the actors involved in the preparation of this plate, because it suggests the trying out of a new dish, following a recipe and accumulating the ingredients needed. Of course, verbal norms are also constraining, but the written recipe exercises rather a different kind of normative pull, both at the level of court recipes (which are seen as the noble, even the right and proper, thing to do) and at the level of the *bourgeois gentilhomme* learning about such behavior from a cookbook. In this respect, peasant cooking is different. Firstly, it relies less on precise quantities, which tend to be specified exactly in the written recipe. Secondly, it tends to be less tied to specific ingredients; one can substitute more easily if one does not think one is preparing *tripe à la mode de Caen*, but simply cooking a dish of tripe for supper. Thirdly, there is more flexibility with regard to procedures.

Perhaps I can interpolate a personal example. When we first came across that Cantal dish, *l'aligot*, we found a recipe printed on the back of a picture postcard (though it is equally available in *La Cuisine occitane*). We took some trouble to accumulate the specified cheese (Tome de Cantal) which was not available in any of the local shops but occasionally at markets, and to follow the instructions given there. Later, upon enquiring about the dish in a restricted area around where we lived, we found several households giving us recipes that differed considerably from one another. There was little "cultural" standardization of the mode of preparation, even though the ingredients remained relatively constant.

While literate cooking is constraining (if one follows the book), it is so partly because, as in the case above, it often provides instruction ("programmed learning") for individuals who do not themselves know how to prepare the dishes. In the town, where children spend a large part of their time at school and are not required to make a great contribution to the house or garden, individuals often learn cooking indirectly from books rather than directly from the familial setting. Such a process necessitates "following a (written) recipe," rather than learning by participation i.e. by oral means. But in situations such as the initial elaboration and later transmission of a courtly cooking, writing appears to have played an important role, firstly, in the accumulation and learning of a wider repertoire of recipes, and secondly, in the retention and organization of that repertoire.

In completely oral societies, at least as far as the ordinary household in Africa is concerned, opportunities for learning and transmission are quite limited. Cooking is usually carried out in the open, and hence generally in the more communal rather than the more private space. Therefore the pressures towards conformity and homogeneity will be greater than in a peasant society in Europe where households are smaller, meals more private. But in any oral culture new recipes have to be learnt by one individual from another in a face-to-face situation. The concrete context would stress the relation of teacher to pupil, e.g. of mother to daughter. For oral learning tends to reduplicate the "initial situation," the process of socialization. With the cooks of courts, once again, to learn is often to place oneself in a subordinate position towards another. But all this is avoided if one can use a written source, which is "objective," "neutral," "impersonal," as far as human relations themselves are concerned.[3] One can learn in privacy as an adult even better than as a child; one can learn to change one's cooking, change one's ways of behaving, without seeking the direct advice of others. Hence the widespread circulation of *printed* works on cooking, etiquette and household management (*manuscript* works tend to be confined to the few who don't need them to the same extent) at a period when the changing socio-economic structure made mobility, social and geographical, a characteristic feature of the life of Western Europe. Indeed one might say that on the domestic level, the publication of manuals of "correct" behavior made possible the rapid assimilation of social climbers. It also contributed, in a wider sense, to the weakening of subcultures in the society, since the "secrets" of one group were being made public to all others, although it took the political change of universal suffrage, the economic change of mechanized industrialization and the communications change of radio and, above all, television, to produce the present pressure towards national, even international, uniformity that characterizes the global village.

The second point concerns the development of a wider repertoire of recipes. In oral cultures, procedures for preparing food can be complex, not so much in terms of what is placed on the table, but in the preparation itself. In West Africa the time taken in preparing millet grains for the porridge, or in making shea-butter, or rendering manioc fit for consumption, all complex processes, is very considerable. But the number of recipes that can be held in oral memory is limited, while those that can be held in written form are unlimited. If one had information upon the different recipes in actual use by a particular woman in a West African society, the total number would be relatively small. Indeed the same is no doubt also true of most peasant and urban households in Europe, especially where the wife is working outside the home as well as within. However, an individual can always consult his own recipe-books, published cookbooks or the columns of the newspaper, in order to extend the number of dishes that he can prepare. Thus the written recipe serves in part to fill the gap created by the absence of Granny, Nanna or Mémé (who has been left behind in the village, or in the town before last), by the feeble communication of culture between mother and daughter (often because of the alternative activities of either or both), and by the possibility of the easy and inexpensive resort to prepared meals, such as fish and chips, delicatessen, T.V. dinners or convenience foods of all kinds. In fulfilling this function the cookbook is also instrumental in extending the range of a society's cuisine as well as that of a particular individual.

While the use of written recipes may emphasize and enshrine differentiation of a hierarchical kind, as well as permitting the accumulation of modes of preparation from many localities, it has limited implications for the growth of knowledge. Such is not the case with the medical recipe mentioned earlier. I have already noted the kind of power over the systematic arrangement of material that writing gave to the study of surgery in the shape of the Edwin Smith Papyrus. Wounds, fractures and dislocations are covered in a relatively systematic way that was to become the accepted one, namely *a capite ad calcem*, from head to toe (Sigerist 1951: 305). Other medical works do not display such an overarching system although the Papyrus Ebers has a definite beginning and end. This papyrus is a compendium devoted to a number of subjects of interest to the physician, usually introduced by an Incipit, such as "The beginning of the physician's secret: knowledge of the heart's movements"; it mostly consists of "collections of recipes with prescriptions for the treatment of internal diseases, of diseases of the eyes, of the skin, of women, and other ailments" (Sigerist 1951: 313). The relation with incantations is close (indeed the boundary is always blurred) and the medical recipe may well be seen as emerging from the spell, just as the prescription in its turn was close both to the provision of cosmetics and to the dietetic or culinary recipe; one required the external application of *materia medica* and related substances, the other the internal consumption of powders, solids and fluids of various kinds.[4]

But the fact that these remedies were set down in the papyrus meant that they could also be made the subject of comment and addition. Whatever had been put down in writing received immense authority, and it is significant that Thoth, the god of writing, was also the patron of physicians. In the introduction to the Papyrus Ebers, Re, the sun god, says: "I will save him from his enemies, and Thoth shall be his guide, he who lets writing speak and has composed the books; he gives to the skillful, to the physicians who accompany him, skill to cure. The one whom the god loves, him he shall keep alive" (Sigerist 1951: 287). While the god had given all knowledge, this knowledge could nevertheless be augmented, by marginal notes and colophons, so that the learning contained in the manuscripts surpassed that of any one individual. Indeed the practice of medicine split into a number of specialisms, and roles proliferated; as Herodotus noted of Egypt, "every physician is for one disease . . . and the whole country is full of physicians" (Herodotus II, 84).

The incorporation of medicine in a written text may well have led to the same split that the arts displayed between composers and reciters, at least in Greek medicine which owed so much to Egypt. For a Greek alchemical treatise defines the difference between the physician on the one hand and the magician on the other; the former acts "mechanically and by books," while the other is a priest, acting "through his own religious feelings" (Gardiner 1947: 268; Sigerist 1951: 364). Here the doctor as "scientist" is being contrasted with the healer as priest; the former has to act by the book, the latter is more free to let his spirit roam, to seek for new remedies, to fill the interstices of the written word with inspired messages from the gods. The cost of book learning was a restriction on spontaneity.

In Mesopotamia, medicine developed along similar lines. Their medical texts were somewhat more carefully arranged, being "collections of cases arranged systematically under a certain heading," the principle of classification being etiological, symptomatic or clinical (Sigerist 1951: 417). These works were not only copied but edited as well; indeed they may have been the least static books in the scribal libraries. Many tablets consisted entirely of recipes, while others listed symptoms and prognoses, but they were added to and changed over time. This gradual process of accumulating, writing down, assessing and augmenting a set of acts designed to relieve the ills of mankind is clearly continuous with the diagnosis and curing of disease in oral cultures. But the systematization of knowledge, combined with the incremental activity involved in the training of literate physicians, meant that a large step had been taken on the road to "modern science", to "rational medicine", to "logico-empirical activity", whatever phrase one deems appropriate. It is the kind of step in the "domestication of the savage mind" that is so ill-described by means of our current binarisms, which smack of the body–soul dichotomy that has for so long been a central feature of human thought. I prefer instead to link these changes with those in the technology of the intellect, the means as well as the modes of communication that enabled man to make these advances in human knowledge.

Notes

1. "Parfois, cependant, une rédaction plus précise undique l'élaboration que cellesci doivent subir, ou bien le produit résultant de cette élaboration, ou encore, peut-être, l'article aquel elles doivent être associés ou mélangées pour obtenir un mets déterminé" (Birot 1964:2).
2. See my piece on "The Sociology of Cooking."
3. See Esther Goody's Malinowski Lecture for 1975 (1977).
4. These substances which, as materia medica, are internalized for curing, may be forbidden as foods, e.g. the flesh of the pig in Ancient Egypt.

References

Birot, M. (1960) *Textes administratifs de la salle 5 du palais* (Archives royals de Mari IX). Paris.
—— (1964) *Textes administratifs de la salle 5 du palais* (2e partie) (Archives royals de Mari XI). Paris.
Bottéro, J. (1956) *Textes administratifs de la salle 110* (Archives royals de Mari VII). Paris.
—— (1958) "Lettres de la salle 110 du palais de Mari", *Revue d'assyriologie et d'archéologie orientale*; 52: 163–176.
Forbes, R. J. (1954) "Chemical, culinary and domestic arts". In C. Singer et al. (eds), *A History of Technology*, Vol. 1, *From Early Times to Fall of Ancient Empires*. Oxford: Clarendon Press.
Gardiner, A. H. (1947) *Ancient Egyptian Onomastica*. London: Oxford University Press.
Miller, G., Gallanter, E. and Pibram, K. (1960) *Plans and the Structure of Behavior*. New York: Holt.
Roden, C. (1968) "Intellectuals". In *Encyclopedia of the Social Sciences*. New York: Nelson.
Rodinson, M. (1949) "Recherches sur les documents Arabes relatifs à la cuisine", *Revue des etudes Islamiques*, 17–18.
Sigerist, H. E. (1951) *A History of Medicine, I: Primitive and Archaic Medicine*. London: Oxford University Press.

8

Time, Sugar, and Sweetness

Sidney W. Mintz

Food and eating as subjects of serious inquiry have engaged anthropology from its very beginnings. Varieties of foods and modes of preparation have always evoked the attention, sometimes horrified, of observant travelers, particularly when the processing techniques (e.g., chewing and spitting to encourage fermentation) and the substances (e.g., live larvae, insects, the contents of animal intestines, rotten eggs) have been foreign to their experience and eating habits. At the same time, repeated demonstrations of the intimate relationship between ingestion and sociality among living peoples of all sorts, as well as the importance attributed to it in classic literary accounts, including the Bible, have led to active reflection about the nature of the links that connect them. Long before students of Native America had invented "culture areas," or students of the Old World had formulated evolutionary stages for pastoralism or semiagriculture, W. Robertson Smith had set forth elegantly the concept of commensality and had sought to explain the food prohibitions of the ancient Semites.[1] But food and eating were studies for the most part in their more unusual aspects—food prohibitions and taboos, cannibalism, the consumption of unfamiliar and distasteful items—rather than as everyday and essential features of the life of all humankind.

Food and eating are now becoming actively of interest to anthropologists once more, and in certain new ways. An awakened concern with resources, including variant forms of energy and the relative costs of their trade-offs—the perception of real finitudes that may not always respond to higher prices with increased production—seems to have made some anthropological relativism stylish, and has led to the rediscovery of a treasure-trove of old ideas, mostly bad, about natural, healthful, and energy-saving foods. Interest in the everyday life of everyday people and in categories of the oppressed—women, slaves, serfs, Untouchables, "racial" minorities, as well as those who simply work with their hands—has led, among other things, to interest in women's work, slave food, and discriminations and exclusions. (It is surely no accident that the best early anthropological studies of food should have come from the pens of women, Audrey Richards[2] and Rosemary Firth.[3]) What is more, the upsurge of interest in meaning among anthropologists has also reenlivened the study of any subject matter that can be treated by seeing the patterned relationships between substances and human groups as forms of communication.

While these and other anthropological trends are resulting in the appearance of much provocative and imaginative scholarship, the anthropology of food and eating remains poorly demarcated, so that there ought still to be room for speculative inquiry. Here, I shall suggest some topics for a study of which the skills of anthropology and history might be usefully combined; and I shall raise questions about the relationship between production and consumption, with respect to some specific ingestible, for some specific time period, in order to see if light may be thrown on what foods mean to those who consume them.

During and after the so-called Age of Discovery and the beginning of the incorporation of Asia, Africa, and the New World within the sphere of European power, Europe experienced a deluge of new substances, including foods, some of them similar to items they then supplemented or supplanted, others not readily comparable to prior dietary components. Among the new items were many imports from the New World, including maize, potatoes, tomatoes, the so-called "hot" peppers (*Capsicum annuum, Capsicum frutescens*, etc.), fruits like the papaya, and the food and beverage base called chocolate or cacao.

Two of what came to rank among the most important post-Columbian introductions, however, did not originate in the New World, but in the non-European Old World: tea and coffee. And one item that originated in the Old World and was already known to Europeans, the sugar cane, was diffused to the New World, where it became, especially after the seventeenth century, an important crop and the source of sugar, molasses, and rum for Europe itself. Sugar, the ingestible of special interest here, cannot easily be discussed without reference to other foods, for it partly supplemented, partly supplanted, alternatives. Moreover, the character of its uses, its association with other items, and, it can be argued, the ways it was perceived, changed greatly over time. Since its uses, interlaced with those of many other substances, expressed or embodied certain continuing changes in the consuming society itself, it would be neither feasible nor convincing to study sugar in isolation. Sweetness is a "taste," sugar a product of seemingly infinite uses and functions; but the foods that satisfy a taste for sweetness vary immensely. Thus, a host of problems arise.

Until the seventeenth century, ordinary folk in Northern Europe secured sweetness in food mostly from honey and from fruit. Lévi-Strauss is quite right to emphasize the "natural" character of honey,[4] for he has in mind the manner of its production. Sugar, molasses, and rum made from the sugar cane require advanced technical processes. Sugar can be extracted from many sources, such as the sugar palm, the sugar beet, and all fruits, but the white granulated product familiar today, which represents the highest technical achievement in sugar processing, is made from sugar cane and sugar beet. The sugar-beet extraction process was developed late, but sugar-cane processing is ancient. When the Europeans came to know the product we call sugar, it was cane sugar. And though we know sugar cane was grown in South Asia at least as early as the fourth century B.C., definite evidence of processing—of boiling, clarification and crystallization—dates from almost a millennium later.

Even so, sugar crudely similar to the modern product was being produced on the southern littoral of the Mediterranean Sea by the eighth century A.D., and thereafter on Mediterranean islands and in Spain as well. During those centuries it remained costly, prized, and less a food than a medicine. It appears to have been regarded much as were spices, and its special place in contemporary European tastes—counterpoised, so to speak, against bitter, sour, and salt, as the opposite of them all—would not be achieved until much later. Those who dealt in imported spices dealt in sugar as well. By the thirteenth century English monarchs had grown fond of sugar, most of it probably from the Eastern Mediterranean. In 1226 Henry III appealed to the Mayor of Winchester to obtain for him three pounds of Alexandrine sugar, if possible; the famous fair near Winchester made it an entrepôt of exotic imports. By 1243, when ordering the purchase of spices at Sandwich for the royal household, Henry III included 300 pounds of *zucre de Roche* (presumably, white sugar). By the end of that century the court was consuming several tons of sugar a year, and early in the fourteenth century a full cargo of sugar reached Britain from Venice. The inventory of a fifteenth-century chapman in York—by which time sugar was beginning to reach England from the Atlantic plantation islands of Spain and Portugal—included not only cinnamon, saffron, ginger, and galingale, but also sugar and "casson sugar." By that time, it appears, sugar had entered into the tastes and recipe books of the rich; and the two fifteenth-century cookbooks edited by Thomas Austin[5] contain many sugar recipes, employing several different kinds of sugar.

Although there is no generally reliable source upon which we can base confident estimates of sugar consumption in Great Britain before the eighteenth century—or even for long after—there is no doubt that it rose spectacularly, in spite of occasional dips and troughs. One authority estimates that English sugar consumption increased about four-fold in the last four decades of the seventeenth century. Consumption trebled again during the first four decades of the eighteenth century; then more than doubled again from 1741–1745 to 1771–1775. If only one-half of the imports were retained in 1663, then English and Welsh consumption increased about twenty times, in the period 1663–1775. Since population increased only from four and one-half million to seven and one-half million, the per capita increase in sugar consumption appears dramatic.[6] By the end of the eighteenth century average annual per capita consumption stood at thirteen pounds. Interesting, then, that the nineteenth century showed equally impressive increases—the more so, when substantial consumption at the start of the nineteenth century is taken into account—and the twentieth century showed no remission until the last decade or so. Present consumption levels in Britain, and in certain other North European countries, are high enough to be nearly unbelievable, much as they are in the United States.

Sugar consumption in Great Britain rose together with the consumption of other tropical ingestibles, though at differing rates for different regions, groups, and classes. France never became the sugar or tea consumer that Britain became, though coffee was more successful in France than in Britain. Yet, the general spread of these substances through the Western world since the seventeenth century has been one of the truly important economic and cultural phenomena of the modern age. These were, it seems, the first edible luxuries to become proletarian commonplaces; they were surely the first luxuries to become regarded as necessities by vast masses of people who had not produced them, and they were probably the first substances to become the basis of advertising campaigns to increase consumption. In all of these ways, they, particularly sugar, have remained unmistakably modern.

Not long ago, economists and geographers, not to mention occasional anthropologists, were in the habit of referring to sugar, tea, coffee, cocoa, and like products as "dessert crops." A more misleading misnomer is hard to imagine, for these were among the most important commodities of the eighteenth- and nineteenth-century world, and my own name for them is somewhat nastier:

> Almost insignificant in Europe's diet before the thirteenth century, sugar gradually changed from a medicine for royalty into a preservative and confectionery ingredient and, finally, into a basic commodity. By the seventeenth century, sugar was becoming a staple in European cities; soon, even the poor knew sugar and prized it. As a relatively cheap source of quick energy, sugar was valuable more as a substitute for food than as a food itself; in western Europe it probably supplanted other food in proletarian diets. In urban centres, it became the perfect accompaniment to tea, and West Indian sugar production kept perfect pace with Indian tea production. Together with other plantation products such as coffee, rum and tobacco, sugar formed part of a complex of "proletarian hunger-killers," and played a crucial role in the linked contribution that Caribbean slaves, Indian peasants, and European urban proletarians were able to make to the growth of western civilization.[7]

If allowance is made for hyperbole, it remains true that these substances, not even known for the most part by ordinary people in Europe before about 1650, had become by 1800 common items of ingestion for members of privileged classes in much of Western Europe—though decidedly not in all—and, well before 1900, were viewed as daily necessities by all classes.

Though research by chemists and physiologists on these substances continues apace, some general statements about them are probably safe. Coffee and tea are stimulants without calories or other food value. Rum and tobacco are both probably best described as drugs, one very high in caloric yield, and the other without any food value at all, though apparently having the effect at times of reducing hunger. Sugar, consisting of about 99.9 percent pure sucrose, is, together with salt, the purest chemical substance human beings ingest and is often labeled "empty calories" by physicians and nutritionists. From a nutritional perspective, all are, in short, rather unusual substances. With the exception of tea, these hunger-killers or "drug foods" destined for European markets were mostly produced in the tropical Americas from the sixteenth century onward until the nineteenth century; and most of them continue to be produced there in substantial amounts. What, one may ask, was the three-hundred-year relationship between the systems of production of these commodities, their political and economic geography, and the steady increase in demand for them?

Though remote from his principal concerns, Marx considered the plantations of the New World among "the chief momenta of primitive accumulation":[8]

> Freedom and slavery constitute an antagonism. . . . We are not dealing with the indirect slavery, the slavery of the proletariat, but with direct slavery, the slavery of the black races in Surinam, in Brazil, in the Southern States of North America. Direct slavery is as much the pivot of our industrialism today as machinery, credit, etc. Without slavery, no cotton; without cotton, no modern industry. Slavery has given their value to the colonies; the colonies have created world trade; world trade is the necessary condition of large-scale machine industry. Before the traffic in Negroes began, the colonies only supplied the Old World with very few products and made no visible change in the face of the earth. Thus slavery is an economic category of the highest importance.[9]

These and similar assertions have been taken up by many scholars, most notably, Eric Williams, who develops the theme in his famous study, *Capitalism and Slavery* (1944). In recent years a lively controversy has developed over the precise contribution of the West India plantations to capitalist growth in the metropolises, particularly Britain. The potential contribution of the plantations has been viewed in two principal ways: fairly direct capital transfers of plantation profits to European banks for reinvestment; and the demand created by the needs of the plantations for such metropolitan products as machinery, cloth, torture instruments, and other industrial commodities. Disputes continue about both of these potential sources of gain to metropolitan capital, at least about their aggregate effect. But there is a third potential contribution, which at the moment amounts only to a hunch: Possibly, European enterprise accumulated considerable savings by the provision of low-cost foods and food substitutes to European working classes. Even if not, an attractive argument may be made that Europeans consumed more and more of these products simply because they were so good to consume. But it hardly seems fair to stop the questions precisely where they might fruitfully begin. Of the items enumerated, it seems likely that sweet things will prove most persuasively "natural" for human consumption—if the word dare be used at all. Hence, a few comments on sweetness may be in order.

Claude Lévi-Strauss in his remarkable *From Honey to Ashes* (1973), writes of the stingless bees of the Tropical Forest and of the astoundingly sweet honeys they produce, which, he says,

> have a richness and subtlety difficult to describe to those who have never tasted them, and indeed can seem almost unbearably exquisite in flavour. A delight more piercing than any normally afforded by taste or smell breaks down the boundaries of sensibility, and blurs its registers, so much so that the eater of honey wonders whether he is savouring a delicacy or burning with the fire of love.[10]

I shall resist an inclination here to rhapsodize about music, sausage, flowers, love and revenge, and the way languages everywhere seem to employ the idiom of sweetness to describe them—and so much else—but only in order to suggest a more important point. The general position on sweetness appears to be that our hominid capacity to identify it had some positive evolutionary significance—that it enabled omnivores to locate and use suitable plant nutrients in the environment. There is no doubt at all that this capacity, which presumably works if the eating experience is coupled with what nutritionists call "a hedonic tone," is everywhere heavily overladen with culturally specific preferences. Indeed, we know well that ingestibles with all four of the principal "tastes"—salt, sweet, sour, and bitter—figure importantly in many if not most cuisines, even if a good argument can be made for the evolutionary value of a capacity to taste sweetness.

Overlaid preferences can run against what appears to be "natural," as well as with it. Sugar-cane cultivation and sugar production flourished in Syria from the seventh century to the sixteenth, and it was there, after the First Crusade, that north Europeans got their first sustained taste of sugar. But the Syrian industry disappeared during the sixteenth century, apparently suppressed by the Turks, who, according to Iban Battuta, "regard as shameful the use of sugar houses." Since no innate predisposition, by itself, explains much about human behavior, and since innate predispositions rarely get studied before social learning occurs—though there is at least some evidence that fetal behavior is intensified by the presence of sucrose, while human newborns apparently show a clear preference for sweetened liquids— how much to weigh the possible significance of a "natural" preference remains moot. For the moment, let it suffice that, whether there exists a natural craving for sweetness, few are the world's peoples who respond negatively to sugar, whatever their prior experience, and countless those who have reacted to it with intensified craving and enthusiasm.

Before Britons had sugar, they had honey. Honey was a common ingredient in prescriptions; in time, sugar supplanted it in many or most of them. (The term "treacle," which came to mean molasses in English usage, originally meant a medical antidote composed of many ingredients, including honey. That it should have come to mean molasses and naught else suggests, in a minor way, how sugar and its byproducts overcame and supplanted honey in most regards.) Honey had also been used as a preservative of sorts; sugar turned out to be much better and, eventually, cheaper. At the time of the marriage of Henry IV and Joan of Navarre (1403), their wedding banquet included among its many courses "Perys in syrippe." "Almost the only way of preserving fruit," write Drummond and Wilbraham, "was to boil it in syrup and flavour it heavily with spices."[11] Such syrup can be made by super-saturating water with sugar by boiling; spices can be added during the preparation. Microorganisms that spoil fruit in the absence of sugar can be controlled by 70 percent sugar solutions, which draw off water from their cells and kill them by dehydration. Sugar is a superior preservative medium—by far.

Honey also provided the basis of such alcohol drinks as mead, metheglin, and hypomel. Sugar used with wine and fruit to make hypocras became an important alternative to these drinks; ciders and other fermented fruit drinks made with English fruit and West Indian sugars represented another; and rum manufactured from molasses represented an important third. Here again, sugar soon bested honey.

The use of spices raises different issues. Until nearly the end of the seventeenth century, a yearly shortage of cattle fodder in Western Europe resulted in heavy fall butchering and the preservation of large quantities of meat by salting, pickling, and other methods. Though some writers consider the emphasis on spices and the spice trade in explanation of European exploration excessive, this much of the received wisdom, at least, seems well founded. Such spices were often used to flavor meat, not simply to conceal its taste; nearly all were of tropical or subtropical origin (e.g., nutmeg, mace, ginger, pepper, coriander, cardamom, turmeric—saffron is an important exception among others). Like these rare flavorings, sugar was a condiment, a preservative, and a medicine; like them, it was sold by Grocers (*Grossarii*) who garbled (mixed) their precious wares, and was dispensed by apothecaries, who used them in medicines. Sugar was employed, as were spices, with cooked meats, sometimes combined with fruits. Such foods still provide a festive element in modern Western cuisine: ham, goose, the use of crab apples and pineapple slices, coating with brown sugar, spiking with cloves. These uses are evidence of the obvious: that holidays preserve better what ordinary days may lose—just as familial crises reveal the nature of the family in ways that ordinary days do not. Much as the spices of holiday cookies—ginger, mace, cinnamon—suggest the past, so too do the brown sugar, molasses, and cloves of the holiday ham. More than just a hearkening to the past, however, such practices may speak to some of the more common ways that fruit was preserved and meat flavored at an earlier time.

Thus, the uses and functions of sugar are many and interesting. Sugar was a medicine, but it also disguised the bitter taste of other medicines by sweetening. It was a sweetener, which, by 1700, was sweetening tea, chocolate, and coffee, all of them bitter and all of them stimulants. It was a food, rich in calories if little else, though less refined sugars and molasses, far commoner in past centuries, possessed some slight additional food value. It was a preservative, which, when eaten with what it preserved, both made it sweeter and increased its caloric content. Its byproduct molasses (treacle) yielded rum, beyond serving as a food itself. For long, the poorest people ate more treacle than sugar; treacle even turns up in the budget of the English almshouses. Nor is this list by any means complete, for sugar turns out to be a flavor-enhancer, often in rather unexpected ways. Rather than a series of successive replacements, these new and varied uses intersect, overlap, are added on rather than lost or supplanted. Other substances may be eliminated or supplanted; sugar is not. And while there are medical concerns voiced in the historical record, it appears that no one considered sugar sinful, whatever they may have thought of the systems of labor that produced it or its effects on dentition. It may well be that, among all of the "dessert crops," it alone was never perceived as an instrument of the Devil.[12]

By the end of the seventeenth century sugar had become an English food, even if still costly and a delicacy. When Edmund Verney went up to Trinity College, Oxford in 1685, his father packed in his trunk for him eighteen oranges, six lemons, three pounds of brown sugar, one pound of powdered white sugar in quarter-pound bags, one pound of brown sugar candy, one-quarter pound of white sugar candy, one pound of "pickt Raisons, good for a cough," and four nutmegs.[13] If the seventeenth century was the century in which sugar changed in Britain from luxury and medicine to necessity and food, an additional statistic may help to underline this transformation. Elizabeth Boody Schumpeter has divided her overseas trade statistics for England into nine groups, of which "groceries," including tea, coffee, sugar, rice, pepper, and other tropical products, is most important. Richard Sheridan points out that in 1700 this group comprised 16.9 percent of all imports by official value; in 1800 it comprised 34.9 percent. The most prominent grocery items were brown sugar and molasses, making up by official value two-thirds of the group in 1700 and two-fifths in 1800. During the same century tea ranked next: The amount imported rose, during that hundred years, from 167,000 pounds to 23 *million* pounds.[14]

The economic and political forces that underlay and supported the remarkable concentration of interest in the West India and East India trade between the seventeenth and nineteenth centuries cannot be discussed here. But it may be enough to note Eric Hobsbawm's admirably succinct summary of the shift of the centers of expansion to the north of Europe, from the seventeenth century onward:

> The shift was not merely geographical, but structural. The new kind of relationship between the "advanced" areas and the rest of the world, unlike the old, tended constantly to intensify and widen the flows of commerce. The powerful, growing and accelerating current of overseas trade which swept the infant industries of Europe with it—which, in fact, sometimes actually

created them—was hardly conceivable without this change. It rested on three things: in Europe, the rise of a market for overseas products for everyday use, whose market could be expanded as they became available in larger quantities and more cheaply; and overseas the creation of economic systems for producing such goods (such as, for instance, slave-operated plantations) and the conquest of colonies designed to serve the economic advantage of their European owners.[15]

So remarkably does this statement illuminate the history of sugar—and other "dessert crops"—between 1650 and 1900 that it is almost as if it had been written with sugar in mind. But the argument must be developed to lay bare the relationships between demand and supply, between production and consumption, between urban proletarians in the metropolis and African slaves in the colonies. Precisely how demand "arises"; precisely how supply "stimulates" demand even while filling it—and yielding a profit besides; precisely how "demand" is transformed into the ritual of daily necessity and even into images of daily decency: These are questions, not answers. That mothers' milk is sweet can give rise to many imaginative constructions, but it should be clear by now that the so-called English sweet tooth probably needs—and deserves—more than either Freud or evolutionary predispositions in order to be convincingly explained.

One of Bess Lomax's better-known songs in this country is "Drill, ye Tarriers, Drill."[16] Its chorus goes:

> And drill, ye tarriers, drill,
>   Drill, ye tarriers, drill,
> It's work all day for the sugar in your tay,
>   Down behind the railway.

As such, perhaps it has no particular significance. But the last two verses, separated and followed by that chorus, are more pointed:

> Now our new foreman was Gene McCann,
>   By God, he was a blamey man.
> Last week a premature blast went off
>   And a mile in the air went Big Jim Goff.
> Next time pay day comes around,
>   Jim Goff a dollar short was found.
> When asked what for, came this reply,
>   You're docked for the time you was up in the sky.

The period during which so many new ingestibles became encysted within European diet was also the period when the factory system took root, flourished, and spread. The precise relationships between the emergence of the industrial workday and the substances under consideration remain unclear. But a few guesses may be permissible. Massive increases in consumption of the drug-food complex occurred during the eighteenth and nineteenth centuries. There also appears to have been some sequence of uses in the case of sugar; and there seems no doubt that there were changes in the use, by class, of sugar and these other products over time, much as the substances in association with which sugar was used also changed. Although these are the fundamentals upon which further research might be based, except for the first (the overall increases in consumption) none may be considered demonstrated or proved. Yet, they are so general and obvious that it would be surprising if any turned out to be wrong. Plainly, the more important questions lie concealed behind such assertions. An example may help.

To some degree it could be argued that sugar, which seems to have begun as a medicine in England and then soon became a preservative, much later changed from being a direct-use product into an indirect-use product, reverting in some curious way to an earlier function but on a wholly different scale. In 1403, pears in syrup were served at the feast following the marriage of Henry IV to Joan of Navarre. Nearly two centuries later, we learn from the household book of Lord Middleton, at Woollaton Hall, Nottinghamshire, of the purchase of two pounds and one ounce of "marmalade" at the astronomical price of 5s. 3d., which, say Drummond and Wilbraham, "shows what a luxury such imported preserved fruits were."[17]

Only the privileged few could enjoy these luxuries even in the sixteenth century in England. In subsequent centuries, however, the combination of sugars and fruit became more common, and the cost of jams, jellies, marmalades, and preserved fruits declined. These changes accompanied many other dietary changes, such as the development of ready-made (store-bought) bread, the gradual replacement of milk-drinking by tea-drinking, a sharp decline in the preparation of oatmeal—especially important in Scotland—and a decrease in the use of butter. Just how such changes took place and the nature of their interrelationship require considerable detailed study. But factory production of jams and the increasing use of store-bought (and factory-made) bread plainly go along with the decline in butter use; it seems likely that the replacement of milk with tea and sugar are also connected. All such changes mark the decline of home-prepared food. These observations do not add up to a lament over the passage of some bucolic perfection, and people have certainly been eating what is now fashionably called "junk food" for a very long time. Yet, it is true that the changes mentioned fit well with a reduction in the time which must be spent in the kitchen or in obtaining foodstuffs, and that they have eased the transition to the taking of more and more meals outside the home. "Only in the worst cases," writes Angeliki Torode of the mid-nineteenth-century English working class, "would a mother hesitate to open her jam jar, because her children ate more bread if there was jam on it."[18] The replacement of oatmeal by bread hurt working-class nutrition; so, presumably, did the other changes, including the replacement of butter by jam. Sugar continues to be used in tea—and in coffee, which never became a lower-class staple in England—but its use in tea is direct, its use in jam indirect. Jam, when produced on a factory basis and consumed with bread, provides an efficient, calorie-high and relatively cheap means of feeding people quickly, wherever they are. It fits well with changes in the rhythm of effort, the organisation of the family, and, perhaps, with new ideas about the relationship between ingestion and time.

"What is wanted," wrote Lindsay, a nutritionist of the early twentieth century, about Glasgow, "is a partial return to the national dish of porridge and milk, in place of tea, bread and jam, which have so universally replaced it in the towns, and which are replacing it even in the rural districts."[19] But why, asks R. H. Campbell, the author of the article in which Lindsay is cited, did people fail to retain the more satisfactory yet cheap diet of the rural areas?"[20] Investigators in Glasgow found a ready answer: "When it becomes a question of using the ready cooked bread or the uncooked oatmeal, laziness decides which, and the family suffers." In the city of Dundee, home of famous jams and marmalades, other investigators made an additional observation: The composition of the family diet appears to change sharply when the housewife goes to work. There, it was noted that such time-consuming practices as broth-making and oatmeal-cooking dropped out of domestic cuisine. Bread consumption increases; Campbell cites a statistic for the nineteenth century indicating that one family of seven ate an average of fifty-six pounds of bread per week.[21] Jam goes with bread. The place of laziness in these changes in diet remains to be established; the place of a higher value on women's labor—labor, say, in jam factories (though women worked mainly in jute factories in Dundee)—may matter more.

The rise of industrial production and the introduction of enormous quantities of new ingestibles occurred during the same centuries in Britain. The relationship between these phenomena is, on one level, fairly straightforward: As people produced less and less of their own food, they ate more and more food produced by others, elsewhere. As they spent more and more time away from farm and home, the kinds of foods they ate changed. Those changes reflected changing availabilities of a kind. But the availabilities themselves were functions of economic and political forces remote from the consumers and not at all understood as "forces." People were certainly not compelled to eat the specific foods they ate. But the range of foods they came to eat, and the way they came to see foods and eating, inevitably conformed well with other, vaster changes in the character of daily life—changes over which they plainly had no direct control.

E. P. Thompson has provided an illuminating overview of how industry changed for working people the meaning—nay, the very perception—of the day, of time itself, and of self within time: "If men are to meet both the demands of a highly-synchronized automated industry, and of greatly enlarged areas of 'free time,' they must somehow combine in a new synthesis elements of the old, and the new, finding an imagery based neither upon the seasons nor upon the market but upon human occasions."[22] It is the special character of the substances described here that, like sugar, they provide calories without nutrition; or, like coffee and tea, neither nutrition nor calories, but stimulus to greater effort, or, like tobacco and alcohol, respite from reality. Their study might enable one to see better how an "imagery based ... upon human occasions" can take shape partly by employing such substances, but not always with much success. Perhaps high tea can one day become a cozy cuppa; perhaps the afternoon sherry can find its equivalent in the grog shop. But a great amount of manufactured sweetness may eventually lubricate only poorly, or even partly take the place of, human relations on all occasions.

The coffee break, which almost always features coffee or tea, frequently sugar, and commonly tobacco, must have had its equivalent before the industrial system arose, just as it has its equivalent outside that system today. I have been accused of seeing an inextricable connection between capitalism and coffee-drinking or sugar use; but coffee and sugar are too seductive, and capitalism too flexible, for the connection to be more than one out of many. It is not that the drug-food habits of the English working classes are the consequence of long-term conspiracies to wreck their nutrition or to make them addicted. But if the changing consumption patterns are the result of class domination, its particular nature and the forms that it has taken require both documentation and specification. What were the ways in which, over time, the changing occupational and class structure of English society was accompanied by, and reflected in, changes in the uses of particular ingestibles? How did those ingestibles come to occupy the paramount place they do in English consumption? Within these processes were, first, innovations and imitations; later, there were ritualizations as well, expressing that imagery based upon human occasions to which Thompson refers. But an understanding of those processes, of those meanings, cannot go forward, I believe, without first understanding how the production of the substances was so brilliantly separated by the workings of the world economy from so-called meanings of the substances themselves.

I have suggested that political and economic "forces" underlay the availabilities of such items as sugar; that these substances gradually percolated downward through the class structure; and that this percolation, in turn, probably fit together social occasion and substance in accord with new conceptions of work and time. And probably, the less privileged and the poorer imitated those above them in the class system. Yet, if one accepts this idea uncritically, it might appear to obviate the research itself. But such "imitation" is, surely, immeasurably more complicated than a bald assertion makes it seem. My research to date is uncovering the ways in which a modern notion of advertising and early conceptions of a large clientele—a mass market, or "target audience" for a mass market—arose, perhaps particularly in connection with sweet things and what I have labeled here "drug-foods." How direct appeals, combined with some tendency on the part of working people to mimic the consumption norms of those more privileged than they, can combine to influence "demand" may turn out to be a significant part of what is meant by meaning, in the history of such foods as sugar.

As anthropologists turn back to the study of food and eating and pursue their interest in meaning, they display a stronger tendency to look at food in its message-bearing, symbolic form. This has resulted in an enlivening of the discipline, as well as in attracting the admiration and attention of scholars in kindred fields. Such development is surely all to the good. But for one interested in history, there is reason to wonder why so few anthropological studies have dealt with long-term changes in such things as food preferences and consumption patterns, to which historians and economic historians have paid much more attention. In part, the relative lack of anthropological interest may be owing to the romanticism of an anthropology once resolutely reluctant to study anything not "primitive." But it appears also to stem from a readiness to look upon symbolic structures as timeless representations of meaning.

Hence, we confront difficult questions about what we take "meaning" to mean and within what limits of space and time we choose to define what things mean. No answers will be ventured here. But if time is defined as outside the sphere of meaning in which we are interested, then certain categories of meaning will remain and may then be considered adequate and complete. In practice, and for the immediate subject-matter, the structure of meaning would in effect be made coterminous with the political economy. For the substances of concern here—plantation products, tropical products, slave products, imported from afar, detached from their producers—the search for meaning can then be confined within convenient boundaries: the boundaries of consumption.

But if one is interested in the world economy created by capitalism from the sixteenth century onward, and in the relationships between the core of that economy and its subsidiary but interdependent outer sectors, then the structure of meaning will not be coterminous with the metropolitan heartland. If one thinks of modern societies as composed of different groups, vertebrated by institutional arrangements for the distribution and maintenance of power, and divided by class interests as well as by perceptions, values and attitudes, then there cannot be a single system of meaning for a class-divided society. And if one thinks that meanings arise, then the separation of how goods are produced from how they are consumed, the separation of colony from metropolis, and the separation of proletarian from slave (the splitting in two of the world economy that spawned them both in their modern form) are unjustified and spurious.

Such substances as sugar are, from the point of view of the metropolis, raw materials, until systems of symbolic extrusion and transformation can operate upon them. But those systems do not bring them forth or make them available; such availabilities are differently determined. To find out what these substances come to mean is to reunite their availabilities with their uses—in space and in time.

For some time now anthropology has been struggling uncomfortably with the recognition that so-called primitive society is not what it used to be—if, indeed, it ever was. Betrayed by its own romanticism, it has sought to discover new subject-matters by imputations of a certain sort—as if pimps constituted the best equivalent of "the primitive" available for study. Without meaning to impugn in the least the scientific value of such research, I suggest that there is a much more mundane modernity equally in need of study, some of it reposing on supermarket shelves. Anthropological interest in things—material objects—is old and highly respectable. When Alfred Kroeber referred to "the fundamental thing about culture . . . the way in which men relate themselves to one another by relating themselves to their cultural material"[23] he meant objects as well as ideas. Studies of the everyday in modern life, of the changing character of such humble matters as food, viewed from the perspective of production and consumption, use and function, and concerned with the emergence and variation of meaning, might be one way to try to renovate a discipline now dangerously close to losing its purpose.

Notes

Versions of this paper were presented during the past few years at the University of Minnesota, Bryn Mawr College, Rice University, Wellesley College, Cornell University, the University of Pennsylvania, and at Johns Hopkins University's Seminar in Atlantic History and Culture. In radically modified form, these materials also formed part of my 1979 Christian Gauss Lectures at Princeton University. I benefited from comments by participants at all of these presentations, and from criticisms from other friends, including Carol Breckinridge, Carol Heim, and Professors Fred Damon, Nancy Dorian, Eugene Genovese, Jane Goodale, Richard Macksey, Kenneth Sharpe, and William Sturtevant.

1. W. Robertson Smith, *Lectures on the Religion of the Semites* (New York, 1889).
2. Audrey I. Richards, *Hunger and Work in a Savage Tribe: A Functional Study of Nutrition Among the Southern Bantu* (London, 1932); *Land, Labour and Diet in Northern Rhodesia: An Economic Study of the Bemba Tribe* (London, 1939).
3. Rosemary Firth, *Housekeeping Among Malay Peasants* (London, 1943).
4. Claude Lévi-Strauss, *From Honey to Ashes* (New York, 1973).
5. Thomas Austin, *Two Fifteenth-Century Cookbooks* (London, 1888).
6. Richard Sheridan, *Sugar and Slavery* (Baltimore, 1974).
7. Sidney W. Mintz, "The Caribbean as a Socio-cultural area," *Cahiers d'Histoire Mondiale*, IX (1966), 916–941.
8. Karl Marx, *Capital* (New York, 1939), I, 738.
9. Karl Marx to P. V. Annenkov, Dec. 28, 1846, *Karl Marx to Friedrich Engels: Selected Works* (New York, 1968).
10. Lévi-Strauss, *From Honey to Ashes*, 52.
11. J. C. Drummond and Anne Wilbraham, *The Englishman's Food* (London, 1958), 58.
12. I am indebted to Professor Jane Goodale of Bryn Mawr College, who first suggested to me that I investigate this possibility.
13. Drummond and Wilbraham, *Englishmen's Food*, 111.
14. Sheridan, *Sugar and Slavery*, 19–20. Statistics on tea are somewhat troublesome. Smuggling was common, and figures on exports are not always reliable. That the increases in consumption were staggering during the eighteenth century, however, is not open to argument. See Elizabeth Schumpeter, *English Overseas Trade Statistics, 1697–1808* (Oxford, 1960).
15. Eric Hobsbawm, *Industry and Empire* (London, 1968), 52.
16. See A. Lomax, *The Folk Songs of North America* (Garden City, N.Y., 1975).
17. Drummond and Wilbraham, *Englishmen's Food*, 54.
18. Angeliki Torode, "Trends in Fruit Consumption," in T. C. Barker, J. C. McKenzie, and John Yudkin, eds., *Our Changing Fare* (London, 1966), 122.
19. R. H. Campbell, "Diet in Scotland: An Example of Regional Variation," in ibid., 57.
20. Ibid.
21. Ibid., 58.
22. E. P. Thompson, "Time, Work Discipline and Industrial Capitalism," *Past and Present*, no. 38 (1967), 96.
23. Alfred Kroeber, *Anthropology* (New York, 1948), 68.

9

Anorexia Nervosa and its Differential Diagnosis

Hilde Bruch[1]

The condition of self-inflicted starvation, without recognizable organic disease and in the midst of ample food, is usually diagnosed as anorexia nervosa; in the German literature it is also called Pubertätsmagersucht. These two names, focusing on the most dramatic aspects of the disorder, actually relate to two essentially different problems. It is of decisive significance whether a patient is preoccupied with his body size, with a relentless pursuit of being thin and a phobic avoidance of being fat, or whether he is preoccupied with the eating function itself and its distorted symbolic meaning, with thinness only an accidental by-product. Somewhat paradoxically, "Magersucht," the pursuit of thinness, appears to be the key issue in the classical anorexia nervosa syndrome, which separates it from other psychiatric conditions associated with loss of weight. This argument will be the topic of this paper.

Concern with the differential diagnosis has pervaded the whole literature since self-starvation was recognized as a rather unique syndrome, approximately 100 years ago (Gull 1873, Lasèque 1873). Exhaustive historical reviews were reported in 3 recent monographs, from the United States by Bliss and Branch (1960), from Germany by Thomae (1961), and from Italy by Selvini (1963).

Bliss and Branch concluded that anorexia nervosa was not a distinct clinical entity but only a symptom found at times in about all psychiatric categories. For their own studies they considered a loss of 25 pounds as a suitable definition of the condition, if the drop in weight was attributable to psychological causes. How unspecific their definition is is well illustrated by their published case histories (22), reflecting a variety of psychiatric conditions with nothing in common except the symptom of weight loss. It is doubtful whether such an undistinguished assortment of patients, ranging in age from 15 to 56 years (only 2 or 3 below 20 at the time of onset), would have inspired Sir William Gull, or anyone else, to recognize an unusual clinical syndrome.

Thomae and Selvini, on the basis of 30 and 26 cases respectively, attempt to delineate anorexia nervosa from other forms of psychological undernutrition. They deal with young patients whose illness appears to be related to the problems of puberty and early adulthood. In a more recent evaluation of over 90 patients, Thomae demonstrated that age is a significant factor (1965). Patients who developed the condition after 18 years of age were "atypical," with symptoms and course distinctly different from the puberty group.

They came, however, to different theoretic and therapeutic conclusions. Thomae offers his theoretic formulation, in traditional psychoanalytic terms, that "obviously it is a drive disturbance," and that "we can say definitely that oral ambivalence underlies the whole symptomatology." He also considers classical analysis the treatment of choice. His results fail to support his pronouncements.

Selvini, on basis of a broader, more up-to-date approach, builds up the concept of anorexia nervosa as a condition in which the body is used as a tool in an individual's struggle for existence. Emaciation is viewed as the key symptom, as an effort to escape passivity and a concrete striving for independence. Therapy under this newly developed orientation was definitely more successful than older approaches.

My own studies have been undertaken under similar considerations. It is essential to delineate the differences in psychological manifestations, dynamic significance and clinical course of various forms of eating disorders. A meaningful grasp of the condition and appropriate treatment is possible only through such clear-cut distinction.

Subjects

This report is based on the study of 43 patients (37 females, 6 males) who had been diagnosed as suffering from anorexia nervosa, and whom I observed between 1942 and the middle of 1964. Twenty-five cases were private patients, whereby my contact ranged from diagnostic consultation to psychoanalytic treatment extending over several years; the others were in-patients at the New York Psychiatric Institute. I supervised their treatment program which in most instances involved intensive psychotherapy, extending in one male patient to a five-year hospitalization plus a four-year follow-up period.

Anorexia nervosa has been thought of as a rare disease, but it appears to be on the increase. I saw only six cases between 1942 and 1954, 12 from 1955 to 1959, and 25 from 1960 to the middle of 1964. A previous report (1961) had been based on the 12 patients seen between 1955 and 1959 (Bruch 1962a). At that time only one patient had been referred for psychiatric consultation within the first year of illness. Of the current group, 16 patients were seen within the first year of illness, six of them within the first six months. The others had been sick, and many had been unsuccessfully in treatment, from 2 to 16 years; in three instances efforts at therapy had been continued for more than 10 years.

The age of onset ranged from 11 to 28 years, with four being over 20 years old when they became ill; three of them had had an earlier episode during adolescence. The female patients, with the exception of the 11-year-old, had had at least one spontaneous menstruation; with two exceptions, they were amenorrheic. The six males had become sick between 12 and 14 years of age, while still in prepuberty. They were markedly retarded in height and in physical and sexual maturation.

The personal and social background information was analyzed according to several descriptive aspects. Ten came from the highest socio-economic class, 4 were daughters of physicians, 23 belonged to the middle class, including 5 male patients, and 6, including one male, were of lower class background. Eight were only children, 17 the oldest of two or more children, 11 were in the middle position, and 7 were youngest children. The age range was that of adolescence and early adulthood. Older patients were not deliberately excluded, but none exhibiting anything resembling the classic picture has come to my attention.

There were 7 Catholics, 11 Protestants, and 25 Jewish patients; all 6 male patients were Jewish. The wide range of ethnic background was characteristic of the population of New York City. Conspicuous is the absence of Negro patients, although there was a proportionate number of Negro patients in the hospital. In most cases the loss of weight was reported as having occurred suddenly in a seemingly well functioning individual who may or may not have been somewhat overweight. The mean weight loss was 45 pounds, with a range from 25 to 85 pounds. The lowest weight, 45 pounds, was observed in an 18½-year-old young man who had been sick for 6½ years. His highest weight, at age 12, had been 91 pounds.

Differential Diagnosis

Not all voluntary abstinence from food is observed in psychiatric patients. As a matter of fact, it has been used effectively as a political weapon in the struggle for independence. Well known examples are the hunger strikes of British suffragettes forcing the "cat and mouse act" on the authorities so that a political prisoner who starved herself was released from jail, and finally resulting in women getting the vote. More recently Ghandi and his followers used food refusal in the service of India's struggle for independence.

These examples illustrate the coercive quality of noneating, effective only in the relation to a responsive partner. I doubt that during the Nazi terror a noneating political prisoner would have been granted any consideration. It would also be an ineffective tool in a family setting of extreme poverty with food scarcity. In our group of patients, even in those of lower economic background, food was available in abundance and its refusal had a disruptive effect on the family.

Voluntary abstinence from food may be a prescribed ritual in many religious traditions. Much has been said about the ascetic significance of food denial in anorexia nervosa. Its more frequent occurrence in Catholic and Jewish patients has been quoted as evidence that a background with ritualistic emphasis on food might predispose to it. Most likely an economic element has been neglected in these surveys. In my series 11 patients were Protestant, all of upper class background. Preoccupation with ritualistic problems played a role in some patients, none of whom exhibited the classic anorexia nervosa syndrome.

This brings me to the core of my discussion, namely, to the question whether there is a clinical entity that deserves and needs to be separated from other psychiatric conditions, and, if yes, what would be its characteristic symptomatology, proper diagnostic classification, prognosis and treatment. To arrive at a meaningful concept of the dynamic significance of this illness, it is necessary to recognize and isolate, in each individual case, the focal point, the crucial problems of the disturbed pattern of living, and to assess the patient's tools for dealing with them.

Such an evaluation demands a clear distinction between the dynamic issues of the developmental impasse resulting in anorexia nervosa, and the secondary, even tertiary problems, symptoms and complications developing in its wake. Refusal to eat in itself provokes marked changes. Many symptoms traditionally included in the clinical description, such as distrustful, negativistic and manipulative behavior, develop as a consequence of the guilt and panic which a "hunger strike" arouses in a family. In long standing cases, every effort was made to reconstruct the patterns of family interaction and concerns before the patient had become sick. This was done with the aid of reports from previous physicians, psychiatrists and hospitals, and also from schools, summer camps or former psychological tests. In addition there was extensive work with the families.

Little attention has been paid, in the psychiatric literature, to the psychic effects of drastic food deprivation, with which the Western World has become only too familiar during the tragic years when millions of people died of externally enforced starvation. Much of what is reported in the psychoanalytic literature, such as infantile regression, obsessive ruminative preoccupation with food, narcissistic self-absorption, etc., etc., has also been described in the physiologic and clinical studies of human starvation (Keys et al. 1950). A meaningful psychological evaluation is often not possible before the worst aspects of the starvation have been corrected.

In the course of these elaborate evaluations it became apparent that the group of 43 cases whom we had considered to represent typical anorexia nervosa consisted of two distinct types. In 30 patients (26 females, 4 males) the main issue was recognized as a struggle for control, for a sense of identity and effectiveness, with relentless pursuit of thinness as a final step in this effort. They were not primarily concerned with eating, although the most bizarre food habits had developed in the course of their illness. Many indulged in enormous eating binges, with subsequent vomiting.

In the smaller group of 13 patients (11 females, 2 males), the primary concern was with the eating function which was used in various symbolic ways. No general picture can be drawn for this group. The loss of weight was incidental to some other problem, and it was often complained of, or valued only secondarily for its coercive effect.

These patients vary considerably in the severity of their illness and accessibility to treatment. In some the conflicts of conversion hysteria and psychoneurosis can be recognized, and in others there are manifest symptoms of schizophrenic reactions. They resemble the patients who had not been included because the noneating was so obviously an incidental symptom. This young group of patients looks deceptively like true anorexia nervosa. Most had been sick for a long time. The superimposed struggles and concerns had created a picture similar to that in long-standing anorexia nervosa. Clarification is possible only through detailed reconstruction of the earlier situation, with focus on the essential basic problems. Only a few examples will be given.

Case 1: The first case is that of a 19-year-old girl who had lost 30 pounds during her first year at college. When she was 14 years old her mother had undergone major surgery, and for a time there was doubt about her recovery. During the following year the patient lost approximately 15 pounds which did not arouse concern; on the contrary, she was admired for being fashionably slim. She was slow to reveal the underlying problem: ever since her mother's operation she was unable to eat unless she could observe the exact amount her mother ate. While living at home she had succeeded in hiding this, but she lost weight rapidly at college, because "I did not know what mother ate." She was unhappy about being so thin. A ludicrous situation developed with the father urging and threatening his wife to eat more so that the daughter would gain weight. Once this problem was in the open, other phobic symptoms became manifest which had been camouflaged as long as she had clung so closely to her mother.

Noneating as part of a psychoneurosis does not necessarily imply a better prognosis than anorexia nervosa. Seven of the 13 noneaters had been sick, or in treatment, for from 5 to 16 years; only 6 of the 30 anorexic patients had been sick, or in treatment, for from 5 to 7 years.

Case 2: OQ was a 32-year-old highly intelligent professional woman who had lost a considerable amount of weight at age 16. At that time she felt that a sickly sister was favored and she was treated as the "cook and bottle washer." There were at least four hospital admissions and an endless number of psychiatrists with whom she had worked in various forms of psychotherapy. She was considered to be a typical case of anorexia nervosa. She presented herself to me with this diagnosis, and was aware that her age made her something of a medical rarity. Evaluation of her long history revealed as the major theme of her life the effort to gain attention, and to dominate through weakness. She exhibited many symptoms and complaints, without underlying physical pathology. In an emergency she would resort to suicidal gestures, one serious enough to land her in a psychiatric hospital. She refused to cooperate until the staff had extracted from her husband a promise not to seek a divorce. The noneating was a nearly accidental symptom in the life of a woman with a pervasive hysterical character structure. She valued it for its coercive effect and had gradually learned every trick to arouse attention and concern, and to keep her weight at a dangerously low level.

More difficult to differentiate from true anorexia nervosa are young schizophrenics in whom noneating is the outstanding feature in a usually not too dramatic picture. The therapeutically most resistant case of my own observation belongs to this group of pseudo-anorexia nervosa.

Case 3: NT was nearly 14 years old when he came to the hospital after he had lost more than 30 pounds during the preceding nine months. The admitting physician described him as looking like a victim of a concentration camp. He was not only emaciated (weighing 60 pounds), but was covered with large bruises, mainly on his forehead. Whenever he did something he felt he was not supposed to do, he would beat his head against the wall. To protect him against injury he was given a football helmet which he wore with pride.

This symptom seemed to represent an imitation of a habit in which his grandmother had indulged, who would beat herself whenever her children or her grandchildren misbehaved, in particular whenever they did not eat all the food she offered. The grandmother had taken care of him because his mother was busy with a feebleminded older brother. The grandmother died while the patient was hospitalized.

He was the first male anorexic patient whom we studied in detail, and the many deviations from the classical picture were explained as different manifestations in the male. In a way the error of our diagnostic impression was clarified by another male anorexic who had survived six years of psychoanalytic treatment when he was admitted at age 18. His reaction to his fellow anorexic was startling: he offered he would rather eat. He explained later, "I took one look at him. Man, that scared me. He was weird!"

By that time, NT had been on the ward for over three years, had been exposed to the whole repertoire of psychiatric regimens (including, when first admitted, to a course of insulin shock therapy) and had been tube fed for over three years, debating at each feeding whether it was "harder" to eat or to be fed.

He was articulate and argumentative in his psychotherapeutic sessions. The leading theme was guilt and atonement, and the noneating was used in this context. There was no concern about his size and appearance. Many other bizarre rituals served the same purpose. He would sleep on the floor because it was "harder" than sleeping in his bed, or he would wet himself instead of going to the toilet with the same explanation. He would feel guilty for causing extra work for the staff and then would devise new rituals to atone for this.

He came from a Jewish family. His illness became manifest following his Bar Mitzvah. To him the outstanding statement in the ceremony was: "Now you are a man," which meant to him that he was now responsible for his sins. Since in Jewish ritual the Day of Atonement is a fast day, he used fasting as the leading means of expiation.

The schizophrenic aspects of his behavior were, of course, not missed, but they were looked upon as part of the anorexia nervosa picture, and not vice versa. Except for the range of his morbid preoccupations, he was rational and precise, kept well informed on current events and was somewhat of the ward philosophizer. A certain competition developed between him and the other anorexic young man, but he decided for continued tube feeding "because it was harder."

One day, two years to the day after his grandmother's death, he decided to leave the hospital. He stated that he would eat because he wanted to finish his education. He did exactly that. When last heard of (September, 1964), he had finished high-school and was enrolled in a business college. He had some "religious" young men as friends, a few years younger than himself, now 23 years old. In the summer he had functioned successfully as a camp counselor. His weight was 135 pounds and his eating habits appeared to be normal; there were no fads and he requested no special diet.

In contrast to the diverse clinical pictures of weight loss that are *not* characteristic of anorexia nervosa (and they can be easily enlarged), the true syndrome is amazingly uniform. In the 30 patients whom we have classified as belonging to this group, *three* areas of disordered psychological functioning were recognized.

The *first* outstanding symptom is a *disturbance in the body image and body concept* of delusional proportions. Cachexia may occur to the same pitiful degree in other patients. Of pathognomic significance for anorexia nervosa is the absence of concern, the vigor and stubbornness with which the often gruesome emaciation is defended as *normal* and *right*. The true anorexic is identified with his skeleton-like appearance, denies its abnormality, and actively maintains it, in contrast to other noneaters who deplore the weight loss.

Recognition of the disturbed body image is important for differential diagnosis and also for appraising treatment progress. Anorexic patients may gain weight for many reasons. Without corrective change in the body image, improvement is apt to be only a temporary remission.

The *second* outstanding characteristic is a *disturbance in the accuracy of perception or cognitive interpretation of stimuli arising within the body.* Most prominent is the failure to interpret enteroceptive signals indicating nutritional need. Awareness of hunger and appetite in the ordinary sense seems to be absent. A patient's sullen statement, "I do not need to eat," is probably an accurate expression of what he experiences most of the time.

There have been endless discussions in the literature whether these patients are truly "anorexic," without desire for food, or whether they "repress" their hunger sensations or obstinately fail to act on them. The crux of the matter is that so little is known about hunger and hunger awareness, and how normal people know why they eat what they eat.

Detailed information may not be forthcoming until considerable therapeutic progress has been made, when bodily sensations begin to be more distinctly recognized. Frequently, the first discovery will not be in the nutritional area but in some other sensory perception.

Not only is there a drastic curtailment of intake, the whole eating habits become disorganized, with peculiar preferences and cravings. Absence of desire for food often alternates with uncontrollable impulses to gorge oneself, which is followed by self-induced vomiting. The lack of control, experienced during these eating binges, is terrifying. Patients express this as, "I do not dare to eat. If I take just one bite I am afraid that I will not be able to stop." The patients considered pseudo-anorexia nervosa did not have such episodes of bulimia.

In the battle against fatness and in an effort to remove unwanted food from the body, many patients resort to the use of laxatives, enemas, or self-induced vomiting. Although the urgent need of keeping the body weight low is given as the motive, other aspects must be considered, namely, that here, too, disturbances in the cognitive awareness of bodily sensations and integration play a role.

Another characteristic manifestation of falsified awareness of the bodily state is *overactivity*, a *denial of fatigue*, which impressed the early writers and was lost sight of in psychoanalytic reports. Overactivity is rarely spontaneously mentioned, certainly not in long-standing cases, but it can be recognized with great regularity if looked for; usually it appears before the noneating phase. Sometimes there has been an intensified interest in athletics and sports; in others, the activities appear to be aimless, e.g., walking by the mile, chinning and bending exercises, refusing to sit down or literally running around in circles. Patients who continue in school will spend long hours on their homework, intent on having perfect grades.

Drive for activity continues until the emaciation is far advanced. By that time, the actual amount of exercise may not be large but appears remarkable, since lassitude, fatigue and avoidance of any effort characteristically accompany chronic food deprivation. In contrast, anorexia nervosa is characterized by a subjective feeling of *not being tired* and *wanting to do things*. This paradoxical sense of alertness must also be considered as expression of conceptual and perceptual disturbances in body awareness.

One may also consider the failure of sexual functioning and the absence of sexual feelings as falling within this area of perceptual and conceptual deficiency. These disturbances too precede the starvation phase and often continue after nutritional restitution.

In addition to the failure of recognizing bodily states, a marked deficiency in identifying emotional states must be mentioned. These patients are limited in recognizing emotional reactions, anxiety and other feelings; even severe depressive episodes may remain masked for a long time.

The *third* outstanding feature is a paralyzing sense of ineffectiveness. Anorexic patients experience themselves as acting only *in response* to demands coming from other people and situations, and *as not* doing anything because they *want to*. While the first two features are readily recognized, the third defect is camouflaged by enormous negativism and stubborn defiance. The indiscriminate nature of their rejecting ordinary demands of living reveals it as a desperate cover-up for an undifferentiated sense of helplessness, a generalized parallel to the fear of eating one bite lest control be lost completely.

The paramount importance of the third characteristic was recognized in the course of extended psychotherapy. Once defined, it can be readily identified early in treatment. If the therapist communicates his awareness of the patient's sense of helplessness without insult to his fragile self esteem, meaningful therapeutic involvement becomes possible, avoiding the exhausting power struggle or futile efforts at persuasion that so often characterize treatment of these patients.

The discovery of this deep-seated ineffectiveness was surprising and seemed to stand in contrast to the whole early development, which, according to the initial reports by the parents, appeared to have been free of difficulties and problems to an unusual degree. The patients were described as having been outstandingly good and quiet children, obedient, clean, eager to please, helpful and precociously dependable, and excelling in school work. They were the pride and joy of their parents, and great things were expected of them. The need for self-reliant independence, which confronts every adolescent, seems to cause an insoluble conflict after a childhood of robot-like obedience. They lack awareness of their own resources, and reliance on their thoughts, feelings and bodily sensations. The obstinate negativistic façade hides a deficit in initiative and autonomy.

Case 4: FT weighed 49 pounds and was 49 inches tall when he entered the hospital at age 18. There were no signs of puberal development. Eating difficulties had begun rather suddenly when he was 12 years old. He had been in treatment, medically and psychiatrically, during the past six years, with several changes of therapist and repeated hospital admissions, at times as a lifesaving measure.

Despair, rage and mutual blame dominated the picture to such an extent that it took considerable perseverance to reconstruct his life history. The patient stayed at the hospital only because he knew that his parents did not want him home. He had ruined their life and they did not want what little there was left completely destroyed. When his mother heard that he was improving and had gained some weight, she reacted, "But I had been led to believe that there was no hope."

And yet he had been a perfect child. His father took pride in his athletic abilities, spending much time with him. His mother had gloried in the way he had responded to her care, growing and developing entirely on schedule. She considered herself an expert on child psychology and was admired and consulted by her neighbors and friends. When he was nine years old she became concerned about his saying "No" so often.

Then it was noted that he became preoccupied with exercise for his own sake, getting up early for long distance running in the park, or doing calisthenics to the point of exhaustion before going to bed. It was also noted that he was exceedingly interested in his fitness, examining not only his biceps for strength but also the muscles of his thighs to see whether they were strong enough.

The patient spoke about this later when he had formed a good relationship in psychotherapy. He explained that he hit on the idea of eating less when the over-exercising became too strenuous. There was the overall feeling of having to *burn up energy* and to get rid of everything that came *from "them."* Staying thin became a goal in itself, the only guarantee that he *owned* his own body.

In another context he spoke about what he had felt he owed his parents. "I am their property, and I owe them willingness to do things for them." In a long discourse he explained, "Everything I have and own comes from them. I feel guilty that I have nothing to show for it. I keep taking and taking from them. At least if *I* could produce, then *they* would be proud. That would be something *they could get from me.* All I do is take and take—food, money, sleep. Everything is theirs—that's why I say 'I am theirs.' Only after I have shown that I can make something out of myself, then I can feel that things are mine."

When early in adolescence he felt rebellious against this conviction of being their property and the obligation of making them proud, he retreated, in a way, to his own body as the only personal reserve for his sense of selfhood and identity. When primitive hunger overpowered him, he would eat voraciously and then throw up so that he would not retain anything that came from "them."

The patient has done amazingly well under a psychotherapeutic approach developed to correct the basic deficits previously described as characteristic for this group. He was the one who had declared, "I'd rather eat" when he met his tube-fed fellow anorexic, but this declaration extended initially only to the avoidance of tube feeding. It took more than a year of intensive work before he recognized the futility of his struggle for identity through manipulating his body, and before he developed some awareness of his inner sensations and trust in his resources and capacities. Only then did he begin to enjoy eating as a normal, biologic function. He even accepted endocrine studies and treatment to enhance his retarded growth and sexual maturation. Formerly, he had rejected any approach to this problem as something external that would "force" his body to change.

When he left the hospital after two years, he was able to live with his family without over-reacting in this perverse way to their admittedly self-centered hopes and expectations. He has since completed college. He has grown, underwent puberty and maintained a normal weight. When last heard of he was concerned about being drafted into the Army.

Discussion

A number of questions offer themselves for discussion. What is the dynamic and genetic background of this syndrome, and how does a family transmit to a child such an ineffective self-concept? To what psychiatric classification shall one assign this condition? What is the prognosis and what is essential for effective treatment?

Intensive family contact permitted detailed reconstruction of the early developmental experiences of our patients. They had been well cared-for children for whom things were done and who had been offered all advantages and privileges of modern child care. They had been exposed to many stimulating influences, in education, in the arts, in athletics, and the like. Yet, there had been a conspicuous deficit in encouragement or reinforcement of self expression. Reliance on their own inner resources, ideas or autonomous decisions had remained undeveloped. Pleasing compliance had become their way of life, which, however, when progressive development demanded more than conforming obedience, turned into indiscriminate negativism, the leitmotif of their illness.

Evidence of disregard of the patient's needs and emotions could be readily recognized in the ongoing family transactions. Uniformly, in giving the histories, the parents focused on the weight loss as the only symptom and on the nuisance value of the illness. They appeared to be impervious to the emotional needs and reactions of the patients, in the past as well as in the present. At times there was a truly shocking discrepancy between the bland and unobservant attitude of the parents, and the evidence of serious physical and emotional illness in the patient.

The question offers itself, under what conditions and in what kind of family setting does such a contradictory development occur, that of perfectionistic performance within a prescribed setting and of paralyzing impotence in the face of responsibility and independence? With widely varying individual features, several common aspects could be recognized. The mothers had often been women of achievement, or career women frustrated in their aspirations. They appeared to be conscientious in their concept of motherhood. They were subservient to their husbands in many details, and yet did not truly respect them. The fathers, despite social and financial success, which was often considerable, felt in some sense "second best." They were enormously preoccupied with outer appearances, in the physical sense of the word, admired fitness and beauty, and expected proper behavior and measurable achievements.

The parents emphasized the normality of their family life, with repeated statements that "nothing was wrong," sometimes with frantic stress on "happiness," directly denying the desperate illness of one member. Often they emphasized the superiority of the now sick child over his siblings. Sometimes one or another child had been a "disappointing" sibling. In four instances there was an older mentally retarded child. In other cases, there was poor school performance, aggressive behavior, allergy, death following a tonsillectomy in a sibling, or loss through an abortion. Only in a few instances had there been any premonition or attention to warning signs that the "well balanced" child who had satisfied his parents' dreams, was heading toward emotional bankruptcy.

In order to visualize how a family, without dramatic signs of discord, fails to transmit an adequate sense of self-effectiveness to a child, I felt it necessary to construct a model of personality development that is simpler, but also more generalized, than that offered in traditional psychoanalytic theories (Bruch 1961). Behavior, from birth on, needs to be differentiated into two forms, namely, that *initiated* in the individual, and that *in response* to external stimuli. This distinction applies both to the biologic and the socio-emotional field, and also to pleasure- and pain-producing states. Considering earliest infancy as a time of complete dependence disregards the fact that the infant has ways of letting his wants and needs be known, and in this way is an active participant in his development.

Behavior in relation to the child can be described as *responsive* and *stimulating*. The interaction between child and environment can be rated as *appropriate* or *inappropriate*, depending on whether it serves organization and development of need awareness, or is retarding or destructive. These elementary distinctions permit the dynamic analysis, regardless of the specific area or content of the behavior, of an amazingly large variety of clinical problems.

For normal development it appears to be essential that *appropriate responses to the clues originating in the child* and *stimulation coming from the outside* are well balanced. Deprivation of the regular sequence of felt discomfort, signal, appropriate response, and felt satisfaction may exert a profound disruption in essential early learning, resulting in disordered awareness of bodily function and body concept (Bruch 1962b). If responses to child initiated clues are continuously inappropriate or contradictory, he will fail to develop a sense of ownership of his own body, and will experience instead that he is *not in control* of its functions, with an overall lack of awareness of living his own life, and a conviction of the ineffectiveness of all efforts and strivings. Anorexia nervosa patients approach life with this basic psychic orientation.

As to the diagnostic classification, efforts were doomed to failure as long as all cases of psychic weight loss were lumped together, or complex external symptoms, such as emaciation, amenorrhea, etc., were simply added to arrive at a diagnosis. In this study an effort was made to grasp the inner significance of the whole syndrome in the dynamic development of each individual. Independently, and working in a different setting and with different theoretical conceptions, Selvini (1963) has come to startlingly similar conclusions, namely that the attitude towards one's own body, its concrete manipulation in the service of otherwise unattainable goals, is the core of the problem.

Even after separating out a functionally and dynamically defined syndrome, meaningful classification remains a difficult and uncertain task. This is related to problems inherent in our current diagnostic terminology, which was developed at the end of the nineteenth century under entirely different conceptions of mental illness. One might reverse the old simile of old wine in new bottles to the opposite, namely, that new wine has been poured into old containers so often that they are now creaking and ready to burst. In spite of all the progress in the field of organic knowledge and dynamic understanding, no new theory has been developed, and no new classification has been proposed that would be generally acceptable.

The nonspecific forms of anorexia offered fewer difficulties in this respect. Their fitting into the one or the other of the current diagnostic categories has been one reason for differentiating them from genuine anorexia nervosa.

In anorexia nervosa much of the old controversy has focused on questions like whether a patient *has* hysteria, schizophrenia, obsessive compulsive neurosis, etc., presupposing a concept of mental illness as a definite disease process which invades an organism. With increasing awareness of psychiatric illness as deviant adaptational patterns, the question needs to be reformulated; under what conditions, and in what area of functioning does a person react in a schizophrenic, hysterical, depressive, etc., fashion?

Viewed under this orientation, the basic psychological issues which were presented for anorexia nervosa closely resemble those that have been described as significant for schizophrenic development. However, as was discussed earlier, there is need to differentiate between anorexia nervosa patients and schizophrenics who refuse to eat for a variety of reasons, but not with the singular goal of achieving autonomy and effectiveness through this bizarre control over the body and its functions.

Not all patients in this group showed other signs of schizophrenic reaction. Some appeared to be neurotic in their overall adaptation; in others, there were definite indications of schizophrenic thinking. Of particular interest are observations on two patients in whom psychological tests had been done repeatedly over several years. The earlier tests failed to show signs of schizophrenic thinking, which became manifest later on, in one case after there was marked clinical improvement and greater freedom of expression.

The significance of depressive features deserves special discussion, namely, whether they indicate depression as the primary illness, or are part of the despair of the schizophrenic reaction. I am inclined to the latter evaluation, whereas Selvini feels that true anorexia nervosa should be classified as standing midway between schizophrenia and depression (1963).

Anorexia nervosa has always been regarded a serious illness, with, at best, a protracted course. The outcome depends, more than has been recognized before, on the appropriateness and effectiveness of the therapeutic approach, something that could not be worked out as long as various forms of psychological undernutrition were lumped together. Also, not sufficient consideration has been given to the self-perpetuating destructive influence of severe nutritional deprivation. Some reports read as if the psychiatrist was so preoccupied with proving the psychological origin of the illness, and the effectiveness of his theoretic concepts, that nothing was done to correct the severe starvation, with its many physiologic and psychic by-effects. Neglecting to do this unnecessarily prolongs the course of the illness.

Traditional psychoanalysis has been conspicuously ineffective in anorexia nervosa. Meyer and Weinroth (1957) expressed doubts about reports of cases "where improvement had been ascribed to the acquisition of insight obtained through psychoanalysis." Although Thomae (1961) openly states that his study was designed to prove the superior value of psychoanalysis, his results fall short of this goal; at best they are equivocal. In Selvini's (1963) group more pertinent understanding of the condition was accompanied by better treatment results.

The same holds true for this group. The report is based on what was learned by modifying and revising psychoanalytic therapy. It was recognized that *giving insight* to these patients through motivational interpretations of "unconscious" processes was not only useless but reinforced a basic defect in their personality structure, namely, the inability to know what they themselves felt; it had always been *mother*, who "knew" how *they* felt. With a change from an interpretative to a more fact finding approach, there was decided improvement in the patients' responses, even in cases that had been sick for long periods (as was illustrated through case 4).

For many, it was the first consistent experience that someone *listened* to what *they had to say* and did not tell them how to feel. The essential therapeutic task in dealing with these patients is to evoke awareness that there are feelings and impulses that originate in themselves, and that they can learn to recognize them. In a later phase of treatment, they need help in evaluating the appropriateness of their impulses and in judging the realistic possibilities of their plans and hopes.

Summary

The literature on anorexia nervosa is characterized by confusion about its proper delineation and diagnostic classification. This appears to be related to failure to discriminate between different types of psychologically determined emaciation, and to the shortcomings of our current psychiatric classification.

This report is based on observations made on 43 patients (37 females, 6 males) who were studied between 1942 and 1964. Instances of under-nutrition secondary to a well defined psychiatric illness were not included.

This group showed two distinct types, 13 cases of refusal to eat in the service of neurotic or schizophrenic conflicts. Although often confused with anorexia nervosa, these patients do not represent a special clinical entity.

True anorexia nervosa, observed in 30 cases, is characterized by body image disturbances of delusional proportions. This is an important diagnostic and prognostic sign.

Disturbance in the accuracy of perception or recognition of bodily states is the second pathognomic sign. Failure to recognize signs of nutritional needs is the outstanding aspect. The paradoxic overactivity of these patients is related to the same cognitive deficit.

The third characteristic feature is an all-pervading sense of ineffectiveness which appears to be the outcome of a personality development with a deficit of confirmation of child-initiated behavior.

"Insight-giving" psychotherapy is ineffective. Evoking awareness of impulses originating within the patient is essential for successful treatment.

Note

1.   From the Department of Psychiatry, College of Physicians and Surgeons and New York State Psychiatric Institute, New York, New York, and the Department of Psychiatry, Baylor University College of Medicine, Houston, Texas.

References

Bliss, E. L. and Branch, C.: *Anorexia Nervosa*. Hoeber, New York (1960).

Bruch, H. Transformation of oral impulses in eating disorders *Psychiat. Quart., 35:* 458 (1961).

Bruch, H. Perceptual and conceptual disturbances in anorexia nervosa. *Psychosom. Med., 24:* 187 (1962a).

Bruch, H. Falsification of bodily needs and body concept in schizophrenia. *Arch. Gen. Psychiat., 6:* 18 (1962b).

Gull, W. W. Anorexia nervosa (apepsia hysterica). *Brit. Med. J., 2:* 527 (1873).

Keys, A. *et al. The Biology of Human Starvation.* Univ. Minnesota Press, Minneapolis (1950).

Lasègue, E. C. De l'anorexie hystérique. *Arch. Gén. Méd., 21:* 385 (1873).

Meyer, B. C. and Weinroth, L. A. Observations on psychological aspects of anorexia nervosa. *Psychosom. Med., 19:* 389 (1957).

Selvini, M. P. *L'Anoressia Mentale.* Feltrinelli, Milano (1963).

Thomae, H. *Anorexia Nervosa.* Huber-Klett, Bern-Stuttgart (1961).

Thomae, H. Verschiedene Typen von Anorexia Nervosa. Presented at the Symposium über Anorexia Nervosa, Goettingen, Germany, April 24/25, 1965. Thieme, Stuttgart (1965).

Gender and Consumption

10 Fast, Feast, and Flesh: The Religious Significance of Food to Medieval
   Women
   Caroline Walker Bynum

11 Appetite as Voice
   Joan Jacobs Brumberg

12 Anorexia Nervosa: Psychopathology as the Crystallization of Culture
   Susan Bordo

13 Feeding Hard Bodies: Food and Masculinities in Men's Fitness Magazines
   Fabio Parasecoli

14 The Overcooked and Underdone: Masculinities in Japanese Food
   Programming
   T. J. M. Holden

15 Japanese Mothers and *Obentōs*: The Lunch-Box as Ideological State
   Apparatus
   Anne Allison

16 Conflict and Deference
   Marjorie DeVault

17 Feeding Lesbigay Families
   Christopher Carrington

10

Fast, Feast, and Flesh: The Religious Significance of Food to Medieval Women

Caroline Walker Bynum

In reading the lives of the [ancients] our lukewarm blood curdles at the thought of their austerities, but we remain strangely unimpressed by the essential point, namely, their determination to do God's will in all things, painful or pleasant.

Henry Suso,[1] German mystic of the fourteenth century

Strange to say the ability to live on the eucharist and to resist starvation by diabolical power died out in the Middle Ages and was replaced by "fasting girls" who still continue to amuse us with their vagaries.

William Hammond,[2] nineteenth-century American physician and founder of the New York Neurological Society

Scholars have recently devoted much attention to the spirituality of the thirteenth, fourteenth, and fifteenth centuries. In studying late medieval spirituality they have concentrated on the ideals of chastity and poverty—that is, on the renunciation, for religious reasons, of sex and family, money and property. It may be, however, that modern scholarship has focused so tenaciously on sex and money because sex and money are such crucial symbols and sources of power in our own culture. Whatever the motives, modern scholars have ignored a religious symbol that had tremendous force in the lives of medieval Christians. They have ignored the religious significance of food. Yet, when we look at what medieval people themselves wrote, we find that they often spoke of gluttony as the major form of lust, of fasting as the most painful renunciation, and of eating as the most basic and literal way of encountering God. Theologians and spiritual directors from the early church to the sixteenth century reminded penitents that sin had entered the world when Eve ate the forbidden fruit and that salvation comes when Christians eat their God in the ritual of the communion table.[3]

In the Europe of the late thirteenth and fourteenth centuries, famine was on the increase again, after several centuries of agricultural growth and relative plenty. Vicious stories of food hoarding, of cannibalism, of infanticide, or of ill adolescents left to die when they could no longer do agricultural labor sometimes survive in the sources, suggesting a world in which hunger and even starvation were not uncommon experiences. The possibility of overeating and of giving away food to the unfortunate was a mark of privilege, of aristocratic or patrician status—a particularly visible form of what we call conspicuous consumption, what medieval people called magnanimity or largesse. Small wonder then that gorging and vomiting, luxuriating in food until food and body were almost synonymous, became in folk literature an image of unbridled sensual pleasure; that magic vessels which forever brim over with food and drink were staples of European folktales; that one of the most common charities enjoined on religious orders was to feed the poor and ill; or that sharing one's own meager food with a stranger (who might turn out to be an angel, a fairy, or Christ himself) was, in hagiography and folk story alike, a standard indication of heroic or saintly generosity. Small wonder too that voluntary starvation, deliberate and extreme renunciation of food and drink, seemed to medieval people the most basic asceticism, requiring the kind of courage and holy foolishness that marked the saints.[4]

Food was not only a fundamental material concern to medieval people; food practices—fasting and feasting—were at the very heart of the Christian tradition. A Christian in the thirteenth and fourteenth centuries was required by church law to fast on certain days and to receive communion at least once a year.[5] Thus, the behavior that defined a Christian was food-related behavior. This point is clearly illustrated in a twelfth-century story of a young man (of the house of Ardres) who returned from the crusades claiming that he had become a Saracen in the East; he was, however, accepted back by his family, and no one paid much attention to his claim until he insisted on eating meat on Friday. The full impact of his apostasy was then brought home, and his family kicked him out.[6]

Food was, moreover, a central metaphor and symbol in Christian poetry, devotional literature, and theology because a meal (the eucharist) was the central Christian ritual, the most direct way of encountering God. And we should note that this meal was a frugal repast, not a banquet but simply the two basic foodstuffs of the Mediterranean world: bread and wine. Although older Mediterranean traditions of religious feasting did come, in a peripheral way, into Christianity, indeed lasting right through the Middle Ages in various kinds of carnival, the central religious meal was reception of the two basic supports of human life. Indeed Christians believed it was human life. Already hundreds of years before transubstantiation was defined as doctrine, most Christians thought that they quite literally ate Christ's body and blood in the sacrament.[7] Medieval people themselves knew how strange this all might sound. A fourteenth-century preacher, Johann Tauler, wrote:

> St. Bernard compared this sacrament [the eucharist] with the human processes of eating when he used the similes of chewing, swallowing, assimilation and digestion. To some people this will seem crude, but let such refined persons beware of pride, which come from the devil: a humble spirit will not take offense at simple things.[8]

Thus food, as practice and as symbol, was crucial in medieval spirituality. But in the period from 1200 to 1500 it was more prominent in the piety of women than in that of men. Although it is difficult and risky to make any quantitative arguments about the Middle Ages, so much work has been done on saints' lives, miracle stories, and vision literature that certain conclusions are possible about the relative popularity of various practices and symbols. Recent work by André Vauchez, Richard Kieckhefer, Donald Weinstein, and Rudolph M. Bell demonstrates that, although women were only about 18 percent of those canonized or revered as saints between 1000 and 1700, they were 30 percent of those in whose lives extreme austerities were a central aspect of holiness and over 50 percent of those in whose lives illness (often brought on by fasting and other penitential practices) was the major factor in reputation for sanctity.[9] In addition, Vauchez has shown that most males who were revered for fasting fit into one model of sanctity—the hermit saint (usually a layman)—and this was hardly the most popular male model, whereas fasting characterized female saints generally. Between late antiquity and the fifteenth century there are at least thirty cases of women who were reputed to eat nothing at all except the eucharist,[10] but I have been able to find only one or possibly two male examples of such behavior before the well-publicized fifteenth-century case of the hermit Nicholas of Flüe.[11] Moreover, miracles in which food is miraculously multiplied are told at least as frequently of women as of men, and giving away food is so common a theme in the lives of holy women that it is very difficult to find a story in which this particular charitable activity does not occur.[12] The story of a woman's basket of bread for the poor turning into roses when her husband (or father) protests her almsgiving was attached by hagiographers to at least five different women saints.[13]

If we look specifically at practices connected with Christianity's holy meal, we find that eucharistic visions and miracles occurred far more frequently to women, particularly certain types of miracles in which the quality of the eucharist as food is underlined. It is far more common, for example, for the wafer to turn into honey or meat in the mouth of a woman. Miracles in which an unconsecrated host is vomited out or in which the recipient can tell by tasting the wafer that the Priest who consecrated it is immoral happen almost exclusively to women. Of fifty-five people from the later Middle Ages who supposedly received the holy food directly from Christ's hand in a vision, forty-five are women. In contrast, the only two types of eucharistic miracle that occur primarily to men are miracles that underline not the fact that the wafer is food but the power of the priest.[14] Moreover, when we study medieval miracles, we note that miraculous abstinence and extravagant eucharistic visions tend to occur together and are frequently accompanied by miraculous bodily changes. Such changes are found almost exclusively in women. Miraculous elongation of parts of the body, the appearance on the body of marks imitating the various wounds of Christ (called stigmata), and the exuding of wondrous fluids (which smell sweet and heal and sometimes are food—for example, manna or milk) are usually female miracles.[15]

If we consider a different kind of evidence—the *exempla* or moral tales that preachers used to educate their audiences, both monastic and lay—we find that, according to Frederic Tubach's index, only about 10 percent of such stories are about women. But when we look at those stories that treat specifically fasting, abstinence, and reception of the eucharist, 30 to 50 percent are about women.[16] The only type of religious literature in which food is more frequently associated with men is the genre of satires on monastic life, in which there is some suggestion that monks are more prone to greed.[17] But this pattern probably reflects the fact that monasteries for men were in general wealthier than women's houses and therefore more capable of mounting elaborate banquets and tempting palates with delicacies.[18]

Taken together, this evidence demonstrates two things. First, food practices were more central in women's piety than in men's. Second, both men and women associated food—especially fasting and the eucharist—with women. There are, however, a number of problems with this sort of evidence. In addition to the obvious problems of the paucity of material and of the nature of hagiographical accounts—problems to which scholars since the seventeenth century have devoted much sophisticated discussion—there is the problem inherent in quantifying data. In order to count phenomena the historian must divide them up, put them into categories. Yet the most telling argument for the prominence of food in women's spirituality is the way in which food motifs interweave in women's lives and writings until even phenomena not normally thought of as eating, feeding, or fasting seem to become food-related. In other words, food becomes such a pervasive concern that it provides both a literary and a psychological unity to the woman's way of seeing the world. And this cannot be demonstrated by statistics. Let me therefore tell in some detail one of the many stories from the later Middle Ages in which food becomes a leitmotif of stunning complexity and power. It is the story of Lidwina of the town of Schiedam in the Netherlands, who died in 1433 at the age of 53.[19]

Several hagiographical accounts of Lidwina exist, incorporating information provided by her confessors; moreover, the town officials of Schiedam, who had her watched for three months, promulgated a testimonial that suggests that Lidwina's miraculous abstinence attracted more public attention than any other aspect of her life. The document solemnly attests to her complete lack of food and sleep and to the sweet odor given off by the bits of skin she supposedly shed.

The accounts of Lidwina's life suggest that there may have been early conflict between mother and daughter. When her terrible illness put a burden on her family's resources and patience, it took a miracle to convince her mother of her sanctity. One of the few incidents that survives from her childhood shows her mother annoyed with her childish dawdling. Lidwina was required to carry food to her brothers at school, and on the way home she slipped into church to say a prayer to the Virgin. The incident shows how girlish piety could provide a respite from household tasks—in this case, as in so many cases, the task of feeding men. We also learn that Lidwina was upset to discover that she was pretty, that she threatened to pray for a deformity when plans were broached for her marriage, and that, after an illness at age fifteen, she grew weak and did not want to get up from her sickbed. The accounts thus suggest that she may have been cultivating illness—perhaps even rejecting food—before the skating accident some weeks later that produced severe internal injuries. In any event, Lidwina never recovered from her fall on the ice. Her hagiographers report that she was paralyzed except for her left hand. She burned with fever and vomited convulsively. Her body putrefied so that great pieces fell off. From mouth, ears, and nose, she poured blood. And she stopped eating.

Lidwina's hagiographers go into considerable detail about her abstinence. At first she supposedly ate a little piece of apple each day, although bread dipped into liquid caused her much pain. Then she reduced her intake to a bit of date and watered wine flavored with spices and sugar; later she survived on watered wine alone—only half a pint a week—and she preferred it when the water came from the river and was contaminated with salt from the tides. When she ceased to take any solid food, she also ceased to sleep. And finally she ceased to swallow anything at all. Although Lidwina's biographers present her abstinence as evidence of saintliness, she was suspected by some during her lifetime of being possessed by a devil instead; she herself appears to have claimed that her fasting was natural. When people accused her of hypocrisy, she replied that it is no sin to eat and therefore no glory to be incapable of eating.[20]

Fasting and illness were thus a single phenomenon to Lidwina. And since she perceived them as redemptive suffering, she urged both on others. We are told that a certain Gerard from Cologne, at her urging, became a hermit and lived in a tree, fed only on manna sent from God. We are also told that Lidwina prayed for her twelve-year-old nephew to be afflicted with an illness so that he would be reminded of God's mercy. Not surprisingly, the illness itself then came from miraculous feeding. The nephew became sick by drinking several drops from a pitcher of unnaturally sweet beer on a table by Lidwina's bedside.

Like the bodies of many other women saints, Lidwina's body was closed to ordinary intake and excreting but produced extraordinary effluvia.[21] The authenticating document from the town officials of Schiedam testifies that her body shed skin, bones, and even portions of intestines, which her parents kept in a vase; and these gave off a sweet odor until Lidwina, worried by the gossip that they excited, insisted that her mother bury them. Moreover, Lidwina's effluvia cured others. A man in England sent for her wash water to cure his ill leg. The sweet smell from her left hand led one of her confessors to confess sins. And Lidwina actually nursed others in an act that she herself explicitly saw as a parallel to the Virgin's nursing of Christ.

One Christmas season, so all her biographers tell us, a certain Catherine, who took care of her, had a vision that Lidwina's breasts would fill with milk, like Mary's, on the night of the Nativity. When she told Lidwina, Lidwina warned her to prepare herself. Then Lidwina saw a vision of Mary surrounded by a host of female virgins; and the breasts of Mary and of all the company filled with milk, which poured out from their open tunics, filling the sky. When Catherine entered Lidwina's room, Lidwina rubbed her own breast and the milk came out, and Catherine drank three times and was satisfied (nor did she want any corporeal food for many days thereafter).[22] One of Lidwina's hagiographers adds that, when the same grace was given to her again, she fed her confessor, but the other accounts say that the confessor was unworthy and did not receive the gift.

Lidwina also fed others by charity and by food multiplication miracles. Although she did not eat herself, she charged the widow Catherine to buy fine fish and make fragrant sauces and give these to the poor. The meat and fish she gave as alms sometimes, by a miracle, went much further than anyone had expected. She gave water and wine and money for beer to an epileptic burning with thirst; she sent a whole pork shoulder to a poor man's family; she regularly sent food to poor or sick children, forcing her servants to spend or use for others money or food she would not herself consume. When she shared the wine in her bedside jug with others it seemed inexhaustible. So pleased was God with her charity that he sent her a vision of a heavenly banquet, and the food she had given away was on the table.

Lidwina clearly felt that her suffering was service—that it was one with Christ's suffering and that it therefore substituted for the suffering of others, both their bodily ills and their time in purgatory. Indeed her body quite literally became Christ's macerated and saving flesh, for, like many other female saints she received stigmata (or so one—but only one—of her hagiographers claims).[23] John Brugman, in the

Vita posterior, not only underlines the parallel between her wounds and those on a miraculous bleeding host she received; he also states explicitly that, in her stigmata, Christ "transformed his lover into his likeness."[24] Her hagiographers state that the fevers she suffered almost daily from 1421 until her death were suffered in order to release souls in purgatory.[25] And we see this notion of substitution reflected quite clearly in the story of a very evil man, in whose stead Lidwina made confession; she then took upon herself his punishment, to the increment of her own bodily anguish. We see substitution of another kind in the story of Lidwina taking over the toothache of a woman who wailed outside her door.

Thus, in Lidwina's story, fasting, illness, suffering, and feeding fuse together. Lidwina becomes the food she rejects. Her body, closed to ordinary intake and excretion but spilling over in milk and sweet putrefaction, becomes the sustenance and the cure—both earthly and heavenly—of her followers. But holy eating is a theme in her story as well. The eucharist is at the core of Lidwina's devotion. During her pathetic final years, when she had almost ceased to swallow, she received frequent communion (indeed as often as every two days). Her biographers claim that, during this period, only the holy food kept her alive.[26] But much of her life was plagued by conflict with the local clergy over her eucharistic visions and hunger. One incident in particular shows not only the centrality of Christ's body as food in Lidwina's spirituality but also the way in which a woman's craving for the host, although it kept her under the control of the clergy, could seem to that same clergy a threat, both because it criticized their behavior and because, if thwarted, it could bypass their power.[27]

Once an angel came to Lidwina and warned her that the priest would, the next day, bring her an unconsecrated host to test her. When the priest came and pretended to adore the host, Lidwina vomited it out and said that she could easily tell our Lord's body from unconsecrated bread. But the priest swore that the host was consecrated and returned, angry, to the church. Lidwina then languished for a long time, craving communion but unable to receive it. About three and a half months later, Christ appeared to her, first as a baby, then as a bleeding and suffering youth. Angels appeared, bearing the instruments of the passion, and (according to one account) rays from Christ's wounded body pierced Lidwina with stigmata. When she subsequently asked for a sign, a host hovered over Christ's head and a napkin descended onto her bed, containing a miraculous host, which remained and was seen by many people for days after. The priest returned and ordered Lidwina to keep quiet about the miracle but finally agreed, at her insistence, to feed her the miraculous host as communion. Lidwina was convinced that it was truly Christ because she, who was usually stifled by food, ate this bread without pain. The next day the priest preached in church that Lidwina was deluded and that her host was a fraud of the devil. But, he claimed, Christ was present in the bread he offered because it was consecrated with all the majesty of the priesthood. Lidwina protested his interpretation of her host, but she agreed to accept a consecrated wafer from him and to pray for his sins. Subsequently the priest claimed that he had cured Lidwina from possession by the devil, while Lidwina's supporters called her host a miracle. Although Lidwina's hagiographers do not give full details, they claim that the bishop came to investigate the matter, that he blessed the napkin for the service of the altar, and that the priest henceforth gave Lidwina the sacrament without tests or resistance.

As this story worked its way out, its theme was not subversive of clerical authority. The conflict began, after all, because Lidwina wanted a consecrated host, and it resulted in her receiving frequent communion, in humility and piety. According to one of her hagiographers, the moral of the story is that the faithful can always substitute "spiritual communion" (i.e., meditation) if the actual host is not given.[28] But the story had radical implications as well. It suggested that Jesus might come directly to the faithful if priests were negligent or skeptical, that a priest's word might not be authoritative on the difference between demonic possession and sanctity, that visionary women might test priests. Other stories in Lidwina's life had similar implications. She forbade a sinning priest to celebrate mass; she read the heart of another priest and learned of his adultery. Her visions of souls in purgatory especially concerned priests, and she substituted her sufferings for theirs. One Ash Wednesday an angel came to bring ashes for her forehead before the priest arrived. Even if Lidwina did not reject the clergy, she sometimes quietly bypassed or judged them.

Lidwina focused her love of God on the eucharist. In receiving it, in vision and in communion, she became one with the body on the cross. Eating her God, she received his wounds and offered her suffering for the salvation of the world. Denying herself ordinary food, she sent that food to others, and her body gave milk to nurse her friends. Food is the basic theme in Lidwina's story: self as food and God as food. For Lidwina, therefore, eating and not-eating were finally one theme. To fast, that is, to deny oneself earthly food, and yet to eat the broken body of Christ—both acts were to suffer. And to suffer was to save and to be saved.

Lidwina did not write herself, but some pious women did. And many of these women not only lived lives in which miraculous abstinence, charitable feeding of others, wondrous bodily changes, and eucharistic devotion were central; they also elaborated in prose and poetry a spirituality in which hungering, feeding, and eating were central metaphors for suffering, for service, and for encounter with God. For example, the great Italian theorist of purgatory, Catherine of Genoa (d. 1510)—whose extreme abstinence began in response to an unhappy marriage and who eventually persuaded her husband to join her in a life of continence and charitable feeding of the poor and sick—said that the annihilation of ordinary food by a devouring body is the best metaphor for the annihilation of the soul by God in mystical ecstasy.[29] She also wrote that, although no simile can adequately convey the joy in God that is the goal of all souls, nonetheless the image that comes most readily to mind is to describe God as the only bread available in a world of the starving.[30] Another Italian Catherine, Catherine of Siena (d. 1380), in whose saintly reputation fasting, food miracles, eucharistic devotion, and (invisible) stigmata were central,[31] regularly chose to describe Christian duty as "eating at the table of the cross the food of the honor of God and the salvation of souls."[32] To Catherine, "to eat" and "to hunger" have the same fundamental meaning, for one eats but is never full, desires but is never satiated.[33] "Eating" and "hungering" are active, not passive, images. They stress pain more than joy. They mean most basically to suffer and to serve—to suffer because in hunger one joins with Christ's suffering on the cross; to serve because to hunger is to expiate the sins of the world. Catherine wrote:

> And then the soul becomes drunk. And after it . . . has reached the place [of the teaching of the crucified Christ] and drunk to the full, it tastes the food of patience, the odor of virtue, and such a desire to bear the cross that it does not seem that it could ever be satiated. . . . And then the soul becomes like a drunken man; the more he drinks the more he wants to drink; the more it bears the cross the more it wants to bear it. And the pains are its refreshment and the tears which it has shed for the memory of the blood are its drink. And the sighs are its food.[34]

And again:

> Dearest mother and sisters in sweet Jesus Christ, I, Catherine, slave of the slaves of Jesus Christ, write to you in his precious blood, with the desire to see you confirmed in true and perfect charity so that you be true nurses of your souls. For we cannot nourish others if first we do not nourish our own souls with true and real virtues. . . . Do as the child does who, wanting to take milk, takes the mother's breast and places it in his mouth and draws to himself the milk by means of the flesh. So . . . we must attach ourselves to the breast of the crucified Christ, in whom we find the mother of charity, and draw from there by means of his flesh (that is, the humanity) the milk that nourishes our soul. . . . For it is Christ's humanity that suffered, not his divinity; and, without suffering, we cannot nourish ourselves with this milk which we draw from charity.[35]

To the stories and writings of Lidwina and the two Catherines—with their insistent and complex food motifs—I could add dozens of others. Among the most obvious examples would be the beguine Mary of Oignies (d. 1213) from the Low Countries, the princess Elisabeth of Hungary (d. 1231), the famous reformer of French and Flemish religious houses Colette of Corbie (d. 1447), and the thirteenth-century poets Hadewijch and Mechtild of Magdeburg. But if we look closely at the lives and writings of those men from the period whose spirituality is in general closest to women's and who were deeply influenced by women—for example, Francis of Assisi in Italy, Henry Suso and Johann Tauler in the Rhineland, Jan van Ruysbroeck of Flanders, or the English hermit Richard Rolle—we find that even to these men food asceticism is not the central ascetic practice. Nor are food metaphors central in their poetry and prose.[36] Food then is much more important to women than to men as a religious symbol. The question is why?

Modern scholars who have noticed the phenomena I have just described have sometimes suggested in an offhand way that miraculous abstinence and eucharistic frenzy are simply "eating disorders."[37] The implication of such remarks is usually that food disorders are characteristic of women rather than men, perhaps for biological reasons, and that these medieval eating disorders are different from nineteenth- and twentieth-century ones only because medieval people "theologized" what we today "medicalize."[38] While I cannot deal here with all the implications of such analysis, I want to point to two problems with it. First, the evidence we have indicates that extended abstinence was almost exclusively a male phenomenon in early Christianity and a female phenomenon in the high Middle Ages.[39] The cause of such a distribution of cases cannot be primarily biological.[40] Second, medieval people did not treat all refusal to eat as a sign of holiness. They sometimes treated it as demonic possession, but they sometimes also treated it as illness.[41] Interestingly enough, some of the holy women whose fasting was taken as miraculous (for example, Colette of Corbie) functioned as healers of ordinary individuals, both male and female, who could not eat.[42] Thus, for most of the Middle Ages, it was only in the case of some unusually devout women that not-eating was both supposedly total and religiously significant. Such behavior must have a cultural explanation.

On one level, the cultural explanation is obvious. Food was important to women religiously because it was important socially. In medieval Europe (as in many countries today) women were associated with food preparation and distribution rather than food consumption. The culture suggested that women cook and serve, men eat. Chronicle accounts of medieval banquets, for example, indicate that the sexes were often segregated and that women were sometimes relegated to watching from the balconies while gorgeous foods were rolled out to please the eyes as well as the palates of men.[43] Indeed men were rather afraid of women's control of food. Canon lawyers suggested, in the codes they drew up, that a major danger posed by women was their manipulation of male virility by charms and potions added to food.[44] Moreover, food was not merely a resource women controlled; it was *the* resource women controlled. Economic resources were controlled by husbands, fathers, uncles, or brothers. In an obvious sense, therefore, fasting and charitable food distribution (and their miraculous counterparts) were natural religious activities for women. In fasting and charity women renounced and distributed the one resource that was theirs. Several scholars have pointed out that late twelfth- and early thirteenth-century women who wished to follow the new ideal of poverty and begging (for example, Clare of Assisi and Mary of Oignies) were simply not permitted either by their families or by religious authorities to do so.[45] They substituted fasting for other ways of stripping the self of support. Indeed a thirteenth-century hagiographer commented explicitly that one holy woman gave up food because she had nothing else to give up.[46] Between the thirteenth and fifteenth centuries, many devout laywomen who resided in the homes of fathers or spouses were able to renounce the world in the midst of abundance because they did not eat or drink the food that was paid for by family wealth. Moreover, women's almsgiving and abstinence appeared culturally acceptable forms of asceticism because what women ordinarily did, as housewives, mothers, or mistresses of great castles, was to prepare and serve food rather than to eat it.

The issue of control is, however, more basic than this analysis suggests. Food-related behavior was central to women socially and religiously not only because food was a resource women controlled but also because, by means of food, women controlled themselves and their world.

First and most obviously, women controlled their bodies by fasting. Although a negative or dualist concept of body does not seem to have been the most fundamental notion of body to either women or men, some sense that body was to be disciplined, defeated, occasionally even destroyed, in order to release or protect spirit is present in women's piety. Some holy women seem to have developed an extravagant fear of any bodily contact.[47] Clare of Montefalco (d. 1308), for example, said she would rather spend days in hell than be touched by a man.[48] Lutgard of Aywières panicked at an abbot's insistence on giving her the kiss of peace, and Jesus had to interpose his hand in a vision so that she was not reached by the abbot's lips. She even asked to have her own gift of healing by touch taken away.[49] Christina of Stommeln (d. 1312), who fell into a latrine while in a trance, was furious at the laybrothers who rescued her because they touched her in order to do so.[50]

Many women were profoundly fearful of the sensations of their bodies, especially hunger and thirst. Mary of Oignies, for example, was so afraid of taking pleasure in food that Christ had to make her unable to taste.[51] From the late twelfth century comes a sad story of a dreadfully sick girl named Alpaïs who sent away the few morsels of pork given her to suck, because she feared that any enjoyment of eating might mushroom madly into gluttony or lust.[52] Women like Ida of Louvain (d. perhaps 1300), Elsbeth Achler of Reute (d. 1420), Catherine of Genoa, or Columba of Rieti (d. 1501), who sometimes snatched up food and ate without knowing what they were doing, focused their hunger on the eucharist partly because it was an acceptable object of craving and partly because it was a self-limiting food.[53] Some of women's asceticism was clearly directed toward destroying bodily needs, before which women felt vulnerable.

Some fasting may have had as a goal other sorts of bodily control. There is some suggestion in the accounts of hagiographers that fasting women were admired for suppressing excretory functions. Several biographers comment with approval that holy women who do not eat cease also to excrete, and several point out explicitly that the menstruation of saintly women ceases.[54] Medieval theology—profoundly ambivalent about body as physicality—was ambivalent about menstruation also, seeing it both as the polluting "curse of Eve" and as a natural function that, like all natural functions, was redeemed in the humanity of Christ. Theologians even debated whether or not the Virgin Mary menstruated.[55] But natural philosophers and theologians were aware that, in fact, fasting suppresses menstruation. Albert the Great noted that some holy women ceased to menstruate because of their fasts and austerities and commented that their health did not appear to suffer as a consequence.[56]

Moreover, in controlling eating and hunger, medieval women were also explicitly controlling sexuality. Ever since Tertullian and Jerome, male writers had warned religious women that food was dangerous because it excited lust.[57] Although there is reason to suspect that male biographers exaggerated women's sexual temptations, some women themselves connected food abstinence with chastity and greed with sexual desire.[58]

Women's heightened reaction to food, however, controlled far more than their physicality. It also controlled their social environment. As the story of Lidwina of Schiedam makes clear, women often coerced both families and religious authorities through fasting and through feeding. To an aristocratic or rising merchant family of late medieval Europe, the self-starvation of a daughter or spouse could be deeply perplexing and humiliating. It could therefore be an effective means of manipulating, educating, or converting family members. In one of the most charming passages of Margery Kempe's autobiography, for example, Christ and Margery consult together about her asceticism and decide that, although she wishes to practice both food abstention and sexual continence, she should perhaps offer to trade one behavior for the other. Her husband, who had married Margery in an effort to rise socially in the town of Lynn and who was obviously ashamed of her queer penitential clothes and food practices, finally agreed to grant her sexual abstinence in private if she would return to normal cooking and eating in front of the neighbors.[59] Catherine of Siena's sister, Bonaventura, and the Italian saint Rita of Cascia (d. 1456) both reacted to profligate young husbands by wasting away and managed thereby to tame disorderly male behavior.[60] Columba of Rieti and Catherine of Siena expressed what was clearly adolescent conflict with their mothers and riveted family attention on their every move by their refusal to eat. Since fasting so successfully manipulated and embarrassed families, it is not surprising that self-starvation often originated or escalated at puberty, the moment at which families usually began negotiations for husbands for their daughters. Both Catherine and Columba, for example, established themselves as unpromising marital material by their extreme food and sleep deprivation, their frenetic giving away of paternal resources, and their compulsive service of family members in what were not necessarily welcome ways. (Catherine insisted on doing the family laundry in the middle of the night.)[61]

Fasting was not only a useful weapon in the battle of adolescent girls to change their families' plans for them. It also provided for both wives and daughters an excuse for neglecting food preparation and family responsibilities. Dorothy of Montau, for example, made elementary mistakes of cookery (like forgetting to scale the fish before frying them) or forgot entirely to cook and shop while she was in ecstasy. Margaret of Cortona refused to cook for her illegitimate son (about whom she felt agonizing ambivalence) because, she said, it would distract her from prayer.[62]

Moreover, women clearly both influenced and rejected their families' values by food distribution. Ida of Louvain, Catherine of Siena, and Elisabeth of Hungary, each in her own way, expressed distaste for family wealth and coopted the entire household into Christian charity by giving away family resources, sometimes surreptitiously or even at night. Elisabeth, who gave away her husband's property, refused to eat any food except that paid for by her own dowry because the wealth of her husband's family came, she said, from exploiting the poor.[63]

Food-related behavior—charity, fasting, eucharistic devotion, and miracles—manipulated religious authorities as well.[64] Women's eucharistic miracles—especially the ability to identify unconsecrated hosts or unchaste priests—functioned to expose and castigate clerical corruption. The Viennese woman Agnes Blannbekin, knowing that her priest was not chaste, prayed that he be deprived of the host, which then flew away from him and into her own mouth.[65] Margaret of Cortona saw the hands of an unchaste priest turn black when he held the host.[66] Saints' lives and chronicles contain many stories, like that told of Lidwina of Schiedam, of women who vomited out unconsecrated wafers, sometimes to the considerable discomfiture of local authorities.[67]

The intimate and direct relationship that holy women claimed to the eucharist was often a way of bypassing ecclesiastical control. Late medieval confessors and theologians attempted to inculcate awe as well as craving for the eucharist; and women not only received ambiguous advice about frequent communion, they were also sometimes barred from receiving it at exactly the point at which their fasting and hunger reached fever pitch.[68] In such circumstances many women simply received in vision what the celebrant or confessor withheld. Imelda Lambertini, denied communion because she was too young, and Ida of Léau, denied because she was subject to "fits," were given the host by Christ.[69] And some women received, again in visions, either Christ's blood, which they were regularly denied because of their lay status, or the power to consecrate and distribute, which they were denied because of their gender. Angela of Foligno and Mechtild of Hackeborn were each, in a vision, given the chalice to distribute.[70] Catherine of Siena received blood in her mouth when she ate the wafer.[71]

It is thus apparent that women's concentration on food enabled them to control and manipulate both their bodies and their environment. We must not under-estimate the effectiveness of such manipulation in a world where it was often extra-ordinarily difficult for women to avoid marriage or to choose a religious vocation.[72] But such a conclusion concentrates on the function of fasting and feasting, and function is not meaning. Food did not "mean" to medieval women the control it provided. It is time, finally, to consider explicitly what it meant.

As the behavior of Lidwina of Schiedam or the theological insights of Catherine of Siena suggest, fasting, eating, and feeding all meant suffering, and suffering meant redemption. These complex meanings were embedded in and engendered by the theological doctrine of the Incarnation. Late medieval theology, as is well known, located the saving moment of Christian history less in Christ's resurrection than in his crucifixion. Although some ambivalence about physicality, some sharp and agonized dualism, was present, no other period in the history of Christian spirituality has placed so positive a value on Christ's humanity as physicality. Fasting was thus flight not so much *from* as *into* physicality. Communion was consuming—i.e., becoming—a God who saved the world through physical, human agony. Food to medieval women meant flesh and suffering and, through suffering, salvation: salvation of self and salvation of neighbor. Although all thirteenth- and fourteenth-century Christians emphasized Christ as suffering and Christ's suffering body as food, women were especially drawn to such a devotional emphasis. The reason seems to lie in the way in which late medieval culture understood "the female."

Drawing on traditions that went back even before the origins of Christianity, both men and women in the later Middle Ages argued that "woman is to man as matter is to spirit." Thus "woman" or "the feminine" was seen as symbolizing the physical part of human nature, whereas man symbolized the spiritual or rational.[73] Male theologians and biographers of women frequently used this idea to comment on female weakness. They also inverted the image and saw "woman" as not merely below but also above reason. Thus they somewhat sentimentally saw Mary's love for souls and her mercy toward even the wicked as an apotheosis of female unreason and weakness, and they frequently used female images to describe themselves in their dependence on God.[74] Women writers, equally aware of the male/female dichotomy, saw it somewhat differently. They tended to use the notion of "the female" as "flesh" to associate Christ's humanity with "the female" and therefore to suggest that women imitate Christ through physicality.

Women theologians saw "woman" as the symbol of humanity, where humanity was understood as including bodiliness. To the twelfth-century prophet, Elisabeth of Schönau, the humanity of Christ appeared in a vision as a female virgin. To Hilde-gard of Bingen (d. 1179), "woman" was the symbol of humankind, fallen in Eve, restored in Mary and church. She stated explicitly: "Man signifies the divinity of the Son of God and woman his humanity."[75] Moreover, to a number of women writers, Mary was the source and container of Christ's physicality; the flesh Christ put on was in some sense female, because it was his mother's. Indeed whatever physiological theory of reproduction a medieval theologian held, Christ (who had no human father) had to be seen as taking his physicality from his mother. Mechtild of Magdeburg went further and implied that Mary was a kind of preexistent humanity of Christ as the Logos was his preexistent divinity. Marguerite of Oingt, like Hildegard of Bingen, wrote that Mary was the *tunica humanitatis*, the clothing of humanity, that Christ puts on.[76] And to Julian of Norwich, God himself was a mother exactly in that our humanity in its full physicality was not merely loved and saved but even given being by and from him. Julian wrote:

> For in the same time that God joined himself to our body in the maiden's womb, he took our soul, which is sensual, and in taking it, having enclosed us all in himself, he united it to our substance. . . . So our Lady is our mother, in whom we are all enclosed and born of her in Christ, for she who is mother of our saviour is mother of all who are saved in our saviour; and our saviour is our true mother, in whom we are endlessly born and out of whom we shall never come.[77]

Although male writers were apt to see God's motherhood in his nursing and loving rather than in the fact of creation, they too associated the flesh of Christ with Mary and therefore with woman.[78]

Not only did medieval people associate humanity as body with woman; they also associated woman's body with food. Woman was food because breast milk was the human being's first nourishment—the one food essential for survival. Late medieval culture was extraordinarily concerned with milk as symbol. Writers and artists were fond of the theme, borrowed from antiquity, of lactation offered to a father or other adult male as an act of filial piety. The cult of the Virgin's milk was one of the most extensive cults in late medieval Europe. A favorite motif in art was the lactating Virgin.[79] Even the bodies of evil women were seen as food. Witches were supposed to have queer marks on their bodies (sort of super-numerary breasts) from which they nursed incubi.

Quite naturally, male and female writers used nursing imagery in somewhat different ways. Men were more likely to use images of being nursed, women metaphors of nursing. Thus when male writers spoke of God's motherhood, they focused more narrowly on the soul being nursed at Christ's breast, whereas women were apt to associate mothering with punishing, educating, or giving birth as well.[80] Most visions of drinking from the breast of Mary were received by men.[81] In contrast, women (like Lidwina) often identified with Mary as she nursed Jesus or received visions of taking the Christchild to their own breasts.[82] Both men and women, however, drank from the breast of Christ, in vision and image.[83] Both men and women wove together—from Pauline references to milk and meat and from the rich breast and food images of the Song of Songs—a complex sense of Christ's blood as the nourishment and intoxication of the soul. Both men and women therefore saw the body on the cross, which in dying fed the world, as in some sense female. Again, physiological theory reinforced image. For, to medieval natural philosophers, breast milk was transmuted blood, and a human mother (like the pelican that also symbolized Christ) fed her children from the fluid of life that coursed through her veins.[84]

Since Christ's body itself was a body that nursed the hungry, both men and women naturally assimilated the ordinary female body to it. A number of stories are told of female saints who exuded holy fluid from breasts or fingertips, either during life or after death. These fluids often cured the sick.[85] The union of mouth to mouth, which many women gained with Christ, became also a way of feeding. Lutgard's saliva cured the ill; Lukardis of Oberweimar (d. 1309) blew the eucharist into another nun's mouth; Colette of Corbie regularly cured others with crumbs she chewed.[86] Indeed one suspects that stigmata—so overwhelmingly a female phenomenon—appeared on women's bodies because they (like the marks on the bodies of witches and the wounds in the body of Christ) were not merely wounds but also breasts.

Thus many assumptions in the theology and the culture of late medieval Europe associated woman with flesh and with food. But the same theology also taught that the redemption of all humanity lay in the fact that Christ was flesh and food. A God who fed his children from his own body, a God whose humanity *was* his children's humanity, was a God with whom women found it easy to identify. In mystical ecstasy as in communion, women ate and became a God who was food and flesh. And in eating a God whose flesh was holy food, women both transcended and became more fully the flesh and the food their own bodies were.

Eucharist and mystical union were, for women, both reversals and continuations of all the culture saw them to be.[87] In one sense, the roles of priest and lay recipient reversed normal social roles. The priest became the food preparer, the generator and server of food. The woman recipient ate a holy food she did not exude or prepare. Woman's jubilant, vision-inducing, inebriated eating of God was the opposite of the ordinary female acts of food preparation or of bearing and nursing children. But in another and, I think, deeper sense, the eating was not a reversal at all. Women became, in mystical eating, a fuller version of the food and the flesh they were assumed by their culture to be. In union with Christ, woman became a fully fleshly and feeding self—at one with the generative suffering of God.

Symbol does not determine behavior. Women's imitation of Christ, their assimilation to the suffering and feeding body on the cross, was not uniform. Although most religious women seem to have understood their devotional practice as in some sense serving as well as suffering, they acted in very different ways. Some, like Catherine of Genoa and Elisabeth of Hungary, expressed their piety in feeding and caring for the poor. Some, like Alpaïs, lay rapt in mystical contemplation as their own bodies decayed in disease or in self-induced starvation that was offered for the salvation of others. Many, like Lidwina of Schiedam and Catherine of Siena, did both. Some of these women are, to our modern eyes, pathological and pathetic. Others seem to us, as they did to their contemporaries, magnificent. But they all dealt, in feast and fast, with certain fundamental realities for which all cultures must find symbols—the realities of suffering and the realities of service and generativity.

Notes

This paper was originally given as the Solomon Katz Lecture in the Humanities at the University of Washington, March 1984. It summarizes several themes that will be elaborated in detail in my forthcoming book, *Holy Feast and Holy Fast: Food Motifs in the Piety of Late Medieval Women*. I am grateful to Rudolph M. Bell, Peter Brown, Joan Jacobs Brumberg, Rachel Jacoff, Richard Kieckhefer, Paul Meyvaert, Guenther Roth, and Judith Van Herik for their suggestions and for sharing with me their unpublished work.

1. Quoted from Suso's letter to Elsbet Stagel in Henry Suso, *Deutsche Schriften im Auftrag der Württembergischen Kommission für Landesgeschichte*, ed. Karl Bihlmeyer (Stuttgart, 1907), 107; trans. (with minor changes) M. Ann Edward in *The Exemplar: Life and Writings of Blessed Henry Suso*, O.P, ed. Nicholas Heller, 2 vols. (Dubuque, Ia., 1962), 1:103.

2. William A. Hammond, *Fasting Girls: Their Physiology and Pathology* (New York, 1879), 6. Quoted in Joan J. Brumberg, " 'Fasting Girls': Nineteenth-Century Medicine and the Public Debate over 'Anorexia,' " paper delivered at the Sixth Berkshire Conference on the History of Women, 1–3 June 1984.

3. On the patristic notion that the sin of our first parents was gluttony, see Herbert Musurillo, "The Problem of Ascetical Fasting in the Greek Patristic Writers," *Traditio* 12 (1956): 17, n. 43. For a medieval discussion, see Thomas Aquinas *Summa theologiae* 2–2.148.3. For several examples of very explicit discussion of "eating God" in communion or of "being eaten" by him, see Augustine of Hippo (d. 430). *De civitate Dei*, in *Patrologiae cursus completus: Series latina*, ed. J.-P. Migne [hereafter *PL*], 41, col. 284c; Hilary of Poitiers (d. 367), *Tractatus in CXXV psalmum*, in PL 9, col. 688b–c; idem, *De Trinitate, PL* 10, cols. 246–47; Mechtild of Magdeburg (d. about 1282 or 1297), *Offenbarungen der Schwester Mechtild von Magdeburg oder Das Fliessende Licht der Gottheit*, ed. Gall Morel (Regensburg, 1869; reprint ed., Darmstadt, 1963), 43; Hadewijch (thirteenth century), *Mengeldichten*, ed. J. Van Mierlo (Antwerp, 1954), poem 16, p. 79; Johann Tauler, sermon 30, *Die Predigten Taulers*, ed. Ferdinand Vetter (Berlin, 1910), 293.

4. See Fritz Curschmann, *Hungersnöte im Mittelalter: Ein Beitrag zur Deutschen Wirtschaftsgeschichte des 8. bis 13. Jahrhunderts* (Leipzig, 1900); Mikhail M. Bakhtin, *Rabelais and His World*, trans. Hélène Iswolsky (Cambridge, Mass., 1968); and Piero Camporesi, *Il pane selvaggio* (Bologna, 1980).

5. For a brief discussion, see P M. J. Clancy, "Fast and Abstinence," *New Catholic Encyclopedia* (New York, 1967), 5:846–50.

6. Lambert, "History of the Counts of Guines," in *Monumenta Germaniae historica: Scriptorum* (hereafter *MGH.SS*), vol. 24 (Hanover, 1879), 615.

7. See Peter Brown, "A Response to Robert M. Grant: The Problem of Miraculous Feedings in the Graeco-Roman World," in *The Center for Hermeneutical Studies* 42 (1982): 16–24; Édouard Dumoutet, *Corpus Domini: Aux sources de la piété eucharistique médiévale* (Paris, 1942).

8. Tauler, sermon 31, in *Die Predigten*, 310; trans. E. Colledge and Sister M. Jane, *Spiritual Conferences* (St. Louis, 1961), 258.

9. André Vauchez, *La Sainteté en Occident aux derniers siècles du moyen âge d'après les procès de canonisation et les documents hagiographiques* (Rome, 1981); Donald Weinstein and Rudolph M. Bell, *Saints and Society: The Two Worlds of Western Christendom, 1000–1700* (Chicago, 1982); Richard Kieckhefer, *Unquiet Souls: Fourteenth-Century Saints and Their Religious Milieu* (Chicago, 1984). See also Ernst Benz, *Die Vision: Erfahrungsformen und Bilderwelt* (Stuttgart, 1969), 17–34.

10. For partial (and not always very accurate) lists of miraculous abstainers, see T E. Bridgett, *History of the Holy Eucharist in Great Britain*, 2 vols. (London, 1881), 2:195ff.; Ebenezer Cobham Brewer, *A Dictionary of Miracles: Imitative, Realistic and Dogmatic* (Philadelphia, 1884), 508–10; Peter Browe, *Die Eucharistischen Wunder des Mittelalters* (Breslau, 1938), 49–54; Thomas Pater, *Miraculous Abstinence: A Study of One of the Extraordinary Mystical Phenomena* (Washington, D.C., 1964); and Herbert Thurston, *The Physical Phenomena of Mysticism* (Chicago, 1952), 341–83 and passim. My own list includes ten cases that are merely mentioned in passing in saints' lives or chronicles. In addition, the following women are described in the sources as eating "nothing" or "almost nothing" for various periods and as focusing their sense of hunger on the eucharist: Mary of Oignies, Juliana of Cornillon, Ida of Louvain, Elisabeth of Spalbeek, Margaret of Ypres, Lidwina of Schiedam, Lukardis of Oberweimar, Jane Mary of Maillé, Alpaïs of Cudot, Elisabeth of Hungary, Margaret of Hungary, Dorothy of Montau (or Prussia), Elsbeth Achler of Reute, Colette of Corbie, Catherine of Siena, Columba of Rieti, and Catherine of Genoa. Several others—Angela of Foligno, Margaret of Cortona, Beatrice of Nazareth, Beatrice of Ornacieux, Lutgard of Aywières, and Flora of Beaulieu—experienced at times what the nineteenth century would have called a "hysterical condition" that left them unable to swallow.

11. Possible male exceptions include the visionary monk of Evesham and Aelred of Rievaulx in the twelfth century and Facio of Cremona in the thirteenth; see *The Revelation to the Monk of Evesham Abbey*, trans. Valerian Paget (New York, 1909), 35, 61; *The Life of Aelred of Rievaulx by Walter Daniel*, ed. and trans. F. M. Powicke (New York, 1951), 48ff.; *Chronica pontificum et imperatorum Mantuana* for the year 1256, *MGH.SS* 24: 216. In general those males whose fasting was most extreme—for example, Henry Suso, Peter of Luxembourg (d. 1386), and John the Good of Mantua (d. 1249)—show quite clearly in their *vitae* that, although they starved themselves and wrecked their digestions, they did *not* claim to cease eating entirely nor did they lose their *desire* for food. See Life of Suso in *Deutsche Schriften*, 7–195; John of Mantua, Process of Canonization, in *Acta sanctorum*, ed. the Bollandists [hereafter *AASS*], October: 9 (Paris and Rome, 1868), 816, 840; Life of Peter of Luxemburg, in *AASS*, July: 1 (Venice, 1746), 513, and Process of Canonization, in ibid., 534–39. James Oldo (d. 1404), who began an extreme fast, returned to eating when commanded by his superiors; life of James Oldo, in *AASS*, April: 2 (Paris and Rome, 1865), 603–4.

12. On food multiplication miracles, see Thurston, *Physical Phenomena*, 385–91. According to the tables in Weinstein and Bell, *Saints and Society*, women are 22 percent of the saints important for aiding the poor and 25 percent of those important for curing the sick, although a little less than 18 percent of the total.

13. Elisabeth of Hungary, Rose of Viterbo, Elisabeth of Portugal, Margaret of Fontana, and Flora of Beaulieu; see Jeanne Ancelet-Hustache, *Sainte Elisabeth de Hongrie* (Paris, 1946), 39–42, n. 1; life of Margaret of Fontana, in *AASS*, September: 4 (Antwerp, 1753), 137; and Clovis Brunel, "Vida e Miracles de Sancta Flor," *Analecta Bollandiana* 64 (1946): 8, n. 4. We owe the story, first attached to Elisabeth of Hungary but apocryphal, to an anonymous Tuscan Franciscan of the thirteenth century; see Paul G. Schmidt, "Die zeitgenössische Überlieferung zum Leben und zur Heiligsprechung der heiligen Elisabeth," in *Sankt Elisabeth: Fürstin, Dienerin, Heilige: Aufsätze, Dokumentation, Katalog*, ed. the University of Marburg (Sigmaringen, 1981), 1, 5.

14. Browe, *Die Wunder*, and Antoine Imbert-Gourbeyre, *La Stigmatisation: L'Extase divine et les miracles de Lourdes: Réponse aux libres-penseurs*, 2 vols. (Clermont-Ferrand, 1894), 2:183, 408–9. The work of Imbert-Gourbeyre is on one level inaccurate and credulous, but for my purposes it provides a good index of stories medieval people were willing to circulate, if not of events that in fact happened. See also Caroline W. Bynum, "Women Mystics and Eucharistic Devotion in the Thirteenth Century," *Women's Studies* 11 (1984): 179–214.

15. See Thurston, *Physical Phenomena*, passim; E. Amann, "Stigmatisation," in *Dictionnaire de théologie catholique*, vol. 14, part I (Paris, 1939), col. 2617–19; Imbert-Gourbeyre, "La Stigmatisation"; and Pierre Debongnie, "Essai critique sur l'histoire des stigmatisations au moyen âge," *Études carmélitaines* 21.2 (1936): 22–59. According to the tables in Weinstein and Bell, women provide 27 percent of the wonder-working relics although only 18 percent of the saints. Women also seem to provide the largest number of myroblytes (oil-exuding saints), although more work needs to be done on this topic. Of the most famous medieval myroblytes—Nicolas of Myra, Catherine of Alexandria, and Elisabeth of Hungary—two were women. On myroblytes, see Charles W. Jones, *Saint Nicolas of Myra, Bari and Manhattan: Biography of a Legend* (Chicago, 1978), 144–53; J.-K. Huysmans, *Sainte Lydwine de Schiedam* (Paris, 1901), 288–91 (which must, however, be used with caution); and n. 85 below.

16. Frederic C. Tubach, *Index Exemplorum: A Handbook of Medieval Religious Tales* (Helsinki, 1969); see entries for "abstinence," "fasting," "bread," "loaves and fishes," "meat," "host," and "chalice."

17. See, for example, *Tractatus beati Gregorii pape contra religionis simulatores*, ed. Marvin Colker, in *Analecta Dublinensia: Three Medieval Latin Texts in the Library of Trinity College, Dublin* (Cambridge, Mass., 1975), 47, 57; and A. George Rigg, " 'Metra de monachis carnalibus': The Three Versions," *Mittellateinisches Jahrbuch* 15 (1980): 134–42, which gives three versions of an antimonastic parody and shows that the adaptation for nuns eliminates most of the discussion of food as temptation.

18. See David Knowles, "The Diet of Black Monks," *Downside Review* 52 [n.s. 33] (1934), 273–90; and Eileen Power, *Medieval English Nunneries, c. 1275 to 1535* (Cambridge, 1922), 161–236.

19. There are four near-contemporary *vitae* of Lidwina, one in Dutch by John Gerlac, two in Latin by John Brugman, and one by Thomas à Kempis. (Both Gerlac and Brugman knew her well: Gerlac was her relative and Brugman her confessor.) See *AASS*, April: 2, 267–360, which gives Brugman's Latin translation of Gerlac's Life with Gerlac's additions indicated in brackets (the *Vita prior*) and Brugman's longer Life (the *Vita posterior*); and Thomas à Kempis, *Opera omnia*, ed. H. Sommalius, vol. 3 (Cologne, 1759), 114–64. See also Huysmans, *Lydwine*.

20. Brugman, *Vita posterior*, 320; see also ibid., 313. It is also important to note that Lidwina at first responded to her terrible illness with anger and despair and had to be convinced that it was a saving imitation of Christ's passion; see *Vita prior*, 280–82, and Thomas à Kempis, Life of Lidwina, in *Opera omnia*, 132–33.

21. For parallel lives from the Low Countries, see Thomas of Cantimpré, Life of Lutgard of Aywières (d. 1246), in *AASS*, June: 4 (Paris and Rome, 1867), 189–210; and G. Hendrix, "Primitive Versions of Thomas of Cantimpré's *Vita Lutgardis*," *Cîteaux: Commentarii cistercienses* 29 (1978): 153–206; Thomas of Cantimpré, Life of Christina Mirabilis (d. 1224), in *AASS*, July: 5 (Paris, 1868), 637–60; Life of Gertrude van Oosten (d. 1358), in *AASS*, January: 1 (Antwerp, 1643), 348–53; and Philip of Clairvaux, Life of Elisabeth of Spalbeek (or Herkenrode) (d. after 1274), *Catalogus codicum hagiographicorum Bibliothecae regiae Bruxellensis*, vol. 1, part 1, ed. the Bollandists, in *Subsidia hagiographica*, vol. 1 (Brussels, 1886), 362–78.

22. *Vita prior*, 283; *Vita posterior*, 344; Thomas à Kempis, Life of Lidwina, 135–36.

23. *Vita posterior*, 331–32, 334–35. See Debongnie, "Stigmatisations," 55–56.

24. *Vita posterior*, 335.

25. See, for example, *Vita prior*, 277, 297. We are told that her relatives and friends benefited especially.

26. See *Vita prior*, 297; see also ibid., 280. And see Thomas à Kempis, Life of Lidwina, 155–56.

27. *Vita prior*, 295–97; *Vita posterior*, 329–35; Thomas à Kempis, Life of Lidwina, 155–56.

28. *Vita posterior*, 330.

29. *Vita mirabile et doctrina santa della beata Caterina da Genova* . . . (Florence, 1568; reprint ed., 1580), 106–7.

30. *Trattato del Purgatorio*, in Umile Bonzi da Genova, *S. Caterina Fieschi Adorno*, 2 vols. (Marietta, 1960), vol. 2, *Edizione critica dei manoscritti cateriniani*, 332–33. Catherine's "works" were compiled after her death, and there

is some controversy about their "authorship" but all agree that the treatise on purgatory represents her own teaching.

31. It is quite easy to establish striking parallels between Catherine's behavior and modern descriptions of anorexia/bulimia. For accounts of Catherine's extended inedia, bingeing, and vomiting, see Raymond of Capua, Life of Catherine, in *AASS*, April: 3 (Paris and Rome, 1866), 872, 876–77, 903–7, 960; and the anonymous *I miracoli di Caterina di Jacopo da Siena di Anonimo Fiorentino a cura di Francesco Valli* (Siena, 1936), 5–9, 23–35. On Catherine's eucharistic devotion, see Raymond, Life of Catherine, 904–5, 907, and 909.

32. See, for example, letter 208, *Le lettere de S. Caterina da Siena, ridotte a miglior lezione, e in ordine nuovo disposte con note di Niccolò Tommaseo*, ed. Piero Misciattelli, 6 vols. (Siena, 1913–22), 3: 255–58; letter 11, ibid., 1:44; letter 340, ibid., 5:158–66; and *Catherine of Siena: The Dialogue*, trans. Suzanne Noffke (New York, 1980), 140.

33. See, for example, *Dialogue*, 170; letter 34, *Le lettere*, 1:157; letter 8, ibid., 1:34–38; letter 75, ibid., 2:21–24.

34. Letter 87, ibid., 2:92.

35. Letter 2* (a separately numbered series), in ibid., 6:5–6. Catherine is fond of nursing images to describe God: see letter 86, ibid., 2:81–88; letter 260, ibid., 4:139–40; letter 1*, ibid., 6:1–4; letter 81 (Tommaseo-Misciattelli no. 239) in *Epistolario di Santa Caterina da Siena*, ed. Eugenio Dupré Theseider, vol. 1 (Rome, 1940), 332–33; *Dialogue*, 52, 179–80, 292, and 323–24; and Raymond of Capua, Life of Catherine, 909.

36. On Hadewijch and Mechtild, see n. 3 above. On Mary of Oignies, see Bynum, "Women Mystics and Eucharistic Devotion." For Elisabeth, see Albert Huyskens, *Quellenstudien zur Geschichte der hl. Elisabeth Landgräfin von Thüringen* (Marburg, 1908); and Ancelet-Hustache, *Elisabeth*. For Colette, see the Lives in *AASS*, March: 1 (Antwerp, 1668), 539–619. For an analysis of the extent of food asceticism and food metaphors in the lives and writings of Francis, Suso, Tauler, Ruysbroeck, and Rolle, see my forthcoming book, *Holy Feast and Holy Fast*. Tauler and Rolle (like Catherine of Siena's confessor, Raymond of Capua) were actually apologetic about their inability to fast. And Tauler and Ruysbroeck, despite intense eucharistic piety, use little food language outside a eucharistic context. See also n. 11 above.

37. See, for example, Thurston, *Physical Phenomena*, passim; Benedict J. Groeschel, introduction to *Catherine of Genoa: Purgation and Purgatory . . .*, trans. Serge Hughes (New York, 1979), 11; J. Hubert Lacey, "Anorexia Nervosa and a Bearded Female Saint," *British Medical Journal* 285 (18–25 December 1982): 1816–17. Rudolph M. Bell is at work on a sophisticated study, the thesis of which is that a number of late medieval Italian religious women suffered from anorexia nervosa.

38. There is good evidence for biological factors in women's greater propensity for fasting and "eating disorders." See Harrison G. Pope and James Hudson, *New Hope for Binge Eaters: Advances in the Understanding and Treatment of Bulimia* (New York, 1984), which argues that anorexia and bulimia are types of biologically caused depression, which have a pharmacological cure. See also "Appendix B: Sex Differences in Death, Disease and Diet" in Katharine B. Hoyenga and K. T. Hoyenga, *The Question of Sex Differences: Psychological, Cultural and Biological Issues* (Boston, 1979), 372–90, which demonstrates that, for reasons of differences in metabolism between men and women, women's bodies tolerate fasting better than men's. For a sophisticated discussion of continuities and discontinuities in women's fasting practices, see Joan Jacobs Brumberg, " 'Fasting Girls': Reflections on Writing the History of Anorexia Nervosa," *Proceedings of the Society for Research on Child Development* (forthcoming).

39. See Browe, *Die Wunder*, 49–50; Jules Corblet, *Histoire dogmatique, liturgique et archéologique du sacrement de l'eucharistie*, 2 vols. (Paris, 1885–86), 1:188–91; and n. 10 above.

40. Even for the modern period there is much evidence that confutes a rigidly biochemical explanation of women's inedia. Most researchers agree that incidents of anorexia and bulimia are in fact increasing rapidly, although recent talk of an "epidemic" may be journalistic over-reaction. See Hilde Bruch, "Anorexia Nervosa: Therapy and Theory," *American Journal of Psychiatry* 139, no. 12 (December 1982): 1531–38; and A. H. Crisp, et al., "How Common Is Anorexia Nervosa? A Prevalence Study," British *Journal of Psychiatry* 128 (1976): 549–54.

41. Lidwina, Catherine of Siena, and Alpaïs of Cudot (d. 1211) apparently saw their inedia as illness, although all three were accused of demonic possession. See n. 20 above; Catherine of Siena, letter 19 (Tommaseo-Misciattelli no. 92), Epistolario, 1:80–82, where she calls her inability to eat an infermità; Raymond of Capua, Life of Catherine, 906; and Life of Alpaïs, in *AASS*, November: 2.1 (Brussels, 1894), 178, 180, 182–83, and 200.

42. Peter of Vaux, Life of Colette, 576, and account of miracles performed at Ghent after her death, ibid., 594–95. See also the two healing miracles recounted in the ninth-century life of Walburga (d. 779) by Wolfhard of Eichstadt, in *AASS*, February: 3 (Antwerp, 1658), 528 and 540–42. Holy women also sometimes cured people who could not fast; see the Life of Juliana of Cornillon, in *AASS*, April: 1 (Paris and Rome, 1866), 475, and Thomas of Cantimpré, Life of Lutgard, 200–201.

43. Barbara K. Wheaton, *Savouring the Past: The French Kitchen and Table from 1300 to 1789* (Philadelphia, 1983), 1–26.

44. Georges Duby, *The Knight, the Lady and the Priest: The Making of Modern Marriage in Medieval France*, trans. Barbara Bray (New York, 1984), 72, 106.

45. On Mary of Oignies, see Brenda M. Bolton, "*Vitae Matrum*: A Further Aspect of the *Frauenfrage*," in *Medieval Women: Dedicated and Presented to Professor Rosalind M. T. Hill* . . . ed. D. Baker, Studies in Church History, subsidia 1 (Oxford, 1978), 257–59; on Clare, see Rosalind B. Brooke and Christopher N. L. Brooke, "St. Clare," in ibid., 275–87. And on this point generally see my essay, "Women's Stories, Women's Symbols: A Critique of Victor Turner's Theory of Liminality," in Frank Reynolds and Robert Moore, eds., *Anthropology and the Study of Religion* (Chicago, 1984), 105–25.

46. Life of Christina *Mirabilis*, 654.

47. On this point, see Claude Carozzi, "Douceline et les autres," in *La religion populaire en Languedoc du XIIIe siècle à la moitié du XIVe siècle*, Cahiers de Fanjeaux, no. 11 (Toulouse, 1976), 251–67; and Martinus Cawley, "Lutgard of Aywières: Life and Journal," *Vox Benedictina: A Journal of Translations from Monastic Sources* 1.1 (1984): 20–48.

48. See Vauchez, *La Sainteté*, 406. For Francesca Romana de' Ponziani (d. 1440) and Jutta of Huy (d. 1228), who found the act of sexual intercourse repulsive, see Weinstein and Bell, 39–40, 88–89.

49. Life of Lutgard, 193–95. To interpret these incidents more psychologically, one might say that it is hardly surprising that Lutgard, a victim of attempted rape in adolescence, should feel anesthetized when kissed, over her protests, by a man. Nor is it surprising that the "mouth" and "breast" of Christ should figure so centrally in her visions, providing partial healing for her painful experience of the mouths of men; see Life of Lutgard, 192–94, and Hendrix, "Primitive Versions," 180.

50. Ernest W. McDonnell, *The Beguines and Beghards in Medieval Culture with Special Emphasis on the Belgian Scene* (Rutgers, 1954; reprint ed., New York, 1969), 354–55.

51. James of Vitry, Life of Mary of Oignies, in *AASS*, June: 5 (Paris and Rome, 1867), 551–56.

52. Addendum to the life of Alpaïs, 207–8.

53. See above n. 29; Life of Ida of Louvain, in *AASS*, April: 2, 156–89, esp. 167; Anton Birlinger, ed., "Leben Heiliger Alemannischer Frauen des XIV–XV Jahrhunderts, 1: Dit erst Büchlyn ist von der Seligen Kluseneryn von Rüthy, die genant waz Elizabeth," *Alemannia* 9 (1881): 275–92, esp. 280–83; Life of Columba of Rieti, in *AASS*, May: 5 (Paris and Rome, 1866), 149*–222*, esp. 159*, 162*–164*, 187*, and 200*. This would also appear to be true of Dorothy of Montau (or of Prussia) (d. 1394); see Kieckhefer, *Unquiet Souls*, 22–33.

54. See, for example, Life of Lutgard, 200; Life of Elisabeth of Spalbeek, 378; Peter of Vaux, Life of Colette, 554–55; Life of Columba of Rieti, 188*. This emphasis on the closing of the body is also found in early modern accounts of "fasting girls." Jane Balan (d. 1603) supposedly did not excrete, menstruate, sweat, or produce sputum, tears, or even dandruff; see Hyder E. Rollins, "Notes on Some English Accounts of Miraculous Fasts," *The Journal of American Folk-lore* 34.134 (1921): 363–64.

55. See Charles T. Wood, "The Doctors' Dilemma: Sin, Salvation and the Menstrual Cycle in Medieval Thought," *Speculum* 56 (1981): 710–27.

56. Albert the Great, *De animalibus libri XXVI. nach der Cölner Urschrift*, vol. 1 (Münster, 1916), 682. Hildegard of Bingen, *Hildegardis Causae et Curae*, ed. P. Kaiser (Leipzig, 1903), 102–3, comments that the menstrual flow of virgins is less than that of nonvirgins, but she does not relate this to diet.

57. See, for example, Jerome, letter 22 *ad Eustochium*, PL 22, cols. 404–5, and *contra Jovinianum*, PL 23, cols. 290–312; Fulgentius of Ruspé, letter 3, PL 65, col. 332; John Cassian, *Institutions cénobitiques*, ed. Jean-Claude Guy (Paris, 1965), 206; and Musurillo, "Ascetical Fasting," 13–19. For a medieval preacher who repeats these warnings, see Peter the Chanter, *Verbum abbreviatum*, PL 205, cols. 327–28.

58. See Life of Margaret of Cortona, in *AASS*, February: 3 (Paris, 1865), 313, and Life of Catherine of Sweden, in *AASS*, March: 3 (Paris and Rome, 1865), 504. On the tendency of male writers to depict women as sexual beings, see Weinstein and Bell, 233–36.

59. *The Book of Margery Kempe*, trans. W. Butler-Bowdon (London, 1936), 48–49.

60. Raymond, Life of Catherine, 869; Life of Rita of Cascia, in *AASS*, May: 5 (Paris and Rome, 1866), 226–28. (There is no contemporary life of Rita extant.)

61. Raymond, Life of Catherine, 868–91 passim; Life of Columba, 153*–161*.

62. On Dorothy, see *Vita Lindana*, in *AASS*, October: 13 (Paris, 1883), 505, 515, 523, and 535–43; see also John Marienwerder, *Vita Latina*, ed. Hans Westpfahl, in *Vita Dorotheae Montoviensis Magistri Johannis Marienwerder* (Cologne, 1964), 236–45, and Kieckhefer, *Unquiet Souls*, 22–33. On Margaret, see Father Cuthbert, *A Tuscan Penitent: The Life and Legend of St. Margaret of Cortona* (London, n.d.), 94–95.

63. See above nn. 36, 53, and 61. On Elisabeth, see the depositions of 1235 in Huyskens, *Quellenstudien*, 112–40, and Conrad of Marburg's letter (1233) concerning her life in ibid., 155–60. See also Ancelet-Hustache, *Elisabeth*, 201–6 and 314–18. The importance of Elisabeth's tyrannical confessor, Conrad of Marburg, in inducing her obsession with food is impossible at this distance to determine. It was Conrad who ordered her not to eat food gained by exploitation of the poor; at times, to break her will, he forbade her to indulge in charitable food distribution.

64. On this point, see Bynum, "Women Mystics and Eucharistic Devotion," and idem, *Jesus as Mother: Studies in the Spirituality of the High Middle Ages* (Berkeley, 1982), chap. 5.

65. See Browe, *Die Wunder*, 34.

66. Life of Margaret of Cortona, 341; see also 343, where she recognizes an unconsecrated host.

67. See, for example, Life of Ida of Louvain, 178–79, and account of "Joan the Meatless" in Thomas Netter [Waldensis], *Opus de sacramentis . . .* (Salamanca, 1557), fols. 111v–112r. See also life of Mary of Oignies, 566, where James of Vitry claims that Mary saw angels when virtuous priests celebrated.

68. See Joseph Duhr, "Communion fréquente," *Dictionnaire de spiritualité, ascétique et mystique, doctrine et histoire*, vol. 2 (Paris, 1953), cols. 1234–92, esp. col. 1260.

69. For Imelda, see Browe, *Die Wunder*, 27–28; for Ida of Léau, see Life, in *AASS*, October: 13, 113–14. See also Life of Alice of Schaerbeke (d. 1250), in *AASS*, June: 2 (Paris and Rome, 1867), 473–74; and Life of Juliana of Cornillon, 445–46.

70. Life of Angela of Foligno, in *AASS*, January: 1, 204. For Mechtild's vision, see Bynum, *Jesus as Mother*, 210, n. 129.

71. See Catherine's own account of such a miracle, *Dialogue*, 239.

72. The work of David Herlihy and Diane Owen Hughes, which argues that, in the thirteenth and fourteenth centuries, the age discrepancy between husband and wife increased and the dowry, provided by the girl's family, became increasingly a way of excluding her from other forms of inheritance and from her natal family, suggests some particular reasons for a high level of antagonism between girls and their families in this period. The antagonism would stem less from the failure of families to find husbands for daughters (as Herlihy suggests) than from the tendency of families to marry girls off early and thereby buy them out of the family when they were little more than children (as Hughes suggests). See David Herlihy, "Alienation in Medieval Culture and Society," reprinted in *The Social History of Italy and Western Europe, 700–1500: Collected Studies* (London, 1978); idem, "The Making of the Medieval Family: Symmetry, Structure and Sentiment," *Journal of Family History* 8.2 (1983): 116–30; and Diane Owen Hughes, "From Brideprice to Dowry in Mediterranean Europe," *Journal of Family History* 3 (1978): 262–96.

73. Vern Bullough, "Medieval Medical and Scientific Views of Women," *Viator* 4 (1973): 487–93; Eleanor McLaughlin, "Equality of Souls, Inequality of Sexes: Women in Medieval Theology," *Religion and Sexism: Images of Woman in the Jewish and Christian Traditions*, ed. R. Ruether (New York, 1974), 213–66; and M.-T d'Alverny, "Comment les théologiens et les philosophes voient la femme?" *Cahiers de civilisation médiévale* 20 (1977): 105–29.

74. See Bynum, *Jesus as Mother*, chap. 4, and idem, " 'And Woman His Humanity': Female Imagery in the Religious Writing of the Later Middle Ages," in *New Perspectives on Religion and Gender*, ed. Caroline Bynum, Stevan Harrell, and Paula Richman, to appear.

75. *Die Visionen der hl. Elisabeth und die Schriften der Aebte Ekbert und Emecho von Schönau*, ed. F W. E. Roth (Brunn, 1884), 60; Hildegard, *Liber divinorum operum*, PL 197, col. 885; and idem, *Scivias*, ed. Adelgundis Führkötter and A. Carlevaris, 2 vols. (Turnhout, 1978), 1:225–306, esp. 231 and plate 15.

76. On Mechtild, see *Jesus as Mother*, 229, 233–34, and 244; and on Hildegard, see Barbara Jane Newman, "O *Feminea Forma*: God and Woman in the Works of St. Hildegard (1098–1179)," Ph.D. diss., Yale, 1981, 131–34. And see Marguerite of Oingt, *Speculum*, in *Les Oeuvres de Marguerite d'Oingt*, ed. and trans. A. Duraffour, R. Gardette, and R. Durdilly (Paris, 1965), 98–99.

77. *Julian of Norwich: Showings*, trans. Edmund Colledge and James Walsh (New York, 1978), long text: 292, 294.

78. See *Jesus as Mother*, chap. 4, and Jan van Ruysbroeck, *The Spiritual Espousals*, trans. E. Colledge (New York, n.d.), 43; idem, "Le Miroir du salut éternel," in *Oeuvres de Ruysbroeck l'Admirable*, trans. by the Benedictines of St.-Paul de Wisques, vol. 3 (3rd ed., Brussels, 1921), 82–83; Francis of Assisi, "Salutation of the Blessed Virgin," trans. B. Fahy, in *St. Francis of Assisi: Writings and Early Biographies: English Omnibus of Sources*, ed. Marion Habig (3rd ed., Chicago, 1973), 135–36; and Henry Suso, *Büchlein der Ewigen Weisheit*, in *Deutsche Schriften*, 264.

79. See P. V. Bétérous, "A propos d'une des légendes mariales les plus répandues: Le 'Lait de la Vierge,' " *Bulletin de l'Association Guillaume Budé* 4 (1975): 403–11; Léon Dewez and Albert van Iterson, "La Lactation de saint Bernard: Légende et iconographie," *Cîteaux in de Nederlanden* 7 (1956): 165–89.

80. See Bynum, " 'And Woman His Humanity.' "

81. On the lactation of Bernard of Clairvaux, see E. Vacandard, *Vie de saint Bernard, abbé de Clairvaux*, 2 vols. (1895; reprint ed., Paris, 1920), 2:78; and Dewez and van Iterson, "La Lactation." Suso received the same vision; see Life, in *Deutsche Schriften*, 49–50. Alanus de Rupe (or Alan de la Roche, d. 1475), founder of the modern rosary devotion, tells a similar story of himself in his Revelations; see Heribert Holzapfel, *St. Dominikus und der Rosenkranz* (Munich, 1903), 21. See also Alb. Poncelet, "Index miraculorum B. V. Mariae quae saec. VI–XV latine conscripta sunt," *Analecta Bollandiana* 21 (1902): 359, which lists four stories of sick men healed by the Virgin's milk. Vincent of Beauvais, *Speculum historiale* (Venice, 1494), fol. 80r, tells of a sick cleric who nursed from the Virgin. Elizabeth Petroff, *Consolation of the Blessed* (New York, 1979), 74, points out that the Italian women saints she has studied nurse only from Christ, never from Mary, in vision. A nun of Töss, however, supposedly received the "pure, tender breast" of Mary into her mouth to suck because she helped Mary rear the Christchild;

Ferdinand Vetter, ed., *Das Leben der Schwestern zu Töss beschrieben von Elsbet Stagel* (Berlin, 1906), 55–56. And Lukardis of Oberweimar nursed from Mary; see her Life in *Analecta Bollandiana* 18 (1899): 318–19.

82. See, for example, Life of Gertrude van Oosten (or of Delft), 350. For Gertrude of Helfta's vision of nursing the Christchild, see *Jesus as Mother*, 208, n. 123. There is one example of a man nursing Christ; see McDonnell, *Beguines*, 328, and Browe, *Die Wunder*, 106.

83. See *Jesus as Mother*, chap. 4, and nn. 35 and 49 above. Clare of Assisi supposedly received a vision in which she nursed from Francis; see "Il Processo di canonizzazione di S. Chiara d'Assisi," ed. Zeffirino Lazzeri, *Archivum Franciscanum Historicum* 13 (1920), 458, 466.

84. See Mary Martin McLaughlin, "Survivors and Surrogates: Children and Parents from the Ninth to the Thirteenth Centuries," *The History of Childhood*, ed. L. DeMause (New York, 1974): 115–18, and *Jesus as Mother*, 131–35.

85. See above nn. 15 and 82. Examples of women who exuded healing oil, in life or after death, include Lutgard of Aywières, Christina *Mirabilis*, Elisabeth of Hungary, Agnes of Montepulciano, and Margaret of Città di Castello (d. 1320). See Life of Lutgard, 193–94; Life of Christina *Mirabilis* 652–54; Huyskens, *Quellenstudien*, 51–52; Raymond of Capua, Life of Agnes, in *AASS*, April: 2, 806, Life of Margaret of Città di Castello, *Analecta Bollandiana* 19 (1900): 27–28, and see also *AASS*, April: 2, 192.

86. Life of Lutgard, 193; Life of Lukardis of Oberweimar, *Analecta Bollandiana* 18 (1899): 337–38; Peter of Vaux, Life of Colette, 563, 576, 585, and 588.

87. For a general discussion of reversed images in women's spirituality, see Bynum, "Women's Stories, Women's Symbols."

11

The Appetite as Voice

Joan Jacobs Brumberg

The symptoms of disease never exist in a cultural vacuum. Even in a strictly bio-medical illness, patient responses to physical discomfort and pain are structured in part by who the patient is, the nature of the care giver, and the ideas and values at work in that society. Similarly, in mental illness, basic forms of cognitive and emotional disorientation are expressed in behavioral aberrations that mirror the deep preoccupations of a particular culture. For this reason a history of anorexia nervosa must consider the ways in which different societies create their own symptom repertoires and how the changing cultural context gives meaning to a symptom such as noneating.[1]

In this chapter I suggest a link between the emergence of anorexia nervosa in the nineteenth century and the cultural predispositions of that era. Just as the incidence of anorexia nervosa today is related to powerful contemporary messages about body image and dieting, there is a cultural context—albeit a somewhat different one—that helps to explain the Victorian anorectic. Again, this is not to say that cultural ideas directly cause the disease. At the outset I acknowledged the etiological complexities and the limitations of historical study; anorexia nervosa is a multidetermined disorder that involves individual biological and psychological factors as well as environmental influences. As a historian, I cannot resolve the problem of causation nor can I chart individual psychopathologies. Historical study does, however, illuminate the larger meanings of food and eating in Victorian society and in that process posits a certain set of cultural preoccupations that had particular impact on adolescent women among the bourgeoisie. In effect, by supplying something of the female "food vocabulary" of a distant era, I hope to explain how there could have been anorexia nervosa before there was Twiggy.

The medical literature of the nineteenth century provides few clues to the meaning of anorexic behavior in that period. Physicians reported the characteristic cry of the anorectic as "I will not eat," but they rarely provided the text of the critical subordinate clause, "I will not eat because. . . ." The medical literature supplied few accounts by Victorian anorectics, *in their own words*, of why they refused their food and why they deprived their bodies as they did. The Victorian anorectic's understanding of her own behavior remains something of a mystery.

Victorian anorectics did present somatic complaints about gastric discomfort and difficulty in swallowing. Because nineteenth-century doctors emphasized physical diagnosis and somatic treatment, they probably reinforced presentations of this type of distress.[2] Parents also found it easier to accept physical rather than emotional reasons as an explanation of their daughter's emaciation and food-refusing behavior. Yet in cases of anorexia nervosa, biomedical reasons for noneating were quickly undercut, since the diagnosis itself meant that there was no organic disease. Most physicians avoided explanations of etiology and concentrated instead on curing the primary symptoms, noneating and emaciation.

Among the few careful medical reports that include any discussion of motivation is Stephen MacKenzie's 1895 account of his anorexic patient's refusing her food "on account of her mother talking to her about being so fat." A decade earlier Jean-Martin Charcot discovered a rose-colored ribbon wound tightly around the waist of a patient with anorexia nervosa. Questioning revealed the girl's preoccupation with the size of her body: the ribbon was a measure that her waist was not to exceed. Max Wallet's 1892 discussion of two cases of anorexia nervosa suggested the same theme and implicated peers. A seventeen-year-old refused her food because of a "fear of being seen as a bit heavy," and a fifteen-year-old stopped eating when she "got the idea that she was too fat after seeing her friends forcing themselves to lose weight."[3]

These behaviors were likely to be dismissed by physicians as the flirtatious "coquetries" and simpleminded "scruples" of female adolescence.[4] In effect, nineteenth-century medicine did not relate anorexia nervosa to the cultural milieu that surrounded the Victorian girl. The ideas of Victorian women and girls about appetite, food, and eating, as well as the cultural categories of fat and thin, were not mentioned as contributing to the disease. Only in the twentieth century has medicine come to understand that society plays a role in shaping the form of psychological disorders and that behavior and physical symptoms are related to cultural systems. Throughout the nineteenth century most doctors gave and accepted formulaic explanations of anorexia nervosa (for example, their patients "craved sympathy" or experienced a "perversion of the will") without providing any substantive discussion of why appetite and food were at issue. These explanations said more about the doctors' general views on adolescence, gender, and hysteria than they did about the specific mentality of patients with anorexia nervosa.

Given the attention paid to anorexia nervosa in late-nineteenth-century Anglo-American medicine, the failure of physicians to document the anorectic's explanations, however mundane or bizarre they might have been, is a provocative omission. This lapse raises a number of questions about the state of doctor-patient relations and the history of diagnostic and therapeutic techniques in the late nineteenth century. What were the dynamics of doctor, patient, and mother within the Victorian examining room? What expectations did the doctor have of his young female patients? A sensitivity to the relationship between culture and symptomatologies prompts additional questions. What motivated young women in the late nineteenth century to persistently refuse food in the face of familial coaxing and professional medical supervision? What role did food and eating play in female identity in the Victorian era?

In the Examining Room

By the 1870s physical examinations included visual observation and manual manipulation of the body, combined with a few rudimentary tests of body temperature, blood, and urine. Because manual examinations were a progressive innovation done only by better-trained physicians, some patients were probably still unfamiliar and uneasy with the latest information-gathering techniques: listening to the body through a stethoscope, manipulation of the body parts, and tactile probing of the body. In cases involving young women, professionally knowledgeable and socially correct doctors did the examination in the presence of the girl's mother as well as a clinical clerk. The clerk recorded information while the physician listened to, poked, and thumped the patient, who remained partially dressed in her underclothing.[5]

The nineteenth-century physician's new faith in the verifiable external signs and sounds of illness shaped the interaction in the examining room. The doctor was more interested in what the body revealed than in anything the patient had to say about her illness. Educated physicians came to regard the process of history taking as secondary to the process of physical examination. Doctors were assured that patient accounts of illness were more often than not prejudiced, ignorant, and unreliable; personal and family narratives were rarely objective and they almost always revealed the ignorance of lay people about medical phenomena. In this atmosphere of suspicion about all patient accounts, volatile adolescent girls were considered particularly unreliable informants.[6]

As a consequence, the professionally correct doctor turned to the girl's mother, in her authoritative role as parent, for information about the patient's medical history and current symptoms. Social convention supported this strategy; as long as an unmarried girl resided at home, her parents unquestionably had authority over her. Consequently, the doctor, who was in the employ of the parents, dealt with the young woman as a child. In the Victorian examining room, the mother was not only a monitor of the physical examination of her daughter's body but a check on the substance of the conversation between doctor and patient.

In this scenario, which assumes that doctor, mother, and patient all played out their expected social roles, the examining room reproduced the situation in the home. The doctor and the mother were the primary conversants; again, two adults, a male and a female, were focused on the girl's wasting body and her refusal of food. Again, she was told that she ought to eat. Her response, shaped by the nature of the medical investigation and parental expectations, was to say that she could not—eating hurt in some vague, nonspecific way. When the examination showed no organic problem to sustain this interpretation, the doctor made his diagnosis: anorexia nervosa.

It is unlikely that the doctor ever dismissed the mother and tried to see the patient alone in order to search out what was troubling the girl and causing her to refuse food. Propriety worked against such a scenario, as did the conception of the patient as a dependent person and the doctor's lack of interest in girlish narratives. Adolescent patients must have sensed the doctors' disinterest in their point of view. A recovered anorectic told her physician that, during treatment, "I saw that you wished to shut me up."[7]

In an era that valued demure behavior in all women, it is not inconceivable that the anorexic girl honored social conventions by respecting her mother's authority and keeping silent. It is also possible that the partially dressed young woman was so embarrassed by her situation and so intimidated by her doctor that she could not speak. Another explanation, culled directly from the medical record, suggests that the patient responded to questions in a diffident manner. Published case reports repeatedly said that girls with anorexia nervosa were "sullen," "sly," and "peevish," implying that they were as parsimonious with their words as with their food.[8] Refusal to sustain conversation with either one's parents or one's doctor went hand in hand with refusal to eat. The anorexic girl used both her appetite and her body as a substitute for rhetorical behavior.

When the doctor had ascertained that the patient had no physical reason not to eat, his forbearance might ebb. At that point an authoritarian regimen of overfeeding, weighing, and isolation was usually instituted. This regimen became the primary basis of the doctor's relationship with his patient. Conversation, when it occurred, centered on the amount of food taken and weight and strength gained. Both doctor and patient acted as if the girl's illness was strictly physical (rather than emotional), despite the fact that differential diagnosis established exactly the opposite. The physician maintained an exclusive focus on the issue of the girl's body and her need to add flesh. To do otherwise was to pander to the sympathies of a hysterical adolescent.

Anorexic patients did sometimes talk to less authoritative female medical personnel and to their peers. Although the evidence is undeniably scanty, a few examples reveal that the Victorian anorectic shared aspects of her compulsion to starve with individuals she perceived as less threatening and more sympathetic than the doctor. If she did not speak to them directly about why she refused her food, she left telltale pieces of evidence that provided some explanation of her behavior. This evidence was rarely uncovered by the supervising doctor.

For example, ward nurses and nursing nuns were in close and intimate contact with patients—feeding them regularly, washing their bodies, and supervising their waking hours. In the 1890s a French physician ousted a high-strung mother from the home where her anorexic daughter was being treated and sent in a nun, *une religieuse*, to care for the emaciated fifteen-year-old. At first, the "new attitude of her caretakers" terrified the girl and she became even more recalcitrant about eating and said that she wished to die. After three months of an enforced dietary regimen of arrowroot, bread, eggs, and beef tea, the girl left the recuperative home "fat" and capable of a normal, active life. At discharge the nun disclosed that she had in her possession a series of letters written by the former patient, which "constituted a peculiarly interesting witness from the point of view of causation of her malady." The letters, addressed to an older male relative, disclosed that the patient's food refusal was generated by her romantic and "singular passion" for this man who, in the young girl's presence, had explicitly admired another woman who was "extremely lean." In the effort to please him, the girl began to starve herself, walk excessively, and lace herself very tightly. Yet she never once told her doctor about her passion for her relative or her desire to be thin.[9]

The nurse functioned as detective in another case involving a twenty-year-old at St. George's Hospital in London. None of the consulting doctors could find an organic reason why their hysterical patient refused to eat, but they continued to work diligently to ease the stomach pains of which she complained. The nurse on the ward, however, regarded the girl as a malingerer and told the doctors, "On December 6th, whilst the girl was apparently suffering . . . the Queen [Victoria] passed the hospital, on her way to open Blackfriar's Bridge; [the girl] rose in bed to watch her out of the window, having been thought utterly unable to move, owing to pain." On yet another occasion, when friends were admitted for hospital visiting hours, the nurse found the supposedly debilitated girl "sitting up in bed, trying on a new coloured frock."

This same patient, who told the doctors that she could not and would not eat, engaged in surreptitious relationships with other patients in order to get bits of their food, which she would eat on the sly. The nurse at St. George's Hospital found a note from the girl indicating that she did eat secretly:

> My dear Mrs. Evans—I was very sorry you should take the trouble of cutting me such a nice peice [sic] of bread and butter yesterday. I would of taken it, but all of them saw you send it, and they would of made enough to have talked about, but I should be very glad if you will cut me a nice peice [sic] of crust and put it in a peice [sic] of paper and send it or else bring it, so as they do not see it, for they all watch me very much.[10]

The nurse's information on this patient provided no real explanation of why the girl would not eat in the presence of her caretakers, but it did confirm the physician's belief that hysterical adolescents were by definition deceptive. This attitude surely affected the doctors' interactions with anorexic patients. If it was assumed that the patient was by nature duplicitous, then any explanation she gave would be suspect.

Sensing the doctor's loyalties to her parents and his suspicious attitude toward her, the anorectic usually chose not to disclose her private preoccupations to an unsympathetic male authority figure. When she spoke, it was almost always of bodily ills: pain after eating, a sour stomach, difficulty in swallowing, flatulence. Deference, fear, and anger all combined to keep her essentially mute. When her bodily pre-occupations were rooted in ideas that the doctor might find childish, inappropriate, or distasteful, her silence became confirmed. Furthermore, there was always the distinct possibility of misunderstanding or embarrassment when girls told personal things to men or boys. The bourgeois world of the nineteenth century was still very much sex segregated.[11] Consequently, enormous emotional risks were involved in baring one's soul to the doctor. Most adult men did not understand the language of girlhood sentiment and knew neither its vocabulary nor its symbols. The silence of the Victorian anorectic was in keeping with her provocative resort to symbolic rather than rhetorical behavior.

The Irrational Appetite

In the late nineteenth century, adolescent girls demonstrated an array of health problems that involved eating and appetite disturbances. These problems lent confusion rather than clarity to the process of making the diagnosis for anorexia nervosa.

In effect, there was a wide spectrum of "picky eating" and food refusal, ranging from the normative to the pathological. Anorexia nervosa was the extreme—but it was not altogether alien, given the range of behaviors that doctors saw in adolescent female patients. As one astute twentieth-century physician wrote about the origins of anorexia nervosa in the era of William Gull, "the conditions of life were well staged for such a disturbance."[12]

The health of young women was definitely influenced by a general female fashion for sickness and debility.[13] The sickly wives and daughters of the bourgeoisie provided the medical profession with a ready clientele. In Victorian society unhappy women (and men) had to employ physical complaints in order to be permitted to take on the privileged "sick role." Because the most prevalent diseases in this period were those that involved "wasting," it is no wonder that becoming thin, through noneating, became a focal symptom. Wasting was in style.

Among women, invalidism and scanty eating commonly accompanied each other. The partnership was familiar enough to become the subject of a satirical novel. In *The Female Sufferer; or, Chapters from Life's Comedy*, Augustus Hoppin satirized the indolent existence of an upper-class invalid who, while ever so ill, managed to run a vigorous social life from her sick chamber. The stylized eating of this "nervous exhaustionist" was central to the author's portrait. Delicate foods such as "tidbits of fruit and jelly," "a snip of a roll," "a wren's leg on toast," were taken only to "appease the cravings of her exhausted nerves"—but not because she was hungry. At times, however, the debilitated patient would become voracious for "dainty" items such as wedding cake, peaches and cream, and freshly cut melon. According to Hoppin, another characteristic type among female sufferers was the woman supposedly perishing of starvation or "pining away." "Well! dying of inanition is doing something, isn't it?" asked one of the admirers who surrounded the sick couch. Another replied, "Inanition being merely action begun, demands too much exertion for [the lady] to finish."[14]

Adolescent girls simply followed and imitated the behavioral styles of adult women. As a consequence, mothers were urged to take action against their daughters' fondness for wasting and debility. In *Eve's Daughters; or, Common Sense for Maid, Wife, and Mother*, Marion Harland told parents:

> Show no charity to the faded frippery of sentiment that prates over romantic sickliness. Inculcate a fine scorn for the desire to exchange her present excellent health for the estate of the pale, drooping, human-flower damsel; the taste that covets the "fascination" of lingering consumption; the "sensation" of early decease induced by the rupture of a blood-vessel over a laced handkerchief held firmly to her lily mouth by agonized parent or distracted lover. All this is bathos and vulgarity . . . Bid her leave such balderdash to the pretender to ladyhood, the low-minded *parvenu*, who, because foibles are more readily imitated than virtues, and tricks than graces, copies the mistakes of her superiors in breeding and sense, and is persuaded that she has learned "how to do it."[15]

Harland, an American, called the "cultivation of fragility" a "national curse."

Of the conditions that affected girls most frequently, dyspepsia and chlorosis both incorporated peculiar eating and both could be confused with anorexia nervosa. Dyspepsia, a form of chronic indigestion with discomfort after eating, was widespread in middle-class adults and in their daughters. Physicians saw the adolescent dyspeptic frequently; advice writers suggested how she should be managed at home; health reformers used her existence to argue for changes in the American diet; and even novelists considered her enough of a fixture on the domestic scene to include her in their portraits of social life.[16] The dyspeptic had no particular organic problem; her stomach was simply so sensitive that it precluded normal eating. Whereas dyspeptic women could be extremely thin, some, according to doctors' reports, were corpulent. Yet dyspepsia sometimes looked much like anorexia nervosa. For example, a physician described his young dyspeptic patients as persons "who enter upon a strict regimen which they follow only too well. By auto-observation and auto-suggestion, by constantly noticing and classifying their foods, and rejecting all kinds that they think they cannot digest, they finally manage to live on an incredibly small amount."[17]

Chlorosis, a form of anemia named for the greenish tinge that allegedly marked the skin of the patient, was the characteristic malady of the Victorian adolescent girl. Although chlorosis was never precisely defined and differentiated, it was unequivocally regarded as a disease of girlhood rather than boyhood. Its symptoms included lack of energy, shortness of breath, dyspepsia, headaches, and capricious or scanty appetite; sometimes the menses stopped. Chlorotic girls tended to lose some weight as a result of poor eating and aversion to specific foods, particularly meat.[18] (Today iron-deficiency anemia corresponds to the older diagnosis of chlorosis.)

Doctors of the Victorian era fostered the notion that all adolescent girls were potentially chlorotic: "Every girl passes as it were through the outer court of chlorosis in her progress from youth to maturity . . . Perhaps, no girl escapes it altogether."[19] In contrast to anorexia nervosa, treatment for this popular disease was relatively easy: large doses of iron salts and a period of rest at home. As a result, parents were not afraid of chlorosis. In fact, it was accepted as a normal part of adolescent development. Many doctors and families were also fond of tonics to stimulate the appetite, restore the blush to the cheek, and cure latent consumption. "Young Girls Fading Away" was the headline of a well-known advertisement for Dr. William's Pink Pills for Pale People, a medicine aimed at the chlorotic market.[20] A vast amount of patent medicine was sold to families that assumed chlorosis in an adolescent whenever her energy, spirits, or appetite waned. In cases that were eventually diagnosed as anorexia nervosa, the patient in the earliest stages may well have been regarded as dyspeptic or chlorotic. Because clinical descriptions of the confirmed dyspeptic, the chlorotic, and the anorectic had many features in common, we must assume that the diagnoses occasionally overlapped.[21]

Taken together, these conditions suggest that young women presented unusual eating and diminished appetite more often than any other group in the population. Apparently, it was relatively normal for a Victorian girl to develop poor appetite and skip her meals, "affect daintiness" and eat only sweets, or express strong food preferences and dislikes.[22] A popular women's magazine told its readership that in adolescence "digestive problems are common, the appetite is fickle, and evidences of poor nourishment abound."[23] Between 1850 and 1900 the most frequent warning issued to parents of girls had to do with forestalling the development of idiosyncrasies, irregularities, or strange whims of appetite because these were precursors of disease as well as signs of questionable moral character.

Ideas about female physiology and sexual development underlay the physician's expectations and his clinical treatment. Doctors believed that women were prone to gastric disorders because of the superior sensitivity of the female digestive system. Using the machine metaphor that was popular in describing bodily functions, they likened a man's stomach to a quartz-crushing machine that required coarse, solid food. By contrast, the mechanisms of a woman's stomach could be ruined if fed the same materials. The female digestive apparatus required foods that were soft, light, and liquid.[24] (Dyspepsia in women could result from the choice of inappropriate foods that required considerable chewing and digestion.)

To the physician's mind, a young woman caught up in the process of sexual maturation was subject to vagaries of appetite and peculiar cravings. "The rapid expansion of the passions and the mind often renders the tastes and appetite capricious," wrote a midcentury physician.[25] Therefore, even normal sexual development had the potential to create a disequilibrium that could lead to irregular eating such as the kind reported in dyspepsia and chlorosis. But physicians reported on eating behavior that was far more bizarre. In fact, the adolescent female with "morbid cravings" was a stock figure in the medical and advice literature of the Victorian period. Stories circulated of "craving damsels" who were "trash-eaters, oatmeal-chewers, pipe-chompers, chalk-lickers, wax-nibblers, coal-scratchers, wall-peelers, and gravel-diggers." The clinical literature also provided a list of "foods" that some adolescent girls allegedly craved: chalk, cinders, magnesia, slate pencils, plaster, charcoal, earth, spiders, and bugs.[26] Modern medicine associates iron-deficiency anemia with eating nonnutritive items, such as pica. For the Victorian physician, nonnutritive eating constituted proof of the fact that the adolescent girl was essentially out of control and that the process of sexual maturation could generate voracious and dangerous appetites.

In this context physicians asserted that even normal adolescent girls had a penchant for highly flavored and stimulating foods. A reputable Baltimore physician, for example, described three girlfriends who constantly carried with them boxes of pepper and salt, taking the condiments as if they were snuff.[27] The story was meant to imply that the girls were slaves of their bodily appetites. Throughout the medical and advice literature an active appetite or an appetite for particular foods was used as a trope for dangerous sexuality. Mary Wood-Allen warned young readers that the girl who masturbated "will manifest an unnatural appetite, sometime desiring mustard, pepper, vinegar and spices, cloves, clay, salt, chalk, charcoal, etc."[28]

Because appetite was regarded as a barometer of sexuality, both mothers and daughters were concerned about its expression and its control. It was incumbent upon the mother to train the appetite of the daughter so that it represented only the highest moral and aesthetic sensibilities. A good mother was expected to manage this situation before it escalated into a medical or social problem. Marion Harland's Mamie, the prototypical adolescent, developed at puberty "morbid cravings of appetite and suffered after eating things that never disagreed with her before." Mamie's mother was cautioned about the possibility that a disturbance of appetite could precipitate an adolescent decline. Mothers were urged to be vigilant: "If Mamie has not a rational appetite, create a digestive conscious [sic] that may serve her instead." Mothers were expected to educate, if not tame, their adolescent daughter's propensity for "sweetmeats, bonbons, and summer drinks" as well as for "stimulating foods such as black pepper and vinegar pickle."[29] "Inflammatory foods" such as condiments and acids, thought to be favored by the tumultuous female adolescent, were strictly prohibited by judicious mothers. Adolescent girls were expressly cautioned against coffee, tea, and chocolate; salted meats and spices; warm bread and pastry; confectionery; nuts and raisins; and, of course, alcohol.[30] These sorts of foods stimulated the sensual rather than the moral nature of the girl.

No food (other than alcohol) caused Victorian women and girls greater moral anxiety than meat. The flesh of animals was considered a heat-producing food that stimulated production of blood and fat as well as passion. Doctors and patients shared a common conception of meat as a food that stimulated sexual development and activity. For example, Lucien Warner, a popular medical writer, suggested that meat eating in adolescence could actually accelerate the development of the breasts and other sex characteristics; at the same time, a restriction on the carnivorous aspects of the diet could moderate premature or rampant sexuality as well as over-abundant menstrual flow. "If there is any tendency to precocity in menstruation, or if the system is very robust and plethoric, the supply of meat should be quite limited. If, on the other hand, the girl is of sluggish temperament and the menses are tardy in appearance, the supply of meat should be especially generous."[31] Meat eating in excess was linked to adolescent insanity and to nymphomania.[32] A stimulative diet of meat and condiments was recommended only for those girls whose development of the passions seemed, somehow, "deficient."

By all reports adolescent girls ate very little meat, a practice that certainly contributed to chlorosis or iron-deficiency anemia. In fact, many openly disdained meat without being necessarily committed to the ideological principles of the health reformers who espoused vegetarianism.[33] Meat avoidance, therefore, is the most apt term for this pattern of behavior. According to E. Lloyd Jones, adolescent girls "are fond of biscuits, potatoes, etc. while they avoid meat on most occasions, and when they do eat meat, they prefer the burnt outside portion." Another doctor confirmed the same problem in a dialogue between himself and a patient. "Oh, I like pies and preserves but I can't bear meat," the young woman reportedly told the family physician. A "disgust for meat in any form" characterized many of the adolescent female patients of a Pennsylvania practitioner of this period.[34]

When it became necessary to eat meat (say, if prescribed by a doctor), it was an event worthy of note. For many, meat eating was endured for its healing qualities but despised as a moral and aesthetic act. For example, eighteen-year-old Nellie Browne wrote to tell her mother that a delicate classmate [Laura] had, like her own sister Alice, been forced to change her eating habits:

> I am very sorry to hear Alice has been so sick. Tell her she must eat meat if she wishes to get well. Laura eats meat *three* times a day.—She says she cannot go without it.—If Laura *can* eat *meat, I am sure Alice can.* If Laura needs it *three* times a day, Alice needs it *six.* (Italics in original.)[35]

After acknowledging the "common distaste for meat" among his adolescent patients, Clifford Allbutt wrote, "Girls will say the entry of a dish of hot meat into the room makes them feel sick."[36]

The repugnance for fatty animal flesh among Victorian adolescents ultimately had a larger cultural significance. Meat avoidance was tied to cultural notions of sexuality and decorum as well as to medical ideas about the digestive delicacy of the female stomach. Carnality at table was avoided by many who made sexual purity an axiom. Proper women, especially sexually maturing girls, adopted this orientation with the result that meat became taboo. Contemporary descriptions reveal that some young women may well have been phobic about meat eating because of its associations:

> There is the common illustration which every one meets a thousand times in a lifetime, of the girl whose stomach rebels at the very thought of fat meat. The mother tries persuasion and entreaty and threats and penalties. But nothing can overcome the artistic development in the girl's nature which makes her revolt at the bare idea of putting the fat piece of a dead animal between her lips.[37]

In this milieu food was obviously more than a source of nutrition or a means of curbing hunger; it was an integral part of individual identity. For women in particular, how one ate spoke to issues of basic character.

"A Woman Should Never be Seen Eating"

In Victorian society food and femininity were linked in such a way as to promote restrictive eating among privileged adolescent women. Bourgeois society generated anxieties about food and eating—especially among women. Where food was plentiful and domesticity venerated, eating became a highly charged emotional and social undertaking. Displays of appetite were particularly difficult for young women who understood appetite to be both a sign of sexuality and an indication of lack of self-restraint. Eating was important because food was an analogue of the self. Food choice was a form of self-expression, made according to cultural and social ideas as well as physiological requirements. As the anthropologist Claude Lévi-Strauss put it, things must be "not only good to eat, but also good to think."[38]

Female discomfort with food, as well as with the act of eating, was a pervasive subtext of Victorian popular culture.[39] The naturalness of eating was especially problematic among upwardly mobile, middle-class women who were preoccupied with establishing their own good taste.[40] Food and eating presented obvious difficulties because they implied digestion and defecation, as well as sexuality. A doctor explained that one of his anorexic patients "refused to eat for fear that, during her digestion, her face should grow red and appear less pleasant in the eyes of a professor whose lectures she attended after her meals."[41] A woman who ate inevitably had to urinate and move her bowels. Concern about these bodily indelicacies explains why constipation was incorporated into the ideal of Victorian femininity. (It was almost always a symptom in anorexia nervosa.) Some women "boasted that the calls of Nature upon them averaged but one or two demands per week."[42]

Food and eating were connected to other unpleasantries that reflected the self-identity of middle-class women. Many women, for good reason, connected food with work and drudgery. Food preparation was a time-consuming and exhausting job in the middle-class household, where families no longer ate from a common soup pot. Instead, meals were served as individual dishes in a sequence of courses. Women of real means and position were able to remove themselves from food preparation almost entirely by turning over the arduous daily work to cooks, bakers, scullery and serving maids, and butlers. Middle-class women, however, could not achieve the same distance from food.[43]

Advice books admonished women "not to be ashamed of the kitchen," but many still sought to separate themselves from both food and the working-class women they hired to do the preparation and cooking. A few women felt the need to make alienation from food a centerpiece of their identity. A young "lady teacher," for example, "regard[ed] it as unbecoming her position to know anything about dinner before the hour for eating arrived . . . [She was] ashamed of domestic work, and graduate[d] her pupils with a similar sense of false propriety."[44] Similarly, in the 1880s in Rochester, New York, a schoolgirl was chastised by her aunt for describing (with relish) in her diary the foods she had eaten during the preceding two weeks.[45]

Food was to be feared because it was connected to gluttony and to physical ugliness. In advice books such as the 1875 *Health Fragments; or, Steps toward a True Life* women were cautioned to be careful about what and how much they ate. Authors George and Susan Everett enjoined: "Coarse, gross, and gluttonous habits of life degrade the physical appearance. You will rarely be disappointed in supposing that a lucid, self-respectful lady is very careful of the food which forms her body and tints her cheeks." Sarah Josepha Hale, the influential editor of *Godey's Lady's Book* and an arbiter of American domestic manners, warned women that it was always vulgar to load the plate.[46]

Careful, abstemious eating was presented as insurance against ugliness and loss of love. Girls in particular were told: "Keep a great watch over your appetite. Don't always take the nicest things you see, but be frugal and plain in your tastes."[47] Young women were told directly that "gross eaters" not only developed thick skin but had prominent blemishes and broken blood vessels on the nose. Gluttony also robbed the eyes of their intensity and caused the lips to thicken, crack, and lose their red color. "The glutton's mouth may remind us of cod-fish—never of kisses." A woman with a rosebud mouth was expected to have an "ethereal appetite." A story circulated that Madam von Stein "lost Goethe's love by gross habits of eating sausages and drinking strong coffee, which destroyed her beauty." Women such as von Stein, who indulged in the pleasures of the appetite, were said to develop "a certain unspiritual or superanimal expression" that conveyed their base instincts. "[We] have never met true refinement in the person of a gross eater," wrote the Everetts.[48]

Indulgence in foods that were considered stimulating or inflammatory served not only as an emblem of unchecked sensuality but sometimes as a sign of social aggression. Women who ate meat could be regarded as acting out of place; they were assuming a male prerogative. In *Daniel Deronda* (1876) George Eliot described a group of local gentry, all men, who came together after a hunt to take their meal apart from the women. As they ate, the men took turns telling stories about the "epicurism of the ladies, who had somehow been reported to show a revolting masculine judgement in venison." Female eating was a source of titillation to men precisely because they understood eating to be a trope for sexuality. Furthermore, women who asked baldly for venison were aggressive if not insatiable. What most bothered the local gentry was the women's effrontery to "ask . . . for the fat—a proof of the frightful rate at which corruption might go in women, but for severe social restraint."[49]

Because food and eating carried such complex meanings, manners at the table became an important aspect of a woman's social persona. In their use of certain kinds of conventions, nineteenth-century novelists captured the crucial importance of food and eating in the milieu of middle-class women. Because they understood the middle-class reverence for the family meal, writers such as Jane Austen and Anthony Trollope saw the meal as an arena for potential individual and collective embarrass-ment. These novelists provided numerous examples of young women whose lives and fortunes hung on the issue of dinner-table decorum. For example, in Austen's *Mansfield Park* (1814) the heroine, Fanny Price, was horrified at the prospect of having her well-bred suitor eat with her family and "see all their deficiencies." Fanny was concerned not only about her family's lower standard of cookery, but about her sister's mortifying tendency to eat "without restraint." In Trollope's *Ralph the Heir* (1871) the family of Mr. Neefit, a tradesman, invited Ralph Newton, a gentleman, to their family table after some degree of preparation and nervousness on the part of the wife and daughter. Newton, who was halfheartedly courting the daughter at the request of her socially ambitious father, ultimately concluded that the young woman was attractive enough but the roughness of her father was unbearable. He found particularly galling the manner in which Mr. Neefit ate his shrimp.[50] Manners at table were often a dead giveaway of one's true social origins. This convention for marking the social distance between the classes was utilized by Mark Twain in the famous scene in *The Prince and the Pauper* (1881) where the prince's impersonator drinks from his fingerbowl.[51]

Women's anxieties about how to eat in genteel fashion were widespread and con-veyed by novelists in a number of different ways. In Elizabeth Gaskell's *Cranford* (1853), the middle-class ladies of the town were made uncomfortable by the presentation of foods that were difficult to eat—in this case, peas and oranges. One woman "sighed over her delicate young peas [but] left them on one side of her plate untasted" rather than attempt to stab them or risk dropping them between the two prongs of her fork. She knew that she could not do "the ungenteel thing"—shoveling them with her knife. So, too, oranges presented difficulties for the decorous middle-class women of Cranford:

> When oranges came in, a curious proceeding was gone through. Miss Jenkyns did not like to cut the fruit; for, as she observed, the juice all ran out nobody knew where; sucking (only I think she used some more recondite word) was in fact the only way of enjoying oranges; but then there was the unpleasant association with a ceremony frequently gone through by little babies; and so, after dessert, in orange season, Miss Jenkyns and Miss Matty used to rise up, possess themselves each of an orange in silence, and withdraw to the privacy of their own rooms to indulge in sucking oranges.[52]

In fact, secret eating was not unknown among those who subscribed to the absurd dictum that "a woman should never be seen eating."[53] This statement, attributed by George Eliot to the famed poet Lord Byron, was the ultimate embodiment of Victorian imperatives about food and gender.

Over and over again, in all of the popular literature of the Victorian period, good women distanced themselves from the act of eating with disclaimers that pronounced their disinterest in anything but the aesthetics of food. "It's very little I eat myself," a proper Trollopian hostess explained, "but I do like to see things nice."[54] Apparently, Victorian girls adopted the aesthetic sensibilities of their mothers, displaying extraordinary interest in the appearance and color of their food, in the effect of fine china and linen, and in agreeable surroundings. A 1904 study of the psychology of foods in adolescence reported that boys most valued companionship at table, whereas girls emphasized "ceremony" and "appointments."[55] Attention to the aesthetics of eating seemed to minimize the negative implications of participating in the gustatory and digestive process.

But Victorian women avoided connections to food for a number of other reasons. The woman who put soul over body was the ideal of Victorian femininity. The genteel woman responded not to the lower senses of taste and smell but to the highest senses—sight and hearing—which were used for moral and aesthetic purposes.[56] One of the most convincing demonstrations of a spiritual orientation was a thin body—that is, a physique that symbolized rejection of all carnal appetites. To be hungry, in any sense, was a social faux pas. Denial became a form of moral certitude and refusal of attractive foods a means for advancing in the moral hierarchy.

Appetite, then, was a barometer of a woman's moral state. Control of eating was eminently desirable, if not necessary. Where control was lacking, young women were subject to derision. "The girl who openly enjoys bread-and-butter, milk, beefsteak and potatoes, and thrives thereby, is the object of many a covert sneer, or even overt jest, even in these sensible days and among sensible people."[57] Given the intensity of concern about control of appetite, it is not surprising that some women found strong attraction in cultural figures whose biographies exemplified the triumph of spirit over flesh. Two figures representing the Romantic and medieval traditions became especially relevant to how young women thought about these issues: Lord Byron and Catherine of Siena. Both spoke to the moral desirability of being thin.

Known to the Victorian reading public as the author of the immensely popular epic poems *Childe Harold* and *Don Juan*, Lord Byron (1788–1824) remained an important cultural figure whose life and work stood, even as late as the third quarter of the century, as a symbol of the power of the Romantic movement.[58] Young women who shared the Romantic sensibility found Byron's poetry inspirational. *Childe Harold*, which detailed a youth's struggle for meaning, spoke to the inner reaches of the soul and helped its readers transcend the "tawdry world." For many, such as Trollope's Lizzie Eustace, Byron was "the boy poet who understood it all."[59]

Although Byron's tempestuous love life served to titillate some and revolt others, the poet's struggles with the relation of his body to his mind were of enormous interest to women. Byron starved his body in order to keep his brain clear. He existed on biscuits and soda water for days and took no animal food. According to memoirs written by acquaintances, the poet had a "horror of fat"; to his mind, fat symbolized lethargy, dullness, and stupidity. Byron feared that if he ate normally he would lose his creativity. Only through abstinence could his mind exercise and improve. In short, Byron was a model of exquisite slenderness and his sensibilities about fat were embraced by legions of young women.[60]

Adults, especially physicians, lamented Byron's influence on youthful Victorians. In addition to encouraging melancholia and emotional volatility, Byronism had consequence for the eating habits of girls. In Britain "the dread of being fat weigh[ed] like an incubus" on Romantic youngsters who consumed vinegar "to produce thinness" and swallowed rice "to cause the complexion to become paler."[61] According to American George Beard, "our young ladies live all their growing girlhood in semistarvation" because of a fear of "incurring the horror of disciples of Lord Byron."[62] Byronic youth, in imitation of their idol, disparaged fat of any kind, a practice which advice writers found detrimental to their good health. "If plump, [the girl] berates herself as a criminal against refinement and aesthetic taste; and prays in good or bad earnest, for a spell of illness to pull her down."[63] Other doctors besides Beard spoke of the popular Romantic association between scanty eating, a slim body, and "delicacy of mind." Beard, however, did not let the blame for modern eating habits rest entirely on the Romantics. He decried the influence of Calvinist doctrine as well. Cultivated people, he said, eat too little because of the old belief that "satiety is a conviction of sin."[64]

Women attuned to the higher senses did find inspiration for their abstemious eating in the austerities of medieval Catholics—particularly Catherine of Siena. Although Protestant Victorian writers presented Catherine's asceticism as a dangerous form of self-mortification, there was also widespread admiration for the spiritual intensity that drove her fasts. Victorian writers used the biography of Saint Catherine to demonstrate how selfhood could be lost to a higher moral or spiritual purpose. This message was considered particularly relevant to girls, in that self-love was supposed to be a distinguishing characteristic of the female adolescent.[65] Saint Catherine's biography was included in inspirational books for girls, and two prominent women of the period demonstrated serious interest in her life. Josephine Butler, an articulate English feminist, published a full-length biography in 1879; and Vida Scudder, a Wellesley College professor of English and a Christian socialist, published her letters in 1895.[66] Because she provided a vivid demonstration of a woman who placed spiritual over bodily concerns, Catherine of Siena was of enormous interest to Anglo-American women.

This lingering ascetic imperative did not go unnoticed by one of the period's most astute observers of religious behavior, William James. In *The Varieties of Religious Experience* the Harvard philosopher and psychologist noted quite correctly that old religious habits of "misery" and "morbidness" had fallen into disrepute. Those who pursued "hard and painful" austerities were regarded, in the modern era, as "abnormal." Yet James noted that young women were the most likely to remain tied to the dying tradition of religious asceticism. Although he understood that ascetic behavior had many sources (what he called "diverse psychological levels"), he did mark girls as the group most likely to embrace "saintliness." "We all have some friend," James wrote, "perhaps more often feminine than masculine, and young than old, whose soul is of this blue-sky tint, whose affinities are rather with flowers and birds and all enchanting innocencies than with dark human passions."[67] Girls seemed to be the most interested in the tenets of what James called saintliness: conquering the ordinary desires of the flesh, establishing purity, and taking pleasure in sacrifice.[68]

Those who were ascetic in girlhood tried to act as well as look like saints. In *The Morgesons* (1862) Veronica, an adolescent invalid and dyspeptic, defined her saintliness through a diet of tea and dry toast. She cropped her hair short in the manner of a penitent; constantly washed her hands with lavender water, as if she were taking a ritual absolution; and on her bedroom wall hung a picture of the martyred Saint Cecilia with white roses in her hair.[69] Although Veronica was a Protestant, she revered Saint Cecilia for her spirituality. Many novelists linked asceticism to physical beauty as well as to spiritual perfection. In short, beautiful women were often "saintlike," a relationship that implied the inverse as well—the "saintlike" were beautiful. Trollope, for example, spoke of a young gentleman who "declared to himself at once that she was the most lovely young woman he had ever seen. She had dark eyes, and perfect eyebrows, and a face which, either for colour or lines of beauty, might have been taken for a model for any female saint or martyr."[70]

By the last decades of the nineteenth century, a thin body symbolized more than just sublimity of mind and purity of soul. Slimness in women was also a sign of social status. This phenomenon, noted by Thorstein Veblen in *The Theory of the Leisure Class*, heralded the demise of the traditional view that girth in a woman signaled prosperity in a man. Rather, the reverse was true: a thin, frail woman was a symbol of status and an object of beauty precisely because she was unfit for productive (or reproductive) work. Body image rather than body function became a paramount concern.[71] According to Veblen, a thin woman signified the idle idyll of the leisured classes.

By the turn of the twentieth century, elite society already preferred its women thin and frail as a symbol of their social distance from the working classes. Consequently, women with social aspirations adopted the rule of slenderness and its related dicta about parsimonious appetite and delicate food. Through restrictive eating and restrictive clothing (that is, the corset), women changed their bodies in the name of gentility.

Women of means were the first to diet to constrain their appetite, and they began to do so before the sexual and fashion revolutions of the 1920s and the 1960s. In the 1890s Veblen noted that privileged women "[took] thought to alter their persons, so as to conform more nearly to the instructed taste of the time."[72] In effect, Veblen documented the existence of a critical gender and class imperative born of social stratification. In bourgeois society it became incumbent upon women to control their appetite in order to encode their body with the correct social messages.[73] Appetite became less of a biological drive and more of a social and emotional instrument.

Historical evidence suggests that many women managed their food and their appetite in response to the notion that sturdiness in women implied low status, a lack of gentility, and even vulgarity. Eating less rather than more became a preferred pattern for those who were status conscious. The pressure to be thin in order to appear genteel came from many quarters, including parents. "The mother, also, would look upon the sturdy frame and ruddy cheeks as tokens of vulgarity."[74] Recall that Eva Williams, admitted to London Hospital in 1895 for treatment of anorexia nervosa, told friends that it was her mother who complained about her rotundity.

A controlled appetite and ill health were twin vehicles to elevated womanhood. Advice to parents about the care of adolescent daughters regularly included the observation that young women ate scantily because they denigrated health and fat for their declassé associations. In 1863 Hester Pendleton, an American writer on the role of heredity in human growth, lamented the fact that the natural development of young women was being affected by these popular ideas. "So perverted are the tastes of some persons," Pendleton wrote, "that delicacy of constitution is considered a badge of aristocracy, and daughters would feel themselves deprecated by too robust health."[75] Health in this case meant a sturdy body, a problem for those who cultivated the fashion of refined femininity. One writer felt compelled to assert: "Bodily health is never pertinently termed 'rude.' It is not coarse to eat heartily, sleep well, and to feel the life throbbing joyously in heart and limb."[76]

Consequently, to have it insinuated or said that a woman was robust constituted an insult. This convention was captured by Anthony Trollope in *Can You Forgive Her?* (1864). After a late-night walk on the grounds of the Palliser estate, the novel's genteel but impoverished young heroine, Alice Vavasour, was criticized by a male guest for her insensitivity to the physical delicacy of her walking companion and host, Lady Glencora Palliser. The youthful and beautiful Lady Glencora caught cold from the midnight romp, but Alice did not. The critical gentleman immediately caught the social implications of the fact that Alice was not unwell from the escapade, and he used her health against her: "Alice knew that she was being accused of being robust . . . but she bore it in silence. Plough-boys and milkmaids are robust, and the accusation was a heavy one."[77] The same associations were relevant thirty years later in the lives of middle-class American girls. Marion Harland observed that the typical young woman "would be disgraced in her own opinion and lost caste with her refined mates were she to eat like a ploughboy."[78]

In the effort to set themselves apart from plowboys and milkmaids—that is, working and rural youth—middle-class daughters chose to pursue a body configuration that was small, slim, and essentially decorative. By eating only tiny amounts of food, young women could disassociate themselves from sexuality and fecundity and they could achieve an unambiguous class identity. The thin body not only implied asexuality and an elevated social address, it was also an expression of intelligence, sensitivity, and morality. Through control of appetite Victorian girls found a way of expressing a complex of emotional, aesthetic, and class sensibilities.

By 1900 the imperative to be thin was pervasive, particularly among affluent female adolescents. Albutt wrote in 1905, "Many young women, as their frames develop, fall into a panic fear of obesity, and not only cut down on their food, but swallow vinegar and other alleged antidotes to fatness."[79] The phenomenon of adolescent food restriction was so widespread that an advice writer told mothers, "It is a circumstance at once fortunate and notable if [your daughter] does not take the notion into her pulpy brain that a healthy appetite for good substantial food is 'not a bit nice,' 'quite too awfully vulgar you know.' "[80]

Because food was a common resource in the middle-class household, it was available for manipulation. Middle-class girls, rather than boys, turned to food as a symbolic language, because the culture made an important connection between food and femininity and because girls' options for self-expression outside the family were limited by parental concern and social convention. In addition, doctors and parents expected adolescent girls to be finicky and restrictive about their food. Young women searching for an idiom in which to say things about themselves focused on food and the body. Some middle-class girls, then as now, became preoccupied with expressing an ideal of female perfection and moral superiority through denial of appetite. The popularity of food restriction or dieting, even among normal girls, suggests that in bourgeois society appetite was (and is) an important voice in the identity of a woman. In this context anorexia nervosa was born.

Notes

1. The best statement in the extensive literature on the relation of culture to obsessive-compulsive disorder and anorexia nervosa is Albert Rothenberg, "Eating Disorder as a Modern Obsessive-Compulsive Syndrome," *Psychiatry* 49 (February 1986), 45–53. Another historian interested in the changing symptomatology of hysteria is Edward Shorter. See his "The First Great Increase in Anorexia Nervosa," *Journal of Social History* 21 (Fall 1987), 69–96.

2. This point is made by Laurence Kirmayer, "Culture, Affect and Somatization, Part II," *Transcultural Psychiatric Research Review* 21 (1984), 254.

3. London Hospital Physician's Casebooks, MS 107 (1897); quoted in Pierre Janet, *The Major Symptoms of Hysteria* (New York, 1907), p. 234; Max Wallet, "Deux cas d'anorexie hystérique," in *Nouvelle iconographie de la Salpêtrière*, ed. J.-M. Charcot (Paris, 1892), p. 278.

4. See Janet, *Major Symptoms*, p. 234, for these terms. Janet did, however, have a more sophisticated interpretation based on his idea of "body shame."

5. My description of late-nineteenth-century examinations is drawn from clinical case records and from Joel Stanley Reiser, *Medicine and the Reign of Technology* (Cambridge, Mass., 1978), and Charles E. Rosenberg, "The Practice of Medicine in New York a Century Ago," *Bulletin of the History of Medicine* 41 (May—June 1967), 223–253; D. W. Cathell, *The Physician Himself and What He Should Add to His Scientific Acquirements* (Baltimore, 1882).

6. I have tried to incorporate, and move beyond, the perspective on male doctor–female patient relations provided in Barbara Ehrenreich and Deirdre English, *Witches, Midwives and Nurses* (New York, n.d.) and *Complaints and Disorders: The Sexual Politics of Sickness* (Old Westbury, N.Y., 1973).
   At the outset I asked, Can the absence of a patient voice be attributed solely to misogyny and authoritarianism on the part of doctors? I think not. Without discounting the relevance of both these well-known attributes of nineteenth-century medical men, I posit that the silence stemmed from a complex of medical and social factors that shaped interactions between doctors and patients in such a way that doctors really had no coherent, firsthand information to give. In a system of medicine that emphasized physical diagnosis, somatic complaints were primary. Traditional ideas about hysteria in women and girls prevailed, providing the rationale for medical and moral therapy.

7. J.-M. Charcot, *Clinical Lectures on Diseases of the Nervous System* (London, 1889), p. 214.

8. See for example John Ogle, "A Case of Hysteria: 'Temper Disease,' " *British Medical Journal* (July 16, 1870), 59; William Gull, "Anorexia Nervosa (Apepsia Hysterica, Anorexia Hysterica)," *Transactions of the Clinical Society of London 7* (1874), 22–28; Thomas Stretch Dowse, "Anorexia Nervosa," *Medical Press and Circular* 32 (August 3, 1881), 95–97, and ibid. (August 17, 1881), 147–148; W.J. Collins, "Anorexia Nervosa," *Lancet* (January 27, 1894), 203.

9. Charles Féré, *The Pathology of Emotions* (London, 1899), pp. 79–80. The home was designated a Maison de Santé.

10. Ogle, "A Case of Hysteria," pp. 57–58.

11. The strongest statement on the separation of male and female spheres is Carroll Smith-Rosenberg, "The Female World of Love and Ritual: Relations between Women in Nineteenth-Century America," *Signs 1* (1975), 1–29. In addition to this generative article see Nancy Cott, *The Bonds of Womanhood* (New Haven, 1977); Ann Douglas, *The Feminization of American Culture* (New York, 1977); Mary Ryan, *Cradle of the Middle Class* (New York, 1982). Ellen Rothman's *Hands and Hearts: A History of Courtship in America* (New York, 1984) provides some rethinking of the Smith-Rosenberg thesis.

12. John Ryle to Parkes Weber, January 27, 1939, PP/FDW F. Parkes Weber Papers, Wellcome Institute, London.

13. Ann Douglas Wood, "The Fashionable Diseases: Women's Complaints and Their Treatment in Nineteenth Century America," *Journal of Interdisciplinary History* 4 (1973), 25–52; John S. Haller, Jr., and Robin Haller, *The Physician and Sexuality in Victorian America* (Urbana, Ill., 1974). See also Judith Walzer Leavitt, ed., *Women and Health in America: Historical Readings* (Madison, Wis., 1984).

14. Augustus Hoppin, *A Fashionable Sufferer; or, Chapters from Life's Comedy* (Boston, 1883), pp. 16–17, 35, 55, 135.

15. Marion Harland, *Eve's Daughters; or, Common Sense for Maid, Wife, and Mother*, with an introduction by Sheila M. Rothman (Farmingdale, N.Y., 1885; reprint ed., 1978), pp. 135, 153.

16. See for example Elizabeth Stoddard, *The Morgesons* (New York, 1862). Stoddard's novel depended in large part on the contrast between two adolescent sisters in a wealthy New England family. Thirteen-year-old Veronica was in a premature adolescent decline and remained at home, isolated from society; sixteen-year-old Cassandra was vigorous and "about in the world." Cassandra served as narrator of the story and presented many intimate details of the behavior of her ailing and dyspeptic sister.

    Stoddard never told the reader what was exactly wrong with the invalid Veronica, but she appeared to combine both dyspepsia and anorexia. Cassandra explained: "Delicacy of constitution the doctor called the disorder. She had no strength, no appetite, and looked more elfish than ever. She would not stay in bed, and could not sit up, so father had a chair made for her, in which she could recline comfortably." Much of the trouble with Veronica revolved around her lack of appetite, complicated by a simultaneous interest in food preparation and in the consumption habits of others. From her chair she directed the family maid to cook elaborate dishes that she would not eat.

    For long periods of time Veronica Morgeson ate only a single kind of food. "As we began our meal," Cassandra recounted, "Veronica came in from the kitchen with a plate of toasted crackers. She set the plate down, and gravely shook hands with me, saying she had concluded to live entirely on toast, but supposed I would eat all sorts of food as usual." Although Veronica ate virtually nothing at the family table, she asked many questions about food and cast aspersions on her sister's normal appetite. Whenever Cassandra said she was hungry, Veronica's eyes "sparkled with disdain." Above and beyond her physical problems with digestion, Veronica Morgeson took some strange emotional delight in her denial of hunger. Her reclusive existence, her listless appetite, and her sensitive stomach all implied that her body was subordinated to other, higher, more spiritual concerns.

17. Bernard Hollander, *Nervous Disorders of Women* (London, 1916), p. 77.

18. For the general history of chlorosis see Frank Panettiere, "What Ever Happened to Chlorosis?" *Alaska Medicine* 15 (May 1973), 68–70; Eugene Stransky, "On the History of Chlorosis," *Episteme* 8: 1 (1974), 26–46; and Ronald E. McFarland, "The Rhetoric of Medicine: Lord Herbert's and Thomas Carew's Poems of Green Sickness," *Journal of the History of Medicine* 30 (July 1975), 250–258. For more perceptive in-depth studies of Great Britain see Karl Figlio, "Chlorosis and Chronic Disease in Nineteenth Century Britain: The Social Constitution of Somatic Illness in a Capitalist Society," *Social History* 3 (1978), 167–197; and Irvine Loudon, "Chlorosis, Anemia, and Anorexia Nervosa," *British Medical Journal* 281 (December 20–27, 1980), 1669–75; and idem, "The Diseases Called Chlorosis," *Psychological Medicine* 14 (1984), 27–36. For the United States see R. P. Hudson, "The Biography of Disease: Lessons Learned from Chlorosis," *Bulletin of the History of Medicine* 51 (1977), 448–463; S. R. Huang, "Chlorosis and the Iron Controversy: An Aspect of Nineteenth-Century Medicine," Ph.D. diss., Harvard University, 1978; Joan Jacobs Brumberg, "Chlorotic Girls, 1870–1920: An Historical Perspective on Female Adolescence," *Child Development* 53 (December 1982), pp. 1468–77; A. C. Siddall, "Chlorosis—Etiology Reconsidered," *Bulletin of the History of Medicine* 56 (1982), 254–260.

19. J. H. Montgomery, *Clinical Observations on Cases of Simple Anemia or Chlorosis Occurring in Young Women in the Decade Following Puberty* (Erie, Pa., 1919), n.p.

20. *Elmira Daily Gazette and Free Press*, March 26, 1898. On women and the patent medicine business see Sarah Stage, *Female Complaints: Lydia Pinkham and the Business of Women's Medicine* (New York, 1979).

21. Loudon, "Chlorosis," posits that after 1850 chlorosis incorporated at least three different kinds of disorders common in young women, each involving loss of weight, anemias, and amenorrhea. He argues that chlorosis was a functional disorder closely related to anorexia nervosa: they were two "closely related conditions, each a manifestation of the same type of psychological reaction to the turbulence of puberty and adolescence" (p. 1675). Loudon is correct that the two diseases had much in common, and his observation probably means that there was more anorexia nervosa than heretofore thought. Anorexia nervosa was usually a class-specific diagnosis, however. Given the family dynamic that was part of the disorder, working-class girls were unlikely to develop anorexia nervosa. Moreover, if a similar pattern of noneating developed in a working-class girl, family reverberations were different and medicine was more likely to call the disorder chlorosis or depression.

22. G. Stanley Hall, *Adolescence*, vol. 1 (New York, 1904), pp. 252–253.

23. Lillie A. Williams, "The Distressing Malady of Being Seventeen Years Old," *Ladies' Home Journal* (May 1909), p. 10.

24. James Henry Bennet, *Nutrition in Health and Disease* (London, 1877), pp. 59–60, 170. Among others who noted that children and women had more easily disturbed digestive systems are Elizabeth Blackwell, *The Laws of Life* (New York, 1852), and Anna Brackett, *The Education of American Girls* (New York, 1874).

25. Edward Smith, *Practical Dietary for Families, Schools, and the Labouring Classes* (London, 1864), p. 141.

26. These descriptions are from Harland, *Eve's Daughters*, pp. 111–113; see also Charles E. Simon, "A Study of Thirty-One Cases of Chlorosis," *American Journal of Medical Sciences* 113 (April 1897), 399–423; Lucien Warner, *A Popular Treatise on the Functions and Diseases of Women* (New York, 1875), p. 70; E.H. Ruddock, *The Lady's Manual of Homeopathic Treatment* (New York, 1869), p. 32.

27. Simon, "Thirty-One Cases," pp. 413–414.

28. Mary Wood-Allen, *What a Young Girl Ought to Know* (Philadelphia, 1905), p. 89.

29. Harland, *Eve's Daughters*, pp. 111–115. The notion of an adolescent decline had some foundation. Among adolescents tuberculosis was a particularly serious threat with a high mortality rate. Most people agreed that once a child passed beyond infancy and early childhood, adolescence stood as the next critical juncture in the life course. Adolescence required caution, whether one was male or female, but physical decline seemed to occur more often among girls. For a summary of a number of different statements about the vulnerability of the female adolescent see Nellie Comins Whitaker, "The Health of American Girls," *Popular Science Monthly* 71 (September 1907), 240.

30. Harvey W. Wiley, *Not By Bread Alone* (New York, 1915), pp. 245, 248–250, 256; Brackett, *Education of American Girls*, pp. 25–26.

31. Warner, *A Popular Treatise*, p. 54.

32. On meat eating and sexual excess see Vern Bullough and Martha Voight, "Women, Menstruation and Nineteenth Century Medicine," *Bulletin of the History of Medicine* 47 (1973), 66–82. According to many physicians, flesh eating contributed to a "neurotic temperament." T. S. Clouston, superintendent of the Edinburgh asylum, espoused a widely held view: "I have found ... a large proportion of the adolescent insane h[ave] been flesh-eaters, consuming and having a craving for much animal food"; see his "Puberty and Adolescence Medico-Psychologically Considered," *Edinburgh Medical Journal* 26 (July 1880), 17. This article was reprinted in the *American Journal of Insanity* in April 1881.

33. See Albert J. Bellows, *The Philosophy of Eating* (New York, 1869), for a typical statement about meat by a health reformer. And see Stephen Nissenbaum, *Sex, Diet and Debility in Jacksonian America: Sylvester Graham and Health Reform* (Westport, Conn., 1980) on nineteenth-century vegetarianism.

34. E. L. Jones, *Chlorosis: The Special Anemia of Young Women* (London, 1897), p. 39; Charles Meigs, *Females and Their Diseases* (Philadelphia, 1848), p. 361; Montgomery, *Clinical Observations*. Susan Williams, *Savory Suppers and Fashionable Feasts: Dining in Victorian America* (New York, 1985), notes that rare or "underdone meat" was "out of fashion" and particularly "disgusting" to women and children (p. 239).

35. Nellie Browne to her mother [April 1859?], Sarah Ellen Browne papers, Schlesinger Library. Jane Hunter called this letter to my attention.

36. J. Clifford Allbutt, *A System of Medicine*, vol. 5 (New York, 1905), p. 517.

37. "The Antagonism between Sentiment and Physiology in Diet," *Current Literature* 42 (February 1907), 222. In reporting on anorexia nervosa, Pierre Janet described the phenomenon of *la crainte d'engraisser*—literally, the fear of taking on grease.

38. Quoted in Claude Fischler, "Food Preferences, Nutritional Wisdom, and Sociocultural Evolution," in *Food, Nutrition and Evolution*, ed. Dwain Watcher and Norman Kretchmer (New York, 1981), p. 58.

39. I borrow the term "subtext" from literary criticism, particularly from the semioticians. Food in nineteenth-century fiction and culture serves as a set of signs and symbols with communicative power. See Roland Barthes, "From Work to Text," in *Textual Strategies: Perspectives in Post-Structuralist Criticism*, ed. J. Harari (Ithaca, N.Y., 1979), pp. 73–81.

40. Williams, *Savory Suppers*, chap. 1, describes a pattern of middle-class concern over eating and correct eating behaviors. See also Jocelyne Kolb, "Wine, Women, and Song: Sensory Referents in the Works of Heinrich Heine," Ph.D. diss., Yale University, 1979, p. 8.

41. Janet, *Major Symptoms*, p. 234. Kolb, "Wine, Women, and Song," writes, "In the early nineteenth century, when Heine was writing, the mere mention of food in a lyrical passage was generally shocking enough to achieve ironic distance" (p. 71). The Romantic conception of food as an emblem of both the positive and negative sides of sensuality continued into the late nineteenth century.

42. Harland, *Eve's Daughters*, p. 81.

43. Although middle-class women were frequently assisted by a household servant, resulting in some reduction of time spent in the kitchen, they spent more and more energy planning meals, purchasing food, and determining ways to make eating an aesthetic experience. See Ruth Schwartz Cowan, *More Work for Mother: The Ironies of Household Technology from the Open Hearth to the Microwave* (New York, 1983).

44. George Everett and Susan Everett, *Health Fragments; or, Steps Toward a True Life* (New York, 1875), p. 35.

45. Fannie Munn Field diary, December 17, 1886, Box 9:18, Munn-Pixley Papers, Department of Rare Books and Special Collections, Rush Rhees Library, University of Rochester, Rochester, N. Y. Susan Williams kindly brought this example to my attention.

46. Everett and Everett, *Health Fragments*, p. 25; Sarah Josepha Hale, *Receipts for the Million* (Philadelphia, 1857), p. 509.

47. [Society for Promoting Christian Knowledge], *Talks to Girls by One of Themselves* (London, 1894), p. 104.

48. Everett and Everett, *Health Fragments*, pp. 26, 29; Leslie A. Marchand, *Byron: A Portrait* (New York, 1970), p. 386.

49. George Eliot, *Daniel Deronda* (New York, n.d.), p. 104.

50. Jane Austen, *Mansfield Park* (New York, 1963), pp. 311–312; Anthony Trollope, *Ralph the Heir* (London, 1871), p. 195.

51. Mark Twain, *The Prince and the Pauper* (New York, 1881).

52. Elizabeth Gaskell, *Cranford* (New York, 1906), pp. 41, 53. Oranges were problematic for American eaters also; see Williams, *Savory Suppers*, pp. 108–109.

53. Eliot, *Daniel Deronda*, p. 104. The actual text of Byron's statement is, "A woman should never be seen eating or drinking, unless it be lobster salad and champagne, the only truly feminine and becoming viands" (Marchand, *Byron*, p. 133). In William Makepeace Thackeray's *History of Pendennis* (London, 1848–50) the character of Blanche is delineated by her peculiar appetite and secret eating: "When nobody was near, our little sylphide, who scarcely ate at dinner more than six grains of rice . . . was most active with her knife and fork, and consumed a very substantial portion of mutton cutlets: in which piece of hypocrisy it is believed she resembled other young ladies of fashion." (Quoted in Ann Alexandra Carter, "Food, Feasting, and Fasting in the Nineteenth Century British Novel," Ph.D. diss., University of Wisconsin, 1978, p. 3.)

54. Anthony Trollope, *Can You Forgive Her?* (New York, 1983), p. 70.

55. Sanford Bell, "An Introductory Study of the Psychology of Foods," *Pedagogical Seminary* 9 (1904), 88–89. In *Adolescence*, vol. 2 (New York, 1907), pp. 14–15, G. Stanley Hall noted that appetite varied a great deal in adolescence in response to "psychic motives." Hall did not differentiate between male and female adolescents, although Bell certainly did.

56. Kolb, "Wine, Women, and Song," suggests that this idea of higher and lower senses was inherited from the eighteenth century—specifically from Chevalier de Jancourt, who wrote on the subject in the famed *Encyclopédie*.

57. Harland, *Eve's Daughters*, p. 153.

58. On Byron's influence among the Victorians see Donald David Stone, *The Romantic Impulse in Victorian Fiction* (Cambridge, Mass., 1980).

59. Lizzie Eustace in Anthony Trollope's *Eustace Diamonds*, quoted in Stone, *Romantic Impulse*, p. 51.

60. On Byron's life and his struggles with food and eating see Edward John Trelawny, *Records of Shelly, Byron and the Author*, ed. David Wright (London, 1973), pp. 11, 35–36, 86, 97–98, 245. Mary Jacobus pointed out to me that Trelawny, notoriously "pro-Shelley," is a less than totally disinterested source of information on Byron's dieting. It is significant that Shelley too was a picky eater and a vegetarian, contributing further to the romance of undereating.

61. J. Milner Fothergill, *The Maintenance of Health* (London, 1874), pp. 80–81. This report of girls swallowing vinegar is not anomalous. See Brumberg, "Chlorotic Girls."

62. George Beard, *Eating and Drinking* (New York, 1871), p. 104.

63. Harland, *Eve's Daughters*, p. 124.

64. Beard, *Eating and Drinking*, p. v. Beard connected the American propensity for scanty eating to health reformers of an evangelical bent. He spoke of the "vast army of Jeremiahs who have gone up and down the land, predicting that gluttony will be our ruin" (p. iv). The links between parsimonious eating and religiosity are also suggested in Anthony Trollope, *Rachel Ray* (London, 1880). The plot of this novel is interesting because of the contrast between Mrs. Ray, a loving mother who likes tea and buttered toast, and her austere daughter, Mrs. Prime, an asexual, pious, churchgoing widow who likes "her tea to be stringy and bitter" and "her bread stale" (p. 5). The older women who influence Rachel Ray are defined by their appetite and eating behavior.

65. See for example Clouston, "Puberty and Adolescence," p. 14.

66. H. Davenport Adams, *Childlife and Girlhood of Remarkable Women* (New York, 1895), chap. 5; Josephine Butler, *Catherine of Siena* (London, 1878); Vida Scudder, *Letters of St. Catherine of Siena* (New York, 1905).

67. William James, *The Varieties of Religious Experience*, with an introduction by Reinhold Niebuhr (New York, 1961), pp. 79, 221, 238–240. James observed that as a consequence of secularization painful austerities or asceticism were not considered abnormal. "A strange moral transformation has within the past century swept over our Western world. We no longer think we are called on to face physical pain with equanimity. It is not expected of a man that he should either endure it or inflict much of it, and to listen to the recitals of cases of it makes our flesh creep morally as well as physically . . . The result of this historical alteration is that even in the Mother Church herself, where ascetic discipline has such a fixed traditional prestige as a factor of merit, it has largely come into desuetude, if not discredit. A believer who flagellates or 'macerates' himself today arouses more wonder and fear than emulation" (p. 238). On asceticism see also Emile Durkheim, *The Elementary Forms of Religious Life*, trans. Robert Nisbet (London, 1976), pp. 299–321.

68. In the twentieth century many ideological followers of G. S. Hall spoke of the idealism of adolescents, that is, their search for moral purity. Asceticism in adolescence is rarely discussed, however. An interesting article that

makes this connection and also distinguishes between adaptive and pathological asceticism is S. Louis Mogul, "Asceticism in Adolescence and Anorexia Nervosa," *Psychoanalytic Study of the Child* 35 (1980), 155–175. Mogul is reliant on Anna Freud, *The Ego and the Mechanisms of Defense* (London, 1937).

69. See note 16 above and Stoddard, *The Morgesons*, pp. 30, 57, 61, 140.
70. Trollope, *Ralph the Heir*, p. 29.
71. Thorstein Veblen, *The Theory of the Leisure Class* (New York, 1967).
72. Ibid., pp. 145–149.
73. My argument about the ways in which cultural and class concerns are encoded in the body follows from Michel Foucault, *History of Sexuality*, vol. 1 (New York, 1980), and *Madness and Civilization* (New York, 1965). The discipline of the body and its relation to social theory is explored by Brian Turner, "The Discourse of Diet," *Theory, Culture and Society* 1: 1 (1982), 23–32.
74. Hester Pendleton, *Husband and Wife; or, The Science of Human Development through Inherited Characteristics* (New York, 1863), p. 66.
75. Ibid., pp. 65–66.
76. Harland, *Eve's Daughters*, p. 134.
77. Trollope, *Can You Forgive Her?* p. 297.
78. Harland, *Eve's Daughters*, p. 111.
79. Allbutt, *A System of Medicine*, vol. 3, p. 485.
80. Harland, *Eve's Daughters*, p. 111.

12

Anorexia Nervosa: Psychopathology as the Crystallization of Culture

Susan Bordo

1996 Prefatory Note

In 1983, preparing to teach an interdisciplinary course called "Gender, Culture, and Experience," I felt the need for a topic that would enable me to bring feminist theory alive for a generation of students that seemed increasingly suspicious of feminism. My sister, Binnie Klein, who is a therapist, suggested that I have my class read Kim Chernin's *The Obsession: Reflections on the Tyranny of Slenderness*. I did, and I found my Reagan-era students suddenly sounding like the women in the consciousness-raising sessions that had first made me aware of the fact that my problems as a woman were not mine alone. While delighted to have happened on a topic that was so intensely meaningful to them, I was also disturbed by what I was reading in their journals and hearing in the privacy of my office. I had identified deeply with the general themes of Chernin's book. But my own disordered relations with food had never reached the point of anorexia or bulimia, and I was not prepared for the discovery that large numbers of my students were starving, binging, purging, and filled with self-hatred and desperation. I began to read everything I could find on eating disorders. I found that while the words and diaries of patients were enormously illuminating, most of the clinical theory was not very helpful. The absence of cultural perspective—particularly relating to the situation of women—was striking.

As a philosopher, I was also intrigued by the classically dualistic language my students often used to describe their feelings, and I decided to incorporate a section on contemporary attitudes toward the body in my metaphysics course. There, I discovered that although it was predominantly my female students who experienced their lives as a perpetual battle with their bodies, quite a few of my male students expressed similar ideas when writing about running. I found myself fascinated by what seemed to me to be the cultural emergence of a set of attitudes about the body which, while not new as *ideas*, were finding a special kind of embodiment in contemporary culture, and I began to see all sorts of evidence for this cultural hypothesis. "Anorexia Nervosa: Psychopathology as the Crystallization of Culture," first published in 1985, was the result of my initial exploration of the various cultural axes to which my students' experiences guided me in my "Gender, Culture, and Experience" and metaphysics courses. Other essays followed, and ultimately a book, *Unbearable Weight: Feminism, Western Culture, and the Body*, further exploring eating disorders through other cultural interconnections and intersections: the historically female disorders, changes in historical attitudes toward what constitutes "fat" and "thin," the structural tensions of consumer society, the post-modern fascination with re-making the self.

Since I began this work in 1983, my then-tentative intuitions have progressively been validated, as I have watched body practices and attitudes that were a mere ripple on the cultural scene assume a central place in the construction of contemporary subjectivity. In 1995, old clinical generalizations positing a distinctive class, race, family and "personality" profile of the woman most likely to develop an eating disorder no longer hold, as images of the slender, tight body become ever-more widely deployed, asserting their homogenizing power over other cultural ideals of beauty, other cultural attitudes toward female appetite and desire. The more generalized obsession with control of the body which I first began to notice in the early eighties now supports burgeoning industries in exercise equipment, diet products and programs, and cosmetic surgery—practices which are engaged in by greater numbers and more diverse groups of people all the time. On television, "infomercials" hawking stomach-flatteners, miracle diet plans, and wrinkle-dissolving cosmetics have become as commonplace as aspirin ads. As the appearance of our bodies has become more and more important to personal and professional success, the incidence of eating disorders has risen, too, among men. All of this has led to an explosion of written material, media attention and clinical study, much of it strongly bearing out my observations and interpretations. I have not, however, incorporated any new studies or statistics into the piece reprinted here. With the exception of a few endnotes, it appears substantially as it did in its original version.

> Historians long ago began to write the history of the body. They have studied the body in the field of historical demography or pathology; they have considered it as the seat of needs and appetites, as the locus of physiological processes and metabolisms, as a target for the attacks of germs or viruses; they have shown to what extent historical processes were involved in what might seem to be the purely biological "events" such as the circulation of bacilli, or the extension of the lifespan. But the body is also directly involved in a political field; power relations have an immediate hold upon it; they invest it, mark it, train it, torture it, force it to carry out tasks, to perform ceremonies, to emit signs.
>
> *Michel Foucault*, Discipline and Punish

> I believe in being the best I can be,
> I believe in watching every calorie . . .
>
> *Crystal Light television commercial*

Eating Disorders, Culture, and the Body

Psychopathology, as Jules Henry has said, "is the final outcome of all that is wrong with a culture."[1] In no case is this more strikingly true than in that of anorexia nervosa and bulimia, barely known a century ago, yet reaching epidemic proportions today. Far from being the result of a superficial fashion phenomenon, these disorders, I will argue, reflect and call our attention to some of the central ills of our culture—from our historical heritage of disdain for the body, to our modern fear of loss of control over our future, to the disquieting meaning of contemporary beauty ideals in an era of greater female presence and power than ever before.

Changes in the incidence of anorexia[2] have been dramatic.[3] In 1945, when Ludwig Binswanger chronicled the now famous case of Ellen West, he was able to say that "from a psychiatric point of view we are dealing here with something new, with a new symptom."[4] In 1973, Hilde Bruch, one of the pioneers in understanding and treating eating disorders, could still say that anorexia was "rare indeed."[5] Today, in 1984, it is estimated that as many as one in every 200–250 women between the ages of thirteen and twenty-two suffer from anorexia, and that anywhere from 12 to 33 percent of college women control their weight through vomiting, diuretics, and laxatives.[6] The New York Center for the Study of Anorexia and Bulimia reports that in the first five months of 1984 it received 252 requests for treatment, as compared to the 30 requests received in all of 1980.[7] Even correcting for increased social awareness of eating disorders and a greater willingness of sufferers to report their illnesses, these statistics are startling and provocative. So, too, is the fact that 90 percent of all anorectics are women, and that of the 5,000 people each year who have part of their intestines removed as an aid in losing weight 80 percent are women.[8]

Anorexia nervosa is clearly, as Paul Garfinkel and David Garner have called it, a "multidimensional disorder," with familial, perceptual, cognitive, and, possibly, biological factors interacting in varying combinations in different individuals to produce a "final common pathway."[9] In the early 1980s, with growing evidence, not only of an overall increase in frequency of the disease, but of its higher incidence in certain populations, attention has begun to turn, too, to cultural factors as significant in the pathogenesis of eating disorders.[10] Until very recently, however, the most that could be expected in the way of cultural or social analysis, with very few exceptions, was the (unavoidable) recognition that anorexia is related to the increasing emphasis that fashion has placed on slenderness over the past fifteen years.[11] This, unfortunately, is only to replace one mystery with another, more profound than the first.

What we need to ask is *why* our culture is so obsessed with keeping our bodies slim, tight, and young that when 500 people were asked what they feared most in the world, 190 replied, "Getting fat."[12] In an age when our children regularly have nightmares of nuclear holocaust, that as adults we should give *this* answer—that we most fear "getting fat"—is far more bizarre than the anorectic's misperceptions of her body image, or the bulimic's compulsive vomiting. The nightmares of nuclear holocaust and our desperate fixation of our bodies as arenas of control—perhaps one of the few available arenas of control we have left in the twentieth century—are not unconnected, of course. The connection, if explored, could be significant, demystifying, instructive.

So, too, we need to explore the fact that it is women who are most oppressed by what Kim Chernin calls "the tyranny of slenderness," and that this particular oppression is a post-1960s, post-feminist phenomenon. In the fifties, by contrast, with middle-class women once again out of the factories and safely immured in the home, the dominant ideal of female beauty was exemplified by Marilyn Monroe— hardly your androgynous, athletic, adolescent body type. At the peak of her popularity, Monroe was often described as "femininity incarnate," "femaleness embodied"; last term, a student of mine described her as "a cow." Is this merely a change in what size hips, breasts, and waist are considered attractive, or has the very idea of incarnate femaleness come to have a different meaning, different associations, the capacity to stir up different fantasies and images, for the culture of the eighties? These are the sorts of questions that need to be addressed if we are to achieve a deep understanding of the current epidemic of eating disorders.

The central point of intellectual orientation for this essay is expressed in its subtitle. I take the psychopathologies that develop within a culture, far from being anomalies or aberrations, to be characteristic expressions of that culture; to be, indeed, the crystallization of much that is wrong with it. For that reason they are important to examine, as keys to cultural self-diagnosis and self-scrutiny. "Every age," says Christopher Lasch, "develops its own peculiar forms of pathology, which express in exaggerated form its underlying character structure."[13] The only aspect of this formulation with which I would disagree, with respect to anorexia, is the idea of the expression of an underlying, unitary cultural character structure. Anorexia appears less as the extreme expression of a character structure than as a remarkably overdetermined *symptom* of some of the multifaceted and heterogeneous distresses of our age. Just as anorexia functions in a variety of ways in the psychic economy of the anorexic individual, so a variety of cultural currents or streams converge in anorexia, find their perfect, precise expression in it.

I will call those streams or currents "axes of continuity": *axes* because they meet or converge in the anorexic syndrome; *continuity* because when we locate anorexia on these axes, its family resemblances and connections with other phenomena emerge. Some of these axes represent anorexia's *synchronicity* with other contemporary cultural practices and forms—bodybuilding and jogging, for example. Other axes bring to light *historical* connections: for instance, between anorexia and earlier examples of extreme manipulation of the female body, such as tight corseting, or between anorexia and long-standing tradition and ideologies in Western culture, such as our Greco-Christian traditions of dualism. The three axes that I will discuss in this essay (although they by no means exhaust the possibilities for cultural understanding of anorexia) are the *dualist axis*, the *control axis*, and the *gender/power axis*.[14]

Throughout my discussion, it will be assumed that the body, far from being some fundamentally stable, acultural constant to which we must *contrast* all culturally relative and institutional forms, is constantly "in the grip," as Foucault puts it, of cultural practices. Not that this is a matter of cultural *repression* of the instinctual or natural body. Rather, there is no "natural" body. Cultural practices, far from exerting their power *against* spontaneous needs, "basic" pleasures or instincts, or "fundamental" structures of body experience, are already and always inscribed, as Foucault has emphasized, "on our bodies and their materiality, their forces, energies, sensations, and pleasures."[15] Our bodies, no less than anything else that is human, are constituted by culture.

Often, but not always, cultural practices have their effect on the body as experienced (the "lived body," as the phenomenologists put it) rather than the physical body. For example, Foucault points to the medicalization of sexuality in the nineteenth century, which recast sex from being a family matter into a private, dark, bodily secret that was appropriately investigated by such specialists as doctors, psychiatrists, and school educators. The constant probing and interrogation, Foucault argues, ferreted out, eroticized and solidified all sorts of sexual types and perversions, which people then experienced (although they had not done so originally) as defining their bodily possibilities and pleasures. The practice of the medical confessional, in other words, in its constant foraging for sexual secrets and hidden stories, actually *created* new sexual secrets—and eroticized the acts of interrogation and confession, too.[16] Here, social practice changed people's *experience* of their bodies and their possibilities. Similarly, as we shall see, the practice of dieting—of saying no to hunger—contributes to the anorectic's increasing sense of hunger as a dangerous eruption from some alien part of the self, and to a growing intoxication with controlling that eruption.

The *physical* body can, however, also be an instrument and medium of power. Foucault's classic example in *Discipline and Punish* is public torture during the Ancien Régime, through which, as Dreyfus and Rabinow put it, "the sovereign's power was literally and publicly inscribed on the criminal's body in a manner as controlled, scenic and well-attended as possible."[17] Similarly, the nineteenth-century corset caused its wearer actual physical incapacitation, but it also served as an emblem of the power of culture to impose its design on the female body.

Indeed, female bodies have historically been significantly more vulnerable than male bodies to extremes in both forms of cultural manipulation of the body. Perhaps this has something to do with the fact that women, besides *having* bodies, are also *associated* with the body, which has always been considered woman's "sphere" in family life, in mythology, in scientific, philosophical, and religious ideology. When we later consider some aspects of the history of medicine and fashion, we will see that the social manipulation of the female body emerged as an absolutely central strategy in the maintenance of power relations between the sexes over the past hundred years. This historical understanding must deeply affect our understanding of anorexia and of our contemporary preoccupation with slenderness.

This is *not* to say that I take what I am doing here to be the unearthing of a longstanding male conspiracy against women or the fixing of blame on any particular participants in the play of social forces. In this I once again follow Foucault, who reminds us that although a perfectly clear logic, with perfectly decipherable aims and objectives, may characterize historical power relations, it is nonetheless "often the case that no one was there to have invented" these aims and strategies, either through choice of individuals or through the rational game plan of some presiding "headquarters."[18] We are not talking, then, of plots, designs, or overarching strategies. This does not mean that individuals do not *consciously* pursue goals that in fact advance their own position. But it does deny that in doing so they are consciously directing the overall movement of power relations or engineering their shape. They may not even know what that shape is. Nor does the fact that power relations involve domination by particular groups—say, of prisoners by guards, females by males, amateurs by experts—entail that the dominators are in anything like full control of the situation or that the dominated do not sometimes advance and extend the situation themselves.[19] Nowhere, as we shall see, is this collaboration in oppression more clear than in the case of anorexia.

The Dualist Axis

I will begin with the most general and attenuated axis of continuity, the one that begins with Plato, winds its way to its most lurid expression in Augustine, and finally becomes metaphysically solidified and scientized by Descartes. I am referring, of course, to our dualistic heritage: the view that human existence is bifurcated into two realms or substances: the bodily or material, on the one hand; the mental or spiritual, on the other. Despite some fascinating historical variations which I will not go into here, the basic imagery of dualism has remained fairly constant. Let me briefly describe its central features; they will turn out, as we will see, to comprise the basic body imagery of the anorectic.

First, the body is experienced as *alien*, as the not-self, the not-me. It is "fastened and glued" to me, "nailed" and "riveted" to me, as Plato describes it in the *Phaedo*.[20] For Descartes, the body is the brute material envelope for the inner and essential self, the thinking thing; it is ontologically distinct from that inner self, is as mechanical in its operations as a machine, is, indeed, comparable to animal existence.

Second, the body is experienced as *confinement and limitation:* a "prison," a "swamp," a "cage," a "fog"—all images that occur in Plato, Descartes, and Augustine—from which the soul, will, or mind struggles to escape. "The enemy ["the madness of lust"] held my will in his power and from it he made a chain and shackled me," says Augustine.[21] In the work of all three philosophers, images of the soul being "dragged" by the body are prominent. The body is "heavy, ponderous," as Plato describes it; it exerts a downward pull.[22]

Third, the body is *the enemy,* as Augustine explicitly describes it time and again, and as Plato and Descartes strongly suggest in their diatribes against the body as the source of obscurity and confusion in our thinking. "A source of countless distractions by reason of the mere requirement of food," says Plato; "liable also to diseases which overtake and impede us in the pursuit of truth; it fills us full of loves, and lusts, and fears, and fancies of all kinds, and endless foolery, and in very truth, as men say, takes away from us the power of thinking at all. Whence come wars, and fightings, and factions? Whence but from the body and the lusts of the body."[23]

And, finally, whether as an impediment to reason or as the home of the "slimy desires of the flesh" (as Augustine calls them), the body is the locus of *all that threatens our attempts at control.* It overtakes, it overwhelms, it erupts and disrupts. This situation, for the dualist, becomes an incitement to battle the unruly forces of the body, to show it who is boss. For, as Plato says, "Nature orders the soul to rule and govern and the body to obey and serve."[24]

All three—Plato, Augustine, and, most explicitly, Descartes—provide instructions, rules, or models of how to gain control over the body, with the ultimate aim—for this is what their regimen finally boils down to—of learning to live without it.[25] By that is meant: to achieve intellectual independence from the lure of the body's illusions, to become impervious to its distractions, and, most important, to kill off its desires and hungers, Once control has become the central issue for the soul, these are the only possible terms of victory, as Alan Watts makes clear:

> Willed control brings about a sense of duality in the organism, of consciousness in conflict with appetite . . . But this mode of control is a peculiar example of the proverb that nothing fails like success. For the more consciousness is individualized by the success of the will, the more everything outside the individual seems to be a threat—including . . . the uncontrolled spontaneity of one's own body . . . Every success in control therefore demands a further success, so that the process cannot stop short of omnipotence.[26]

Dualism here appears as the offspring, the by-product, of the identification of the self with control, an identification that Watts sees as lying at the center of Christianity's ethic of anti-sexuality. The attempt to subdue the spontaneities of the body in the interests of control only succeeds in constituting them as more alien and more powerful, and thus more needful of control. The only way to win this no-win game is to go beyond control, to kill off the body's spontaneities entirely—that is, to cease to *experience* our hungers and desires.

This is what many anorectics describe as their ultimate goal. "[I want] to reach the point," as one puts it, "when I don't need to eat at all."[27] Kim Chernin recalls her surprise when, after fasting, her hunger returned: "I realized [then] that my secret goal in dieting must have been the intention to kill off my appetite completely."[28]

It is not usually noted, in the popular literature on the subject, that anorexic women are as obsessed with *hunger* as they are with being slim. Far from losing her appetite, the typical anorectic is haunted by it—in much the same way that Augustine describes being haunted by sexual desire—and is in constant dread of being overwhelmed by it. Many describe the dread of hunger, "of not having control, of giving in to biological urge," to "the craving, never satisfied thing,"[29] as the "original fear" (as one puts it),[30] or, as Ellen West describes it, "the real obsession." "I don't think the dread of becoming fat is the real . . . neurosis," she writes, "but the constant desire for food . . . [H]unger, or the dread of hunger, pursues me all morning . . . Even when I am full, I am afraid of the coming hour in which hunger will start again." Dread of becoming fat, she interprets, rather than being originary, served as a "brake" to her horror of her own unregulatable, runaway desire for food.[31] Bruch reports that her patients are often terrified at the prospect of taking just one bite of food, lest they never be able to stop.[32] (Bulimic anorectics, who binge on enormous quantities of food—sometimes consuming up to 15,000 calories a day[33]—indeed *cannot* stop.)

These women experience hunger as an alien invader, marching to the tune of its own seemingly arbitrary whims, disconnected from any normal self-regulating mechanisms. Indeed, it would not possibly be so connected, for it is experienced as coming from an area *outside* the self. One patient of Bruch's says she ate breakfast because "my stomach wanted it," expressing here the same sense of alienation from her hunger (and her physical self) as Augustine's when he speaks of his "captor," "the law of sin that was in my member."[34] Bruch notes that this "basic delusion," as she calls it, "of not owning the body and its sensations" is a typical symptom of all eating disorders. "These patients act," she says, "as if for them the regulation of food intake was outside [the self]."[35] This experience of bodily sensations as foreign is, strikingly, not limited to the experience of hunger. Patients with eating disorders have similar problems in identifying cold, heat, emotions, and anxiety as originating in the self.[36]

While the body is experienced as alien and outside, the soul or will is described as being trapped or confined in this alien "jail," as one woman describes it.[37] "I feel caught in my body," "I'm a prisoner in my body":[38] the theme is repeated again and again. A typical fantasy, evocative of Plato, imagines total liberation from the bodily prison: "I wish I could get out of my body entirely and fly!"[39] "Please dear God, help me . . . I want to get out of my body, I want to get out!"[40] Ellen West, astute as always, sees a central meaning of her self-starvation in this "ideal of being too thin, of being *without a body*."[41]

Anorexia is not a philosophical attitude; it is a debilitating affliction. Yet, quite often a highly conscious and articulate scheme of images and associations—virtually a metaphysics—is presented by these women. The scheme is strikingly Augustinian, with evocations of Plato. This does not indicate, of course, that anorectics are followers of Plato or Augustine, but that the anorectic's metaphysics makes explicit various elements, historically grounded in Plato and Augustine, that run deep in our culture.[42] As Augustine often speaks of the "two wills" within him, "one the servant of the flesh, the other of the spirit," who "between them tore my soul apart," so the anorectic describes a "spiritual struggle," a "contest between good and evil," often conceived explicitly as a battle between mind or will and appetite or body.[43] "I feel myself, quite passively," says West, "the stage on which two hostile forces are mangling each other."[44] Sometimes there is a more aggressive alliance with mind against body: "When I fail to exercise as often as I prefer, I become guilty that I have let my body 'win' another day from my mind. I can't wait 'til this semester is over . . . My body is going to pay the price for the lack of work it is currently getting. I can't wait!"[45]

In this battle, thinness represents a triumph of the will over the body, and the thin body (that is to say, the nonbody) is associated with "absolute purity, hyper-intellectuality and transcendence of the flesh. My soul seemed to grow as my body waned; I felt like one of those early Christian saints who starved themselves in the desert sun. I felt invulnerable, clean and hard as the bones etched into my silhouette."[46] Fat (that is to say, becoming *all* body) is associated with the taint of matter and flesh, "wantonness,"[47] mental stupor and mental decay.[48] One woman describes how after eating sugar she felt "polluted, disgusting, sticky through the arms, as if something bad had gotten inside."[49] Very often, sexuality is brought into this scheme of associations, and hunger and sexuality are psychically connected. Cherry Boone O'Neill describes a late-night binge, eating scraps of leftovers from the dog's dish:

> I started slowly, relishing the flavor and texture of each marvelous bite. Soon I was ripping the meager remains from the bones, stuffing the meat into my mouth as fast as I could detach it.
>   [Her boyfriend surprises her, with a look of "total disgust" on his face.]
>   I had been caught red-handed . . . in an animalistic orgy on the floor, in the dark, alone. Here was the horrid truth for Dan to see. I felt so evil, tainted, pagan . . . In Dan's mind that day, I had been whoring after food.[50]

A hundred pages earlier, she had described her first romantic involvement in much the same terms: "I felt secretive, deceptive, and . . . tainted by the ongoing relationship" (which never went beyond kisses).[51] Sexuality, similarly, is "an abominable business" to Aimee Liu; for her, staying reed-thin is seen as a way of avoiding sexuality, by becoming "androgynous," as she puts it.[52] In the same way, Sarah, a patient of Levenkron connects her dread of gaining weight with "not wanting to be a 'temptation' to men."[53] In Liu's case, and in Sarah's, the desire to appear unattractive to men is connected to anxiety and guilt over earlier sexual abuse. Whether or not such episodes are common to many cases of anorexia,[54] "the avoidance of any sexual encounter, a shrinking from all bodily contact," is, according to Bruch, characteristic of anorectics.[55]

The Control Axis

Having examined the axis of continuity from Plato to anorexia, we should feel cautioned against the impulse to regard anorexia as expressing entirely modern attitudes and fears. Disdain for the body, the conception of it as an alien force and impediment to the soul, is very old in our Greco-Christian traditions (although it has usually been expressed most forcefully by male philosophers and theologians rather than adolescent women!).

But although dualism is as old as Plato, in many ways contemporary culture appears *more* obsessed than previous eras with the control of the unruly body. Looking now at contemporary American life, a second axis of continuity emerges on which to locate anorexia. I call it the *control axis*.

The young anorectic, typically, experiences her life as well as her hungers as being out of control. She is a perfectionist and can never carry out the tasks she sets herself in a way that meets her own rigorous standards. She is torn by conflicting and contradictory expectations and demands, wanting to shine in all areas of student life, confused about where to place most of her energies, what to focus on, as she develops into an adult. Characteristically, her parents expect a great deal of her in the way of individual achievement (as well as physical appearance), yet have made most of the important decisions for her.[56] Usually, the anorexic syndrome emerges, not as a conscious decision to get as thin as possible, but as the result of her having begun a diet fairly casually, often at the suggestion of a parent, having succeeded splendidly in taking off five or ten pounds, and then having gotten hooked on the intoxicating feeling of accomplishment and control.

Recalling her anorexic days, Aimee Liu recreates her feelings:

> The sense of accomplishment exhilarates me, spurs me to continue on and on. It provides a sense of purpose and shapes my life with distractions from insecurity. . . . I shall become an expert [at losing weight]. . . . The constant downward trend [of the scale] somehow comforts me, gives me visible proof that I can exert control.[57]

The diet, she realizes, "is the one sector of my life over which I and I alone wield total control."[58]

The frustrations of starvation, the rigors of the constant physical activity in which anorectics engage, the pain of the numerous physical complications of anorexia: these do not trouble the anorectic. Indeed, her ability to ignore them is further proof to her of her mastery of her body. "This was something I could control," says one of Bruch's patients. "I still don't know what I look like or what size I am, but I know my body can take anything."[59] "Energy, discipline, my own power will keep me going," says Liu. "Psychic fuel, I need nothing and no one else, and I will prove it . . . Dropping to the floor, I roll. My tailbone crunches on the hard floor . . . I feel no pain. I will be master of my own body, if nothing else, I vow."[60] And, finally, from one of Bruch's patients: "*You make of your own body your very own kingdom where you are the tyrant, the absolute dictator.*"[61]

Surely we must recognize in this last honest and explicit statement a central modus operandi for the control of contemporary bourgeois anxiety. Consider compulsive jogging and marathon-running, often despite shin splints and other painful injuries, with intense agitation over missing a day or not meeting a goal for a particular run. Consider the increasing popularity of triathlon events such as the Iron Man, whose central purpose appears to be to allow people to find out how far they can push their bodies—through long-distance swimming, cycling, and running—before they collapse. Consider lawyer Mike Frankfurt, who runs ten miles every morning: "*To run with pain is the essence of life.*"[62] Or consider the following excerpts from student journals:

The best times I like to run are under the most unbearable conditions. I love to run in the hottest, most humid and steepest terrain I can find. . . . For me running and the pain associated with it aren't enough to make me stop. I am always trying to overcome it and the biggest failure I can make is to stop running because of pain. Once I ran five of a ten-mile run with a severe leg cramp but wouldn't stop—it would have meant failure.[63]

When I run I am free . . . The pleasure is closing off my body—as if the incessant pounding of my legs is so total that the pain ceases to exist. There is no grace, no beauty in the running—there is the jarring reality of sneaker and pavement. Bright pain that shivers and splinters sending its white hot arrows into my stomach, my lung, but it cannot pierce my mind. I am on automatic pilot—there is no remembrance of pain, there is freedom—I am losing myself, peeling out of this heavy flesh. . . . Power surges through me.[64]

None of this is to dispute that the contemporary concern with fitness has non-pathological, nondualist dimensions as well. Particularly for women, who have historically suffered from the ubiquity of rape and abuse, from the culturally instilled conviction of our own helplessness, and from lack of access to facilities and programs for rigorous physical training, the cultivation of strength, agility, and confidence clearly has a positive dimension. Nor are the objective benefits of daily exercise and concern for nutrition in question here. My focus, rather, is on a subjective stance, become increasingly prominent, which, although preoccupied with the body and deriving narcissistic enjoyment from its appearance, takes little pleasure in the *experience* of embodiment. Rather, the fundamental identification is with mind (or will), ideals of spiritual perfection, fantasies of absolute control.

Not everyone, of course, for whom physical training is a part of daily routine exhibits such a stance. Here, an examination of the language of female body-builders is illustrative. Body-building is particularly interesting because on the surface it appears to have the opposite structure to anorexia: the body-builder is, after all, building the body *up*, not whittling it down. Body-building develops strength. We imagine the body-builder as someone who is proud, confident, and perhaps most of all, conscious of and accepting of her physicality. This is, indeed, how some female body-builders experience themselves:

I feel . . . tranquil and stronger [says Lydia Cheng]. Working out creates a high everywhere in my body. I feel the heat. I feel the muscles rise, I see them blow out, flushed with lots of blood. . . . My whole body is sweating and there's few things I love more than working up a good sweat. That's when I really feel like a woman.[65]

Yet a sense of joy in the body as active and alive is *not* the most prominent theme among the women interviewed by Trix Rosen. Many of them, rather, talk about their bodies in ways that resonate disquietingly with typical anorexic themes.

There is the same emphasis on will, purity, and perfection: "I've learned to be a stronger person with a more powerful will . . . pure concentration, energy and spirit." "I want to be as physically perfect as possible." "Body-building suits the perfectionist in me." "My goal is to have muscular perfection."[66] Compulsive exercisers—whom Dinitia Smith, in an article for *New York* magazine, calls "The New Puritans"—speak in similar terms: Kathy Krauch, a New York art director who bikes twelve miles a day and swims two and a half, says she is engaged in "a quest for perfection." Mike Frankfurt, in describing his motivation for marathon running, speaks of "the purity about it." These people, Smith emphasizes, care little about their health: "They pursue self-denial as an end in itself, out of an almost mystical belief in the purity it confers."[67]

Many body-builders, like many anorectics, unnervingly conceptualize the body as alien, not-self:

> I'm constantly amazed by my muscles. The first thing I do when I wake up in the morning is look down at my "abs" and flex my legs to see if the "cuts" are there. . . . My legs have always been my most stubborn part, and I want them to develop so badly. Every day I can see things happening to them . . . I don't flaunt my muscles as much as I thought I would. I feel differently about them; they are my product and I protect them by wearing sweaters to keep them warm.[68]

Most strikingly, body-builders put the same emphasis on *control*: on feeling their life to be fundamentally out of control, and on the feeling of accomplishment derived from total mastery of the body. That sense of mastery, like the anorectic's, appears to derive from two sources. First, there is the reassurance that one can overcome all physical obstacles, push oneself to any extremes in pursuit of one's goals (which, as we have seen, is a characteristic motivation of compulsive runners, as well). Second, and most dramatic (it is spoken of time and again by female body-builders), is the thrill of being in total charge of the shape of one's body. "Create a masterpiece," says *Fit* magazine. "Sculpt your body contours into a work of art." As for the anorectic— who literally cannot *see* her body as other than her inner reality dictates and who is relentlessly driven by an ideal image of ascetic slenderness—so for the body-builder a purely mental conception comes to have dominance over her life: "You visualize what you want to look like . . . and then create the form." "The challenge presents itself: to rearrange things." "It's up to you to do the chiseling; you become the master sculptress." "What a fantasy, for your body to be changing! . . . I keep a picture in my mind as I work out of what I want to look like and what's happened to me already."[69] Dictation to nature of one's own chosen design for the body is the central goal for the body-builder, as it is for the anorectic.

The sense of security derived from the attainment of this goal appears, first of all, as the pleasure of control and independence. "Nowadays," says Michael Sacks, associate professor of psychiatry at Cornell Medical College, "people no longer feel they can control events outside themselves—how well they do in their jobs or in their personal relationships, for example—but they can control the food they eat and how far they can run. Abstinence, tests of endurance, are ways of proving their self-sufficiency."[70] In a culture, moreover, in which our continued survival is often at the mercy of "specialists," machines, and sophisticated technology, the body acquires a special sort of vulnerability and dependency. We may live longer, but the circumstances surrounding illness and death may often be perceived as more alien, inscrutable, and arbitrary than ever before.

Our contemporary body-fetishism expresses more than a fantasy of self-mastery in an increasingly unmanageable culture, however. It also reflects our alliance *with* culture against all reminders of the inevitable decay and death of the body. "Everybody wants to live forever" is the refrain from the theme song of *Pumping Iron*. The most youth-worshipping of popular television shows, *Fame*, opens with a song that begins, "I want to live forever." And it is striking that although the anorectic may come very close to death (and 15 percent do indeed die), the dominant experience throughout the illness is of *invulnerability*.

The dream of immortality is, of course, nothing new. But what is unique to modernity is that the defeat of death has become a scientific fantasy rather than a philosophical or religious mythology. We no longer dream of eternal union with the gods; instead, we build devices that can keep us alive indefinitely, and we work on keeping our bodies as smooth and muscular and elastic at forty as they were at eighteen. We even entertain dreams of halting the aging process completely. "Old age," according to Durk Pearson and Sandy Shaw, authors of the popular *Life Extension*, "is an unpleasant and unattractive affliction."[71] The mega-vitamin regime they prescribe is able, they claim, to prevent and even to reverse the mechanisms of aging.

Finally, it may be that in cultures characterized by gross excesses in consumption, the "will to conquer and subdue the body" (as Chernin calls it) expresses an aesthetic or moral rebellion.[72] Anorectics initially came from affluent families, and the current craze for long-distance running and fasting is largely a phenomenon of young, upwardly mobile professionals (Dinitia Smith calls it "Deprivation Chic").[73] To those who are starving *against* their wills, of course, starvation cannot function as an expression of the power of the will. At the same time, we should caution against viewing anorexia as a trendy illness of the elite and privileged. Rather, its most outstanding feature is powerlessness.

The Gender/Power Axis

Ninety percent of anorectics are women. We do not, of course, need to know that particular statistic to realize that the contemporary "tyranny of slenderness" is far from gender-neutral. Women are more obsessed with their bodies than men, less satisfied with them,[74] and permitted less latitude with them by themselves, by men, and by the culture. In a 1984 *Glamour* magazine poll of 33,000 women, 75 percent said they thought they were "too fat." Yet by Metropolitan Life Insurance Tables, themselves notoriously affected by cultural standards, only 25 percent of these women were heavier than their optimal weight, and a full 30 percent were *below* that weight.[75] The anorectic's distorted image of her body—her inability to see it as anything but too fat—although more extreme, is not radically discontinuous, then, from fairly common female misperceptions.

Consider, too, actors like Nick Nolte and William Hurt, who are permitted a certain amount of softening, of thickening about the waist, while still retaining romantic-lead status. Individual style, wit, the projection of intelligence, experience, and effectiveness still go a long way for men, even in our fitness-obsessed culture. But no female can achieve the status of romantic or sexual ideal without the appropriate *body*. That body, if we use television commercials as a gauge, has gotten steadily leaner since the mid-1970s.[76] What used to be acknowledged as an extreme required only of high fashion models is now the dominant image that beckons to high school and college women. Over and over, extremely slender women students complain of hating their thighs or their stomachs (the anorectic's most dreaded danger spot);

often, they express concern and anger over frequent teasing by their boyfriends. Janey, a former student, is 5′ 10″ and weighs 132 pounds. Yet her boyfriend Bill, also a student of mine, calls her "Fatso" and "Big Butt" and insists she should be 110 pounds because (as he explains in his journal for my class) "that's what Brooke Shields weighs." He calls this "constructive criticism" and seems to experience extreme anxiety over the possibility of her gaining any weight: "I can tell it bothers her yet I still continue to badger her about it. I guess that I think that if I continue to remind her things will change faster."[77] This sort of relationship, in which the woman's weight has become a focal issue, is not at all atypical, as I have discovered from student journals and papers.

Hilde Bruch reports that many anorectics talk of having a "ghost" inside them or surrounding them, "a dictator who dominates me," as one woman describes it; "a little man who objects when I eat" is the description given by another.[78] The little ghost, the dictator, the "other self" (as he is often described) is always male, reports Bruch. The anorectic's *other* self—the self of the uncontrollable appetites, the impurities and taints, the flabby will and tendency to mental torpor—is the body, as we have seen. But it is also (and here the anorectic's associations are surely in the mainstream of Western culture) the *female* self. These two selves are perceived as at constant war. But it is clear that it is the male side—with its associated values of greater spirituality, higher intellectuality, strength of will—that is being expressed and developed in the anorexic syndrome.[79]

What is the meaning of these gender associations in the anorectic? I propose that there are two levels of meaning. One has to do with fear and disdain for traditional female roles and social limitations. The other has to do, more profoundly, with a deep fear of "the Female," with all its more nightmarish archetypal associations of voracious hungers and sexual insatiability.

Adolescent anorectics express a characteristic fear of growing up to be mature, sexually developed, and potentially reproductive women. "I have a deep fear," says one, "of having a womanly body, round and fully developed. I want to be tight and muscular and thin."[80] Cherry Boone O'Neill speaks explicitly of her fear of womanhood.[81] If only she could stay thin, says yet another, "I would never have to deal with having a woman's body; like Peter Pan I could stay a child forever."[82] The choice of Peter Pan is telling here—what she means is, stay a *boy* forever. And indeed, as Bruch reports, many anorectics, when children, dreamt and fantasized about growing up to be boys.[83] Some are quite conscious of playing out this fantasy through their anorexia; Adrienne, one of Levenkron's patients, was extremely proud of the growth of facial and body hair that often accompanies anorexia, and especially proud of her "skinny, hairy arms."[84] Many patients report, too, that their father had wanted a boy, were disappointed to get "less than" that, or had emotionally rebuffed their daughter when she began to develop sexually.[85]

In a characteristic scenario, anorexia develops just at the outset of puberty. Normal body changes are experienced by the anorectic, not surprisingly, as the takeover of the body by disgusting, womanish fat. "I grab my breasts," says Aimee Liu, "pinching them until they hurt. If only I could eliminate them, cut them off if need be, to become as flat-chested as a child again."[86] The anorectic is exultant when her periods stop (as they do in *all* cases of anorexia[87] and as they do in many female runners as well). Disgust with menstruation is typical: "I saw a picture at a feminist art gallery," says another woman. "There was a woman with long red yarn coming out of her, like she was menstruating. . . . I got that *feeling*—in that part of my body that I have trouble with . . . my stomach, my thighs, my pelvis. That revolting feeling."[88]

Some authors interpret these symptoms as a species of unconscious feminist protest, involving anger at the limitations of the traditional female role, rejection of values associated with it, and fierce rebellion against allowing their futures to develop in the same direction as their mothers' lives.[89] In her portrait of the typical anorexic family configuration, Bruch describes nearly all of the mothers as submissive to their husbands but very controlling of their children.[90] Practically all had had promising careers which they had given up to care for their husbands and families full-time, a task they take very seriously, although often expressing frustration and dissatisfaction.

Certainly, many anorectics appear to experience anxiety about falling into the lifestyle they associate with their mothers. It is a prominent theme in Aimee Liu's *Solitaire*. Another woman describes her feeling that "[I am] full of my mother . . . she is in me even if she isn't there" in nearly the same breath as she complains of her continuous fear of being "not human . . . of ceasing to exist."[91] And Ellen West, nearly a century earlier, had quite explicitly equated becoming fat with the inevitable (for an elite woman of her time) confinements of domestic life and the domestic stupor she associates with it:

> Dread is driving me mad . . . the consciousness that ultimately I will lose everything; all courage, all rebelliousness, all drive for doing; that it—my little world—will make me flabby, flabby and fainthearted and beggarly.[92]

Several of my students with eating disorders reported that their anorexia had developed after their families had dissuaded them from choosing or forbidden them to embark on a traditionally male career.

Here anorexia finds a true sister-phenomenon in the epidemic of female invalidism and "hysteria" that swept through the middle and upper-middle classes in the second half of the nineteenth century.[93] It was a time that, in many ways, was very like our own, especially in the conflicting demands women were confronting: the opening up of new possibilities versus the continuing grip of the old expectations. On the one hand, the old preindustrial order, with the father at the head of a self-contained family production unit, had given way to the dictatorship of the market, opening up new, nondomestic opportunities for working women. On the other hand, it turned many of the most valued "female" skills—textile and garment manufacture, food processing—out of the home and over to the factory system.[94] In the new machine economy, the lives of middle-class women were far emptier than they had been before.

It was an era, too, that had been witnessing the first major feminist wave. In 1840 the World Anti-Slavery Conference had been held, at which the first feminists spoke loudly and long on the connections between the abolition of slavery and women's rights. The year 1848 saw the Seneca Falls Convention. In 1869, John Stuart Mill published his landmark work "On the Subjection of Women." And in 1889 the

Pankhursts formed the Women's Franchise League. But it was an era, too (and not unrelatedly, as I shall argue later), when the prevailing ideal of femininity was the delicate, affluent lady, unequipped for anything, but the most sheltered domestic life, totally dependent on her prosperous husband, providing a peaceful and comfortable haven for him each day after his return from his labors in the public sphere.[95] In a now famous letter, Freud, criticizing John Stuart Mill, writes:

> It really is a still-born thought to send women into the struggle for existence exactly as men. If, for instance, I imagine my gentle sweet girl as a competitor it would only end in my telling her, as I did seventeen months ago, that I am fond of her and that I implore her to withdraw from the strife into the calm uncompetitive activity of my home.[96]

This is exactly what male doctors *did* do when women began falling ill, complaining of acute depression, severe headaches, weakness, nervousness, and self-doubt.[97] Among these women were such noted feminists and social activists as Charlotte Perkins Gilman, Jane Addams, Elizabeth Cady Stanton, Margaret Sanger, British activist Josephine Butler, and German suffragist Hedwig Dohm. "I was weary myself and sick of asking what I am and what I ought to be," recalls Gilman,[98] who later went on to write a fictional account of her mental breakdown in the chilling novella *The Yellow Wallpaper*. Her doctor, the famous specialist S. Weir Mitchell, instructed her, as Gilman recalls, to "live as domestic a life as possible. Have your child with you all the time . . . Lie down an hour every day after each meal. Have but two hours intellectual life a day. And never touch pen, brush or pencil as long as you live."[99]

Freud, who favorably reviewed Mitchell's 1887 book and who advised that psychotherapy for hysterical patients be combined with Mitchell's rest cure ("to avoid new psychical impressions"),[100] was as blind as Mitchell to the contribution that isolation, boredom, and intellectual frustration made to the etiology of hysteria. Nearly all of the subjects in *Studies in Hysteria* (as well as the later *Dora*) are acknowledged by Freud to be unusually intelligent, creative, energetic, independent, and, often, highly educated. (Berthe Pappenheim—"Anna O."—as we know, went on after recovery to become an active feminist and social reformer.) Freud even comments, criticizing Janet's notion that hysterics were "psychically insufficient," on the characteristic coexistence of hysteria with "gifts of the richest and most original kind."[101] Yet Freud never makes the connection (which Breuer had begun to develop)[102] between the monotonous domestic lives these women were expected to lead after they completed their schooling, and the emergence of compulsive daydreaming, hallucinations, dissociations, and hysterical conversions.

Charlotte Perkins Gilman does make that connection. In *The Yellow Wallpaper* she describes how a prescribed regime of isolation and enforced domesticity eventuates, in her fictional heroine, in the development of a full-blown hysterical symptom, madness, and collapse. The symptom, the hallucination that there is a woman trapped in the wall paper of her bedroom, struggling to get out, is at once a perfectly articulated expression of protest and a completely debilitating idée fixe that allows the woman character no distance on her situation, no freedom of thought, no chance of making any progress in leading the kind of active, creative life her body and soul crave.

So too for the anorectic. It is indeed essential to recognize in this illness the dimension of protest against the limitations of the ideal of female domesticity (the "feminine mystique," as Betty Friedan called it) that reigned in America throughout the 1950s and early 1960s—the era when most of their mothers were starting homes and families. This was, we should recall, the era following World War II, an era during which women were fired en masse from the jobs they had held during the war and shamelessly propagandized back into the full-time job of wife and mother. It was an era, too, when the "fuller figure," as Jane Russell now calls it, came into fashion once more, a period of "mammary madness" (or "resurgent Victorianism," as Lois Banner calls it), which glamorized the voluptuous, large-breasted woman.[103] This remained the prevailing fashion tyranny until the late 1960s and early 1970s.

But we must recognize that the anorectic's protest, like that of the classical hysterical symptom, is written on the bodies of anorexic women, not embraced as a conscious politics—nor, indeed, does it reflect any social or political understanding at all. Moreover, the symptoms themselves function to preclude the emergence of such an understanding. The idée fixe—staying thin—becomes at its farthest extreme so powerful as to render any other ideas or life-projects meaningless. Liu describes it as "all encompassing."[104] West writes: "I felt all inner development was ceasing, that all becoming and growing were being choked, because a single idea was filling my entire soul."[105]

Paradoxically—and often tragically—these pathologies of female protest (and we must include agoraphobia here, as well as hysteria and anorexia) actually function as if in collusion with the cultural conditions that produced them.[106] The same is true for more moderate expressions of the contemporary female obsession with slenderness. Women may feel themselves deeply attracted by the aura of freedom and independence suggested by the boyish body ideal of today. Yet, each hour, each minute spent in anxious pursuit of that ideal (for it does not come naturally to most mature women) is in fact time and energy taken from inner development and social achievement. As a feminist protest, the obsession with slenderness is hopelessly counterproductive.

It is important to recognize, too, that the anorectic is terrified and repelled, not only by the traditional female domestic role—which she associates with mental lassitude and weakness—but by a certain archetypal image of the female: as hungering, voracious, all-needing, and all-wanting. It is this image that shapes and permeates her experience of her own hunger for food as insatiable and out of control, that makes her feel that if she takes just one bite, she will not be able to stop.

Let us explore this image. Let us break the tie with food and look at the metaphor: hungering . . . voracious . . . extravagantly and excessively needful . . . without restraint . . . always wanting . . . always wanting too much affection, reassurance, emotional and sexual contact, and attention. This is how many women frequently experience themselves, and, indeed, how many men experience women. "Please, God, keep me from telephoning him," prays the heroine in Dorothy Parker's classic "A Telephone Call,"[107] experiencing her need for reassurance and contact as being as out of control and degrading as the anorectic does her desire for food. The male counterpart to this is found in Paul Morel in Lawrence's *Sons and Lovers*: "Can you never like things without clutching them as if you wanted to pull the heart out of them?" he accuses Miriam as she fondles a flower. "Why don't you have a bit more restraint, or reserve, or something. . . . You're always begging things to love you, as if you were a beggar for love. Even the flowers, you have to fawn on them."[108] How much psychic authenticity do these images carry in 1980s America? One woman in my class provided a stunning insight into the connection between her perception of herself and the anxiety of the compulsive dieter. "You know," she said, "the anorectic is always convinced she is taking up too much space, eating too much, wanting food too much. I've never felt that way, but I've often felt that I was *too much*—too much emotion, too much need, too loud and demanding, too much *there*, if you know what I mean."[109]

The most extreme cultural expressions of the fear of woman as "too much"—which almost always revolve around her sexuality—are strikingly full of eating and hungering metaphors. "Of woman's unnatural, *insatiable* lust, what country, what village doth not complain?" queries Burton in *The Anatomy of Melancholy*.[110] "You are the true hiennas," says Walter Charleton, "that allure us with the fairness of your skins, and when folly hath brought us within your reach, you leap upon us and *devour* us."[111]

The mythology/ideology of the devouring, insatiable female (which, as we have seen, is the image of her female self the anorectic has internalized) tends historically to wax and wane. But not without rhyme or reason. In periods of gross environmental and social crisis, such as characterized the period of the witch-hunts in the fifteenth and sixteenth centuries, it appears to flourish.[112] "All witchcraft comes from carnal lust, which is in women *insatiable*," say Kramer and Sprenger, authors of the official witch-hunters handbook, *Malleus Malificarum*. For the sake of fulfilling the "*mouth* of the womb . . . [women] consort even with the devil."[113]

Anxiety over women's uncontrollable hungers appears to peak, as well, during periods when women are becoming independent and are asserting themselves politically and socially. The second half of the nineteenth century, concurrent with the first feminist wave discussed earlier, saw a virtual flood of artistic and literary images of the dark, dangerous, and evil female: "sharp-teethed, devouring" Sphinxes, Salomes, and Delilahs, "biting, tearing, murderous women." "No century," claims Peter Gay, "depicted woman as vampire, as castrator, as killer, so consistently, so programmatically, and so nakedly as the nineteenth."[114] No century, either, was so obsessed with sexuality—particularly female sexuality—and its medical control. Treatment for excessive "sexual excitement" and masturbation in women included placing leeches on the womb,[115] clitoridectomy, and removal of the ovaries (also recommended for "troublesomeness, eating like a ploughman, erotic tendencies, persecution mania, and simple 'cussedness' ").[116] The importance of female masturbation in the etiology of the "actual neurosis" was a topic in which the young Freud and his friend and colleague Wilhelm Fliess were especially interested. Fliess believed that the secret to controlling such "sexual abuse" lay in the treatment of nasal "genital spots"; in an operation that was sanctioned by Freud, he attempted to "correct" the "bad sexual habits" of Freud's patient Emma Eckstein by removal of the turbinate bone of her nose.[117]

It was in the second half of the nineteenth century, too, despite a flurry of efforts by feminists and health reformers,[118] that the stylized "S-curve," which required a tighter corset than ever before, came into fashion.[119] "While the suffragettes were forcefully propelling all women toward legal and political emancipation," says Amaury deRiencourt, "fashion and custom imprisoned her physically as she had never been before."[120] Described by Thorstein Veblen as a "mutilation, undergone for the purpose of lowering the subject's vitality and rendering her permanently and obviously unfit for work," the corset indeed did just that.[121] In it a woman could barely sit or stoop, was unable to move her feet more than six inches at a time, and had difficulty in keeping herself from regular fainting fits. (In 1904, a researcher reported that "monkeys laced up in these corsets moped, became excessively irritable and within weeks sickened and died"!)[122] The connection was often drawn in popular magazines between enduring the tight corset and the exercise of self-restraint and control. The corset is "an ever present monitor," says one 1878 advertisement, "of a well-disciplined mind and well-regulated feelings."[123] Today, of course, we diet to achieve such control.

It is important to emphasize that, despite the practice of bizarre and grotesque methods of gross physical manipulation and external control (clitoridectomy, Chinese foot-binding, the removal of bones of the rib cage in order to fit into the tight corsets), such control plays a relatively minor role in the maintenance of gender/power relations. For every historical image of the dangerous, aggressive woman there is a corresponding fantasy—an ideal femininity, from which all threatening elements have been purged—that women have mutilated themselves *internally* to attain. In the Victorian era, at the same time that operations were being performed to control female sexuality, William Acton, Richard von Krafft-Ebing, and others were proclaiming the official scientific doctrine that women are naturally passive and "not very much troubled with sexual feelings of any kind."[124] Corresponding to this male medical fantasy was the popular artistic and moral theme of woman as ministering angel; sweet, gentle, domestic, without intensity or personal ambition of any sort.[125] Peter Gay suggests, correctly, that these ideals must be understood as a reaction-formation to the era's "pervasive sense of manhood in danger," and he argues that few women actually fit the "insipid goody" (as Kate Millett calls it) image.[126] What Gay forgets, however, is that most women *tried* to fit—working classes as well as middle were affected by the "tenacious and all-pervasive" ideal of the perfect lady.[127]

On the gender/power axis the female body appears, then, as the unknowing medium of the historical ebbs and flows of the fear of woman as "too much." That, as we have seen, is how the anorectic experiences her female, bodily self: as voracious, wanton, needful of forceful control by her male will. Living in the tide of cultural backlash against the second major feminist wave, she is not alone in constructing these images. Christopher Lasch, in *The Culture of Narcissism*, speaks of what he describes as "the apparently aggressive overtures of sexually liberated women" which convey to many males the same message—that women are *voracious, insatiable,*" and call up "early fantasies of a possessive, suffocating, *devouring* and castrating mother."[128]

Our contemporary beauty ideals, by contrast, seemed purged, as Kim Chernin puts it, "of the power to conjure up memories of the past, of all that could remind us of a woman's mysterious power."[129] The ideal, rather, is an "image of a woman in which she is not yet a woman": Darryl Hannah as the lanky, newborn mermaid in *Splash*; Lori Singer (appearing virtually anorexic) as the reckless, hyperkinetic heroine of *Footloose*; The Charley Girl; "Cheryl Tiegs in shorts, Margaux Hemingway with her hair wet; Brooke Shields naked on an island";[130] the dozens of teenage women who appear in Coke commercials, in jeans commercials, in chewing gum commercials.

The images suggest amused detachment, casual playfulness, flirtatiousness without demand, and lightness of touch. A refusal to take sex, death, or politics too deadly seriously. A delightfully unconscious relationship to her body. The twentieth century has seen this sort of feminine ideal before, of course. When, in the 1920s, young women began to flatten their breasts, suck in their stomachs, bob their hair, and show off long, colt-like legs, they believed they were pursuing a new freedom and daring that demanded a carefree, boyish style. If the traditional female hourglass suggested anything, it was confinement and immobility. Yet the flapper's freedom, as Mary McCarthy's and Dorothy Parker's short stories brilliantly reveal, was largely an illusion—as any obsessively cultivated sexual style must inevitably be. Although today's images may suggest androgynous independence, we need only consider who is on the receiving end of the imagery in order to confront the pitiful paradox involved.

Watching the commercials are thousands of anxiety-ridden women and adolescents (some of whom may well be the very ones appearing in the commercials) with anything *but* an unconscious relation to their bodies. They are involved in an absolutely contradictory state of affairs, a totally no-win game: caring desperately, passionately, obsessively about attaining an ideal of coolness, effortless confidence, and casual freedom. Watching the commercials is a little girl, perhaps ten years old, whom I saw in Central Park, gazing raptly at her father, bursting with pride: "Daddy, guess what? I lost two pounds!" And watching the commercials is the anorectic, who associates her relentless pursuit of thinness with power and control, but who in fact destroys her health and imprisons her imagination. She is surely the most startling and stark illustration of how cavalier power relations are with respect to the motivations and goals of individuals, yet how deeply they are etched on our bodies, and how well our bodies serve them.

Notes

This essay was presented as a public lecture at Le Moyne College, was subsequently presented at D'Youville College and Bennington College, and was originally published in the *Philosophical Forum* 17, no. 2 (Winter 1985). I wish to thank all those in the audiences at Le Moyne, D'Youville, and Bennington who commented on my presentations, and Lynne Arnault, Nancy Fraser, and Mario Moussa for their systematic and penetrating criticisms and suggestions for the *Forum* version. In addition, I owe a large initial debt to my students, particularly Christy Ferguson, Vivian Conger, and Nancy Monaghan, for their observations and insights.

1. Jules Henry, *Culture Against Man* (New York: Alfred A. Knopf, 1963).

2. When I wrote this piece in 1983, the term *anorexia* was commonly used by clinicians to designate a general class of eating disorders within which intake-restricting (or abstinent) anorexia and bulimia-anorexia (characterized by alternating bouts of gorging and starving and/or gorging and vomiting) are distinct subtypes (see Hilde Bruch, *The Golden Cage: The Enigma of Anorexia Nervosa* [New York: Vintage, 1979], p. 10; Steven Levenkron, *Treating and Overcoming Anorexia Nervosa* [New York: Warner Books, 1982], p. 6; R. L. Palmer, *Anorexia Nervosa* [Middlesex: Penguin, 1980], pp. 14, 23–24; Paul Garfinkel and David Garner, *Anorexia Nervosa: A*

Multidimensional Perspective [New York: Brunner/Mazel, 1982], p. 4). Since then, as the clinical tendency has been increasingly to emphasize the differences rather than the commonalities between the eating disorders, bulimia has come to occupy its own separate classificatory niche. In the present piece I concentrate largely on those images, concerns, and attitudes shared by anorexia and bulimia. Where a difference seems significant for the themes of this essay, I will indicate the relevant difference in a footnote rather than over-complicate the main argument of the text. This procedure is not to be taken as belittling the importance of such differences, some of which I discuss in "Reading the Slender Body."

3. Although throughout history scattered references can be found to patients who sound as though they may have been suffering from self-starvation, the first medical description of anorexia as a discrete syndrome was made by W. W. Gull in an 1868 address at Oxford (at the time he called the syndrome, in keeping with the medical taxonomy of the time, *hysteric apepsia*). Six years later, Gull began to use the term *anorexia nervosa*; at the same time, E. D. Lesegue independently described the disorder (Garfinkel and Garner, *Anorexia Nervosa*, pp. 58–59). Evidence points to a minor "outbreak" of anorexia nervosa around this time (see Jacobs Brumberg, *Fasting Girls*, Cambridge: Harvard University Press, 1988), a historical occurrence that went unnoticed by twentieth-century clinicians until renewed interest in the disorder was prompted by its reemergence and striking increase over the past twenty years (see note 11 of "Whose Body Is This?" for sources that document this increase). At the time I wrote the present piece, I was not aware of the extent of anorexia nervosa in the second half of the nineteenth century.

4. Ludwig Binswanger, "The Case of Ellen West," in Rollo May, ed., *Existence* (New York: Simon and Schuster, 1958), p. 288. He was wrong, of course. The symptom was not new, and we now know that Ellen West was not the only young woman of her era to suffer from anorexia. But the fact that Binswanger was unaware of other cases is certainly suggestive of its infrequency, especially relative to our own time.

5. Hilde Bruch, *Eating Disorders* (New York: Basic Books, 1973), p. 4.

6. Levenkron, *Treating and Overcoming Anorexia Nervosa*, p. 1; Susan Squire, "Is the Binge-Purge Cycle Catching?" *Ms.* (Oct. 1983).

7. Dinitia Smith, "The New Puritans," *New York Magazine* (June 11, 1984): 28.

8. Kim Chernin, *The Obsession: Reflections on the Tyranny of Slenderness* (New York: Harper and Row, 1981), pp. 63, 62.

9. Garfinkel and Garner, *Anorexia Nervosa*, p. xi. Anorectics characteristically suffer from a number of physiological disturbances, including amenorrhea (cessation of menstruation) and abnormal hypothalamic function (see Garfinkel and Garner, *Anorexia Nervosa*, pp. 58–59, for an extensive discussion of these and other physiological disorders associated with anorexia; also Eugene Garfield, "Anorexia Nervosa: The Enigma of Self-Starvation," *Current Contents* [Aug. 6, 1984]: 8–9). Researchers are divided, with arguments on both sides, as to whether hypothalamic dysfunction may be a primary cause of the disease or whether these characteristic neuroendocrine disorders are the result of weight loss, caloric deprivation, and emotional stress. The same debate rages over abnormal vasopressin levels discovered in anorectics. Touted in tabloids all over the United States as the "explanation" for anorexia and key to its cure. Apart from such debates over a biochemical predisposition to anorexia, research continues to explore the possible role of biochemistry in the self-perpetuating nature of the disease, and the relation of the physiological effects of starvation to particular experiential symptoms such as the anorectic's preoccupation with food (see Bruch, *The Golden Cage*, pp. 7–12; Garfinkel and Garner, *Anorexia Nervosa*, pp. 10–14).

10. Initially, anorexia was found to predominate among upper-class white families. There is, however, widespread evidence that this is now rapidly changing (as we might expect; no one in America is immune to the power of popular imagery). The disorder, it has been found, is becoming more equally distributed, touching populations (e.g., blacks and East Indians) previously unaffected, and all socioeconomic levels (Garfinkel and Garner, *Anorexia Nervosa*, pp. 102–3). There remains, however, an overwhelming disproportion of women to men (Garfinkel and Garner, *Anorexia Nervosa*, pp. 112–13).

11. Chernin's *The Obsession*, whose remarkable insights inspired my interest in anorexia, remains *the* outstanding exception to the lack of cultural understanding of eating disorders.

12. Chernin, *The Obsession*, pp. 36–37. My use of the expression "our culture" may seem overly homogenizing here, disrespectful of differences among ethnic groups, socioeconomic groups, subcultures within American society, and so forth. It must be stressed here that I am discussing ideology and images whose power is *precisely* the power to homogenize culture. Even in pre-mass-media cultures we see this phenomenon: the nineteenth-century ideal of the "perfect lady" tyrannized even those classes who could not afford to realize it. With television, of course, a massive deployment of images becomes possible, and there is no escape from the mass shaping of our fantasy lives. Although they may start among the wealthy elite ("A woman can never be too rich or too thin"), media-promoted ideas of femininity and masculinity quickly and perniciously spread their influence over everyone who owns a TV or can afford a junk magazine or is aware of billboards. Changes in the incidence of anorexia among lower-income groups (see note 10, above) bear out this point.

13. Christopher Lasch, *The Culture of Narcissism* (New York: Warner Books, 1979), p. 88.

14. I choose these three primarily because they are where my exploration of the imagery, language, and metaphor produced by anorexic women led me. Delivering earlier versions of this essay at colleges and conferences, I discovered that one of the commonest responses of members of the audience was the proffering of further axes; the paper presented itself less as a statement about the ultimate meaning or causes of a phenomenon than as an invitation to continue my "unpacking" or anorexia as a crystallizing formation. Yet the particular axes chosen have more than a purely autobiographical rationale. The dualist axes serve to identify and articulate the basic body imagery of anorexia. The control axis is an exploration of the question "Why now?" The gender/power axis continues this exploration but focuses on the question "Why women?" The sequence of axes takes us from the most general, most historically diffuse structure of continuity—the dualist experience of self—to ever narrower, more specified arenas of comparison and connection. At first the connections are made without regard to historical context, drawing on diverse historical sources to exploit their familiar coherence in an effort to sculpt the shape of the anorexic experience. In this section, too, I want to suggest that the Greco-Christian tradition provides a particularly fertile soil for the development of anorexia. Then I turn to the much more specific context of American fads and fantasies in the 1980s, considering the contemporary scene largely in terms of popular culture (and therefore through the "fiction" of homogeneity), without regard for gender difference. In this section the connections drawn point to a historical experience of self common to both men and women. Finally, my focus shifts to consider, not what connects anorexia to other general cultural phenomena, but what presents itself as a rupture from them, and what forces us to confront how ultimately opaque the current epidemic of eating disorders remains unless it is linked to the particular situation of women.

    The reader will notice that the axes are linked thematically as well as through their convergence in anorexia: the obsession with control is linked with dualism, and the gender/power dynamics discussed implicitly deal with the issue of control (of the feminine) as well.

15. Michel Foucault, *The History of Sexuality*. Vol. 1: *An Introduction* (New York: Vintage, 1980), p. 155.

16. Foucault, *History of Sexuality*, pp. 47–48.

17. Hubert L. Dreyfus and Paul Rabinow, *Michel Foucault: Beyond Structuralism and Hermeneutics* (Chicago: University of Chicago Press, 1983), p. 112.

18. Foucault, *History of Sexuality*, p. 95.

19. Michel Foucault, *Discipline and Punish* (New York: Vintage, 1979), p. 26.

20. Plato, *Phaedo*, in *The Dialogues of Plato*, ed. and trans. Benjamin Jowett, 4th ed., rev. (Oxford: Clarendon Press, 1953), 83d.

21. St. Augustine, *The Confessions*, trans. R. S. Pine-Coffin (Middlesex: Penguin, 1961), p. 164.

22. *Phaedo* 81d.

23. *Phaedo* 66c. For Descartes on the body as a hindrance to knowledge, see *Conversations with Burman* (Oxford: Clarendon Press, 1976), p. 8, and *Passions of the Soul* in *Philosophical Works of Descartes*, 2 vols., trans. Elizabeth S. Haldane and G. R. T. Ross (Cambridge: Cambridge University Press, 1969), vol. 1, p. 353.

24. *Phaedo* 80a.

25. Indeed, the Cartesian "Rules for the Direction of the Mind," as carried out in the *Meditations* especially, are actually rules for the transcendence of the body—its passions, its senses, the residue of "infantile prejudices" of judgment lingering from that earlier time when we were "immersed" in body and bodily sensations.

26. Alan Watts, *Nature, Man, and Woman* (New York: Vintage, 1970), p. 145.

27. Bruch, *Eating Disorders*, p. 84.

28. Chernin, *The Obsession*, p. 8.

29. Entry in student journal, 1984.

30. Bruch, *The Golden Cage*, p. 4.

31. Binswanger, "The Case of Ellen West," p. 253.

32. Bruch, *Eating Disorders*, p. 253.

33. Levenkron, *Treating and Overcoming Anorexia Nervosa*, p. 6.

34. Bruch, *Eating Disorders*, p. 270; Augustine, *Confessions*, p. 164.

35. Bruch, *Eating Disorders*, p. 50.

36. Bruch, *Eating Disorders*, p. 254.

37. Entry in student journal, 1984.

38. Bruch, *Eating Disorders*, p. 279.

39. Aimee Liu, *Solitaire* (New York: Harper and Row, 1979), p. 141.

40. Jennifer Woods, "I Was Starving Myself to Death," *Mademoiselle* (May 1981): 200.

41. Binswanger, "The Case of Ellen West," p. 251 (emphasis added).

42. Why they should emerge with such clarity in the twentieth century and through the voice of the anorectic is a question answered, in part, by the following two axes.

43. Augustine, *Confessions*, p. 165; Liu, *Solitaire*, p. 109.

44. Binswanger, "The Case of Ellen West," p. 343.

45. Entry in student journal, 1983.

46. Woods, "I Was Starving Myself to Death," p. 242.

47. Liu, *Solitaire*, p. 109.

48. "I equated gaining weight with happiness, contentment, then slothfulness, then atrophy, then death." (From case notes of Binnie Klein, M.S.W., to whom I am grateful for having provided parts of a transcript of her work with an anorexic patient.) See also Binswanger, "The Case of Ellen West," p. 343.

49. Klein, case notes.

50. Cherry Boone O'Neill, *Starving for Attention* (New York: Dell, 1982), p. 131.

51. O'Neill, *Starving for Attention*, p. 49.

52. Liu, *Solitaire*, p. 101.

53. Levenkron, *Treating and Overcoming Anorexia Nervosa*, p. 122.

54. Since the writing of this piece, evidence has accrued suggesting that sexual abuse may be an element in the histories of many eating-disordered women (see note 2 in "Whose Body Is This?").

55. Bruch, *The Golden Cage*, p. 73. The same is not true of bulimic anorectics, who tend to be sexually active (Garfinkel and Garner, *Anorexia Nervosa*, p. 41). Bulimic anorectics, as seem symbolized by the binge-purge cycle itself, stand in a somewhat more ambivalent relationship to their hungers than do abstinent anorectics. See "Reading the Slender Body," in this volume, for a discussion of the cultural dynamics of the binge-purge cycle.

56. Bruch, *The Golden Cage*, p. 33.

57. Liu, *Solitaire*, p. 36.

58. Liu, *Solitaire*, p. 46. In one study of female anorectics, 88 percent of the subjects questioned reported that they lost weight because they "liked the feeling of will power and self-control" (G. R. Leon, "Anorexia Nervosa: The Question of Treatment Emphasis," in M. Rosenbaum, C. M. Franks, and Y. Jaffe, eds., *Perspectives on Behavior Therapy in the Eighties* [New York: Springer, 1983], pp. 363–77).

59. Bruch, *Eating Disorders*, p. 95.

60. Liu, *Solitaire*, p. 123.

61. Bruch, *The Golden Cage*, p. 65 (emphasis added).

62. Smith, "The New Puritans," p. 24 (emphasis added).

63. Entry in student journal, 1984.

64. Entry in student journal, 1984.

65. Trix Rosen, *Strong and Sexy* (New York: Putnam, 1983), p. 108.

66. Rosen, *Strong and Sexy*, pp. 62, 14, 47, 48.

67. Smith, "The New Puritans," pp. 27, 26.

68. Rosen, *Strong and Sexy*, pp. 61–62.

69. Rosen, *Strong and Sexy*, pp. 72, 61. This fantasy is not limited to female body-builders. John Travolta describes his experience training for *Staying Alive*: "[It] taught me incredible things about the body . . . how it can be reshaped so you can make yourself over entirely creating an entirely new you. I now look at bodies almost like pieces of clay that can be molded." ("Travolta: 'You Really Can Make Yourself Over,'" *Syracuse Herald-American* Jan. 13, 1985.)

70. Smith, "The New Puritans," p. 29.

71. Durk Pearson and Sandy Shaw, *Life Extension* (New York: Warner, 1982), p. 15.

72. Chernin, *The Obsession*, p. 47.

73. Smith, "The New Puritans," p. 24.

74. Sidney Journard and Paul Secord, "Body Cathexis and the Ideal Female Figure," *Journal of Abnormal and Social Psychology* 50: 243–46; Orland Wooley, Susan Wooley, and Sue Dyrenforth, "Obesity and Women—A Neglected Feminist Topic," *Women's Studies Institute Quarterly* 2 (1979): 81–92. Student journals and informal conversations with women students have certainly borne this out.

75. "Feeling Fat in a Thin Society," *Glamour* (Feb. 1984): 198.

76. The same trend is obvious when the measurements of Miss America winners are compared over the past fifty years (see Garfinkel and Garner, *Anorexia Nervosa*, p. 107). Some evidence has indicated that this tide is turning and that a more solid, muscular, athletic style is emerging as the latest fashion tyranny.

77. Entry in student journal, 1984.

78. Bruch, *The Golden Cage*, p. 58.

79. This is one striking difference between the abstinent anorectic and the bulimic anorectic: in the binge-and-vomit cycle, the hungering female self refuses to be annihilated, is in constant protest. And, in general, the rejection of femininity discussed here is *not* typical of bulimics, who tend to strive for a more "female"-looking body as well.

80. Entry in student journal, 1983.

81. O'Neill, *Starving for Attention*, p. 53.

82. Entry in student journal, 1983.

83. Bruch, *The Golden Cage*, p. 72; Bruch, *Eating Disorders*, p. 277. Others have fantasies of androgyny: "I want to go to a party and for everyone to look at me and for no one to know whether I was the most beautiful slender

woman or handsome young man" (as reported by therapist April Benson, panel discussion, "New Perspectives on Female Development," third annual conference of the Center for the Study of Anorexia and Bulimia, New York, 1984).

84. Levenkron, *Treating and Overcoming Anorexia Nervosa*, p. 28.

85. See, for example, Levenkron's case studies in *Treating and Overcoming Anorexia Nervosa*, esp. pp. 45, 103; O'Neill, *Starving for Attention*, p. 107; Susie Orbach, *Fat Is a Feminist Issue* (New York: Berkley, 1978), pp. 174–75.

86. Liu, *Solitaire*, p. 79.

87. Bruch, *The Golden Cage*, p. 65.

88. Klein, case study.

89. Chernin, *The Obsession*, pp. 102–3; Robert Seidenberg and Karen DeCrow, *Women Who Marry Houses: Panic and Protest in Agoraphobia* (New York: McGraw-Hill, 1983), pp. 88–97; Bruch, *The Golden Cage*, p. 58, Orbach, *Fat Is a Feminist Issue*, pp. 169–70. See also my discussions of the protest thesis in "Whose Body Is This?" and "The Body and the Reproduction of Femininity" in this volume.

90. Bruch, *The Golden Cage*, pp. 27–28.

91. Bruch, *The Golden Cage*, p. 12.

92. Binswanger, "The Case of Ellen West," p. 243.

93. At the time I wrote this essay, I was unaware of the fact that eating disorders were frequently an element of the symptomatology of nineteenth-century "hysteria"—a fact that strongly supports my interpretation here.

94. See, among many other works on this subject, Barbara Ehrenreich and Deirdre English, *For Her Own Good* (Garden City: Doubleday, 1979), pp. 1–29.

95. See Martha Vicinus, "Introduction: The Perfect Victorian Woman," in Martha Vicinus, ed., *Suffer and Be Still: Women in the Victorian Age* (Bloomington: Indiana University Press, 1972), pp. x–xi.

96. Ernest Jones, *Sigmund Freud: Life and Work* (London: Hogarth Press, 1956), vol. 1, p. 193.

97. On the nineteenth-century epidemic of female invalidism and hysteria, see Ehrenreich and English, *For Her Own Good*; Carroll Smith-Rosenberg, "The Hysterical Woman: Sex Roles and Conflict in Nineteenth-Century America," *Social Research* 39, no. 4 (Winter 1972): 652–78; Ann Douglas Wood, "The 'Fashionable Diseases': Women's Complaints and Their Treatment in Nineteenth Century America," *Journal of Interdisciplinary History* 4 (Summer 1973).

98. Ehrenreich and English, *For Her Own Good*, p. 2.

99. Ehrenreich and English, *For Her Own Good*, p. 102.

100. Sigmund Freud and Josef Breuer, *Studies on Hysteria* (New York: Avon, 1966), p. 311.

101. Freud and Breuer, *Studies on Hysteria*, p. 141; see also p. 202.

102. See especially pp. 76 ("Anna O."), 277, 284.

103. Marjorie Rosen, *Popcorn Venus* (New York: Avon, 1973); Lois Banner, *American Beauty* (Chicago: University of Chicago Press, 1983), pp. 283–85. Christian Dior's enormously popular full skirts and cinch-waists, as Banner points out, are strikingly reminiscent of Victorian modes of dress.

104. Liu, *Solitaire*, p. 141.

105. Binswanger, "The Case of Ellen West," p. 257.

106. This is one of the central themes I develop in "The Body and the Reproduction of Femininity," the next essay in this volume.

107. Dorothy Parker, *Here Lies: The Collected Stories of Dorothy Parker* (New York: Literary Guild of America, 1939), p. 48.

108. D. H. Lawrence, *Sons and Lovers* (New York: Viking, 1958), p. 257.

109. This experience of oneself as "too much" may be more or less emphatic; depending on such variables as race, religion, socioeconomic class, and sexual orientation. Luise Eichenbaum and Susie Orbach (*Understanding Women: A Feminist Psychoanalytic Approach* [New York: Basic Books, 1983]) emphasize, however, how frequently their clinic patients, nonanorexic as well as anorexic, "talk about their needs with contempt, humiliation, and shame. They feel exposed and childish, greedy and insatiable" (p. 49). Eichenbaum and Orbach trace such feelings, moreover, to infantile experiences that are characteristic of all female development, given a division of labor within which women are the emotional nurturers and physical caretakers of family life. Briefly (and this sketch cannot begin to do justice to their rich and complex analysis): mothers unwittingly communicate to their daughters that feminine needs are excessive and bad and that they must be contained. The mother does this out of a sense that her daughter will have to learn the lesson in order to become properly socialized into the traditional female role of caring for others—of feeding others, rather than feeding the self— and also because of an unconscious identification with her daughter, who reminds the mother of the "hungry, needy little girl" in herself, denied and repressed through the mother's *own* "education" in being female: "Mother comes to be frightened by her daughter's free expression of her needs, and unconsciously acts toward her infant daughter in the same way she acts internally toward the little-girl part of herself. In some ways

the little daughter becomes an external representation of that part of herself which she has come to dislike and deny. The complex emotions that result from her own deprivation through childhood and adult life are both directed inward in the struggle to negate the little-girl part of herself and projected outward onto her daughter" (p. 44). Despite a real desire to be totally responsive to her daughter's emotional needs, the mother's own anxiety limits her capacity to respond. The contradictory messages she sends out convey to the little girl "the idea that to get love and approval she must show a particular side of herself. She must hide her emotional cravings, her disappointments and her angers, her fighting spirit. . . . She comes to feel that there must be something wrong with who she really is, which in turn must mean that there is something wrong with what she needs and what she wants. . . . This soon translates into feeling unworthy and hesitant about pursuing her impulses" (pp. 48–49). Once she has grown up, of course, these feelings are reinforced by cultural ideology, further social training in femininity, and the likelihood that the men in her life will regard her as "too much" as well having been schooled by their own training in masculine detachment and autonomy.

(With boys, who do not stir up such intense identification in the mother and who, moreover, she knows will grow up into a world that will meet their emotional needs [that is, the son will eventually grow up to be looked after by his future wife, who will be well trained in the feminine arts of care], mothers feel much less ambivalent about the satisfaction of needs and behave much more consistently in their nurturing. Boys therefore grow up, according to Eichenbaum and Orbach, with an experience of their needs as legitimate, appropriate, worthy of fulfillment.)

The male experience of the woman as "too much" has been developmentally explored as well, in Dorothy Dinnerstein's ground-breaking *The Mermaid and the Minotaur: Sexual Arrangements and Human Malaise* (New York: Harper and Row, 1976). Dinnerstein argued that it is the woman's capacity to call up memories of helpless infancy, primitive wishes of "unqualified access" to the mother's body, and "the terrifying erotic independence of every baby's mother" (p. 62) that is responsible for the male fear of what he experiences as "the uncontrollable erotic rhythms" of the woman. Female impulses, a reminder of the autonomy of the mother, always appear on some level as a threatening limitation to his own. This gives rise to a "deep fantasy resentment" of female impulsivity (p. 59) and, on the cultural level, "archetypal nightmare visions of the insatiable female" (p. 62).

110. Quoted in Brian Easlea, *Witch-Hunting, Magic, and the New Philosophy* (Atlantic Highlands, N. J.: Humanities Press, 1980), p. 242 (emphasis added).

111. Quoted in Easlea, *Witch-Hunting*, p. 242 (emphasis added).

112. See Peggy Reeve Sanday, *Female Power and Male Dominance* (Cambridge: Cambridge University Press, 1981), pp. 172–84.

113. Quoted in Easlea, *Witch-Hunting*, p. 8.

114. Peter Gay, *The Bourgeois Experience: Victoria to Freud*. Vol. 1: *Education of the Senses* (New York: Oxford University Press, 1984), pp. 197–201, 207.

115. Chernin, *The Obsession*, p. 38.

116. Ehrenreich and English, *For Her Own Good*, p. 124.

117. See Jeffrey Masson's controversial *The Assault on Truth: Freud's Suppression of the Seduction Theory* (Toronto: Farrar Strauss Giroux, 1984) for a fascinating discussion of how the operation (which, because Fliess failed to remove half a meter of gauze from the patient's nasal cavity, nearly killed her) may have figured in the development of Freud's ideas on hysteria. Whether or not one agrees fully with Masson's interpretation of the events, his account casts light on important dimensions of the nineteenth-century treatment of female disorders and raises questions about the origins and fundamental assumptions of psychoanalytic theory that go beyond any debate about Freud's motivations. The quotations cited in this essay can be found on p. 76; Masson discusses the Eckstein case on pp. 55–106.

118. Banner, *American Beauty*, pp. 86–105. It is significant that these efforts failed in large part because of their association with the women's rights movement. Trousers like those proposed by Amelia Bloomer were considered a particular badge of depravity and aggressiveness, the *New York Herald* predicting that women who wore bloomers would end up in "lunatic asylums or perchance in the state prison" (p. 96).

119. Banner, *American Beauty*, pp. 149–50.

120. Amaury deRiencourt, *Sex and Power in History* (New York: David McKay, 1974), p. 319. The metaphorical dimension here is as striking as the functional, and it is a characteristic feature of female fashion: the dominant styles always decree, to one degree or another, that women *should not take up too much space*, that the territory we occupy should be limited. This is as true of cinch-belts as it is of foot-binding.

121. Quoted in deRiencourt, *Sex and Power in History*, p. 319.

122. Kathryn Weibel, *Mirror, Mirror: Images of Women Reflected in Popular Culture* (New York: Anchor, 1977), p. 194.

123. Christy Ferguson, "Images of the Body: Victorian England," philosophy research project, Le Moyne College, 1983.

124. Quoted in E. M. Sigsworth and T. J. Wyke, "A Study of Victorian Prostitution and Venereal Disease," in Vicinus, ed., *Suffer and Be Still*, p. 82.

125. See Kate Millett, "The Debate over Women: Ruskin vs. Mill," and Helene E. Roberts, "Marriage, Redundancy, or Sin: The Painter's View of Women in the First Twenty-Five Years of Victoria's Reign," both in Vicinus, ed., *Suffer and Be Still*.

126. Gay, *The Bourgeois Experience*, p. 197; Millett, "Debate over Women," in Vicinus, ed., *Suffer and Be Still*, p. 123.

127. Vicinus, "Introduction," p. x.

128. Lasch, *The Culture of Narcissism*, p. 343 (emphasis added).

129. Chernin, *The Obsession*, p. 148.

130. Charles Gaines and George Butler, "Iron Sisters," *Psychology Today* (Nov. 1983): 67.

13

Feeding Hard Bodies: Food and Masculinities in Men's Fitness Magazines

Fabio Parasecoli

Body image has become relevant to many men, as the flourishing of male fitness magazines and the consequent development of theoretical studies on the subject demonstrate. In this article, I will examine the close connections between food, masculinities, and body image in male fitness magazines, a booming sector of the fitness and wellness publishing industry. I will attempt a qualitative analysis of editorial features and advertising pages of the October 2002 issues of *Men's Health, Men's Fitness,* and *Muscle and Fitness.* I will also analyze *Men's Health Guide to Women,* a supplement to *Men's Health* published in the same month, designed as a sort of a how-to manual for men who want to increase their success with women. Since I will deal with communication and media, I will adopt a semiotic approach, emphasizing how the frequent references to food and eating, which clearly constitute a deep concern, actually construct a coherent discourse about masculinities, using words, images, and metaphors as signs within a complex code that needs to be deciphered.

Plural Masculinities

All over the Western world, a new breed of leisure publications that deal with various intimate aspects of men's lives have recently invaded kiosks: men's fitness magazines. These publications have now abandoned the closets of gay men and the lockers of professional body builders to be conspicuously displayed in dentist's waiting rooms or on coffee tables next to football magazines.

Amongst these, *Men's Health, Men's Fitness,* and *Muscle and Fitness* have proved to be particularly successful over time, in terms of popularity and sheer sales.[1] They constitute a particularly significant segment in this kind of literature and deserve closer examination. I will examine the differences in both editorial and advertising materials that assume a certain range of diversity within the target readership. *Muscle and Fitness* is more specifically geared toward professional body builders, while the other two are geared toward an audience of men who, although conscious of their bodies, are also interested in other aspects of their masculinities. Launched in 1988, *Men's Health,* owned by Rodale, is definitely the most popular, claiming a circulation of 1.7 million, with 22 editions in 30 countries.[2] Rodale, a family owned company, publishes magazines, such as Organic Gardening, Prevention, Bicycling, Runner's World, and Backpacker, and has established a public image of health consciousness and commitment against tobacco and hard liquor. At the end of 2002, both *Men's Fitness* and *Muscles and Fitness,* originally owned by Weider Publications Inc. and founded by the bodybuilder Joe Weider in the 1940s, were bought by American Media, which already owns many tabloids, such as *National Enquirer* and *Star*, and also operates Distribution Services to place these periodicals in supermarkets. In other words, bodybuilding has gone mainstream.

The enthrallment with the body image, previously imposed mostly on women, is now becoming a common feature in masculine practices and identification processes, to the point that the expression "the Adonis complex" has been created, referring to the more pathological, obsessive forms of this phenomenon.[3] Recent literature on body images has developed in the frame of theories that consider multiple masculinities as constructed collectively in culture and sustained in all kinds of institutions (the school, the gym, the army, the workplace). Connell writes,

> Men's bodies are addressed, defined and disciplined, and given outlet and pleasures, by the gendered order of society. Masculinities are neither programmed in our genes, nor fixed by social structure, prior to social interaction. They come into existence as people act. They are actively produced, using the resources and strategies available in a given social setting.[4]

Masculinities are not fixed or defined once and for all; they do not represent embodiments of discrete states of being. They vary in time and place, in different historical, social, and cultural environments. As practices, they sometimes articulate contradictory desires, emotions, and ideals, denying the very notion of a static and defined identity. These concurrent masculinities are not equivalent: some tend to be considered more desirable than others, even when they are not the most common, and thus become "hegemonic," a standard against which men embodying other kinds of masculinities assess their self perception and often also their self-esteem.[5] Body images play a fundamental role in defining what the dominant masculinity model should look like. Nevertheless, men's bodies are not blank pages that become the receptacle for all kinds of power and social determinations: they are actual agents, and they interact with other aspects of the social practices determining masculinities. As Connell emphasizes,

> To understand how men's bodies are actually involved in masculinities we must abandon the conventional dichotomy between changing culture and unchanging bodies . . . Through social institutions and discourses, bodies are given social meaning. Society has a range of "body practices" which address, sort and modify bodies. These practices range from deportment and dress to sexuality, surgery and sport.[6]

The growing attention to the male body—it has sometime been argued—is, at least partially, a result of the mainstreaming and the normalization of gay culture.[7] Nevertheless, also in heterosexual contexts, male strong bodies have traditionally served as metaphors for sexual potency, power, productivity, dominance, independence, and control. Both discourses are somehow articulated in the contemporary hegemony of the muscular body type (also known as mesomorphic, as opposed to ectomorphic, slim, and endomorphic, overweight), often in connection with a phenomenon sometimes defined as "re-masculinization."[8] Until a few decades ago, the aspiration for a muscular build was a prerogative of a small circle of professional and amateur bodybuilders, who were also involved in different forms of competition, giving to the whole scene the veneer of a sport. In time, after large sections of the gay community embraced the muscular body as desirable and prestigious, the same attitude became more and more visible, also in heterosexual—or should we now say metrosexual—circles. A renewed attention to the body and its appearance goes well beyond the concern for its athletic potential, which was a normal element of the sport subculture, uniting all men, gay and straight, in the same awe for the bulging muscle. These phenomena explain the growing success of men's fitness magazines, which, at any rate, carefully avoid dealing with issues of sexual preference, and ban any hint of homoeroticism, which is, nevertheless, always lurking behind the glossy pages of the magazines.

The growing prestige granted to the muscular body places increasing pressure on men to take greater care of their looks.[9] Men seem to adopt different strategies to make sense of their bodies when they do not meet the hegemonic expectations; the three predominant ways to adjust the discrepancy between the ideal and the real body have been defined as reliance, where the individual works on his body to reach the model; reformulation, where each individual adjusts his conception of hegemonic masculinity to meet his abilities; rejection, where the individual totally refuses the hegemonic model.[10] In the case of reliance, usually great amounts of energy, money, and time are invested in gaining the desired body image, often with anxious undertones that reveal a certain preoccupation with control over one's body. In this context, food plays a fundamental, though often concealed, role. Diets and eating habits are interpreted as a key element in the construction of a fit body.

Advertising to the Hardliners

Needless to say, the food and supplement industry has tapped into these trends to acquire new consumers for highly processed products that ensure growing revenues for a sector structurally plagued by intense competition.[11] Many of the advertising pages in these magazines often play with a sense of inadequacy, or with a desire for emulation in order to increase sales, proposing behaviors and values. These constructs hinge on dedication and effort that help to form the constructs of hegemonic masculinities.[12]

On the first page of *Muscle and Fitness*, we find an ad page by Animal for "Hardcore Training Packs." Next to a hip bodybuilder, wearing a woolen hat and a grungy sweater, we read the sentence: "Shut up and train," followed by a small print copy text:

> Every day you train is judgment day. Each rep, each plate matters. You don't make time for talk. All you care about is moving weight. Nothing else. This is hardcore. This is Animal. Can you handle it?[13]

A few pages later, the same man is shown, while lifting heavy weights, with an expression of near pain on his face. The text states:

> Go hard . . . or go home. Balls-to-the-wall training. You sweat. You push. It hurts. In here, there's no room for crybabies, no place for talking trash. Just raw lifting. This is the real deal. This is Animal. Can you handle it?[14]

In the same magazine, another ad shows a weight lifting bench with pillows and a blanket in an empty room. The text reads, "Obsessed is just a rod the lazy use to describe the dedicated."[15] Amen, one would fervently respond. What we face here is a fullblown cult. The new religion requires total dedication, in anticipation of the final ordeal when all believers will present themselves to the Big Trainer to have their thighs and deltoids carefully measured. Good results can be attained only by severe, unrelenting, and even painful workouts.

Nevertheless, followers are offered ways to cheat a little, as in any religion. The magazine pages are full of ads for protein drinks, integrators, engineered nutrition items, and dietary supplements that can help adepts to reach their goals. Exercising is necessary, but science can help to the point that one can grow muscle while sleeping, as with products such as SomnaBol-PM, Night charger, or NitroVarin-PLS.[16] None-theless, the ad specifies that there is absolutely no need to stop taking regular proteins during the day.

As a matter of fact, most advertising in *Muscle & Fitness* is related to nutrition, meeting the readers' needs to ensure the correct intake of substances required to bulk up, and expressing the food and supplement industry's necessity to increase sales. The technical terminology it uses aims itself at readers who are supposedly familiar with choline bitartrate (for serious mental energy and acuity, we are told) or glutamine peptides, who are extremely aware of their daily protein intake, and who would never ingest the wrong kind of proteins. We are made privy of "the ugly truth behind the collagen."

> Pigs feet, cattle hide, crushed bone, fish skin. How did they ever end up in your chocolate protein bar? Easy. After those bones and skins get soaked in lime to remove all hair and grease, seared with acid until they disintegrate and then molded into edible gelatin or collagen, they become part of countless protein bars, even the best sellers.[17]

The text, advertising the VHT 100% real Protein Chocolate Rum Cake, is placed next to a full page picture of a meat grinder where a cow skull, a not-at-all-bad-looking fish, a bone, and a pig foot are being cranked in what is clearly an impossible fashion. The effect is quite overwhelming, even if for a non-American, the pig foot and the fish actually look appetizing. But the correlation with death, clearly expressed by the skull and the bone, would make even the most adventurous eater flinch. To balance the disturbing effect of the image, the text is followed by a very reassuring chart that explains how the protein efficiency ratio is much lower in collagen than in, say, milk protein, beef protein, or best of all, in whole eggs. The advertisement clearly tries to convince the potential consumers to buy a product that provides the necessary nutrients, all the while avoiding any contamination with the less appealing aspects of food, namely death and corruption.

The assumed goal for this magazine's target readers, it is clear, is gaining muscle mass. The three page advertisement for Cytodine, "The Single Most Revolutionary Advancement in Bodybuilding History," recurs to the old routine of the before and after pictures, except that the before pictures portray bodies that would make the average male seethe in sheer envy. The same trick is deployed a few pages later, where we are told that an Idrise Ward-El was able to lose 20 pounds of fat and gain 25 pounds of muscle with the help of Muscle Tech supplements, like Cell-tech.[18] The before picture of Mr. Ward-El shows us a stocky but muscular and good looking black man, who, in the after picture, actually boasts huge muscles (and no longer wears glasses). At this point, it is necessary to note that Asian men are conspicuously absent from these magazines, both in the editorial and the advertising material. The underlying but unexpressed assumption is that Asian men are either not interested in achieving a muscular body for cultural reasons, or physically not able to do so.

Buy a Better Body

Muscle & Fitness readers are hardliners and dedicated body-builders. The advertising is overwhelmingly concentrated on nutritional items. The few exceptions are pages for sports gear. On the other hand, advertising in both *Men's Fitness* and *Men's Health*, with a greater emphasis on fitness and general well-being, also promotes colognes, after shave moisturizer, razors, a juice processor, trekking shoes, even cars, fashion and hi-tech. The advertising in *Men's Health* definitely appeals to needs other than fitness or muscle building. As we will see, even women become a center of interest. It is relevant to point out that in all three magazines, the advertising pages for nutritional products can be classified in two typologies: the ones where each item is promoted exclusively for its nutritional value without any reference to taste, and others where flavor plays a certain, if peculiar, role.

Only one advertisement in the three magazines refers to the idea of actually cooking food. The headline is: "Trim, tone, define, & sculpt! Your complete Diet, Training, Nutrition and Fat Free Cook Book Collection." We are informed that the cook, a black woman who smilingly displays her amazing biceps, besides being a master chef, was a contender in the Miss Olympia body building competition. She's not alone: the other author is a white nutrition consultant with a health science degree. The ad interestingly uses a woman, as muscled as she is, to evoke the idea of the kitchen, while the scientific side of the cookbooks, guaranteeing good results, is the work of a man.[19] This element—as we will see—often recurs when the preparation of food is mentioned. The fact that the cook is black and the scientist is white adds further layers to this advertisement, revealing race biases that are otherwise invisible in this kind of literature, probably based on the assumption that part of the readership is composed of black males.

The scientific seal that seems fundamental to these ads is particularly evident in those belonging to the first typology, advertising nutritional supplements. Long lists of components are given in uncompromising terminology, aimed at convincing readers that the products are actually systematically engineered to improve their muscle mass. Pictures and details of muscled persons, and images of the advertised items, usually accompany written texts. Since their names are difficult to remember, and in the end they all sound the same, it is important to display the actual box or flacon in the ad, to make sure that readers will be able to recognize it when they are shopping at their local health shop. Most of the products in this typology do not make any reference to the act of eating. Sometimes a flavor is given to them, especially when it comes to drinks and bars, but it is not relevant in the economy of most ads.

On the other hand, the second typology employs the element of taste to make the advertised products more appealing, even if only to ensure a certain variety of flavors for the same product. The important elements are still nutritional, but consumers can turn to taste to avoid boredom. Even in this second category, there is little, if any, reference to actual food. Some ads display fruit, milk, or chocolate next to the supplement. Designer Whey promotes a peach flavored bar with the catching title "Finally! Because we're so sick and tired of chocolate." The paradox is that the ad does not refer to real chocolate, of course, but the ersatz flavor that can be found in other bars.[20] The Maxxon bar ad boldly refers to the actual act of eating and munching, usually not even considered, almost repressed. We see four images of the bar. In the first one, it has just been bitten, then more and more chunks of it disappear, supposedly in the watering mouth of a reader. Under each image, we read "Crunch . . . crunch . . . crunch . . ." and finally "Mmmmm!" In this unusual case, the supplement is not mysteriously incorporated in the body to increase muscle mass, but it is actually chewed.[21]

Labrada Nutrition promotes a Carb Watchers Lean Body Banana Split Meal Replacement Bar. In the ad, the bar is actually dipped in a banana split, with whipped cream, cherries, and even chocolate prominently displayed. This is probably the most risqué image in the three magazines, but the impact is, of course, balanced by the copy that points out that the bar provides 30 g of proteins and just 4 g of sugar. The suggestion is that, by choosing that product, health conscious consumers are allowed to sin with their minds and mouths, but their bodies will not even notice it.[22] Pages promoting everyday food products that one would find at a corner store or a supermarket are absent in *Muscle & Fitness*. While in *Men's Fitness*, we find only a few representations, *Men's Health* publishes ads for Heinz Tomato Ketchup, Wendy's, V8 vegetable juice, and Stouffer's frozen meals, and nutrition supplements become much less frequent. The readers of the latter magazine appear to be considered by the advertisers as full blown eating creatures, even if the products, not necessarily flat-belly friendly, refer to a kind of food consumption that does not include any actual cooking.

From this semiotic analysis of the advertising pages, we can already draw a few conclusions about the relationships among food, nutrition, masculinities, and muscle building, with body images playing a paramount role in the choices of what these men—for whom the magazines are meant—eat. Taste and sensual appreciation don't play a central role. Different flavorings are added only to ensure variety to meals that otherwise would be always the same. The key element here is nutrition: how much protein the body absorbs, and how much fat it is able to burn. Readers seem to show a certain need to be reassured in their quest for fitness, and the agency that guarantees this sense of protection is science, connoted as exact, matter of fact, serious, and above all, masculine.

Science, considered as a legitimate masculine way of thinking and approaching reality, also plays a very important role in the editorial features concerning food and nutrition, second in importance only to the stories about fitness and muscle building. In the three magazines, we find sections with nutrition advice: all of them quote the sources the news is taken from, and in *Men's Fitness*, we even see the face of the experts that give the tips.[23] Despite the continuous reference to science, the general discourse is far from adopting an actual scientific approach: tips are just tips. Readers are offered bits and pieces of unrelated news about this specific nutrient or the other, without any systematic connection.[24] As it often happens in diet and nutrition communication, editorial staffs opt for clear, simple, and ready-to-apply pieces of advice, avoiding any difficulty intrinsic in the subject matter.

In *Muscle & Fitness*, readers are told to decrease their consumption of Java coffee for a leaner diet, since in an experiment made in Norway, "the intake of chocolate, sweets, cakes, sweet biscuits, pastry, and jams increased when coffee intake was higher, and decreased when coffee intake was lower. The opposite effect occurred with fish dishes, other drinks, as well as with physical activity."[25] No explanation is given about the phenomenon, not even tentatively. On the other hand, in *Men's Fitness*, we readers are told to take caffeine before a workout, because "caffeine stimulates the release of free fatty acids, which are utilized for energy when you exercise, leading to fat loss."[26]

Another feature in *Men's Fitness* is dedicated to soy, endorsed for its muscle-building and protecting capacity.[27] The whole story is an attempt to explain that eating soy and soy-derived products will not transform readers into "bandy-legged yogis with torsos the shape of Coke bottles." As a matter of fact, after reassuring bodybuilders that the phytoestrogens, contained in soy and similar to estrogens, do not constitute a problem for masculinity, the author affirms, "real men eat tofu, in moderation." In the Nutrition Bulletin section, *Men's Health* invites readers to add fish oil to their diet to avoid the risk of developing both an irregular heartbeat and insulin resistance, to eat strawberry and black raspberries to reduce risks of cancer, to drink soda to help quench appetite, milk to reduce the risk of colon cancer, and wine for better breath.[28] Readers are presented a mass of unrelated information that seems to play with their health fears. "Smart strategies" are proposed to fight any kind of problem connected with food.

In the Weight Loss Section of *Men's Health*, readers are taught what to order at an Italian restaurant by two juxtaposed pictures.[29] One is a serving of lasagna with antipasto and two breadsticks (959 calories), the other is eggplant parmigiana with a side of spaghetti with marinara sauce and two slices of garlic bread (1,246 calories). Frankly, both look quite unappetizing. The text states the obvious, underlying the necessity of limiting caloric intake, without providing the reader with any actual advice or information about better and healthier eating.

Lose Fat, Build Muscle, Eat Smart

As in the case of the advertising pages, the goal of editorial features dealing with nutritional models and attitudes is usually not to follow a balanced and constant diet that can ensure body fitness in the long run, but rather to obtain fast results that are immediately visible on the body. The focus is on food as building material for a better-looking, longer-living body, rather than as a source of pleasurable experiences or a marker of cultural identity, let alone a cherished and hated instrument for caring and nurturing. Each magazine presents some sort of diet. For example, *Muscle & Fitness* introduces King Kamali, a 6 meal, 7 snack daily program that has, at its core, heavy and hard training. The program mixes fresh food, such as fruit or chicken breast, with protein powders and meal replacements.

The same issue of *Muscle & Fitness* also dedicates a long feature to a Crash Course in Nutrition for college students, with a strong caveat against consuming fast food and, in general, "sweet, salty, and fatty temptations."[30] Fast food is not bad, per se, as it often solves time and budget, problems common to many students. As a matter of fact, the editor lists many snacks, categorized into "crunchy," "fruity," and "substantial," that can help youngsters fight their hunger without fattening them up. The aim is, once again, to bulk up, to add muscle, and to lose fat. Male students are not supposed to waste time cooking, which might make them look effeminate. They are advised to "stock up on canned beans, frozen vegetables, bagged pasta, and even canned tuna to provide a basis for many nutritious meals." One single recipe is given, and that is for batches of pasta "made ahead of time and refrigerated in a sealed container until you're ready to assemble the meal." As a matter of fact, all that students are supposed to do is toss some precooked or canned food in the pasta.

Men's Fitness presents us with the 15/21 Quickstart Plan to lose up to 15 pounds in 21 days.[31] For each week, a daily diet is prescribed (not proposed) that has to be repeated, unchanged, for seven days. The diet is composed of 6 meals, with two main intakes and 4 snacks. Readers are also given eight "ground rules for maximal fat loss," such as "Eat the meals in order," "Eat all the food prescribed," "Don't cheat." The language transpires sheer severity: If you want to lose fat, you have to be totally dedicated, even ready to sacrifice yourself on the kitchen stove, for you are required to boil eggs, steam vegetables and fish, and bake potatoes. The feature is illustrated with pictures of fresh, if unappetizing, food, in light yellow dishes and bowls.

The same magazine, nevertheless, deals with the issue of daily food intake, in an article aimed at teaching readers to maintain the muscle they have earned in months of training without losing definition.[32] The story features appetizing pictures of simple and nourishing dishes, just like those one could see in a food magazine. Readers are given another series of clear rules. "The idea is to control as much as possible of what goes into what you eat." The control issue is paramount: If one does not keep one's otherwise wild appetites in check, one can neither lose fat and build muscle, nor maintain one's body frame. Once again, most of the dishes proposed in the maintenance menu require little or no preparation. If they do, it's not mentioned. So readers are told to eat lean turkey, chicken breasts, or salmon, but no information is given about how to fix them.

Do Real Men Cook?

When actual food is advertised, it is ready-made or fast food. Potential readers are not supposed to have any connection with buying, storing, and cooking food, all activities apparently belonging to the feminine. Male subjects cannot perform activities related to the preparation of food without affecting their masculine traits and the inscription of these in a cultural order that is deeply gendered. By reiteration, the norms intrinsic to these practices and processes, highly regulated and ritualized, are likely to be incorporated in the very body of the individual, which thus enters the domain of cultural intelligibility.[33] They constitute the conditions of emergence and operation, the boundaries and stability of the gendered subject. Thus, the production or materialization of a masculine subject is also its subjection, its submission to rules and norms, including the ones regulating the kitchen. The embodiment of gender through reiterated practices, including food related activities, reveals the influence of power structures that, in Foucaultian fashion, are omnipresent and pervasive, not necessarily connected with specific institutions, and not always imposed on the subject from the outside.[34] At any rate, the food and supplement industry seems eager to and capable of exploiting these elements to encourage readers to buy products that reinforce their perceived masculinities by avoiding gendered activities, such as buying food and cooking, while turning them into better consumers of high added value, hence more expensive, products.

Usually, no cooking is required from readers. In *Muscle & Fitness*, we find a feature about eating fish.[35] "Bodybuilders, grab your can openers—try these 10 scrumptious, high-protein seafood recipes you can prepare in minutes." Readers are given alternatives to the boring old tuna, such as canned or preserved salmon, clams, crab, and shrimp. Again, these dishes can be assembled more than cooked, saving time and sparing the readers' self-esteem. On the pages in *Men's Fitness*, triple-deckers become "a tower of muscle-building power."[36] The title is reinforced by the picture of a never-ending series of superposed layers of sandwich, filled with a scrumptious bounty of food. The abundance is so overwhelming that the sandwich passes the limits of the page and continues on the next one. Then, of course, limitations are given about what kind of bread, cheese, and condiments are to be used. Again, there is no cooking involved. This is real men's food, even if it gingerly avoids mayonnaise, full-fat cheese, and fattening bread (described as "bullets to be dodged").

Men's Health has a section called, "A man, a can, a plan," that gives a recipe for "Beer-n-sausage bake: a tasty, filling, and alcohol-tinged meal."[37] This is the quintessential macho meal: cans, beer, and no cooking. Well, actually there is some cooking involved. A sausage has to be grilled in a skillet and then put into the oven with the other ingredients, but we are still at a manageable level of assemblage. Anyway, everybody knows that cooking meat, as on the occasion of barbecues, is a man's thing. Exceptions are rare. In a story aimed at promoting pork, after reassuring readers of its leanness, *Muscle & Fitness* entices its readers with tips for making it tastier ("fruits go well with pork") and even gives instructions for a tropical stir-fried pork tenderloin, re-inforced by an inviting picture of the dish. It goes without saying, we are given the exact quantity of calories and nutrients per serving.[38]

In another section of *Muscle & Fitness*, Laura Creavalle (the author of the above mentioned cookbooks advertised in the same magazine) gives the recipe for a home-baked and healthy fruit and nut bread.[39] The recipe is simple and quick, and while the text refers to "moist, delicious cakes that warm the hearts of young and old alike," it also warns against the excess of fat in baked goods. It is interesting that the cook is a woman, and while she refers to the emotional connotations of food, she also puts herself under the banners of the fight against inordinate consumption of victuals.

Again, dedicating time and effort in fixing meals is perceived as connected to the nurturing role that is considered typical of females. Women re-affirm their nature by performing their role of caregiver. They are responsible with feeding not only their own, but also others' bodies, ensuring their survival, but submitting them to the constant temptation of the unchecked, always invading flesh. This short examination of food preparation practices, as explained in the feature articles, reconfirms the conclusions deducted from the analysis of the advertising pages. Cooking is perceived as one of the most identifiable performative traits of femininity. Men should avoid participating in the transformation of food at the stove, an almost alchemical activity dangerously close to the growth of flesh, inherently difficult to keep in check.

How to Feed a Naked Woman

As a matter of fact, according to *Men's Health Guide to Women*, there is an occasion when a man is supposed to cook: When he wakes up after a night of sex, and he wants to fix breakfast for the naked woman dozing in his bed.[40] "Once you get her into bed, these breakfasts will keep her there," affirms the writer, who, being a woman, knows what she is talking about. The five recipes she gives are simple and quite fast. The point is to nourish the body in ways that are propitious to more sex, obviously the sole goal of all the cooking. So readers are told that the crustless quiche they are taught to prepare is "high in protein to help control SHBG, a substance that makes it harder for your body to use testosterone and arginine. Argigine is an amino acid that improves the blood flow to your penis and may also improve sexual stamina for both men and women." In the introduction to Eggs Benedict with Broiled Grapefruit the author states:

> Eggs, bacon, English muffin: the combination provides extra zinc, a mineral that she needs to stay well lubricated and that you need to keep producing semen; You'll pick up niacin, too—a B vitamin essential for the secretion of histamine, a chemical that helps trigger explosive orgasms. And since studies have found that too much or too little dietary fat can decrease levels of libido-boosting testosterone, the recipe has been tweaked to provide an optimal 28 percent of calories from fat. The pink grapefruit? It's for her; women like pink stuff.

Is that not a known fact? Nevertheless, readers are clearly advised to avoid thinking about women, except as objects of sexual desire, waiting in bed to be fed. "Remember when your mom used to make French toasts on Saturday mornings? Try not to think about it. Instead, think about thiamin and riboflavin." Thank goodness science is there to help men achieve their goals and, to some extent, to reassure their apparently wobbling self-confidence. Recipes have more the function of a placebo that works like Viagra, rather than a sensual, sexual experience to be shared with one's partner. Taste is not mentioned once.

Herbs and other natural ingredients are also supposed to help with one's sex life.[41] They surely will not hurt, and they might actually help, "whether you're an old man battling occasional bouts of impotence or a younger man whose sex drive is sometimes slowed by stress." A whole list is given, with specific explanations for each substance. A similar approach is evident in a feature in *Men's Health*, "The Sex for Life Diet."[42]

The article you are about to read is based on the simple notion that a) men like food and b) men like sex, so c) wouldn't it be great if you could actually eat your way to more fun in the bedroom? Grunt if you agree. Or maybe just sharpen your knife and fork. With help from nutritionists and the latest research, we've discovered 10 superfoods that can help you at every age and stage of sex life—whether it's seducing women in your 20s, producing Mini-Me's in your 30s, or inducing your equipment to keep working in your 40s and beyond.

Science is invoked not only in the content of the story, but the whole argument is presented like a robust, if actually fake, Aristotelian syllogism. Men act according to logic, even when their bodies crave sex and food, and live their lives according to neat plans. In your 20s, you have fun; in your 30s, you concentrate on reproduction (not out of love, but with the goal of creating replicas of your narcissistic self); in your 40s, there is nothing much left to do but worry about your faltering pleasure tools.

In the pictures, we see a handsome couple engaged in various activities. In one, the smiling, blond, thin, and cute woman feeds pizza to her muscle-bound, tight T-shirt wearing partner. In the second, he has just dipped a strawberry in whipped cream, while she coyly averts her gaze from him, with a finger in her mouth. In the third, she holds a glass of red wine in her hand, while he smiles and leans on her. The images make a clear distinction between male food (pizza) and female food (strawberries, whipped cream, wine). Wine seems to haunt men as a dangerous world, often perceived as intrinsically feminine, that they are obliged to cope with and that causes performance anxieties. Men wander in uncharted territories, far from the reassuring haven of beer. In the article, "22 Ways to Make an Impression," readers are given a few basic tips:[43] "11. Never bring out a half-consumed bottle of wine sealed with aluminum foil. 12. When choosing a bottle of wine to take to a dinner party, spend between $10 and $15. That's for a bottle, not a gallon. 13. When a wine steward gives you a cork, sniff it. Don't check it by plunging it rapidly in and out of your pursed lips."

In other words, never mimic fellatio in front of your date, as allusive to future pleasure as that may be. Why is the "wine steward" supposed to give you the cork to sniff? If that operation needs to be done, it should be his task, and he will pour some wine in your glass to swirl and, yes, sniff. The story is written for men who clearly are not used to ordering wine, and prefer eating at the local diner, or even greasy spoon, rather than at a restaurant where a somehow menacing and mysterious "wine steward" might embarrass them. We find another hint to the anxiety provoked by dining experiences in "24 Rules for a Successful Relationship," which interestingly mentions dinner in the same sentence with the choice of living quarters or, even, reproduction.[44]

Put your foot down the next time both of you are making plans for dinner—or, heck, deciding where to live and whether to have children. Women rate agreeable men as more attractive than stubborn ones, but only if the nicer guys have a dominant streak. If strength and decisiveness are missing, nice guys come off as meek.

In only one instance taste is mentioned together with sex. In "9 Tricks of Domestic Bliss," giving tips on where to have sex when at home, we read:

> Use a sturdy wooden table, which is more comfortable than the floor and a better height than the kitchen counter, says Louanne Cole Weston, Ph.D., a sex therapist in Fair Oaks, California. Have your partner lie back on the table with her pelvis near the edge. Then reach for some food—anything that can be licked off is fair game. Giving your tongue something tasty to aim for can help you dwell in one spot longer—and she'll love that.[45]

Cunnilingus can be fun, but it is better to give it more taste. Here, women are equaled to food, put on a table, and eaten. They don't necessarily taste good though.

Balanced Diets, Controlled Bodies

In all the examined material, a strong desire to control not only one's body appearance, but also to curb one's desires and appetites is evident. A fit male body becomes the material expression of one's dominion over the self. *Muscle & Fitness* features a story about "taming the craving for sugary treats."[46] Sweets are interestingly connected with women, either because they supposedly yearn for them more than men do, "especially during that week of the month," or because they control the administration of sweets in the household, and not only for children. The professional bodybuilder, Garrett Downing, tells readers,

> Since I have problems with portion control, I put my wife in charge. She's the keeper of the sweets. She'll give me one or two cookies and hide the rest from me. Otherwise, I could actually sit down and eat an entire box . . . I think if you try to eat clean all the time, you get to a point where you get a little insane. Having that bit of a sweet treat brings a little sanity back into your life. If you're training consistently and the rest of your diet is intact, occasional treats won't do damage. When you don't train consistently but cheat frequently, that's when they catch up. Even when you're dieting, you can allow yourself a treat—a couple of cookies, a slice of pie or cake.[46]

The woman as the "keeper of the sweet" is quite a powerful, if involuntary, metaphor. The phrase reveals an ongoing struggle between a "clean" diet and the unrelenting desire for sensory satisfaction that can lead the more dedicated man to "cheating." Sweets—and appetite in general—are clearly perceived as feminine. This is a recurrent—though often latent and disguised—element that plays an important role in structuring the nutritional discourse in these magazines. As a matter of fact, only women appear capable of keeping men's desire in check, probably because they are supposed to deal constantly with their own. The agency of desire, thus, becomes the agency controlling desire. This ambivalence reminds us of Melanie Klein's theories on the ambivalent desires and fears of devouring and being devoured in infants.[47] According to the famed psychoanalyst, the mother's breast, and any other source of nourishment, that ensures satisfaction and then disappears, is perceived as both good and bad, creating frustrations in infants and cannibalistic drives aiming at the destruction and ingestion of the desired objects. Since infants experience their own cannibalistic drives as dangerous, they protect themselves by externalizing and projecting them outside, on to the breast, from which they fear retaliation in the form of ingestion. Thus, desire and hunger are perceived as potentially destructive and, for that reason, projected on the outside. This anxiety ridden dualism that characterizes the first months of life—defined by Klein as "paranoid-schizoid"—appears to be successively sublimated in the dichotomy that, according to the feminist theorist Susan Bordo, haunts Western civilization from Plato on. She writes, "the construction of the body as something apart from the true self (whether conceived as soul, mind, spirit, will, creativity, freedom . . .) and as undermining the best efforts of that self."[48] As it happens, when gender is applied to this dualism, women are on the side of the lower bodily drives, embodying appetites and desires that weigh down men in their attempt to achieve freedom from materiality. In the passage from the feature, "Taming the Craving for Sugary Treats," we quoted, women are the keepers of the occasional but controlled treats that allow men to attain the perfect trained body by freeing them of excessive stress about food.[46] If we follow the Kleinian hypothesis, sweet treats, of which the writer acknowledges the irresistible appeal, materialize the schizoid attitude of infants desiring their mothers' breast and being frustrated at them when they do not get total satisfaction. The desire for limitless enjoyment cannot be met; the craved symbiosis between the child and the mother, where the infant fantasizes to get rid of all individuality, is unattainable and intrinsically impossible, because the body is always already inscribed as singular and autonomous in the cultural order, despite its inherent fragmentation.[49] Similarly, the alluring and threatening sweets, symbolizing unbridled pleasure, must be kept at a distance, submitted to the woman who knows how to deal with this kind of danger.

In the development of these dynamics leading to a loss of the individual distinction that is clearly perceived as a masculine trait, women coincide with the dimension that has been defined by Julia Kristeva as the abject, "a threat that seems to emanate from an exorbitant outside or inside, ejected beyond the scope of the possible, the tolerable, the thinkable . . . The abject has only one quality of the object—that of being opposed to I."[50] As Mary Douglas has shown in her study on purity and danger, the abject is what subverts order, codified systems, and stable identities. In this specific context, sweets endanger the whole effort to build a distinctively masculine, muscular body. They belong to a dimension both wanted and, precisely for this reason, demonized, in that it could condemn the male body to lose its frontiers that were gained with such great difficulty.

As Susan Bordo has pointed out, the body is identified "as animal, as appetite, as deceiver, as prison of the soul and confounder of its projects." It is always in opposition with the spiritual self and rationality that mirror the divine.[51] Historically, Bordo argues, women have been identified with the debasing dimension of the body, chaotic and undisciplined. Men, on the other side, are supposed to reflect the spirit. Masculinity is embodied in their control over the flesh, a metaphorical equivalent of their dominion over female unchecked carnality. The dichotomy is inscribed in the male body as a series of clear oppositions between hard and soft, thick and thin, and, of course, fit and flabby. Food is a temptation that can make man fall, unless it is stripped down to its nutritional components, purified by the intervention of scientific rationality. Within these nutrients, the main contest is between protein, the building material for good muscle, and fat, the symbol of the uncontrollable flesh. Carbohydrates are in a middle ground; they are the fuel that allows us to work out, but they can easily fall in the realm of the enemy if they are consumed in excessive quantity. The battle is fought in every man's body, and it takes strenuous efforts. Food and nutrition often play an important role in the discourse proposed in this literature, deploying scientific language to reassure readers of the effectiveness of the advice given. This rhetoric proposes a strong desire to control not only one's body appearance, but also to curb one's desires and appetites. A fit body becomes the material expression of one's dominion over the self, over the flesh and appetites that often appear as tainted by a definite feminine character. Control clearly does not imply cooking; most of the dishes proposed in those magazines require little or no preparation. Cooking food seems to constitute a threat to the reader's masculinity; men consume, they do not get involved with the chores related to food. Men's fitness magazines present themselves as scientific weapons, offering practical advice and helping readers to discern when food is a friend and when, more often, a foe.

Notes

1. See also Connell, R. W. (1995) *Masculinities*. University of California Press. Connell, R.W. (2000) *The Men and the Boys*. University of California Press. Bordo, S. *The Male Body: A New Look at Men in Public and in Private*. New York: Farrar, Straus and Giroux, 1999.
2. See the company's official website, www.rodale.com
3. Harrison, G., Pope, Katharine, A., Phillips, R. O. (2000) *The Adonis Complex*. Touchstone Books, New York.
4. Connell, R. W. (2000), *The Men and the Boys*. Polity, Cambridge, p. 12.
5. Connell, R. W. (1995) *Masculinities*. University of California Press, Berkeley CA.
6. Connell, R. W. (2000) *The Men and the Boys*. Polity, Cambridge, p. 57.
7. Bronki, M. (1998) "The Male Body in the Western Mind." *Harvard Gay and Lesbian Review*, 5(4):28.
8. Jeffords, S. (1989) *The Remasculinization of America: Gender and the Vietnam War*. Indiana University Press, Bloomington.
9. Wienke, C. (1998) Negotiating the male body: Men, masculinity, and cultural ideas. *The Journal of Men's Studies*, 6(3):255.
10. Gerschick T. and Miller, A. (1994) Gender Identities at the Crossroads of Masculinity and Physical Disability. *Masculinites*, 2:34–55.
11. Marion Nestle (2002) *Food Politics*. Berkeley: University of California Press, pp. 1–30.
12. Naomi Wolf (1991) *The Beauty Myth: How images of beauty are used against women*. New York: W. Morrow.
13. *Muscle & Fitness*, November 2002, p. 1.
14. *Muscle & Fitness*, November 2002, pp. 138–139.
15. *Muscle & Fitness*, November 2002, p. 87.
16. *Muscle & Fitness*, November 2002, pp. 58–60.
17. *Muscle & Fitness*, November 2002, pp. 20–21.
18. *Muscle & Fitness*, November 2002, p. 204.
19. *Muscle & Fitness*, November 2002, p. 17.
20. *Muscle & Fitness*, November 2002, p. 65.
21. *Muscle & Fitness*, November 2002, p. 70.
22. *Muscle & Fitness*, November 2002, p. 57.
23. *Men's Fitness*, "10 ways to leave your blubber," October 2002, pp. 46–50.
24. For the reasons behind this approach, see Marion Nestle (2002). *Food Politics*. University of California Press.
25. *Muscle & Fitness*, "Cut back on Java for a leaner, healthier diet," November 2002, p. 44.
26. *Men's Fitness*, "Expert's tip: Cut the bad stuff in half," October 2002, p. 50.
27. *Men's Fitness*, "The power of soy," October 2002, pp. 134–136.
28. *Men's Health*, "Nutrition bulletin," October 2002, p. 52.
29. *Men's Health*, "Eat this not that," October 2002, p. 72.
30. *Muscle & Fitness*, "Crash course in nutrition," October 2002, pp. 158–162.
31. *Men's Fitness*, "Quick start three week program," October 2002, p. 77.
32. *Men's Fitness*, "Going the distance," October 2002, pp. 118–121.
33. Judith Butler, *Bodies that Matter*. Routledge, New York (1993), pp. 1–23. Butler limits herself to sex, but I do think that also food can be approached in the same way. She wonders: "Given that normative heterosexuality is clearly not the only regulatory regime operative in the production of bodily contours or setting the limits of bodily intelligibility, it makes sense to ask what other regimes of regulatory production contour the materiality of bodies." p. 17.

34. "Power must be understood in the first instance as a multiplicity of force relations, as the process which, through ceaseless struggles and confrontations, transforms, strengthens or reverses them: as the support which these force relations find in one another, thus forming a chain or a system, or on the contrary, the disjunction and contradictions which isolate them from one another; and lastly, as the strategies in which they take effect, whose general design or institutional crystallization is embodied in the state apparatus, in the formulations of the law, in the various social hegemonies." Michel Foucault (1990). *The History of Sexuality Vol. I.* Vintage Books, New York, p. 93.
35. *Muscle & Fitness*, "Go fish," November 2002, pp. 168–174.
36. *Men's Fitness*, "Science of sandwich," October 2002, 96–98.
37. *Men's Health*, "A man, a can, a plan," October 2002, pp. 66.
38. *Muscle & Fitness*, "Pork slims down," November 2002, p. 68.
39. *Muscle & Fitness*, "Home-baked and healthy," November 2002, p. 206.
40. *Men's Health Guide to Women*, "How to feed a naked woman," October 2002, pp. 56–59.
41. *Men's Health Guide to Women*, "Better sex naturally," October 2002, pp. 146–151.
42. *Men's Health*, "The sex for life diet," October 2002, pp. 156–159.
43. *Men's Health Guide to Women*, "22 ways to make an impression," October 2002, pp. 4–5.
44. *Men's Health Guide to Women*, "24 rules for a successful relationship," October 2002, pp. 166–167.
45. *Men's Health Guide to Women*, "9 tricks of domestic bliss," October 2002, pp. 208–209.
46. *Men's Fitness*, "Sweet tooth?" November 2002, p. 210.
47. *The Selected Melanie Klein* (1986) Edited by Juliet Mitchell, The Free Press.
48. Bordo, S. (1993) *Unbearable weight.* University of California Press, p. 3.
49. Jacques Lacan, "Le stade du miroir comme formateur de la fonction du Je." In *Ecrits*, Editions du Seuil, Paris (1966) See also Slavoj Zizek (1989) *The Supreme Object of Ideology.* Verso, London, pp. 121–129.
50. Kristeva, J. (1982) *Powers of Horror.* Columbia University Press, New York, p. 1.
51. Bordo, S. (1993) *Unbearable Weight.* University of California Press, p. 3.

14

The Overcooked and Underdone: Masculinities in Japanese Food Programming

T. J. M. Holden

This paper explores the representation of masculinity in Japanese television cooking shows. It does this for a number of reasons: first, because television is the most consumed medium in Japan today, present in every household and viewed, on average, 3.5 hours a day; second, because food is present on nearly every channel, in some form, on and in between virtually every program, all day, every day; third, because gender representations, especially masculinity, are a major component of these communications; fourth, because the version of masculinity that is communicated is a very narrowly constructed, univocal type. Following Ito (1996), this type is shown to be "masculine hegemony," a construct with three essential elements: authority, power, and possession. Working through an array of on-screen data, I show that, regardless of the "kind" of male present in food shows, men invariably embody one or all of these masculine characteristics. Importantly, women also reinforce these qualities, either by facilitating manifestation or adopting these traits themselves. Few, if any, deviations from these depictions can be located, and when they are, they can be explained in terms of the corporate structure of television in contemporary Japan. A key observation is that, despite the prevalence of hegemonic masculinity, it is not played out through the iconic Japanese male, the salaried worker. In fact, in contrast with the pervading socio-economic reality in the "real world," this male archetype is wholly absent inside the food-show screen.*

To the Western ear, the phrase "Japanese food show" will likely conjure images of teams of smocked chefs hustling through a rangy in-studio kitchen, racing the clock, concocting ingenious ways to prepare a particular ingredient, thereby pleasing a panel of judges and defeating a crafty culinary rival. In fact, though, food battles are only one genre of food show in Japan; a genre, itself, that is widely represented on television. Moreover, battles are but one way that gender and, in this particular case, masculinities are expressed in Japanese culinary TV. Stated another way, on Japanese television, food shows are manifest—even ubiquitous—and food is a dominant means by which identity discourse transpires (Holden 2003).[1] While masculinity is but one component of identity, it is a major one. It is a discursive formation that emerges prominently at various turns in TV food shows, in multiple ways.

This article's purpose is to demonstrate the degree to which discourse about masculinity courses through Japanese food shows. So, too, does it seek to open for

* I wish to thank Takako Tsuruki for the invaluable assistance she provided throughout this research. In particular, her wise counsel, cultural and linguistic interpretations, and apt examples immeasurably improved the original paper. So, too, am I indebted to the advice of two anonymous reviewers, as well as the continued support of the editors of this special issue.

Address correspondence to T.J.M. Holden, Department of Multi-Cultural Studies, Graduate School of International Cultural Studies (GSICS), Tohoku University, Sendai 980–8576, Japan. E-mail: holden@intcul.tohoku.ac.jp consideration the communication architecture and set of codes through which masculinities are expressed. This is important for at least two reasons: first, because it has not been done before; and second, because (not unlike the false perception that *Iron Chef* is representative of the universe of Japanese food shows) prevailing assumptions about Japanese masculinity are similarly truncated. The paltry range of masculinities depicted on culinary TV must be said to play a part in that. For the most part, masculinity is a narrow, repetitive discourse; hence, the "overcooked" in this article's title. What is underdone is both ironic and intriguing. First, the salary-man—the prototypical version of Japanese masculinity—is virtually invisible; secondly, although a wide range of male characterizations (that is, fashions and mannerisms and lifestyles) may be on display, the actual range of masculinities represented on TV is close to nil. Despite the fact that a more protean set of representations concerning masculinity exist as social text out in the world beyond the screen, inside the box, these masculinities are, like the salaryman, incapable of being found.

Japanese Masculinity in the Academic Literature

What are the prevailing assumptions about Japanese masculinity? Until recently, discourse was nearly univocal, confined to the social type, "salaryman." The urban, middle class, white-collar worker has remained a relatively uncontested figure in both academic literature and public consciousness. Emblematic of the "typical" Japanese male since the 1960s (e.g., Vogel 1963; Plath 1964), this caricature persisted relatively unabated into the 1990s (e.g., Rohlen 1974; Allison 1994). Now, however, that image is beginning to change. As Roberson and Suzuki (2003:8) recently observed, the salaryman is but an idealized version of Japanese masculinity. Its wide currency may be explained because it articulates other powerful "discursive pedagogies," such as the capitalist employee, state taxpayer and family provider. The authors cite Ito (1996), who has argued that past conceptions of Japanese masculinity have been driven by views of hegemony and, in particular, three "inclinations" characterize the dominant discourse concerning masculinity.[2] These inclinations, identified as (interpersonal) authority, power, and possession (especially in relation to women), obviously align easily with conceptions of men as workers, members of power structures, protectors, and "bread-winners."

Importantly, Roberson and Suzuki assert, the salaryman is not the sole version of masculine identity in contemporary Japan. Indeed, there are numerous discourses available regarding what is "male" in contemporary Japan.[3] Such a critique is consistent with a general, quiet revisionism that has transpired in Japanese studies over the past two decades; one that has alleged greater heterogeneity in Japanese identity.[4]

The unitary image of masculine identity in the form of salaryman reflects an association of masculinity with particular institutional sites (for instance, *inside corporation* or *outside home*). This has been a standard, unreflecting, academic trope during the post-Pacific War era. It is also (coincidentally) consistent with the way Hall (1994) has theorized that identity *ought* to be decoded (i.e., within institutional contexts). For most researchers of Japanese masculine identity, those institutional contexts have centered on the state, the workplace, and the school

(Connell 1995). By inversion (i.e., reflecting a relative absence or exclusion), the institution of family (and its locus, the home) can also be included in gender-identity discourse.[5]

Perhaps in reaction to the institutional emphasis, Roberson and Suzuki's volume is rich with alternatives: civil movements, transnational information flows, transgender practices, day-laborers. Non-institutional theorization of identity is an important maneuver, but does not minimize the importance of institutions in bounding, framing, and providing meaning to contemporary identity. This is particularly true in an era of "reflexive modernization" (Robertson 1992; Beck 1994), constituted by "late modern" or "post-traditional" societies (Giddens 1994), such as Japan. In this article, in particular, it is the media institution (generally) and television (specifically) through which masculine identity is found to flow. As shown in my previous work on "mediated identity" (2003), such formal institutional sites are heavily implicated in the gender-identity calculus.[6] In a word, media (such as television) are institutions— no different than the state, corporation, or family—that provide the ideational and "physical" context within which masculinity is represented and through which it is reproduced.

In this article, I explore one genre within this institutional site of television communication in Japan: food discourse. Surveying this content, one soon learns that Japanese masculinities are both on-message and beyond-message vis-à-vis past academic framings. Consistent with what has heretofore been alleged about masculine identity in Japan, there is a widespread hegemonic masculinity on display. At the same time (and significantly), that hegemonic masculinity is *not* played out through the aegis of the corporate worker. Despite the pervasive expression of masculine identity through food talk, and despite the fact that such identity tends to be hegemonic in nature, there is nary a salaryman to be found. Japan's televisual masculinity is singularly hegemonic, yet it is not confined to a particular model or "type" of person. It is communicated through any number of people—both male *and* female (as we shall see), people who are both conventional and unconventional in appearance, job designation, or background.

Japanese Television and Food Discourse: A Précis

All of this is important because television is the preeminent medium of communication in Japan. It has a diffusion rate of 100%,[7] is viewed by virtually every Japanese person every day,[8] and outpaces other popular forms of information processing, such as newspapers (86%), cell phones (73%), and the internet (27%). It has been reported that, on average, at least one TV set plays 7 to 8 hours a day in each Japanese dwelling, with personal viewing rates per day approaching 225 minutes.[9] A recent European survey ranks Japan second worldwide in terms of daily TV viewership.[10]

While television is dominant, one might wish to argue that food is not. A conservative accounting—based on genres reported in television guides—suggests that TV food shows comprise but 5% of programming between 6 a.m. and 6 p.m. The reality, though, is quite different. Begin with the fact that, unlike other countries (in which food shows are generally confined to specialty cable channels, or else a particular hour on a particular day), Japan's food shows can be found on at least one commercial station during "golden time"[11] on multiple days of the week. In past years, there has been either a food-themed show or a show with a regular food segment every day of the week in prime time.

The best current embodiment is *"Dochi no Ryāri Syou"* (*Which?! Cooking Show*). Now in its seventh season, it is a highly rated Thursday night offering from 9 to 10 p.m. In this show, seven entertainers must choose between two dishes prepared before their eyes by rival chefs from a prestigious cooking academy. The guests are allowed to sample the food and are given a chance to change "sides" if or when their preferences for the respective dishes shift. Their decisions are often influenced by two hosts—both popular male TV fixtures, one in his late 40s, the other in his early 60s—who interview chefs, cajole the guests, and make impassioned appeals for their support. *Dochi* also serves as a window on the world, with segments on the people and places associated with one of the key ingredients in each dish: a fisherman, for instance, a dairy farmer, or cabbage grower, all toiling away in their respective remote corners of Japan.

In addition to shows that are exclusively about food, a number of golden hour variety shows feature regular segments built around food. For instance, SMAPXS-MAP—now in its sixth year and hosted by Japan's premier "boys band" (SMAP)—includes a "Bistro" segment in which an invited guest (generally a female entertainer) is welcomed into the bistro by the "owner" (generally SMAP's lead singer Masahiro Nakai), interviewed about her life and career, then eats (and judges) rather elaborate, multi-course meals prepared by competing teams (which are comprised of SMAP pairs).[12]

The popular, Thursday night variety show *Tonnerus' Minasan no Okage Deshita* (Tunnel's: Because of Everyone's Good Will) offers a weekly segment in which two guests—usually one male and one female—are invited to sit alongside one member of the (male) comedy duo, *Tonnerus*. Both guests are served four dishes, which they must consume while being interviewed about the food as well as their life histories. Free discussion and casual banter co-mingle with on-camera consumption. At the end of the show, all four participants vote as to which dish the guests consumed and pretended to enjoy, but in actuality detested.

Gotchi Battaru is a third show in which food plays an important, entertaining role. It is actually an elaborate segment of another show, the Friday evening variety hour, *Guru Guru Ninety-Nine*. Four regulars (comedians, usually, and all men) travel each week to a different top-rated (and pricey) restaurant and try to guess the price of a set of dishes prepared for them. An invited guest from the entertainment world accompanies them.[13] After all individual estimates are summed, the guest farthest from the price of the entire meal must pay for everyone. Cumulative, weekly totals are also kept and posted on the show's web site, listing how many times each regular has lost, and how far in arrears he is. Discussion during the show is balanced between good-natured ribbing of individual guestimates, information about how the food is prepared, and comments about how each dish tastes.

These are four examples of food discourse on Japanese TV, reflective of a larger pool of shows in which food plays either a primary or secondary role. Factor in the number of shows in which food appears in an ancillary role (for instance, during morning "wake-up" programs that discuss urban culinary trends or local village festivals, or else travel shows that present the foods of target destinations that can be consumed) and the percentage of food-related discourse on Japanese television increases exponentially. This description does not even begin to tap the great reservoir of "inadvertent food discourse" in which food serves as an incidental, but prominent background feature during dramas, quiz shows, newscasts, sporting events, and the like. Finally, one must not forget the ubiquitous presence of food advertising on TV, which has been found to account for as much as 20% of all ads broadcast in a one month period (Holden 2001).[14]

All considered, it is *impossible* to view food discourse as a trivial or negligible element in Japanese televisual communication. Food is present on virtually every channel, every hour, every day of the week, throughout the broadcast day.

Characteristics of Televisual Masculinities

What then, of gender, in general, and masculinity, in particular, is in these televisual culinary productions? First, these elements are neither invisible, nor insignificant. Furthermore, scrutiny of the content of food shows supports recent theorization on gender. To wit, rather than simple sets of stereotypical differences between classes tagged as "male" and "female," masculinity and femininity clearly emerge as social constructions, i.e., sets of reproduced practices and performances that mimic and support a system of power.[15] In fact, the ways in which gender identities (in general) and masculinities (in particular) are communicated in these televisual productions faithfully reflect Ito's (1996) trinity of authority, power, and possession. Similarly, there are cases in which femininities are constructed and communicated in such a way as to embody and buttress Ito's hegemonic masculinity. Let's consider concrete examples of these elements, in turn.

Power: Masculinity as Competition

To begin, let's return to *Dochi*, previously introduced, in which rival dishes are hawked by two male hosts. These front-men are combatants who do whatever they can to secure victory: interviewing the competing chefs (who are almost always men), sampling the food, cajoling the guests to join their side, and making impassioned appeals for support. At the end of the contest, one exults in victory, the other despairs in loss. Their win-loss record is updated weekly on *Dochi*'s website. At the close of each show, the victorious host holds court center stage, consuming the favored meal with the winning guests. He gloats and needles the losing host as well as those unfortunate guests who voted incorrectly. These minority members are made to observe and, sometimes, even serve the winners. *Dochi*'s discourse, in short, is one of contestation, of dominance achieved, and of subordination suffered; it operates in the vernacular of power.[16] Its conflictual, competitive discourse is one normally associated with games—not unlike the *Iron Chef* show, with its clock, rival

combatants, teams of specialists, sideline announcer, play-by-play and color commentators, and final judges. Such competitive shows adopt the rhetoric, the visual,
contextual, and practical tropes of sport, "an institution created by and for men,"
(Messner and Sabo 1990), whose practices service the reproduction of hegemonic
masculinity.[17]

Viewed in this way, shows like *Dochi* and *Iron Chef* support a sporting, contentious
masculinity. They conjure constructions of gender in terms of combat—not coincidentally performed almost exclusively by men. And lest one wonder whether this is
but an aberration, we must note that this discursive practice is not confined to one
or two shows. *Bistro SMAP*, after all, is a competition between teams. The results are
not simply points on a weekly chart, but kisses acquired from the female guests. And
even in shows where food is used to measure intellect, sophistication, and judgment
(e.g., *Gotchi Battaru*), the discursive frame centers on competition to avoid pecuniary
loss and, thus, public "face."

In keeping with the notion that gender is not simply reducible to male/female
categorizations, there are those Japanese food shows in which women battle one
another for judges' approval. When they do, these females adopt the vernacular of
(hegemonic) male discourse.[18] They are operating within an authoritative structuration of power, working against rivals for a favorable personal result. In a word,
women in the context of Japanese mediated identity are not immune from operating
in the rhetoric, manifesting the core trait of hegemonic masculinity.[19]

Authority (I): Masculinity as Executive Function

Among the categories that Goffman identified in his qualitative assessment of *Gender
Advertisements* (1976) was "executive function," the role (of elevated position, control, and authoritative action) that men adopt when paired in ads with women. This
function was patent in my own content analysis of gender in Japanese television ads
(1999); it also seems widely replicated in Japanese television food shows.

Men are executives insofar as they are accorded the lead and the power to direct.
All activity flows through them, or else beneath their commanding gaze. In food
shows, masculine guidance can take the form of two guises: host and chef.

Host   As *Dochi*'s description suggests, men often appear in the role of host. This is
not a hard and fast rule—*Emiko no O-shaberi Kukingu* (Emiko's Cooking Talk)—
features a female host. Importantly, though, Emiko defers to the chef who is a man.
As is true of all food shows, in the matter of food preparation, culinary direction, and
advice, the chef operates as chief executive. For the host, the role is clearly defined
and circumscribed: hosts greet guests, interview them about their lives, solicit their
opinions about life and food, ensure that attention is accorded to the chef's (often
backgrounded) work in the kitchen, and facilitate the flow between and balance these
various elements. Important among the latter is timekeeping and scheduling; hosts
determine when final judgments will be rendered. As guardians of continuity,
they also verbally validate results tendered by chefs or guests. In a word, they exert
administrative control over the communication event.

In cases where there are multiple hosts of differing gender, executive function adheres to a "gender order," with males invariably reigning over women. Consider the show *Chūbō Desu Yo* (This is the Kitchen). Airing at 11:30 p.m. on Saturdays, the hosts' job is not only to welcome guests and make them feel at home, but also to prepare a meal with them in an in-studio kitchen. Like *Dochi*, the guest offers judgment on the food prepared, and, like *Dochi*, that decision has the power to make the hosts exult in triumph or deflate in defeat. Unlike *Dochi*, however, the hosts are not rivals, and, importantly, unlike *Dochi*, they stand in a particular (power) relation to one another. The female host of *Chūbō* (Ikumi Kimura) is introduced at the outset of each show as an "announcer." Moreover, she wears the same green and yellow sticker on her apron that all newly minted drivers in Japan affix to their car windows—signifier of a beginner. Tellingly, Ms. Kimura has worn that sticker for over three years. By contrast, the male host (formerly a popular singer named Masaaki Sakai) introduces himself as possessor of "three stars"—the highest rating that can be awarded to a prepared dish on the show—and "Master Chef."

Gender ranking does not end there. During the course of the half hour, Sakai provides directions to Kimura in the kitchen, during the interview segments, and at the dinner table when the time comes to ask the guest for his or her evaluation of the completed meal. Often Sakai will interrupt his work in the kitchen to engage the guest in conversation, leaving Kimura to toil on her own, making sure that the preparation moves toward completion. In addition, during the critical moments of the show, when a segment has to be concluded or a result announced, it is Sakai who takes the lead. "This dish is *finished*!" he will intone after the casserole comes out of the oven. Or, as they consume the food, he will suggest, "Ms. Kimura, please ask the guest for his final evaluation." Kimura-san will then dutifully inquire, "For this dish, how many stars will you give?" Once the guest has responded, it is left to Sakai to affirm the judgment. As the camera focuses tight on his face, he shouts theatrically toward the rafters, "for this dish . . . one and a half stars!"

Chef   There are numerous shows in which the chef also adopts the executive function by instructing the host and/or guests in the ways of food preparation. *Chūbō* is emblematic of this, introducing three chefs at the outset, who perform on their own premises. Having viewed three variations on the show's selected meal, the three amateur cooks now follow one of the demonstrated recipes. At various stages of the preparation, the three loosely discuss the method they are following. In particular, though, once the meal has been completed and is being consumed, they discuss where they may have improved on the meal—what ingredient was in too little or too great supply, where the oven or stove was used for too long or too short a time. In short, the amateurs note their deviations from the chef's instructions and chastise themselves for failing to conform to his direction.

In addition, *Chūbō* features a short segment introducing a resident apprentice in one of the chef's kitchens. Almost always, these chefs-in-training are young men in their early twenties. In every case to date, these young men are depicted receiving commands from elder male employers. Here again, then, Japan's food shows cast men and masculinity in a discourse of authority.

With but two exceptions, the featured chefs in all of Japan's food show genres are men—the exceptions being the case of desserts and *katei ryōri* (home cooking).[20] When these dishes are featured, female chefs consistently appear. However, because neither category of food is widely represented on food shows, the female presence tends to be overshadowed by that of the male. As a consequence, viewers are apt to perceive "chef" as a male role and, logically, see men as culinary authorities. It is not a stretch to assert that, on the other side of the gender equation, the significance that flows out of the two areas reserved for female expertise (desserts and home cooking) communicates that women are "sweet," soft, peripheral or decentered (i.e., not associated with main courses), less sophisticated or elaborate, and also are specialists in meals served in private rather than out in the public sphere.

Authority (II): Masculinity in Profession

The executive function is not the only way in which status and authority are communicated in cooking shows. Another is the provision of expert knowledge. And like the direction of stage and culinary activity, this is another function that is performed predominantly by men. The recognition of a chef as an "expert" occurs in numerous ways in food shows.

First, and most obviously, is the invocation of the title "chef" to those who are called upon to perform in these TV productions. The deference hosts show to these culinary workers—soliciting their opinion about preparation, allowing them to explain the peculiarities and secrets of each ingredient—goes a long way toward elevating cooks to a position of authority. Clothes, too, serve as markers of professional association, and guest chefs never fail to appear in the starched white aprons and *toques* of those who cook for a living. Finally, and most importantly, is the chef's resume. In Japan, where organizational affiliation is one of the significant markers of legitimacy, food shows take pains to introduce their kitchen authorities not simply by name or age, but by pedigree. For example, they name the schools in which they have trained, the countries in which have they apprenticed, and under which banner they now wield a spatula. In a word, this discursive formation is framed institutionally, in terms of economy and social sanction.

Dochi serves as exemplar of this intellectual construction, drawing its chefs exclusively from one corporate group, *Tsuji* in Osaka—arguably Japan's most prestigious professional cooking academy. The title "*Tsuji*" (or its offshoot "*Ecole Culinaire Nationale*") flashed beneath the in-studio chef or else on the food show's web page is enough to communicate "expert." To Japanese media consumers, "*Tsuji*" connotes "rigorous training," "knowledge," "competence," "professionalism," and "quality-control."

As mentioned earlier, *Dochi* (not unlike most other food shows) calls upon its professionals to provide advice in between segments of host/guest repartee. Culinary experts explain the "dos and don'ts" associated with particular foods and tricks for preparing a meal to perfection. Hosts are careful to respond with affirmative noises, such as "Oh, I see" or "That makes sense," or even "Incredible!"—clearly stressing the presence of a knowledge hierarchy. In this way, the message is communicated that chefs are "professionals," not merely because they have a title and an impressive uniform, but because they are experts and leaders in their field.

Status in a Binary Universe: The Comparison with Women   One of the major areas of contestation in gender studies—appropriated from structuralism and ushered in large part, by Judith Butler (1990)—is the issue of language as totality, a closed system in which signs give rise, by inference, to (often invisible) paired opposites. On these terms, "man" begets "woman" as "feminine" conjures "masculine." As Hughes (2002:15) has observed, "In the male-female binary, to be a woman requires us to have a corresponding concept of man. Without this relation, the terms alone would have no reference point from which to derive their meaning."

Butler's influence—along with Foucault's (1980)[21]—was to move analysis beyond simplistic binaries. At the same time, the structure of meaning in Japanese televisual productions is predominantly dualistic, creating sign-pairs of male/chef and female/not chef. In this way, what is present, what is communicated, and what "exists," is an absence of females in the role of "chief cook," and the banishment of women from public kitchens as either professional or apprentice. All of this can produce the view that women are *not* cooking authorities, and that "chef" is a male identification, rather than a female one.

Probing this possibility further, we find that women who appear as cooks in Japanese cooking shows are often featured in one of two ways: in the primary guise as "*talento*" (entertainer), or else in the capacity as "housewife." In the case of the former, women seldom, if ever, offer culinary advice. Their cooking duties are mere props to their true identity, star, singer, sex symbol, or actress. In the case of the latter, women prepare foods and engage in activities associated with the private domain of the household.

An example of the former is found in the Sunday afternoon show, "*Iron Shufu.*"[22] A spin-off of *Iron Chef*, this variety show features female guests, all former entertainers who are now married. The show has a number of components: two rounds of quizzes (one centering on food customs, another concerning ingredients, nutrition, and calories), then a round in which kitchen skills are on display. For instance, housewives might have to run an obstacle course while flipping stir-fry in a wok, or grasp slippery *konyaku* with chopsticks. One week, there aired a task involving whipping cream, after which sticky hands and quivering fingers were made to thread three needles in succession. Following this ordeal, contestants were asked a battery of personal questions regarding life with their husbands (e.g., where was their first date, what was the first present they received from their husband, when is their wedding anniversary).

Once all these tasks are completed, the two highest scoring guests (measured in terms of fastest time through the obstacle course and most correct quiz answers) are pitted against one another in a cook-off. They are given thirty minutes to prepare a meal in the *katei* (or "home-cooking") style. Like its namesake, *Iron Chef*, one featured ingredient must be integrated into the menu. An additional stipulation (since it is *katei* style) is that one of the courses must be served with rice. A panel of celebrity judges—along with the president of a cooking school (i.e., a professional/expert)—offers evaluative comments and scores the two contestants. In numerous ways, then,

Iron Shufu embodies elements of the masculine hegemonic discourse: competition, expert evaluation, and female cooks associated with private (home-made) food. It also casts women in overtly-domesticated roles that differ in multiple, stereotypical ways from those accorded to men. In this way, patriarchal gendered discourse is reproduced.[23]

Markers of Masculine Identity   It should be observed that there *are* a few food shows in which female chefs prepare foods other than desserts or home cooking. In these cases, however, an interesting designation is attached to the cooks; an appellation that appears to undercut their status as authority. Their title is "*riyōri kenkyu ka*"—literally "food researchers." One tangible effect of this title is that it tends to soften the impression left when a woman is offering advice to a male announcer or host.

This is not so in reverse, of course. Where women are being instructed and a man is in the tutelary role, there is no shying away from affixing the title "chef" or "*sensei*" (teacher), providing his professional affiliation, and clothing him in the garb of the professional cook. A prime example is the after-hours entertainment (Saturday 12:30 a.m.), *Ai no Ēpuron 3* (The Love of Three Aprons), in which three young (generally sexy) *talento* are assigned the task of preparing a particular dish (for instance, apple pie) without the benefit of a recipe. The final product is then presented to a panel of (generally) male entertainers. The program's website explains that the "women must make the dishes for these men with love."

The bulk of the show involves the heaping of (generally critical) judgments by the male hosts upon each of the women's food productions. Thereafter, the dishes are assessed by a professional (male) chef. His comments, though generally respectful, aim at improving the women's effort next time out (with the implicit assumption that there *will* be a next time). Due to the deference paid to him by the guests and hosts, as well as his uniform and title, he comes across as an authority possessing special knowledge; his words are treated as insights beyond reproach.

Possession: Masculinity as Ownership

Punctuating and possibly stoking the go-go era of Japanese socioeconomic development were distinct epochs in which particular trinities of goods were sought. Thus, there were the three Ss of the late 1950s and early 1960s: *senpūki*, *sentakuki*, and *suihanki* (fan, washing machine, and electric rice cooker); the three Cs of the late 1960s: *kā*, *kūr ā*, and *karā terebi* (car, air conditioner, and color TV); and the three Js of the late 1970s: *jūeru*, *jetto*, and *jūtaku* (jewels, jetting, and house).[24] Aside from travel by jet, all of these items were goods to be owned. They were statuses secured through acquisition and were communicated via conspicuous display.

Of course, these trinities center on consumption; however, they also reflect a discourse of possession. It is this rhetoric that can also be spied in Japanese food productions, particularly in relation to the chefs who appear. In numerous shows, the chefs are introduced on the premises of restaurants they have founded, manage, and maintain. Cameras capture them either outside the door of their business or else inside, in the dining area. Invariably, they proudly bow in greeting and offer some remarks of invitation. Viewers are treated not only to tours of their kitchens, but are shown menus, sample the décor, drink in the ambiance, and even watch the chef as he prepares and then consumes his product.

Chūbō, previously mentioned, is noted for such excursions to the owner-chef's domain. So, too, though, are the numerous shows in which hosts travel to a particular locale (perhaps in a village off the beaten path) or else seek out a particularly special dish. In such cases, the chef becomes something more than a food preparer; he becomes host in his own right, commander of a world of his own invention, and interviewee. His status as owner lends an additional power to his countenance. He is not only executive, not only employer, not only expert, he is also landholder, proprietor, and business owner. In Japan, for historical (social class-based) reasons, these are quite powerful statuses to hold. And, of course, it goes without saying, these are roles that are almost exclusively held by men, at least in the Japanese televisual universe.

Alternative Conceptions of Televisual Masculinity

Thus far, we have explored how masculinity in TV food programming is consistent with past conceptions of Japanese masculinity; in a word, it embodies a hegemonic discourse of authority, power, and possession. Here, I wish to briefly identify two elements that suggest alternative, though not necessarily inconsistent, conceptions of gender identity.

Creation: Masculinity as Production

When Sherry Ortner offered the now-famous assertion (1974) that women are nature and men are culture, she was referring to the notion that the male world is "made"; it is a world invented, produced, rendered, and controlled. Certainly, this is the message from Japan's food shows—where the key producers are generally all male. Production transpires within an institutional context (media) and, within that context, an (generally) organizational structure. Such a structure is "man-made"; it is a humanly constructed, artificial environment, configured to confer status and facilitate the expression of power. The tools wielded and the products crafted on these shows, may or may not belong to the cooks, but the fact that they are produced in audio-visual spaces generally presided over by men and filtered through rhetorical strategies that are often regarded as "hegemonic masculinity" suggests that these productions are, in fact, male; they are possessions of the male producer world and, hence, can be associated with masculine identity.

By contrast, for women—who are so often associated with the "natural realm"— their televisual role is generally one of nurturer or consumer. As such, their job is to *facilitate* food production (as hosts) or else serve as end-users (as guests).[25] Certainly, exceptions can be located as in the case of the *Three Aprons* or the *Iron Shufu* shows, previously described. In each case, however, production is for purposes supportive of a patriarchal frame, namely, satisfying the dictates of male hosts or else proving one's wherewithal in providing an amenable home for a husband. Because competition is involved, the women in these shows subrogate themselves to and adopt the logic of hegemonic male discourse. Even when they are not governed by the male world, they seek to uphold and reproduce the logic of that world.

Freedom: Masculinity as Agency

If the message of some TV food shows is that women exist within a clearly delineated, bounded structure, the same could obviously be said of men. As previously mentioned, chefs are often depicted as members of organizations (as in the case of the *Tsuji* performers) or else (as in the case of the *Chūbō* chefs and a wealth of other shop proprietors) as proud possessors (creators, owners, executives) of structures that, incidentally, are "man-made."

At the same time, this image of attachment must be counterbalanced with the impressions of independence often communicated by Japanese media productions. As Gill (2003:145) has written, "Japanese male fantasies frequently stress the mobile: the sportsman, the traveler, the man of action, the magically endowed superhero." For men, and especially television viewers, the majority are tied to structures of "permanence and stasis" (Ibid:146) and, so, pine for an alternative model of existence—a model offered by the television shows. This is not so much embodied by the chefs who have hung out their shingle and run their own businesses; rather, it is in the aegis of the entertainers and guests who saunter onto the food show stage seemingly unencumbered and free of institutional affiliation or organizational layering.

This is a version of Japanese masculinity that is less well known—one that has few exemplars out in the free world, one that is often relegated to the realm of wish fulfillment (for instance, movies centering on the vagabond peddler "*Tora-san*," leaderless *samurai*, like the "*47 Ronin*," or meandering monks like "*Zatoichi*," or, more recently, daily news about highly-publicized "free agents" who have migrated to play baseball in America). It is a version of masculinity that, far from the quintessential salaryman, views male identity in terms of autonomy and individually-oriented existence. It is a disparate image of masculinity, one which may have little referent in reality, but is, nonetheless, persistently cropping up in televisual productions.

Alternative Masculinities

While the general argument on these pages has been that, with regard to Japan's televisual food shows, little alternative discourse circulates concerning masculinity, this is not completely the case. As we saw in the previous passage, discrepant masculinities *do* exist. And, in fact, these discrepant versions are greater—more extensive and farther reaching—than simply that of the autonomous agent just described. Here, I'd like to consider a few of these deviations, and also what that may tell us about contemporary Japanese society.

TV's Widest Angle: Masculinity's "Multiplicity"

It is not infrequent that alternative genders—transgendered men and female mascu-linities—surface on Japanese TV.[26] Food shows and food advertising, in particular, often feature performances of multiple genders.[27] Consider, for example, *Dochi*. Generally, six of the seven invitees rotate weekly,[28] often striking a numerical balance between men and women. Among the former, past episodes have included a trans-vestite, numerous *rikishi* (sumo wrestlers), retired baseball and tennis players, actors, singers, comedians, writers, and producers.

The transvestite, in particular, warrants mention here. His name is Akihiro Miwa, and he is a cultural icon. A former cabaret singer, Miwa is as famous for his elegant gowns as he is for his silky singing voice and his romantic involvement with a famous novelist, the late Yukio Mishima. A writer and TV personality, as well as a regular on variety shows, Miwa is accorded respect, with little hint of derision or disdain. The same can also be said for the homosexual twins "Píco" and "Osugi," the "new half," "Pítā," and the ubiquitous and enormously popular transvestite, Ken'ichi Mikawa. While one would be hard-pressed to claim that transgendered men are widely represented on Japanese television, it would also be impossible to deny their presence. Rarely does a day pass without the appearance of a person embodying an alternative conception of gender on mass-distributed, mass-consumed Japanese television.

Alongside these versions of masculinity are also other "models." On *Dochi* alone, one encounters an obese wrestler from Hawaii; a waif-like singer from Japan's long-est-running boy's band, SMAP; a forty-something producer in scruffy beard, blue jeans, signature cowboy boots and ten-gallon hat; a Japan-raised, blond-haired, grungy, earring-studded Canadian; an elderly actor with assiduously trimmed goatee, adorned in *yukata* (traditional male *kimono*). In short, one can hardly claim that what is broadcast is the narrow, repetitive discourse of masculinity embodied by salarymen in gray suits and conservative ties. So, too, could one hardly assert that this motley mélange of free agents fits the profile of power wielding, authoritative, possessive hegemonists—at least on the surface.

The Illusion of Freedom

It must be recognized, however, that while such "models" of masculinity may materi-alize on-screen, good reasons exist to view their social impact with caution. As guests, these men stand in an asymmetric relationship to those who manage the show, specifically, the hosts and chefs in front of the cameras. For these latter groups, invariably, action is wrapped in the vernacular of masculine hegemony. Significantly, no matter what model of masculinity hosts and chefs may *appear* to communicate via their appearance, they uniformly manage to channel food talk into discourse concerning authority, power, and possession.

It also must be noted that a disjuncture exists between the televisual and the real worlds. Food shows place a plethora of free agents on display and communicate alternative masculinities and femininities in far greater measure than the stereotypical types comprising the world beyond the screen. To wit, in Japan today, organizational work still accounts for upwards of 70% of those employed;[29] day laborers and casual or part-time workers also comprise a significant sector of workers. Nonetheless, in show after show, from the food-centered *Dochi, Chubo Desu Yo,* and *Kakurea Gohan,* to the weekly cooking segments on SMAPXSMAP, *Tonnerus'* (*Minasan no Okage Deshita*) and *Gotchi Battaru,* workers—both within and on the margins of "organizational society"—are *never* invited to sit at the TV table.

What's more, while the actors and actresses, athletes, singers, comedians, and the like *appear* to be "free agents," it is also apparent that this is mere illusion; they are far from free. Almost all of the food consuming-performers on screen belong to invisible corporate structures that book them onto these shows, not only to reap money, but more importantly, to gain further exposure for them, their popular cultural product. As such, the consumer-performers on food shows offer the illusion of independence, reproducing a myth of masculine and feminine freedom that in actually doesn't exist. In its stead stands the more hegemonic, structured model of masculinity that pervades almost all of Japanese society today.

The Absence of Vision

In the same way, although transgendered and alternative masculinities are represented on these shows, it is generally only through the aegis of a handful of prominent entertainers, the few, established well-known, accepted "others" who make the perpetual rounds in what is a finite, hermetic, televisual universe. Today, these performers who began as public curiosities rotate from show to show, appearing in a variety of genres equally distributed across the four major channels and spread throughout the seven day viewing week. The consequence is that the message that they might embody of alternative versions of masculinity stands a very real likelihood of being absorbed into, and even overshadowed by, the intimacy cultivated through repetitive exposures of star and host and encounters between viewer and performer.[30] It is this affective bond, I would aver, that may easily lead to the emotional embrace of the one or three or five alternatively masculine "regulars," without having to inculcate the ontological potentials they actually embody. The result may be that viewers become desensitized to, or even come to ignore, the performativities that these personalities signify, the various transgender potentials of "transvestite," "drag queen," "new half," or "homosexual."[31]

The Tight Focus of Televisual Masculine Identity

There is no end to food shows on Japanese television. No two are exactly identical, but all are broadcast for a purpose. To be sure, they exist to educate and entertain. Occasionally they may carry some deep unspoken or less motivated purpose, for instance, the mediation of identity. When this occurs it might be identity defined in terms of the nation, interest group, or individual (Holden 2001, 2003). Or, as shown here, it may be identity cast in terms of gender.

Televisual food shows clearly play a powerful role in communicating masculine identity in contemporary Japan.[32] Clearly, too, such shows are not amenable to the representation of all aspects of masculinity. Beyond the gender performativities previously mentioned, a number of contexts are absent from the screen in which masculinities are generally reproduced. For instance, save for the simulated kitchens in which chefs toil, workplaces are almost entirely absent. Also missing are homes, sites where parenting occurs.[33] Class also is invisible, as are men who are unemployed or else under-employed. Not surprisingly, the homeless are non-existent. In short, there is so much that bears on masculinity that televisual productions ignore, deny, or banish from public view.

The discourse that does appear in these productions serves to present, interpret, translate, and/or modify masculinities. Interestingly, as pervasive as gendered iden-tifications are, the emblematic masculinity for Japan, the salaryman, is entirely pulled from the frame. In his place are other figures—numerous tropes, codes, characters, social processes, institutions, organizational structures, and human agents—both visible and invisible, who are employed to communicate masculine identity. It is a certain kind of identity, a singular kind of identity that is consistently organized and communicated in terms of authority, power, possession, production, and—only seemingly—autonomy.

Notes

1. This work, like two precursors (Holden 1999, 2001), is based on a systematic sampling of the universe of Japanese TV shows. As explained in that earlier work, recording transpired over the course of an entire month and was supplemented with new programming as some shows were retired and others debuted. Analysis was based on the construction of three distinguishable data sets: (1) an "ideal week" of prime time food shows, (2) food-related advertising, and (3) regular programming in which "inadvertent food discourse" was regularly introduced. Especially considering the extensive amount of air-time accorded to the last category, it was con-cluded that food discourse is ubiquitous on Japanese television, playing virtually every hour, on every channel, every day.
2. Ito's concept was developed in association with his introduction to men's studies—written in Japanese. Clearly, however, the concept is not culturally bound and can be applied to other contexts.
3. Presaging this work, perhaps, was McLelland's (2000), which argues that homosexuality in Japan does not reduce to a neat, unitary discourse.
4. Among a chorus of writers Lebra and Lebra (1986), Moeur and Sugimoto (1986), Harootunian (1989), and Befu (2000) have observed that there is no homogenous Japan, comprised of a single class, gender, geography, ethnicity, occupation, or generation.
5. Iwao (1993:271), while arguing that Japan has witnessed a dramatic opening up of the public sphere (and, attendant institutional sites) for women, discusses how family has remained one institution which an earlier generation of women use to define their identity.
6. In my conceptualization, "mediated identity" is interactive and institutional, involving: (1) significations, (2) conveyed through representations of sameness and difference, (3) by media, and (4) brought into relief by: (a) references to (socially constructed) group-based traits, and (b) the depiction of relationships between: (i) individuals and/or (ii) groups. Even more recent work on cell phone users (Holden forthcoming) suggests that the above definition requires modification to allow for the power of users to communicate representations of themselves and actively construct identities by consciously utilizing media.
7. Japan: Profile of a Nation, Kodansha (1995:247). See also Kazuo Kaifu, "Japan's Broadcasting Digitization Enters the Second Stage: Its present state and prospects," NHK Culture Broadcasting Institute, No.11 (New Year, 2000). In fact, the diffusion rate as early as 1965 was 95%.
8. 95% of the population according to Shuichi Kamimura and Mieko Ida. See: "Will the Internet Take the Place of Television?: From a Public Opinion Survey on 'The Media in Daily Life,' " NHK Culture Broadcasting Institute, No. 19 (New Year, 2002); url: http://www.nhk. or.jp/bunken/bcri-news/bnls-feature.html.

9. NHK's Research Institute reported a figure of 3 hours and 44 minutes in a 2001 survey. This statistic has consistently topped three hours since 1960. Kato (1998:176), reports that viewer rates averaged three hours and eleven minutes in 1960 and three hours and twenty-six minutes in 1975.

10. Bosnia is the only country to rank ahead of Japan in terms of daily average viewing. See: " '2002'; Une Année de Télévision dans Le Monde; analyse les paysages télévisuels et les programmes préférés de 1.4 milliard de téléspectateurs dans 72 territoires audiovisuals."

11. The Japanese equivalent of "prime time," running from 8 p.m. to 10 p.m., Monday through Sunday.

12. The Bistro SMAP website can be accessed at the following url: http://www.fujitv.co.jp/b_hp/smapsmap/bistro.html. Pages such as the following (http://www.eonet.ne.jp/~smapy/SMAP-DATA.htm) feature cumulative data on guests, meals prepared, team combinations, winning teams, and number of victories amassed by each SMAP member. Awards are given to the chef (among the 5 SMAP members) who has recorded the most victories in a season.

13. Generally Japanese, although in the first season Chris Carter and Jackie Chan were invited. Carter won and Chan finished last of the four contestants.

14. The actual numbers were 681 out of 3,656 ads. Food was the second most advertised product category—behind "events" and about equal to "cars" and "sundries". This is significant insofar as Japan boasts the world's second largest advertising market, amounting to $223,250,000 just for television. This translates into 957,447 ads, consuming 6,016 broadcasting hours per year. (*Source*: Dentsu Koukoku Nenkan, '02–'03 [Dentsu Advertising Yearbook, 2002–2003], Tokyo: Dentsu, 2002; pp. 57, 90, 89 [respectively]). Advertising serves not only a major motor for consumption-based capitalist societies; it also works as one of the major means by which cultural communication occurs. Through ads, television exerts both a socializing and ideological power, narrowly and repetitiously re/producing images of gender, cultural values, history, nationalism, and political, social and personal identity (among others). I have explored these manifestations in, among other works, studies of gender advertisements (1999) and "adentity" (2000a) in Japan.

15. See, for instance, Hearn and Morgan (1990) and Broad and Kaufman (1994).

16. One might wish to have a "representative" example of masculine hegemony, in the form of one particular host. However, there are any number of "male figures" or "characters" across the TV spectrum who adopt the traits of hegemonic masculinity. The number of cases is so large, and any particular TV personality may accentuate a certain masculine trait over another, that it is best left to work through various cases in which these traits are expressed. In this way, the characteristics constituting hegemony take precedence over the particular "hegemonic representative"; so, too, does this afford us the freedom to recognize cases in which, for instance, women adopt those characteristics, thereby, serving to "enrich" and strengthen the spread of this masculine hegemonic discursive formation.

17. Indeed, Messner (1992) has argued that sport is one of the primary areas in which hegemonic masculinity is learned and perpetuated. To the degree that food shows adopt sporting rhetoric, then, they serve as such communication vehicles as well.

18. To underscore this point, consider the case of the female chef who finally prevailed on *Dochi*. Having appeared and lost a number of times, when her dessert finally won she gushed: "That's the first time I won!" Reflecting on past efforts, which resulted in her having to sit for the cameras, in a room off center stage, eating the food (by herself) that she had prepared, she explained, "When I lost, that (i.e. sitting alone eating my food) was the hardest thing . . ."

19. Importantly, it is not "female masculinity" (Halberstam 1998) that they manifest, rather a male model of masculinity.

20. Home cooked meals are those served everyday, featuring a soy sauce base and/or rice.

21. See, for instance, Fausto-Sterling (1993) and Halberstam (1998).

22. "Shufu" is the word for "housewife" in Japanese.

23. An example of what Sugimoto (1997) decries as the overarching ideology of the male-centered family in Japan; one that, in his opinion, is deeply sexist and patriarchal.

24. These triplets have, over the years, held currency in the popular culture, from marketers to journalists to everyday citizens, and, hence were widely discussed. The recitation here, however, is from Kelly (1992). In the early 1990s, just prior to the bursting of the "economic bubble," young women talked about searching for mates who possessed the "three kos (highs)": physical height, job status, and salary. In the late 1990s, after half a decade of economic downturn, these same women complained that their salarymen husbands embodied the "three Ks": kitanai, kusai, kirai (dirty, smelly, hateful). Mathews (2003:116), more optimistically, speaks of a new set of three Cs for men: bringing home a *comfortable* income, being *communicative*, and *cooperating* with childcare and housework.

25. In Tobin's (1992:10) words, "consumption is associated with the sphere of women." At the same time, Tobin is careful to observe that, while such associations may exist, they amount to critically under-assessed stereotypes. In fact, in Japan today, women also produce and men consume. Too much emphasis has been accorded to this artificial (and inaccurate) dichotomy—a point to bear in mind when applying Ortner to contemporary Japan.

26. Here I refer to the women of Takarazuka theatre in which an all-female cast plays roles both male and female. These stars have, on occasion, crossed over into TV. According to Nakamura and Matsuo (2003:59), Takarazuka is a "special type of asexual, agendered space," one allowing "both female and male fans, regardless of their sexual orientations, (to) temporarily transcend their everyday gender expectations and roles." A recent, quite different, example of female masculinity was the 2003–04 Georgia (canned coffee) TV ad campaign in which women, dressed Takarazuka-style, as salarymen, wreak havoc in their office space, tormenting their (male) boss, while dancing, rapping, taunting, and laughing with delight.

27. Here I invoke Lunsing (2003:20) who employs " 'transgender' in the broadest sense possible". In his words, "the majority of people have at least some attributes ascribed to the opposite gender and thereby can be seen to engage in transgender activity."

28. One guest holds near-permanent status: Tsuyoshi Kusanagi, singer for the J-pop mega-group, SMAP. Kusanagi's appeal may lie in his asexual, if not effeminate, countenance. As a character with a blurry gender identity, he appears to comfortably rest between the weekly groups of three males and three females. It is not uncommon to see Kusanagi among the winning side in a 4 to 3 split, often in cases when the food choice cleaves panelists along traditional gender lines.

29. Precise statistical confirmation can be illusory. One such study—from a Marxist/worker's perspective (Voice of Electricity Workers) can be found at the following url: http://www.eefi.org/0702/070215.htm. Its data is culled from numerous official and unofficial studies in the mid-1990s which, combined, work the figure toward the 70% threshold. More recent reports from the Ministry of Health, Labour and Welfare (1999) speak to dramatic decline in the number of employees in blue collar jobs (due to economic downturn), as well as a glut of professional and technical workers.

30. English-language studies of Japanese television are rather sparse. One of the best, by Painter (1996), argues that TV productions seek to produce "quasi-intimacy" by "emphasiz(ing) themes related to unity (national, local, cultural, or racial) and unanimity (consensus, common sense, identity) . . ." (198). I would go further. Theorization I am currently completing suggests that affective bonds are formed among a national community via the hermetic circulatory process I described earlier. The result, in the first instance, is a fusing of affective ties between viewer, performer, host, production team, and TV station. In the second instance, it is the dislodging and transfer of such bonds (as the performers and viewers jump from program to program and station to station, day after day). In the final instance, it is the creation of a sort of seamless, floating family locked in an on-going communal conversation. Although Painter doesn't suggest as much, it would seem that the intimacy he first described today serves increasingly as a stepping stone in the forging of solidarity among TV's national viewers; a solidarity that in contemporary Japan, exerts greater unifying power than that of corporate, legal, religious or ritualistic activities and formations.

31. On these categories—"the various tropes constructing transgender" in contemporary Japan—see Lunsing (2003).

32. Female identity, to be sure, is just as rich an area of inquiry. Femininities are more variegated than masculinities. There are so many "types" of women on-screen; at the same time, ultimately, the range of female motion is less extensive. For, as we have seen in this research (at least when it comes to food shows) so much of feminine discourse is subrogated to expressions of masculine hegemony. Women end up operating within those terms, either employing such tools of power, themselves or deferring to them.

33. In fact, men are rarely, if ever, depicted cooking in homes. Male culinary acts are: always for show (i.e., in the studio); associated with work (i.e., in the role of chef); or else at play (i.e., during seasonal picnics, generally in commercials).

References

Allison, A. (1994) *Nightwork: Sexuality, Pleasure and Corporate Masculinity in a Tokyo Hostess Club.* Chicago: University of Chicago Press.

Baudrillard, J. (1994) (1981) *Simulacra and Simulation*, S.F. Glaser (ed.). Ann Arbor: The University of Michigan Press.

Beck, U. (1994) "The Reinvention of Politics: Towards a Theory of Reflexive Modernization." In U. Beck, A. Giddens and S. Lash (eds.). *Reflexive Modernization: Politics, Tradition and Aesthetics in the Modern Social Order*, Oxford: Polity Press, pp. 1–55.

Befu, H. (2001) *Hegemony of Heterogeneity: An Anthropological Analysis of* Nihonjinron. Melbourne: Trans Pacific Press.

Brod, H., and Kaufman, M. (eds.) (1994) *Theorizing Masculinities*, Thousand Oaks: Sage Publications.

Butler, J. (1990) *Gender Trouble: Feminism and the Subversion of Identity.* London and New York: Routledge.

Clammer, J. (2000) "Received Dreams: Consumer Capitalism, Social Process, and the management of the emotions in contemporary Japan." In J.S. Eades, T. Gill, and H. Befu (eds.). *Globalization and Social Change in Contemporary Japan*. Melbourne: Trans Pacific Press. pp. 203–223.

Connell, R. W. (1995) *Masculinities*. Berkeley: University of California Press.

Fausto-Sterling, A. (1993) "The Five Sexes: Why Male and Female are not Enough." *The Sciences* (March–April): pp. 20–25.

Foucault, M. (1980) *Power/Knowledge: Selected Interviews and Other Writings, 1972–77*. C. Gordeon, (ed.) Brighton: Harvester.

Giddens, A. (1994) "Living in a Post-Traditional Society," in U. Beck, A. Giddens and S. Lash (eds.). *Reflexive Modernization: Politics, Tradition and Aesthetics in the Modern Social Order*. Oxford: Polity Press, pp. 56–109.

Gill, T. (2003) "When Pillars Evaporate: Structuring Masculinity on the Japanese Margins." In J.E. Roberson and N. Suzuki (eds.). *Men and Masculinities in Contemporary Japan: Dislocating the Salaryman Doxa*. London and New York: RoutledgeCurzon, pp. 144–161.

Goffman, E. (1976) *Gender Advertisements*. New York: Harper and Row Publishers.

Halberstam, J. (1988) *Female Masculinity*. Durham, N.C. and London: Duke University Press.

Hall, S. (1994) "Introduction: Who Needs 'Identity'?" In S. Hall and P. du Gay (eds.). *Questions of Cultural Identity*. London and Thousand Oaks: Sage Publishing, pp. 1–17.

—— (1980) "Encoding/Decoding in Television Discourse." In S. During (ed.). *The Cultural Studies Reader*. London and New York: Routledge, pp. 90–103.

Harootunian, H. D. (1989) "Visible Discourses/Invisible Ideologies." In M. Miyoshi and H. D. Harootunian (eds.). *Postmodernism and Japan*. Durham and London: Duke University Press.

Hearn, J., and Morgan, D. (eds.) (1990) *Men, Masculinities and Social Theory*. London: Unwin Hyman.

Holden, T.J.M. (1999) " 'And Now for the Main (Dis)course . . .' or: Food as Entree in Contemporary Japanese Television." *M/C: A Journal of Media and Culture*. Volume 2, Issue 7 (October 20).

—— (2000a) "*Adentity*: Images of Self in Japanese Television Advertising." *The International Scope Review*. Volume 2, Issue 4 (Winter).

—— (2000b) " 'I'm Your Venus'/You're a Rake: Gender and the Grand Narrative in Japanese Television Advertising." *Intersections: Gender, History and Culture in the Asian Context*. Issue 3 (February).

—— (2001) "Food on Japanese Television: The Entree and the Discourse." Presented at the 3rd Annual Meeting of *Anthropology of Japan in Japan*, National Museum of Ethnology, Osaka, Japan. May 12th and 13th.

—— (2003) "Japan's 'Mediated' Global Identities." In T.J. Scrase, T.J.M. Holden, and S. Baum (eds.). *Globalization, Culture and Inequality in Asia*. Melbourne: Trans Pacific Press, pp. 144–167.

—— Forthcoming. "The Social Life of Japan's Adolphenic." In P. Nilan and C. Feixa (eds.). *Global Youth? Hybrid Identities: Plural Worlds*. London: Routledge.

Hughes, C. (2002) *Key Concepts in Feminist Theory and Research*. London and Thousand Oaks: Sage Publications.

Ito, K. (1996) *Danseigaku Nyumon* (Introduction to Men's Studies). Tokyo: Sakuhinsha.

Iwao, S. (1993) *The Japanese Woman: Traditional Image and Changing Reality*. New York: The Free Press.

Kelly, W.W. (1992). "Tractors, Television and Telephones: Reach out and Touch Someone in Rural Japan." In J.J. Tobin (ed.). *Re-Made in Japan: Everyday Life and Consumer Taste in a Changing Society*. New Haven and London: Yale University Press, pp. 77–88.

Lebra, T.S., and Lebra, W.P. (1986) (1974) *Japanese Culture and Behavior: Selected Readings*. Honolulu: University of Hawaii Press.

Lunsing, W. (2003) "What Masculinity?: Transgender Practices Among Japanese 'Men.' " In J.E. Roberson and N. Suzuki (eds.). *Men and Masculinities in Contemporary Japan: Dislocating the Salaryman Doxa*. London and New York: RoutledgeCurzon, pp. 20–36.

Mathews, G. (2003) "Can a 'Real Man' Live for his Family?: *Ikigai* and Masculinity in Today's Japan." In J.E. Roberson and N. Suzuki (eds.). *Men and Masculinities in Contemporary Japan: Dislocating the Salaryman Doxa*. London and New York: RoutledgeCurzon, pp. 109–125.

McLelland, M. (2000) *Male Homosexuality in Modern Japan: Cultural Myths and Social Realities*. Richmond, Surrey: Curzon.

Messner, M. (1992) *Power at Play: Sports and the Problem of Masculinity*. Boston: Beacon Press.

Messner, M., and Sabo, D. (1990) "Introduction: Toward a Critical Feminist Reappraisal of Sport, Men and the Gender Order." In M. Messner and D. Sabo (eds.). *Sport, Men and the Gender Order: Critical Feminist Perspectives*. Champagne, Il: Human Kinetics, pp. 1–15.

Moeur, R., and Sugimoto, Y. (1986) *Images of Japanese Society*. London and New York: Kegan Paul International.

Nakamura, K., and Matsuo, H. (2003) "Female Masculinity and Fantasy Spaces: Transcending Genders in the Takarazuka Theatre and Japanese Popular Culture." In J.E. Roberson and N. Suzuki (eds.). *Men and Masculinities in Contemporary Japan: Dislocating the Salaryman Doxa*. London and New York: RoutledgeCurzon, pp. 59–76.

Ortner, S. (1974) "Is Female to Male as Nature is to Culture?" In *Woman, Culture, and Society*. Michelle Zimbalist Rosaldo and Louise Lamphere (eds.). Stanford University Press.

Painter, A.A. (1996) "Japanese Popular Daytime Television, Popular Culture, and Ideology," In J.W. Treat (ed.). *Contemporary Japan and Popular Culture*. Surrey: Curzon Press.

Plath, D. (1964) *The After Hours: Modern Japan and the Search for Enjoyment*. Berkeley: University of California Press.

Roberson, J.E., and N. Suzuki (eds.) (2003). *Men and Masculinities in Contemporary Japan: Dislocating the Salaryman Doxa*. London and New York: RoutledgeCurzon.

Robertson, R. (1992) *Globalization: Social Theory and Global Culture*. London and Thousand Oaks: Sage Publications.

Rohlen, T.P. (1974) *For Harmony and Strength: Japanese White-Collar Organization in Anthropological Perspective*. Berkeley: University of California Press.

Sugimoto, Y. (1997) *An Introduction to Japanese Society*. Cambridge: Cambridge University Press.

Tobin, J.J. (ed.) (1992) *Re-made in Japan: Everyday Life and Consumer Taste in a Changing Society*. New Haven and London: Yale University Press.

Vogel, E.F. (1963) *Japan's New Middle Class: The Salary Man and His Family in a Tokyo Suburb*. Berkeley: University of California Press.

15

Japanese Mothers and *Obentōs*: The Lunch-Box as Ideological State Apparatus

Anne Allison

Obentōs are boxed lunches Japanese mothers make for their nursery school children. Follow-ing Japanese codes for food preparation—multiple courses that are aesthetically arranged—these lunches have a cultural order and meaning. Using the *obentō* as a school ritual and chore—it must be consumed in its entirety in the company of all the children—the nursery school also endows the *obentō* with ideological meanings. The child must eat the *obentō*, the mother must make an *obentō* the child will eat. Both mother and child are being judged; the subjectivities of both are being guided by the nursery school as an institution. It is up to the mother to make the ideological operation entrusted to the *obentō* by the state-linked institution of the nursery school, palatable and pleasant for her child, and appealing and pleasurable for her as a mother.

Introduction

Japanese nursery school children, going off to school for the first time, carry with them a boxed lunch *(obentō)* prepared by their mothers at home. Customarily these *obentōs* are highly crafted elaborations of food: a multitude of miniature portions, artistically designed and precisely arranged, in a container that is sturdy and cute. Mothers tend to expend inordinate time and attention on these *obentōs* in efforts both to please their children and to affirm that they are good mothers. Children at nursery school are taught in turn that they must consume their entire meal according to school rituals.

Food in an *obentō* is an everyday practice of Japanese life. While its adoption at the nursery school level may seem only natural to Japanese and unremarkable to out-siders, I will argue in this article that the *obentō* is invested with a gendered state ideology. Overseen by the authorities of the nursery school, an institution which is linked to, if not directly monitored by, the state, the practice of the *obentō* situates the producer as a woman and mother, and the consumer as a child of a mother and a student of a school. Food in this context is neither casual nor arbitrary. Eaten quickly in its entirety by the student, the *obentō* must be fashioned by the mother so as to expedite this chore for the child. Both mother and child are being watched, judged, and constructed; and it is only through their joint effort that the goal can be accomplished.

I use Althusser's concept of the Ideological State Apparatus (1971) to frame my argument. I will briefly describe how food is coded as a cultural and aesthetic apparatus in Japan, and what authority the state holds over schools in Japanese society. Thus situating the parameters within which the *obentō* is regulated and structured in the nursery school setting, I will examine the practice both of making and eating *obentō* within the context of one nursery school in Tokyo. As an anthropologist and mother of a child who attended this school for fifteen months, my analysis is based on my observations, on discussions with other mothers, daily conversations and an interview with my son's teacher, examination of *obentō* magazines and cookbooks, participation in school rituals, outings, and Mothers' Association meetings, and the multifarious experiences of my son and myself as we faced the *obentō* process every day.

I conclude that *obentō* as a routine, task, and art form of nursery school culture are endowed with ideological and gendered meanings that the state indirectly manipulates. The manipulation is neither total nor totally coercive, however, and I argue that pleasure and creativity for both mother and child are also products of the *obentō*.

Cultural Ritual and State Ideology

As anthropologists have long understood, not only are the worlds we inhabit symbolically constructed, but also the constructions of our cultural symbols are endowed with, or have the potential for, power. How we see reality, in other words, is also how we live it. So the conventions by which we recognize our universe are also those by which each of us assumes our place and behavior within that universe. Culture is, in this sense, doubly constructive: constructing both the world for people and people for specific worlds.

The fact that culture is not necessarily innocent, and power not necessarily transparent, has been revealed by much theoretical work conducted both inside and outside the discipline of anthropology. The scholarship of the neo-Marxist Louis Althusser (1971), for example, has encouraged the conceptualization of power as a force which operates in ways that are subtle, disguised, and accepted as everyday social practice. Althusser differentiated between two major structures of power in modern capitalist societies. The first, he called (Repressive) State Apparatus (SA), which is power that the state wields and manages primarily through the threat of force. Here the state sanctions the usage of power and repression through such legitimized mechanisms as the law and police (1971: 143–5).

Contrasted with this is a second structure of power—Ideological State Apparatus(es) (ISA). These are institutions which have some overt function other than a political and/or administrative one: mass media, education, health and welfare, for example. More numerous, disparate, and functionally polymorphous than the SA, the ISA exert power not primarily through repression but through ideology. Designed and accepted as practices with another purpose—to educate (the school system), entertain (film industry), inform (news media), the ISA serve not only their stated objective but also an unstated one—that of indoctrinating people into seeing the world a certain way and of accepting certain identities as their own within that world (1971: 143–7).

While both structures of power operate simultaneously and complementarily, it is the ISA, according to Althusser, which in capitalist societies is the more influential of the two. Disguised and screened by another operation, the power of ideology in ISA can be both more far-reaching and insidious than the SA's power of coercion. Hidden in the movies we watch, the music we hear, the liquor we drink, the textbooks we read, it is overlooked because it is protected and its protection—or its alibi (Barthes 1957: 109–111)—allows the terms and relations of ideology to spill into and infiltrate our everyday lives.

A world of commodities, gender inequalities, and power differentials is seen not therefore in these terms but as a naturalized environment, one that makes sense because it has become our experience to live it and accept it in precisely this way. This commonsense acceptance of a particular world is the work of ideology, and it works by concealing the coercive and repressive elements of our everyday routines but also by making these routines of everyday familiar, desirable, and simply our own. This is the critical element of Althusser's notion of ideological power: ideology is so potent because it becomes not only ours but us—the terms and machinery by which we structure ourselves and identify who we are.

Japanese Food as Cultural Myth

An author in one *obentō* magazine, the type of medium-sized publication that, filled with glossy pictures of *obentōs* and ideas and recipes for successfully recreating them, sells in the bookstores across Japan, declares, "the making of the *obentō* is the one most worrisome concern facing the mother of a child going off to school for the first time" (*Shufunotomo* 1980: inside cover). Another *obentō* journal, this one heftier and packaged in the encyclopedic series of the prolific women's publishing firm, *Shufunotomo*, articulates the same social fact: "first-time *obentōs* are a strain on both parent and child" (*"hajimete no obentō wa, oya mo ko mo kinchoshimasu"*) (*Shufunotomo* 1981: 55).

An outside observer might ask: What is the real source of worry over *obentō*? Is it the food itself or the entrance of the young child into school for the first time? Yet, as one look at a typical child's *obentō*—a small box packaged with a five- or six-course miniaturized meal whose pieces and parts are artistically arranged, perfectly cut, and neatly arranged—would immediately reveal, no food is "just" food in Japan. What is not so immediately apparent, however, is why a small child with limited appetite and perhaps scant interest in food is the recipient of a meal as elaborate and as elaborately prepared as any made for an entire family or invited guests?

Certainly, in Japan much attention is focused on the *obentō*, investing it with a significance far beyond that of the merely pragmatic, functional one of sustaining a child with nutritional foodstuffs. Since this investment beyond the pragmatic is true of any food prepared in Japan, it is helpful to examine culinary codes for food preparation that operate generally in the society before focusing on children's *obentōs*.

As has been remarked often about Japanese food, the key element is appearance. Food must be organized, reorganized, arranged, rearranged, stylized, and restylized to appear in a design that is visually attractive. Presentation is critical: not to the extent that taste and nutrition are displaced, as has been sometimes attributed to Japanese food, but to the degree that how food looks is at least as important as how it tastes and how good and sustaining it is for one's body.

As Donald Richie has pointed out in his eloquent and informative book *A Taste of Japan* (1985), presentational style is the guiding principle by which food is prepared in Japan, and the style is conditioned by a number of codes. One code is for smallness, separation, and fragmentation. Nothing large is allowed, so portions are all cut to be bite-sized, served in small amounts on tiny individual dishes, and arranged on a table (or on a tray, or in an *obentō* box) in an array of small, separate containers.[1] There is no one big dinner plate with three large portions of vegetable, starch, and meat as in American cuisine. Consequently the eye is pulled not toward one totalizing center but away to a multiplicity of de-centered parts.[2]

Visually, food substances are presented according to a structural principle not only of segmentation but also of opposition. Foods are broken or cut to make contrasts of color, texture, and shape. Foods are meant to oppose one another and clash; pink against green, roundish foods against angular ones, smooth substances next to rough ones. This oppositional code operates not only within and between the foodstuffs themselves, but also between the attributes of the food and those of the containers in or on which they are placed: a circular mound in a square dish, a bland-colored food set against a bright plate, a translucent sweet in a heavily textured bowl (Richie 1985: 40–41).

The container is as important as what is contained in Japanese cuisine, but it is really the containment that is stressed, that is, how food has been (re)constructed and (re)arranged from nature to appear, in both beauty and freshness, perfectly natural. This stylizing of nature is a third code by which presentation is directed; the injunction is not only to retain, as much as possible, the innate naturalness of ingredients—shopping daily so food is fresh and leaving much of it either raw or only minimally cooked—but also to recreate in prepared food the promise and appearance of being "natural." As Richie writes, "the emphasis is on presentation of the natural rather than the natural itself. It is not what nature has wrought that excites admiration but what man has wrought with what nature has wrought" (1985: 11).

This naturalization of food is rendered through two main devices. One is by constantly hinting at and appropriating the nature that comes from outside—decorating food with season reminders, such as a maple leaf in the fall or a flower in the spring, serving in-season fruits and vegetables, and using season-coordinated dishes such as glass-ware in the summer and heavy pottery in the winter. The other device, to some degree the inverse of the first, is to accentuate and perfect the preparation process to such an extent that the food appears not only to be natural, but more nearly perfect than nature without human intervention ever could be. This is nature made artificial. Thus, by naturalization, nature is not only taken in by Japanese cuisine, but taken over.

It is this ability both to appropriate "real" nature (the maple leaf on the tray) and to stamp the human reconstruction of that nature as "natural" that lends Japanese food its potential for cultural and ideological manipulation. It is what Barthes calls a second-order myth (1957:114–17): a language that has a function people accept as only pragmatic—the sending of roses to lovers, the consumption of wine with one's dinner, the cleaning up a mother does for her child—which is taken over by some interest or agenda to serve a different end—florists who can sell roses, liquor companies that can market wine, conservative politicians who campaign for a gendered division of labor with women kept at home. The first order of language ("language-object"), thus emptied of its original meaning, is converted into an empty form by which it can assume a new, additional, second order of signification ("metalanguage" or "second-order semiological system"). As Barthes points out, however, the primary meaning is never lost. Rather, it remains and stands as an alibi, the cover under which the second, politicized meaning can hide. Roses sell better, for example, when lovers view them as a vehicle to express love rather than the means by which a company stays in business.

At one level, food is just food in Japan—the medium by which humans sustain their nature and health. Yet under and through this code of pragmatics, Japanese cuisine carries other meanings that in Barthes' terms are mythological. One of these is national identity: food being appropriated as a sign of the culture. To be Japanese is to eat Japanese food, as so many Japanese confirm when they travel to other countries and cite the greatest problem they encounter to be the absence of "real" Japanese food. Stated the other way around, rice is so symbolically central to Japanese culture (meals and *obentōs* often being assembled with rice as the core and all other dishes, multifarious as they may be, as mere compliments or side dishes) that Japanese say they can never feel full until they have consumed their rice at a particular meal or at least once during the day.[3]

Embedded within this insistence on eating Japanese food, thereby reconfirming one as a member of the culture, are the principles by which Japanese food is customarily prepared: perfection, labor, small distinguishable parts, opposing segments, beauty, and the stamp of nature. Overarching all these more detailed codings are two that guide the making and ideological appropriation of the nursery school *obentō* most directly: 1) there is an order to the food: a right way to do things, with everything in its place and each place coordinated with every other, and 2) the one who prepares the food takes on the responsibility of producing food to the standards of perfection and exactness that Japanese cuisine demands. Food may not be casual, in other words, nor the producer casual in her production. In these two rules is a message both about social order and the role gender plays in sustaining and nourishing that order.

School, State, and Subjectivity

In addition to language and second-order meanings I suggest that the rituals and routines surrounding *obentōs* in Japanese nursery schools present, as it were, a third order, manipulation. This order is a use of a currency already established—one that has already appropriated a language of utility (food feeds hunger) to express and implant cultural behaviors. State-guided schools borrow this coded apparatus: using the natural convenience and cover of food not only to code a cultural order, but also to socialize children and mothers into the gendered roles and subjectivities they are expected to assume in a political order desired and directed by the state.

In modern capitalist societies such as Japan, it is the school, according to Althusser, which assumes the primary role of ideological state apparatus. A greater segment of the population spends longer hours and more years here than in previous historical periods. Also education has now taken over from other institutions, such as religion, the pedagogical function of being the major shaper and inculcator of knowledge for the society. Concurrently, as Althusser has pointed out for capitalist modernism (1971: 152, 156), there is the gradual replacement of repression by ideology as the prime mechanism for behavior enforcement. Influenced less by the threat of force and more by the devices that present and inform us of the world we live in and the subjectivities that world demands, knowledge and ideology become fused, and education emerges as the apparatus for pedagogical and ideological indoctrination.

In practice, as school teaches children how and what to think, it also shapes them for the roles and positions they will later assume as adult members of the society. How the social order is organized through vectors of gender, power, labor, and/or class, in other words, is not only as important a lesson as the basics of reading and writing, but is transmitted through and embedded in those classroom lessons. Knowledge thus is not only socially constructed, but also differentially acquired according to who one is or will be in the political society one will enter in later years. What precisely society requires in the way of workers, citizens, and parents will be the condition determining or influencing instruction in the schools.

This latter equation, of course, depends on two factors: 1) the convergence or divergence of different interests in what is desired as subjectivities, and 2) the power any particular interest, including that of the state, has in exerting its desires for subjects on or through the system of education. In the case of Japan, the state wields enormous control over the systematization of education. Through its Ministry of Education (Monbusho), one of the most powerful and influential ministries in the government, education is centralized and managed by a state bureaucracy that regulates almost every aspect of the educational process. On any given day, for example, what is taught in every public school follows the same curriculum, adheres to the same structure, and is informed by textbooks from the prescribed list. Teachers are nationally screened, school boards uniformly appointed (rather than elected), and students institutionally exhorted to obey teachers given their legal authority, for example, to write secret reports (*naishinsho*) that may obstruct a student's entrance into high school.[4]

The role of the state in Japanese education is not limited, however, to such extensive but codified authorities granted to the Ministry of Education. Even more powerful is the principle of the "*gakureki shakkai*" (lit., academic pedigree society), by which careers of adults are determined by the schools they attend as youth. A reflection and construction of the new economic order of post-war Japan,[5] school attendance has become the single most important determinant of who will achieve the most desirable positions in industry, government, and the professions. School attendance itself is based on a single criterion: a system of entrance exams which determines entrance selection, and it is to this end—preparation for exams—that school, even at the nursery-school level, is increasingly oriented. Learning to follow directions, do as one is told, and "*ganbaru*" (Asanuma 1987) are social imperatives, sanctioned by the state, and taught in the schools.

Nursery School and Ideological Appropriation of the *Obentō*

The nursery school stands outside the structure of compulsory education in Japan. Most nursery schools are private; and, though not compelled by the state, a greater proportion of the three- to six-year-old population of Japan attends pre-school than in any other industrialized nation (Tobin 1989; Hendry 1986; Boocock 1989).

Differentiated from the *hoikuen,* another pre-school institution with longer hours which is more like daycare than school,[6] the *yochien* (nursery school) is widely perceived as instructional, not necessarily in a formal curriculum but more in indoctrination to attitudes and structure of Japanese schooling. Children learn less about reading and writing than they do about how to become a Japanese student, and both parts of this formula—Japanese and student—are equally stressed. As Rohlen has written, "social order is generated" in the nursery school, first and foremost, by a system of routines (1989: 10, 21). Educational routines and rituals are therefore of heightened importance in *yochien,* for whereas these routines and rituals may be the format through which subjects are taught in higher grades, they are both form and subject in the *yochien.*

While the state (through its agency, the Ministry of Education) has no direct mandate over nursery-school attendance, its influence is nevertheless significant. First, authority over how the *yochien* is run is in the hands of the Ministry of Education. Second, most parents and teachers see the *yochien* as the first step to the system of compulsory education that starts in the first grade and is closely controlled by Monbusho. The principal of the *yochien* my son attended, for example, stated that he saw his main duty to be preparing children to enter more easily the rigors of public education soon to come. Third, the rules and patterns of "group living" (*shudanseikatsu*), a Japanese social ideal that is reiterated nationwide by political leaders, corporate management, and marriage counselors, is first introduced to the child in nursery school.[7]

The entry into nursery school marks a transition both away from home and into the "real world," which is generally judged to be difficult, even traumatic, for the Japanese child (Peak 1989). The *obentō* is intended to ease a child's discomfiture and to allow a child's mother to manufacture something of herself and the home to accompany the child as s/he moves into the potentially threatening outside world. Japanese use the cultural categories of *soto* and *uchi; soto* connotes the outside, which in being distanced and other, is dirty and hostile; and *uchi* identifies as clean and comfortable what is inside and familiar. The school falls initially and, to some degree, perpetually, into a category of *soto.* What is ultimately the definition and location of *uchi,* by contrast, is the home, where family and mother reside.[8] By producing something from the home, a mother both girds and goads her child to face what is inevitable in the world that lies beyond. This is the mother's role and her gift; by giving of herself and the home (which she both symbolically represents and in reality manages[9]), the *soto* of the school is, if not transformed into the *uchi* of the home, made more bearable by this sign of domestic and maternal hearth a child can bring to it.

The *obentō* is filled with the meaning of mother and home in a number of ways. The first is by sheer labor. Women spend what seems to be an inordinate amount of time on the production of this one item. As an experienced *obentō* maker, I can attest to the intense attention and energy devoted to this one chore. On the average, mothers spend 20–45 minutes every morning cooking, preparing, and assembling the contents of one *obentō* for one nursery school-aged child. In addition, the previous day they have planned, shopped, and often organized a supper meal with left-overs in mind for the next day's *obentō*. Frequently women[10] discuss *obentō* ideas with other mothers, scan *obentō* cook-books or magazines for recipes, buy or make objects with which to decorate or contain (part of) the *obentō*, and perhaps make small food portions to freeze and retrieve for future *obentōs*.[11]

Of course, effort alone does not necessarily produce a successful *obentō*. Casualness was never indulged, I observed, and even mothers with children who would eat anything prepared *obentōs* as elaborate as anyone else's. Such labor is intended for the child but also the mother: it is a sign of a woman's commitment as a mother and her inspiring her child to being similarly committed as a student. The *obentō* is thus a representation of what the mother is and what the child should become. A model for school is added to what is gift and reminder from home.

This equation is spelled out more precisely in a nursery school rule—all of the *obentō* must be eaten. Though on the face of it this is petty and mundane, the injunction is taken very seriously by nursery school teachers and is one not easily realized by very small children. The logic is that it is time for the child to meet certain expectations. One of the main agendas of the nursery school, after all, is to introduce and indoctrinate children into the patterns and rigors of Japanese education (Rohlen 1989; Sano 1989; Lewis 1989). And Japanese education, by all accounts, is not about fun (Duke 1986).

Learning is hard work with few choices or pleasures. Even *obentōs* from home stop once the child enters first grade.[12] The meals there are institutional: largely bland, unappealing, and prepared with only nutrition in mind. To ease a youngster into these upcoming (educational, social, disciplinary, culinary) routines, *yochien obentōs* are designed to be pleasing and personal. The *obentō* is also designed, however, as a test for the child. And the double meaning is not unintentional. A structure already filled with a signification of mother and home is then emptied to provide a new form: one now also written with the ideological demands of being a member of Japanese culture as well as a viable and successful Japanese in the realms of school and later work.

The exhortation to consume one's entire *obentō*[13] is articulated and enforced by the nursery school teacher. Making high drama out of eating by, for example, singing a song; collectively thanking Buddha (in the case of Buddhist nursery schools), one's mother for making the *obentō*, and one's father for providing the means to make the *obentō*; having two assigned class helpers pour the tea, the class eats together until everyone has finished. The teacher examines the children's *obentōs*, making sure the food is all consumed, and encouraging, sometimes scolding, children who are taking too long. Slow eaters do not fare well in this ritual, because they hold up the other students, who as a peer group also monitor a child's eating. My son often complained about a child whose slowness over food meant that the others were kept inside (rather than being allowed to play on the playground) for much of the lunch period.

Ultimately and officially, it is the teacher, however, whose role and authority it is to watch over food consumption and to judge the person consuming food. Her surveillance covers both the student and the mother, who in the matter of the *obentō* must work together. The child's job is to eat the food and the mother's to prepare it. Hence, the responsibility and execution of one's task is not only shared but conditioned by the other. My son's teacher would talk with me daily about the progress he was making finishing his *obentōs*. Although the overt subject of discussion was my child, most of what was said was directed to me: what I could do in order to get David to consume his lunch more easily.

The intensity of these talks struck me at the time as curious. We had just settled in Japan and David, a highly verbal child, was attending a foreign school in a foreign language he had not yet mastered; he was the only non-Japanese child in the school. Many of his behaviors during this time were disruptive: for example, he went up and down the line of children during morning exercises hitting each child on the head. Hamadasensei (the teacher), however, chose to discuss the *obentōs*. I thought surely David's survival in and adjustment to this environment depended much more on other factors, such as learning Japanese. Yet it was the *obentō* that was discussed with such recall of detail ("David ate all his peas today, but not a single carrot until I asked him to do so three times") and seriousness that I assumed her attention was being misplaced. The manifest reference was to boxed lunches, but was not the latent reference to something else?[14]

Of course, there was another message for me and my child. It was an injunction to follow directions, obey rules, and accept the authority of the school system. All of the latter were embedded in and inculcated through certain rituals: the nursery school, as any school (except such nonconventional ones as Waldorf and Montessori) and practically any social or institutional practice in Japan, was so heavily ritualized and ritualistic that the very form of ritual took on a meaning and value in and of itself (Rohlen 1989: 21, 27–28). Both the school day and the school year of the nursery school were organized by these rituals. The day, apart from two free periods, for example, was broken by discrete routines—morning exercises, arts and crafts, gym instruction, singing—most of which were named and scheduled. The school year was also segmented into and marked by three annual events—sports day (*undokai*) in the fall, winter assembly (*seikatsu happyokai*) in December, and dance festival (*bon odori*) in the summer. Energy was galvanized by these rituals, which demanded a degree of order as well as a discipline and self-control that non-Japanese would find remarkable.

Significantly, David's teacher marked his successful integration into the school system by his mastery not of the language or other cultural skills, but of the school's daily routines—walking in line, brushing his teeth after eating, arriving at school early, eagerly participating in greeting and departure ceremonies, and completing all of his *obentō* on time. Not only had he adjusted to the school structure, but he had also become assimilated to the other children. Or, restated, what once had been externally enforced now became ideologically desirable; the everyday practices had moved from being alien (*soto*) to being familiar (*uchi*) to him, that is, from being someone else's to being his own. My American child had to become, in some sense, Japanese, and where his teacher recognized this Japaneseness was in the daily routines such as finishing his *obentō*. The lesson learned early, which David learned as well, is that not adhering to routines such as completing one's *obentō* on time results not only in admonishment from the teacher, but in rejection from the other students.

The nursery-school system differentiates between the child who does and the child who does not manage the multifarious and constant rituals of nursery school. And for those who do not manage, there is a penalty, which the child learns to either avoid or wish to avoid. Seeking the acceptance of his peers, the student develops the aptitude, willingness, and in the case of my son—whose outspokenness and individuality were the characteristics most noted in this culture—even the desire to conform to the highly ordered and structured practices of nursery-school life. As Althusser (1971) wrote about ideology: the mechanism works when and because ideas about the world and particular roles in that world that serve other (social, political, economic, state) agendas become familiar and one's own.

Rohlen makes a similar point: that what is taught and learned in nursery school is social order. Called *shudanseikatsu* or group life, it means organization into a group where a person's subjectivity is determined by group membership and not "the assumption of choice and rational self-interest" (1989: 30). A child learns in nursery school to be with others, think like others, and act in tandem with others. This lesson is taught primarily through the precision and constancy of basic routines: "Order is shaped gradually by repeated practice of selected daily tasks . . . that socialize the children to high degrees of neatness and uniformity" (p. 21). Yet a feeling of coerciveness is rarely experienced by the child when three principles of nursery-school instruction are in place: (1) school routines are made "desirable and pleasant" (p. 30), (2) the teacher disguises her authority by trying to make the group the voice and unity of authority, and (3) the regimentation of the school is administered by an attitude of "intimacy" on the part of the teachers and administrators (p. 30). In short, when the desire and routines of the school are made into the desires and routines of the child, they are made acceptable.

Mothering as Gendered Ideological State Apparatus

The rituals surrounding the *obentō*'s consumption in the school situate what ideological meanings the *obentō* transmits to the child. The process of production within the home, by contrast, organizes its somewhat different ideological package for the mother. While the two sets of meanings are intertwined, the mother is faced with different expectations in the preparation of the *obentō* than the child is in its consumption. At a pragmatic level the child must simply eat the lunch box, whereas the mother's job is far more complicated. The onus for her is getting the child to consume what she has made, and the general attitude is that this is far more the mother's responsibility (at this nursery school, transitional stage) than the child's. And this is no simple or easy task.

Much of what is written, advised, and discussed about the *obentō* has this aim explicitly in mind: that is making food in such a way as to facilitate the child's duty to eat it. One magazine advises:

The first day of taking *obentō* is a worrisome thing for mother and *boku* (child[15]) too. Put in easy-to-eat foods that your child likes and is already used to and prepare this food in small portions. (*Shufunotomo* 1980:28)

Filled with pages of recipes, hints, pictures, and ideas, the magazine codes each page with "helpful" headings:

- First off, easy-to-eat is step one.
- Next is being able to consume the *obentō* without leaving anything behind.
- Make it in such a way for the child to become proficient in the use of chopsticks.
- Decorate and fill it with cute dreams (*kawairashi yume*).
- For older classes (*nencho*), make *obentō* filled with variety.
- Once he's become used to it, balance foods your child likes with those he dislikes.
- For kids who hate vegetables . . .
- For kids who hate fish . . .
- For kids who hate meat . . . (pp. 28–53)

Laced throughout cookbooks and other magazines devoted to *obentō*, the *obentō* guidelines issued by the school and sent home in the school flier every two weeks, and the words of Japanese mothers and teachers discussing *obentō*, are a number of principles: 1) food should be made easy to eat: portions cut or made small and manipulated with fingers or chopsticks, (child-size) spoons and forks, skewers, toothpicks, muffin tins, containers, 2) portions should be kept small so the *obentō* can be consumed quickly and without any left-overs, 3) food that a child does not yet like should be eventually added so as to remove fussiness (*sukikirai*) in food habits, 4) make the *obentō* pretty, cute, and visually changeable by presenting the food attractively and by adding non-food objects such as silver paper, foil, toothpick flags, paper napkins, cute handkerchiefs, and variously shaped containers for soy sauce and ketchup, and 5) design *obentō*-related items as much as possible by the mother's own hands including the *obentō* bag (*obentōfukuro*) in which the *obentō* is carried.

The strictures propounded by publications seem to be endless. In practice I found that visual appearance and appeal were stressed by the mothers. By contrast, the directive to use *obentō* as a training process—adding new foods and getting older children to use chopsticks and learn to tie the *furoshiki*[16]—was emphasized by those judging the *obentō* at the school. Where these two sets of concerns met was, of course, in the child's success or failure completing the *obentō*. Ultimately this outcome and the mother's role in it, was how the *obentō* was judged in my experience.

The aestheticization of the *obentō* is by far its most intriguing aspect for a cultural anthropologist. Aesthetic categories and codes that operate generally for Japanese cuisine are applied, though adjusted, to the nursery school format. Substances are many but petite, kept segmented and opposed, and manipulated intensively to achieve an appearance that often changes or disguises the food. As a mother insisted to me, the creation of a bear out of miniature hamburgers and rice, or a flower from an apple or peach, is meant to sustain a child's interest in the underlying food. Yet my child, at least, rarely noticed or appreciated the art I had so laboriously contrived. As for other children, I observed that even for those who ate with no obvious "fussiness," mothers' efforts to create food as style continued all year long.

Thus much of a woman's labor over *obentō* stems from some agenda other than that of getting the child to eat an entire lunch-box. The latter is certainly a consideration and it is the rationale as well as cover for women being scrutinized by the school's authority figure—the teacher. Yet two other factors are important. One is that the *obentō* is but one aspect of the far more expansive and continuous commitment a mother is expected to make for and to her child. "*Kyoiku mama*" (education mother) is the term given to a mother who executes her responsibility to oversee and manage the education of her children with excessive vigor. And yet this excess is not only demanded by the state even at the level of the nursery school; it is conventionally given by mothers. Mothers who manage the home and children, often in virtual absence of a husband/father, are considered the factor that may make or break a child as s/he advances towards that pivotal point of the entrance examinations.[17]

In this sense, just as the *obentō* is meant as a device to assist a child in the struggles of first adjusting to school, the mother's role is generally perceived as that of support, goad, and cushion for the child. She will perform endless tasks to assist in her child's study: sharpen pencils and make midnight snacks as the child studies, attend cram schools to verse herself in subjects her child is weak in, make inquiries as to what school is most appropriate for her child, and consult with her child's teachers. If the child succeeds, a mother is complimented; if the child fails, a mother is blamed.

Thus, at the nursery-school level, the mother starts her own preparation for this upcoming role. Yet the jobs and energies demanded of a nursery-school mother are, in themselves, surprisingly consuming. Just as the mother of an entering student is given a book listing all the pre-entry tasks she must complete—for example, making various bags and containers, affixing labels to all clothes in precisely the right place and of precisely the right size—she will be continually expected thereafter to attend Mothers' Association meetings, accompany children on field trips, wash her child's clothes and indoor shoes every week, add required items to her child's bag on a day's notice, and generally be available. Few mothers at the school my son attended could afford to work in even part-time or temporary jobs. Those women who did tended either to keep their outside work a secret or be reprimanded by a teacher for insufficient devotion to their child. Motherhood, in other words, is institutionalized through the child's school and such routines as making the *obentō* as a full-time, kept-at-home job.[18]

The second factor in a woman's devotion to over-elaborating her child's lunch box is that her experience doing this becomes a part of her and a statement, in some sense, of who she is. Marx writes that labor is the most "essential" aspect to our species-being and that the products we produce are the encapsulation of us and therefore our productivity (1970: 71–76). Likewise, women are what they are through the products they produce. An *obentō* therefore is not only a gift or test for a child, but a representation and product of the woman herself. Of course, the two ideologically converge, as has been stated already, but I would also suggest that there is a potential disjoining. I sensed that the women were laboring for themselves apart from the agenda the *obentō* was expected to fill at school. Or stated alternatively, in the role that females in Japan are highly pressured and encouraged to assume as domestic manager, mother, and wife, there is, besides the endless and onerous responsibilities, also an opportunity for play. Significantly, women find play and creativity not outside their social roles but within them.

Saying this is not to deny the constraints and surveillance under which Japanese women labor at their *obentō*. Like their children at school, they are watched not only by the teacher but by each other, and they perfect what they create, at least partially, so as to be confirmed as a good and dutiful mother in the eyes of other mothers. The enthusiasm with which they absorb this task, then, is like my son's acceptance and internalization of the nursery-school routines; no longer enforced from outside, it is adopted as one's own.

The making of the *obentō* is, I would thus argue, a double-edged sword for women. By relishing its creation (for all the intense labor expended, only once or twice did I hear a mother voice any complaint about this task), a woman is ensconcing herself in the ritualization and subjectivity (subjection) of being a mother in Japan. She is alienated in the sense that others will dictate, inspect, and manage her work. On the reverse side, however, it is precisely through this work that the woman expresses, identifies, and constitutes herself. As Althusser pointed out, ideology can never be totally abolished (1971: 170); the elaborations that women work on "natural" food produce an *obentō* that is creative and, to some degree, a fulfilling and personal statement of themselves.

Minami, an informant, revealed how both restrictive and pleasurable the daily rituals of motherhood can be. The mother of two children—one aged three and one a nursery school student—Minami had been a professional opera singer before marrying at the relatively late age of 32. Now, her daily schedule was organized by routines associated with her child's nursery school: for example, making the *obentō*, taking her daughter to school and picking her up, attending Mothers' Association meetings, arranging daily play dates, and keeping the school uniform clean. While Minami wished to return to singing, if only on a part-time basis, she said that the demands of motherhood, particularly those imposed by her child's attendance at nursery school, frustrated this desire. Secretly snatching only minutes out of any day to practice, Minami missed singing and told me that being a mother in Japan means the exclusion of almost anything else.[19]

Despite this frustration, however, Minami did not behave like a frustrated woman. Rather she devoted to her mothering an energy, creativity, and intelligence I found to be standard in the Japanese mothers I knew. She planned special outings for her children at least two or three times a week, organized games that she knew they would like and would teach them cognitive skills, created her own stories and designed costumes for afternoon play, and shopped daily for the meals she prepared with her children's favorite foods in mind. Minami told me often that she wished she could sing more, but never once did she complain about her children, the chores of child-raising, or being a mother. The attentiveness displayed otherwise in her mothering was exemplified most fully in Minami's *obentōs*. No two were ever alike, each had at least four or five parts, and she kept trying out new ideas for both new foods and new designs. She took pride as well as pleasure in her *obentō* handicraft; but while Minami's *obentō* creativity was impressive, it was not unusual.

Examples of such extraordinary *obentō* creations from an *obentō* magazine include: 1) ("donut *obentō*"): two donuts, two wieners cut to look like a worm, two cut pieces of apple, two small cheese rolls, one hard-boiled egg made to look like a rabbit with leaf ears and pickle eyes and set in an aluminum muffin tin, cute paper napkin added, 2) (wiener doll *obentō*): a bed of rice with two doll creations made out of wiener parts (each consists of eight pieces comprising hat, hair, head, arms, body, legs), a line of pink ginger, a line of green parsley, paper flag of France added, 3) (vegetable flower and tulip *obentō*): a bed of rice laced with chopped hard-boiled egg, three tulip flowers made out of cut wieners with spinach precisely arranged as stem and leaves, a fruit salad with two raisins, three cooked peaches, three pieces of cooked apple, 4) (sweetheart doll *obentō*—*abekku ningyo no obentō*): in a two-section *obentō* box there are four rice balls on one side, each with a different center, on the other side are two dolls made of quail's eggs for heads, eyes and mouth added, bodies of cucumber, arranged as if lying down with two raw carrots for the pillow, covers made of one flower—cut cooked carrot, two pieces of ham, pieces of cooked spinach, and with different colored plastic skewers holding the dolls together (*Shufunotomo* 1980: 27, 30).

The impulse to work and re-work nature in these *obentōs* is most obvious perhaps in the strategies used to transform, shape, and/or disguise foods. Every mother I knew came up with her own repertoire of such techniques, and every *obentō* magazine or cookbook I examined offered a special section on these devices. It is important to keep in mind that these are treated as only flourishes: embellishments added to parts of an *obentō* composed of many parts. The following is a list from one magazine: lemon pieces made into butterflies, hard-boiled eggs into *daruma* (popular Japanese legendary figure of a monk without his eyes), sausage cut into flowers, a hard-boiled egg decorated as a baby, an apple piece cut into a leaf, a radish flaked into a flower, a cucumber cut like a flower, a *mikan* (nectarine orange) piece arranged into a basket, a boat with a sail made from a cucumber, skewered sausage, radish shaped like a mushroom, a quail egg flaked into a cherry, twisted *mikan* piece, sausage cut to become a crab, a patterned cucumber, a ribboned carrot, a flowered tomato, cabbage leaf flower, a potato cut to be a worm, a carrot designed as a red shoe, an apple cut to simulate a pineapple (pp. 57–60).

Nature is not only transformed but also supplemented by store-bought or mother-made objects which are precisely arranged in the *obentō*. The former come from an entire industry and commodification of the *obentō* process: complete racks or sections in stores selling *obentō* boxes, additional small containers, *obentō* bags, cups, chopsticks and utensil containers (all these with various cute characters or designs on the front), cloth and paper napkins, foil, aluminum tins, colored ribbon or string, plastic skewers, toothpicks with paper flags, and paper dividers. The latter are the objects mothers are encouraged and praised for making themselves: *obentō* bags, napkins, and handkerchiefs with appliqued designs or the child's name embroidered. These supplements to the food, the arrangement of the food, and the *obentō* box's dividing walls (removable and adjustable) furnish the order of the *obentō*. Everything appears crisp and neat with each part kept in its own place: two tiny hamburgers set firmly atop a bed of rice; vegetables in a separate compartment in the box; fruit arranged in a muffin tin.

How the specific forms of *obentō* artistry—for example, a wiener cut to look like a worm and set within a muffin tin—are encoded symbolically is a fascinating subject. Limited here by space, however, I will only offer initial suggestions. Arranging food into a scene recognizable by the child was an ideal mentioned by many mothers and cook-books. Why those of animals, human beings, and other food forms (making a pineapple out of an apple, for example) predominate may have no other rationale than being familiar to children and easily re-produced by mothers. Yet it is also true that this tendency to use a trope of realism—casting food into realistic figures—is most prevalent in the meals Japanese prepare for their children. Mothers I knew created animals and faces in supper meals and/or *obentō*s made for other outings, yet their impulse to do this seemed not only heightened in the *obentō* that were sent to school but also played down in food prepared for other age groups.

What is consistent in Japanese cooking generally, as stated earlier, are the dual principles of manipulation and order. Food is manipulated into some other form than it assumes either naturally or upon being cooked: lines are put into mashed potatoes, carrots are flaked, wieners are twisted and sliced. Also, food is ordered by some human rather than natural principle; everything must have neat boundaries and be placed precisely so those boundaries do not merge. These two structures are the ones most important in shaping the nursery school *obentō* as well, and the inclination to design realistic imagery is primarily a means by which these other culinary codes are learned by and made pleasurable for the child. The simulacrum of a pineapple recreated from an apple therefore is less about seeing the pineapple in an apple (a particular form) and more about reconstructing the apple into something else (the process of transformation).

The intense labor, management, commodification, and attentiveness that goes into the making of an *obentō* laces it, however, with many and various meanings. Overarching all is the potential to aestheticize a certain social order, a social order that is coded (in cultural and culinary terms) as Japanese. Not only is a mother making food more palatable to her nursery-school child, but she is creating food as a more aesthetic and pleasing social structure. The *obentō*'s message is that the world is constructed very precisely and that the role of any single Japanese in that world must be carried out with the same degree of precision. Production is demanding; and the producer must both keep within the borders of her/his role and work hard.

The message is also that it is women, not men, who are not only sustaining a child through food but carrying the ideological support of the culture that this food embeds. No Japanese man I spoke with had or desired the experience of making a nursery-school *obentō* even once, and few were more than peripherally engaged in their children's education. The male is assigned a position in the outside world, where he labors at a job for money and is expected to be primarily identified by and committed to his place of work.[20] Helping in the management of home and the raising of children has not become an obvious male concern or interest in Japan, even as more and more women enter what was previously the male domain of work. Females have remained at and as the center of home in Japan, and this message too is explicitly transmitted in both the production and consumption of entirely female-produced *obentō*.

The state accrues benefits from this arrangement. With children depending on the labor women devote to their mothering to such a degree, and women being pressured as well as pleasurized in such routine maternal productions as making the *obentō*—both effects encouraged and promoted by institutional features of the educational system, which is heavily state-run and at least ideologically guided at even the nursery-school level—a gendered division of labor is firmly set in place. Labor from males, socialized to be compliant and hardworking, is more extractable when they have wives to rely on for almost all domestic and familial management. And females become a source of cheap labor, as they are increasingly forced to enter the labor market to pay domestic costs (including those vast debts incurred in educating children) yet are increasingly constrained to low-paying part-time jobs because of the domestic duties they must also bear almost totally as mothers.

Hence, not only do females, as mothers, operate within the ideological state apparatus of Japan's school system, which starts semi-officially with the nursery school, they also operate as an ideological state apparatus unto themselves. Motherhood *is* state ideology, working through children at home and at school and through such mother-imprinted labor that a child carries from home to school as the *obentō*. Hence the post-World War II conception of Japanese education as egalitarian, democratic, and with no agenda of or for gender differentiation, does not in practice stand up. Concealed within such cultural practices as culinary style and child-focused mothering is a worldview in which the position and behavior an adult will assume has everything to do with the anatomy she/he was born with.

At the end, however, I am left with one question. If motherhood is not only watched and manipulated by the state but made by it into a conduit for ideological indoctrination, could not women subvert the political order by redesigning *obentō*? Asking this question, a Japanese friend, upon reading this paper, recalled her own experiences. Though her mother had been conventional in most other respects, she made her children *obentōs* that did not conform to the prevailing conventions. Basic, simple, and rarely artistic, Sawa also noted, in this connection, that the lines of these *obentōs* resembled those by which she was generally raised: as gender-neutral, treated as a person not "just as a girl," and being allowed a margin to think for herself. Today she is an exceptionally independent woman who has created a life for herself in America, away from homeland and parents, almost entirely on her own. She loves Japanese food, but the plain *obentōs* her mother made for her as a child, she is newly appreciative of now, as an adult. The *obentōs* fed her, but did not keep her culturally or ideologically attached. For this, Sawa says today, she is glad.

Notes

The fieldwork on which this article is based was supported by a Japan Foundation Postdoctoral Fellowship. I am grateful to Charles Piot for a thoughtful reading and useful suggestions for revision and to Jennifer Robertson for inviting my contribution to this issue. I would also like to thank Sawa Kurotani for her many ethnographic stories and input, and Phyllis Chock and two anonymous readers for the valuable contributions they made to revision of the manuscript.

1. As Dorinne Kondo has pointed out, however, these cuisinal principles may be conditioned by factors of both class and circumstance. Her *shitamachi* (more traditional area of Tokyo) informants, for example, adhered only

casually to this coding and other Japanese she knew followed them more carefully when preparing food for guests rather than family and when eating outside rather than inside the home (Kondo 1990: 61–2).

2. Rice is often, if not always, included in a meal; and it may substantially as well as symbolically constitute the core of the meal. When served at a table it is put in a large pot or electric rice maker and will be spooned into a bowl, still no bigger or predominant than the many other containers from which a person eats. In an *obentō* rice may be in one, perhaps the largest, section of a multi-sectioned *obentō* box, yet it will be arranged with a variety of other foods. In a sense rice provides the syntactic and substantial center to a meal yet the presentation of the food rarely emphasizes this core. The rice bowl is refilled rather than heaped as in the preformed *obentō* box, and in the *obentō* rice is often embroidered, supplemented, and/or covered with other foodstuffs.

3. Japanese will both endure a high price for rice at home and resist American attempts to export rice to Japan in order to stay domestically self-sufficient in this national food *qua* cultural symbol. Rice is the only foodstuff in which the Japanese have retained self-sufficient production.

4. The primary sources on education used are Horio 1988; Duke 1986; Rohlen 1983; Cummings 1980.

5. Neither the state's role in overseeing education nor a system of standardized tests is a new development in post-World War II Japan. What is new is the national standardization of tests and, in this sense, the intensified role the state has thus assumed in overseeing them. See Dore (1965) and Horio (1988).

6. Boocock (1989) differs from Tobin *et al.* (1989) on this point and asserts that the institutional differences are insignificant. She describes extensively how both *yochien* and *hoikuen* are administered (*yochien* are under the authority of Monbusho and *hoikuen* are under the authority of the Koseisho, the Ministry of Health and Welfare) and how both feed into the larger system of education. She emphasizes diversity: though certain trends are common amongst pre-schools, differences in teaching styles and philosophies are plentiful as well.

7. According to Rohlen (1989), families are incapable of indoctrinating the child into this social pattern of *shundanseikatsu* by their very structure and particularly by the relationship (of indulgence and dependence) between mother and child. For this reason and the importance placed on group structures in Japan, the nursery school's primary objective, argues Rohlen, is teaching children how to assimilate into groups. For further discussion of this point see also Peak 1989; Lewis 1989; Sano 1989; and the *Journal of Japanese Studies* issue [15(1)] devoted to Japanese pre-school education in which these articles, including Boocock's, are published.

8. For a succinct anthropological discussion of these concepts, see Hendry (1987: 39–41). For an architectural study of Japan's management and organization of space in terms of such cultural categories as *uchi* and *soto*, see Greenbie (1988).

9. Endless studies, reports, surveys, and narratives document the close tie between women and home, domesticity and femininity in Japan. A recent international survey conducted for a Japanese housing construction firm, for example, polled couples with working wives in three cities, finding that 97 percent (of those polled) in Tokyo prepared breakfast for their families almost daily (compared with 43 percent in New York and 34 percent in London); 70 percent shopped for groceries on a daily basis (3 percent in New York, 14 percent in London), and only 22 percent of them had husbands who assisted or were willing to assist with housework (62 percent in New York, 77 percent in London) (quoted in *Chicago Tribune* 1991). For a recent anthropological study of Japanese housewives in English, see Imamura (1987). Japanese sources include *Juristo zokan sogo tokushu* 1985; *Mirai shakan* 1979; *Ohirasori no seifu kenkyukai* 3.

10. My comments pertain directly, of course, to only the women I observed, interviewed, and interacted with at the one private nursery school serving middle-class families in urban Tokyo. The profusion of *obentō*-related materials in the press plus the revelations made to me by Japanese and observations made by other researchers in Japan (for example, Tobin 1989; Fallows 1990), however, substantiate this as a more general phenomenon.

11. To illustrate this preoccupation and consciousness: during the time my son was not eating all his *obentō*, many fellow mothers gave me suggestions, one mother lent me a magazine, my son's teacher gave me a full set of *obentō* cookbooks (one per season), and another mother gave me a set of small frozen-food portions she had made in advance for future *obentōs*.

12. My son's teacher, Hamada-sensei, cited this explicitly as one of the reasons why the *obentō* was such an important training device for nursery-school children. "Once they become *ichinensei* [first-graders], they'll be faced with a variety of food, prepared without elaboration or much spice, and will need to eat it within a delimited time period."

13. An anonymous reviewer questioned whether such emphasis placed on consumption of food in nursery school leads to food problems and anxieties in later years. Although I have heard that anorexia is now a phenomenon in Japan, I question its connection to nursery-school *obentōs*. Much of the meaning of the latter practice, as I interpret it, has to do with the interface between production and consumption, and its gender linkage comes from the production end (mothers making it) rather than the consumption end (children eating it). Hence, while control is taught through food, it is not a control linked primarily to females or bodily appearance, as anorexia may tend to be in this culture.

14. Fujita argues, from her experience as a working mother of a daycare (*hoikuen*) child, that the substance of these daily talks between teacher and mother is intentionally insignificant. Her interpretation is that the mother is not to be overly involved in nor too informed about matters of the school (1989).

15. "*Boku*" is a personal pronoun that males in Japan use as a familiar reference to themselves. Those in close relationship with males—mothers and wives, for example—can use *boku* to refer to their sons or husbands. Its use in this context is telling.

16. In the upper third grade of the nursery school (the *nencho* class; children aged five to six) that my son attended, children were ordered to bring their *obentō* with chopsticks rather than forks and spoons (considered easier to use) and in the traditional *furoshiki* (piece of cloth that enwraps items and is double-tied to close it) instead of the easier-to-manage *obentō* bags with drawstrings. Both *furoshiki* and chopsticks (*o-hashi*) are considered traditionally Japanese, and their usage marks not only greater effort and skills on the part of the children but their enculturation into being Japanese.

17. For the mother's role in the education of her child, see, for example, White (1987). For an analysis, by a Japanese, of the intense dependence on the mother that is created and cultivated in a child, see Doi (1971). For Japanese sources on the mother-child relationship and the ideology (some say pathology) of Japanese motherhood, see Yamamura (1971); Kawai (1976); Kyutoku (1981); *Sorifu seihonen taisaku honbuhen* (1981); *Kadeshobo shinsha* (1981). Fujita's account of the ideology of motherhood at the nursery-school level is particularly interesting in this connection (1989).

18. Women are entering the labor market in increasing numbers, yet the proportion who do so in the capacity of part-time workers (legally constituting as much as thirty-five hours per week but without the benefits accorded to full-time workers) has also increased. The choice of part-time over full-time employment has much to do with a woman's simultaneous and almost total responsibility for the domestic realm (Juristo 1985; see also Kondo 1990).

19. As Fujita (1989: 72–79) points out, working mothers are treated as a separate category of mothers, and non-working mothers are expected, by definition, to be mothers full-time.

20. Nakane's much-quoted text on Japanese society states this male position in structuralist terms (1970). Though dated, see also Vogel (1963) and Rohlen (1974) for descriptions of the social roles for middle-class, urban Japanese males. For a succinct recent discussion of gender roles within the family, see Lock (1990).

References

Althusser, Louis (1971) *Ideology and ideological state apparatuses (Notes toward an investigation in Lenin and philosophy and other essays)*. New York: Monthly Review Press.

Asanuma, Kaoru (1987) *"Ganbari" no kozo (Structure of "Ganbari")*. Tokyo: Kikkawa Kobunkan.

Barthes, Roland (1957) *Mythologies*. Trans. Annette Lavers. New York: Noonday Press.

Boocock, Sarane Spence (1989) Controlled diversity: An overview of the Japanese preschool system. *The Journal of Japanese Studies* 15(1): 41–65.

Chicago Tribune (1991) Burdens of working wives weigh heavily in Japan. January 27, section 6, p. 7.

Cummings, William K. (1980) *Education and equality in Japan*. Princeton, NJ: Princeton University Press.

Doi, Takeo (1971) *The anatomy of dependence: The key analysis of Japanese behavior*. Trans. John Becker. Tokyo: Kodansha International, Ltd.

Dore, Ronald P. (1965) *Education in Tokugawa Japan*. London: Routledge and Kegan Paul.

Duke, Benjamin (1986) *The Japanese school: Lessons for industrial America*. New York: Praeger.

Fallows, Deborah (1990) Japanese women. *National Geographic* 177(4): 52–83.

Fujita, Mariko (1989) "It's all mother's fault": Childcare and the socialization of working mothers in Japan. *The Journal of Japanese Studies* 15(1): 67–91.

Greenbie, Barrie B. (1988) *Space and spirit in modern Japan*. New Haven, CT: Yale University Press.

Hendry, Joy (1986) *Becoming Japanese: The world of the pre-school child*. Honolulu: University of Hawaii Press.

—— (1987) *Understanding Japanese society*. London: Croom Helm.

Horio, Teruhisa (1988) *Educational thought and ideology in modern Japan: State authority and intellectual freedom*. Trans. Steven Platzer. Tokyo: University of Tokyo Press.

Imamura, Anne E. (1987) *Urban Japanese housewives: At home and in the community*. Honolulu: University of Hawaii Press.

Juristo zokan Sogotokushu (1985) Josei no Gensai to Mirai (The present and future of women). 39.

Kadeshobo shinsha (1981) *Hahaoya (Mother)*. Tokyo: Kadeshobo shinsha.

Kawai, Jayao (1976) *Bosei shakai nihon no Byori (The pathology of the mother society—Japan)*. Tokyo: Chuo koronsha.

Kondo, Dorinne K. (1990) *Crafting selves: Power, gender, and discourses of identity in a Japanese workplace.* Chicago, IL: University of Chicago Press.

Kyutoku, Shigemori (1981) *Bogenbyo (Disease rooted in motherhood).* Vol II. Tokyo: Sanma Kushuppan.

Lewis, Catherine C. (1989) From indulgence to internalization: Social control in the early school years. *Journal of Japanese Studies* 15(1): 139–157.

Lock, Margaret (1990) Restoring order to the house of Japan. *The Wilson Quarterly* 14(4): 42–49.

Marx, Karl and Frederick Engels (1970) (1947). *Economic and philosophic manuscripts,* ed. C. J. Arthur. New York: International Publishers.

Mirai shakan (1979) Shufu to onna (Housewives and women). Kunitachishi Komininkan Shimindaigaku Semina—no Kiroku. Tokyo: Miraisha.

Mouer, Ross and Yoshio Sugimoto (1986) *Images of Japanese society: A study in the social construction of reality.* London: Routledge and Kegan.

Nakane, Chie (1970) *Japanese society.* Berkeley: University of California Press.

Ohirasori no Seifu kenkyukai (1980) Katei kiban no jujitsu (The fullness of family foundations). (Ohirasori no Seifu kenkyukai—3). Tokyo: Okurasho Insatsukyoku.

Peak, Lois (1989) Learning to become part of the group: The Japanese child's transition to preschool life. *The Journal of Japanese Studies* 15(1): 93–123.

Richie, Donald (1985) *A taste of Japan: Food fact and fable, customs and etiquette, what the people eat.* Tokyo: Kodansha International Ltd.

Rohlen, Thomas P. (1974) *The harmony and strength: Japanese white-collar organization in anthropological perspective.* Berkeley: University of California Press.

—— (1983) *Japan's high schools.* Berkeley: University of California Press.

—— (1989) Order in Japanese society: attachment, authority, and routine. *The Journal of Japanese Studies* 15(1): 5–40.

Sano, Toshiyuki (1989) Methods of social control and socialization in Japanese day-care centers. *The Journal of Japanese Studies* 15(1): 125–138.

Shufunotomo Besutoserekushon shiri-zu (1980) Obentō 500 sen. Tokyo: Shufunotomo Co., Ltd.

Shufunotomohyakka shiri-zu (1981) 365 nichi no obentō hyakka. Tokyo: Shufunotomo Co.

Sorifu Seihonen Taisaku Honbuhen (1981) Nihon no kodomo to hahaoya (Japanese mothers and children): koku-saihikaku (international comparisons). Tokyo: Sorifu Seishonen Taisaku Honbuhen.

Tobin, Joseph J., David Y. H. Wu, and Dana H. Davidson (1989) *Preschool in three cultures: Japan, China, and the United States.* New Haven CT: Yale University Press.

Vogel, Erza (1963) *Japan's new middle class: The salary man and his family in a Tokyo suburb.* Berkeley: University of California Press.

White, Merry (1987) *The Japanese educational challenge: A commitment to children.* New York: Free Press.

Yamamura, Yoshiaki (1971) *Nihonjin to haha: Bunka toshite no haha no kannen ni tsuite no kenkyu (The Japanese and mother: Research on the conceptualization of mother as culture).* Tokyo: Toyo-shuppansha.

16

Conflict and Deference

Marjorie DeVault

Studies of contemporary couples suggest that equal sharing of family work is quite rare (e.g. Hood 1983; Hertz 1986; Hochschild 1989). In most families, women continue to put family before their paid jobs, and take primary responsibility for housework and child care. Indeed, some studies suggest that husbands require more work than they contribute in families: Heidi Hartmann (1981a) shows that working women with children and husbands spend more time on housework than single mothers, whether their husbands "help" with the work or not. Some analysts suggest that women's responsibility for housework persists because men's careers are typically more lucrative than women's, but Berk's study of the determinants of couples' family work patterns (1985) shows that these decisions do not depend on economics alone. In two-paycheck families, women continue to do more household work than men, even when such a pattern is not economically rational. Berk's explanation is that the "production of gender"—of a sense that husband and wife are acting as "adequate" man and woman—takes precedence over the most economically efficient production of household "commodities." Her conclusion, arrived at through statistical analysis of a large sample of households, is consistent with clues from the speech of informants in this study, who connected cooking with "wife" and asserted the importance of "a meal made by (a) mother."

This chapter explores the effect of these gendered expectations on the expression of conflict and on relations of service and deference. In order to take account of the diversity of families in this period of change, we must consider several different patterns, ranging from families attempting to share household work to those where relations of male dominance and female submission are enacted and enforced physically, through violence and abuse. I will suggest that expectations of men's entitlement to service from women are powerful in most families, that these expectations often thwart attempts to construct truly equitable relationships and sometimes lead to violence. I do not mean to argue that husbands are all tyrannical or that marriages aiming at egalitarianism do not represent significant change. Rather, my aim is to identify the varied but powerful effects of taken-for-granted beliefs and expectations about gender, and to begin to confront these expectations as barriers to change.

Problems of Sharing

Even when husbands and wives are committed to the idea of sharing household responsibilities, the character of family work contributes to an asymmetry in effort and attention to household needs. Perhaps because the household routine is such a coherent whole, it often seems easiest for one person to take responsibility for its organization, even when others share the actual work. This person—traditionally, the housewife—is the one who keeps an entire plan in mind. In the households I studied, three men shared cooking with their wives, and all three reported that for the most part they take directions from their wives. Ed, a psychologist who has begun to do more and more cooking for his family, still commented:

> It's not my domain. I have the minute to minute decisions, because I'm the one who's here, but she's really the one who decides things. I just carry out the decisions. It feels more like my domain now, because I've been doing so much more than I used to, but she's still in charge. She's the organized one.

He perceives her as "organized"; in fact, it is her activity that produces this perception, since even with his increased participation in carrying out tasks, she is the one who does the organizing. Similarly, Robin, whose husband Rick does virtually all the cooking, reported that she was "the supervisor":

> He handles the greater portion of, you know, taking care of stuff around the house. I tell him. I say, "OK, we need this, this, and this done," and that's what he does. Like I say, "You better go to the store, we need some milk," or "We need the laundry done," and he'll do it. He's the employee, and I'm the supervisor.

Rick agreed. When I asked if "keeping track of things" was a large part of his work, he responded:

> You're talking to the wrong person. Robin tells me when I've got to remember stuff. She's—if she could be a computer, she'd have the greatest memory bank in the world. I'm scatter-brained. And I don't remember a lot of things, I'll let things run down and just let things go completely out . . . She'll keep a list in her head for a week.

Though he is quite comfortable and skilled at doing the discrete tasks of cooking, he seems not to have learned the practices associated with responsibility for feeding; she, rather than he, does the work of keeping things in mind.

The fact that standards and plans for housework are typically unarticulated also makes it difficult for husbands and wives to share work in the family. The houseworker defines the work as part of an overall design for the household routine, but this design is only partially conscious. Phyllis, a white single mother, complained, "Once you've got the whole system in your head, it's very hard to translate that into collective work, I think." She has tried to share the work with her daughters, but finds it difficult:

> We once made schedules. There was probably something in the paper about that. And I tried to make—you know, we would all take a chore, and write it down and do it. And you know something? When you work all day, and come home, it's almost easier to do it than to have to supervise other people doing something they don't want to do.

Like a manager, Phyllis is responsible for planning and overseeing a range of activities involving others. At first glance, the problem—to get the children to do some of the work—seems similar to the problem of supervision in the workplace. However, a wife or mother lacks the formal authority of a manager. Further, since family work is mostly invisible, household members learn not to expect to share it, and the woman in charge of a household may not even be able to specify fully the tasks to be divided. The invisibility of monitoring, for example, can make it difficult to share the work of provisioning, because one never knows whether another is thinking about what is needed. A woman whose husband shares some housework commented:

> I'll be surprised. Like just the other day, we had just a little bit of milk left. And I know he saw that, because he handled the milk carton. But he went to the store and he didn't buy milk.

Tasks such as planning and managing the sociability of family meals are also invisible, and since maintaining their invisibility is part of doing the work well, people are often unable, or reluctant, to talk explicitly about them.

Serious attempts to share housework require a great deal of communication among household members, because the overall design for the work must be developed, and then continually revised, collectively. Sometimes these attempts at equality provoke conflict over the definition of the work. The problems involved were evident in the comments of a black professional couple, Ed and Gloria, who have been moving slowly toward a more equal division of household work during the course of their marriage. They have had to redefine some activities. Gloria enjoys gardening and yardwork, for example, and used to consider them part of her housework. Her husband Ed is not so interested, and argues that these activities are "hobbies"; he explained, "I say, 'Look, don't be dumping on me because I'm not doing that, because it's not my hobby.' " When they told me about their routine, both of their accounts emphasized that their activities and decisions depended on a variety of factors, and had to be made practically from moment to moment—both kept repeating, "It depends." They must talk about these things, or find some other way of making decisions together. For example, he reported:

> She does more cooking, but every now and then she'll say that she wants me to do the cooking. And if I'm really uptight, I'll say, "Look, I'll go out and buy it [laughing]. You're working, we can spend the money." Otherwise I'll go ahead and do it.

She described unspoken decisions about cleaning up after dinner:

> That depends on how tired I am or how tired I feel he is. If I've had a hard week or a hard day I just leave, and I don't care, I just walk out. And either Ed does it or it's there in the morning. And then it depends on how much time I have. If I have to go straight out, then I don't know what happens. But if I have some time then I might clean up some dishes. And then if I'm feeling up, or if I think he's down, then I'll clean up after supper.

Making these decisions requires sensitivity, and when there are problems, explicit talk:

We just feel each other out. We know each other so well that we can read each other. I know
when he's uptight and he knows when I'm uptight. We don't really talk about it. The only time
we talk is when we're not reading each other very well, and then one person starts to feel
dumped on. Then we talk, we say "What's going on?" and we try to work it out.

Such comments illustrate the extent to which housework is typically hidden from
view. When one person takes responsibility for the work, others rarely think about
it. Even the one who does it—because so much of her thought about it is never
shared—may not be fully aware of all that is involved. Her work can come to seem
like a natural expression of caring.

When couples begin to share the work of care, its "workful" unnaturalness—the
effort behind care—is necessarily exposed. The underlying principles of housework
must be made visible. The work must be seen as separable from the one who does it,
instead of in the traditional way as an expression of love and personality. Some
couples do attempt to discuss the effort required to produce the kind of family life
they desire. But as the next section will show, many people accept with little thought
pervasive cultural expectations that connect relations of service with the definitions
of "husband" and "wife," and "mother" and "father."

Husbands and Wives

When I talked with women about their household routines, many of them spoke of
their partners' preferences as especially compelling considerations. In discussions
about planning meals, several women mentioned their husbands' wage work as
activity that conferred a right to "good food": "He works hard," or "He is in a very
demanding work situation." (And their more detailed comments indicate that they
use the phrase "good food" here to mean "food that he likes.") These comments are
consistent with the findings from studies of power within marriage, which suggest
that paid employment brings power and influence within the family (Blood and
Wolfe 1960). Family members easily recognize the importance of paid work, and
Charles and Kerr (1988) found that many women rationalized their sole responsibility for cooking in terms of their husbands' wage work. However, beliefs about
work and domestic service do not operate in the same way for men and women.
While men who work are said to deserve service, women who work for pay are (at
most) excused from the responsibility of providing that service; they are rarely
thought entitled to service themselves (Murcott 1983). Studies of the impact of
women's employment on family patterns suggest that men's work at home increases
only slightly when their wives take jobs. Employed wives typically manage household
tasks by redefining the work and doing less than they would if they were home. I will
argue here that women's service for husbands is based on more than the importance
of men's paid work. Women's comments about feeding reveal powerful, mostly
unspoken beliefs about relations of dominance and subordination between men and
women, and especially between husbands and wives. They show that women learn to
think of service as a proper form of relation to men, and learn a discipline that
defines "appropriate" service for men.

Rhian Ellis (1983) suggests that many incidents of domestic violence are triggered by men's complaints about the preparation and service of meals. She notes, for example, the expectation in some working-class subcultures that a wife will serve her husband a hot meal immediately when he returns from work, and she cites examples from several studies of battered women who report being beaten when they violated this expectation. Typical accounts portray husbands who return home hours late, sometimes in the middle of the night, and still expect to be served well and promptly. In other studies, researchers report that conflict can arise from men's complaints about the amount or quality of food they are served. Some of the researchers who report such incidents remark on the fact that violence can be triggered by such "trivial" concerns, but Ellis suggests that the activities of cooking and serving food in particular ways are in fact quite significant because they signal a wife's acceptance of a subservient domestic role and deference to her husband's wishes. In these situations, men insist on enforcing exaggerated relations of dominance and subordination within the family. We will see below that some of the patterns taken for granted in families without explicit violence are based on similar assumptions about women's deference to men's needs: assumptions that women should work to provide service and that men are entitled to receive it. Though battering represents an extreme version of inequity between husbands and wives, it highlights the significance of the observations that will follow, and suggests a vicious circle: the idea that some version of womanly deference is "normal" may contribute to an ideology of male entitlement that supports violence against wives and mothers.

The households I studied were not, to my knowledge, ones in which violence was prevalent. Only a few of the people I interviewed spoke of any discord with their spouses. But their reports of daily routines suggested that implicit marital "bargains" were often based on taken-for-granted notions of men's entitlement to good food and domestic service. It was clear, for example, that in most households, wives are very sensitive to husbands' evaluations of their cooking. Teresa, a young Chicana woman, described the pressure of learning to cook when she was first married and her husband and brother-in-law, who ate with them regularly, were both "judging" her:

> It was really a lot of strain, to make two men happy, who were judging you. You know, "You don't make it half as good as my mother did." So that kind of pushed me a little to learn fast.

After seven years of marriage, her husband still has a great deal of influence over her routine. He buys most of the meat for their household, since they both agree that he knows more about butchering and can purchase better meat than she does. He has high standards and communicates them clearly. When he goes shopping with her, Teresa reported, "I go through the canned food aisle, and he'll say, 'Why are you taking so many cans?' " Teresa does not think of her husband as particularly demanding. She not only accepts her husband's preferences, but also thinks of her occasional failures to satisfy them as "cheating":

> I do cheat a little and I—like beans take about an hour to make—so if I forget that I've run out of beans and I don't make any, I'll just open a can of beans and just warm them up. But when he tastes them—I don't know, there must be something about the taste—he'll say, "These are canned beans, right?" No matter how hard I try to hide the can.

Teresa told me proudly that he likes the way she cooks and that he has gained so much weight that people tease him about what a good cook she must be. But she is still anxious about judgments of her cooking:

> Maybe that's why I still don't like cooking, because I know that every time I serve something on the table I'm going to be judged and criticized, and you know, "This tastes awful."
> [MD: Do you really get that?]
> No, no. But I'm afraid of it, right.

She describes a more relaxed kind of meal on the days when her husband works a second job and she eats casually with the children:

> Now Saturday, my husband works on Saturdays, and that's the day that my kids are home also. So Saturday would be hot dog day. . . . To them it's a big treat to have hot dogs. And to me it's another treat, because I don't have to go to the whole trouble of doing or preparing a whole dinner. . . . [She explains that her husband works until 10 or 11 at night.] So that means the whole day, we don't have a father to yell over us. That's what Felix says, he says, "Oh, we don't have a dad to yell over us." So we'll have some kind of Campbell's soup, some kind of vegetable or chicken or something, out of a can.

Some writers have suggested that men are especially authoritarian in Hispanic families, and the pattern that Teresa describes—with "a dad to yell over" the family— is consistent with such an idea. However, I found instances of a similar attitude among women in all of the class and ethnic groups included in this study. While none of these women—including Teresa—described their own attitudes in terms of service or deference to husbands (in fact, many took some trouble to explain to me that they were not mere servants in their own households), they spoke of accepting husbands' demands in a matter-of-fact tone that illustrates the force of male preference. For example, one affluent white woman explained:

> My husband doesn't like a prepared salad dressing. So I make my own. And he now is on a kick of having me make orange juice, rather than buying the frozen concentrate. So I'm going to have to go out and get an orange juice squeezer.

Her tone suggests that she takes for granted that his preferences should determine her work: that he is "on a kick" means that she will buy new equipment and adopt a new method for preparing juice. The same idea appears in one of Donna's comments, as she talked about the foods that she and her husband like:

> I used to have pork chops three or four times a week. And then he just said, "I don't want pork chops and that's it." And I haven't bought them since.

She tells a small story here that conveys the drama involved in such mundane matters. By casting her report in this way, she indicates how easily (and thus, we assume, how legitimately in her mind) he puts forward his claims: "he just said . . . And I haven't bought them since." Susan, remembering times when she had to scrimp and save, explains how things are different now: "If the man wants a steak, he gets it. Period. No questions asked." Again, the sense is of men's entitlement to make claims and have them met within the family.

A number of women referred to husbands as the sources for elaborated standards. For example, a recently divorced, white professional woman commented on how the change had affected her cooking; before the divorce, meals for her husband and children had been rather elaborate:

> The expectations of supper, you know, the big meal of the day. I knew that I had to have meat and potatoes, and you know, the usual fanfare. But you know, that really required me to be home, by five o'clock in the afternoon, to get all of this ready by six. But I knew that was expected of me.

After the divorce, food was "not a priority," and her routines became "very simple." She is concerned about nutrition, and carefully monitors her children's eating, but she no longer prepares elaborate meals:

> After my husband left, things got very simple with food. I found, you know, I didn't have to be in the kitchen, and shopping all of the time. I had always flirted with the idea. But you know, being married, you're just a slave to the kitchen. And once I got out of that, I just had more choices. I mean I had more flexibility in what I could do and couldn't do.

She does not explain exactly why her choices expanded; we do not know precisely how her husband influenced her decisions before the divorce. But she describes a shift from "the usual fanfare" to a "very simple" routine, and her language is clearly that of constraint and permission: her husband's presence—whether explicitly or through her own expectation—dictated what she "could do and couldn't do."

The fact that many women seem unaware of this tone of deference toward husbands, or at least are unable to articulate its basis, does not detract from its force. Indeed, for those who value some form of egalitarianism in their marriages, the requirement to serve a husband—which might be resisted if it were more explicit—is not necessarily diminished by its invisibility. One white woman, whose husband is an executive, provides an example. When she is home alone during the day, she eats very casually, and she talked about how difficult it is, when her husband occasionally works at home, to prepare a "really decent lunch." When I asked why he could not eat the same lunch that she does every day, she was unable to explain except on the basis of an inarticulate "feeling." She attempts to explain her concern as an issue of nutrition, but her talk about non-fattening food is peppered with references to meals that are "really decent" in some other way:

> It would be hard, if I had to feed him every day, to think up really decent lunches. I eat yogurt, and peanut butter and jelly, all kinds of fattening things that I shouldn't, and every day I say I shouldn't be doing this. But it's hard to think of a well-rounded, man-sized meal.

I asked what she might prepare for him, and she gave an example:

> All right. Yesterday, we thought our girls were coming out the night before and I had bought some artichokes for them, so I cooked them anyway. So I scooped out the center and made a tuna salad and put it in the center, on lettuce, tomatoes around. And then, I had made zucchini bread. . . . So that was our lunch. If I had to do it every day I would find it difficult. [MD: When you make a distinction between the kind of lunch you would have and what you'd fix for him, what's involved in that? Is it because of what he likes, or what?]

> I just feel he should have a really decent meal. He would not like—well, I do terrible things and
> I know it's fattening. Like I'll sit down with yogurt and drop granola into it and it's great. Well,
> I can't give him that for lunch.
> [MD: Why not?]
> He doesn't, he wouldn't like it, wouldn't appreciate it. Or peanut butter and jelly, for instance,
> it's not enough of a lunch to give him.

This woman's discussion is quite confusing: when she is alone, she does "terrible things" that her husband "wouldn't appreciate." Yet she went on to report that in fact, he likes peanut butter and jelly and would probably enjoy a peanut butter sandwich. Still, it is "not enough of a lunch to give him." In fact, her explanations—the references to nutrition concerns and what he "appreciates" or not—obscure what seems the real point, that "not enough of a lunch," means simply that she has not done enough to prepare it. Her example shows how much trouble she takes to prepare a "decent" meal and serve it attractively. A "man-sized meal" may not be so much larger in quantity, but should be a meal she has worked on for him.

What we see in all of these comments are specific versions of a socially produced sense of appropriate gender relations, a sense that certain activities are associated with the very fundamental cultural categories "man"/"husband"/"father" and "woman"/"wife"/ "mother." (Haavind [1984] discusses similar interpretive frames for marital exchanges among Norwegian couples.) These associations are learned early and enforced through everyday observation of prevailing patterns of gender relations; they are rarely justified or even articulated explicitly, but explicit statement is hardly necessary. For most people, these understandings have become part of a morally charged sense of how things should be, so that even those who strive for some version of equity are prey to their pervasive effects.

Men Who Cook

Perhaps the one who cooks tries, "naturally," to please the one he or she cooks for, regardless of gender. That is, in homes where men did as much cooking as women, perhaps the inequalities of service and deference found in typical family settings would be less pronounced. The possibility is one that is difficult to assess, precisely because gender is so strongly associated with activity. In the sample of households I studied, only three men cooked more than occasionally. Of these three, only one (Rick, a white transportation worker) has taken on almost all of the cooking, as women have typically done in traditional families; one (Ed, a black psychologist) prepares the dinners most evenings, but usually by finishing the preparation of foods his wife organizes during the previous weekend; and the third (a white professional worker) reported cooking about twice each week. Since even Rick reported that his wife is the one who organizes their routine, reminding him what needs to be done each day, none of these men has taken on the kind of sole responsibility for family work that has been the most common pattern for women, and none seems to do even half of the coordinative work of planning. In short, they cook, but they are only beginning to share the work of "feeding the family" in the broader sense I have been developing. These few men's reports cannot provide any definitive answers to questions about gender and household activity, but they are worth examining along-side their wives' accounts if only as suggestive pointers toward understanding how men's and women's understandings of family work might differ. Their comments indicate that they have begun to learn and practice household skills such as preparing specific foods, juggling schedules in order to bring the family together for a meal, and improvising with the materials at hand. However, their reports also suggest that they do not feel the force of the morally charged ideal of deferential service that appears in so many women's reports.

When women talked about what they cook, they frequently referred to husbands' and children's preferences as the fixed points around which they designed the meals. Such comments are mostly absent from the reports of these men, except for occasional references to particular foods that young children will not accept. Rick, for example, likes cooking partly because he can be inventive; his explanation emphasizes his own creativity more than the tastes of those he serves:

> I don't use any recipes. It's just by what I want to put in the thing. And if I like it . . . It's just whatever I have up here, what can I think of now?

He reported that his wife and child like his cooking, but only late in the interview, in the context of a longer exchange. When I asked him what kinds of cooking he did best, his answer again suggested freedom from external standards even as he mentioned his attention to his wife's tastes; he explained that he has not mastered baking, and then continued:

> Other than the meats and stuff, it's just my imagination, whatever I want to do with it. And whatever Robin, you know, I might ask her, like the taste of this, do you like it? And if she doesn't like it I won't put it in.

Picking up on this last comment, I attempted to probe for some elaboration of his feelings about the evaluations of others:

> [MD: So what about Robin and Kate? Do they think you're a pretty good cook?]
> Yeah. I surprise them. I surprise myself sometimes [laughing].
> [MD: Yeah. They don't have too many complaints, then?]
> No. If so, I don't know . . . [laughing and shaking his head as if the thought had never occurred to him].

Rick takes on more responsibility than the other men I talked with, and he does so even in the face of some pointed teasing from some of his working-class friends who disapprove. But even though he understands the urgent demands of children who need to be fed, he seems able to set limits for his efforts more comfortably than most of the women I talked with, and has more success saying no to the tasks he simply does not enjoy. Both he and Robin reported that she cooks when they entertain large groups; he explained that she cooks "if it's gonna be any effort out on cooking—you know, let's go, go for it—instead of just fixing one or two things like I do." And in contrast to the many women who spoke of their own "laziness" and "bad habits," his matter-of-fact reporting of his faults, while perhaps more reasonable, is striking in its lack of shame:

I don't remember a lot of things, I'll let things run down and just let things go completely out. I do that quite frequently. Keeping things in your head—well, I know what I really need. But it's going out and getting it—and having the money. Or just going out and getting it, that you know, if I'm extra tired and I don't feel like going out, I'm just not going to go out. The heck with that. I'll think of something else, or do something else with it, or not use it, whatever.

Like the women I have discussed, Rick takes advantage of the flexibility built into housework to avoid work he dislikes; unlike many women, he seems to do so guiltlessly.

A somewhat different situation provides another version of this attitude. Ed took up cooking not because he enjoys it but because he had to, when his wife's new job and long commute meant that she returned home too late to prepare their meals. As he explained the routine, she plans meals for each day, but he decides when he begins the cooking whether or not to follow her plan, and she accepts whatever he decides. I saw the system in operation when I observed a dinner at their house. Just as he started to serve, Ed stopped and said, "Oh, I forgot—Gloria would have liked us to have a salad." But then, shrugging, "Well, I didn't get to it." The ability simply to dismiss the work that cannot be completed, without the anxieties that plague so many women, springs at least in part from differential cultural expectations: the notion that caring work is optional or exceptional for men while it is obligatory for women.

These comments can only be taken as suggestive, since so few men were interviewed, and since even these represent only a few of the variety of household arrangements that are possible. Still, the language of these few men points toward some rather different understandings of what in the work is "burdensome" or "convenient." Ed, for example, sees as a "burden" an aspect of planning that the women I talked with took so much for granted that few even mentioned it. He reported:

We occasionally get some fresh vegetables, but we usually have frozen vegetables. Because the fresh stuff, somebody has to be there to consume it, you can't delay. Then it becomes my burden, you know, I have to be thinking, and orchestrating, how to use such a variety of food.

And another man explained to me that he and his wife never use "convenience foods," but then defined as a "convenience" the extra work that she puts in on the weekends:

One convenience that we will use sometimes is this. You know when you come home from work and you have to cook you don't really have much time. You don't have time to simmer sauces or anything like that. So sometimes on the weekend Katherine will make up something, or if she has time maybe two or three things, something that can be heated up later in the week and will actually taste better then.

Here, we see how thoroughly women's household work is obscured from view and thus framed as not requiring discussion or negotiation. Even as they began to share this work, all three of these men continued to attribute imbalances in the division of labor to their wives' "personalities," defining much of the extra work that wives did in terms of fortuitous propensities for organization or planning.

Men who care for children alone probably take the work of feeding more seriously, simply because they are forced to take sole responsibility for this work that most men are unaccustomed to doing. In addition to the men who served as informants in this study, I talked briefly with two single fathers who do all of the work to care for their families. Both emphasized the terrific worries produced by the necessity of feeding their children. One talked of an "overwhelming anxiety" that began each day as he finished work and realized that he had to get a meal on the table for his three demanding children; the other wrote me that I should emphasize "the STRUGGLE involved in this feeding work," and the fact that "behind closed doors, dinner is often a nightmare." These are the comments of men who cannot rely on partners even to help out or fill in (much less to plan menus and prepare meals during the weekend), who are forced to learn, as most women do, that "if I don't do it, no one else will." Such comments may also reflect the lack of guidance—and resulting panic—for men who must unlearn cultural expectation, disentangling work and gender, separating "care" from "woman"; they must learn to provide service instead of receiving it.

Struggle and Silence

Overt conflict about who will do housework is surprisingly rare.[1] Informants in only two of the thirty households I studied discussed any sustained, explicit conflict about who would do the work or how it should be done, and results from a large survey are similar: over half of the wives responding reported no difference of opinion over who should do what, and only seven percent reported "a lot" of difference of opinion (Berk 1985:188; see also Haavind 1984). Fairly quickly, most houseworkers develop adjustments that are satisfactory enough to mute potential complaints from household members. The boundaries that produce complaints become the givens of the work, and those who do it find ways to manage around these givens. The perception that routine is chosen provides an interpretive frame for redefining the adjustments that are made.

Susan, who is quite happy with her household routine, told a story from the early days of her marriage, of her first definitions of her work, an episode of resentment, and its resolution:

> When we first got married, I played "Suzy Homemaker." I was young and stupid—what did I know? We lived in the suburbs and I worked in the city, and I had to get up at five every morning to get to work. And then on my days off, I'd get up to fix him breakfast, and you know, put on makeup, all that kind of thing. After a while my sister-in-law kind of pulled me aside, and told me I'd better cool it, or he'd get used to that kind of thing.
>
> I still remember, once I came home after a grueling day, and there was my old man, sitting in front of the TV with his potato chips. I said, "God, in my next life I hope I come back as a 26-inch Zenith, I'd get more attention!" That was probably our first fight. But it had been brewing for about four months. You know, we were just getting used to our differences.
>
> I like the way I do things, I'm used to it. I just get it all done on Monday and then I don't have to worry about it. If I don't do it, I'm a wreck by Wednesday. It's not that I like this kind of work, but you have to do it.

She thinks of this as something other than conflict over work: they were "getting used to [their] differences." She knows that she "has to do" housework, so she has found a way of doing it that she "likes," or at least, that she is "used to." Now, her husband goes to work before she is up; he has coffee and a doughnut, or buys something to eat on the job, and she sleeps until her daughter wakes up. She says, "If I'm in a pinch, my husband's not beyond doing the laundry, or washing a floor." But it is clear that she accepts responsibility for the housework: it is her domain, and though she would not say that she likes the work, she has accepted what seems to her a satisfactory compromise.

In many households, like this one, such compromises are negotiated with little overt struggle. Some men accept more responsibility, or are less demanding than others; some women are satisfied to take on the family work with little help. But accommodations cannot always be found. When conflict about housework does arise, it can be quite painful, at least partly because it carries so much emotional significance. Women who resist doing all of the work, or resist doing it as their husbands prefer, risk the charge—not only from others, but in their own minds as well—that they do not care about the family. When I talked with Jean, for example, she was engaged in an ongoing struggle to get her husband to share the housework. She spoke of her continuing frustration in two interviews, a year apart, and I felt her ambivalence when she told me:

> In spite of all this, I love him. [Laughing, but then serious] No, I do love him, and I'm willing to make some sacrifices, but there are times when I really just go off half-crazy. Because the pressure is just too much sometimes. I just feel it's not fair. It's not a judicious way to live, a fair and equal way to live.

She seems afraid that any complaint will be heard as her lack of feeling for her husband. She insists, indeed, that she is willing to make "sacrifices," as a loving wife should. His resistance to helping with family work is apparently not subject to such an interpretation.

Since feeding work is associated so strongly with women's love and caring for their families, it is quite difficult for women to resist doing all of the work. Bertie, for example, had been married for over twenty years when I talked with her, and had experienced a long period of difficult change. She told the following story:

> There was a time when I was organized, did things on time, on a schedule. I cooked because I felt a responsibility to cook. I felt guilty if I didn't give my husband a certain kind of meal every day. . . . When I made the transition it was hard. For me and him. And he's still going through some problems with objecting to it. But I felt that I had put undue pressure on myself, by trying to do what people used to do, you know . . . when the husband could pay the bills, and the wife took care of the house . . .
>
> I told my family that there were certain things that I needed, which went neglected for many years. And when I recognized my own needs, there was a problem . . . I had given so much of my life to my husband and children, that he thought that I was wrong, not to give them that much time anymore. But I needed to go back to school, needed to improve myself, I needed time to myself . . . They've come to accept it now. Five, six years ago it was really rough. But now they accept, that you know, I'm a person. I am to be considered a person. I have rights, you know, to myself. It was a rough ride for a while. And I suppose it could have gone in another direction. But it didn't.

Things had changed by the time I talked with Bertie: she was pursuing a degree and spent less time cooking elaborate meals. Her husband and children have not taken on much of the work burden, and typically do without meals when she is away from home in the evening—sometimes the girls will prepare sandwiches—so Bertie continues to be responsible for the bulk of the work, simply doing less than in the past. Yet when she talks about their struggles, she still worries about being "selfish":

> I do take that time now for myself. But I count my study time, and my class time, as my own, you know. You know, so that I don't—I try not to be selfish.

When I asked what "being selfish" meant, she replied:

> I make sure that I have time with them. If my stuff gets to be too much, whatever is necessary, whatever, whatever is important, I try to do. Because we still have the children to raise. So there must be some sharing. They're still there, I can't treat them as though they're not there. Even though they're pretty independent, on their own, they still require a lot of attention. So I have to be careful not to give too much to myself. Because you can fall into that. You know, studying too much. It's hard to describe.

It is, indeed, hard to describe. She claims that she has "rights," like any other person. Yet she will do "whatever is necessary" for her family, and must be "careful not to give too much" to herself. Her talk reveals quite different standards for evaluating her own needs and those of others. Raising such issues within the family requires this intense scrutiny of a woman's own desires.

Bertie's system for accounting her time was not unique. Jean, the other woman who was struggling with these issues, reported considering her situation in similar terms. She identified two blocks of time as "hers," but both were hers in a rather ambiguous sense:

> I feel the only real time I have to myself is usually my lunch hour [at her job]. I consider that my time for myself.

She has to be firm to maintain even this break officially sanctioned by her employer. Her husband, who works at night and wakes up around noon, would like her to come home so that they could spend time together. But, she reported:

> I do kind of resist that, on any kind of a regular basis. Although, in order to keep a marriage going I should maybe not do that. It's hard for me. Because I know there's a need there, I feel that too. But this gets into all kinds of other issues, about him not helping. I feel that if he would help out more around the house, I'd be more willing to come home and spend more time with him. But you know, I feel like this is my own time for me.

On Sundays they go to church, partly for the children, and partly because she and her husband enjoy the "social aspect," and she counts this activity as her own time as well:

> So Sunday morning is never a time when I can do chores or anything. And in a way, I mean, I count that as time for me, in a different sort of way.

These women have asserted to their families that they are people too, with rights (and it is surely striking that they feel a need to put forward such a claim). However, they still must calculate which time to claim as "theirs," and the logic of caring for others labels any too-active exercise of their rights as "selfish."

These women's stories help to show why there were only two of them in the group I studied, why conflict about housework is infrequent. They illustrate how, in the family context, a mother's claims for time to pursue her own projects can so easily be framed as a lack of care, and a mother's claim even to be "a person" may be taken as "selfish." If the act of pressing a claim for time off or help from others is so fraught with interpersonal danger, it is perhaps not surprising that so many women choose to accommodate to inequitable arrangements instead of resisting them.

The Language of Choice

Many women spoke of doing work they did not enjoy in order to please their husbands. However, very few of them expressed explicit discomfort about these efforts, and only those described above reported any sustained, overt conflict with their husbands. I was puzzled and a bit dismayed by their complacency about what I saw as inequity. But as I analyzed their reports, I began to see how the organization of family work contributes to their responsiveness to husbands' demands. As I showed in the last chapter, individuals have considerable flexibility in designing household routines, and they choose routines shaped to the idiosyncrasies of those in the family. They find ways to adjust to special demands, and then take their adjustments for granted, often describing them as no particular trouble. These choices, and the sense of autonomy that comes with making them, combine to hide the fact that they are so often choices made in order to please others.

Both deference and a sense that her deferential behavior has been freely chosen can be seen in Donna's comments, for example. Her husband, a mail carrier, is moody and unpredictable, difficult to please and quite openly critical when he does not like the meal she prepares. Although she told me several times that he is "not fussy," she also reported checking with him about every evening's menu before she begins to cook. I remarked that his preferences seemed quite important, and she responded:

> Yeah. I like to satisfy him, you know, because a lot of times I'll hear, "Oh, you don't cook good," or something like that.

The possibility of such criticism becomes part of the context within which she plans her work. She thinks ahead about what to prepare, but final decisions depend on his responses. When I asked what she would prepare the night of our interview, she could not answer:

> I haven't talked to him today, so I really don't know . . .
> [MD: Does he always know what he wants?]
> Well—I give him choices. Or he'll say "I don't care." So then it's up to me and I just take out something. Hopefully tonight—I would like to have the pot roast—so maybe he'll say yes. Because he actually bought it the other day, so he might want it.

Donna finds ways to build a routine that provides some shape for her work and still allows accommodation to his day-to-day tastes. She explained, for instance, how she plans her shopping:

> Like I'll ask him, what do you want me to pick up? you know, what kind of meat do you want me to pick up? And he'll go through the paper, and he'll tell me, do this, get this. But as far as really making it out [a menu], I just don't. Because sometimes he might not be in the mood for it, he might not want it, or something like that. So I just leave it up in the freezer.

Her scheme sometimes involves an extra trip to the store:

> Then if he wants something, then I'll just go to the store and get what he wants. It's really kind of day by day. I find it easier that way. I couldn't sit there and write what I'm having for dinner every day. I just can't do that.

When I asked why not:

> I don't know, I figure maybe it's just me. I just can't sit there and write, well, we're going to have this and this and this. And then that day you might not have a taste for it. And then you'll want something else. That's the way I look at it.

She "finds it easier" to plan meals day by day, and she presents this as just her way, a personal inclination. It seemed clear to me that her strategy was shaped by her husband's demands, in response to his moodiness and in order to avoid his sharp words of criticism. Within the constraints of their relationship, she does make choices in order to avoid trouble. Interpreting her accommodations as choices freely made, she translates his peculiarities into a general observation: "you might not have a taste for it. And then you'll want something else." And thus she presents the result of her strategizing as her own belief: "the way I look at it."

Even when family members are not so demanding, the pattern of choosing to adjust to others is common. Another woman explained how she has chosen a breakfast routine that lets her sleep a bit later instead of eating with the family:

> During the week I usually get their food to the table and then I make a lunch [for her husband]. It's more pragmatic. I could get up earlier and do that, but I choose to stay in bed and avoid sitting at the table.

Such comments stress autonomy and choice; however, it is clear that these women's decisions are not so freely made as they suggest. When husbands decide to press their claims, these become the fixed points around which adjustments and "choices" are made. One white woman, married to a journalist, reported a more conflictual negotiation over the breakfast routine, and explained how she has adjusted her morning schedule to accommodate her husband's ideas about breakfast:

> Breakfast has turned into more of a social occasion than I perhaps would care for. For my husband it's a real social affair, and we got into huge fights years ago. He always from the day we got married expected me to get up and fix his breakfast, no matter what time he was going anywhere. Then we lived overseas and we had two maids, and I couldn't see any point in

getting up just to sit with him—I didn't even have to *make* the breakfast. Well, that was a "dreadful, dreadful thing." Finally he got over that, and I don't mind getting up, you see, all right, that's a personal thing . . . So I usually fix breakfast for the two of us. Which is nice—but I would like to be able to read the newspaper, myself.

On this issue, her husband is adamant—it is a "dreadful, dreadful thing" not to have breakfast together—and she has adjusted to this "personal" preference. But she also describes her adjustment in somewhat contradictory terms: she "doesn't mind getting up," it's "all right," even "nice," but still, she would rather read the newspaper.

The choices that women talk about are not entirely illusory: in many ways, houseworkers can choose to do the work as they like. They adopt different general strategies: some maintain that they "couldn't live with" a regular routine, while others describe themselves as "disciplined" and "big on rules." To some extent, people even choose not to do the kinds of work they dislike. One woman, who would like to "just forget about" cooking, has simplified her food routine so that her work is quite automatic: she prepares meals that are "very easy to cook, and very quick also." And Sandra, who enjoys cooking and prepares elaborate meals for her husband (in addition to simpler, early meals for the children), thinks of her efforts as "compensation" for the cleaning that she does not enjoy, and often does not do. Still, these real choices—some of which certainly do ease the burdens of housework—seem also to provide a rationale for deference: women emphasize their freedoms and minimize their adjustments to others.

As women make choices about housework, their decisions include calculations about when to press their own claims and when to defer to others. The choice to do something in the way one prefers oneself is made to fit among the more compelling demands of others, especially husbands. The houseworker comes to understand her work in terms of a compromise that seems fair: since she is free to choose in some ways, it is only fair to defer in others. Most women seem only partly conscious of this logic. They, and others, notice the choices but not the deference. In their talk about decision making they tend to conflate benefits for the household group all told with their own more specific interests and preferences.

Such calculation can be seen in this affluent white woman's comment about going out to dinner. She explained that she does not really enjoy cooking and "always looks forward to" going out. But she often thinks that her husband is not so interested, and "being sensitive" to him can interfere with her enjoyment:

> If it's just family, and we wait a real long time, I keep thinking of all the things I could be doing instead of waiting, and then I wonder if it's really worth going out for dinner. And I think part of that is being sensitive to my husband's thinking. Because he has to eat out every noon anyway. And sometimes a couple of times a week he has to take people out for dinner. So it's not a great pleasure for him, to go out again just to get me out of the kitchen. So if it takes a real long time, I just feel, why did I do this, you know?

She does not feel this way—and can relax and enjoy herself—when she eats out with her son; knowing her husband's feeling changes her own, to the point that she wonders, "Why did I do this, you know?" Donna, managing a household with little extra money, worries about the expense of eating out, but talks about the decision to go out in much the same way:

> To me, I can go to the store and pay for two days', meals, or go to a restaurant and pay for one meal. And a lot of times the kids don't finish their meal. . . . I was going to suggest for Easter, going out to eat. But maybe I'm better off just getting something and having it here.

Her own preferences disappear into those of the household group: though she might enjoy a holiday, the children might not eat and money spent would be wasted. In the end she concludes, "I'm better off" doing the work of cooking at home.

These are issues that are difficult for many women to discuss; their talk is often hesitant, sometimes contradictory. But my point is not that they are confused or disingenuous about their positions relative to others. Rather, I mean to show that they are in a situation not easily described in terms of either autonomy or control. These women do make choices about their work, though many of those choices are made within a structure of constraints produced by others. The work itself is defined in terms of service to others, and husbands' demands are given special force through cultural assumptions about appropriate relations between husbands and wives. What a husband insists on typically becomes a requirement of the work, and a woman who arranges the routine to satisfy her own preferences as well as his may simply be making her work more difficult. The fact that so many women frame these accommodations as "choices" means that they are less likely to make choices more obviously in their own behalf when the interests of family members conflict. In such situations, women seem to assume that they have made enough choices, and often come to define deference as equity.[2]

Dominance and Subordination in Everyday Life

I have argued throughout this book that the work of "feeding a family" is skilled practice, a craft in which many women feel pride and find much satisfaction. Such a view suggests that wives care for family members not because they are coerced or compelled by despotic husbands, but because they believe the work of care is valuable and important. The discussion in this chapter, however, suggests as well that the work of caring—however valuable or valued by those who do it—is implicated in subtle but pervasive ways in relations of inequality between men and women. Some husbands insist quite explicitly that their wives display subordination by providing domestic service. For most men, however, such coercion is as unnecessary as it would be unpleasant. In most households, wives display deference to husbands simply because catering to a man is built into a cultural definition of "woman" that includes caring activity and the work of feeding. For many women and men, patterns of "womanly" service for men simply "feel right." In some cases, the recognition of a husband's claim to service is quite direct ("Like I said, whatever he wants, I'll make it the way he likes, and everything he likes with it."); in other reports, references to "really decent," "man-sized" meals point toward a more diffuse sense that a husband, because he is a man, deserves special service. Both kinds of statements show how the everyday activities of cooking, serving, and eating become rituals of dominance and deference, communicating relations of power through nonverbal behavior (Henley 1977).

Many women take pleasure in preparing food that pleases, in serving family members, in rewarding a husband for his work at a difficult job. Many think of the craft of attentive service as work they choose. But few women are themselves the recipients of a similarly attentive service in return. We might assume that men who cook like to please the ones they cook for. But they do not talk about preparing a meal that is "enough for a woman." Indeed, they cannot talk (or think) this way: the idea of a "woman-sized meal" is so dissonant with prevailing cultural meanings that it sounds quite wrong. Here, we see how categories of expression interact with people's everyday family activity. The gender inequalities inherent in language and in a multitude of nonverbal behaviors are woven into the fabric of social relations produced as people go about the mundane affairs of everyday life. Even when fathers cook, their activity—however similar to that of mothers who cook—is framed differently. There are no terms within which men think of cooking as service for a woman, no script suggesting that husbands should care for wives through domestic work. Some women are beginning to insist on more equal relations, and some husbands are beginning to strūggle at taking equal responsibility for family work. But these attempts are made without a cultural imagery to support them, and in opposition to established understandings about appropriate activities for men and women.

Because of the expectation that women will be responsible for caring work, their own independent activities are likely to conflict with requirements for family service. When wives and mothers assert their rights to pursue individual projects, they often discover the limits of choice and the force of cultural expectation. When women resist—by demanding help with housework or a respite from serving others—they challenge a powerful consensual understanding of womanly character by suggesting that women's care for others is effort rather than love. Many have trouble speaking plainly about the limits of caring work, and many find that in the long run it is easier to do all the work required than to press claims for an equitable division of labor.

The invisibility of the work that produces "family," the flexibility underlying perceptions of "choice" about the work, and the association between caring work and the supposedly "natural" emotions of a loving wife and mother all tend to suppress conflict over housework. Since the work itself is largely unrecognized, and often misidentified as merely "love" rather than also effort, redefinition is required before questions of dividing the work can be discussed. Those who have benefited from the work often have trouble recognizing it as such, and indeed, have little incentive to do so. Further, many women find that they can make enough choices and adjustments in some areas that accommodation in others seems preferable to sustained conflict. Those who insist on negotiating new household patterns must confront their own and others' sense that they do so out of "selfishness" or insufficient concern for others. Even as they struggle for more equitable arrangements, these women carefully ration ("count") the time and attention they give to their own needs, while attempting to provide "whatever" their families require. Their demands for themselves are painfully visible within the family, while their accommodations to others remain largely unacknowledged.

The claim that caring is work, or that this work should be shared by all those who are able to do it, must be made against powerful beliefs about the naturalness and importance of family life, and about men's and women's dispositions and roles. For a woman to provoke and sustain conflict in this area is to risk the charge that she is unnatural or unloving. The costs of conflict are high. Conversely, when a husband complains, or even hints at complaint, his claims carry with them the weight of generations of traditional practice and a body of expert advice about housekeeping and family life based on the assumption that women will serve others. As women adjust and accommodate, choosing deference to others and fitting their own projects into frames established by others, their actions contribute to traditional assumptions about woman's "nature." Thus, we see that in addition to its constructive, affiliative aspect, the work of care—as presently organized—has a darker aspect, which traps many husbands and wives in relations of dominance and subordination rather than mutual service and assistance.

Notes

A portion of this chapter first appeared in "Conflict over Housework: A Problem That (Still) Has No Name," in *Research in Social Movements, Conflicts and Change: A Research Annual*, edited by Louis Kriesberg (Greenwich, CT: JAI Press, 1990), pp. 189–202. Reprinted by permission.

1. This is not to say that other kinds of family conflicts are not sometimes expressed through the provision of food or reactions to it. I was less interested in these emotional dynamics associated with food than with the practices of organizing, preparing, and serving it, and I did not ask interviewees to speculate about the significance of food in conflicts other than those focused on family work issues. Charles and Kerr (1988) provide somewhat more information about food practices as expressions of interpersonal struggles, especially between parents and young children.
2. Haavind, in an analysis of "Love and Power in Marriage," based on studies of Norwegian couples, makes a similar observation about understandings of "choice" that accompany marriage based on romantic love: "If you have the right to marry whom you please, the responsibility for how it works out is also yours. Therefore, it is difficult to share our disappointment with anyone outside our own marriage" (1984:161).

References

Berk, Sarah Fenstermaker (1985) *The Gender Factory: The Apportionment of Work in American Households*. New York: Plenum Press.

Blood, Robert O. and Donald M. Wolfe (1960) *Husbands and Wives*. New York: Free Press.

Charles, Nickie and Marion Kerr (1988) *Women, Food and Families*. Manchester: Manchester University Press.

Ellis, Rhian (1983) The Way to a Man's Heart: Food in the Violent Home. In Anne Murcott, ed., *The Sociology of Food and Eating*, 164–171. Aldershot: Gower Publishing.

Haavind, Hanne (1984) Love and Power in Marriage. In Harriet Hotter, ed. *Patriarchy in a Welfare Society*, 136–167. Oslo: Universitetsforlaget.

Hartmann, Heidi I. (1981) The Family as the Locus of Gender, Class, and Political Struggle: The Example of Housework. *Signs* 6:366–94.

Hertz, Rosanna (1986) *More Equal than Others: Women and Men in Dual Career Marriages*. Berkeley: University of California Press.

Hochschild, Arlie, with Anne Machung (1989) *The Second Shift: Working Parents and the Revolution at Home*. New York: Viking.

Hood, Jane C. (1983) *Becoming a Two-Job Family*. New York: Praeger.

Henley, Nancy M. (1977) *Body Politics: Power, Sex and Nonverbal Communication*. New York: Simon and Schuster.

Murcott, Anne (1983) "It's a Pleasure to Cook for Him": Food, Mealtimes and Gender in some South Wales Households. In Eva Garmarnikow et al. eds. *The Public and the Private*, 78–90. London: Heinemann.

17

Feeding Lesbigay Families

Christopher Carrington

> To housekeep, one had to plan ahead and carry items of motley nature around in the mind and at the same time preside, as mother had, at the table, just as if everything, from the liver and bacon, to the succotash, to the French toast and strawberry jam, had not been matters of forethought and speculation.
>
> *Fannie Hurst*, Imitation of Life

> Life's riches other rooms adorn. But in a kitchen home is born.
>
> *Epigram hanging in the kitchen of a lesbian family*

Preparing a meal occurs within an elaborate set of social, economic, and cultural frameworks that determine when and with whom we eat, what and how much we eat, what we buy and where we go to buy it, and when and with what tools and techniques we prepare a meal. Many people associate the activities of preparing and sharing meals together with family. As sociologist Marjorie DeVault convincingly argues in *Feeding the Family* (1991), the work of preparing and sharing meals creates family. Many lesbigay families point to the continuous preparation of daily meals and/or the occasional preparation of elaborate meals as evidence of their status as families. The labor involved in planning and preparing meals enables family to happen in both heterosexual and in lesbigay households. However, both the extent and the character of feeding activities can vary dramatically from one household to the next and often reflects the influence of socioeconomic factors like social class, occupation, and gender, among others.

In this chapter I pursue two objectives in the investigation of feeding work in lesbigay families. First, I explore feeding work through analyzing its character and revealing its often hidden dimensions. This entails some discussion of how families conceive of and articulate the work of feeding. For instance, participants use a number of rhetorical strategies to portray the organization of feeding in their households. Many participants use two distinctions: cooking/cleaning and cooking/shopping. I will show how these distinctions function to create a sense of egalitarianism and to obfuscate rather than clarify the process of feeding. This inevitably leads to questions about the division of feeding work in lesbigay households. While an investigation of feeding work reveals that a less than egalitarian division often develops within many lesbigay households, I refrain from any explanation of why until chapter 5. Second, in this chapter I explore how socioeconomic differences among lesbigay households influence feeding activities, and vice-versa. Therein, just as feeding work can create family identity for participants, so too can feeding work create gender, ethnic, class, and sexual identities.

The Character of Feeding Work

As DeVault (1991) so aptly describes in her study of feeding work in heterosexual families, the people who do feeding work often find it difficult to describe the task. Commonly held definitions and most sociological investigations of domestic labor often reduce feeding work to cooking, shopping, and cleaning up the kitchen—the most apparent expressions of feeding. But when interviewing and engaging in participant observation with people who perform these functions, it becomes clear that cooking and shopping refer to a wide range of dispersed activities that punctuate the days of those who feed. It includes things like knowledge of what family members like to eat, nutritional concerns, a sense of work and recreation schedules, a mental list of stock ingredients in the cupboard, a mental time line of how long fruits and vegetables will last, etc. Frequently, these activities go unnoticed because they often happen residually and unreflectively. For instance, the way that one comes to know about the character and qualities of food stuffs—through experimentation, through conversations with colleagues, through browsing in a cookbook at a bookstore, through reading the food section in the newspaper—these activities often appear as recreation or as an expression of personal interest and not as forms of work. Yet, to successfully feed a family, such activities must occur and consume the energy and time of those who do them. In order to illuminate the full character of feeding work in lesbigay families, I want to look behind the traditional typologies of cooking, shopping, and cleaning and reveal the dynamic and invisible character of much of the work involved in feeding these families.

Planning Meals

Feeding actually consists of a number of distinct processes including planning, shopping, preparation, and management of meals. Planning presumes the possession of several forms of knowledge about food, about the household, about significant others, and about cultural rules and practices toward food. In most of the lesbigay families in this study, one person emerges as a fairly easily identified meal planner; hereafter I refer to such persons as planners. Planning for most families means thinking ahead, perhaps a day or two or even a week, but in many cases just a few hours before a meal. For those who decide what to eat on a day-to-day basis, they often decide and plan meals while at work. Matthew Corrigan, an office administrator, put it like this:

> Usually we decide something at the last minute. Or sometimes we go out with someone. We rarely go out with just us two, but with others as well. If Greg is home in time and has an inspiration, he will make something, but the general pattern is for me to throw something together when I get home. I usually decide at work what to make.

A retail clerk, Scott McKendrick, reports: "We decide right before we eat. We go out to the store and buy enough brown rice for several days or a package of chicken breasts or broccoli. We shop every three days." "We" actually refers to Scott's partner, Gary Hosokawa, a thirty-six-year-old bookkeeper who works for a small hotel. Earlier in the interview Scott reports that his partner cooks 75 percent of the time, and later in the interview he indicates that his partner stops at the store several days a week. In fact, Scott makes few planning decisions. His partner Gary makes most of them. Beth Wilkerson, who works as a nonprofit fundraiser, reports: "I think about it in the morning and stop and pick it up on my way home from work. But, sometimes, I get ahead of myself and will cook on a weekend for the next week."

Many times the partner who pulls together meals on a daily basis does not consider what they do as planning. One computer engineer, Brad O'Neil, explains: "I hardly ever plan. It just happens at the last minute." Further questioning reveals he often decides at work what to make and often stops at the store to buy missing ingredients. Like the other planners, Brad knows the foodstuffs available at home, he knows where to go to get what he needs, and he knows how to prepare the food. Mentally, he draws the connections between things at home, the things he needs, and a potential meal. He plans, though he fails to recognize his efforts as such.

In some instances, someone plans for a longer period of time, often a week. Those partners who plan by the week more readily recognize the planning they do. While some planners find the effort enriching and pleasant, many others express a certain amount of frustration with the process, particularly in deciding what to make. Randy Ambert, an airline flight attendant, expresses the frustration this way:

> I find that he doesn't give me any input. I rarely make things he doesn't like, but he doesn't tell me what he wants to have, I have to do that every single week. I am constantly searching for clues as to what he likes and doesn't like. I don't think he truly appreciates how much effort it takes.

Sucheng Kyutaro, an office manager for a real estate agency, explains:

> I also find it hard to figure out what she likes to eat, I think. It's a pain to get her to tell me what she will eat and then she becomes annoyed when I forget it the next time. I think about her every time I try to come up with some dinner items.

These comments illustrate one of the hidden forms of work involved in feeding: learning what others will and will not eat and learning to predict their responses. Hochschild (1983) uses the terms *emotion work* or *emotion management* to refer to this kind of empathetic activity, quite often performed by women, but as my research suggests, by many gay men as well. When thinking about domestic work, most people conjure up images of cleaning bathrooms, buying groceries and cooking meals. Emotion management involves the process of establishing empathy with another, interpreting behavior and conducting yourself in a way that "produces the proper state of mind in others" (Hochschild 1983, 7). Emotion management involves the management of feelings, both of your own and those of others. For example, it involves efforts to soothe feelings of anger in another or to enhance feelings of self-worth when someone "feels down." Sucheng's effort to "think about her partner every time" she plans a meal, in order to avoid producing "annoyance" in her partner, constitutes emotion work. Sucheng hopes to create an emotional state of satisfaction and happiness in her partner through her feeding efforts.

Most lesbigay families, just like most heterosexual families, do not engage in the emotion work of feeding in an egalitarian way. For example, partners in lesbigay relationships do not share equal knowledge of the food tastes and preferences of each other. Queries about the food preferences of partners reveal a highly differentiated pattern, where the planners possess extensive and detailed knowledge of their partner's preferences and food concerns, while their partners know comparatively little about the planner's tastes. In response to a question about his partner's food preferences, Steven Beckett, a retired real estate agent, reports that "he will not eat 'undercooked' or what the rest of us call normally cooked chicken. If he finds any red near the bone he will throw it across the room. Milk has to be low fat. Pork has to be quite cooked. He likes things quite spiced. He doesn't like peas or Brussels sprouts." Steven's partner, Anthony Manlapit, answers, "I think he likes a lot of things, I know he likes to eat out a lot." In response to the same question, Robert Bachafen, a school librarian, responds, "Yes. I stay away from radishes, shellfish, and certain soups. It's basically trial and error. He won't touch barbecued meats. I have learned over the years what he will and will not eat." His partner, Greg Sandwater, an architect, says, "I can't think of anything. He is not too fussy." Emily Fortune, a homemaker and mother of infant twins, as well as an accountant who works at home replies, "She doesn't care for pork. She has a reaction to shrimp. She doesn't like fish that much. Used to be that she wouldn't eat chicken. She doesn't like bell peppers. She doesn't like milk." Her partner, Alice Lauer, a rapid transit driver, says, "There aren't too many things she doesn't like." A finance manager for a savings and loan company, Joan Kelsey, replies, "Liver, brussels sprouts, she doesn't like things with white sauces. She is not as fond of junk food as I am." Her partner, Kathy Atwood, an accountant, responds, "I can't think of much of anything she doesn't like."

Steven, Robert, Alice, and Joan all plan meals and hold responsibility for the lion's share of feeding work. The partners of these meal planners confidently assert that their partners tend to like most things. Not true. In interviews I asked explicit questions about food likes and dislikes and found that planners indeed hold food preferences though their partners often do not readily know them. The ease with which planners cite detailed accounts of their partner's food likes and dislikes suggests that they use such knowledge with some frequency. Coming to know a partner's food preferences takes work: questioning, listening, and remembering. Successful feeding depends on this effort.

Learning about Food and Food Preparation

While learning the food preferences of others constitutes one dimension of the planning process, learning about food and preparation techniques constitutes another. The planners turn to a number of sources to inspire and guide their decisions. About half of the planners turn to cookbooks every few weeks. A part-time real estate agent, Lawrence Sing, provides the following insight into his process:

> It depends on what I am reading lately. I read a lot of food magazines and cookbooks, and we have a garden, which produces a lot of fresh greens and vegetables, that will have some bearing on it. It depends on what is seasonal. I am currently enamored with Dean Ornish's book *(Eat More, Weigh Less)*, which focuses on high-carbohydrate, low-fat, low-protein diet with lots of fiber. And I think that this would be salubrious for all of us so I am directing our eating in that direction by making sure that more of those kinds of foods appear.

The nutritional comments in this passage deserve comment, and I will return to them shortly. In addition to cookbooks, many planners read the food section of the local paper or magazines about food. Margarita Lopez, a graduate student in the social sciences, tells me:

> I read the food section of the *Chronicle* every week. I often pick up recipes from there. She complains if we have the same thing all the time, so I am always looking for more creative ways to cook things. When we go on vacation, I will buy a new cookbook, and I always look to the cooking section when we go into a bookstore, sort of searching for ideas, you know.

Lance Converse, a forty-year-old healthcare administrator, remarks:

> I make up a menu. Most of it is in my head. I make things up. He voices things that he likes and dislikes. I especially note when he says he doesn't like something. We eat lots of chicken and fish and avoid red meats. I read *Gourmet* and *Better Homes and Gardens* quite often to come up with ideas.

Others learn about food in conversations with friends and family. Michael Herrera, an office administrator, speaks of the importance of his mother to learning about food:

> We tend to make the same things over and over. I make enchiladas a lot. I make a lot of pasta, with a salad and bread. But I try to include new things. Karen [a third member of the household] doesn't do as much of the cooking. I tend to keep it all together. I turn to my mother for help often. She gives me ideas about different things to make and I will call her in the middle of preparing something if I get confused about how to do something.

Others let the marketplace guide their process. Mary Ann Callihan, an artist and craftsperson, speaks of weekly trips to the farmer's market:

> I try to decide many of the week's meals based on what is available. I often get ideas from other people I see there each Saturday. I have some acquaintances there, who I really only ever see at the farmer's market. We talk to the farmers and to their wives mostly and to one another, and I come away with ideas of how to prepare things differently.

In Mary Ann's case, shopping becomes much more than purchasing food stuffs; it becomes an occasion for learning what to prepare and how to prepare it.

Nutritional Concerns as Feeding Work

While the planners need to learn the food preferences of family members and continuously learn about food and its preparation from a variety of sources, they must
also take into consideration a whole set of concerns about nutrition. Such concerns
seem omnipresent in our society, though my research indicates some significant
variation by both gender and age regarding this issue. Many of the planners work
within fairly stringent guidelines regarding the nutritional content of the meals they
prepare. As we earlier saw with Lawrence's effort to promote a healthy diet for his
family through "directing our eating in that direction by making sure that more of
those kinds of foods appear," the planners regularly oversee the nutritional strategies
and dietary regimes. Many lesbian and gay-male (in particular) families fight over
the nutritional content of food. These concerns cast a long shadow over the entire
feeding process for some of the male planners. Joe McFarland, an attorney, states:

> We constantly fight about it. I am more conscious of fat and calorie content and seem to have
> to remind Richard constantly about it. He prepares great meals, but I am trying desperately to
> stay in shape. He gets upset because sometimes I just won't eat what he made or very much of
> it. I don't see why it's so hard for him to make low-fat stuff.

His partner, Richard Neibuhr, who does the feeding work, perceives of it this way:

> Yes. Well, we try to cut fat. We eat a lot less red meat. We eat fish and poultry and we always
> skin the poultry because that's not too good for you. We eat a lot of rice and potatoes and
> avoid white cheese and butter. It all boils down to him trying to sustain his sexual attractive-
> ness, I think. It would be easier for me to tell you things he will eat. He will not eat pork, no
> sausage. He eats one or two types of fish, he eats chicken and pasta. Anything other than that,
> it's a battle royal to get him to eat it, it's too fat! He won't eat Greek, Mexican, Chinese—forget
> it. He will eat Italian, but only with light marinara sauces, sea bass, sole, skinless chicken, and
> that's about it. He is very picky and it all comes down to his effort to look beautiful. It's a lot of
> work to come up with meals that meet his dietary standards and yet still taste good and don't
> bore you to death.

Note the extent of Richard's knowledge about his partner's food preferences. Also
note that Richard carries the burden of making sure the meals remain nutritionally
sound.

Lesbian households also report conflicts over the nutritional quality of meals.
Deborah James, a daycare worker, shares the following thoughts:

> Yeah, she doesn't want me to fry things because it makes such a mess and she has to clean it
> up. She would rather that I stir-fry and make more vegetables. She likes my cooking because it
> tastes good, but she would rather eat healthier than I would. I think my cooking sort of
> reminds her of her growing up. She partly likes that and she partly doesn't because it reminds
> her of being poor and I think the food I make sometimes, she thinks she's too good for it, that
> she should eat like rich people eat. I try to keep her happy, though.

Emily Fortune, a work-at-home accountant in her early thirties who recently became
the mother of twins, maintains concern over both her partner's nutrition and her
newborn children's nutrition. Emily, in response to a question about conflict over
food, states:

> Yes, I am still nursing, so I watch out for what the babies are going to get. We are no longer vegetarians. We both were at one time, but I eat a lot of protein for the babies. I think Alice would prefer a more vegetarian diet that was better for her. I try to think up meals that are healthy for all of them, both Alice and the babies. It's hard, though.

In most cases the meal planner becomes responsible for preparing meals that conform to dietary preferences and nutritional regimes.

In sum, planning meals, learning about foodstuffs and techniques, considering the preferences and emotions of significant others, and overseeing nutritional strategies frame the essential yet invisible precursor work to the actual daily process of preparing a meal. However, before the preparation begins, one must shop.

Provisioning Work

Shopping includes much more than the weekly trips to buy food products. DeVault (1991) recommends the use of the term *provisioning* to capture the character of the work involved in shopping. Provisioning assumes several forms of mental work that precede the actual purchase of food, including determining family members' food preferences, dietary concerns, as well as culturally specific concerns about food. Further, provisioning depends upon the following additional activities: developing a standard stock of food, learning where to buy the "appropriate" food, monitoring current supplies, scheduling grocery trips, making purchases within particular financial constraints and building flexibility into the process. Each of these components appears in lesbigay households and most often fall to the planner to orchestrate and perform.

Quite often these dimensions of provisioning go unrecognized and get subsumed in the rhetorical strategies participants use to describe the division of feeding work. More than half of the lesbigay family members use the distinction cooking/shopping to describe the division of tasks in meal preparation. This distinction creates an egalitarian impression, as in the phrase "She cooks and I shop." But the distinction conceals. In most cases the person with responsibility for cooking either did the actual shopping himself or herself or they prepared a list for the other person to use at the market. Responses to queries about who writes grocery lists illustrate this dynamic, as Carey Becker, a part-time radiologist, put it: "I write the list for major shopping for the most part, as well. She knows what brand to buy, so I don't tell her that. But I am the one who knows what we need and I make the list up for her." Daniel Sen Yung, a health educator, speaks of a similar pattern: "During the week when *we* are cooking, *we* write things on the list. He is more likely to do that, I guess, because he is cooking and will run out of stuff, so I get it at the store. He knows what we need." Note the recurrence of the phrase "knows what we need." In both instances the phrase refers to the possession of a stock knowledge of foodstuffs. Planners develop and possess an extensive mental list of standard ingredients used in their kitchens. In the research I asked to see the current grocery list, should one exist. I saw thirty-two such lists. In the wide majority of cases, just one person wrote the list or wrote more than three-quarters of the items on the list. In those relationships where participants make a distinction between shopping and cooking meals, lists become longer and more detailed. For instance, many planners specify brand names or write down terms like *ripe avocado* instead of just *avocado*. The cumulative effect of such detailed list writing greatly simplifies the work of shopping and undermines the seeming egalitarianism of the cooking/shopping distinction.

The Significance of Small Grocery Trips

In most households, shopping includes a number of smaller trips to supplement throughout the week. In the wide majority of lesbigay families, one person makes these supplemental trips. In most cases, the person who bears responsibility for meal preparation does this type of shopping. They often stop at the store at lunch or, more often, on their way home from work. While many families initially indicate that they split cooking from shopping, in reality the person who cooks often shops throughout the week while the other partner makes "major" shopping trips, usually on weekends and often using a shopping list prepared by the planner at home. At first glance the intermittent shopping trips during the week appear ad hoc and supplemental in character. Yet for many lesbigay households, particularly in the lesbian and gay enclaves of San Francisco, these little trips constitute the essential core of feeding. Many planners shop at corner markets near their homes or places of work, frequently purchasing the central ingredients for the meal that evening—fresh meats and vegetables, breads, and pastas. In many respects the weekend shopping actually looks more supplemental. Again, the grocery lists prove instructive. They contain many more entries for items like cereals, granola bars, sugar, mustard, yogurt, and soda than for the central elements in evening meals: vegetables, fresh pasta, fish, chicken, bread, potatoes, corn, prepared sauces, and often milk.

The intermittent shopper often operates with a great degree of foresight. Alma Duarte, a bookkeeper for a small business, belies the ad hoc characterization of daily grocery trips:

> Every couple of days, I go to the store. I do the in-between shopping. Every couple of days, I run down and pick up stuff we need or are running out of. I buy vegetables, fish, and bread. I buy the heavy items at Calla Market because she has a back problem, so I buy like soda and detergent and kitty litter, but she does the big shopping on weekends. I almost always go and buy fish on Monday afternoons at a fish market near where I work and then I stop at the produce market near home. And I have to go and get fresh vegetables every couple days over there.

Alma speaks of at least three destinations, and she obviously organizes her schedule to accommodate these different trips. One might think that this kind of shopping appears rather routine, after all, she describes it as "in-between." Note that she does not consider this effort major, rather her partner does "big" shopping. Yet each of Alma's little shopping trips consists of a rather large number of choices about feeding. She must decide on what kind of fish and conceive of other items to serve with it. She decides when to go for vegetables and chooses among myriad varieties, making sure not to buy the same ones over and over yet also measuring the quality of the produce. Interestingly, she identifies her partner as the shopper, though she actually makes most of the feeding and provisioning decisions for her family.

Deciding Where to Shop

Deciding where to shop often appears as a matter of routine, even to those who make such choices. Yet these choices mirror complex thought processes and a variety of social-structural considerations. When asking participants where they shop, most offered the name of a large grocery store nearby. Most participants indicate that they chose grocery stores based on two factors: proximity and selection. However, when asking about and going with planners to where they shop, a much more elaborate answer emerges. To begin with, planners, like Alma, often shop at more than one store. Lily Chin, an office manager, describes her shopping strategy:

> I shop at Food for Less, Trader Joe's, Costco, and a corner market. It depends on what I need and how much time I have. If I have a lot of time, I might go over to Trader Joe's, but there it's hit and miss. Sometimes they have what I need and sometimes not. Usually, to go there, you take a lot of time and you thoroughly check out their selection. You let their products decide what you buy. If I am picking up fish for that night, then I stop at Andronico's because they have such a nice fish selection. If I just need some fresh vegetables I will go home and park the car and walk down to the corner produce market. Sometimes I call Carol and ask her to stop. She stops at Food for Less because she can get in and out quickly. She doesn't like to go to places where you have to find a parking space.

Her partner, Carol Len, a social worker, says that they shop at Food for Less because of the convenience. She said little else. In contrast, planners report thinking about a wide array of factors in deciding where to shop. Lily makes decisions based on a constellation of factors: menu, parking, time availability, and quality. Lily traverses this constellation of factors on a routine basis. The weighing of these factors and the creation of a daily strategy to achieve a successful meal remains mostly invisible, yet essential. Lily's partner portrays the shopping effort as one consisting of mostly convenience, a central factor in her own experience of shopping, but just one concern among many in an effort to feed a lesbigay family.

Moreover, the decision of where to shop involves concerns about social class, ethnic identity, and the sexual identities of lesbigay family members. Delineating the significance of each of these social-structural factors poses great difficulties, yet their import appears clear. For instance, concerns about the healthfulness of food reflect class-related concerns, but it also reflects concerns for many gay men about physical appearance. DeVault points to the centrality of food style as a marker of class identity in contemporary society (1991, 203–26). She notes the middle- and upper-middle-class concern with producing interesting and entertaining meals, the preference for elaborate and exotic meals, the preoccupation with the healthfulness of meals—all symbolic indicators of class standing in our society. These matters of style appear in lesbigay households as well, with more affluent, more professional families emphasizing such concerns. Working/service-class lesbigay families report much more routine in meal preparation, and they go to fewer shops. Partly this occurs because such families do not own automobiles, relying instead on public transportation, which of course is far less convenient. They emphasize the closeness of stores as central to their shopping habits. But beyond this, such families report a standard stock of meals cooked quite frequently. For instance, a lesbian family where one woman works as a retail clerk and the other as an accounting clerk, often eat a product called Chicken Tonight at least twice a week. Letty Feldmen, who does much of the provisioning, remembers in her weekly shopping to buy chicken and two jars of the product. Such families report engaging in less feeding work in terms of the variety and elaborateness of the meals, although, as I will show shortly, working/ service-class lesbigay families eat more meals at home and spend more time managing the cost of meals. For affluent families, the meal planner takes into consideration the class-related concerns about style in deciding where to shop. John Chapman, a homemaker partnered with a successful attorney, implies these class-related concerns in his shopping routine:

> I buy most things out of my head. I go down the aisle and it's like a major deal to go. I like to look and find out what's new. I go over the grocery store with a fine-tooth comb. I am always looking for something new and interesting to make. I try to get away from routine and keep things constantly new. I will pick up prepared meals sometimes from a specialty deli. And whenever we have someone over, well, then I really make sure it's something that people will enjoy and enjoy talking about. That often means that I will have to shop somewhere much nicer, like Petrini's Market or Andronico's in the inner Sunset.

In addition to these class-related concerns, many lesbigay families also take ethnic identity concerns into account in their shopping routines. Henry, a Latino social worker for a Bay Area county, pointed to the importance of maintaining his and his partner's Latino identity in his decision of where to shop:

> I go on Mission Street to the open markets a lot. I like the markets because the fruit and vegetables are good, and I can speak Spanish if I don't understand things. I think it's impor-tant to support the merchants in the community. Food is a part of our heritage, and I don't want us to lose that. I try to keep us connected, in part, through serving meals that reflect our identity. I can buy tortillas and cheese for cheaper prices at Costco, but I don't because that undermines the Mission merchants. It's more of an effort sometimes to find what you are looking for, but you also find things you can't find in white stores.

Sandy Chao, a director of a daycare center, sounded a similar theme:

> I try to get over to Clement Street when I can. There are many Chinese grocers and produce markets over there. I also try to go out on Ocean Avenue, where there is a new Vietnamese store. I try to support the Asian businesses as much as I can. It's just a part of being who we are, and without those things, we lose some of our identity.

While class and ethnicity shape the provisioning choices of meal planners, concerns about lesbian and gay identity and community also influence provisioning work. Many participants spoke of the importance of shopping in lesbigay-owned businesses, or at least in lesbigay-friendly territory. Bill, an artist and painter, expresses this sentiment:

> I try to shop in the gay community. A lot of times that means paying more. The "gay tax," you know. But I feel a lot safer; not that most of the city isn't safe, it's just that I like to be with my own kind. It makes it less stressful. Sometimes it means waiting in longer lines. Like Market Street Safeway, for instance. I could shop elsewhere, I have a car, but I like to go there because that's where the lesbigay people are. I often stop at Harvest Market in Castro, too. They have

great produce and a lot of healthy foods, even though it's expensive, it's gay and it's healthy—
both of those things are pretty important to me a lot of the time.

As the above quotations make clear, deciding where to shop involves a vast array
of considerations. Balancing the various concerns mostly falls onto the shoulders of
the planners in lesbigay families.

Monitoring: Supplies, Schedules, and Finances

Another component of feeding work revolves around the efforts of planners to
monitor the supply of foods and other household products. DeVault suggests the
complexity of monitoring work:

> Routines for provisioning evolve gradually out of decisions that are linked to the resources
> and characteristics of particular households and to features of the market. . . . Monitoring also
> provides a continual testing of typical practices. This testing occurs as shoppers keep track of
> changes on both sides of the relation: household needs and products available. (1991, 71)

Lesbigay family members, both planners and others, attest to clear patterns of
specialization when it comes to who keeps tabs on both food products and other
household products like cleaning supplies, toiletries, and items like dinner candles.
Very few families reported splitting this effort up equally, and even those that did, did
so gingerly. Susan Posner, an employment recruiter in the computer industry,
recounts: "Neither one of us keeps track. It just kind of surfaces that we need something. We don't have a list or anything. I guess whoever runs out of toilet paper first.
And I guess I run out of toilet paper a lot. *[Laughter]* Okay, so maybe I do it." Susan's
comments should give pause to students of domestic labor. Her comments reveal not
only that she does the work of monitoring supplies, but also that she seems either
unaware or perhaps to be attempting to deny that she does it. The planners, who do
much of the monitoring, frequently speak of the dynamic character of the work. Tim
Cisneros, a registered nurse, describes how he needed to change his routine in order
to get the right deodorant for his partner:

> I mostly shop at Diamond Heights Safeway, though now I go over to Tower Market as well, at
> least once a month. I started doing that because Paul is hyperallergic to most deodorants, and
> he needs to use one special kind, and Safeway stopped carrying it. I tried to get it at the
> drugstore next to Safeway, but they don't carry it, either. So now, I go to Tower to get it. I buy
> other stuff while I'm there, so it's not really a big deal.

Tim captures the dynamic quality of provisioning. As demands change in the
household or products change in the market, he comes up with new strategies to
maintain equilibrium. Among roughly half of the families reporting a shopping/
cooking division, more discussion of products occurs. Narvin Wong, a financial
consultant in healthcare, comments:

> We sometimes get our wires crossed. I buy what he puts on the list, and that's almost always
> what we usually buy. I mean, I know what we need, but sometimes he changes his mind about

what he needs and I don't always remember him telling me. Like a few weeks ago, he put olive oil on the list, and I bought olive oil. He says he told me that he wanted to start using extra-virgin olive oil. Well, I didn't hear that, and he yelled and steamed about it when he unpacked the groceries.

Narvin's comments suggest that perhaps less of this kind of conflict takes place in households where one person performs both the actual shopping and the provisioning work behind the trips to the store. Further, Narvin's partner, Lawrence Shoong, says that he often tries to go with Narvin to the store. Why?

Because it gives me a chance to see what's out there. Narvin doesn't look for new things. Even if he does, he doesn't tell me about them. I like to know what's in season and just to see what's new. And inevitably, he forgets stuff. I know I should write it on the list, which is what he says, but when I go to the store, I can go up and down the rows and remember what we have and what we need.

Again, this points to the interdependent and dynamic character of feeding work. To do it successfully, given the way our society distributes foodstuffs and defines appropriate eating, the meal planner needs to stay in contact with the marketplace. Many planners do this through reading grocery flyers in the paper or in the mail, but many also try to stay in contact with the store itself. As Lawrence's comments suggest, much of provisioning work takes place in one's mind, the place where much of the hidden work of monitoring takes place.

Just as the work of monitoring the household and the marketplace come into view as highly dynamic processes, so too does the planner monitor the dynamic schedules of family members as a part of provisioning work. This means that many planners shop with the goal of providing a great deal of flexibility in meal options. For instance, many planners report selecting at least some dinner items that they can easily move to another night of the week should something come up. Sarah Lynch, a graphic artist who works in a studio at home, captures the dynamic circumstances under which she provisions meals:

I never have any idea when Andrea will get here. She may stay at the bank until 11:00 at night. Sometimes, she doesn't know up until right before 5:00 whether she will be able to come home. So I still want us to eat together and I want her to get a decent meal, so I try to buy things that I can make quickly and that still taste good. She will call me from her car phone as soon as she is leaving the city. That gives me about an hour. I often will make something like a lasagna that I can then heat up when she calls, or I buy a lot of fresh pastas, in packages, you know the ones, and I will start that when she calls. I also try to buy a lot of snacklike items, healthy ones, but things like crackers and trail mix and dried fruit so that she can eat those things if she is really hungry and went without lunch or something, and so I can eat while I am waiting for her to get here.

Similarly, Matthew Corrigan, an office administrator, provisions meals to accommodate the schedule of his partner:

I sometimes find it hard to keep a handle on things. Greg is active in a number of voluntary things—our church and a hospice for PWAs—and he serves on a City task force on housing issues. I am never completely sure he will be here for the meals I plan. So I try to have a lot of food around that we can make quickly and easily, like soups, veggie burgers, and pastas. If it gets too tight, and it often does, we will eat out, or just he will eat out and then I eat something at home, and hopefully I have something here to make.

Both Matthew and Sarah provision their households in light of the need to build in flexibility around the work and social schedules of their partners and themselves. The effort they put into this kind of dynamic provisioning frequently goes unnoticed and often appears routine, but they clearly think about these scheduling concerns in the work of provisioning for their families.

In addition to the efforts planners make in monitoring schedules, supplies, and markets, many also report concerns about monitoring finances. Participants refer to financial concerns in deciding what to eat and where to shop. The cutting of coupons both illustrates financial concerns and demonstrates the de facto division of provisioning work in many lesbigay families. Rarely do all family members report cutting coupons. For the most part, one partner, the planner/provisioner, cuts coupons. Tim Reskin, a clerk in a law firm, describes his use of coupons:

> Over time I have developed a sense of the brands that I know that we prefer. Sometimes I have coupons that I use. It's not like I will use any coupon, but if there's something that seems interesting or we don't have an opinion about, I might use that to decide. I always go through the Sunday paper and cut out the usable coupons. Cost is a big criterion for us.

In addition to cutting coupons, the less affluent households more often report that they compare prices, watch for sales and buy large-portioned products at discount stores like Costco and Food for Less. Lower-income planners also report spending more time reading grocery advertisements and going to different stores to buy sale items with the purpose of saving money in mind.

The estimates provided by participants regarding food expenditures provide additional insight into the division and character of feeding work. Weekly grocery expenditures vary significantly for lesbigay families, ranging from thirty dollars a week in the less affluent households to over two hundred and fifty dollars a week in the wealthier ones. Not that family members always agree on the cost of groceries. Planners estimate spending roughly thirty dollars more per week on groceries than their partners estimate. The thirty-dollar figure functions both as the mean and the median among the one hundred and three adult participants.

This knowledge gap in food expenditures points to several interesting dynamics. First, it reflects planners' knowledge of the cost of the many small trips to the store during the week. Second, it indicates planners' greater attentiveness to the cost of food items. Finally, it often points to the expectation of family members that planners should monitor and limit the cost of food. Consider the following examples. Tim Reskin and Phillip Norris live in a distant East Bay suburb in a modest apartment. While they both work in the city, they live in the suburbs to avoid the high cost of housing in the city. Phillip performs much of the work of feeding their family. He plans the meals and creates much of the grocery list. In explaining why Tim does the larger shopping on weekends, they both speak about financial concerns. Tim puts it this way: "Money, that's a major part of the reason why he doesn't go to the store. For him, he doesn't take price into consideration as much as he should. I use coupons and don't get distracted by advertising gimmicks at the store. I am more conscious of money." Phillip sees things somewhat differently, but points to the issue of cost as well:

He shops. He feels he has more control at the store. He feels he's a smarter shopper. I tend to look for high quality, whereas he tends to look for the best price. He does seem sharper. Well, he thinks he has a better handle on excessive spending. I don't know, though. You know, he does limit things during the major shopping, but then I have to go out and get things during the week. I am very, very careful about watching the cost. But you know what? Partly I have to go out and get things because of his complaints. He says that I cook blandly, like an Englishman. But the fact is, I work with what he brings home, and if he won't buy spices or sauces or whatever, in order to save money, then the food will taste bland. He denies that's what happens, but it is.

Phillip actually bears a significant part of the responsibility for monitoring food costs. Note the phrase "I am very, very careful about watching the cost." Phillip's comments illustrate the interdependent character of monitoring costs and planning meals, but further, he also must consider his partner's satisfaction with the meals.

A strikingly similar example emerges within a lesbian household. Marilyn Kemp and Letty Bartky live in one of the lesbigay neighborhoods of San Francisco. They both work in lower-level administrative jobs and find themselves struggling financially. They talk about the high cost of housing and how to cut corners in order to stay living in San Francisco. Letty, who performs much of the feeding work in the family, comments on Marilyn's approach to weekend shopping:

Marilyn likes to shop like a Mormon—you know, be prepared for six months. She buys these huge boxes of stuff. I think it's silly. I am more into going two or three times a week. Also, I like to take time to make up my mind, while she just wants to get through the store as fast as possible. She bitches that I don't make interesting things to eat, but what does she expect given our financial constraints and her shopping regimen.

Marilyn sees it differently: "We don't disagree much at all about shopping. I do it because I am more cost-conscious than she is."

Both of the above families suggest that it is one thing to manage the cost of groceries while shopping, but another matter entirely to manage the cost of groceries in the broader context of feeding the family.

Preparing Meals

The actual physical work of preparing meals each day requires thorough analysis. In some instances, the physical preparation of the meal occasionally begins in the morning when meal planners take items from the freezer to defrost for the evening meal. Some planners report other early morning efforts such as marinating meats, vegetables, or tofu. Most planners begin the meal preparation shortly after arriving home from work. Many report emptying the dishwasher or putting away dry dishes from the rack as one of the first steps in getting ready to prepare the evening meal. This points to the ambiguity of the cooking/cleaning distinction offered by many participants. The majority of meal planners arrive home from work earlier than their partners, more than an hour earlier in most cases. Depending on the menu items, which vary widely, participant's estimate that meal preparation takes approximately an hour. The preparation of the meal involves mastering a number of different tasks, including coordinating the completion time of different elements of the meal, managing unexpected exigencies like telephone calls or conversations with family members, coping with missing ingredients or short supplies, and engaging in all of the techniques of food preparation, from cutting vegetables to kneading pizza dough to deboning fish to barbecuing meats.

Many meal preparers find it difficult to capture the character of the process involved in creating meals. Even those who seem well versed in cooking find it difficult to characterize the process in its true fullness and complexity. Clyde Duesenberry, who prepares most of the meals in the house, comments:

> We make pesto quite a bit. We have that with fried chicken or some sausage or whatever. We like breaded foods. We like Wienerschnitzel. We will have bleu cheese on burgers. We roast chickens quite often. We watch cooking shows quite a bit. The story about Mike, he can do it if he wants to, he knows the basics of cooking. One time I got called out on a call on a Sunday evening and he took over and we never had chicken as good as he made it. He cooks the broccoli. He knows how to do that. It isn't the recipe that makes a cook, it's the mastery of techniques. I can't even begin to cover all the territory of cooking you are asking about.

The meal preparer realizes the complexity of the work involved and struggles to put it into words. Usually their words belie the full extent of their effort as they struggle to express what they do. Another participant, Daniel Sen Yung, says:

> I tend to steam things a lot. I use dressings. A lot of salads and stuff. I don't know [said with exasperation], it's hard to describe, I just do it. Each time it seems like there are different things to do. I call someone and ask them, or I look at a cookbook, or I just experiment and hope for the best. I make some things over and over and each thing has its own routine.

The daily physical process of meal preparation takes on a highly dynamic and thoughtful quality. While some meals take on a routine quality in some households, for most the process appears much more vigorous and multifaceted. It requires the constant attention, the knowledge, and the physical labor of meal preparers.

Feeding and Cleaning

In contrast to meal preparation, the cleanup of the evening meal appears much more routine, requiring less mental effort, less time, less knowledge, and less work. As previously indicated, many lesbigay households use a cooking/cleaning distinction to explain the organization and division of feeding work. However, in close to one-third of the sample, the person who prepares the meal also cleans up after the meal. The cleaning component deserves closer analysis. Those who clean up the kitchen estimate a median time of thirty minutes. They talk about clearing the dishes, loading the dishwasher, putting away leftover food, and wiping the counters and the stovetop. Some include taking out trash or wiping floors, but not most. Among the less affluent families, participants include washing and drying the dishes. As I briefly noted earlier, meal preparers often empty the dishwasher or put away dry dishes when they begin preparing the evening meal. The work of cleaning up is highly routinized in most lesbigay households, requiring little decision making and little emotional work. Barbara Cho, a shift supervisor for a hotel, notes that the cleanup actually allows her time to think and unwind from her day: "It's not a big deal. I clean off the table. Sandy helps bring in the dishes. I load the dishwasher, wipe off the counter, and put stuff away. It's a great time for me to think about things, I often reflect on my day or decide what I'm going to do that night. It helps me unwind." Rarely do meal planners/preparers conceive of their feeding work in these terms. Most spoke of the importance of staying focused on meal preparation in order to avoid burning meats or overcooking vegetables, and of coordinating meal items so they reach completion at the same time.

Several meal preparers also note that they make some effort to limit the mess caused by meal preparation and actually do a lot of cleaning as they go. Sucheng Kyutaro, who prepares most meals eaten at home, notes:

> She complains a lot if I make too much of a mess during cooking. So I kind of watch it as I go. I try to clean the major things as I cook. I will rinse out pans, like if I make spaghetti sauce, I will run all the remains from the vegetables down the garbage disposal and rinse out the sauce pan and I always wipe off the stove. She really doesn't like cleaning that up at all.

Gary Hosokawa, a payroll supervisor who does much of the feeding work in his family, remarks: "Usually I cook and he cleans. Although I am really anal about keeping a clean kitchen, so I clean a lot while I am cooking. There is not that much for him to do."

The preceding comments demonstrate the limited character of cleaning up after meals in many lesbigay households, and they further undermine the salience of the cooking/cleaning distinction employed by many participants to indicate the egalitarianism of their household arrangements.

Feeding Work and the Creation of Gender, Class, Ethnic, and Family Identities

Feeding and the Production of Gender Identity

Recent empirical and theoretical work on the sociology of gender conceives of the production or achievement of gender identity as resulting from routine and continuous engagement in certain kinds of work and activities socially defined as gendered (Berk 1985; Coltrane 1989; West and Zimmerman 1987). This perspective emerges from a school of thought in sociology that understands gender as a dynamic and purposeful accomplishment: something people produce in social interaction (Cahill 1989; Goffman 1977; Kessler and McKenna 1978). West and Zimmerman point to the significance of action, interaction, and display in the process of "doing gender": "a person's gender is not simply an aspect of what one is, but, more fundamentally, it is something that one does, and does recurrently, in interaction with others" (1987, 140). Gender is not the product of socialized roles in which individuals continually recast themselves. Rather, gender requires continual effort to reproduce in everyday life. Since in this society and many others gender constitutes an essential component of the system of social classification, "doing gender" results in keeping social relationships orderly, comprehensible, and stratified. Frequently, individuals possess an awareness of doing gender while deciding how to conduct themselves in daily life. How masculine should one appear while observing a sports event? How feminine should one appear in a television interview? Berk demonstrates how household tasks function as occasions for creating and sustaining gender identity (1985, 204). Coltrane (1989), in his study of fathers who become extensively involved in the work of childcare, shows how such men must manage the threats to gender identity that such work poses to them. To violate the gendered expectations of others often leads to stigma and to challenges to the gender identity of the violator. Coltrane found that men who care for (feed, clean, teach, hold) infants often face stigma from coworkers and biological relatives, and that oftentimes the men hide their caring activities from these people to avoid conflict and challenges to their masculine identity. Men performing domestic labor—or women who fail to—produces the potential for stigma, a matter of great significance for gay and lesbian couples, where the reality of household life clashes with cultural gender expectations.

Accordingly, managing the gendered identity of members of lesbigay families becomes a central dynamic in the portrayal of feeding work both within the family and to outsiders. In general, feeding work in the household constitutes women's work, even when men engage in the work. That link of feeding with the production of womanly status persists and presents dilemmas for lesbigay families. Let me begin my analysis of this dynamic by pointing to a rather odd thing that happened in interviews with lesbigay family members. I interviewed family members separately to prevent participants from constructing seamless accounts of household activities. In so doing, inconsistencies occurred in the portrayal of domestic work, including feeding work. In six of the twenty-six male families, both claim that they last cooked dinner. In four of the twenty-six female couples, both claim that the other person last cooked dinner. How can one explain this? Were the participants simply confused? Why a persistent gendered pattern of confusion? Lesbian families do more often report sharing in the tasks of meal preparation than do gay-male families, so here the confusion may reflect the presence of both partners in the kitchen. Participant observation confirms such a pattern. For instance, in two of the four female households observed in depth, both women spent the majority of time during meal preparation together in the kitchen. I do not mean to imply that they share every task or divide meal preparation equally. Frequently, they engage in conversation and one person assists by getting things out the refrigerator, chopping celery, or pulling something out of the oven. Mostly, one person prepares and manages the meal while the other helps. The question remains, Why do men who did not prepare a meal claim to have done so, and why do women report that their partners who assisted actually prepared the meal?

The answer looks different for female and male households. In some lesbian couples the partner who performs much of the feeding work seems to also concern herself with preventing threats to the gender identity of her less domestically involved partner. This pattern seems most persistent among lesbian couples where one of the partners pursues a higher-paying, higher-status occupation. Consider the following examples. Cindy Pence and Ruth Cohen have been together for eleven years. Cindy works as a nurse, and she does much of the feeding work in the family.

Ruth works as a healthcare executive. Ruth works extensive hours, and it often spills over into their family life, something Cindy dislikes. Ruth acknowledges that Cindy does much of the domestic work in their relationship, including the feeding work. Ruth says that she tries to help out when she can, and she tries to get home to help with dinner. They each claim that the other person cooked the last meal at home. During my interview with Ruth I asked about work and family conflicts, and how work might impinge on family life. Ruth comments:

RUTH: Cindy's great about my work. She does so much. I don't think I could handle it all without her. She sort of covers for me, I guess and I feel guilty about it, but I also know that she appreciates how hard I work for us.
CC: What do you mean, she covers up for you?
RUTH: I mean, she gives me credit for doing a lot of stuff at home that I don't really do. I mean, I help her, but it's not really my show. She does it and I really appreciate it. But I feel terrible about it.

The same kind of feelings emerged in an interview with Dolores Bettenson and Arlene Wentworth. Both women work as attorneys for public entities, though Dolores's job requires less overtime and allows for a more flexible schedule. Dolores reports working for wages forty hours per week, while Arlene reports working for wages around sixty hours per week, including frequent trips to the office on Saturdays and on Sunday evenings. Dolores handles much of the feeding work of the family, while the couple pays a housekeeper to do much of the house cleaning. Both Dolores and Arlene initially report that they split responsibility for cooking. Arlene, in response to a question about conflicts over meal preparation, says:

> I think things are pretty fifty-fifty, we are pretty equal. I guess she does things more thoroughly than I do, and she complains about that, but she always gives me a lot of credit for the stuff I do. Sometimes, I think she gives me too much credit, though, and I feel guilty about it, because, as I said, she takes that kind of stuff more seriously than I do, I just don't have as much time.

Both Ruth's and Arlene's comments reveal a pattern of the more domestically involved partner assigning credit for completing domestic work that the less domestically involved partner did not do. I suspect that this occurs in part to provide "cover" for women who spend less time doing domestic work, less time "doing gender."

In a similar vein, the pattern among gay males appears the opposite, and more intensely so, in some respects. Engaging in routine feeding work violates gendered expectations for men. I emphasize routine because men can and do participate in ceremonial public cooking such as the family barbecue. Yet the reality of household life requires that someone do feeding work. In heterosexual family life, men are usually capable of avoiding feeding work, but in gay-male families, someone must feed. Only in a very few, quite affluent households did feeding work seem particularly diminished and replaced by eating most meals in the marketplace. Most gay-male couples eat at home, and among many of them I detect a pattern of men colluding to protect the masculine status of the meal planner/preparer. The conflicted claims of who last prepared an evening meal illustrate this dynamic and also reveal the ambiguous feelings held by men who feed about their status and their work. Bill Fagan and Rich Chesebro have been together for three years. Rich works at a large software company and Bill works as an artist. Bill often works at home and carries much of the responsibility for domestic work, including feeding work. Both men claim that they last prepared the evening meal. Other questions in the interview reveal that, in fact, Bill prepared the last meal. When I asked about the last time they invited people over for dinner, Bill replied that it was two days before the interview, on Saturday night. When asking about what he prepared, he responded: "Well, let me see, *last night I made lasagna.* Oh, and that night, Saturday, I broiled tuna." It turns out that Bill made the last meal, though Rich claims that he did. Why should Rich make such a claim? Part of the answer lies in Rich's concern, expressed several times, that Bill not become overly identified with domestic work. When asking Rich about who last went to the grocery store, Rich replies: "I think that Bill might have, but it's not that big of a deal really. He really likes his work as an artist and that's where *his true interest lies*" (emphasis added). This response initially confused me. I ask about a trip to the store, and I receive an answer emphasizing Bill's work as an artist. I let this response pass, but later in the interview, I ask Rich about who last invited someone over to dinner, and who typically does this. In a similar rhetorical move, Rich says: "Well, I suspect Bill might be the one to do that, but I don't think it's that significant to him, really. *His real love* is his work as an artist, that's where he puts most of his energy" (emphasis added). At this point in the interview I decide to pursue Rich's intent in moving us from matters of domesticity to Bill's status as an artist. I ask Rich why he brought up Bill's work as an artist in the context of who most likely invites people to dinner. "Well, because I worry that people will get the wrong idea about Bill. I know that he does a lot of stuff around here, but he really wants to become an artist, and I don't want people to think of him as a housewife or something. He has other interests." As these comments disclose, Rich attempts here to manage the identity of his partner. This interpretation receives further confirmation on the basis of Rich's answer to a question about how he would feel about his partner engaging in homemaking full-time and working for wages only partly or not at all:

> Well, I wouldn't like it at all. I don't see how that could be fair, for one person to contribute everything and the other to give little or nothing to the relationship. Plus, what about one's self-respect? I don't see how one could live with oneself by not doing something for a living. I would not be comfortable at all telling people that Bill is just a housewife. If he wanted to do his artwork and do more of the housework, that would be okay, I guess, but that's kind of how we do it already.

While Rich attempts to shield Bill from identification with domestic work, both in order to protect Bill from the status of "a housewife or something" but also to confirm his own belief that domestic work holds little value, the reality remains that Bill does much of the domestic work, including feeding work. Doing feeding work ties Bill to a more feminine identity. Bill put it like this in response to a question about whether the roles of heterosexual society influence the character of his relationship with Rich:

> I think that the functions all need to be handled. There is a certain amount of mothering that is required and whether that is done by a man or a woman does not matter. But mothering per se is an important function. And there is a certain amount of fathering having to do with setting goals and directions and creating focus. I guess people do think of me as more of the mother in our relationship, because I cook and invest a lot in our home, but that's their problem. Sometimes I feel strange about it, but I like to do it and I like the family life that we have together.

Bill's words capture the ambiguity of feelings about feeding work and other domestic work that I heard frequently in many lesbigay households. On the one hand, Bill recognizes the importance of such work (mothering) to creating a family life. On the other hand, he feels strange about his participation in domestic work. His partner worries about people identifying Bill with domestic work and emphasizes Bill's identity as artist.

Another gay-male household illustrates a similar set of dynamics. Nolan Ruether and Joe Mosse have been together just under two decades. They live in an affluent suburban community outside the city. They both work in healthcare, though in very different settings and with quite different responsibilities. Nolan is a registered nurse and reports working forty hours per week. Joe works in a medical research lab and reports working closer to fifty hours per week. He also has a part-time job on the weekends. Nolan handles much of the domestic life of the family, including much of the feeding work. Both partners claim that they last cooked an evening meal. They actually eat separately more often than not, with Nolan eating a meal at home in the early evening that he cooks for himself and Joe either eating something on the run or eating something late when he arrives home. Joe indicates that he does the major shopping on weekends, though Nolan makes frequent trips to the market during the week and says that "I often pick up things that will be easy for him to prepare when he gets home from work, and I frequently will make something that he can simply warm up when he gets in." Nolan actually cooked the last meal, while Joe warmed up the meal when he arrived home late. Throughout the interview Joe emphasizes the egalitarian character of their relationship and diminishes the amount and significance of domestic work in the household. When I ask about conflicts over meals or meal preparation, Joe responds, "Um, well, we hardly ever eat meals at all. There is no work to conflict about. I eat out and he tends after his own." Nolan reports that they eat at home half the week, while they go out the other half. Nolan reports that they do not plan meals, though he says that they do communicate on the subject: "We don't plan meals, really. We either are both at home and I ask him what he wants or I call him at work and then I just go to the store and buy it. We don't keep a lot of food here, because I tend to run out to the store most every day." Nolan does much of the feeding work. He engages in routine provisioning work for the family, and he plans many of the family's social occasions that involve food. He also does most of the emotional labor related to food. Now consider Nolan's responses to questions about his feelings toward traditional gender roles for men and women in American society:

> I certainly see the value in it, in ways I never did before I was in a relationship. There's a lot of work to be done to keep a house nice and to make life pleasant. I get pretty tired sometimes, I don't think Joe has any sense of it, really. He is off so much doing work, but he works so much by choice. You know, I think I said, we don't really need the money, but he couldn't imagine being around here doing this stuff.

Does he think that traditional gender roles influence the pattern of domestic life in your relationship?

> Well, in the sense that I do everything and he does very little, yes, I think it resembles the traditional pattern. He would of course deny it and get angry if I pushed the topic, so I don't bring it up, and I feel it's kind of difficult to talk to anyone about it because, well, because they might think of me as a complaining housewife or something, and I don't think most people can understand a man doing what I do. So when he says there isn't that much to do around here, I just sort of let him believe what he wants. It isn't worth the trouble.

Notice the ambiguity of feeling Nolan expresses about talking to others about his situation. Nolan's restraint (emotion work) actually enables Joe to diminish the importance of the work that Nolan does to maintain family life. Joe's approach to his household life closely resembles the pattern of need reduction detected by Hochschild (1989, 202) among some men in her study of heterosexual couples. Consider the following excerpts from our interview:

CC: Tell me about your feelings toward traditional roles for men and women in the family in American society.

JOE: I think those roles have declined a lot. It's more diverse now. I am really glad that such roles have declined. I feel that there should be two people out earning incomes. I don't think that people should stay at home. I can't see the value in it. Everyone should have outside interests. And I especially don't think that a man should be stuck in the home, cooking and stuff like that. Nolan works full-time, and we mostly eat out.

CC: Do you feel like the prescriptions for such roles influence or shape your relationship? Why or why not?

JOE: Definitely not. We are both men and we both work for a living, and so we don't really fit those images. We don't have much domestic work here, especially since I am not here that much.

CC: What would you think of your partner or yourself engaging in homemaking full-time and working for wages only partly or not at all?

JOE: I would not be too pleased with it. There has to be a common goal that you both work toward. For one to contribute everything would not be fair. I don't make that much of a mess, and so I don't think that there would be anything for him to do.

Throughout this exchange Joe not only diminishes the presence of domestic work through emphasizing how often he is gone but also expresses his concern that his partner not become overly identified with domestic work. Nolan's own feelings of ambiguity about doing domestic work, and the threat it poses to his gender identity, actually keep him from talking about his circumstances.

One observes another example of the salience of gender to the portrayal of feeding work in dinner parties. Using Goffman's conceptions of "frontstage" and "backstage" work, sociologist Randall Collins suggests that cooking meals for dinner parties constitutes a frontstage activity that "generally culminates with the housewife calling the family or guests to the table and presiding there to receive compliments on the results of her stage (or rather table) setting" (1992, 220). The backstage work, much more arduous and time intensive, consists of a wide array of different kinds of invisible work: planning, provisioning, and monitoring. The work is often invisible in the sense that these forms of work receive little public recognition during dinner parties. In most lesbigay families the responsibility for both the elaborateness and the exoticness of foods for dinner parties becomes the responsibility of the meal planner/preparer, and often this person takes front stage at the dinner party.

However, in ten of the lesbigay households, and in contrast to the normative pattern in heterosexual families described by Collins, the person who cooks for dinner parties often only engages in the frontstage work while the other partner performs much of the backstage or more hidden forms of work. These ten households share a common pattern. In the male families the person who performs routine feeding work and performs the backstage work for dinner parties also works for wages in somewhat female-identified occupations: two nurses, a primary education teacher, a legal secretary, a social worker, and an administrative assistant. The frontstage males work in male-identified occupations: two accountants, an engineer, an attorney, a physician, and a midlevel manager. In the female families exhibiting a split between frontstage and backstage feeding, frontstage women work in male-identified occupations: two attorneys, a higher-level manager, and a college professor. The occupations of the backstage females include two retail sales workers, a nurse, and an artist. Taking the front stage in such dinner parties may well function as a strategy on the part of these lesbigay couples to manage threats to the gender identity of the domestically engaged men or the less domestically engaged women.

Confirming this pattern in the words of participants proves somewhat elusive. None of the women who do backstage feeding work for dinner parties expressed dissatisfaction about this. And while several of the men complain that they do not receive credit for the backstage work they do in preparation for such dinners, they also seem reluctant to make much of a fuss about it. Tim Cisneros, who works as a nurse and does much of the routine feeding work, responds to a question about who last prepared a meal for dinner guests, and why, by saying, "Well, I guess you would have to say he did. Though, I am the one who did most of the prep work for it. He gets a lot of pleasure out of cooking fancy stuff for others, and while I think I should get some credit, I don't make a big deal about it. I would feel kind of weird pointing it out, so I just let him take the credit. It's easier that way." Tim's observations convey his awareness of an inconsistency between frontstage and backstage, and his assertion of the ease of maintaining that inconsistency makes it plausible to think that there is little to gain, and there may even be a cost, in disclosing the inconsistency.

As evidenced by the above cases, gender operates as a continuous concern for lesbigay families, but in ways more complex than many accounts of lesbigay family life indicate. The gender strategies deployed by the different participants suggest an abiding concern about maintaining traditional gender categories, and particularly of avoiding the stigma that comes with either failing to engage in domestic work for lesbian families or through engaging in domestic work for gay-male families. The portrayal of feeding work by lesbigay households conforms to these gender-related concerns, and partners tend to manage the identity of their respective partners.

Feeding Work and the Production of Class Identity

Feeding work in lesbigay families both reflects and perpetuates social-class distinctions. Patterns of meal preparation and patterns of sociability forged through the sharing of meals across families reflect the presence of social-class distinctions among lesbigay families. These social-class distinctions seem quite apparent but, historically, sociologists have found it difficult to find such class differences among lesbigay people. Two decades ago, when sociologist Carol Warren conducted a study of gay life, she concluded:

> It is clear that members tend to think of themselves, no matter what the abstract criteria, as members of an elite class. . . . since an elegant upper-middle-class lifestyle is one of the status hallmarks of the gay community, it is quite difficult to tell, and especially in the context of secrecy, what socioeconomic status people actually have. (1974, 85)

The families in this study do not lead secret lives: only five of the 103 adults interviewed completely hid their identities from coworkers, and only six hid their identities from biological relatives. It seems that as the lesbigay community becomes more visible, so too do differences among lesbigay people become more visible. Gender and racial distinctions pervade lesbigay life. Social-class distinctions also pervade lesbigay life and patterns of feeding often reflect and reproduce such distinctions.

The organization, preparation, and hosting of dinner parties plays a significant role in the production of class distinctions among lesbians and gays. Upper-middle-class lesbigay families report organizing and participating in dinner parties for friends and coworkers with much greater frequency than middle- and working/service-class families, except for some ethnically identified families, to whom I will return shortly. In terms of household income, the top 25 percent of families report either holding or attending a dinner party at least two times per month. In the bottom 25 percent, families rarely report any such occasion.

These meals function to reproduce social-class alliances and identities. For example, these meals become occasions for professionals to identify potential clients, learn of potential job opportunities, learn of new technologies, or stay abreast of organizational politics. During election cycles, these dinner parties among the affluent can take on political significance. Many lesbian and gay politicians in the city of San Francisco use such occasions as fundraisers. The lesbigay politician attends the gathering and the campaign charges between $50 and $500 per person. Among the wealthier participants living in San Francisco proper and earning household incomes of over $80,000, nearly every household reports either attending or hosting such a meal. These dinner parties provide access to power and influence on policy, and they play a crucial role in the political order of San Francisco lesbigay politics.

In a wider sense, dinner parties contribute to the creation of social-class identities. As DeVault comments on the function of hosting dinner parties, "it also has significance in the mobilization of these individuals as actors in their class: it brings together 'insiders' to a dominant class, and marks their common interests" (1991, 207). Given DeVault's observation, what does it mean to claim that these shared meals "mark their common interests"?

Beyond the more obvious career advantages and the sheer social enjoyment of such occasions, part of the answer lies in the symbolic meaning people attach to cuisine and the style in which hosts present it. Collins conceives of the symbolic meanings people attach to such occasions as an example of "household status presentation" (1992, 219). In other words, the choice of cuisine and the style of serving constructs social-class identity for the participants. Further, dinner parties function as occasions for the display and sharing of "cultural capital" (Bourdieu 1984). Bourdieu conceives of cultural capital as knowledge and familiarity with socially valued forms of music, art, literature, fashion, and cuisine, in other words, a sense of "class" or "good taste." Bourdieu distinguishes cultural capital from economic capital, therein arguing that some members of society may possess higher levels of cultural capital yet possess less wealth, and vice-versa. The household often functions as the site where cultural capital or good taste finds expression. Dinner parties often operate as a stage upon which the hosts display and call attention to various forms of cultural capital, including everything from works of art to home furnishings, from musical selections to displays of literature, from table settings to the food itself. The elaborate and the exotic quality of meals plays a central role in the upper-middle-class lesbigay dinner party. The higher-status participants often speak of specializing in a particular cuisine. Joe Mosse speaks of his interest in Indian cuisine:

> We often have Indian food when guests come over. I started cooking Indian food as a hobby many years ago. We've collected lots of Indian cookbooks, and I do a lot of different dishes. People generally love it. It's unusual and people remember that. You can get Indian food when you go out, especially in San Francisco, but how many people actually serve it at home?

Such dinner parties become elaborate both through the featuring of exotic menus and through the serving of a succession of courses throughout the meal: appetizers, soups, salads, main courses, coffee and tea, desserts, and after-dinner drinks. These dinner parties often feature higher-quality wines, and the higher-status participants often know something about the wine; this becomes part of the conversation of the evening. In sum, the dinner party serves as an opportunity for the creation and maintenance of class distinction.

Class differences impact the character and extent of everyday meals as well. More affluent families spend less time in preparing meals for everyday consumption than do the less affluent. A number of factors contribute to this. First, the more affluent eat out quite frequently. One in five lesbigay families report eating four or more meals per week in restaurants. Those who eat out more often earn higher incomes. Second, the more affluent use labor-saving devices like microwaves and food processors more frequently. Third, they often purchase prepared meals from upscale delis and fresh pasta shops. All of this purchasing feeding work in the marketplace enables more affluent couples to achieve a greater degree of egalitarianism in their relationships. These couples resemble the dual-career heterosexual couples studied by Hertz (1986). To the outside observer, and to the participants, affluent lesbi-gay families are more egalitarian in terms of feeding work, and they purchase that equality in the marketplace. As Hertz so eloquently argues:

On the surface, dual-career couples appear to be able to operate as a self-sufficient nuclear family. Nonetheless, they are dependent as a group and as individuals on a category of people external to the family. Couples view their ability to purchase this service as another indication of their self-sufficiency (or "making it"). Yet, appearances are deceptive. . . . What appears to be self-sufficiency for one category of workers relies on the existence of a category of less advantaged workers.

Accordingly, scanning the lesbian and particularly the gay-male enclaves of San Francisco one discovers a preponderance of food service establishments offering relatively cheap and convenient meals: taquerias, Thai restaurants, pasta shops, hamburger joints, Chinese restaurants, and delis. Most of these establishments employ women, ethnic and racial minorities, and less educated, less affluent gay men. The low wages earned by these workers enable more affluent lesbigay families to purchase meals in the marketplace and to avoid conflicts over feeding work. Many affluent lesbigay families report deciding to simply eat out rather than face the hassle of planning, preparing, and cleaning up after meals. Less affluent families may not make that choice, and thus they spend much more time and effort in the production of routine meals.

Feeding Work and the Production of Ethnic Identity

While one's social-class status influences whether one attends dinner parties with much frequency, those families with strong ethnic identifications do report more shared meals. Among these groups the sharing of food, and the ethnic character of that food, becomes an important expression of ethnic heritage and cultural identity. Gary Hosokawa, an Asian-American of Hawaiian descent, speaks of the centrality of his *hula* group to his social life and understands that group in familial terms. In response to a question about how he thinks about family, Gary replies:

Its really strange, hard to explain. In Hawaiian culture, your *hula* group is family. In ancient times, the *hula* teacher would choose students to become *hula* dancers and they would live together and become family. They ate together, slept together. They were picked from their own family groups and became a part of another family group. *My hula* group is my family, we eat together, we dance together. It is a very deep, spiritual thing for me.

For Gary, the sharing of meals functions to create and sustain ethnic identity.

In like manner, Michael Herrera and Federico Monterosa, a Latino couple together for three years who live with a young lesbian woman, Jenny Dumont, consider themselves a family. They also speak of a larger family, consisting of other Latino and non-Latino friends, as well as Federico's cousins. Michael and Federico report recurrent dinner and brunch gatherings, often featuring Mexican foods. Michael remarks:

We like to make enchiladas a lot, and sometimes we have meat, like beef or something, but always with a Mexican soup for our family gatherings. We get together every couple of weeks. It's very important to me. It's one thing that I think a lot of my Anglo friends feel really envious about, we sort of have a family and many of them don't.

When comparing the appeal of ethnically identified foods to the upper strata of the lesbigay community (mostly Euro-Americans) with the appeal of such foods to the Asian and Latino participants it becomes clear that the food symbolizes very different things for each group. For the affluent lesbigay families the food represents creativity and contributes to the entertaining atmosphere of the dinner party. It carries status due to its exoticness and the difficulty of its preparation.

In contrast, among Asian and Latino lesbigay participants, food expresses ethnic heritage and symbolizes ethnic solidarity, and sometimes resistance to cultural assimilation. Many Asian and Latino participants pride themselves not on the variety of cuisine but on the consistent replication of the same cuisine and even the same meals.

Feeding Work and the Production of Family

Feeding work plays a pivotal role in the construction of lesbigay families. The comments of meal planners/preparers suggest a conscious effort to create a sense of family through their feeding work. For instance, Kathy Atwood and Joan Kelsey, a lesbian couple in their midthirties and living in a sort of lesbian enclave in the Oakland neighborhood of Rockridge, both speak of sharing meals as constitutive of family. Kathy, talking about why she considers some of her close friends as family, says, "Well, we eat with them and talk to them frequently. I have known one of them for a very long time. They are people we could turn to in need. They are people who invite us over for dinner and people with whom we spend our fun times and because of that, I think of them as family."

Other participants point to sharing meals, as well as jointly preparing the meals, as evidence of family. Fanny Gomez and Melinda Rodriguez have been together for nine years. Fanny does much of the feeding work for the family. Fanny tells of how she and a close friend, Jenny, whom she considers a part of the family, actually get together to prepare meals for holidays and birthdays:

> I certainly think of Jenny as family. She is the partner of the couple friend that I mentioned earlier. She and I get together to plan meals and celebrations. It feels like family to me when we talk, go shop together, and then cook the meals. I mean, it's like family when we eat the meal together, too, it's just that preparing the meal, I guess, it reminds me of working with my mother in her kitchen.

For Fanny, the planning, provisioning, and preparation of the meal constitutes family. The feeding work of Fanny and Kathy links material and interpersonal needs together and results in the creation of family.

We have seen that feeding work within lesbigay families is neither inconsequential nor simple. Strangely, much conversation and academic analysis concerning feeding work reduces the complexity of the enterprise, minimizes its significance, and legitimates the view held by many participants that they don't really do very much feeding work—a view held by those who do it as well as by those who don't. In part we can explain this sentiment by remembering that those who feed often lack the vocabulary to articulate their efforts to others. Few people will tell others that they spent part of their day monitoring the contents of their refrigerator, but they do. Such work remains invisible. We must also understand that many participants diminish feeding because they don't want to face the conflict that a thorough accounting, as I have just provided, might produce in their relationship. Moreover, given the potential for stigma that exists for the men who feed, and the women who don't, it becomes even clearer why the work of feeding remains particularly hidden in lesbigay households.

Finally, concealing the labors of feeding reflects the cultural tendency to romanticize domestic activities as well as to romanticize the relationships such activities create. Spotlighting the labor involved tarnishes the romantic luster that people attach to domesticity. I can't count how many times I have heard people who feed respond to compliments by saying, "Oh, it was nothing, really." Perhaps this is just a matter of self-deprecation, but it might also suggest a cultural cover-up of the laborious character of such efforts. The dinner guests don't really want to hear about the three different stores one went to in search of the ingredients, nor the process of planning and preparing the meal, nor the fight one had with one's spouse about whom to invite or what to serve. A thorough investigation of the labors involved in feeding the family reveals that feeding is work. Recognizing feeding as work raises the impertinent question of why the effort often goes uncompensated, a question that leads directly to issues of exploitation and inequality, issues ripe with the potential for social and family conflict.

Given the social precariousness of lesbian and gay relationships, mostly due to the lack of social, political, and economic resources, the tendency of the participants to avoid such conflicts is probably essential to their long-term survival. When some resources exist, as in the case of economic resources, assuaging such conflicts becomes easier. When ample economic resources exist, feeding becomes less arduous with affluent families turning to the marketplace for meals and preparing meals at home teeming with creativity, quality, symbolic meaning, and nutritional content. When lesbigay families lack economic resources, as is the case among many of the working/service-class families, feeding looks different: routine, fatiguing, nutritionally compromised, and symbolically arid (in the sense that the capacity of feeding to produce a sense of family is compromised). Participants rarely conceive of eating ramen noodle soup on the couch as constitutive of their claim to family status, but they frequently conceive of eating a nutritionally complete meal at a dining-room table as constitutive of such a claim.

References

Berk, S. Fenstermacher (1985) *The Gender Factory: The Apportionment of Work in American Households*. New York: Plenum Press.

Bourdieu, P. (1984) *Distinction: A Social Critique of the Judgment of Taste*. Cambridge, Mass.: Harvard University Press.

Cahill, S. (1989) Fashioning males and females: Appearance management and the social reproduction of gender. *Symbolic Interaction*, 12 (2): 281–98.

Collins, R. (1992) Women and the production of status cultures. In M. Lamont and M. Fournier, eds., *Cultivating Differences: Symbolic Boundaries and the Making of Inequality*. Chicago: University of Chicago Press.

Coltrane, S. (1989) Household labor and the routine production of gender. *Social Problems*, 36 (5): 473–90.

DeVault, M. (1991) *Feeding the Family: The Social Organization of Caring as Gendered Work*. Chicago: University of Chicago Press.

Goffman, E. (1977) The arrangement between the sexes. *Theory and Society*, 4 (Fall): 301–31.

Hochschild, A. (1983) *The Managed Heart: Commercialization of Human Feeling*. Berkeley: University of California Press.

Kessler, S. and W. McKenna (1978) *Gender: An Ethnomethodological Approach*. Chicago: University of Chicago Press.

Warren, C. (1974) *Identity and Community in the Gay World*. New York: John Wiley.

West, C. and D. Zimmerman (1987) Doing Gender. *Gender and Society*, 1 (2): 125–51.

Food and Identity Politics

18 How to Make a National Cuisine: Cookbooks in Contemporary India
Arjun Appadurai

19 "Real Belizean Food": Building Local Identity in the Transnational Caribbean
Richard Wilk

20 Let's Cook Thai: Recipes for Colonialism
Lisa Heldke

21 More than Just the "Big Piece of Chicken": The Power of Race, Class, and
Food in American Consciousness
Psyche Williams-Forson

22 *Mexicanas'* Food Voice and Differential Consciousness in the San Luis
Valley of Colorado
Carole Counihan

23 Rooting Out the Causes of Disease: Why Diabetes is So Common Among
Desert Dwellers
Gary Paul Nabhan

24 Slow Food and the Politics of Pork Fat: Italian Food and European Identity
Alison Leitch

25 Taco Bell, Maseca, and Slow Food: A Postmodern Apocalypse for Mexico's
Peasant Cuisine?
Jeffrey M. Pilcher

26 The Raw and the Rotten: Punk Cuisine
Dylan Clark

27 Salad Days: A Visual Study of Children's Food Culture
Melissa Salazar, Gail Feenstra, and Jeri Ohmart

18

How to Make a National Cuisine: Cookbooks in Contemporary India*

Arjun Appadurai

Cookbooks, which usually belong to the humble literature of complex civilizations, tell unusual cultural tales. They combine the sturdy pragmatic virtues of all manuals with the vicarious pleasures of the literature of the senses. They reflect shifts in the boundaries of edibility, the proprieties of the culinary process, the logic of meals, the exigencies of the household budget, the vagaries of the market, and the structure of domestic ideologies. The existence of cookbooks presupposes not only some degree of literacy, but often an effort on the part of some variety of specialist to standardize the regime of the kitchen, to transmit culinary lore, and to publicize particular traditions guiding the journey of food from marketplace to kitchen to table. Insofar as cookbooks reflect the kind of technical and cultural elaboration we grace with the term *cuisine*, they are likely, as Jack Goody has recently argued, to be representations not only of structures of production and distribution and of social and cosmological schemes, but of class and hierarchy (1982). Their spread is an important sign of what Norbert Elias has called "the civilizing process" (1978). The increased interest of historians and anthropologists in cookbooks should therefore come as no surprise (Chang 1977; Cosman 1976; Khare 1976a, 1976b).

This essay discusses cookbooks produced by a particular type of society at a particular moment in its history. The last two decades have witnessed in India an extremely significant increase in the number of printed cookbooks pertaining to Indian food written in English and directed at an Anglophone readership. This type of cookbook raises a variety of interesting issues that are involved in understanding the process by which a national cuisine is constructed under contemporary conditions. Language and literacy, cities and ethnicity, women and domesticity, all are examples of issues that lie behind these cookbooks. In examining these issues in the

* Earlier versions of this article were presented at the University of California, Santa Barbara; the University of California, Berkeley; Bryn Mawr College; the Center for Advanced Study in the Behavioral Sciences, Stanford; the Oriental Club of Philadelphia, and the University of Houston. I am grateful for comments and criticism made by participants at each of these occasions, but must especially thank Burton Benedict, Gerald Berreman, Stanley Brandes, Donald Brown, Shelly Errington, Nelson Graburn, Ulf Hannerz, Alan Heston, Joanna Kirkpatrick, Del Kolve, Robert Krauss, Paul Rabinow, Renato Rosaldo, Corine Schomer, Judith Shapiro, Brian Spooner, Dennis Thompson, and Aram Yengoyan. I owe special thanks to my wife, colleague, and fellow cook, Carol A. Breckenridge. The staff of the Center for Advanced Study in the Behavioral Sciences made the task of completing this paper far less painful than it would otherwise have been.

Indian case, we can begin to sharpen our comparative instincts about how cuisines are constructed and about what cookbooks imply and create. But before I begin to describe and interpret these books, I need to introduce a comparative problem regarding culinary traditions to which they draw our attention.

Cookbooks appear in literate civilizations where the display of class hierarchies is essential to their maintenance, and where cooking is seen as a communicable variety of expert knowledge.[1] Cookbooks in the preindustrial world are best documented in the agrarian civilizations of Europe, China, and the Middle East. In these cases, the historical impetus for the production of the earliest cookbooks seems to have come from royal or aristocratic milieus, because these were the ones that could afford complex cuisines and had access to the special resources required for the production and consumption of written texts.

The evolution of a high cuisine, to use Goody's term, does not follow exactly the same form or sequence in each of these locales. But with the possible exceptions of China and Italy, there is in every case a powerful tendency to emphasize and reproduce the difference between "high" and "low" cuisines, between court food and peasant food, between the diet of urban centers and that of rural peripheries. Imperial cuisines always drew upon regional, provincial, and folk materials and recipes. Preindustrial elites often displayed their political power, their commercial reach, and their cosmopolitan tastes by drawing in ingredients, techniques, and even cooks from far and wide. Yet these high cuisines, with their emphasis on spectacle, disguise, and display, always seek to distance themselves from their local sources. The regional idiom is here decisively subordinated to a central, culturally superior, idiom. French *haute cuisine* is exemplary of this type of high cuisine. In the cases of China and Italy, by contrast, regional cuisines *are* the *hautes cuisines*, and no imperial or metropolitan culinary idiom really appears to have achieved hegemony, even today. In the Chinese case, to the degree that a civilizational standard has emerged, it appears to be the colorless common denominator of the complex regional variants. In Italy, at least until very recently, it appears to be impossible to speak of a high, transregional cuisine.

In India, we see another sort of pattern, one that is, in some respects, unique. In this pattern the construction of a national cuisine is essentially a postindustrial, postcolonial process. But the traditional Indian picture has some parallels with those of the other major culinary regions of the world. Like the cooking of ancient and early-medieval Europe, preindustrial China, and the precolonial Middle East, cooking in India is deeply embedded in moral and medical beliefs and prescriptions. As in the Chinese and Italian cases, the premodern culinary traditions are largely regional and ethnic. As in Ottoman Istanbul in the seventeenth century, court cuisines drew on foods and recipes from great distances (Sharar 1975). But in contrast to all these preindustrial cases, in India before this century, the emergence of a gustatory approach to food (that is, one that is independent of its moral and medical implications), the related textualization of the culinary realm, and the production of cookbooks seem to have been poorly developed (Khare 1976a).

In the Indian case, the cuisine that is emerging today is a national cuisine in which regional cuisines play an important role, and the national cuisine does not seek to hide its regional or ethnic roots.[2] Like their counterparts in England and France in the early eighteenth century, the new Indian cookbooks are fueled by the spread of print media and the cultural rise of the new middle classes. As in all the other cases, but notably later, food may finally be said to be emerging as a partly autonomous enterprise, freed of its moral and medical constraints. The Indian pattern may well provide an early model of what might be expected to occur with increasing frequency and intensity in other societies having complex regional cuisines and recently acquired nationhood, and in which a postindustrial and postcolonial middle class is constructing a particular sort of polyglot culture. This pattern, which is discussed in the rest of this essay, might well be found, with the appropriate cultural inflections, in places like Mexico, Nigeria, and Indonesia.

The Social World of the New Indian Cuisine

The audience as well as the authors of the English-language cookbooks produced in India in the last two decades are middle-class urban women. But the middle class in India is large and highly differentiated. It includes civil servants, teachers, doctors, lawyers, clerks, and businessmen, as well as film stars, scientists, and military personnel. Some of these persons belong to the upper middle class and a few to the truly wealthy. This middle class is largely to be found in the cities of India, which include not only the international entrepôts such as Bombay and Calcutta, and the traditional capitals like Delhi and Madras, but also smaller industrial, railroad, commercial, and military towns of varying orders of size, complexity, and heterogeneity.

The women who read the English-language women's magazines, such as *Femina* and *Eve's Weekly*, as well as the cookbooks in English that are closely allied sociologically to these magazines, are not only members of the super-elite of the great Westernized cities, they also belong to the professional and commercial bourgeoisie of smaller towns throughout India. As more and more public organizations (such as the army, the railroads, and the civil service), as well as more and more business corporations, circulate their professional personnel across India, increasing numbers of middle-class families find themselves in cities that harbor others like themselves, who are far from their native regions. This spatially mobile class of professionals, along with their more stable class peers in the cities and towns of India, creates a small but important class of consumers characterized by its multiethnic, multicaste, polyglot, and Westernized tastes. This class is linked in particular towns by a network of clubs, social committees, children's schools, cookery classes, and residential preferences. They are nationally linked by their tastes in magazines, clothing, film, and music, and by their interpersonal networks in many cities. Though this class has some very wealthy persons in it, along with some who can barely afford to belong to it, its core consists of government servants, middle-rung professionals, owners of medium-sized businesses, and middle-rank corporate employees. It is this class, rather than the sophisticated super-elites of Bombay, Delhi, Calcutta, Madras, and a few other cities, that is constructing a new middle-class ideology and consumption style for India, which cuts across older ethnic, regional, and caste boundaries. Cookbooks are an important part of the female world of this growing urban middle class.

The interplay of regional inflection and national standardization reflected in the new cookbooks is the central preoccupation of this essay. It represents the culinary expression of a dynamic that is at the heart of the cultural formation of this new middle class. Cookbooks allow women from one group to explore the tastes of another, just as cookbooks allow women from one group to be represented to another.

In the social interaction that characterizes these urban middle-class families, women verbally exchange recipes with one another across regional boundaries and are eager to experiment with them. The oral exchange of recipes is, from the technical point of view, the elementary process that underlies the production of these cookbooks. In many of the introductions to these cookbooks, the authors thank women they have known in various metropolitan contexts for sharing recipes and skills. In some cases, it is possible to discern a progression from orally exchanged recipes to full-fledged ethnic or "Indian" cookbooks. The terseness of many of the recipes in the new cookbooks may testify to the fact that they are intended only as references and aids in a largely oral form of urban interaction.

But the exchange of recipes also has other implications, for it is frequently the first stage in a process that leads to carefully controlled interethnic dining. In urban kitchens, there is often a good deal of informal culinary interaction between female householders that by-passes the question of formal dining. In a society where dining across caste or ethnic boundaries is still a relatively delicate matter, recipes sometimes move where people may not. In traditional India, as we shall see, commensal boundaries were central to the edifice of the caste system. But the movement of recipes in the new urban middle-class milieu is one sign of the loosening of these boundaries. In many cases, the movement of recipes across caste, language, and ethnic boundaries is accompanied by an increase in formal (and informal) entertaining and dining across these boundaries.

In turn, the exchange of recipes, oral in origin but aided and intensified by the new cookbooks, clearly reflects and reifies an emerging culinary cosmopolitanism in the cities and towns of India, which is reflected in other consumption media as well. The trend setters as well as followers in this process are women who often-times work in multiethnic job settings (as their husbands do), whose children are acquiring broader tastes in school lunchrooms and street-vendors' stalls, and whose husbands feel the pressure to entertain colleagues and contacts at home. In all these contexts, what are created, exchanged, and refined are culinary stereotypes of the Other, stereotypes that are then partly standardized in the new cookbooks.

The predicament of these middle-class women is quite complex, however, for the homogenization of a certain middle-class life style calls for diversification of consumption patterns in many domains, including clothing, interior decoration, and cuisine. In the domain of food, the push to diversify the housewife's culinary skills comes from a variety of sources: the push of guests who want to taste your regional specialties (as they have constructed them in the course of their own interactions, travels, and readings of cookbooks), the push of children who are tired of "the same old thing," and the push of ambitious husbands to display the metropolitan culinary ranges of their wives. At the same time that she is dealing with these pressures to diversify her skills and add to her inventory of ethnic food specialties, the typical middle-class housewife also has another clientele, composed of her husband (in another, more primordial guise), her more traditional in-laws and other relatives, and important country cousins who crave food in the specialized mode of the region, caste, and community from which they originally come. This clientele is either simply conservative in its tastes or, worse still, has acquired new-fangled urban notions of authenticity regarding their own natal cuisine. Many middle-class housewives are thus on a perpetual seesaw that alternates between the honing of indigenous culinary skills (and if they have lost them, there are books to which they can turn) and the exploration of new culinary regions. It is the tacit function of the new cookbooks to make this process seem a pleasant adventure rather than a tiring grind.

What is tiring is not only the acquisition, refinement, and display of constantly new culinary wares, but other, less subtle pressures. In the middle-class world I am describing, the budget is a central instrument, for husbanding money as well as time. Many of the new cookbooks emphasize that they are specifically designed to resolve shortages of time and money in urban settings. They therefore frequently offer menus, shortcuts, and hints on how to get more out of less. Some of them explicitly recognize the dual pressure on working women to earn part of the family's livelihood and simultaneously to cater to the culinary sophistication of their families and friends. While authenticity, attractiveness, and nutritional value remain the dominant values of the new cookbooks, efficiency, economy, and utility are becoming increasingly respectable themes.

One very striking example of how this new metropolitan pragmatic begins to erode traditional concepts can be seen in the role of leftover foods in the new cookbooks. Leftovers are an extremely sensitive category in traditional Hindu thought (Khare 1976b; Marriott 1968; Appadurai 1981) and, though in certain circumstances they are seen as positively transvalued, most often the eating of leftovers or wastes carries the risk of moral degradation, biological contamination, and loss of status. Their treatment and the etiquette that surrounds them stand very near the moral center of Hindu social thought. Yet the new cookbooks, which are in other respects hardly iconoclastic, frequently suggest ways to use leftovers and wastes intelligently and creatively. While the traditional prohibitions concerned food contaminated by human saliva rather than by the cooking or serving process, all waste products customarily bear some of the aura of risk associated with leftovers in the narrow sense. Several books contain chapters on the treatment of leftovers. There is even one cookbook, *Tasty Dishes from Waste Items* (Reejhsinghani 1973a), that is built entirely around this principle. Its author goes so far as to say in her introduction that she is "taking these discarded articles of food out of the wastebin and [making] interesting and delightfully different dishes from them." As caste differences come increasingly to be perceived as differences between ethnic entities (Dumont 1970), so food differences come to be seen as consumption issues divorced from the realm of taboo and prohibition. Of course, as food emerges from its traditional moral and social matrix, it becomes embedded in a different system of etiquette—that of the drawing room, the corporate gathering, the club event, and the restaurant.

The history of food consumption outside the domestic framework has yet to be written for India, but there is little doubt that traditional nondomestic commensality was confined to religious and royal milieus, where traditional social or religious boundaries could be maintained even in public eating places. To some extent, public eating places in modern India still seek to maintain boundaries among castes, regions, and food preferences. But restaurants, both humble and pretentious, have increasingly become arenas for the transcendence of ethnic difference and for the exploration of the culinary Other. Restaurant eating has become a growing part of public life in Indian cities, as wealthy families begin to socialize in restaurants and as working men and women find it easier to go out for their main meals than to bring food to work with them. These restaurants tend to parallel, in their offerings, the dialectic of regional and national logics to be noted in the new cookbooks. These twin developments sustain each other.

In addition to the homes and restaurants of the new middle classes, where the new cuisine (in both its provincial and its national forms) is being practiced, transmitted, and learned, a variety of public arenas offer versions of it: food stands in train stations, dining cars of the trains themselves, army barracks, and clubs, student hostels, and shelters of all kinds. Although each of these public arenas contributes to the new interethnic and transregional cuisine in a different way and to a different degree, they all represent the heightened importance of institutional, large-scale, public food consumption in India. The efflorescence of increasingly supralocal and transethnic culinary arenas explains why the pace of change in traditional commensal boundaries (so critical to the caste system) is so much greater than in the realm of marriage, a matter on which there has recently been a lively exchange (Khare 1976b; Goody 1982). Food boundaries seem to be dissolving much more rapidly than marriage boundaries because eating permits a variety of registers, tied to particular contexts, so that what is done in a restaurant may be different from what is seen as appropriate at home, and each of these might be different in the context of travel, where anonymity can sometimes be assured. This kind of compartmentalization, to use Milton Singer's felicitous phrase, is not a realistic option in the domain of marriage, though it might well be in the domain of sexual relations. The new cuisine permits the growing middle classes of Indian towns and cities to maintain a rich and context-sensitive repertoire of culinary postures, whereas in the matter of marriage, there is the stark and usually irreversible choice between staying within the ambit of caste rules or decisively, permanently, and publicly breaking them.

The symbiotic differentiation of both class and cuisine that is flourishing in Indian cities is supported by changes in the technology and economy of cooking. The food blender, spice grinder, and refrigerator are seen in more and more homes. There is a large and growing food industry, selling both ingredients and instant foods of many varieties. The commercialization of agriculture and the increasing sophistication of transport, marketing, and credit make it possible to obtain a wide variety of fruits, vegetables, grains, and spices in most major Indian cities. Major food companies advertise prominently in the women's magazines, sponsor specialized cookbooks, and advertise the glamour of culinary ethnicity. As in the contemporary West, the modern machinery and techniques alleged to be labor-saving devices are in fact agents in the service of an ideology of variety, experimentation, and elaboration in cuisine that puts middle-class housewives under greater pressure than in the past. Thus the *seductiveness of variety* (discussed later in this essay), as an important part of the ideological appeal of the new cookbooks, masks the pressures of social mobility, conspicuous consumption, and budgetary stress for many middle-class wives. Regarded from this point of view, the publishing industry, catering industry, food industries, and the commercial sector in agriculture all have something to gain from the new culinary developments in Indian cities. But the majority of housewives see it as an exciting process of culinary give and take in which they are both contributors and beneficiaries. Before looking closely at the rhetoric of the recent spate of cookbooks, it is necessary to examine briefly the historical and cultural back-drop against which they have materialized.

Culinary Texts and Standards in Indian History

In this section, we consider two distinct but interrelated questions. One is the question of why Indian history has not, until recently, witnessed the same degree of textualization of the culinary realm as several other complex civilizations. The other is the question of the historical forces that until this century have militated against the formation of a civilizational culinary standard in India. These questions involve brief excursions into aspects of the Hindu, Islamic, and colonial contexts of Indian history.

The historical example of India runs counter to our instincts, for here is a case of a highly differentiated, literate, text-oriented, and specialist civilization that has not produced a high cuisine on the French, Chinese, or Middle Eastern model. The puzzle is deepened when we consider that food is a central trope in classical and contemporary Hindu thought, one around which a very large number of basic moral axioms are constructed and a very large part of social life revolves (Appadurai 1981; Khare 1976a, 1976b; Marriott 1968). Food in India is closely tied to the moral and social status of individuals and groups. Food taboos and prescriptions divide men from women, gods from humans, upper from lower castes, one sect from another. Eating together, whether as a family, a caste, or a village, is a carefully conducted exercise in the reproduction of intimacy. Exclusion of persons from eating events is a symbolically intense social signal of rank, of distance, or of enmity. Food is believed to cement the relationship between men and gods, as well as between men themselves. Food is never medically or morally neutral. Whatever the perception of the purely gustatory aspect of particular foods, the issue of their implications for the health, the purity, and the moral and mental balance of the consumer are never far out of sight. Feasting is the great mark of social solidarity, as fasting is the mark of asceticism or piety. Each of these patterns is to be seen in other societies, but the case could be made that the convergence of the moral, social, medical, and soteriological implications of food consumption is nowhere greater than in traditional Hindu India. We are therefore left with the question: Why did Hindu India, so concerned with food as a medium of communication on the one hand and with matters of hierarchy and rank on the other, not generate a significant textual corpus on cuisine?

If we take the long view of food in Hindu thought, which has left ample textual deposits, it is possible to assert that while *gastronomic* issues play a critical role in the Hindu texts, *culinary* issues do not. That is, while there is an immense amount written about *eating* and about feeding, precious little is said about *cooking* in Hindu legal, medical, or philosophical texts. Even a cursory examination of the secondary literature that bears on the subject (Zimmerman 1982; Kane 1974; Khare 1976b) is sufficient to show that food is principally either a moral or a medical matter in traditional Hindu thought. It should also be pointed out that these two dimensions, as in early European and Chinese thought, are deeply intertwined. But the vast body of rules, maxims, prescriptions, taboos, and injunctions concerning food virtually nowhere contains what we would call recipes. Ingredients and raw materials are sometimes mentioned (often in connection with perceptions of balance, seasonality, and the humors), and cooked foods also appear frequently, in connection with special ritual observances. But the processes that transform ingredients into dishes are invariably offstage. Recipes, the elementary forms of the culinary life, are missing in the great tradition of Hinduism.

Yet it is clear that cooking is a highly developed art in Hindu India. How are we to account for the absence of recipes and cookbooks from the otherwise omnivorous tendency of the Hindu elite to codify every sector of life?

The answer must be sought on two levels. The first has already been hinted at: Hindu thought is deeply instrumental, though in a specific cultural mode. Its burning concern is always, however indirectly, to break the epistemological and ontological bonds of this world. Food becomes relevant to this concern as a matter of managing the moral risks of human interactions, or as a matter of sustaining the appetites of the gods (who in turn bestow grace and protection), or as a matter of cultivating those bodily or mental states that are conducive to superior gnosis. In each case, food prescriptions and food taboos are two sides of the same coin. Food thus stays encompassed within the moral and medical modes of Hindu thought, and never becomes the basis of an autonomous epicurean or gustatory logic.

Let us now consider the question of why a pan-Indian Hindu cuisine did not emerge in India. Two possible explanations must be rejected despite their intuitive appeal. The first is the idea that the host of prohibitions and taboos surrounding food in Hindu India so impoverished the dietary base, especially of the upper castes, that the elementary conditions for the emergence of a complex gastronomic culture could not be met. Though it is true that food is surrounded by a large range of prescriptions and proscriptions in Hindu India, this clearly did not prevent the development of fairly elaborate regional and courtly high cuisines. Further, the existence of a very large set of medical and moral do's and don'ts in the Chinese case had no such repressive effect (Goody 1982:111–12). This leads to a second plausible hypothesis that must also be rejected, and that is the explanation which says that Hindu India was not a unified political entity before colonial rule, and thus the institutional framework for standardization, communication, and the hegemony of some culinary center was absent until the formation of the modern nation-state. The problem with this suggestion is that it does not account for the quite high degree of pan-Indian standardization in other social and cultural forms and expressions, not least the so-called caste system, its ritual accompaniments, and the Hindu religious axioms on which it is founded.

Though the problem deserves more extended research and analysis, I suggest that there are two specific cultural factors that have made it difficult for a premodern Hindu high cuisine to emerge. The first is that there was a deep assumption in Hindu thought that local variation in custom (ācāra) must be respected by those in power, and that royal duty consists in protecting such variation unless it violates social and cosmic law (*dharma*). When, in addition, we bear in mind that the producers, distributors, and guardians of the major textual traditions, the Brahmans, did not particularly care (from a religious point of view) about the culinary or gastronomic side of food, we can begin to see why a poorly developed culinary textual tradition in premodern Hindu India and the nonemergence of a *Hindu* culinary standard for all of India might be related phenomena. What little we do know of the Hindu science of cooking—*pāka sāstra*—(see Prakash 1961) suggests that the cook-book tradition, both in Sanskrit and in the vernacular, was informal, fragmentary, and minor. Whether this is the result of a small number of texts or of indifferent preservation and transmission, the impression of a minor genre is unmistakable.

Like other humble traditions that do not enter the ambit of high Hindu thought, Hindu culinary traditions stayed oral in their mode of transmission, domestic in their locus, and regional in their scope. This does not, of course, mean that they were static, insulated from one another, or immune to changes in method or in raw materials. What it does mean is that there was no powerful impetus toward the evolution of a pan-Indian Hindu cuisine. The regional cuisines each had their festive foods, their royal elaborations, and their luxury dishes interacting with plainer, peasant diets keyed to ecological and seasonal factors (Breckenridge 1986). Though it is hard to tell much in retrospect about how this interaction worked, it is plain that, as regards cuisine, traditional Hindu India was thoroughly Balkanized.

With the arrival of the Mughals in India in the first half of the sixteenth century, the textualization of culinary practice took a significant step forward. The famous Mughal administrative manual, the *Ain-I-Akbari*, contains a recipe section, though the text as a whole is devoted to various aspects of statecraft. It is very likely that the culinary traditions of the princely houses of early modern North India were influenced by the practices of the Mughal court. It is also probable that the current pan-Indian availability (particularly in restaurants) of what is called Mughlai cuisine is closely tied to the political spread of Mughal hegemony through most of the subcontinent.

Mughlai cuisine is a royal cuisine that emerged from the interaction of the Turko-Afghan culinary traditions of the Mughal rulers with the peasant foods of the North Indian plains. Because of its diffusion through the royal courts of North India, and because it is the cuisine of reference for the great restaurateurs of northern and western India, Mughlai cuisine has become synonymous, particularly for foreigners, with Indian food. Though it represents an important step toward an Indian cuisine, its Indic base is restricted to the north and west of the subcontinent. It derives nothing of significance from the cuisines of Maharashtra, Bengal, Gujarat, or of any of the southern states. Though some version of Mughlai food is available throughout contemporary India, it cannot be considered an Indian cuisine if by that designation we mean a cuisine that draws on a wide set of regional traditions. It is the limiting case of a tradition that is "high" without being a civilizational standard.

The textualization of culinary traditions was intensified by the arrival of the printing press. The proliferation of presses, journals, and books in the nineteenth-century colonial context did, among other things, usher in the prototypes of the modern cookbook. Thus, in Maharashtra in the nineteenth century, there are books on household management published in Marathi that contain recipes. There is every reason to suppose that this was happening in the other major linguistic regions of India. In the first half of this century, magazines and newspapers began to address the urban housewife by carrying recipe columns. There is also evidence that the modern vernacular periodical press nurtured the popular taste for cookbooks and cooking skills, both prerequisites for the recent rise of the English-language cookbook. One example of these genetic links is a book by Kala Primlani called *Indian Cooking*, first published in 1968. Primlani's book began its career as a series of recipe columns in the Sindhi-language daily *Hindvasi*; it was then published in book form in Sindhi before it achieved its English incarnation. An even more famous example of the shift from an Indian- to an English-language book is *Samaiththu Pār* (Meenakshi Ammal, 1968), whose title was literally translated into the English *Cook and See*, since many of the young Tamil women to whom the original was addressed were functional illiterates in their mother tongue and could not read Tamil. These examples suggest that the vernacular cookbooks and the English-language cookbooks are not wholly discrete genres.[3]

Though the colonial version of Indian cuisine is the most significant precursor of the emergent national cuisine of the last two decades, it was not confined to the homes of the colonial elite and it did not end with colonialism. Some of its content, and a good deal of its ethos, provided the basis of the culinary manuals and procedures of the Indian army, which even today represents a rather specialized subcontinental culinary standard. And with broader reach, there are certain clubs, restaurants, and hotels that carry on the colonial culinary tradition. The other enclave in which some of the Anglo-Indian ethos of this colonial cuisine is preserved is the Parsi community (Mehta 1979).[4]

In the national cuisine that has emerged in the last two decades, Mughlai cuisine (to a considerable extent) and colonial cuisine (to a lesser extent) have been incorporated into a broader conception of Indian food. The shape of this new national repertoire can be seen in the recent proliferation of cookbooks, whose structure and rhetoric are the topic of the next two sections.

Proliferation of Genres and the Culinary Other

The most striking characteristic of English-language books on Indian cooking is the rapid specialization that has occurred within this young field. There are already cookbooks directed toward special audiences, such as *The Working Woman's Cookbook* (Patil 1979) and *Cooking for the Single Person* (Reejhsinghani 1977). There are also entire cookbooks devoted to specific food categories, such as chutneys and pickles (Jagtiani 1973), snacks (Currim and Rahimtoola 1978), vegetable dishes (Lal 1970), and leftovers (Reejhsinghani 1973a). There are books about Indian cooking produced in the West and oriented partly to Euro-American cooks in search of adventure and partly to expatriate Indians in search of their culinary pasts (D. Singh 1970; Aziz 1983; B. Singh 1961; Hosain and Pasricha 1962; Kaufman and Lakshmanan 1964; Jaffrey 1981; Time-Life 1969). There are cookbooks to guide vegetarians (Srivaran 1980), cookbooks produced by large companies that revolve entirely around a single industrially produced food product (Narayan 1975), and cookbooks that reflect the penetration of the large-scale frameworks of catering schools and restaurant kitchens into the domestic milieu (Philip 1965; Bisen 1970). There are books on Indian cooking that were first published in the United States or England and subsequently republished in India (Attwood 1972; Day 1963). Finally, authentic tokens of a flourishing publishing industry, there are cookbooks based entirely on sales gimmicks, like *Film Stars' Favourite Recipes* (Begum 1981). This inventory is representative but by no means exhaustive.

The proliferation of subspecies suggests that a possible index for the emergence of an authentic high cuisine is precisely the emergence of such crosscutting functional classifications. My assumption here is that a complex culinary repertoire underlies and facilitates the type of deconstruction and recombination to which these specialized books testify. Such specialization is very different from the regional and oral Balkanization that characterized premodern Indian cuisine. This argument leads me to suggest that the surest sign of the emergence of an authentically Indian cuisine is the appearance of cookbooks that deal with special audiences and special types of food. To dissolve this seeming paradox, I turn now to the emphasis of the new books on specialized regional or ethnic cuisines.

An historian of China has suggested that among the prerequisites to the emergence of a full-fledged cuisine is a widely based variety of recipes, so that "cuisine does not develop out of the cooking traditions of a single region" (Freeman 1977:144). The Indian material suggests a further refinement of this observation. In the Indian case, perhaps the central categorical thrust is the effort to define, codify, and publicize regional cuisines. There are books now on virtually every major regional cuisine, as well as on several ethnic minority cuisines, such as the Parsi and Moplah. It is difficult to imagine a book such as Rachel Mutachen's *Regional Indian Recipes* (1969) being published much earlier than it was.

These regional and ethnic cookbooks do two things: Like tourist art (Graburn 1976), they begin to provide people from one region or place a *systematic* glimpse of the culinary traditions of another, and they also represent a growing body of food-based characterizations of the ethnic Other.[5] These two functions are distinct but intimately connected. A few examples will serve to capture the texture of this communicative mode. In a book called *Delicious Bengali Dishes*, Aroona Reejhsinghani says of Bengalis: "Besides sweets they are also very fond of rice and fish, especially fresh-water fish: a true Bengali will eat fish at least once daily and no celebration is said to be complete unless there are a few dishes of fish served in it" (Reejhsinghani 1975:1). Or consider the following characterization of Gujaratis and Gujarati cuisine by Veena Shroff and Vanmala Desai in *100 Easy-To-Make Gujarati Dishes* (1979:i–ii):

> Few states in India have such a variety of savory dishes as Gujarat, or a tradition of making and storing snacks. And in a Gujarati home, sweets and snacks are always waiting to be offered to a welcome guest. The use of condiments (*vaghar*) is a practice peculiar to the region. There is a widespread use in Gujarat of mustard seed, fenugreek, thymol, asafoetida and other additives that both make the food tastier and help digestion.

Examples of such ethnic cameos could be multiplied. They play an important part in the introductory sections of the regional and ethnic cookbooks. It is worth noticing that their authors are either transplanted and uprooted professionals (like Premila Lal, a Sindhi born in Tanzania who returned to India) characterizing cuisines that they have themselves learned in a cosmopolitan context, or they are self-advertisements by articulate urban members of a particular ethnic group who seek to publicize its culinary wares, as in the case of Shroff and Desai, both Gujaratis living in Delhi. It should also be noted that these small ethnological cameos hark back to the potted portraits that are the stuff of government gazetteers and ethnographic encyclopedias from the colonial period, where tribes, castes, and linguistic groups were often metonymously captured through the use of the telling custom or the distinctive piece of material technology.

What we see in these many ethnic and regional cookbooks is the growth of an anthology of naturally generated images of the ethnic Other, a kind of "ethno-ethnicity," rooted in the details of regional recipes, but creating a set of generalized gastroethnic images of Bengalis, Tamils, and so forth. Such representations, produced by both insiders and outsiders, constitute reflections as well as continuing refinements of the culinary conception of the Other in contemporary India.

But not only do these constructions build on long-standing and distinct regional cuisines. They also invent and codify new, overarching categories which make sense only from a cosmopolitan perspective. Perhaps the best example of such a process is the growing number of books on "South Indian" cuisine (e.g., Reejhsinghani 1973b; Skelton and Rao 1975) which, taking a distinctly northern perspective, collapse the distinctions between Tamil, Telugu, Kannada, and Malayali cuisines and lump them together as South Indian cuisine. These books divide their recipes into functional classes organized around basic ingredients or courses (thus crosscutting the internal regional categories), though the names of the various dishes do reveal their specific local origins to those who recognize the names. Of course, all cuisines, however local, reflect the aggregation upward of more humble and idiosyncratic cuisines from as far down as individual household culinary styles. Such telescoping and recategorization is also doubtlessly a slow and constant feature of history in complex societies. But certain regional forms and levels are relatively stable and well formed in the Indian case, and it is these that are now being vigorously articulated in print, juxtaposed, and reaggregated.

One consequence of the bustling marketplace of regional and ethnic culinary images is the sense of advocacy that animates many of the authors, who seem aware that there is a good deal of crowding in the gallery of regional or ethnic cuisines and some danger of exclusion from it. Thus Ummi Abdullah, who has produced a specialized book on *Malabar Muslim Cookery*, states: "I would consider my efforts recompensed if at least some of the traditional Moplah recipes find a permanent place on the Indian menu" (1981:4). A slightly different strategy is exemplified by Shanta Ranga Rao, who boldly calls her book *Good Food From India* (1968), though it is exclusively a collection of recipes from a rather small subcommunity from a microregion in South India.

Books like the one by Shanta Ranga Rao remind us that Indian regional or ethnic cookbooks in English are the self-conscious flip side of books that are engaged in constructing a national cuisine. In this, they differ markedly from vernacular cook-books, which take their regional context and audience largely for granted.

The new cookbooks are not simple or mechanical replicas of existing oral reper-toires. The transition to print in this particular social and cultural context results in a good deal of editing. Most of the ethnic or regional books are selective in specific ways. When written by insiders, they represent fairly complex compromises between the urge to be authentic and thus to include difficult (and perhaps, to the outsider, disgusting) items and the urge to disseminate and popularize the most easily under-stood and appreciated items, and to promote those already popular, from one's special repertoire. Outsiders who write these books, on the other hand, end up including the easy-to-grasp and more portable examples from alien ethnic or regional cuisines, partly because their own tastes for the exotic are first nurtured in restaurants or other public eating contexts, where the subtleties of that cuisine (which are often domestic) have already been pared down. In both cases, one of the results of the exchange of culinary images is the elimination of the most exotic, peculiar, distinctive, or domestic nuances in a particular specialized cuisine. In national or "Indian" cookbooks, of course, the selective process is much more obtrusive, and whole regional idioms are represented by a few "characteristic" dishes, which frequently are not, from the insider's perspective, the best candidates for this role.

In the jostling of the various local and regional traditions for appreciation and mutual recognition, certain linguistic and regional traditions with greater access to urban resources, institutions, and media are pushing humbler neighbors out of the cosmopolitan view: Thus Telugu cuisine is being progressively pushed out of sight by Tamil cuisine, Oriya by Bengali cuisine, Kannada by Marathi, Rajasthani by Gujarati, and Kashmiri by Punjabi. This does not mean that the humbler traditions have no cookbooks (theirs are frequently in the relevant vernacular), but they are losing in the struggle for a place in the cultural repertoire of the new national (and international) middle classes.

The construction of, and traffic in, culinary representations of the ethnic, regional, or linguistic Other has one dimension that is not reflected in the new cookbooks. These books, whether national or regional, uniformly contain positive ethnic stereotypes; but the orally communicated images of the culinary Other are often less than complimentary, as in other parts of the world throughout history. Thus, South Indians are said to eat (and enjoy) excessively runny food that trickles down their arms to the elbows, Gujaratis are said to eat "sickeningly sweet" food, Punjabi food is said to be heavy and greasy, Telugu food to be inedibly hot, Bengali food to be smothered in pungent mustard oil, and so forth. The new cookbooks, therefore, represent the friendly end of a traffic in interethnic images that has its seamy side.

I turn now to the question of how, at the same time as cookbooks in India are generating an anthology of specialized culinary representations, they are also increasingly responsible for constructing the idea of a national or "Indian" cuisine.

The Ingredients of a National Cuisine

In the contemporary Indian situation, and to some degree generically, cookbooks appear to belong to the literature of exile, of nostalgia and loss. These books are often written by authors who now live outside India, or at least away from the subregion about which they are writing. Sometimes they are intended for Indians abroad, who miss, in a vague and generalized way, what they think of as Indian food. Sometimes they are written to recollect and reconstruct the colonial idea of Indian food, and in such cases their master trope is likely to be curry, a category of colonial origin. The nostalgia for the glow of empire, in which recipes are largely a Proustian device, is the underlying rationale of at least one book, *The Raj Cookbook* (1981), which has a few "colonial" recipes, squeezed between sundry photographs, advertisements, and newspaper clippings from the twilight of the raj.

How do modern-day cookbooks go about constructing conceptions of a national cuisine in the context of an increasingly articulated gallery of specialized ethnic and regional cuisines? Although each book has a characteristically different strategy, there are a few standard devices. The first is simply to inflate and reify an historically special tradition and make it serve, metonymously, for the whole. I have already mentioned that Shanta Ranga Rao flatly asserts that hers is a book of Indian recipes when in fact it is much more local in its scope. There is also the widespread assumption, referred to earlier, in cookbooks and restaurants both in India and abroad, that conflates Mughlai food with Indian food.

Another strategy for constructing a national cuisine is inductive rather than nominal: The author assembles a set of recipes in a more or less subjective manner and then, in the introduction to the book, gropes for some theme that might unify them. For many books this theme is found, not surprisingly, in the spices and spice combinations, which are often discussed in loving detail. But even here, since regional variation is so great, the search for universals is often forgotten. Other authors discuss processes, such as roasting, frying, basting, etcetera, in the Indian context in an effort to tie together the diversity of regional cuisines. Yet others take an encyclopedic approach and list a set of implements and ingredients (on the model of the French cookbook's *batterie de cuisine*). Finally, and also in the inductive and encyclopedic mode, there are many books that focus on a particular kind of food, such as pickles, and simply provide a set of recipes from many regions. More cautious authors assume nothing general at all, but content themselves with some diluted ideas and comments about Hindu festivals and customs, where again they create a relatively weak sense of the Indian by juxtaposing specialized observances. In one way or another, many of the prefaces to these collections are inductive, intuitive, and encyclopedic in their approach to what constitutes an Indian cuisine.

But beneath the superficial inductivism lies a deeper set of assumptions concerning the structure of an Indian meal that seems shared by many of these authors. These assumptions can be represented as a structural model of an Indian meal and are reflected in the organization of chapters in many of the books. The structure may ideal-typically be represented in terms of the following sets of items, usually each given a separate chapter: rice-based preparations; breads (usually made of wheat flour, but also including rice and lentil-based pancakes); lentil preparations; vegetable preparations, sometimes subdivided; sweets and savories, which laps over into the contemporary Western domain of the "snack"; pickles and chutneys; and sometimes beverages. Salient sets of regional recipes are then brought together under the appropriate headings. This organization seems to reflect a fairly natural (that is, cultural) ordering of the range of preparations that emphasizes their distinctive properties in terms of the base ingredient (grain, lentil, vegetable, etcetera), or of the process used to prepare them (thus pickles, though based on vegetables and fruits, are processed differently from regular vegetable dishes), or of the mode of consumption (thus snacks and savories are largely set apart by the context—either time of day or of year—in which they are consumed). Though this structure creates strange regional and ethnic bedfellows (the Tamil *dosai* is placed along with the Punjabi *chapati* under the rubric of "breads" even though the first is a snack food and the second a basic meal item), it facilitates the collection and dissemination of regionally and ethnically variable recipes.

What suggests that what is emerging is more than an arbitrary hodgepodge of regional recipes is the increasingly widespread invoking of the menu idea. Many recent cookbooks have suggested menus, based on a series of slots of the sort I have discussed above, which are then filled with items from different regional or ethnic traditions. The interesting thing about this process is that while, in European and some other cuisines, the idea of a menu is associated with a succession of courses, Indian meals do not normally have a significant sequential dimension. Everything arrives more or less at once in most everyday contexts, although certain key liquid accompaniments to the base grains and certain key condiments may appear at different points. In festive contexts, the temporal dimension is greatly elaborated, and French-style courses are more prominent. Routinely, however, the Indian menu is a synchronic set of discriminations and does not display the diachronic discrimination associated with the idea of courses. Variety is not so closely tied to rhythm and pace as it is in other complex culinary traditions. But the idea of a menu is clearly a way to organize the proliferation of specialized regional and ethnic traditions and to subordinate them to the counterweight of an Indian culinary idiom. The concept of the menu is sufficiently well developed that at least one book, *The Working Woman's Cookbook* (Patil 1979), is a collection of recipes organized entirely as a sequence of suggested menus which combine regional items in extremely interesting ways. What such a book suggests is the availability of an Indian meal structure, as well as an inventory of regional and ethnic options that can be combined and recombined on this scaffolding.

The appearance of structural devices for organizing a national cuisine is accompanied by the development of a sometimes fairly explicit nationalist and integrationist ideology. Thus, for example, a newspaper review of *Indian Recipes* (Lal 1980) says: "Hindi may or may not help in unifying the country; while it is trying hard, there may be no harm in letting an Uttar Pradesh snack win over a Tamil Nadu heart."

But nationalist exhortations are of limited rhetorical value in the arena of the dining room, the kitchen, and the grocery, and the more subtle and effective ploy of many of the transregional cookbooks involves the seductiveness of variety. Thus, Thangam Philip, a major Indian authority on cooking and nutrition, says in her introduction to another author's cookbook that "if you wish to move out from the traditional and classic recipes of your own community to a wider repertoire, you will find Malini Bisen's *Vegetable Delights* a delightful aid." Or listen to Aruna Sheth, who says in her introduction to *The New Indian Cookbook* (1968), "Can it be that we are not aware of variety in India? It was with this thought in mind that I started to look for variety in the form of dishes from different provinces in India." She goes on to present the following revealing anecdote: "Even the Indians are unfamiliar with many dishes of different provinces. I have made *chakalis*, which are a favorite in Maharashtra and sent a plateful to a friend of mine from the North. Next day when she met me she thanked me for the *masala jalebis* that were so delicious." This little anecdote contains a good deal of information. It shows, among other things, something of the cross-ethnic urban interaction, at the nodes of which many authors of the new cookbooks stand. The author made a savory snack item that she has rendered a Maharashtrian favorite, a judgement that exemplifies the culinary stereotyping mentioned previously. Her friend ("from the North"), unfamiliar with this item, categorized it as a type of sweet with which she is familiar (a *jalēbi* is a deep-fried sweet that in its pretzel-like quality and in some of its ingredients resembles the *chakali*), but by adding the prefatory *masala* (savory spicing) she created something of a culinary oxymoron. This sort of linguistic misclassification is a constant accompaniment of the social interaction associated with the construction of these new cuisines.

There is one final sign that the idea of national Indian cuisine is now taken for granted—though its structure and logic are by no means standardized—and that is the proliferation of cookbooks that subsume and absorb "Indian" recipes into other, more transcendent, categories. Examples abound. Thangam Philip's *Modern Cookery* (1965), produced with an eye to nutritional benefits, restaurant cooking, and extremely Europeanized urban audiences, makes Indian recipes "modern" by looking at them from the perspective of the nutritionist, the food technologist, and the caterer. Madhur Jaffrey's brilliant *Vegetarian Cooking* (1981), like several others, juxtaposes Indian vegetarian recipes with those from the Middle and Far East, thus appealing mainly, in this case, to a particular audience in the United States. Others stick to Indian vegetarian recipes. The third volume of Harvey Day's multivolume *Curries of India* runs the reverse operation and includes dishes from Indonesia, Iran, Iraq, Turkey, Malaya, and Thailand under the label *curry*, illustrating a kind of gastronomic imperialism under the colonial trope of curry. There are also books like the recent *Appetizingly Yours* (Currim and Rahimtoola 1978) that are clearly directed to wealthy urban audiences in India, where Indian and Western snacks are promiscuously combined, with the Indian side of the book drawing on a trans-regional inventory. Thus not only is a national cuisine being constructed from regional or local traditions, but access to this national repertoire permits it to be subordinated to the purposes of other, more general classifications.

The idea of an "Indian" cuisine has emerged because of, rather than despite, the increasing articulation of regional and ethnic cuisines. As in other modalities of identity and ideology in emergent nations, cosmopolitan and parochial expressions enrich and sharpen each other by dialectical interaction. Especially in culinary matters, the melting pot is a myth.

Conclusion

The emergence of a national cuisine in contemporary India suggests a processual model that needs to be tested comparatively in other postcolonial situations in the contemporary world. The critical features of this model are the twin processes of regional and ethnic specialization, on the one hand, and the development of over-arching, crosscutting national cuisines, on the other. These processes are likely to be reflected and reproduced in cookbooks designed by and for the urban middle classes, and particularly their female members, as part of the larger process of the construction of complex public cultures involving media, travel, and entertainment.

Of equal comparative interest are the historical and cultural contexts against which the new national cuisines are appearing, contexts that are likely to vary considerably. In the Indian case, a national cuisine has developed recently in spite of a relative historical disinterest in gastronomic issues in classical (Hindu) traditions, so that both the textualization of the culinary realm and the creation of a civilizational culinary standard are recent processes. The final question that deserves further com-parative investigation is whether the long-term historical and cultural idiosyncrasies of each case make the culinary dynamics of contemporary societies different, in spite of certain broad processual similarities. To answer these questions, we need to view cookbooks in the contemporary would as revealing artifacts of culture in the making.

Notes

1. The single most important comparative treatment of cuisine from a sociological point of view is found in Goody (1982). In addition to that study, which has provided a good deal of the comparative perspective in this essay, I have also consulted the following sources for my understanding of non-Indian culinary traditions: Ahsan (1979); Austin (1888); Chang (1977); Cosman (1976); Forster and Ranum (1979); Furnivall (1868); Revel (1979); Roden (1972); Rodinson (1950); Root (1977); Vehling (1977). Goody (1982) contains an excellent and extensive bibliography.

2. Goody's discussion of the Indian material (Goody 1982: 114–26) takes issue at several points with the approach and arguments of R. S. Khare (1976a, 1976b). On the question of whether a pan-Indian high cuisine existed in premodern India, Goody appears to have confused the question of regional and courtly *high* cuisines with the matter of a *national* cuisine. For the latter, there is little evidence until the second half of this century. I am also inclined to support Khare's view that the cultural significance of cooking within the Hindu system remains incidental. More exactly, it might be said that Brahmanical normative thought gives short shrift to cooking, but royal practices as well as the divine cuisines of the great temples show highly differentiated, though regional, styles of cooking. On sacred cuisine in premodern South India, see Breckenridge (1986). Even here, collections of recipes are hard to find, though lists of ingredients, dishes, and meals frequently appear.

3. The question of vernacular cookbooks deserves separate treatment. In general, the new cookbooks raise a series of interesting linguistic and epistemological issues that have been omitted from this essay because of limitations of space. It should be noted that the shift to English is only the most obvious of a series of changes in the sociology of language reflected in the new cookbooks.

4. This manual itself appears to be modelled on the important Victorian treatise on household management published by Isabella Beeton (1861). I am grateful to Justin Silber of the University of Houston for drawing this work to my attention. The spread of European ideologies of household management to the colonies in the nineteenth and twentieth centuries is an important topic for comparative research.

5. For a sociobiological treatment of ethnic cuisine in contemporary Western contexts, see van den Berghe (1984).

References

Abdullah, U. (1981) *Malabar Muslim Cookery*. New Delhi: Orient Longman Limited.

Ahsan, M. M. (1979) *Social Life under the Abbasids*. London: Longman.

Appadurai, A. (1981) "Gastro-Politics in Hindu South Asia." *American Ethnologist*, 8:3 (August 1981), 494–511.

Attwood, M. S. (1972) *Adventures in Indian Cooking*. Bombay: Jaico. (Originally published as *A Taste of India* (Boston: Houghton Mifflin, 1969).)

Austin, T. (ed.) (1888) *Two Fifteenth-Century Cookbooks* (Early English Text Society, no. 91). London: Early English Text Society.

Aziz, K. (1983) *Indian Cooking*. New York: Putnam.

Beeton, I. (1861) *The Book of Household Management*. London: S. O. Beeton.

Begum, S. (1981) *Film Stars' Favourite Recipes*. Bombay: India Book House.

Bisen, M. (1970) *Vegetable Delights: 650 Original Indian Recipes*. Bombay: Wilco.

Breckenridge, Carol A. (1986) "Food, Politics, and Pilgrimage in South India, A.D. 1350–1650," in *Aspects in South Asian Food Systems: Food, Society, and Culture*, R. S. Khare and M. S. A. Rao, eds., 21–53. Durham: Carolina Academic Press.

Chang, K. C. (ed.) (1977) *Food in Chinese Culture: Anthropological and Historical Perspectives*. New Haven and London: Yale University Press.

Cosman, M. P. (1976) *Fabulous Feasts: Medieval Cookery and Ceremony*. New York: Braziller.

Currim, M., and Rahimtoola, M. (1978) *Appetizingly Yours: Snacks for All Occasions*. Delhi: Macmillan.

Day, H. (1963) *Curries of India*. Bombay: Jaico.

Dumont, L. (1970) *Homo Hierarchicus: The Caste System and Its Implications*. Chicago: University of Chicago Press.

Elias, N. (1978) *The Civilizing Process*. Oxford: Oxford University Press.

Forster, R., and Ranum, O. (1979) *Food and Drink in History*. Vol. 5 of *Selections from the Annales: Economies, Sociétiés, Civilisations*. Baltimore: Johns Hopkins University Press.

Freeman, M. (1977) "Sung," in *Food in Chinese Culture: Anthropological and Historical Perspectives*, K. C. Chang, ed., 141–76. New Haven and London: Yale University Press.

Furnivall, F. J. (1868) *Early English Meals and Manners* (Early English Text Society, no. 32). London: Early English Text Society.

Goody, J. (1982) *Cooking, Cuisine, and Class: A Study in Comparative Sociology*. Cambridge: Cambridge University Press.

Graburn, N. (1976) *Ethnic and Tourist Arts: Cultural Expressions from the Fourth World*. Berkeley: University of California Press.

Hosain, A., and Pasricha, S. (1962) *Indian Cooking: A Practical Introduction to Authentic Indian Food*. London: Paul Hamlyn Ltd.

Jaffrey, M. (1981) *World of the East: Vegetarian Cooking*. New York: Knopf.

Jagtiani, D. (1973) *Chutneys and Pickles of India*. Bombay: India Book House.

Kane, P. V. (1974) *History of Dharmasastra*. 2d ed. Poona: Bhandarkar Oriental Research Institute.

Kaufman, W. I., and Lakshmanan, S. (1964) *The Art of India's Cookery*. Garden City, New York: Doubleday and Co.

Khare, R. S. (1976a) *Culture and Reality: Essays on the Hindu System of Managing Foods*. Simla: Indian Institute of Advanced Study.

—— (1976b) *The Hindu Hearth and Home*. Durham: Carolina Academic Press.

Lal, P. (1970) *Vegetable Dishes*, Bombay: India Book House.

—— (1980) *Indian Recipes*. Calcutta: Rupa and Company.

Marriott, M. (1968) "Caste Ranking and Food Transactions: A Matrix Analysis," in *Structure and Change in Indian Society*, M. Singer and B. S. Cohn, eds., 133–71. Chicago: Aldine.

Meenakshi Ammal, S. (1968) *Samaiththu Pār*. 3 vols. Madras: S. Meenakshi Ammal Publications.

Mehta, J. (1979) *101 Parsi Recipes*. Bombay: J. Mehta.

Mutachen, R. (1969) *Regional Indian Recipes*. Bombay: Jaico.

Narayan, P. (1975) *The New Dalda Cookbook*. Delhi: Vikas.

Patil, V. (1979) *The Working Woman's Cookbook*. Bombay: India Book House.

Philip, T. E. (1965) *Modern Cookery for Teaching and the Trade*. 2 vols. Bombay: Orient Longman.

Prakash, Om (1961) *Food and Drinks in Ancient India*. Delhi: Munshi Ram Manohar Lal.

Primlami, K. (1968) *Indian Cooking*. Bombay: India Book House.

The Raj Cookbook. 1981. Delhi: Piper Books.

Ranga Rao, S. (1968) *Good Food from India*. Bombay: Jaico.

Reejhsinghani, A. (1973a) *Tasty Dishes from Waste Items*. Bombay: Jaico.

—— (1973b) *The Art of South Indian Cooking*. Bombay: Jaico.

—— (1975) *Delicious Bengali Dishes*. Bombay: Jaico.

—— (1977) *Cooking for the Single Person*. New Delhi: Vikas.

Revel, J.-F. (1979) *Un festin en paroles*. Paris: Pauvert.

Roden, C. (1972) *A Book of Middle-East Food*. New York: Knopf.

Rodinson, M. (1950) *Recherches sur les documents arabes relatifs a la cuisine*. Paris: Librarie Orientaliste P. Geuthner.

Root, W. (1977) *The Food of Italy*. New York: Random.

Sharar, A. M. (1975) *Lucknow: The Last Phase of an Oriental Culture*, E. S. Harcourt and F. Husain, eds. and trans. London: Paul Elek.

Sheth, A. (1968) *The New Indian Cookbook*. Delhi: Hind Pocket Books.

Shroff, V., and Desai, V. (1979) *100 Easy-to-Make Gujarati Dishes*. Delhi: Vikas.

Singh, B. (1961) *Indian Cookery*. London: Mills and Boon.

Singh, D. (1970) *Indian Cookery*. Harmondsworth: Penguin Books.

Skelton, M. L., and Rao, G. G. (1975) *South Indian Cookery*. Delhi: Orient Paperbacks.

Srivaran, M. L. (1980) *Unusual Vegetarian Cookery*. Delhi: Hind Pocket Books.

Steel, F. A., and Gardiner, G. (1888) *The Complete Indian Housekeeper and Cook*. London: William Heinemann.

Time-Life. (1969) *Recipes: The Cooking of India*. New York: Time-Life Books.

van den Berghe, P. L. (1984) "Ethnic Cuisine: Culture in Nature." *Ethnic and Racial Studies*, 7:3 (July 1984) 387–97.

Vehling, J. D., trans. (1977) *Apicius: Cooking and Dining in Imperial Rome*. New York: Peter Smith.

Zimmerman, Francis. (1982) *La jungle et le fumet des viandes*. Paris: Editions du Seuil.

19

"Real Belizean Food": Building Local Identity in the Transnational Caribbean

Richard Wilk

Food and cooking can be an avenue toward understanding complex issues of cultural change and transnational cultural flow. Using examples from Belize, I discuss the transformation from late colonial times to the present in terms of hierarchies of cuisine and changes in taste. In recent Belizean history, food has been used in personal and political contexts to create a sense of the nation at the same time that increased political and economic dependency has undercut national autonomy. I suggest several possible ways to conceptualize the complex and contradictory relationship between local and global culture.

It is an anthropological truism that food is both substance and symbol, providing physical nourishment and a key mode of communication that carries many kinds of meaning (Counihan and Van Esterik 1997). Many studies have demonstrated that food is a particularly potent symbol of personal and group identity, forming one of the foundations of both individuality and a sense of common membership in a larger, bounded group. What is much less well understood is how such a stable pillar of identity can also be so fluid and changeable, how the seemingly insurmountable boundaries between each group's unique dietary practices and habits can be maintained, while diets, recipes, and cuisines are in a constant state of flux (Warde 1997:57–77).

In a world of constant cultural contact, international media, and marketing, the process of change in diets seems to have accelerated, but the boundaries that separate cultures have not disappeared. The difficult conundrum of stability and change, of borrowing and diffusion, without growing similarity or loss of identity, which we find in the world's food consumption, appears in many other realms of culture too (Appadurai 1996; Friedman 1994). It is clear that older modernization and acculturation theories that predicted a growing homogenization and Westernization of the world's cultures are inadequate in a world that seems to constantly generate new diversities, new political and social divisions, and a host of new fundamentalisms.

Many social scientists have pointed to the resurgence of nationalism and ethnicity in the last two decades, and some argue that strengthened local identities are a direct challenge to globalizing diffusion of consumer culture (Jenkins 1994; Tobin 1992). Popular discourse also opposes the authentic local or national culture or cuisine with an anonymous artificial mass-marketed global culture of McDonalds and Disney. These arguments reproduce a diffusionary model of history as Westernization or cultural imperialism. In this paper I continue to build an alternative approach, recognizing that the strengthening of local and national identities and global mass-market capitalism are not contradictory trends but are in fact two aspects of the same process (Beckett 1996; Miller 1997). I will further this argument through an analysis of recent changes in the cuisine of a tiny and marginal place, a mere dot on the globe, which may in some ways be unique, but which is in other ways typical of many tiny and unique places that are increasingly integrated into the world market for consumer goods.

I have worked in the Central American and Caribbean nation of Belize since 1973. Belize is a wonderful place to study the relationship between food and national identity because nationhood is such a recent construct there. Belize only attained independence in 1981, and until that time nationhood was primarily a matter of political rhetoric, government commissions, and debate among the educated elite. Since that time foreign media, tourism, and migration have spread a broad aware-ness that Belize needs a national culture, cuisine, and identity to flesh out the bare institutional bones of nationhood provided by the British. Belizeans know their flag, anthem, capital, and great founding father, and they now know they should have a culture to go with them.[1]

Belizean nationalism contends with a variety of other forms of identification, including regional, familial, class, language, and ethnicity. Many, but not all, Belizeans identify with a language or ethnic group, what are usually called "races" or "cul-tures." The "Creole" category was once applied to local Europeans, then was extended to those of mixed European and African ancestry, and now it has become a general term for people with multiple or overlapping ethnic backgrounds (see Stone 1994). Diverse groups of migrants from neighboring Spanish-speaking countries are usually labeled "Spanish," though many are of Amerindian origins. In addition there are relatively bounded ethnic groups defined by language, residence, and liveli-hood, including Mopan, Kekchi, East Indians, Mennonites, Chinese, and Garifuna. Of course, in a country of only 200,000 people, all "ethnic" boundaries are crosscut by multiple personal and familial relationships.

Most of the puzzles and problems of uniqueness and sameness, of boundaries and open flows, of imperialism and resistance, which we find in other parts of the developing world, are also found in Belize. While it does not always occupy center stage, food and cooking are an important and sometimes dramatic entry into under-standing this uneven process. To show how rapidly Belize has been transformed, and to argue for the crucial importance of foreign influence in the growth of local culture, I begin with a contrast between two meals, separated by 17 years.

Eating Culture: A Tale of Two Meals

Meal One: Tea with the Gentles. May 1973, Orange Walk Town (a predominantly Hispanic community in the northern sugar zone).

Henry and Alva Gentle, both schoolteachers in the Creole lower middle class, invited a 19-year-old archaeology student to their home for a meal. I showed up one evening about 5 p.m., just as the family sat down to the evening meal. "Dinner" at midday was the main meal of the day in Belize; evening "Tea" was usually leftovers from dinner, with some bread or buns and coffee or tea.

Everyone was excited about having a foreign visitor in the house and stopped eating when I entered. I was ushered into the uninhabited and unused front "parlor" for a few minutes while furious activity took place in the kitchen, and children were sent running out of the back door to the local shops. Then we all went back to the kitchen table to finish tea. Everyone sat before small plates of fish and plantain dumplings simmered in coconut milk, with homemade coconut bread, and either fresh fruit juice or tea.

After six weeks on archaeological camp food (most of which had been sent out in a large crate from England), my mouth watered at the prospect of tasting something authentically Belizean. Instead, as an honored guest I was treated to the best food the house could afford, food they thought I would be comfortable with: a plate of greasy fried canned corned beef (packed, as I later found out, in Zimbabwe), accompanied by six slices of stale Mexican "Pan Bimbo" white bread, a small tin of sardines in tomato sauce, and a cool Seven-up with a straw. When this was presented, the whole family paused in their own meal to smile shyly and expectantly, waiting for my pleased reaction. I did as well as I could, given that I dislike both sardines and canned meat.

Despite the food, the meal was relaxed and fun, with a lot of joking between family members. There was no formal beginning or end to the meal; late arrivals were seated and served. People got up and went to the kitchen to fill their plates when they wanted more food. Gradually people drifted out to the verandah to sit and discuss the day, and see who was walking by on the street.

Meal Two: Dinner with the Lambeys. August 1990, Belmopan (a small town of about 3,000, also the nation's capital, with a mixed population).

Lisbeth and Mike Lambey lived across the street from us in a nice part of town, populated mostly by middle-rank civil servants. Mike worked in the refugee office, and Lisbeth was an assistant to a permanent secretary. A bit more educated than the Gentles and better paid, they still belonged to the same salaried middle class. Lisbeth was born into an old Creole family in Belize City, while Mike's family included Creole and East Indians from the southern part of Belize. All of Mike's family live in various parts of Belize, but all of Lisbeth's seven siblings have emigrated to the United States.

When Mike invited us to dinner, over the front wall of his yard one morning, he said, "I want you to taste some good local food." By 1990, many younger Belizeans, especially those with both spouses working for wages, had shifted to what was seen as an "American" practice of eating a light lunch and the main family meal in the evening.

When my wife and I arrived, we were seated in front of the TV, handed large glasses of rum & Coke (made with local rum and "local" Coke), and were shown a videotape about the refugee situation in Belize, produced by the United Nations. Then we sat down to a formal set table; all the food was laid out on platters, and it was passed from hand to hand. Everything we ate, both Mike and Lisbeth said, was produced in Belize and cooked to Belizean recipes. We had tortillas (from the

Guatemalan-owned factory down the road), stewed beans (which I later found were imported from the United States, but sold without labeling), stewed chicken (from nearby Mennonite farms), salad (some of the lettuce was from Mexico) with bottled French dressing (Kraft from England), and an avocado with sliced white cheese (made locally by Salvadoran refugees). We drank an "old-fashioned" homemade pineapple wine.

The meal was formal, with a single common conversation about local politics to which each of us contributed. When everyone was finished, Lisbeth circulated a small dish of imported chocolate candies, and we all rose together and went back to the TV. Mike proudly produced a videocassette on which a friend had taped Eddie Murphy's "Coming to America" from a satellite broadcast. Both Lisbeth and Mike found the film hilarious and especially liked the parts where the two unsophisticated bumbling Africans show their ignorance of New York and American ways. We had seen the film before, and pleading the need to put our daughter to bed, went home before it was over.

That same week Radio Belize carried advertisements for the grand opening of the first self-proclaimed "Belizean Restaurant."[2] Owned by a Belize-born couple who had recently returned from a 20-year sojourn in the United States, the advertisement asked customers to "Treat yourself to a Belizean Feast. Authentic Belizean dishes— Garnachas, Tamales, Rice and Beans, Stew Chicken, Fried Chicken." All these foods were already served in numerous restaurants all over the country; the only other kind of food available in most places is Chinese. But this was the first time they had been granted the public distinction of being the national cuisine.[3]

In the next two years the notion of Belizean restaurants and Belizean cuisine became commonplace, and most people accepted that there was indeed a national and traditional lexicon of recipes. In fact, as an anthropologist running around the country asking questions about Belizean food, I often found that my very efforts loaned legitimacy to the idea of a national cuisine. How is it that British and Mexican dishes, and global standards like stewed and fried chicken, emerged so quickly as an emblematic Belizean cuisine? This is a clear example of nation building, as a contextualized and nonreflective set of practices was codified and labeled as characteristic of the nation as a whole (Lofgren 1993).

In Belize, cuisine has been nationalized in a process quite distinct from that described by Appadurai in India (1988). There, cookbooks were crucial instruments, and regional diversity has been the main theme. National cuisine explicitly incorporates and crosscuts local traditions, which are simultaneously codified as each local group finds a significant contrastive "other" in neighboring areas and in the superordinate national melange. In Belizean cuisine the internal contention over how different ethnic and regional groups will be incorporated in the national has been quite muted. Instead, Belizean cooking has emerged through an explicit contrast with an externalized "other." The crucible of Belizean national cooking has been the *transnational* arena: the flow of migrants, sojourners, tourists, and media that increasingly links the Caribbean with the United States. Caribbean nationalism and identity is now problematized and contested, debated and asserted, in this shifting transnational terrain (Basch et al. 1994; Olwig 1993).

Belizean National Culture

The emergence of Belizean cuisine is just a small part of a general process, which has unevenly and imperfectly established and legitimized Belizean national culture (Everitt 1987; Medina 1997; Wilk 1993b). Just twenty years ago, the concept of Belizean Culture" was no more than politicians' rhetoric and a project for a small group of foreign-trained intellectuals. Official national dance troupes, endless patriotic speeches, history textbooks, and border disputes with neighboring Guatemala have helped establish the existence of a category of "Belizean Culture," but they have not filled it with meaning. It does not take a social scientist to define or observe the ways Belizeans contest the meaning of the category of local culture; this is the stuff of daily conversation and debate. The process is contrastive, defining the self through defining difference. Belizeans speak of "floods" and "invasions" of foreign goods, television, preachers, tourists, money, entrepreneurs, music, language, drugs, gangs, tastes, and ideas (Wilk 1993a). Contrasts between the local and the foreign are on everyone's lips, though there are many shades of opinion about which is good or bad, and about where it will all end up. People constantly talk about authenticity and tradition, contrasting the old "befo' time" with everything new, foreign, and "modan."

Popular ideas about how the foreign is affecting "little Belize" are dramatized in jokes and stories about Belizeans who mimic or affect foreign ways. The traveler who returns after a few weeks in the States with an American accent is a common figure of fun. So is the returned cook who no longer recognizes a catfish, and the shopper who buys pepper sauce from an American supermarket to bring home, and pays a hefty import duty before looking at the label and seeing that it was made in Belize.

Opportunities for drawing contrasts appear often. Estimates of the number of Belizeans living in the United States range widely, but 60,000 seems a reasonable low estimate, or 30% of the total living in Belize (Vernon 1990). Los Angeles is the second-largest Belizean settlement. There is constant flow of people, goods, and money between domestic and foreign communities (Wilk and Miller 1997). At the same time, more than 140,000 foreign tourists visit Belize every year. Belize is flooded with American media, from books and magazines to a barrage of cable and satellite television.

Elsewhere I have argued that one effect of the increasing prominence of the foreign in Belize has been the objectification of the local (Wilk 1993a, 1995). Many issues that were once seen as localized, ethnic, and even familial are now interpreted in a global context. The problems of youth, social welfare, ethnicity, and gender roles, for example, are now placed in a global contrast of "our way" or "our Belizean traditions" with "those in the States." This has led to the emergence of a political and cultural discourse about otherness and sameness. During colonial times, foreign culture was received indirectly, with the expatriate and local colonial elite acting as selective agents and gatekeepers. Now all classes have direct access to foreign culture, and the foreign is no longer as closely associated with wealth and power (Wilk 1990).

This suggests an important way of reading the differences between the two meals I discussed above. While class differences between my two hosts account for some of the variation, the most dramatic differences between the Gentles and the Lambeys have little to do with changes in the content of Belizean culture or identity. Instead, they result from changing knowledge about foreigners and increased consciousness of culture itself. The Gentles, despite (or maybe even because of) their education really knew very little about Americans or American culture. Travelers were rare; personal contact with foreigners was sought for the very purpose of learning. The Gentles wanted to please me—they just did not know how Americans were different from the British. They could not know that young Americans wanted something "local," "cultural," and "authentic"—the very things they knew educated and rich people looked down upon. No wonder that our relationship did not flourish; I felt like I was a disappointment to them, and they probably felt the same way about me. My zeal to teach them to value their indigenous culture probably sounded false and condescending, if not just incoherent and weird.

Seventeen years later, the Lambeys know how to play this game properly. They are Belizean nationalists who know that they are supposed to have something authentic and local to offer. They have been abroad and have learned to perceive and categorize differences as "national" and "cultural." They have learned that foreigners expect them to be Belizean, and they know how to do the job. They are as busy creating traditions and national culture as the itinerant Belizean woodcarvers who now tell tourists that their craft was handed down from their African ancestors (rather than taught by Peace Corps volunteers). Serving an authentic Belizean meal, for the Lambeys, is a performance of modernity and sophistication. The emotion it evokes for them is closer to pride or defiance than to nostalgia, the warmth of memory, or the comfort of repeated family habits. On the contrary, the meal expresses a sense of distance. A Belizean student, home for the summer from Jamaica, expressed this distanced, critical stance when, discussing his feeling about Belizean nationalism and ethnic identity, he told me "you can respect something without believing in it." Belizeans often express a similar sophisticated tolerance toward the stereotypes of their country that appear constantly in travel guides and tourist magazines—"Belize the unspoiled wilderness," etc. (Munt and Higinio 1993).

To sneer at this accomplishment, at the Lambeys' dinner, by laughing at its shallowness or inauthenticity, is to miss its point (Friedman 1992). Mastering the performance and the role asserts a claim to categorical equality, to knowledge and power. I did not really understand how important a step this was until I talked with older Belizeans about their first trips to America.

A newspaper editor told me that as a child he and his friends looked at the names of cities on the labels of American products, at the goods advertised on the pages of magazines plastered on their house walls, and fantasized about what the United States was like. He thought the advertisements depicted a real world. So on his first trip to the United States, he was excited about going to the places where the products came from. And it was a shock to find how ignorant he had been—and that so many of the things people ate and used in the United States actually came from Japan and other countries. He felt humiliated and foolish for having confused fantasy with reality.

He related that even recently Belizeans were relatively unsophisticated consumers and were "easy marks" for advertising—bad loans and credit terms and shoddy goods—because they lacked the experience and sophistication of Americans. As an ardent and radical nationalist, he decried the effects of television on local culture, but in the same breath praised the way TV has "raised Belizeans' consciousness," making them knowledgeable and aware of the rest of the world. No longer were they blinded by surface appearances. He, and other Belizeans I spoke with, expressed an optimistic hope that in seeing more clearly the problems of the rest of the world, Belizeans would learn to value what they had at home (Salzman [1996] finds this same effect on the island of Sardinia).[4]

The invasion of foreign media and goods and increased knowledge and sophistication are two sides of the same coin, two elements of the same process.[5] Outsiders tend to focus attention on the former, on the ways that certain meanings, messages, and practices are imparted and forced on powerless consumers of media and advertising (e.g., Lundgren 1988). We may see this as seduction or as a transformation of consciousness. But at the same time, it leads to an accumulation of knowledge about the world. By putting this knowledge into play in their everyday consuming lives, by performing and enacting and using unfamiliar goods, Belizean consumers transform abstract images, words, and names into the familiar appliances of life in Belize. Through consumption the foreign is made part of local existence, and it therefore comes under the same sorts of (albeit limited) control.

Belizeans are becoming sophisticated consumers, and in the process they are gaining a form of power that was previously denied, or was rationed and controlled by the local elite. Now Belizeans can penetrate deceit and "comparison shop," by playing one message off against another, and break away from the "brand loyalty" of the colonial consumer who had limited information. Furthermore, Belizeans have become more aware that they use and manipulate goods to their own social ends; they have acquired the distance necessary to view goods as tools to be manipulated rather than signs to be accepted or rejected. This sort of distance, of course, does not necessarily diminish desire or emotional longing for the foreign. For example, when in my 1990 survey I asked 1,136 high school students what they would buy with $50, 28% chose imported tennis shoes, usually by brand, color, and type of lacing.

When I first went to Belize I was asked many strange questions about the United States—was it true that people in California had sex in public? Did Indians still attack travelers in the West? Today I am more likely to be engaged in debate about Los Angeles street gangs or Bill Clinton's troubles with his cabinet. Through media, Belizeans tap more levels of American discourse—they get the "official" word on CNN news, and the gossip from the *National Enquirer, Life-Styles of the Rich and Famous* (which has featured a Belizean resort twice), and *A Current Affair*.

Belize has always had a lively network of gossip and rumor about local politics and goings-on, but for England and America there was only the official word. While Belizeans do not usually feel equal to Americans in most political and material ways, they now know more of the dimensions of their perceived inferiority. This is very different from the more pervasive, generalized, and threatening inferiority that people felt during colonial times. Their new depth of knowledge of the world does not in and of itself create tastes for foreign things. Knowledge "of" the world is not the same as knowledge of how it works, but it lends new textures of significance to foreign things and imparts a much richer field of meaning to them. The taste for foreign goods over local equivalents so often observed in developing countries (e.g., Orlove 1997) can then be seen as a consequence of the desire to know more about the world, to become more sophisticated, to acquire new forms of knowledge, and to make that knowledge material.

I am not arguing that having access to the American tabloid press has been a great boon to Belize and has changed the global balance of culture and power. But this access does indicate a series of shifts in the relationship between Belize and the rest of the world, which transforms Belizeans as consumers. In the process, a colonial-era hierarchical discourse that opposed the backward local against the modern and foreign has begun to crumble. One way to describe this process is through the metaphor of drama.

The Drama of Local vs. Global

The many manifestations of the "global ecumene" in different places often appear to be chaotic, a pastiche of shreds and patches (Foster 1991:251). One place to seek structure in the flow of goods and meanings is to look for an underlying power and interest groups, focusing for example on the interests of the state, or of multinational corporations (see critique in Miller 1994). A complementary approach can instead seek an emerging structure in the form of a narrative or story, played over and over in different settings with new characters and variations—a repetitious drama instead of a social or spatial order.

Observers of global cultural process find tendencies toward both homogenization and heterogeneity (Arizpe 1996; Friedman 1990; Hannerz 1987; Howes 1996; Lofgren 1993; Tobin 1992). If we think of the global culture as constituted by drama, we can perhaps locate the homogeneity in a common dramatic structure of encounter, while the local actors, symbols, and performances of the drama proliferate in splendid diversity. In this way, the global ecumene becomes a unifying drama, rather than a uniform culture, a constant array of goods or a constellation of meanings.

One unifying dramatic theme is a struggle that reads as "the local against the foreign," or *our* culture vs. the powerful and dangerous *other*. This drama is played out in many permutations, at many different levels, often nested within each other. What is universal is the drama itself, not its outcome. And this is in itself a significant opening for both cross-cultural communication and misunderstanding, for example, in the trade disputes between the United States and Japan, or the long-standing cultural ambivalence of the French toward Americans (Kuisel 1993).

If we follow this argument in Belize, we see that participation in the global ecumene is not so much a matter of acquiring goods or building a nation. Instead, there is a process of learning to share and participate in a core drama, in which national identity is essential costume. Social scientists studying consumption have been actors—sometimes scriptwriters—for this performance. We tend to force debate into two positions—hegemony and resistance. On one hand there is capitalism's need for expansion of markets, the breakdown of community and local economies, and the capacity of advertising to create envy, social composition, and new needs. Consumer goods are essential components of new market-oriented systems of ranking and hierarchy.

On the other hand, we can emphasize the ways objects are used to resist capitalism, to maintain local systems, to forge links with an authentic past, to build identities defined by local systems of meaning and nonmarket social relationships. The moral of this story is that the technological apparatus of capitalism, including television and other media, has been turned to very local and anti-hegemonic purposes. New gardens of cultural diversity will continue to spring up from the leveled and furrowed fields of international capitalism; even corporate icons of homogeneity like McDonalds can become local institutions (Watson 1997).

This contrast of seductive globalism and authentic localism is an extremely potent drama because it has no solution—it is an eternal struggle, where each pole defines its opposite, where every value carries its own negation. The players in the drama are always taking positions as advocates for one pole or the other, but they are actually locked in a dance—dependent on each other for the support of opposition. Anthropologists and folklorists tend to embrace this drama, attached to both its tragedy (ethnocide) and its comedic moments (bumbling development bureaucrats outwitted by canny natives). The key development in the last 20 years is that the drama has escaped from academic confinement and is no longer the province of the educated. Everyone in Belize is now concerned with foreign influence, local authenticity, and the interpretation of various kinds of domination and resistance (Wilk 1993a).

Colonial to Global Drama: An Example

To bring the discussion back to earth in Belize, we need to begin with the colonial regime of consumption and describe the way it set the stage and provided scenery for the drama of local vs. global. To stretch my metaphor, this section is a brief look "backstage" at local political economic reality. My data come from extensive archival work with newspapers and documents and from numerous personal histories. For contemporary consumption practices and tastes I draw on a survey of 1,136 high school students from four diverse institutions (Babcock and Wilk 1998) and a door-to-door survey of 389 people in Belize City, Belmopan, and a large village in the Belize District, both conducted in 1990. The latter survey replicated parts of Bourdieu's French work that was the basis of *Distinction* (1984), with a focus on likes and dislikes for a wide variety of foods, music, television, and reading material.[6] These data and my interviews were focused on people's likes and dislikes for particular dishes. I have much less material on the presentation, serving, and context of consumption of foods, though this topic is also very important.

The Colonial Taste for Imports

During the nineteenth century Belize was a logging colony, the source of famous "Honduras mahogany" that was used to build European railway cars and furniture. The colony was dependent on imported foods of all kinds. While scattered rural subsistence farmers produced most of their own food, the standard diet of mahogany cutters and working-class urban dwellers was imported flour and salt meat. A weekly ration for workers was four pounds of pork and seven of flour, eaten as dishes like "pork and doughboys."[7] The managerial and mercantile European and Creole middle and upper classes consumed a wide variety of imported foods and drink and limited amounts of a few local vegetables, fish, poultry, and game. Given the very uneven quality of imported goods, consumers who could afford packaged and branded foodstuffs became highly "brand loyal" to established lines and companies. Brands were ranked according to both price and quality, with the highest ranks from Britain and lower ranks from the United States and Latin America. Access to the best brands was tightly controlled both through price and a strict system of exclusive distributorships that kept them in only the "best" shops. Branding was a key element of cultural capital, and it came to connote quality (see Burke 1996 on soap in Zimbabwe). The poor bought generic goods, often dipped out of barrels, and had little choice when compelled to purchase in company stores through various forms of debt-servitude and payments of wages in goods.

Diet was highly class stratified. A single scale of values placed local products at the bottom and increasingly expensive and rare imports at the top. Imports were available to anyone who had the money, but in practice people did not usually consume above their class, whatever their economic resources. These strictures were only relaxed during the Christmas season, when the most exclusive products circulated at every kind of festivity. At this time the elite placed special emphasis on rare delicacies obtained directly from England through private networks of friends and relatives. Among the laboring class, almost any kind of store-bought food was considered superior to the rural diet based on root crops, rice, game, fruits, and vegetables.

This relatively simple hierarchy put the greatest pressure on the thin middle class of local petty merchants, low-level officials, tradespeople, and clerical workers. They did not have the resources for an exclusively imported diet but had to work hard to distance themselves from the kind of cheap and local foods that were the rural and working-class staple. One consequence was that they shunned local foods like fresh fish and game meat. These were common fare for the rural poor (for whom they were part of a subsistence lifestyle) and the upper elite (for whom selected varieties were considered exotic delicacies as long as they were prepared according to European recipes and smothered with imported sauces).

While the middle class depended most heavily on imports, menus from elite ceremonial meals included local snapper fillets in fish courses, garnished with oysters imported fresh from New Orleans. Venison, duck, and a few other local game animals with European analogs also appeared on the tables of the local gentry. Lobster is a good example of a stratified taste "sandwich": eaten by the poor because it was cheap, by the elite because it was prized in Europe, but shunned by the middle class as a "trash fish." Older Belizeans from middle-class backgrounds say that their mothers would not allow a lobster in the house; some of these people never ate lobster until it became a major export commodity in the 1970s, and then they had to stop when the price skyrocketed in the 80s.

The middle class built dietary diversity by borrowing foods from the Hispanic mercantile and managerial elites in the northern part of the country. "Spanish" food, especially festive dishes like tamales, relleno (a stuffed chicken stew), and tacos, entered the middle-class diet as a safely exotic option—associated neither with the class below nor the class above. "Spanish" food quickly became naturalized as part of the middle-class Creole diet. In this case, the "foreign" quickly became "local" and authentic.

There was some local resistance to or evasion of colonial food hierarchy. Foreign dishes were often localized or made affordable by substituting ingredients, renaming, and recombination. Kin ties cut across class and ethnic boundaries, and local produce therefore circulated between classes through networks of extended kin. Certain kinds of rare "country foods," especially wild game, honey, hearts of palm and the like, came to carry a connotation of familism and embeddedness; of belonging to place, even for the urban middle class. Some rustic dishes were enjoyed only in privacy, or on special occasions or during visits to country relatives. Festivals and celebrations also provided sites for the legitimate consumption of local products as holiday foods, especially fruit wines and snacks like cashew and pumpkin seed. But while neighboring countries like Mexico had a substantial indigenous population whose dishes could be adapted and co-opted by an educated elite, Mayan cooking entered Belize only indirectly, mainly through the influence of the refugees from Yucatan's caste war in the 1850s.

The colonial regime of consumption in Belize, therefore, was similar to that seen by Bourdieu in modern France (1984). A relatively stable social hierarchy was defined by differential access to economic and cultural capital, which takes the form of "taste" and thoroughly naturalized predisposition and preference. Goods were positional markers within the hierarchy, both the means by which culture is internalized as taste and an external symbolic field through which groups identify boundaries and define differences among classes. There was a slow flow downward, as lower ranks emulated the elite and the elite found new markers. There was also a stable degree of resistance, as some local products and practices were regenerated and adopted upwards through kin ties.

Of course, in France the hierarchy is much more elaborate, with a clear division between economic and cultural elites and a dynamic and innovative upper middle class. In Belize the fashion system was more firmly controlled by an elite that tended to combine political, economic, and cultural power through the media, government, schools, and shops. They managed and censored the flow of goods and information into and around the colony and enforced quite uniform standards within the elite class through many of the same exclusive social practices used in London at the time (Mennell 1984:200–228). Boundaries with lower classes were policed through many forms of class and racial discrimination. Up until the 1970s there were racially exclusive clubs, and working-class people were not welcome and subject to humiliation in the few shops that catered to the elite (Conroy 1997).

Elite power was embodied in practices of consumption, and through roles as cultural gatekeepers, the elite were arbiters of taste in everything. The choices of the middle and working classes were limited to accepting or rejecting what was offered. They could not find alternative sources of goods, information, or taste, except in the immediately neighboring Hispanic republics.

Pupsi and Crana; The Beginning of the End

This hierarchy of taste remained remarkably stable until the early 1960s. An incident, where food entered national political debate, illustrates one of the ways the colonial order began to unravel. In 1963 the British finally granted Belize limited local self-government. The anticolonial Peoples United Party was promptly elected, led by George Price, who had been imprisoned for nationalist activity in the 1950s (Shoman 1985). With limited legislative power, Price began to make symbolic changes in flags, official dress, and the names of towns and landmarks. He chose to de-emphasize some colonial holidays and changed the name of the country from British Honduras to Belize. There was a lot of popular support for most of these measures.

Then he gave a speech about the local economy, which exposed the country's ambivalence about the depth of the decolonizing project. In the speech he suggested that it was time to stop aping the food standards of the colonial masters. Belizeans would have to become self-sufficient in food and value the "traditional" local foods instead of copying foreign models and continuing to depend on imported foodstuff. He told his audience they should eat less imported wheat bread and more of their own products, drinking fever-grass tea and sweet potato wine, and eating pupsi and crana (abundant local river fish) instead of imported sardines. Like a number of colonial agricultural officers, he argued that it was unreasonable for a country rich in fertile soil, surrounded by abundant sea life, to import grain and fish from Europe.

The pro-British opposition, which had unsuccessfully fought Price's other cultural initiatives, now found an issue that aroused popular support that threatened Price's whole nationalist project.

> When the PUP started they promised you ham and eggs, etc., if you put them in power. They also promised you self government. But today when they get Self Government, they tell you to boil fever grass and eat pupsi and other river fishes. What will they tell you to eat when they get independence? [Belize Billboard, January 5, 1964]
>
> The human body is like a machine, and it must have fuel to keep it running. And the fuel of the human machine is food, protective and sustaining foods such as milk and other dairy products, eggs, vegetables, fruits, whole grain and enriched bread and other cereals, just to name a few. Food must supply the vitamins needed, along with other essential nutrients such as proteins to keep the body running in high gear. . . . It is obvious that pupsi and crana, which live mostly in polluted swamps[,] cannot replace, as our premier advises, our sources of vitamin rich food and proteins, the most important ingredients in our diet. Conditions such as pellagra and ariboflavinosis (disease dues [sic] to lack of vitamin B) occur in people who live continuously on restricted diets such as corn and salt pork only. [Belize Billboard, January 11, 1964]

While the PUP's previous nationalist program challenged British political authority, the suggestion about diet turned the entire edifice of colonial cultural values and hierarchy literally upside-down. The words *pupsi* and *crana* became a rallying point for the formation of an opposition party. Price's suggestion was also unpopular among working- and middle-class people, who felt that they were being told to be satisfied with poverty instead of "improving" their lot. Many who had supported Price had seen his goal in terms of equity—a society where everyone would eat "high table" imported foods, sardines, and wine, not one where even the elite would be eating "bush food." Surprised by the reaction, Price moderated his position to one of import substitution, particularly supporting local rice and bean production, and he never publicly again called for a change in the national diet.[8] The violent reaction to his speech shows just how resistant to change the cultural order of colonial taste had become. It brought home to Belizeans the realization that the end of colonialism would mean more than a new flag and new street names. The "Pupsi and Crana" speech was the point where the colonial regime of taste was no longer part of the taken-for-granted of everyday life.

The Royal Rat

Queen Elizabeth's 1985 visit to Belize marked a major symbolic recognition of Belize's independence, at a time when Belizeans were increasingly worried about the veracity of British defense guarantees against Guatemalan aggression (the result of a long-standing border dispute). A major event during the visit, the first by a reigning Monarch, was a state banquet at the residence of the British High Commissioner in Belmopan, the new capital city built by the British. A selection of Belizean delicacies was prepared for the royal party by the best local cooks.

One of the tastiest wild mammals of the Belizean rain forest is a large rodent called a *gibnut* or *paca (Agouti paca)*. Highly prized in the rural diet, it was never widely eaten by the urban middle class. At the suggestion of local cooks and officials, but with the approval of the High Commissioner, roast gibnut was given the place of national honor as a main meat course at the Queen's banquet. She did not eat very much of it, but as a graceful veteran of hundreds of inedible feasts of local specialties, she still praised it to the cook. There the story would have ended, but for the British tabloid press. The Sun and other British newspapers produced a slew of outraged headlines, variations on the theme of "Queen Served Rat by Wogs." Angry letters were printed in the British press by citizens who were enraged by this assault on HRH's dignity.

The press reports were quickly transmitted back to Belize, where they provoked outrage and widespread anger, even among those who had never eaten gibnut. A few conciliatory conservative writers tried to explain to the British that the gibnut was *not* a rat and suggested that the incident was merely a misunderstanding. But most Belizeans saw this as an example of British arrogance and racism. For the first time, a Belizean dish became a matter of public pride. Nationalist chefs and nutritionists defended the Belizean gibnut as tasty, healthful, and nutritious. Reinterpreted as a national delicacy, today it often appears on restaurant menus as "Royal Rat," and its high price and legitimacy in national cuisine place heavy hunting pressure on remaining populations.

The incident of the royal rat came at a crucial time, just after political independence was granted in 1981. A legitimate category of Belizean food was beginning to emerge. The government, interested in cutting down food imports, had halfheartedly sponsored several campaigns through the Ministry of Education aimed at promoting production and consumption of local foods with help from CARE and the Peace Corps. During the 1970s the thrust was one of substituting local products for imports in familiar recipes: making bread with plantain flour, jams with local fruits. During the eighties the emphasis shifted to rediscovering (or reinventing) traditional foods; eating pupsi and crana was no longer unthinkable.

Belizean restaurants in the United States, cookbooks, public festivities where food is served, and the expensive dining rooms of foreign-owned luxury hotels were all crucial stages where ideas about Belizean food were tried out. By 1990, many dishes that were once markers of rural poverty had been converted into national cuisine. Others had quietly disappeared. Foreigners, expatriates, tourists, and emigrants were crucial agents in formulating and valuing the local.

Taste and Hierarchy Today

Colonial Belize had a clear hierarchy of social and economic strata, marked by their food practices and preferences. As Bourdieu points out (1984), these class tastes are bound together into systems, with internal logic and structure, by sentiments and dispositions rooted in childhood and a lifetime of learning. He focuses on schools as crucial institutions where these tastes are transmitted and ordered.

The social and economic stratification of Belize has not changed drastically since the 1950s. There are still quite exclusive and cosmopolitan mercantile economic and bureaucrat-technocrat elites, a diverse petit-bourgeoisie and functionary middle class, and a large and partially destitute working class. This hierarchy is crosscut in complex ways by ethnicity, family ties, political alliances, regional loyalties, and rural/urban differences.

Under the colonial regime, diversity was managed and regulated through the flows of fashion and taste, which entered through the gatekeepers at the top who were legitimate models for emulation or resistance. Today this hierarchy has been drastically undercut by new flows of information and goods. Travel, once the province of the privileged elite, is now practical for most Belizeans. My survey shows that 73% of Belizean adults have traveled out of their country, while 34% have lived abroad for three months or more. Variation by class is not high. For example, 45% of skilled manual workers have lived abroad, compared to 54% of business managers. The average Belizean has 2.6 immediate consanguineal relatives living abroad, again with little difference broken down by class, education, or wealth.

The worlds these travelers encounter are quite diverse; some end up in middle-class suburbs, others in urban ghettos. Some bring back luxury cars, high fashion clothing, and silverware, some display diplomas and stereos, others bring the latest rap tunes, BK basketball shoes, Gumby haircuts, and dread belts.

Similarly, tourists who once concentrated in a few oceanside resorts are now diffusing through the countryside and cities as ecotourists in search of unspoiled nature, ancient ruins, and authentic folk culture. A growing portion of the population has direct contact with foreigners; Belizeans now make sophisticated distinctions between different kinds of foreigners, and see them as representing diverse cultural backgrounds.

Finally, greater access to electronic and print media has vastly broadened the images that Belizeans have of the world and has destroyed much of the gatekeeping role of the elite. Satellite-fed cable television had reached 35% of the urban population by 1990, and broadcast stations served almost all of the rest. Forty-six percent of high school students reported that they watched TV every day. Belizeans now have a diversity of models, fantasies, and dramas to choose from, in a kind of "global menu" (Petch 1987; Wilk 1993a, 1995). People are now faced with the need to contrast, weigh, and choose. What seems to be emerging is both a clearer definition of the national and local, and a less hierarchical diversification of lifestyles.

In one survey I asked 389 Belizeans to rate 21 main course dishes on a four-point scale from love to hate (Wilk 1997). Eight were clearly Belizean Creole food, and the other 13 were foreign in varying ways (Hispanic, Chinese, and Indian dishes represented Belizean minority ethnic groups, while dishes like macaroni and cheese or pizza have no local constituency). The responses were striking in their lack of clear order or hierarchy; tastes did not cluster together, nor did they help disaggregate the population by class or education.

In a correlation matrix of preferences for each dish with various socioeconomic measures, differences in gender and age were as strongly correlated with food preferences (r^2 ranging from .08 to .12) as were differences in wealth, job status, and education. Ethnicity was a surprisingly weak variable in explaining differences in taste, especially so among the young. All ethnic, age, and income groups showed a high degree of agreement in their preferences for basic nationalized dishes like rice and beans or tamales. 41% of high school students, for example, volunteered rice and beans with stewed chicken as their favourite dish on an open-ended question. Most important, there was little difference between ethnic/language categories.

The neat orderings of taste that Bourdieu found in France are absent from the Belizean data on food or other kinds of preferences for art, clothing, and music. I tried numerous measures of cultural and economic capital, but could not make simple directional patterns out of taste. There was no consensus on "highbrow," "middlebrow," and "lowbrow" that was more pronounced than the differences based on basic demographic categories like age and gender. Belize is a mosaic when it comes to consumer preferences, not a simple hierarchy. Recent investigation in developed countries also finds the connection between taste and class to be attenuated (Land 1998; Warde 1997; though also see Featherstone 1990).

Conclusion: Cultural Capital Revisited

When Bourdieu uses the metaphor of capital he implies something stored up, acquired, and kept. The stability of high culture is perhaps this kind of capital— knowledge of Shakespeare remains a class signifier for a whole lifetime. But in Belize this kind of stable regime of taste is literally a thing of the past, either in the form of indigenous cultures that are increasingly marginalized as "traditional," or as an imperfectly translated and assimilated colonial high culture. What remains is fluid and changeable; this year's imported high culture (e.g., Lambada) is next year's street music. In such a small population, where kin ties crosscut wealth and class, exclusivity is very difficult to maintain. Belize may simply not be big enough for the exclusive economic and cultural elites or the challenging and mobile middle class that maintain the fashion system in France. The constant drain of emigration, the flow of media, and arrival of expatriates and returned migrants do not eliminate local tastes or fashions, but do prevent these tastes from crystallizing and forming a fashion system of their own.

If consumption and taste are reflections of and constitutive of power, what are the sources of power in Belize? First and foremost is economic capital in the form of ownership of business, next is ownership of land (though this has been very uneven historically). Another important source of power is education, both schooling and the knowledge and practices of a class. In Belize these knowledges and practices have usually been of two kinds: access to foreign goods, objects, styles, and knowledge; and family connections.

But now these are far from the only sources of power in Belize. Today knowledge of the foreign is no longer the monopoly of either the economic or cultural elite; it is accessible to many people directly through travel and indirectly through television and the movement of relatives back and forth. The knowledge of things foreign can be seen as a kind of cultural capital, but its source lies only partially within the family or the educational systems, and it has no legitimizing institution like a university or a social register. It can be obtained in many ways, and so loses some of its power to make social distinctions.

Bourdieu's analysis concentrates on the cultural means that reflect, constitute, and enforce the power of dominant classes. In his scheme, the rest of society belongs to the dominated classes. "It must never be forgotten that the working-class 'aesthetic' is a dominated 'aesthetic' which is constantly obliged to define itself in terms of the dominant aesthetic" (1984:41). In colonial Belize this dominance was complete among the small Bourgeoisie, but the scheme excluded the working classes to a degree that prevented them from completely accepting its dominance. Alternative systems of power grew in the working-class majority milieu that have been seen in other parts of the Caribbean as "reputation" or "the transient mode" (Austin 1983; Miller 1990; Wilson 1973). Power in this working-class context emerges from personal ability, personal knowledge, and a personal history. There are many arenas in which this kind of power can be gained through particular forms of competence: in verbal play, sexual prowess, violence, dancing, making music, having babies (Abrahams 1983). These forms of cultural capital are obtained through shared and lived experience, informal education, and social and economic commitment. Forms of consumption that legitimize and reflect this kind of power are not the simple opposites of the dominant forms. They include: foreign kinds of music (dub, rap); forms of dress (today derived from U.S. street-gang fashion), drinks, and foods; and local speech styles, vocabulary, and music.

Is this kind of capital totally antithetical to the "dominant" mode? In the past colonial society, perhaps. Today, however, it is possible to blend and meld the two in new and creative ways. For example, the educated middle class have now adopted local, "Caribbean" or pan-African styles and practices that are self-consciously similar to working-class fashions (though not dangerously similar). They can eat "roots food," listen to Caribbean music, and wear dreadlocks, but they are neat and clean dreadlocks. The hierarchy of power and capital is not gone, but it is no longer mirrored by simple hierarchies of taste.

Tastes and preferences are therefore always polysemic in Belize; there is no over-whelming order imposed by a strict hierarchy of capital. Fashion exists not in Bourdieu's two-dimensional space, linked to underlying variation in class, but in a multidimensional space tied to a series of different sources of power inside and outside of Belizean society. These other kinds of power include access to foreign culture through relatives, visits, tourism, or temporary emigration. As Basch et al. (1994) point out, transnational migration is now at least partially motivated by what the emigrant can *bring* and *send home*. Foreign goods create local identity on a global stage.

The paradoxical result is that in an increasingly open, global society like Belize, tastes and preferences are now more deeply localized than ever before. Local knowledge of history, people, personalities, and politics determine taste, much more than they ever did under the protective boundaries of the British empire.

Notes

Acknowledgments. An early version of this paper was presented at the conference "Defining the National," organized by Ulf Hannerz and Orvar Lofgren, in Lund, Sweden during April of 1992. I thank the participants in that conference for their comments. I also thank James Carrier, Beverly Stoeltje, Danny Miller, Anne Pyburn, and four excellent anonymous reviewers for comments that have improved this paper in many ways. Small portions of this paper previously appeared in *Ethnos* (Wilk 1993b). The research for this paper was supported by a grant from the Wenner-Gren Foundation and a Fulbright research fellowship. Special thanks to Inez Sanchez and Gloria Crawford who taught me most of what I know and appreciate about Belizean cuisine.

1. The national flag and symbols were chosen in 1973 through a public contest; the struggle over them is a good story in itself. Medina (1997) and Judd (1989) provide excellent material on the historical emergence of ethnic categories in Belize, while Bolland (1988) defines the broader political economic context of ethnicity and class in the country.

2. Belizean restaurants appeared first in New York, then Los Angeles; Belizean cuisine is a concept invented almost entirely outside of Belize. Under a national label it submerges ethnic distinctions between what Belizeans call "Spanish" food (tamales, garnachas, chirmole), Garifuna dishes like cassava bread and fish stew (sere), and Creole foods like boil-up and rice-and-beans.

3. There was, at this time, a "Belizean dish of the day" on the menu at the most expensive and exclusive hotel in town; but this was clearly a performance for tourists, since few Belizeans could afford to eat there.

4. To illustrate we have the following quotes from the 1990 Miss Universe Belize, over nationwide radio at the opening of the annual agricultural show. "While foreigners rush to take advantage of our resources, exhibited here today, fellow Belizeans, the prosperity you seek is right before your eyes." The columnist "Smokey Joe," in the Amandala newspaper, Feb. 23, 1990, says, "I wonder why it is that everyone who comes to this country can see the beauty, but we who live here can't. They see the same garbage that we lovingly put all over the place. They see the beauty that we refuse to see. They bless us; we curse ourselves. They praise us: we condemn ourselves. Is this a better land for them, and a plague to us?"

5. Many Belizeans tell jokes and stories about ignorant bumpkins who went to the States full of wild dreams. But in 30 interviews and discussions with high school students in 1990, I never found such innocence, though Lundgren (1988) did find many naive fantasies among Belizean elementary school children.

6. The full presentation of the data from my national and high school surveys would require far more space than is available here. Extensive correlation, partial correlation, and multivariate analysis was done on this large data set; full results will be reported elsewhere.

7. This diet of flour and salt meat was common among nonagricultural hand laborers in many parts of the world in the nineteenth century. Like sugar and dried codfish, these were cheap commodities produced by relatively standardized means, which could be shipped long distances and provided cheaply by employers where there was no ready local source of foodstuff (see Mennell 1984). The early cuisine of Belize is clearly derived from the preserved rations served to sailors on board ships, whereas more agricultural parts of the Caribbean developed dishes based on local products (Mintz 1996).

8. It is interesting that today most government appeals for people to consume local foods are couched in the language of health once again. Now Belizean foods are touted as fresh and natural, as opposed to preserved and processed imports.

References

Abrahams, Roger (1983) *The Man-of-Words in the West Indies*. Baltimore: Johns Hopkins University Press.

Appadurai, Arjun (1988) How to Make a National Cuisine: Cookbooks in Contemporary India. *Comparative Studies in Society and History* 30(1):3–24.

—— (1996) *Modernity at Large*. Minneapolis: University of Minnesota Press.

Arizpe, Lourdes (ed.) (1996) *The Cultural Dimensions of Global Change: An Anthropological Approach*. Vendôme, France: UNESCO.

Austin, Diane (1983) Culture and Ideology in the English-Speaking Caribbean: A View from Jamaica. *American Ethnologist* 10(2): 223–240.

Babcock, Elizabeth, and Richard Wilk (1998) International Travel and Consumer Preferences among Secondary School Students in Belize, Central America. *Caribbean Geography* 8(1):32–45.

Basch, Linda, Nina Schiller, and Cristina Blanc (1994) *Nations Unbound*. Langhorne, PA: Gordon & Breach.

Beckett, Jeremy (1996) Contested Images: Perspectives on the Indigenous Terrain in the Late 20th Century. *Identities* 3(1–2):1–13.

Bolland, Nigel (1988) *Colonialism and Resistance in Belize*. Belize City: Cubola Productions.

Bourdieu, P. (1984) *Distinction: A Social Critique of the Judgment of Taste*. Cambridge, MA: Harvard University Press.

Burke, Timothy (1996) *Lifebuoy Men, Lux Women*. Durham, NC: Duke University Press.

Conroy Richard (1997) *Our Man in Belize: A Memoir*. New York: St. Martin's Press.

Counihan, C., and P. Van Esterik (1997) Introduction. In *Food and Culture: A Reader*. C. Counihan and P. Van Esterik, eds. Pp. 1–7. New York: Routledge.

Everitt, J. C. (1987) The Torch Is Passed: Neocolonialism in Belize. *Caribbean Quarterly* 33(3&4):42–59.

Featherstone, M. (1990) Perspectives on Consumer Culture. *Sociology* 24(1): 5–22.

Foster, Robert J. (1991) Making National Cultures in the Global Ecumene. *Annual Review of Anthropology* 20:235–260.

Friedman, Jonathan (1990) Being in the World: Globalization and Localization. *Theory, Culture and Society* 7:311–328.

—— (1992) The Past in the Future: History and the Politics of Identity. *American Anthropologist* 94(4):837–859.

—— (1994) *Cultural Identity and Global Process*. London: Sage.

Hannerz, Ulf (1987) The World in Creolization. *Africa* 57(4):546–559.

Howes David (1996) Introduction: Commodities and Cultural Borders. In *Cross Cultural Consumption: Global Markets, Local Realities*. David Howes, ed. Pp. 1–16. London: Routledge.

Jenkins, Richard (1994) Rethinking Ethnicity: Identity, Categorization and Power. *Ethnic and Racial Studies* 17(2):197–223.

Judd, Karen (1989) Cultural Synthesis or Ethnic Struggle? Creolization in Belize. *Cimarron* 1–2:103–118.

Kuisel, Richard (1993) *Seducing the French: The Dilemma of Americanization*. Berkeley: University of California Press.

Land, Birgit (1998) *Consumer's Dietary Patterns and Desires for Change*. Working Paper 31, Center for Market Surveillance, Research and Strategy for the Food Sector, Aarhus, Denmark.

Lofgren, Orvar (1993) Materializing the Nation in Sweden and America. *Ethnos* 58(3–4):161–196.

Lundgren, Nancy (1988) When I Grow Up I Want a Trans Am: Children in Belize Talk about Themselves and the Impact of the World Capitalist System. *Dialectical Anthropology* 13:269–276.

Medina, Laurie (1997) Defining Difference, Forging Unity: The Construction of Race, Ethnicity and Nation in Belize. *Ethnic and Racial Studies* 20(4):757–780.

Mennell, S. (1984) *All Manners of Food: Eating and Taste in England and France from the Middle Ages to the Present*. Oxford: Blackwell.

Miller, Daniel (1990) Fashion and Ontology in Trinidad. *Culture and History* 7:49–78.

—— (1994) *Modernity: An Ethnographic Approach*. London: Berg Publishers.

—— (1997) *Capitalism: An Ethnographic Approach*. Oxford: Berg Publishers.

Mintz, Sidney (1996) *Tasting Food, Tasting Freedom*, Boston: Beacon Press.

Munt, Ian, and Egbert Higinio (1993) Eco-Tourism Waves in Belize. In *Globalization and Development: Challenges and Prospects for Belize. Speareports 9*. Society for the Promotion of Education and Research, Belize City. Pp. 34–48.

Olwig, Karen (1993) Defining the National in the Transnational: Cultural Identity in the Afro-Caribbean Diaspora. *Ethnos* 58(3–4): 361–376.

Orlove, Benjamin (ed.) (1997) *The Allure of the Foreign: Imported Goods in Post-Colonial Latin America*. Ann Arbor: The University of Michigan Press.

Petch, T. (1987) Television and Video Ownership in Belize. *Belizean Studies* 15(1):12–14.

Salzman, Philip (1996) The Electronic Trojan Horse: Television in the Globalization of Paramodern Cultures. In *The Cultural Dimensions of Global Change: An Anthropological Approach*. Lourdes Arizpe, ed. Pp. 197–216. Vendôme, France: UNESCO.

Shoman, Assad (1985) *Party Politics in Belize*. Benque Viejo, Belize: Cubola.

Stone, Michael (1994) Caribbean Nation, Central American State: Ethnicity, Race, and National Formation in Belize, 1798–1990. Ph.D. dissertation, University of Texas, Austin.

Tobin, Joseph (1992) Introduction: Domesticating the West. In *Re-made in Japan: Everyday Life and Consumer Taste in a Changing Society*. James Tobin (ed.), pp. 1–41. New Haven, CT: Yale University Press.

Vernon, Dylan (1990) Belizean Exodus to the United States: For Better or Worse. *In Speareport 4*, Society for the Promotion of Education and Research, Belize City. Pp. 6–28.

Warde, Alan (1997) *Consumption, Food and Taste*. London: Sage Publications.

Watson, James, (ed.) (1997) *Golden Arches East: McDonald's in East Asia*. Stanford: Stanford University Press.

Wilk, Richard (1990) Consumer Goods as Dialogue about Development: Research in Progress in Belize. *Culture and History* 7: 79–100.

—— (1993a) "It's Destroying a Whole Generation": Television and Moral Discourse in Belize. *Visual Anthropology* 5:229–244.

—— (1993b). Beauty and the Feast: Official and Visceral Nationalism in Belize. *Ethnos* 53(3–4):1–25.

—— (1995) Colonial Time and TV Time: Television and Temporality in Belize. *Visual Anthropology Review* 10(1): 94–102.

—— (1997) A Critique of Desire: Distaste and Dislike in Consumer Behavior. *Consumption, Markets and Culture* 1(2):175–196.

Wilk, Richard, and Stephen Miller (1997) Some Methodological Issues in Counting Communities and Households. *Human Organization* 56(1):64–71.

Wilson, Peter (1973) *Crab Antics*. New Haven, CT: Yale University Press.

20

Let's Cook Thai: Recipes for Colonialism

Lisa Heldke

I think I've finally figured out why I like Thanksgiving dinner so much, why I enjoy having it at my house, cooking all the food myself, and eating it—sometimes for days afterward. It's because I never wonder what to fix. I prepare virtually the same meal every year. It's a ritual for me; turkey, stuffing, mashed potatoes, gravy, squash, and pumpkin and mince pie appear every year. I like it this way. It's comfortable. It's delicious. I do it only once a year. And my mom does it that way. And there, perhaps, lies the crux of the matter. I have been eating this meal one day a year for my entire life, and over the years, it has come to be virtually the only full meal that my mother and I cook in common.

When I went away to graduate school some fifteen years ago, I entered a world of experimental cooking and eating, a world heavily populated with academics and people with disposable incomes who like to travel. It's a world in which entire cuisines can go in and out of vogue in a calendar year. Where lists of "in" ingredients are published in the glamorous food magazines to which some of us subscribe. In which people whisper conspiratorially about this place that just opened serving Hmong cuisine. It's a wonderful world, full of tastes I never tasted growing up in Rice Lake, Wisconsin, textures I never experienced in the land of hot dish. I love cooking and eating in that world.

However, I never know what to cook when I invite people over for dinner. Sometimes I get paralyzed with indecision. The night before the event, the floor of my living room is covered with cookbooks bristling with book-marks. There are cookbooks by my bed and next to the bathtub, even some actually in the kitchen. I've sketched out five possible menus, each featuring foods of a different nationality, most of them consisting of several dishes I've never cooked before. My mom doesn't do this. When she invites guests for dinner, she selects a menu from among her standards, preparing foods she's prepared and enjoyed countless times before, knowing that once again they will turn out well and everyone will enjoy the meal. I miss that. I envy that—especially when I spend three hours trying to decide on a menu, or when I try a new dish for company and it turns out to be awful and everyone at the meal has to try to pretend they are enjoying it.

So why do I do it? Surely no one holds a freshly sharpened carving knife to my throat and says "cook Indonesian next week when those people you barely know come over for dinner!" What's my motivation, anyway? Excellent question.

And, as it turns out, disturbing answer—an answer of which I've come to be deeply suspicious.

After years of adventurous eating in graduate school and now as a professor, I have come to be seriously uncomfortable about the easy acquisitiveness with which I approach a new kind of food, the tenacity with which I collect eating adventures—as one might collect ritual artifacts from another culture without thinking about the appropriateness of removing them from their cultural setting. Other eating experiences have made me reflect on the circumstances that conspired to bring such far-flung cuisines into my world. On my first visit to an Eritrean restaurant, for example, I found myself thinking about how disturbing and how complicated it was to be eating the food of people who were in the middle of yet another politically and militarily induced famine. On another occasion, an offhand remark in a murder mystery I was reading started me thinking about the reasons there were so many Vietnamese restaurants in Minneapolis/St. Paul, reasons directly connected to the U.S. war in Vietnam and the resultant dislocation of Vietnamese, Laotian, and Hmong people.[1]

Cultural Colonialism

Eventually, I put a name to my penchant for ethnic foods—particularly the foods of economically dominated cultures. The name I chose was "cultural food colonialism." I had come to see my adventure cooking and eating as strongly motivated by an attitude bearing deep connections to Western colonialism and imperialism. When I began to examine my tendency to go culture hopping in the kitchen, I found that the attitude with which I approached such activities bore an uncomfortable resemblance to the attitude of various nineteenth- and early twentieth-century European painters, anthropologists, and explorers who set out in search of ever "newer," ever more "remote" cultures they could co-opt, borrow from freely and out of context, and use as the raw materials for their own efforts at creation and discovery.[2]

Of course, my eating was not simply colonizing; it was also an effort to play and to learn about other cultures in ways that I intended to be respectful. But underneath, or alongside, or over and above these other reasons, I could not deny that I was motivated by a deep desire to have contact with—to somehow own an experience of—an exotic Other as a way of making myself more interesting. Food adventuring, I was coming to decide, made me a participant in cultural colonialism, just as surely as eating Mexican strawberries in January made me a participant in economic colonialism.

This chapter is part of a larger work, *Let's Eat Chinese*, which explores the nature of cultural food colonialism: What is it? What are its symptoms, its manifestations, its cures? Who does it? Where? Why? In that work, I consider a range of activities in which food adventurers participate—everything from dining out to cooking to food journalism—and how these activities manifest and reproduce cultural food colonialism. Here I look specifically at ethnic cookbooks. Cookbooks, like restaurant reviews, and like dining in ethnic restaurants, manifest cultural food colonialism in two ways: first, they speak to the food adventurer's never-ending quest for novel eating experiences—where novelty is also read as exoticism, and second, they turn the ethnic Other into a resource for the food adventurer's own use.

Before plunging further into an exploration of these two features of cultural colonialism, I pause to situate my project within the field of feminist thought. I see it as a feminist project on at least two levels. At the first, most banal level, my field of inquiry—food and cooking—is something traditionally regarded as "women's arena." For at least several decades now, feminist theorists have been exploring those domains of human experience traditionally identified as "belonging" to women. Reproductive issues, childbirth, the work of mothering, sexuality, sex work, pornography, women's health, and any number of other features of human life have come to be examined by scholars because of the efforts of feminists who have seen these "women's issues" as relevant *theoretical* issues. Feminist theorists have now begun to turn serious attention to food and eating—for nearly the first time since that original food theorist Plato took it up.[3]

I also understand this as a feminist project because of my theoretical approach; my work attempts to take up challenges posed by various strands of feminist theory, most notably those strands developed by feminists of color and Third World feminists (for example, bell hooks, Trinh T. Min-ha, Joanna Kadi). One of the most important lessons white feminists learned from the work of feminists of color in the 1980s was that oppression—women's oppression—always exists along multiple axes simultaneously. Feminists must therefore take racism and classism seriously as central features of *women's* oppression—not as add-ons that can be considered after the "real" challenges of "women's" oppression have been met.

In the 1990s that lesson further evolved to emphasize the importance of investigating one's own privilege within systems of oppression—consider, for example, Ruth Frankenburg's analyses of the nature of whiteness. My work takes up the feminist project of interrogating my own location in systems of privilege and oppression—systems that variously privilege and marginalize me.[4] I explore cultural food colonialism, in part, in an attempt to understand my racial/ethnic and class privilege.

Let's Cook Thai

Take a walk through the cookbook section of your local book supermarket and you will confront a gigantic subsection of books promising to teach their readers how to cook some ethnic cuisine. The shelves will hold several works that have become classics in the field, such as Claudia Roden's *A Book of Middle Eastern Food* and *An Invitation to Indian Cooking* by Madhur Jaffrey. It will also include a significant number of new arrivals—new both in the sense of their publication dates and in terms of the cuisine they tout. (In the past fifteen years, for example, mainstream America has "discovered" the cuisines of Southeast Asia—especially Thailand, Vietnam, and Indonesia—and, even more recently, the foods of both East and West Africa.) You will also find a number of books that are the culinary equivalent of *If This Is Tuesday, It Must Be Belgium*—cookbooks that give you a smattering of recipes from every region of the globe, along with a winsome anecdote or two about the people of that region. The ethnic cookbook market has exploded in the United

States, as has the market for the equipment, ingredients, and spices to cook the foods of the world. How does this explosion of interest in ethnic cooking feed into the phenomenon of cultural food colonialism?

Consider one example, *The Original Thai Cookbook* by Jennifer Brennan. Brennan opens her book with these lines: "It is dusk in Bangkok and you are going out to dinner. The chauffeured Mercedes 280 sweeps you from your luxury hotel through streets lined with large, spreading trees and picturesque tile-roofed wooden shops and houses." Brennan goes on to describe "your" arrival at an elegant Thai home—where you are greeted by an "exquisite, delicately boned Thai woman, youthful but of indeterminate age"—and also your meal—a "parade of unfamiliar and exotic dishes" (3–4).

Renato Rosaldo has coined the phrase "imperialist nostalgia" to describe the longing of the colonizer for that which he perceives to be destroyed by imperialism. Brennan here evokes what might be called nostalgia *for* imperialism when she invites her readers to imagine themselves as wealthy and privileged visitors in a culture not their own, and in which they are treated with great deference and respect by some of the wealthiest, most important people in the culture.

Brennan invites her readers to *be* the protagonist of this colonialist story. Her descriptions invite those readers to luxuriate in a fantasy of wealth and also beauty; she suggests that her readers—who are primarily women—should see themselves as "tall and angular" (5), a description that manifests long-standing Euro-American standards of feminine beauty emphasizing long, thin limbs. Reading this description, I find myself seduced by the glamorous role she has assigned me, a middle-class Euro-American woman who has never even been in a Mercedes, let alone traveled to Thailand, and whose body, while fairly tall, would never be described as angular. Just throw in some high cheekbones and a tousled mop of thick, blond hair while you're at it, and I'll sign up for this fantasy tour.

Brennan's description also effectively reduces the identity of the imaginary Thai hostess to her relationship with her guests; Brennan has invented this woman expressly to provide us Western "dinner guests" with pleasure, both visual and gustatory. The "exotic woman" provides just the right touch of beauty, mystery, and servility to get us Western gals into the spirit of imagining ourselves as the heroines of this colonialist culinary tale.[5]

Although she eventually gets down to the business of telling her readers how to cook their own food (a detail that acknowledges that we do not in fact have a Thai cook of our own), Brennan never completely dismisses the colonialist fantasy she has created. Her introduction to this book is just one illustration of the ways that ethnic cookbooks manifest and foster cultural colonization—in this case, by perpetuating a view of the Other as existing to serve and please the reader, and by creating a vision of this Other culture as exotic and alluring.

The Quest for the Exotic

Modern Western colonizing societies have been characterized in part by an obsessive attraction to the new, the unique, the obscure, and the unknown, where "new" is understood in relation to the colonizing society. Desire for new territory, new goods, new trade routes, and new sources of slaves sent European colonizers out to capture and control the rest of the world. The desire to understand the essence of human nature sent European and American anthropologists on a quest to find new, primitive societies not yet exposed to (Western) culture. The desire for new, unadulterated inspiration prompted European painters to move to places far from home. And today, desires for new flavors, new textures, and new styles of dining send us adventuring eaters flipping through the Yellow Pages, scouring the ads in ethnic and alternative newspapers, and wandering down unfamiliar streets in our cities, looking for restaurants featuring cuisines we've not yet experienced—"exotic" cuisines. For the food adventurer, the allure and attraction of such cuisines often consists quite simply in their unfamiliarity and unusualness.

Why does the novel hold such fascination for the food adventurer in American culture? We adventurers come to demand a continual supply of novelty in our diets in part simply to remain entertained. We crave the new just because it is unusual, unexpected, different; differentness is something we have come to expect and require. Of course, it is in the nature of novelty that it is quickly exhausted; if what we crave is novelty per se, our quest will be never-ending. Food magazines often feature articles informing their readers (in all seriousness) about which cuisines and ingredients are now "out" and which have come "in." In a single article, the daily diet of the people of Thailand can be declared passé in the United States.

Novelty is also attractive to adventuring food colonizers because it marks the presence of the exotic, where exotic is understood to mean not only "not local" but also "excitingly unusual." The exotic, in turn, we read as an indication of authenticity. Exotic food is understood as authentic precisely *because* of its strangeness, its novelty. Because it is unfamiliar to me, I assume it must be a genuine or essential part of that other culture; it becomes the marker of what distinguishes my culture from another. Whatever is so evidently not a part of my own culture must truly be a part of this other one. So, in a three-step process, that which is novel to me ends up being exotic, and that which is exotic I end up defining as most authentic to a culture.

How does the quest for the novel-exotic-authentic show up in ethnic cookbooks? Ethnic cookbooks teach their readers how to make the strange familiar by teaching them how to replicate unknown dishes. But how can the cookbook writer achieve this goal without sacrificing the exoticism of the food, given that exoticism has its roots in novelty—in unfamiliarity?

One answer is that for the cook, casual familiarity with a cuisine still radically unfamiliar to most of "us" represents a relationship to the exotic that is itself worth considerable cultural capital, in Pierre Bourdieu's term. A person who achieves such familiarity in a sense becomes the exotic—or at least the exotic once removed. If I can make Indonesian dishes that other food adventurers can only eat in restaurants, I become a kind of exotic myself. Jennifer Brennan—the cookbook author who took us for a ride in Bangkok in her Mercedes—approaches novelty this way in her cookbook.

Jennifer Brennan: The Exotic as Familiar

In the preface to *The Original Thai Cookbook*, Brennan writes that although there are now "Oriental" and Thai markets in "nearly every town," they are filled with "a dazzling and, sometimes, baffling array of foodstuffs: native herbs and spices . . . unusual species of fish; unlabeled cuts of meat; vegetables you might consider weeding from your garden; assortments of strange canned foods and sauces—all with exotic names, sometimes foreign language labels—all purveyed by shopkeepers unfamiliar with English."[6] In other words, although ingredients for Thai food are readily available to non-Thai cooks, availability does not automatically spell familiarity. Brennan emphasizes that language is of little help to the cook here; things are unlabeled, or labeled in a "foreign language," and the people in the stores speak another language too.[7]

Recall Brennan's invitation to imagine yourself visiting an elegant Thai home and being served a banquet. Brennan's lengthy description of this imaginary event highlights the glamorous novelty of everything from the street scenes to the clothing to the way the foods are presented. Her "reassurance of exoticism" serves two related purposes. First, it assures the nervous home cook, perhaps preparing a meal with which to impress her coworkers, that she is not simply naive or ignorant or overly cautious. This food really *is* strange! You don't have to be embarrassed about finding it so, because it really is! Second, it may confirm for the cook that the food she will learn to make is, in some apparently objective sense, exotic, and even familiarity with it cannot alter that fact. Its exoticism means that a cook will definitely earn cultural capital if she serves it to her dinner guests, for whom home-cooked Thai food is still likely to be a novelty.

Another example illustrates both purposes. It comes from a cookbook published twenty years before Brennan's, titled *Japanese Food and Cooking*. In the foreword, author Stuart Griffin describes the respective experiences of "Mrs. American Housewife" and "Mr. American Husband," who have moved to Japan. Mr. American Husband's arrival predates that of his wife, so he has had time to explore Japanese cuisine and to determine that he "could leave a lot of Japanese food alone," specifically the "big, briny tubs of pickles, the fish stands where every species eyed him, and the small stool-and-counter shops with the stomach-turning cooking-oil smells. . . . But he found lots of things that he wanted to eat and did like" (xii). When Mrs. American Housewife arrives, her husband and her Japanese cook enter into a conspiracy to get her to try Japanese dishes. By the end of the foreword, Mrs. American Housewife is hosting dinner parties for her (American) friends, featuring an "entirely Japanese" menu (xiv)—prepared, of course, by her cook. In presenting such foods to her guests, Mrs. A. becomes to her friends a kind of exotic herself.

When Brennan and Griffin describe food as "strange" and "stomach-turning," to whom are they speaking? Brennan identifies her audience as English-speaking people in the United States. (Notably, the gender references have disappeared by the time Brennan published her cookbook; nevertheless, it is still safe to assume that the person wielding the cookbook in a kitchen is a woman.) But in emphasizing the "unfamiliarity" of the foods, Brennan actually specifies her audience much further.

Presumably, Thai Americans would find many of the ingredients in Thai food quite familiar. The same would likely be true of many Vietnamese Americans, Chinese Americans, Indian Americans, Malaysian Americans—any people whose heritage foods have influenced and been influenced by Thai foods. The ingredients Brennan describes would, in fact, be deeply unfamiliar only to *certain* English speakers in the United States. But Brennan's description of the Asian grocery recognizes no such distinctions; "you" will find things strange in such a store, she notes. "Strange," like "exotic," comes to mean strange in principle because strange to us.

Even authors who are insiders to the cuisines about which they write come to use words evincing novelty and exoticism to describe their own cuisines. Claudia Roden, for example, evokes notions of the exotic Middle East when she variously describes certain salads as "rich and exotic" (59), baba ghanoush as "exciting and vulgarly seductive" (46), and Turkish Delight as a food "no harem film scene could be without" (423). That such descriptions also often employ sexual imagery is, of course, no accident; as we saw earlier in Brennan's description, linking food with sexuality or the sexual attractiveness of women is one way to emphasize the exoticism of a food. Women reading this cookbook may feel as though we are being invited to see ourselves as "vulgarly seductive" by extension, when we cook this food.

The Other as Resource

Middle-class members of a colonizing society such as the United States inhabit an atmosphere in which it becomes customary to regard members of a colonized culture as "resources," sources of materials to be extracted to enhance one's own life. In the case of cultural colonialism, the materials are cultural ones. It is no coincidence that the cultures most likely to undergo such treatment at the hands of food adventurers are those described as Third World or nonwhite. There is a tangled interconnection in Euro–American culture between those cultures defined as exotic and Other, and those identified as Third World.

In the world of the ethnic cookbook, the cooking techniques of the Other become marvelous resources that can be scooped up, "developed," and sold to Us, without giving much attention or credit to the women actually responsible for preserving and expanding this cuisine. Recipes become commodities we are entitled to possess when they are taken up into the Western cookbook industry; foods become "developed" when they can be prepared in the West.

In her book *Imperial Eyes*, Mary Louise Pratt suggests that to treat the Other as a resource for one's own use can take many forms—even veneration and admiration.[8] This observation is well worth keeping in mind when one examines ethnic cookbooks, because many of them exhibit appreciation for a food tradition even as they preserve an "essential colonized quality" (163) in the relationship between the cookbook writer and her cook/informant. The case of recipe collecting provides one excellent illustration of this.

Borrowing or Stealing?

Where do recipes come from? And when is it proper to say that a recipe was "stolen," or inadequately credited? When it comes to cookbooks in general, and "ethnic" cookbooks in particular, the definitions of these terms decidedly favor the interests of colonizers. A cookbook author is described as having "stolen" recipes only if they have previously appeared in *published* form—a form of communicating that privileges people on the basis of class and education as well as race, and often sex. Consider the following case: Ann Barr and Paul Levy in *The Foodie Handbook* praise Claudia Roden for her careful "anthropological" work to credit sources of her recipes in *A Book of Middle Eastern Food* (110). But although Roden is careful to identify the sources of the recipes she reproduces when those sources are cookbooks, she acknowledges unpublished sources by name only in the acknowledgments to her book—and then only in a brief, general list of those women to whom she is "particularly indebted."

Barr and Levy's praise for the integrity of "scholar cooks"[9] such as Roden rests on the unstated assumption that only published sources require crediting—an assumption validated by and codified in copyright laws and institutional policies regarding plagiarism. This assumption allows them to regard as highly principled Roden's practice of sometimes describing, but almost never naming, the Middle Eastern women cooks from whom she receives these previously unpublished recipes. But when coupled with her careful crediting of previously published recipes, her practices actually create cookbooks that reflect and reinforce the colonialist and classist societies into which they are received and from which they come. (By "societies from which they come," I mean primarily colonialist Western societies. Their work comes from these societies in a complex manner because Roden is an Egyptian Jew who was educated and has spent much of her adult life in Europe; she is a kind of "insider outsider" who, in part because of her class position, did not in fact learn to cook until she went to Europe.)

In Roden's cookbook, the unpublished women who contributed recipes become interchangeable parts, relevant only for the (universalizable) quality of their being "native cooks." She tells "colorful" stories about some of them in the body of the book, but the reader can match their names to their stories (or their recipes) in only a few cases—and then only with assiduous detective work. They need not be identified definitively, because they cannot be stolen from; they do not own their creations in any genuine (read: legally binding) sense of the word. On the other hand, the creations of cookbook authors, who have access to the machinery of publishing, must be respected and properly attributed.

Barr and Levy's praise of Roden is situated in the context of their discussion of a case of recipe plagiarism. For cookbook writers, the ethics of recipe "borrowing" versus "theft" seem to follow the rules governing plagiarism. According to these, borrowing only becomes theft if a recipe has already appeared in print and one fails explicitly to acknowledge it. Cookbook writers express shock, dismay, or anger when another writer reproduces one of their published recipes without citing its source. However, they waste no emotion over the writer who reproduces recipes gathered "in the field" from unpublished "sources" who go unnamed and uncredited.[10]

Taking up the legal issue, Barr and Levy argue that it is both "mad" and "unenforceable" to suggest that an individual ought to have the right to copyright the directions for an omelet or a traditional French casserole (108); how could any individual "own" the procedure for making dishes so ubiquitous? On the other hand, they favorably report that in a 1984 case Richard Olney successfully sued Richard Nelson for copyright infringement, claiming that Nelson had reproduced thirty-nine of Olney's published recipes in one of his own cookbooks. Thus, while they are uncomfortable with the idea of copyrighting some kinds of recipes, Barr and Levy suggest that justice was served in the Olney case, because Olney is an "originator" of recipes (110)—as opposed to an anthologist (like Nelson) or an anthropologist (like Roden). In support of their view, Barr and Levy quote passages of the relevant recipes to show that Nelson copied ingredients, procedures, even stylistic touches from Olney.

Olney "owned" his recipes in a way that no one can own the omelet recipe because he both "invented" and published them. The latter step apparently is necessary; Barr and Levy have no pity for the author who prints recipes on index cards and distributes them to her friends and then cries "thief."[11] But in the end, it doesn't matter; just because you receive in the mail an unsigned copy of *On the Road*, written in pencil on paper torn from a wide-ruled spiral notebook, you cannot publish Jack Kerouac's words under your own name.

But what of other cases, in which the recipes in question are not originals, but are the "ethnic" equivalents of the omelet? How are we to understand the "ownership rights" of an "anthropologist cook" who publishes the recipes she has "collected in the field," only to have someone else republish those recipes in their own book? Does the anthropologist have the right to complain about theft? Barr and Levy suggest that she does, in their sympathetic consideration of Claudia Roden, whose work has been the site of much borrowing by other cookbook authors. Barr and Levy report that Roden is pleased to see people using the recipes she gathered but not so pleased to see those recipes reappearing in print. In particular, she is "hurt and angry that Arto der Haroutunian, in his books . . . has a great many of the same recipes as hers (*some of which had never been in print before*), similarly described and including some of the mistakes. . . . As a writer who gathered her material physically, Mrs. Roden feels 'He has stolen my shadow' " (112, emphasis added). Roden's anger here suggests that it is not only the originator of a recipe whose work can be stolen; you are also a victim of theft if the recipes you collect and publish are subsequently published by someone else.

That Roden is the victim of a theft of "original material" seems obvious on one level. In a context defined by copyright law, der Haroutunian's acts do constitute a kind of plagiarism of Roden's original work. But consider the matter again; what he stole were, for the most part, recipes she gathered from other women—along with published texts she excerpted and organized in a particular way. Publishing, in this context, comes to be its own kind of originality—or comes to mean originality.

Furthermore, publishing a thing seems to make it one's own, regardless of who "owned" it in the first place. Roden does not claim that she created the recipes; indeed she expressly describes them as belonging to the particular towns, villages, communities, countries in which she located them (Barr and Levy 112). Nevertheless, she says that *her* shadow has been stolen by der Haroutunian.

While I agree that some kind of harm has been done to Roden by those who have republished parts of her book, I want to redirect the discussion to the kinds of harm that her explanation obscures—namely, harm done to the Middle Eastern women at whose stoves Roden stood and from whom she learned the recipes she reproduced in her cookbook. This harm does not fall neatly under the category of theft, because the women cannot be regarded as the owners of recipes in the sense required. The language of property does not help us to understand such harm for at least two reasons: first, these cooks have not laid any claim to the recipes (say, by publishing them themselves or by cooking them in a restaurant for paying customers), nor will they likely do so, because, second, the recipes from which they cook are often as common to them as the omelet is to a French cook, and thus not the sort of thing they would be inclined to think could be owned. It is not appropriate to describe Claudia Roden as "stealing" recipes from the women with whom she studied. She could not rectify the harm done to them simply by documenting the "originators" of her recipes.

Roden erases or generalizes the identities of most of the women who give her recipes.[12] She does identify various of her relatives as the sources of recipes; she notes, for example, that "my mother discovered [this recipe] in the Sudan, and has made it ever since" (43). But in most cases, she mentions only the primary region in which a dish is served and says nothing about the particular woman or women from whom she got the recipe. (Recipes are peppered with phrases such as "A Greek favorite," and "Found in different versions in Tunisia and Morocco.") By contrast, she carefully notes the dishes she found in particular published cookbooks. (For example, with respect to a chicken recipe, she notes: "A splendid dish described to me by an aunt in Paris, the origin of which I was thrilled to discover in al-Baghdadi's medieval cooking manual" [184].) The effect of this differential treatment is to blur the "ordinary" women who contributed to this cookbook into a mass of interchangeable parts. She renders invisible the work done by members of this mass to create, modify, adapt, and compile recipes; it does not matter which individual was responsible for which modification. Only her own work on these tasks is visible in her text.

A critic could reasonably respond that, in fact, these recipes have no originators: "You have already pointed out that many of them are as ubiquitous in the Middle East as the omelet is in France. Are you advocating that Roden give credit to particular women for their contributions of a particular recipe? Why would that make sense, given their ubiquity?" My first answer would be that it makes as much sense to credit these conduits as it does to credit Roden herself; their participation in making these recipes available is certainly as relevant as hers. Her interaction with the publishing industry should not alone give her special, superior claims to ownership.

Furthermore, I do not advocate making particular women, or even their communities, the owners of recipes any more than I advocate allowing Roden to make that claim. The problem emerges from the fact that Roden and other anthropologist cooks transform "traditional" recipes into commodities. They treat the recipes they gather as resources—raw materials onto which they put their creative stamp—by surrounding the recipes with scholarly background information, a personal anecdote, or a relevant quotation from a work of poetry or literature. With this creative transformation, the recipes become property that can be stolen.

Roden has gathered the creative productions of various peoples to make her own cultural creation. Other women's (often other cultures') recipes are the raw materials she harvested and "refined" into a work of "genuine culture." In her case, this refinement involved situating the recipes alongside erudite quotations from Middle Eastern texts, stories she collected about the various dishes, or accounts of the processes by which she came into possession of the recipes. Because she regards her work as a genuine cultural product—not just a "natural outgrowth" of a culture, as is the case with the recipes she collects—she expresses outrage at its being plagiarized.

I want to look at the issue of recipe originality, borrowing, and theft in one final way before leaving it. Another factor contributing to the complexity of this issue is that everyday recipe creation has traditionally been women's work done in the home, whereas the harms that are identified and codified by law tend to be those that befall the creative work men have traditionally done in more public arenas.

Like other kinds of women's creative work, such as quilting and weaving, recipe creation tends to be social. By this I mean not only that the physical work may be done in groups, but also that the creation is frequently the result of many women's contributing their own ideas to the general plan, often over considerable spans of time. A recipe that passes from one cook to another may undergo slight modifications to accommodate differences in taste or unavailability of an ingredient, to streamline or complicate a process, or for unidentifiable reasons.[13] We might say that such a recipe is "original" to everyone and no one; its beginning is unknown, but the contributions of particular cooks may be read in it by someone who knows how to decipher such an "evolutionary" record. (I once heard a food expert analyze the Thanksgiving dinners of several families living in different parts of the United States. On the basis of the foods present in the meal, and the way that those foods were prepared, she was able to identify, with accuracy, the areas of the country in which that family had lived over the past generations.) Cooks who have contributed to the evolution of a recipe may well be pleased and proud when someone else takes up their modification—whether or not the other cook knows who is responsible for it.

The categories originality and plagiarism pertain to more individualistic art forms, such as novel writing and painting—which have been regarded as "high art" and have primarily been the purview of a privileged minority of men. These terms cannot be applied so aptly to other art forms, particularly collective and cumulative ones. However, they often are (awkwardly) applied to these forms, perhaps as an attempt to gain legitimacy for them. For example, women often have (in reality and in fiction) appealed to claims of originality and ownership, to accuse others of "stealing" their recipes. Of course, the women who have done so are often subjected to mockery. (My morning radio station regularly plays the song "Lime Jello, Marshmallow, Cottage Cheese Surprise," in which a woman discusses the dishes that have been brought to a "ladies' " potluck luncheon. The singer, who brought the title dish to the potluck, at one point exclaims, "I did not steal that recipe, it's lies, I tell you, lies!" The line is greeted with guffaws of laughter from the audience.) I suggest that such mockery reflects both the pervasive sense that an individual's recipes are not original to them and thus cannot be owned by them and the sense that, even if recipes could be stolen, their theft would be no crime since recipes simply aren't important enough to be the objects of moral concern.[14]

Recipes for Anticolonialism

When I began exploring cultural food colonialism, I naively believed that cooking was an unambiguously anticolonialist activity, one that could be employed by anyone who wished to develop a way of living in the world that resisted colonialism. Although I no longer think of cooking as magically resistant and am deeply critical of much that goes on in the pages of various ethnic cookbooks, I still believe that cooking has important potential as a site for anticolonialist activity. Food is a wonderful medium for this because culinary diversity is already so much a part of the daily life of many Americans—and because food in general is an essential part of everyone's daily life. Because we must eat, opportunities for becoming anticolonialist in the kitchen present themselves to us with tremendous frequency. Most of us don't go to art museums or concerts every day of the week—but we do eat dinner every night, and we often cook it ourselves.

The question for me is, How can one enact anticolonialist resistance in the kitchen, the grocery story, the cookbook? How can I transform my ethnic cooking into what bell hooks calls a critical intervention in the machinery of colonialism? We who would be anticolonialists must learn how to engage with cuisines, cooks, and eaters from cultures other than our own—not as resources but as conversation partners. We must also recognize that our privilege (racial, ethnic, class, gender) is something food adventurers cannot simply give up, that while I have the luxury of experimenting with other cuisines in the kitchen, my privilege—and my guilt about it—will not be banished simply by my forsaking this luxury.

Writing on the subject of white racism against blacks, bell hooks argues, "Subject to subject contact between white and black which signals the absence of domination . . . must emerge through mutual choice and negotiation . . . [S]imply by expressing their desire for 'intimate' contact with black people, white people do not eradicate the politics of racial domination" (28). Cookbook author Jennifer Brennan does not magically neutralize the colonialist dynamic between herself and her private cook simply by making herself at home in the kitchen and cozying up to the cook to ask sincere, well-meaning questions about whether the Thai really use ketchup in their *paad thai*. Such presumed intimacy can in fact reinforce the dynamic, because it highlights the unreciprocated ways Brennan has access to her cook.

What would mutual choice and negotiation require in such an interaction, what must happen in order to change this into a subject-to-subject exchange? At the very least, Brennan and her "changing parade of household cooks" (preface) would have to discuss the terms under which it would be justifiable for her to publish their work—to take their skills and recipes and market them under her own name. It would require that the cooks be able to make an informed choice about whether or not to participate in this cookbook-making enterprise in the first place—that they understand the larger, long-term consequences of participating in it. Would they still want to participate in the project if they knew how angry Claudia Roden was about Middle Eastern cuisine's unfortunate transformation in the pages of Euro-American cookbooks? They may; my point is that mutual choice would require that both parties making the choices have sufficient information on which to base their choices. Given that Brennan has the publishing power—and more access to information about how these recipes will be used in the United States—she has a particular obligation to exchange that information with her cooks.

"Mutual recognition of racism," hooks continues, "its impact both on those who are dominated and those who dominate, is the only standpoint that makes possible an encounter between races that is not based on denial and fantasy" (28). Such a requirement would transform the way in which Euro-American cookbook writers such as Brennan collect their materials for ethnic cookbooks; it would also result in cookbooks with a very different format. Perhaps ethnic cookbooks written by outsiders to a cuisine could be constructed in ways that actually acknowledge and grapple with the fact of continued colonialist domination by Western cultures. Perhaps cookbooks could be written as genuine collaborations, as opposed to the de facto collaborations they so often are now. (It's worth nothing, in this regard, that collaboration is already highly developed and appreciated in one arena—namely, the community fund-raising cookbook. In such cookbooks, one often finds multiple, nearly identical, copies of a recipe, each credited to a different cook. Such repetition may seem ridiculous to someone who just wants to know how to make a dish, but when it is considered as a record of how a community cooks, it becomes a valuable source of information.)

hooks' message, translated into the realm of food adventuring, is that our only hope for becoming anticolonialists lies in our placing the colonizing relationship squarely in the center of the dining table; only by addressing colonialism directly through our cooking and eating can we possibly transform them into activities that resist exploitation. If "eating ethnic" cannot remain pleasurable once we acknowledge how domination shapes our exchanges with the Other, then we must acknowledge that it is a pleasure well lost.[15]

The anticolonialist aims to disengage from an attitude and a way of life that exploit and oppress and to develop alternatives that subvert the colonizing order. We need to learn how to participate in anticolonialist exchanges of food. We need to find useful, anticolonialist ways to make dinner.

Notes

This article is an earlier version of an argument that eventually appeared in *Exotic Appetites: Ruminations of a Food Adventurer* (Routledge 2004).

1. Sara Paretsky, in her 1985 novel *Killing Orders*, writes, "I stopped for a breakfast falafel sandwich at a storefront Lebanese restaurant. The decimation of Lebanon was showing up in Chicago as a series of restaurants and little shops, just as the destruction of Vietnam had been visible here a decade earlier. If you never read the news but ate out a lot you should be able to tell who was getting beaten up around the world" (36).

2. Explorers Richard Burton and Henry Schoolcraft, for example, "discovered" the headwaters of the Nile and the Mississippi, respectively—with the help of local folks who already knew what Burton and Schoolcraft had come to discover. For an analysis of Burton's much-aided journey to the headwaters of the Nile, see Mary Louise Pratt, *Imperial Eyes*. For an analysis of Schoolcraft's use of Ojibwe experts to locate the headwaters of the Mississippi, see Gerald Vizenor, *The People Named the Chippewa*.

   The painter Paul Gauguin went to Tahiti to "immerse [him]self in virgin nature, see no one but savages, live their life" so that he might make "simple, very simple art"—using their lives and art as his raw material (qtd. in Guerin 48).

3. Plato makes such frequent use of food to illustrate his claims that one is forced to conclude that the references

are anything but accidental. For two considerations of Plato's conceptions of food, see my "Foodmaking as a Thoughtful Practice" and "Do You Really Know How to Cook?"

4. In what may seem an ironic turn, it is the centrality of feminist theory to my own way of doing theory—the centrality of challenges, questions, and critiques from feminists of color—that makes it sometimes appear as if my work "isn't about women at all." My work is *not* about women, when that phrase is understood to mean "about women by not being about men, about women by talking about women and gender exclusively." But it is precisely this notion of feminism—and of being "about women"—that I wish to undermine, following in the path of Third World feminist theorists. Such "aboutness" necessarily brackets or erases race, class, and other markers of difference, as not being central to "women's" identities.

5. Trinh T. Min-ha notes: "Today, the unspoiled parts of Japan, the far-flung locations in the archipelago, are those that tourism officials actively promote for the more venturesome visitors. Similarly, the Third World representative the modern sophisticated public ideally seeks is the *unspoiled* African, Asian, or Native American, who remains more preoccupied with her/his image of the *real* native—the *truly different*—than with the issues of hegemony, racism, feminism, and social change (which s/he lightly touches on in conformance to the reigning fashion of liberal discourse)" (88).

6. There is something more than a little odd about the name of this cookbook. What does it mean for a Westerner to lay a claim to the "territory" of Thai food, by describing as "original" a book that records a culture not her own? The book jacket explains the meaning *original* is to have in this context: this is the first Thai cookbook published in English in the United States. This explanation of the word *original* tends to invite the conclusion that something comes into existence only when it does so in the United States.

7. It is worth noting that the situation with respect to Thai foods has changed dramatically since Brennan published this book in 1981. Now, not only are there more Asian groceries in the United States, but also one can buy many of the ingredients for Thai foods in food cooperatives and upscale supermarkets. One can even buy various premixed "Thai spices" in foil packets and glass jars, with directions clearly labeled in idiomatically flawless American English. They are considerably more expensive and less accessible than the ubiquitous taco seasoning packets, but I have little doubt that these spice packets will one day be every bit as common, as Thai food becomes a part of the mainstream U.S. consumer economy.

8. Discussing Maria Graham Callcott's *Journal of a Residence in Chile during the Year 1822*—a section of the book in which Graham describes learning to make pottery—Pratt writes: "Rather than treating the artisanal pottery works as a deplorable instance of backwardness in need of correction, Graham presents it in this episode almost as a utopia, and a matriarchal one at that. The family-based, non-mechanized production is presided over by a female authority figure. Yet even as she affirms non-industrial and feminocentric values, Graham also affirms European privilege. In relation to her, the potters retain the essential colonized quality of *disponibilité*—they unquestioningly accept Graham's intrusion and spontaneously take up the roles Graham wishes them to" (163). As Pratt suggests, to treat the Other as a resource for one's own use can take many forms, some of which even involve veneration and admiration. This observation is well worth keeping in mind when one examines ethnic cookbooks; many of them exhibit appreciation for a food tradition, even as they preserve the "essential colonized quality" of the relationship between the cookbook writer and her cook/informant.

9. Levy describes Elizabeth David as the inspiration for—and the original example of—a group of people he names "scholar cooks" (31). Among the scholar cooks, presumably, the anthropologist cooks are just one subspecies. Other scholar cooks include diplomat-turned-fish-specialist Alan Davidson, and Jane Grigson.

10. If we consider cooking itself—rather than cookbook writing—we may locate another, similar definition of theft, one very much rooted in class and gender. Chefs can be thought of as stealing one another's dishes, particularly "signature dishes" they have invented, even if those dishes have never been published. (They cannot, so far as I know, sue for such theft, their primary recourse is probably ridicule.) Famous chefs, who get paid for their work, can most easily make this claim, because they can produce the most evidence for it. Unknown chefs—like unknown songwriters—will have more difficulty proving that they've been robbed of a culinary idea. And women who are not chefs for pay but simply cook at home for their families will have insurmountable difficulty; a recipe must become a commodity before it can be stolen. The thing to note is that with cooking, as with cookbooks, proving originality is important—but having the power to reinforce a claim to originality is crucial.

11. Indeed, the prevalence of recipe-card exchanges serves to temper some of their outrage over the Olney-Nelson case; it seems that Nelson got his recipes not out of Olney's book but from a set of recipe cards he had received in the mail and used for years in his cooking class with no idea of their origin.

12. Lutz and Collins, in *Reading National Geographic*, examine the similar ways that *National Geographic* transforms the individuals in its photographs into "types," nearly interchangeable members of the group known as Other (see esp. chap. 4).

13. Not all recipes change, of course, some foods are temperamental enough that cooks feel disinclined to change them, for fear that they will fail. Not all cooks feel comfortable modifying recipes either. In my own family, for example, my mother and I are much more likely to tinker with a recipe than are my two sisters. (Interestingly

enough, both of them are trained as scientists.) For one philosophical discussion of the processes of recipe creation and exchange, see my "Recipes for Theory Making."

14. I would contrast this to the respect, bordering on reverence, with which recipes used in restaurants are often treated, the deference with which a customer asks for the recipe for a particular dish and the gratitude they heap on the cook/chef willing to pass it along. In food magazines, this difference is sometimes manifested in the presence of two separate recipe columns. In one, readers ask other readers to share a recipe for some particular food ("I'm looking for a good recipe for pumpkin bread. Does anyone have one that uses orange juice?"). In the other, readers write in to ask for specific recipes they have tasted in restaurants (recipes they perhaps were too intimidated to request in person), and almost invariably, they couch their requests in the language of a supplicant: "Do you think you could ever possibly get them to release the recipe for this chicken dish I had?"

15. Wendell Berry, in "The Pleasures of Eating," describes an extensive pleasure in one's food, which "does not depend on ignorance" of the conditions under which that food is grown, harvested, and brought to you. For him, this pleasure is not "merely aesthetic" but ethical, political, and environmental as well (378).

References

Barr, Ann, and Paul Levy. *The Official Foodie Handbook: Be Modern—Worship Food*. New York: Timber House (1984).

Berry, Wendell. "The Pleasures of Eating." *Cooking, Eating, Thinking: Transformative Philosophies of Food*. Ed. Deane Curtin and Lisa Heldke. Bloomington: Indiana UP (1992) 374–79.

Bourdieu, Pierre. *Distinction: A Social Critique of the Judgment of Taste*. Trans. Richard Nice. Cambridge: Harvard UP (1984).

Brennan, Jennifer. *The Original Thai Cookbook*. New York: Perigee (1981).

Frankenberg, Ruth. *White Women, Race Matters: The Social Construction of Whiteness*. Minneapolis: U of Minnesota P (1993).

Griffin, Stuart. *Japanese Food and Cooking*. Rutland, Vt.: Charles E. Tuttle (1959).

Guerin, Daniel, ed. *The Writings of a Savage Paul Gauguin*. Trans. Eleanor Levieux. New York: Viking (1978).

Heldke, Lisa. "Do You Really Know How to Cook? A Discussion of Plato's Gorgias." Unpublished paper.

—— . "Foodmaking as a Thoughtful Practice." *Cooking, Eating, Thinking: Transformative Philosophies of Food*. Ed. Deane Curtin and Lisa Heldke. Bloomington: Indiana UP (1992).

—— . "Recipes for Theory Making." *Hypatia* 3.2 (1988): 15–31.

hooks, bell. *Black Looks: Race and Representation*. Boston: South End (1992).

Iyer, Pico. *Video Night in Kathmandu*. New York: Knopf (1988).

Jaffrey, Madhur. *An Invitation to Indian Cooking*. New York: Vintage (1973).

Kadi, Joanna. *Thinking Class*. Boston: South End (1996).

Lutz, Catherine A., and Jane L. Collins. *Reading National Geographic*. Chicago: U of Chicago P (1993).

Min-ha, Trinh T. *Woman Native Other: Writing Postcoloniality and Feminism*. Bloomington: Indiana UP (1989).

Paretsky, Sara. *Killing Orders*. New York: Ballantine (1985).

Pratt, Mary Louise. *Imperial Eyes: Travel Writing and Transculturation*. New York: Routledge (1992).

Roden, Claudia. *A Book of Middle Eastern Food*. New York: Vintage (1974).

Rosaldo, Renato. *Culture and Truth*. Boston: Beacon (1989).

Vizenor, Gerald. *The People Named the Chippewa: Narrative Histories*. Minneapolis: U of Minnesota P (1984).

21

More than Just the "Big Piece of Chicken". The Power of Race, Class, and Food in American Consciousness

Psyche Williams-Forson

In 1999 HBO premiered Chris Rock's stand-up comedy routine *Bigger and Blacker*. One of the jokes deals with what Rock humorously calls the "big piece of chicken."[1] Using wit, Chris Rock delivers a semi-serious treatise on parenting and marriage. First, he admonishes the audience for not recognizing that "a real daddy" receives little praise for "making the world a better place . . ." A man, or "daddy", according to Rock, pays bills, provides food, and all of a family's other necessities. Despite his efforts, he rarely receives any praise for his "accomplishments." Although these tasks are clearly part and parcel of adult responsibilities, Rock ignores this truism in an effort to set up his commentary on the intersection of race, class, gender, and food. Continuing, he argues, "Nobody appreciates daddy . . ."

By way of illustrating why a father needs and deserves such concern Rock points out that fathers work hard all day fighting against the stresses of life. Then a father—particularly an African American father—comes home to more stress:

> And what does daddy get for all his work? The *big piece of chicken*. That's all daddy get is the *big piece of chicken*. That's right. And some women don't want to give up the big piece of chicken. Who the fuck is you to keep the *big piece of chicken*? How dare you keep *the big piece of chicken*! A man can't work for 12 hours and come home to a wing! When I was a kid, my momma [would] lose her mind if one of us ate the *big piece of chicken* by accident. "What the fuck? You ate the *big piece of chicken*. Oh Lawd, no, no, no! Now I got to take some chicken and sew it up. Shit! Give me two wings and a poke chop. Daddy'll never know the difference."[2]

Chris Rock's kind of humor has an extensive history as a form of black expressive culture. Physically, he walks back and forth on stage, bobbing and weaving as he shares different versions of his comic narrations, turning out stories from "everyday conversational talk."[3] Rock uses this form of performance or narrativizing to wage social commentary on a variety of issues including stereotypes of black people and chicken. When an artist uses stereotypes there are a number of factors that have to be considered including the purposes to which such oversimplifications are put. Stated more plainly, the humor of Chris Rock makes us wonder about the subversive ways in which objects like food can be used to contest hegemonic representations of blackness and the ways in which performances of blackness reveal complicated aspects of identity.

Investigating Intersections

As more or less correctly stated, there are roughly two methodological schools of thought when talking about African American foodways. There are those that focus on the food itself and its connections to the African Diaspora. Among them are historians of the American South (e.g., Karen Hess, Joe Gray Taylor and Sam Hilliard) and African American studies (e.g., Tracy Poe and Robert Hall), archeologists (e.g., Theresa Singleton and Anne Yentsch), geographer Judith Carney, anthropologist Tony Whitehead, and independent foodways scholars (e.g., Jessica Harris, Howard Paige, Joyce White, and Diane Spivey). Those who focus generally on the intersections of food and identity, representation, and/or contestation are literary scholars Anne Bower, Kyla Wazana Tompkins, Doris Witt, and Rafia Zafar, sociologist William Whit, anthropologist Charles Joyner, and folklorist Patricia Turner; media specialist Marilyn Kern-Foxworth, and historians Kenneth Goings and M. M. Manring.[4] My research into the realm of African American foods is not only about locating, identifying, and understanding the connections between foods but also the people who consume them. This approach goes beyond the theories that argue we are what we eat and the ways our foods reflect our cultural identity. Rather, the method I employ asks us to consider what we learn about African American life and culture by studying the intersections of food, gender, race, class, and power. How do African American historical, socioeconomic, and political spaces influence the foods that are consumed? How is this consumption a part of the performance of black class? Furthermore, what do we learn about African Americans when black people willingly engage in perpetuating the oversimplified images or ideas that are sometimes held by the larger American society?

Black people have long been engaged in ideological warfare involving food, race, and identity. Most commonly known are the stereotypes concerning black people's consumption of fried chicken and watermelon. Though these stereotypes have been around for centuries they are still pervasive in the contemporary American psyche. Consider, for instance, the numerous postcards, invitations, and other ephemera that illustrate African American men, women, and children with watermelon.

Black feminist scholar Patricia Hill Collins suggests the need to be attuned to the ways in which processes of power underlie social interactions and are involved in the process of external definition. These definitions can be challenged, however, through the process of "self-definition." The acts of "challenging the political knowledge-validation process that result[s] in externally defined stereotypical images . . . can be unconscious or conscious acts of resistance."[5] One engages in the process of self-definition by identifying, utilizing, and more importantly, redefining symbols—like chicken or watermelon—that are commonly affiliated with African Americans. By doing this, black people refuse to allow the wider American culture to dictate what represents their expressive culture and thereby what represents blackness. But this process of defining one's self is fraught with complications and complexities particularly if the group fails to understand or acknowledge that there is a power structure at work behind the creation of common affiliations, labels, or stereotypes.

Collins explains these complications further in her delineation of self-valuation or the replacement of negative images with positive ones. This process of replacement can be equally as problematic as the original external definition if we fail to understand and to recognize the stereotype as a controlling image. This concept is perhaps best illustrated by the example of Chris Rock's comedy that opens this essay. Though I will return to Rock's funny side later, Collins' caution is registered here. The exchange of one set of controlling images for another does little to eradicate the defining image itself. Consequently, black people need to be clear about the ways in which historical, social, political, and economic contexts have established reductionist narratives and how these accounts are embedded in food.

One way that blacks can both deal with these narratives and gain independence from them is to begin by taking a close look at the historical basis of various food stereotypes. These stereotypes tend to be distorted portrayals of those cultural behaviors that are and have been used in order to diminish black personal and collective power.

Stereotypes Abound

Stereotypes involving black people have been around for years. Indeed, they continue to exist.[6] Elsewhere I argue extensively for the partial evolution of some of these stereotypes as ideologies shaped from laws and ordinances passed during the seventeenth and eighteenth centuries.[7] It was and continues to be my contention that these depictions partly emerged as a way to control the economic gains of enslaved and free men and women who bartered and traded in the marketplace. Historians often cite newspaper articles, court documents, and travelers' accounts among other critical sources detailing information on early African and African American entrepreneurs of food. Nineteenth-century travelers' diaries, for example, indicate "flocks of poultry [were] numerous" and, "there are very few [slaves] indeed who are denied the privilege of keeping dunghill fowls, ducks, geese, and turkeys." Moreover, some black people would often sit by the wharf for days on end waiting to buy foods like chicken and then sell them for exorbitant prices.[8] Historian Philip Morgan notes a similar practice whereby some travelers would instruct their stewards to hold in reserve various foods like bacon so they would have bartering power with "the Negroes who are the general Chicken Merchants [sic]."[9]

As with any encroachment, the bartering and trading by African Americans ushered in a slew of regulations that sought to limit items being sold door-to-door and in the market. To be sure the ambiguous ownership of goods prior to sale was one of the many reasons for stalling and halting the sale of goods. Foods were not supposed to be sold prior to passing through the town gates, and in particular customers were not supposed to purchase goods whose ownership might be difficult to trace. This included items such as chickens, which were often sold outside the market. Archeologist Anne Yentsch maintains that foods such as oysters, salted fish in large barrels or casks, cattle, sheep, and hogs that were alive could easily be traced because they were by-and-large produced by small farmers.[10]

Chickens, on the other hand were far harder to pinpoint. Even though several blacks had chickens their masters and neighboring farms had them as well. Sometimes these birds roamed freely and thus were traded or sold in an effort to obtain more favorable goods. Often times, especially during the colonial era, it was difficult to ascertain the exact origins of a bird. Except among the wealthy, most chickens during that time were not kept in hen houses. Chicken and fowl were free to roam finding food and shelter wherever possible, an issue that easily lends support to the charge of theft. Additionally, there was no widespread formalized system of breeding in early America when many Africans and Native Americans were engaged in bartering. Consequently, it was difficult to distinguish most common fowl from one another with the exception of certain kinds of partridges, pheasant, and hens. This reality, however, did little to hinder the accusations of theft, which were not only levied against slaves but also free blacks and fugitives.

These claims were fueled by black people's use of trading practices like forestalling, which legal ordinances did little to reduce. According to the *South Carolina Gazette*, one writer complained that almost on a daily basis, black women could be found huckstering and forestalling "poultry, fruit, eggs," and other goods "in and near the Lower Market . . . from morn till night," buying and selling what and how they pleased to obtain money for both their masters and themselves. Often times their prices were exorbitant and they would use all kinds of marketing strategies to choose which white people to sell to and for how much.[11] Robert Olwell captures this point when he explains: "as slaveholders, Carolina whites felt that slaves should be generally subordinate, but as property holders and capitalists they also had to recognize the legitimacy of the market in which sellers had the right to seek the highest price for their goods."[12]

Many whites viewed blacks with "great prejudice" when they sought to engage in capitalist enterprises. Under slavery's oppression, blacks, regardless of their status, were to be subordinate at all times. Any deviation from this norm was a threat to the social order that had been systematically and institutionally constructed over time. Consequently, any element of freedom recognized and enjoyed by black people, and particularly women, was an affront to white social power. Lawrence McDonnell explains it this way: "The marketplace . . . is a neutral zone, a threshold between buyer and seller. . . . Master and slave confronted each other at the moment of exchange as bearers of commodities, stripped of social dimensions . . . [this] linked black sellers with White buyers, and hence with White society, not only by assertion of black humanity but through White objectification. Slaves appeared here equally purposeful as Whites."[13]

Money and a small measure of market power assaulted the charade played out during slavery that sought to convince black people that freedom would never come. Attributing black economic gain to theft helped to perpetuate the travesty. By attributing stealing by slaves to an inherent nature rather than a condition of their circumstances (or even to a performance of sorts), slave owners were able to deflect attention from their own participation in this aspect of slave victimization. Morally, it was much better to believe that slaves were natural thieves than to believe that the institution of enslavement contributed to their larceny. Clearly there is some truth to the claim that slaves engaged in thievery; the extent to which this was the case, however, is rooted in white patriarchal ideology.[14]

Though devoid of a disposition toward theft, some slaves did engage in pilfering and stealing. Some scholars however, have referred to these acts as skill and cunning. Eugene Genovese's study of African American life and culture, suggests this when he writes, "for many slaves, stealing from their own or other masters became a science and an art, employed as much for the satisfaction of outwitting Ole' Massa as anything else."[15] *In Weevils in the Wheat*, for example, ex-slave Charles Grandy tells that hunger was a motivating factor for stealing food. He says, "I got so hungry I stole chickens off de roos'. . . . We would cook de chicken at night, eat him an' bu'n de feathers. . . . We always had a trap in de floor fo' de do' to hide dese chickens in."[16] This is just one example of African American trickster heroism that not only reflects a kinship to African traditions but also views this type of behavior as both morally acceptable and necessary for survival. At the same time, it is a subversive cultural form that uses humor in its expression.

John Roberts' point about early African Americans should be registered here: "Given the desperate and oppressive circumstances under which they lived, enslaved Africans could not be overly concerned with the masters' definition of 'morality' of behaviors that enhanced their prospects for physical survival and material well-being. The task that they confronted, however, was how to make such individually devised solutions to a collective problem function as a behavior strategy for the group without endangering their adaptability or the physical well-being of members of their community."[17] Although the oppressive circumstances of today are nowhere near those of enslavement, the delicate balance of performing individual behavior and yet not suffering collective consequences is still applicable. Teasing out this sense of balance and its complications might become more apparent as I discuss African American performances of stereotypes during the late nineteenth and early twentieth century.

The South suffered a devastating loss of free labor with the end of the Civil War and migrations of newly freed blacks; it found itself in a precarious situation. Its infrastructure was suffering economically, politically, socially, culturally, and physically. Suddenly, the millions of freed blacks became an overwhelming problem. What about their rights? Would they be given rights? How and to what end? How would white Southerners keep their subordinates in line? Was this even possible anymore? These and many other questions played themselves out on the political landscapes of the day. But they were also played out on cultural playing fields as well. According to historian Kenneth Goings, the loss of control over black people registered such a blow among white Southerners that they began using emerging technology as one means of reasserting control and reclaiming power.[18]

Advancing technology, namely the camera, was useful for depicting African Americans—men, women, boys, and girls—as visually conciliatory. As Grant McCracken intimates, such illustrations were useful for alleviating some of the "nervous prostration" brought on by the rapid changes of the time. Goods and commodities were used in an effort to alleviate some of the distress caused by the social, political, economic, and cultural transformations.[19] Goods were particularly

useful for helping individuals contemplate the "possession of an emotional condition, a social circumstance, even an entire lifestyle" by making desires concrete.[20]
These illustrations, or commodities of racism, were coveted possessions. They
enabled their owners not only to possess the physical object but also to mentally
covet the pastoral image of the gallant South that whites wished to maintain. This
interpretation is certainly not the only reason that people might have purchased
these kinds of photos. But for sure these images and their owners were complicit in
spreading the network of racial power.

What quickly emerged through this visual communication was an ideology of
black inferiority, which assisted in the formulation of racist stereotypes. These stereotypes were perpetuated by advertisements, trading cards, sheet music and stereoviews
like that which illustrates an African American baby in a buggy, caption reading:
"When I Dit Big, 'Oo'll Have to Roost Hiah." This, and countless other images are
clearly staged as if to appear natural. More than likely it was the case of African
Americans performing to stay alive. From the thieving child, to the salacious lover of
white hens, African Americans—particularly men and boys—were constantly
ridiculed; more often than not it was centered on the stereotypical image of the coon.

Kenneth Goings, whose study *Mammy and Uncle Mose* historicizes the cultural and
political economy of black collectibles, maintains that the coon image was one of the
most offensive stereotypes. M. L. Graham used this coon motif as the mascot for his
little-known "Coon Chicken Inn" restaurants. The emblem, a black-faced man with
large, extended red lips, was typically symbolic of how whites would stereotype black
people with food to endorse various products like fried chicken. Considered a most
effective advertising technique, images like these reinforced the stereotypical Old
South/New South myth of the loyal, happy servant just waiting to be used by the
master—and now the consumer.

The restaurant with all of its accoutrements became a metaphor for whites using
and discarding black service. When the meal is complete, the napkins, plates and
utensils bearing the black-faced logo are discarded and along with it any remnant of
the serviceable "darker" that is no longer needed or desired. This act of symbolic and
physical disposal provided whites with what Goings describes in a similar discussion
as a sense of "racial superiority" and a "therapeutic sense of comfort."[21] Manipulating these objects of material culture enabled white Americans not only to forge an
alliance across class lines, but also to more collectively subjugate and vilify black
people. The ideology of black inferiority provided a safeguard for white America
during a time when their racial, economic, and political balance was perceived as
unstable and threatened.

Unfolding against this backdrop, are the numerous ways that food becomes interlaced with discourses of power, race, class, and gender in American consciousness.
Chicken, for example, which was once championed as a celebrated food of the South
prepared by some of the best culinary talent turns into an object of ridicule and
defacement. Chicken—both the bird and the food—is fraught then, with paradoxes
in the contexts of the historical and economic circumstances of the South. On the
one hand, black consumption of chicken was seen as normative; on the other hand,
this consumption was also perceived as negative. The issue is made more complex
when we read chicken—the food—as a cultural text.

Fried chicken, a largely southern food that emerged out of social institutions shaped by racial complexities, is one of many foods that blurs the lines between the "symbolic separations [of] those who prepare the food and those who consume it."[22] Black women were widely credited with lining "Southern groaning boards." This was their rightful place as loyal cooks—a cultural demarcation that became necessary for symbolically separating the domestic rituals of the South. Black women prepared and cooked fried chicken for white families but they did not consume it; and, thereby they maintained the purity of southern cuisine. Mentally, this belief was important for reinforcing the necessary symbolic distance between cook and consumer. This configuration is made problematic and complicated, however, by the insistence that black people are zealous about their consumption of fried chicken.

What becomes necessary then are carefully coded words and messages. Namely, the word Southern becomes coded for white, while "soul food" is decoded as black. Diane Spivey has labeled this coding phenomenon, "Whites Only Cuisine." She says:

> The end of the [Civil] war also signaled the beginning of the redefining of southern White heritage. Food was a factor in the efforts of southern White elites to hold on to their old way of life. Cooking and cuisine were remade to look uniquely southern. . . . Asserting that the recipes were "southern" made [their] cookbooks exclusionary, and therefore racist, because the cookbooks and recipes contained therein were heralded as the creations of the elite southern White women. In an attempt to promote southern White culture, therefore, the concept of "southern cooking" started out as *Whites Only Cuisine.*[23]

Given the mass exchange of foods and food habits that occurred between early Africans, Europeans, and Native Americans it is almost impossible for one group or another to claim any recipe as original or native to their culture. With the ebb and flow of people across continents, regions, and lands come vast amounts of mutual exchange resulting in multi-amalgamations between and among cultures and foods changing and evolving over time.

The intersections of food with power and other variables enable a reading of the ways in which the idea of blackness as performance boldly emerges. As I have discussed thus far, since their arrival in this country, Africans and Africans born in America have been performing race in myriad ways. Long after the auction block performance African Americans engaged in other racial acts like participating in staged photographs and witnessing their recipes being usurped. A good many of these performances of racial roles for survival involved food. Part of understanding the food and foodways of African Americans asks that we also question what all of these performances had to do with blackness and issues of identity? And how has agency been a part of the performance of this blackness? Turning back to Chris Rock's comedic discussion of the "big piece of chicken" helps us to think a bit more about these questions.

In the vignette that opens this essay Chris Rock is explaining how the children of an African American family have eaten the "big piece of chicken" even though they are aware that this piece of meat belongs to their father. In the comedy, Rock makes manhood and fatherhood synonymous with the right to have the largest piece of chicken not simply as a reward but as a right. Rock argues that this is the father's just portion because "daddy can't work all day and come home to a wing." Implied, of course, is the fact that because the father leaves the home to work and engages in a number of anxieties outside the home, he therefore deserves all of the praise—including culinary recompense.

On the surface one could argue that this routine is simply another of Rock's treatises on the ills of society. Every race of people can identify with this scenario—one of the many aspects of Rock's performances that endear him to diverse audience members. However, my contention is that this scene is multifaceted. Rock is, in effect, performing blackness in ways that can be described as both subversive as well as oppressive, rendering this piece to be about more than "the big piece of chicken." Rock is dissident in that, more or less, he follows the basic formula of delivering an African American trickster folktale. Consider, for example, Jacob Stoyer's slave narrative, *Sketches of My Life in the South*, wherein he tells a story of man named Joe and how he outsmarted the master's wife, Mrs. King. According to the story, Joe killed and dressed a turkey that belonged to the King family. In his haste to get the bird into a pot without being caught, he neglected to cut it leaving its knees to stick out of the pot. To hide his thievery, Joe threw one of his shirts over the pot. When Joe failed to respond to the calling of Mr. King, Mrs. King came into the kitchen to inquire of his whereabouts. Discovering the theft, which Joe declined to know anything about, she saw to it that Joe was punished for "allowing the turkey to get into the pot."[24] The point here is the way in which Joe was able to dupe, if only briefly, the King family. The larger issue is the momentary reversal of power executed by Joe in his performance as a "dumb slave."

Similarly, Rock manages to dupe both white and black audience-goers who usually have paid a somewhat hefty price to enjoy a laugh. By performing racist, sexist, and otherwise problematic comedies Rock proffers the illusion that he buys into these notions as truisms. In doing so, he is a relative trickster, perpetuating the racist perception of black people as chicken lovers.[25] But as E. Patrick Johnson argues, "blackness does not only reside in the theatrical fantasy of the White imaginary that is then projected onto black bodies. Nor is blackness always consciously acted out. It is also the inexpressible yet undeniable racial experience of black people—the ways in which the 'living of blackness' becomes a material way of knowing."[26] Among black audiences then, it is not surprising that Rock's performances are laudatory and celebratory. Many watch the performances of him and other comedians and laugh uproariously knowing that much of what is being performed has all kinds of negative implications. Yet, there is something to be said for these dramatic interludes, which often make audiences momentarily forget about their troubles. The very fact of the matter is that these comedians are enjoyed precisely because they engage in the slipperiness of black cultural politics.

Part of this slipperiness derives from another suggestion offered by Johnson: "The interanimation of blackness and performance and the tension between blackness as 'play' and material reality further complicates the notion of what . . . 'playing black' is and what 'playing black' ain't." Rock engages this question of "black is/black ain't" with his audiences. With white audiences he leads them into thinking that he is performing what "black is" as he mocks, mimics, and ridicules black people. With black audiences, he relies upon a number of "in-group" techniques to offer black audiences comic relief while simultaneously playing to a number of 'truth claims.'[27]

Later, using a similar coded performance as the trickster hero in African American folktales, Rock turns the tables on this segment of the audience using the rhetoric of race, class, and gender.

The art of verbal play has always been a vehicle of self-expression for black men and women, although women have only recently been recognized as engaging in such. Rock understands the role of signifying in the black community and employs it well in his routines. From Rock's references to the fact that daddy experiences stress all day from working in a "white world," we can assume that daddy feels little or no economic power. Consequently, in order to establish his manhood, he needs to assert his authority at home. One of the ways he is able to affirm his household status is by eating the "big piece of chicken." Here the chicken functions metaphorically and literally as a source and a reflection of masculine power. Rock's subtle explication of this power enables him to dupe his black audience-goers—particularly the women.

Using children as the catalyst, Rock creates the scenario of mama as a culinary artist. After her children consume the forbidden big piece of chicken, she is able to flawlessly recreate it by expertly sewing together two wings and a pork chop. In addition to all of the other work mama has done during the day—caring for children, running a household, cooking and other chores—she now has to make up for the fact that one of her children has eaten the wrong piece of chicken: "Oh Lawd, no, no, no! Now I got to take some chicken and sew it up. Shit! Give me two wings and a poke chop. Daddy'll never know the difference."[28] The challenge posed by this situation is perhaps the cause for Rock acknowledging, "Now mama got the roughest job, I'm not gonna front." Denying daddy his rightful portion is a measure of disrespect that will surely bring wrath upon the children. To avoid this mama tries to make amends. Mama then has not only procured, prepared, and presented the food, she now has to alter and re-prepare the meal while simultaneously protecting her children. After all of this, mama will undoubtedly have to "do a jig" so that daddy does not recognize her necessary handiwork. She then will have to placate him if he discovers the ruse.

The discussion of mama's incredible talents is double-edged because while plentiful, her culinary and household ingenuity must not be celebrated because to do so would reveal that daddy is eating something less than the "big piece of chicken." Equally problematic is that it has gone unnoticed that while mama is not in the paid work force, she is nonetheless very much involved in a system of work. Her work, unfortunately, is largely domestic, economically undervalued, and from the standpoint of this example, aesthetically unappreciated. Because even though mama has created another large portion for daddy she cannot speak of it because it will only make daddy feel that he gets little for all his "hard work." Unspoken are the stories of mama's day of work, her troubles, and her battles—many of which are represented symbolically by the chicken.

It is not surprising that Rock would gender his discussion to include some kind of praise of a mother's culinary abilities. As Pamela Quaggiotto notes, "the mother determines when, what, and how much family members will eat. . . . She controls the symbolic language of food, determining what her dishes and meals will say about herself, her family, and world."[29] And yet his depictions of "mama" are both enlightening and baffling for what they seem to reveal/hide about Rock's gender and racial agenda. Clearly the parody and humor of this situation are evident. Though the audience knows it is a joke, there is uncertainty over whether mama's work is being praised or ridiculed. Moreover, there are the questions of whether or not Rock is waging some sort of commentary on racial stereotypes involving black people and chicken. For example, it is significant to note that Rock never specifies whether the chicken is fried or baked. In fact, he does not have to because he relies upon a certain amount of a priori knowledge that assumes that chicken eating in the company of black people means "fried." Comforted by the fact that black audience members bring to bear their own life experiences and cultural memories surrounding food and thus know what "black is" and "black ain't" he is able to launch into his dramatization.

Maybe it is a similar comfort and ease that Rock attends to when at the end of the routine he admonishes women to remember their "proper" place in dealing with men. Undoubtedly many in the audience see mama's work as a labor of love that is taken for granted, not needing any particular recognition. In fact, Rock half-heartedly suggests this when he implies that daddy has the primary responsibilities in the household. As if rethinking this assertion, Rock soon after backtracks by supplying his one line of praise for mama. Despite all of this backpedaling, by the end of his show Rock is clear about his direction as he definitively reinstates his masculinist stance. He closes his performance with: "Women talk too much. They always want you to be listenin.' Let a man get situated! Let me get my other foot in the door! Let me get somethin' to eat. Let me get somethin' to drink! Let me take a shit! Go in the fuckin' kitchen and get me my *big piece of chicken!*" Having said this, Rock drops the mike and struts off the stage amid the cheers and shouts of approval from men—and women.[30]

Food objects are useful for elucidating the type of obscurities revealed by this kind of close reading. Additionally, they are politicized by the meanings inscribed in their uses and associations historically and contemporarily. This is particularly salient to an article like chicken that is perceived to be generic in its uses among all races and ethnicities of people. The meanings that chicken holds for black people are as diverse as its members. But when chicken is placed in various contexts alongside performances of power and race then it is plentiful for what it reveals beyond being a portion of food.

This essay has attempted to illustrate the importance of moving beyond studying merely the foods of various cultures to include the behaviors, actions, contexts, and histories that involve them. As this article has also suggested foods like chicken, that have been used to stereotype African American people, are often actually under-girded by intersecting variables of race, gender, class, and power. This fact, perhaps more than any other, lends credence to the notion that food is always about more than what it seems.

Notes

1. For a more detailed analysis of this particular routine of Chris Rock's see Williams-Forson, *Building Houses Out of Chicken Legs: Black Women, Food, and Power* (Chapel Hill: UNC Press, 2006), 178–185.
2. Chris Rock, *Bigger and Blacker*.

3. Geneva Smitherman, "The Chain Remain the Same: Communicative Practices in the Hip Hop Nation." *Journal of Black Studies* 28, 1 (September 1997): 3–25.

4. Though in no way this dichotomous, most of these scholars can be roughly divided into these categories, as Krishnendu Ray observes. See book review by Ray, *Building Houses out of Chicken Legs*, in *Food & Foodways*, 15:1–6, 2007. Also see *African American Foodways: Explorations of History & Culture*, Ed. Anne Bower (Urbana: University of Illinois Press, 2007).

5. Patricia Hill Collins, "Learning from the Outsider Within: The Sociological Significance of Black Feminist Thought," *Social Problems* 33, no. 6 (1986): 516–517.

6. African Americans have long been caricatured as brand mascots for various food and household products. For example, a grinning black chef named Rastus was used to represent Cream of Wheat hot cereal and a pair of black children who were known as the Gold Dust Twins, were used to advertise soap powder. In addition to the now infamous Aunt Jemima, who sold pancake mix, there have been numerous other grinning black women who were "Jemima-like" that were used to sell fried chicken, shortening, and cookware. It also should be noted that other races and ethnicities have also been stereotyped where food is concerned. First there was Frito Bandito, who spoke in an exaggerated Mexican accent and then there was the Chihuahua who muttered ¡Yo Quiero Taco Bell! In March 2007 Masterfoods USA, a unit of Mars Foods attempted to hoist the stereotypical depiction of "Uncle Ben" from servant to chairman of the board. The attempt was met with mixed success. See Stuart Elliott, "Uncle Ben, Board Chairman." *The New York Times*. 30 March 2007, C1.

7. For a more lengthy discussion see Williams-Forson, *Building Houses Out of Chicken Legs*, in particular chapters 1 and 2. See also Williams-Forson, " 'Suckin' the Chicken Bone Dry': African American Women, Fried Chicken, and the Power of a National Narrative." In *Cooking Lessons: The Politics of Gender and Food*. Ed. Sherrie Inness. (Lanham, MD: Rowman & Littlefield, 2000): 200–214.

8. Anne E. Yentsch, *A Chesapeake Family and Their Slaves: A Study in Historical Archaeology* (Great Britain: Cambridge University Press, 1994), 242.

9. Philip D. Morgan, *Slave Counterpoint: Black Culture in the Eighteenth-Century Chesapeake and Lowcountry* (Chapel Hill: University of North Carolina Press for the Omohundro Institute of Early American History and Culture, Williamsburg, Virginia, 1998), 359.

10. Yentsch, 245.

11. Quoted in Robert Olwell, " 'Loose, Idle, and Disorderly': Slave Women in the Eighteenth-Century Charleston Marketplace." In *More Than Chattel: Black Women and Slavery in the Americas*, Eds., David Barry Gaspar and Darlene Clark Hine. (Bloomington: Indiana University Press, 1996): 97–110. See also, Yentsch, 242–243; Phillip Morgan, 368–372.

12. Olwell, 102.

13. McDonnell, "Money Knows No Master: Market Relations and the American Slave Community." In *Developing Dixie: Modernization in a Traditional Society*, ed., Winfred B. Moore, Jr., Joseph F. Tripp, and Lyon G. Tyler, Jr. (Westport: Greenwood Press, 1988): 31–44.

14. Lichtenstein, " 'That Disposition to Theft, With Which They Have Been Branded': Moral Economy, Slave Management, and the Law." *Journal of Social History* 21 (1989): 413–40.

15. Eugene Genovese, *Roll, Jordan, Roll: The World the Slaves Made* (New York: Vintage Books, Inc., 1976), 606.

16. Charles L. Perdue, Jr., Thomas E. Barden, and Robert K. Phillips, Eds. *Weevils in the Wheat* (Charlottesville: University of Virginia Press, 1976), 116.

17. Roberts, *From Trickster to Badman: The Black Folk Hero in Slavery and Freedom*. (Philadelphia: University of Pennsylvania Press, 1989): 33. Lawrence Levine also establishes connections between the African American trickster and the acquisition of food. See "The Slave as Trickster," in *Black Culture, Black Consciousness*. (New York: Oxford University Press, 1977), 121–133.

18. Kenneth W. Goings, *Mammy and Uncle Mose: Black Collectibles and American Stereotyping* (Bloomington: Indiana University Press, 1994), 4–7.

19. Grant McCracken's discussion is a good one on the ways in which consumer goods helped to preserve hopes and ideals during the Victorian era. See "The Evocative Power of Things," *Culture and Consumption* (Bloomington: Indiana University Press, 1990), 104.

20. McCracken, 110.

21. Ibid., 47.

22. Mary Titus, " 'Groaning Tables and Spit in the Kettles': Food and Race in the Nineteenth-Century South," *Southern Quarterly* 20, no. 2–3 (1992),15.

23. Diane M. Spivey, "Economics, War, and the Northern Migration of the Southern Black Cook," in *The Peppers, Crackling, and Knots of Wool Cookbook: The Global Migration of African Cuisine* (New York: State University of New York Press, 1999), 263.

24. See Jacob Stroyer, *Sketches of My Life, Sketches of My Life in the South*. Part I. 1849–1908. Salem: Salem Press (1879). *Documenting the American South: The Southern Experience in Nineteenth-Century America*, eds. Lee

Ann Morawski and Natalia Smith (2001). Academic Affairs Lib., U. of North Carolina, Chapel Hill. ⟨http://docsouth.unc.edu/neh/storyer/stroyer.html⟩ (June 1, 2007).

25. Williams-Forson, *Building Houses Out of Chicken Legs*, 176–181.
26. E. Patrick Johnson, *Appropriating Blackness: Performance and the Politics of Authenticity* (Durham: Duke University Press, 2003), 8.
27. Patricia Turner and Gary Alan Fine suggest that when rumors and/or stereotypes are based on information that *could be* correct it is considered a truth-claim. Truth claims contain a certain amount of "cultural logic" because they make "cultural sense" (i.e. all black people eat fried chicken) even though no systemic, definitive evidence exists in which to substantiate them. *Whispers on the Color Line: Rumor and Race in America.* (Berkeley: University of California Press, 2001), 18.
28. Chris Rock, *Bigger and Blacker.* Videocassette. HBO Studios (1999).
29. Pamela Quaggiotto as quoted in Carole M. Counihan, "Female Identity, Food, and Power in Contemporary Florence," *Anthropological Quarterly* 61 (1988): 52.
30. It is quite easy to become overcome with laughter by Rock's prose and delivery. The immediacy and dramatic nature of the moment invite this response. It is only later, once you have had a chance to relive the scene that one might realize the sexism inherent in both the rhetoric and the performance as Rock leaves the stage seemingly in command, having said all that he has had to say.

22

Mexicanas' Food Voice and Differential Consciousness in the San Luis Valley of Colorado[1]

Carole Counihan

I never cooked you know. I was always a bookworm. Ever since I was a growing up. When it was time for the dishes, they couldn't find me, so my poor sister had to do them by herself. . . . We had an outhouse—a soldiers', a government toilet outside—and I'd take a book, you know, and I'd go there, and they'd say, "Where's Helen?" And somebody would pop up and say, "Oh she's in the toilet reading, she could be." And when I thought the dishes were half done or done I'd pop up. I never was responsible for them, they never depended on me, and my sister was such a good cook. She was a good cook and she griped about me not taking turns on the dishes but she didn't fight, she didn't mind. She was grown up on the job, you know, it was natural for her.[2]

These words of Helen Ruybal, a ninety-nine-year-old widow, mother of two, and former teacher, are part of a long-term ethnographic project I have been conducting since 1996 in the Mexican-American town of Antonito in rural Southern Colorado. I collected food-centered life histories from nineteen women, including Ruybal, and suggest that they reveal women's voice, identity, and worldview.

Antonito is six miles north of the New Mexico state line in the San Luis Valley, an eight thousand square-mile cold desert valley lying at approximately eight thousand feet above sea level between the San Juan and Sangre de Cristo mountains. Antonito is located in Conejos County in the Upper Rio Grande region on the northern frontier of greater Mexico.[3] The population of Antonito is 90% Hispanic and thus it is an excellent site to study the contemporary experience of rural Chicanas and Chicanos. My forthcoming book, tentatively titled *Mexicanas' Stories of Food, Identity and Land in the San Luis Valley of Colorado*,[4] gives a full exposition of how my nineteen interviewees described land and water, defined food and meals, and enacted family, gender, and community relations. In this paper, I use excerpts from one woman's interviews to make two points—first, to affirm the value of the food-centered life history methodology, and second, to suggest how women can display differential consciousness through their practices and beliefs surrounding food.

Food Voice, Feminist Anthropology and *Testimonios*

For over two decades, I have been using a food-centered life history methodology in Italy and the United States to present women's food voices.[5] Food-centered life histories are semi-structured tape-recorded interviews with willing participants, on their beliefs and behaviors surrounding food production, preparation, distribution, and consumption. I developed this methodology out of a feminist goal of fore-grounding the words and perspectives of women who have long been absent in recorded history. In the interviews women describe material culture as well as their subjective remembrances and perceptions. Topics include gardening, preserving food, past and present diets, recipes, everyday and ritual meals, eating out, foods for healing, eating in pregnancy, breast-feeding, and many other subjects (see Appendix 1 for a list of interview themes). For many women (and some men), food is a significant voice of self-expression. In the meals they cook, the rituals they observe, and the memories they preserve, women communicate powerful meanings and emotions.

Like other feminist ethnographers, I have grappled with how to present an authentic picture of my respondents, one that is as much theirs as possible.[6] I used a tape-recorder so I could begin the process of representation with their words. Before doing interviews, I established informed consent, telling people in Antonito who I was and what I was doing there, promising confidentiality, and giving them the choice to participate or not.[7] I asked for their permission to tape-record, explaining that I wanted to have their verbatim descriptions of their experiences, but I also assured them that they could turn the tape recorder off at any time and decline to answer any questions, both of which they did on occasion. While I tried eventually to address all of the topics on my list (see Appendix 1), interviews were conversations with their own momentum and wandered into many non-food topics.

I have not followed the practice of some ethnographers of citing transcriptions verbatim, but at the urging of participants, I have edited the transcriptions to achieve readability, while staying as close to their original language as possible. I eliminated repetition and filler expressions (e.g. "like," "you know"), edited lightly for grammar, and moved around phrases and sections to achieve greater coherence.

My methodology coheres with two linked intellectual traditions—feminist anthropology and *testimonios*. It shares feminist anthropology's goals of placing women at the center, foregrounding women's diversity, challenging gender oppres-sion, and reconstructing theory based on women's experiences (Moore 1988). My use of food-centered life histories to give voice to marginalized women links with the research of other feminists who have examined food as women's voice[8]—particularly Hauck-Lawson's (1998) research on immigrants in New York City, Pérez's (2004) "Kitchen Table Ethnography" with women on both sides of the U.S.-Mexico border, and Abarca's (2001, 2004, 2006) "culinary chats" with Mexican and Mexican-American working class women—all of which use women's food stories to theorize about their identity, agency, and power.[9]

Like the *testimonios* gathered by the Latina Feminist Group, food-centered life histories are about "telling to live." *Testimonios* involve participants speaking for themselves about events they have witnessed, events centered on "a story that *needs* to be told—involving a problem of repression, poverty, subalternity, exploitation, or simply survival" (Beverly 1993: 73). *Testimonios* are personal narratives that reveal individual subjectivity while calling attention to broad political and economic forces. They grew out of Latin American liberation movements at the same time that feminism was emerging in politics and scholarship. Sternbach (1991) highlighted the fact that these movements shared "breaking silences, raising consciousness, envision-ing a new future, and seeking collective action" (Sternbach 1991: 95). Both feminism and *testimonios*, she said, linked the personal and the political, the "private, domestic or intimate sphere" with the "public, historic or collective one" (Sternbach 1991: 97). This is also the aim of food-centered life histories: to thrust the traditionally private sphere of cooking and feeding into the public arena and show the impact of women's experiences on culture and history. The experiences and voices of rural Colorado *Mexicanas* have been left out of the historical record for too long,[10] and recuperating them enhances understanding of the diversity of Mexican and Mexican American women. It fulfills a central goal in feminist ethnography, enriches our understanding of American culture, and makes possible more inclusive political policy.

Food and Differential Consciousness

Across cultures and history, food work can represent drudgery and oppression but also power and creativity. My second goal in this article is to show how women can challenge subordination and strive for agency through their food-centered life histories by evincing what Chela Sandoval (1991) calls "differential consciousness." Differential consciousness is a key strategy used by dominated peoples to survive demeaning and disempowering structures and ideologies. It is the ability to acknowledge and operate within those structures and ideologies but at the same time to generate alternative beliefs and tactics that resist domination.[11] Differential con-sciousness is akin to Scott's (1990) idea of the "hidden transcripts" developed by oppressed peoples to undermine public discourses upholding power structures. Ruybal and other women in her community took diverse stances towards food and were able, in Sandoval's words, to function "within yet beyond the demands of the dominant ideology" (1991: 3).

Women can develop differential consciousness in their relationship to food, as Ruybal did, by challenging the dichotomy between production and reproduction that has been so detrimental to women's social status.[12] As Engels originally pointed out in *The Origin of the Family, Private Property and the State*, and feminist anthropolo-gists have elaborated upon,[13] the splitting of production and reproduction led to the privatization and devaluation of women's labor both inside and outside the home, and, to quote Engels (1972: 120), the "world historical defeat of the female sex." Interpreting Engels, Eleanor Leacock (1972: 41) argued that a major force in the subjugation of women has been "the transformation of their socially necessary labor into a private service." This process has characterized much of women's food work with the global decline of subsistence farming and the separation of production and consumption, but women in Antonito have resisted it in several ways.

Ruybal pursued three strategies throughout her life that displayed differential con-sciousness and enabled her to overcome the production–reproduction dichotomy surrounding food. First, she rejected cooking as pillar of her own identity yet respected women who did it—especially her sister Lila. Second, she welcomed and legitimized her husband's cooking, and thus reduced the dichotomization of male and female labor. Third, she produced and sold *queso*—a fresh, white cow's milk cheese—and, thus, transformed kitchen work into paid, productive labor. In these ways, Ruybal minimized food's oppressive dimensions and enhanced its empowering ones. I focus on Ruybal's experience but place it into the broader cultural context by referring to other women I have interviewed in Antonito, some of whom shared Ruybal's strategies of publicly valuing women's domestic labor, enlisting men's help in the home, and making money from food. In contrast, other women in Antonito found cooking to be a symbol and channel of oppression (Counihan 2002, 2005). Food work offered diverse and conflicting avenues of self-realization for *Mexicanas* in Antonito as it has for women everywhere.

Antonito, Colorado

Antonito is a small town running six blocks east to west and twelve blocks south to north along U.S. Route 285 and State Highway 17 in Southern Colorado. Several Indian groups, especially the Ute, Navaho, and Apache, originally inhabited the region around Antonito in what is today Conejos County. This area was claimed by Spain until Mexican independence in 1821 and by Mexico until the Treaty of Guadalupe Hidalgo in 1848, when it became part of the United States. In the mid-1850s, the earliest Hispanic settlers immigrated from Rio Arriba County, New Mexico and settled in the agricultural hamlets of Conejos, Guadalupe, Mogote, Las Mesitas, San Rafael, San Antonio, Ortiz, La Florida, and Lobatos on the Conejos, San Antonio, and Los Pinos rivers. When the Denver and Rio Grande Railroad tried to build a depot in 1881 in the county seat of Conejos, landowners refused to sell their property, so the railroad established its station and a new town in Antonito.[14]

The town grew steadily due to its commercial importance, saw mills, perlite mines, ranching, and agriculture through and after World War II, with its population peaking at 1255 in 1950 and then dropping steadily to 873 in 2000. In the 2000 census, ninety percent of residents declared themselves "Hispanic." Today the town hosts a pharmacy, a locally owned supermarket, three restaurants, a seasonal food stand, two gas stations, a video store, a hair salon, and several gift and used-goods stores. The climate is cold, dry, and dusty with average annual rainfall a meager eight inches and only two frost-free months a year. Traditional agriculture and ranching depended on the complex ditch or *acequia* system that channeled water from the rivers into the fields, but today commercial agriculture relying on center-pivot irrigation is increasingly common.

Today, poverty is widespread in Antonito, Conejos County, and the predominantly Hispanic rural region of Northern New Mexico and Southern Colorado that Martínez (1998: 70) calls the *siete condados del Norte,* the seven rural counties of northern New Mexico and southern Colorado with Chicano/a majorities.[15] In the Antonito area, important employers are the town, the county, the perlite mine, the schools, the hospitals in La Jara and Alamosa, and the service economy in Alamosa thirty miles north. Many people get by on odd jobs, baby-sitting, trading in used goods, home health care work, public assistance jobs, and welfare. In the summer and fall there is a small tourist economy due to the popular Cumbres and Toltec Scenic Railroad, and to hunting, fishing, and vacationing in the nearby San Juan Mountains.

Helen Ruybal

Helen Gallegos Ruybal grew up with her parents, two brothers, and one sister in the small farming and ranching hamlet of Lobatos, four miles east of Antonito. They owned a modest five acres of land that Helen's mother inherited from her parents. Helen's father used the land to raise some crops and farm animals to provide for their subsistence. She said, *My father used to milk four or five cows, to get around, to get going. And we had two or three pigs and he took care of them and butchered them at the right time, and we had lots of pork.* Helen was not born into the local elite, called *ricos,* whom she defined as those having *money and ranches and animals and cows and water,* but she did achieve membership in the Hispanic elite through education, work, and accumulation of wealth.

Ruybal's parents followed the traditional division of labor: *he provided and she raised the children.* When her father was young, Ruybal said, *He was a common laborer.* But later he opened a small store and also taught school for a while. She remarked: *My father opened up a little convenience store just in a room of the house because the school was there and the kids would go buy candy and go buy peanuts. And they had cigarettes and tobacco and all those things . . . And my mother just cooked and sewed and raised the kids and put up the garden food.* Ruybal's mother was like most women in the community, including her sister Lila, whose primary work was gardening, preserving food, cooking for the family, caring for children, sewing, quilting, cleaning, washing clothes, and other forms of reproductive labor.

Rejecting Cooking, Respecting Women Who Cooked

But Ruybal diverged from the norm represented by her mother and sister and spent as little time as possible on domestic chores throughout her life, while maximizing her productive paid work outside the home. Her food-centered life history revealed both the tactics she followed and the ideologies she developed to support her choices—ideologies grounded in differential consciousness. Ruybal eschewed the housewife role and cooked as little as possible, but valued and benefited from the help of her mother, sister, and husband. Her strategy minimized the subordinating dimensions of reproductive labor.

Even as a girl Ruybal had ambitions beyond the traditional female role: *I wanted to be different. I wanted to go my own way,* she said. She aspired to *education, earning money, and doing some good to people.* Assistance from her sister and parents was critical to Ruybal's ability to study and work: *I had my likes and dislikes supported at home. . . . My parents were interested in education for all of us . . . and my folks believed in going without so we could have supplies and go to school and we never missed it.* By running a store and raising their own food, Ruybal's parents were able to send her to Loretto Academy, a Catholic boarding school in Santa Fe, New Mexico, where she completed high school and teacher certification. She returned home and began teaching while she went part-time and summers to Adams State College in nearby Alamosa. She achieved a BA in 1954, which enhanced her credentials and earning power. She was employed steadily, first in several different hamlets around Antonito, and later in the better paying Chama, New Mexico public schools forty miles away over the San Juan Mountains.

Ruybal's employment gave her financial independence, which meant that she did not have to marry for economic reasons as many girls did. Helen's future husband, Carlos Ruybal, courted her for years and both families supported the match, but Helen resisted marriage: *I wanted to be free to do what I wanted. . . . I didn't want to be tied down. . . . I didn't want to get married, and I refused to all the time, for the sake of not having a family to keep. . . . I wanted to work, and I felt like if I had children, I wasn't going to be able to work. I skipped marriage for a long time.*

Not only did she avoid marriage, but Ruybal also rejected cooking and the prominent role it played in many women's identity: *I never cooked you know. I was always a bookworm. Ever since I was a growing up.* For Ruybal food production, preparation, and clean-up were marginal activities she avoided if she could: *I'm not really a kitchen guy, you know what I mean, a provider in the kitchen.* Nonetheless, she could not escape cooking entirely but made it clear that she was a haphazard and indifferent cook. For example, one day she visited me and brought a gift of bread she had just made, saying, *Is it good? I thought it was kind of good. Sometimes it doesn't come out right. . . . I'm not a good cook [laughs], I'm not a steady cook.* Another time Ruybal spoke about making home-made tortillas, which she and everyone else in town thought were superior to store-bought ones, but she acknowledged her own uncertain skill: *Sometimes I make tortillas. And sometimes they come out good and sometimes they don't, not so good. And oh well.*

Ruybal eventually succumbed to cultural pressure to marry and she had to manage the household and the two children who came soon after her marriage. Crucially important was the support of her mother and sister: *My sister Lila was my right hand; she raised my kids. I'd come from my home, one mile, and I'd leave my kids there. What they didn't have, they had it there, and what they had, well they used it. She took care of them, fed them, and cleaned them up, and when I came in the evening I visited with her, and I picked them up, and I went home.*

Ruybal respected her sister's domestic identity, proficiency in the home, and accomplished cooking: *Lila had six children. . . . She used to sew and crochet and knit and make quilts, pretty ones. . . . And she cooked and she baked. . . . Her children still remember the jelly rolls, and they came out perfect like the ones in the store. . . . She used to make pies, a table full of pies, apple pies. . . . And she had such a good heart, and she was a good cook, she was a good housekeeper.*

Throughout her life Ruybal valued her sister and worked hard to stay on good terms with her: *We got along fine until she died. We were in favor of each other always, since we were growing up. . . . And we never got mad at each other, and we never got into a fight. . . . If it was for my side, she'd go out of her way to do it, and I'd go out of my way to appreciate it. I gave her a lot of things. . . . If she needed twenty dollars, I gave it to her. . . . I always would give her every gift like that, any amount. And she would accept it. . . . I had a good job in the first place, and I had less children, and more money, more money coming in. I was working and I couldn't miss a day and she never earned money. She just cooked, and washed and ironed, and took care of her kids and my kids.*

Even though she said her sister "just" cooked, Ruybal was able to appreciate and benefit from her sister's assumption of traditional female duties while at the same time she rejected them for herself. Not all women in Antonito were able to forge mutual respect out of difference, and public criticism of other women's choices was not uncommon. But Ruybal and her sister displayed differential consciousness by valuing and benefiting from each other's different choices vis-à-vis domestic labor and public work.

Blending Gender Roles: Involving Husband in Cooking

Ruybal's food-centered life history revealed how she improved her status by involving her husband Carlos in domestic chores and thus challenged the splitting of male and female labor so instrumental to women's subordination.[16] Ruybal did not marry until she was sure that Carlos would support her career as a teacher. She said, *He thought of me. If I was going to work, he didn't want to put any objections, just go ahead and work. And . . . the first thing, [my daughter] Carla came. And Carlos helped me a lot and I helped him a lot. . . . And then, not even two years later, [my son] Benito came. I wondered how far I was going that way. And I didn't want a large family. . . . So after that, well, we just didn't let our family grow bigger. . . . We were both combined. We both wanted the same thing.*

Because of her economic contributions to the marriage and busy work schedule, Ruybal was able to secure her husband's help at home and to skirt some of the domestic labor that fell to most women: *I never had to cook. . . . I had kids, but they went to boarding school. I had them in the summer and Carlos used to help me a lot. In fact, they'd be with him at the ranch. . . . He would [cook] when I wasn't home. On Fridays when I came from school he had supper ready. He did fried potatoes, he did fried beans, and he did everything fried quickly, because he didn't want to be at the stove watching it. . . . And he cooked and he had a good meal and I helped him too, we both cooked.* Carlos learned to cook as many men did during summers in the all-male sheep and cattle camps, but unlike most, he utilized his skills at home.

Ruybal described the prevailing gender ideology based on clear and separate male/female work and the differential strategies and consciousness that she—and some others—upheld: *People just wanted to go that a lady's job is a lady's job. . . . They didn't expect the wife to go out and plow the garden or to pick up the plants or brush. They didn't like for them to do men's jobs. . . . And a man's job would be a man's job. . . . But I knew husbands that did all the housework . . . and they took care of the babies, put them to bed and fed them, changed them, dressed them, and changed the diapers, and people would laugh at that, for me they did too. . . . People would be nasty about it— some would—they were jealous. . . . They didn't want men to be that soft and kind-hearted. . . . But others said, well, she deserves it, that he be considerate. She deserves that help. . . . They would consider it right, she deserves it if she works and earns the bread and butter, why not do the dishes for her and do the floor, and make the beds and things like that? And others would think that that's ladies work—make the beds and wash the floor. . . . Some men didn't do anything but eat and provide—provide flour, provide money, provide salt and pepper. All those things but they wouldn't do anything in the kitchen.*

Carlos liked to be a helper always. . . . I always had some other little thing to do, understand? I didn't have it to sit down here and watch. I did other things that had to be done, even little things and bigger things in the home, or in my job, my duty. Because I had to be prepared for that every day and I saw that I was before I tackled anything else. In the mornings, when I went to Taos [to teach], he'd get up early in the morning and run my car, warm and ready to go and he'd come in and prepare breakfast for me. . . . He saw me out and the dishes were left on the table and buying more bread was left on the table and he'd get those things ready for the week, he'd do it.

Ruybal's relatively egalitarian marriage went against the publicly stated value of men controlling family and budget. One man told me that he knew several marriages that fell apart when the women bettered themselves through education, attained jobs outside the home, and gained financial independence. Husbands did not always define their wives' economic success as a boon, but according to Ruybal, Carlos respected her brains and business acumen, and they worked together as successful business partners, with Helen bringing in a steady salary and Carlos managing the growing ranch. Their cooperation allowed them to maximize their economic position and accumulate land and cattle, attaining the status of *ricos*.

Ruybal's economic power outside the home raised her value in the home. Involving her husband in the family cooking and admiring him for it improved the status of food work, reduced her domestic workload, established reciprocity, and challenged the subordination implicit in the expectations that women feed and serve men (cf. DeVault 1991). In contrast, another Antonito woman, Bernadette Vigil, described how her Puerto Rican husband humiliated her by forcing her to cook rice his way, and threw her creations against the wall until she "*got it right*" (Counihan 2002, 2005). Vigil's was an extreme situation, and most women fell somewhere in between Ruybal and Vigil, cooking—sometimes willingly and sometimes not—and spending much of their time on domestic chores, especially feeding men and children.

Transforming Reproduction to Production: Making and Selling Cheese

Ruybal's food-centered life history detailed how she transformed food work from "private service" to public gain by selling cheese she made from the milk of the family's cows. She said, *For ten years, at least ten years, maybe more, I made cheese, white cheese. My husband and my son used to milk at the ranch and bring it from there to town. . . . Cheese was a luxury item, like ice cream on a cake. . . . Oh, that used to be my job, and I'd use that money for a lot of little things. Even big things, I'd just put it with the rest of the money. . . . That was a job, but I liked the idea, I didn't work hard. Even when I went to school, I'd leave the cheese hanging and I'd go away and come back and it was all ready to take it out and put it in the pan in the refrigerator. . . . As long as I had the milk, instead of throwing it away, I made cheese. That's what I did it for more than for an income. But I loved to get the money that I got from my work.*

Converting reproductive labor to productive work enhanced Ruybal's pride, money, and power—in her culture and in her marriage—and enabled her to develop differential consciousness towards food work as she simultaneously minimized its importance in her identity and valued its economic contribution. Many other women in Antonito used their food preparation skills to make money. For example, Ramona Valdez grew up on a ranch in Guadalupe with her parents and two siblings, and from the 1930s through the 1950s, she regularly made cheese and butter, which she sold for fifty cents a pound. Valdez also raised and sold turkeys. Through these activities, she was able to accumulate $800, a lot of money in the 1950s. Pat Gallegos made and sold cheese in the 1990s. Flora Romero was renowned for planning and cooking the food for funeral dinners and weddings. Gloria Garcia and Dora Sandoval both owned and ran local restaurants, and they catered weddings and parties as well. Several women made and sold tortillas, burritos, *empanaditas*, or tamales. Selling food in public gave what they ate in the home a monetary value and transformed food work from undervalued "reproduction" into remunerated "production." By holding differential attitudes towards these diverse forms of food preparation women were able to value their own work and that of others.

Conclusion

Ruybal's food-centered life history showed how she used food as a path to dignity and power, key issues in women's mediation of gender roles. She was among a minority of women in her community who achieved a college degree, a steady career, and a reduced domestic workload; nonetheless she was not unique but rather fell on a continuum of acceptable roles for women. Indeed almost all of the nineteen women I interviewed worked for money for varying periods of time. As they went in and out of the work force, their domestic roles contracted and expanded. Their experiences showed the permeability of the boundaries between public and private, production and reproduction, a permeability that some women, like Ruybal, were able to exploit to gain social prestige and economic power.

Ruybal's food-centered life history revealed her differential consciousness. She functioned "within yet beyond" dominant beliefs about women's food roles—*within* by valuing domestic labor and those who did it, *beyond* by curtailing cooking and using food to further her identity as a worker: *within* by recognizing the gender-dichotomized power structure of her culture, *beyond* by transcending gender oppositions. She, more than many women in her community, managed to shape "the relationship between women's reproductive and productive labor" (Moore 1988: 53), an essential step toward gender equality. Ruybal's flexible attitudes and activities surrounding food enabled her to be economically empowered and socially valued, and to attain the sense of belonging and respect that are hallmarks of what scholars have called cultural citizenship (Flores & Benmayor 1997).

Food-centered life histories are a valuable means to gather information that may otherwise be inaccessible (Hauck-Lawson 1998). They can reveal women's nutritional status, economic realities, psycho-emotional states, social networks, family concerns, and even spouse abuse (Ellis 1983). This information can buttress public policies relevant to women's needs, such as the WIC program, food stamps, and meals for senior citizens; small-business loans for women to start up food-based enterprises; and publicly funded child-care programs to permit women to work and attain parity with men.

I have used food-centered life histories to project Ruybal's voice into the public arena and counter the silencing that has been a central weapon in women's oppression (hooks 1989). Ruybal and other *Mexicanas'* food stories are *testimonios* that counteract erasure and affirm the value of women's labor, memory, and resourcefulness. They increase understanding of Chicanas' diversity in the United States, and challenge universalizing and demeaning portrayals (Zavella 1991).

Appendix 1 Food Centered Life Histories

Food-centered life histories consist of tape-recorded semi-structured interviews with willing participants, focusing on behaviors, experiences, beliefs, and memories centered on food production, preservation, preparation, cooking, distribution, and consumption. The following is a list of key topics presented
in condensed form. In an interview, questions are not nearly so condensed and they follow the flow of
conversation. Many lead naturally to further topics without prompting.

(1) Consumption
What are people/you/your family eating?
    (a)   Where do foods come from, local vs. imported foods?
    (b)   Diversity of cuisines?
    (c)   Vegetarianism?
    (d)   Picky eaters?
    (e)   Processed foods?
    (f)   Fast vs. slow foods?
    (g)   Dietary make-up?
    (h)   Nutritional composition?
    (i)   Seasonal, weekly variation?

Describe the quality of food.
Describe your meals: names, when, what, where, with whom.
    (a)   Describe eating at home.
    (b)   Describe eating out—school, daycare, restaurants, fast food, etc.

Describe the atmosphere and social relations of the eating experience.
Describe the current practices about outsiders eating in the home.
Describe the most important holidays and the role of food and commensality.
Do you know anyone with fussy eating habits, eating disorders, body image issues?
Do you know anyone suffering hunger, malnutrition, or food-related health problems?
How do individuals and the community deal with hunger and malnutrition?
Describe the relationship between food and health. Are foods used in healing?
Describe beliefs and practices surrounding eating in pregnancy.
Describe beliefs and practices surrounding eating during the post-partum period.
Describe the beliefs and practices surrounding infant feeding.
Over your lifetime, what are the most important changes in foodways? Their causes? Effects?
Describe outstanding food memories, good or bad.
Describe symbolic foods and their meanings.

(2) Production
How and by whom are foods produced, processed, and prepared?
Who cooks with what principal foods, ingredients, spices, and combinations?
What are some key recipes?
How are singles, couples or families handling the division of labor at home and at work?
    (a)   Who does the meal planning, shopping, cooking, serving, clearing, dish-washing?
    (b)   Who does other chores—bathroom, floors, clothes-washing, ironing, child-care?

How are boy and girl children being raised vis-à-vis food chores?
Describe the kitchen, place in the home, appliances, cooking tools, and technology.
Is there home gardening, canning, drying, freezing, brewing, baking, etc.? Recipes?
Describe the garden, layout, plants, labor, yearly cycle.

(3) Distribution
Describe your food acquisition.
Who procures food, by what means, where, when, and at what cost?
Do you shop in a grocery store, supermarket, farmers' market, coop, or CSA?
Is food exchanged or shared? How, with whom, when, why?
Is there a food bank, food pantry or soup kitchen in the community? Describe.

(4) Ideology
Describe food uses in popular culture, literature, films, art, advertising, music, etc.
Is food used in religion, magic, or witchcraft?

(5) Demographic Data:
Describe date of birth, marriage, children, parents, occupations, residences, etc.

Notes

1. An earlier version of this paper was published as "Food as Mediating Voice and Oppositional Consciousness for Chicanas in Colorado's San Luis Valley," in *Mediating Chicana/o Culture: Multicultural American Vernacular*, ed. Scott Baugh, Cambridge, England: Cambridge Scholars Press, 2006, pp. 72–84. I thank Scott Baugh for offering new ways of thinking about my work. I thank Penny Van Esterik for helpful comments and my husband, Jim Taggart, for his support and insights. I thank the people of Antonito for their hospitality and all the women who participated in my research for their generosity, especially Helen Ruybal, an extraordinary woman who passed away at the age of 100 in 2006. This paper is dedicated to her memory.

2. Direct quotations from my interviews with Helen Ruybal appear in italics throughout the paper.

3. Paredes (1976: xiv) defined Greater Mexico as "all the areas inhabited by people of a Mexican culture" in the U.S. and Mexico. See also Limón (1998). "*Mexicano*" is one of the most common terms that the people of Antonito use to describe themselves, along with Hispanic, Spanish, and Chicana/o. The life story of Helen Ruybal, like the lives of other women living in rural areas of Greater Mexico, differs in many ways from those of the urban Mexican and Mexican-American women of her generation explored by Ruiz (1993).

4. This book is under contract with the University of Texas Press.

5. See especially my book *Around the Tuscan Table: Food, Family and Gender in Twentieth Century Florence* (2004) as well as Counihan (1999, 2002, 2005).

6. See Behar and Gordon (1995), Gluck and Patai (1991), and Wolf (1992).

7. The American Anthropological Association Code of Ethics guided my research: http://www.aaanet.org/committees/ethics/ethcode.htm

8. On the food voice see Brumberg (1988), especially in her chapter "Appetite as Voice" reprinted in this volume. See also Avakian (1997) who collected personal accounts of cooking and eating from women of various class and ethnic groups. Thompson (1994) collected stories from eighteen women of color and lesbians who used food to cry out against racism, poverty, abuse, and injustice. Hauck-Lawson (1998) showed how one Polish American woman expressed through food her social isolation, depression, and declining self-image—issues that she was unable to speak about directly and that affected her health and diet.

9. Historically, the production, preservation and preparation of food were central to women's roles and identity in the Hispanic Southwest (Deutsch 1987). See Cabeza de Baca (1949, 1954), Gilbert (1942) and Jaramillo (1939, 1955) on the recipes, cooking, and culture of Hispanic New Mexico. Many of the Mexican American women interviewed by Elsasser et al. (1980) in New Mexico and by Patricia Preciado Martin (1992, 2004) in Arizona described foodways and dishes similar to those of Antonito. Williams (1985) and Blend (2001a, b) used a feminist perspective to uncover both the liberating and oppressive dimensions of women's food work and responsibility. Abarca (2001, 2004, 2006) used "culinary chats" and Pérez used "kitchen-table ethnography" to explore Mexican and Mexican-American women's diverse lives. See also Bentley (1998), Montaño (1992), Taggart (2002, 2003), and Taylor and Taggart (2003). For fascinating analyses of literary representations of Chicanas and food, see Ehrhardt (2006), Goldman (1992), and Rebolledo (1995).

10. Deutsch (1987: 11) wrote, "Written history of female minorities or 'ethnics' is rare, that of Chicanas or Hispanic women rarer though increasing, and of Chicanas or Hispanic women in Colorado virtually non-existent."

11. See Segura and Pesquera (1999) on diverse oppositional consciousness among Chicana clerical workers in California. Gloria Anzaldúa's "*Oyé como ladra: el lenguaje de la frontera*" is a wonderful example of differential consciousness expressed through language use (1987: 55–6).

12. Two recent discussions of Latinas' role in transnational food production that undermine the production/reproduction, male/female dichotomy are Barndt (2002) and Zavella (2002).

13. See Lamphere (2000), Leacock (1972), Moore (1988), Rosaldo (1974), Sacks (1974), Sargent (1981).

14. On the history, culture, and land use of the San Luis Valley, see Aguilar (2002), Bean (1975), Deutsch (1987), García (1998), Gutierrez and Eckert (1991), Martínez (1987, 1998), Peña (1998), Simmons (1979), Stoller (1982), Swadesh (1974), Taggart (2002, 2003), Taylor and Taggart (2003), Tushar (1992), Weber (1991), and Weigle (1975).

15. The *siete condados del norte* are Costilla and Conejos Counties in Colorado, and Taos, Río Arriba, San Miguel, Mora, and Guadalupe Counties in New Mexico (Martínez 1998: 70).

16. See Ybarra (1982) and Pesquera (1993) on the relationship of Chicanas' earning power and work to husbands' sharing of household labor.

References

Abarca, Meredith (2001) "*Los Chilaquiles de mi 'ama*: The Language of Everyday Cooking." In Sherrie A. Inness, ed., *Pilaf, Pozole and Pad Thai: American Women and Ethnic Food*. Amherst: University of Massachusetts Press, pp. 119–44.

—— (2004) "Authentic or Not, It's Original," *Food and Foodways* 12, 1: 1–25.

—— (2006) *Voices in the Kitchen: Views of Food and the World from Mexican and Mexican American Working-Class Women*. College Station: Texas A & M UP.

Aguilar, Louis (2002) "Drying Up: drought pushes 169-year-old family-owned ranch in southern Colorado to the edge of extinction." *The Denver Post*, July 29, 2002.

Anzaldúa, Gloria (1987) *Borderlands/La Frontera: The New Mestiza*. San Francisco: Aunt Lute.

Avakian, Arlene Voski (ed.) (1997) *Through the Kitchen Window: Women Explore the Intimate Meanings of Food and Cooking*. Boston: Beacon Press.

Barndt, Deborah (2002) *Tangled Routes: Women, Work, and Globalization on the Tomato Trail*. Lanham: Rowman and Littlefield.

Bean, Luther E. (1975) *Land of the Blue Sky People*. Alamosa, CO: Ye Olde Print Shoppe.

Behar, Ruth, and Deborah A. Gordon (eds) (1995) *Women Writing Culture*. Berkeley: University of California Press.

Bentley, Amy (1998) "From Culinary Other to Mainstream America: Meanings and Uses of Southwestern Cuisine." *Southern Folklore* 55, 3: 238–52.

Beverly, John (1993) *Against Literature*. Minneapolis: University of Minnesota Press.

Blend, Benay (2001a) "I am the Act of Kneading: Food and the making of Chicana Identity." In Sherrie A. Inness, ed., *Cooking Lessons: The Politics of Gender and Food*. New York: Rowman and Littlefield, pp. 41–61.

—— (2001b) " 'In the Kitchen Family Bread is Always Rising!' Women's Culture and the Politics of Food." In Sherrie A. Inness, ed., *Pilaf, Pozole and Pad Thai: American Women and Ethnic Food*. Amherst: University of Massachusetts Press, pp. 119–144.

Brumberg, Joan Jacobs (1988) *Fasting Girls: The Emergence of Anorexia Nervosa as a Modern Disease*. Cambridge: Harvard UP.

Cabeza de Baca, Fabiola (1982 [1949]) *The Good Life*. Santa Fe: Museum of New Mexico Press.

—— (1994 [1954]) *We Fed Them Cactus*. Albuquerque: University of New Mexico Press.

Counihan, Carole (1999) *The Anthropology of Food and Body: Gender, Meaning and Power*. New York: Routledge.

—— (2002) "Food as Women's Voice in the San Luis Valley of Colorado." In Carole Counihan ed., *Food in the USA: A Reader*. New York: Routledge, pp. 295–304.

—— (2004) *Around the Tuscan Table: Food, Family and Gender in Twentieth Century Florence*. New York: Routledge.

—— (2005) "Food as Border, Barrier and Bridge in the San Luis Valley of Colorado." In Arlene Voski Avakian, and Barbara Haber, eds, *From Betty Crocker to Feminist Food Studies: Critical Perspectives on Women and Food*. Amherst: University of Massachusetts Press.

Deutsch, Sarah (1987) *No Separate Refuge: Culture, Class, and Gender on an Anglo-Hispanic Frontier in the American Southwest, 1880–1940*. New York: Oxford.

DeVault, Marjorie (1991) *Feeding the Family: The Social Organization of Caring as Gendered Work*. Chicago: University of Chicago Press.

Ehrhardt, Julia C. (2006) "Towards Queering Food Studies: Foodways, Heteronormativity, and Hungry Women in Chicana Lesbian Writing." *Food and Foodways* 14, 2: 91–109.

Ellis, Rhian (1983) "The Way to a Man's Heart: Food in the Violent Home." In Anne Murcott, ed., *The Sociology of Food and Eating*. Aldershot: Gower Publishing, pp. 164–171.

Elsasser, Nan, et al. (eds) (1980) *Las Mujeres: Conversations from a Hispanic Community*. New York: Feminist Press.

Engels, Frederick (1972 [1942]) *The Origin of the Family, Private Property and the State*. New York: International.

Flores, William V., and Rina Benmayor (eds) (1997) *Latino Cultural Citizenship: Claiming Identity, Space and Rights*. Boston: Beacon.

García, Reyes (1998) "Notes on (Home) Land Ethics: Ideas, Values and the Land." In Devon Peña, ed., *Chicano Culture, Ecology, Politics: Subversive Kin*. Tucson: University of Arizona Press, pp. 79–118.

Gilbert, Fabiola Cabeza de Baca (1970 [1942]) *Historic Cookery*. Santa Fe: Ancient City.

Gluck, Sherna Berger, and Daphne Patai (1991) *Women's Words: The Feminist Practice of Oral History*. New York: Routledge.

Goldman, Anne (1992) " 'I Yam What I Yam': Cooking, Culture and Colonialism." In Sidonie Smith and Julia Watson, eds, *De/Colonizing the Subject: The Politics of Gender in Women's Autobiography*. Minneapolis: University of Minnesota Press, 169–95.

Gutierrez, Paul, and Jerry Eckert (1991) "Contrasts and Commonalities: Hispanic and Anglo Farming in Conejos County, Colorado." *Rural Sociology* 56, 2: 247–63.

Hauck-Lawson, Annie (1998) "When Food is the Voice: A Case-Study of a Polish-American Woman." *Journal for the Study of Food and Society* 2, 1: 21–8.

hooks, bell (1989) *Talking Back: Thinking Feminist, Thinking Black*. Boston: South End.

Jaramillo, Cleofas (1981) (orig. 1939) *The Genuine New Mexico Tasty Recipes*. Santa Fe: Ancient City Press.

—— (2000) (orig. 1955) *Romance of a Little Village Girl*. Albuquerque: University of New Mexico Press.

Lamphere, Louise (2000) "The Domestic Sphere of Women and the Public World of Men: The Strengths and Limitations of an Anthropological Dichotomy." In Caroline B. Brettell and Carolyn F. Sargent, eds, *Gender in Cross-Cultural Perspective*. Upper Saddle River: Prentice, pp. 100–109.

Latina Feminist Group (2001) *Telling to Live: Latina Feminist Testimonios*. Durham: Duke UP.

Leacock, Eleanor Burke (1972) " 'Introduction.' Frederick Engels." *The Origin of the Family, Private Property and the State*. New York: International, pp. 7–67.

Limón, José E. (1998) *American Encounters: Greater Mexico, the United States, and Erotics of Culture*. Boston: Beacon.

Marsh, Charles (1991) *People of the Shining Mountains: The Utes of Colorado*. Boulder: Pruett.

Martin, Patricia Preciado (1992) *Songs My Mother Sang to Me: An Oral History of Mexican American Women*. Tucson: University of Arizona Press.

—— (2004) *Beloved Land: An Oral History of Mexican Americans in Southern Arizona*. Collected and edited by Patricia Preciado Martin. With photographs by José Galvez, Tucson: University of Arizona Press.

Martínez, Rubén O. (1987) "Chicano Lands: Acquisition and Loss." *Wisconsin Sociologist* 24, 2–3: 89–98.

—— (1998) "Social Action Research, Bioregionalism, and the Upper Rio Grande." In Devon Peña, ed., *Chicano Culture, Ecology, Politics: Subversive Kin*. Tucson: University of Arizona Press, pp. 58–78.

Montaño, Mario (1992) *The History of Mexican Folk Foodways of South Texas: Street Vendors, Offal Foods, and Barbacoa de Cabeza*. Ph.D. Dissertation, University of Pennsylvania.

Moore, Henrietta (1988) *Feminism and Anthropology*. Minneapolis: University of Minnesota Press.

Paredes, Américo (1976) *A Texas-Mexican Cancionero*. Urbana: University of Illinois Press.

Peña, Devon (ed) (1998) *Chicano Culture, Ecology, Politics: Subversive Kin*. Tucson: University of Arizona Press.

Pérez, Ramona Lee (2004) "Kitchen Table Ethnography and Feminist Anthropology." Paper presented at the Association for the Study of Food and Society Annual Conference, Hyde Park, NY, June 12.

Pesquera, Beatríz M. (1993) " 'In the Beginning, He Wouldn't Lift Even a Spoon': The Division of Household Labor." In Adela Torre and Beatríz M. Pesquera, eds, *Building with Our Hands: New Directions in Chicana Studies*. Berkeley: University of California Press, 181–95.

Rebolledo, Tey Diana (1995) *Women Singing in the Snow: A Cultural Analysis of Chicana Literature*. Tucson: University of Arizona Press.

Rosaldo, Michelle Zimbalist (1974) "Women, Culture and Society: A Theoretical Overview." In Michelle Zimbalist Rosaldo and Louise Lamphere, eds, *Women, Culture and Society*. Stanford: Stanford University Press, pp. 17–42.

Ruiz, Vicki L. (1993) " 'Star Struck': Acculturation, Adolescence, and the Mexican American Woman, 1920–1950." In Adela Torre and Beatríz M. Pesquera, eds, *Building with Our Hands: New Directions in Chicana Studies*. Berkeley: University of California Press, 109–29.

Sacks, Karen (1974) "Engels Revisited: Women, the Organization of Production, and Private Property." In Michelle Zimbalist Rosaldo and Louise Lamphere, eds, *Women, Culture and Society*. Stanford: Stanford University Press, pp. 207–22.

Sandoval, Chela (1991) "U.S. Third World Feminism: The Theory and Method of Oppositional Consciousness in the Postmodern World." *Genders* 10, 1: 1–24.

Sargent, Lydia (ed.) (1981) *Women and Revolution: The Unhappy Marriage between Marxism and Feminism*. Boston: South End.

Scott, James C. (1990) *Domination and the Arts of Resistance: Hidden Transcripts*. New Haven: Yale UP.

Segura, Denise A., and Beatríz M. Pesquera (1999) "Chicana Political Consciousness: Re-negotiating Culture, Class, and Gender with Oppositional Practices." *Aztlán* 24, 1: 7–32.

Simmons, Virginia McConnell (1979) *The San Luis Valley: Land of the Six-Armed Cross*. Boulder: Pruett.

Sternbach, Nancy Saporta (1991) "Re-membering the Dead: Latin American Women's 'Testimonial' Discourse." *Latin American Perspectives* 70, 18, 3: 91–102.

Stoller, Marianne L. (1982) "The Setting and Historical Background of the Conejos Area." In Marianne L. Stoller and Thomas J. Steele, eds, *Diary of the Jesuit Residence of Our Lady of Guadalupe Parish, Conejos, Colorado, December 1871–December 1875*. Colorado Springs, CO: Colorado College, pp. xvii–xxviii.

Swadesh, Frances León (1974) *Los Primeros Pobladores: Hispanic Americans of the Ute Frontier*. Notre Dame: University of Notre Dame Press.

Taggart, James M. (2002) "Food, Masculinity and Place in the Hispanic Southwest." In Carole Counihan ed., *Food in the USA: A Reader*. New York: Routledge, pp. 305–13.

—— (2003) "José Inez Taylor." *Slow* 31: 46–49.

Taylor, José Inez, and James M. Taggart (2003) *Alex and the Hobo: A Chicano Life and Story*. Austin: University of Texas Press.

Thompson, Becky W. (1994) *A Hunger So Wide and So Deep: American Women Speak Out on Eating Problems.* Minneapolis: University of Minnesota Press.

Tushar, Olibama López (1992) *The People of El Valle: a History of the Spanish Colonials in the San Luis Valley.* Pueblo: Escritorio.

Weber, Kenneth R. (1991) "Necessary but Insufficient: Land, Water, and Economic Development in Hispanic Southern Colorado." *Journal of Ethnic Studies* 19, 2: 127–142.

Weigle, Marta (ed.) (1975) *Hispanic Villages of Northern New Mexico.* A Reprint of vol. II of the 1935 Tewa Basin Study, with Supplementary Materials. Edited, with introduction, notes and bibliography by Marta Weigle. Santa Fe, NM: The Lightning Tree.

Williams, Brett (1985) "Why Migrant Women Feed Their Husbands Tamales: Foodways as a Basis for a Revisionist View of Tejano Family Life." In Linda Keller Bown and Kay Mussell, eds, *Ethnic and Regional Foodways in the United States.* Knoxville: University of Tennessee Press, pp. 113–126.

Wolf, Margery (1992) *A Thrice Told Tale: Feminism, Postmodernism, and Ethnographic Responsibility.* Stanford: Stanford University Press.

Ybarra, L. (1982) "When Wives Work: The Impact on the Chicano Family." *Journal of Marriage and the Family* 44: 169–178.

Zavella, Patricia (1991) "Reflections on Diversity among Chicanas." *Frontiers* 12, 2: 73–85.

—— (2002) "Engendering Transnationalism in Food Processing: Peripheral Vision on Both Sides of the U.S.-Mexican Border." In C. G. Vélez-Ibañez, and Anna Sampaio, eds, *Transnational Latina/o Communities: Politics, Processes, and Cultures.* Lanham: Rowman. pp. 225–45.

23

Rooting Out the Causes of Disease: Why Diabetes is So Common Among Desert Dwellers

Gary Paul Nabhan

From the land of the Navajo, let us go southward into Mexico once again, to a coastal community of another indigenous people. Although genetically unrelated, the Navajo of the United States and the Seri of Mexico share a problem that has both a genetic and a nutritional component: adult-onset diabetes. This nutrition-related disease is one of the three top causes of death among these two Native American groups and among many other indigenous communities as well. Ironically, a half century ago, its presence as a health risk was so minor in these communities that more Indians were dying each year of accidental snake bite than of diabetes. To understand why that change occurred, and what it means for all of us, we must listen not just to epidemiologists, but to the native peoples themselves.

It was in a small, run-down health clinic on a beach of Mexico's Sea of Cortés that an Indian elder gave me a memorable lesson about gene-food interactions. It was a lesson nested in place—the hot desert coastline studded with giant cactus; that particular Indian village, where people cooked most of their food on small campfires in the sandy spaces between shabby government-built houses; and in that clinic, with no windows and no equipment, so rarely frequented by a doctor that we had planted a garden of healing herbs around it in case there was ever a medical emergency. It was in this place that Seri Indian Alfredo López Blanco challenged me—and Western-trained scientists in general—to pay protracted attention to diet change and its role in disease.

I had accompanied my wife Laurie Monti, nurse-practitioner turned medical anthropologist, who was screening Seri families for adult-onset diabetes. The disease was already running rampant through neighboring tribes, but because the Seri are the last culture in Mexico to have retained hunting, fishing, and foraging traditions instead of adopting agriculture, there was some hope that they could stave it off. Only a few of the some 650 tribal members had ever been screened for the non-insulin-dependent form of diabetes, and that smaller, earlier sample had suggested that only 8 percent of the tribe suffered from chronically high blood-sugar levels and low insulin sensitivity.

While Laurie was screening Seri families in the sole office that contained any semblance of sanitary surfaces, I was in the "waiting room"—a sort of stripped-down echo chamber full of barking dogs and crying babies—trying to interview the elders of each family about their genealogical histories. I was attempting to ascertain whether the genetic susceptibility to diabetes of individuals with 100 percent Seri ancestry might be different from those who claimed that some of their ancestors came from among the neighboring Pima and Papago (O'odham) tribes in Arizona, the ethnic populations reputed to have the highest incidence of diabetes in the world.

Alfredo López Blanco returned to the waiting room after Laurie confirmed that his blood-sugar levels were unusually high. Alfredo, who had worked as a fisherman since he was a boy, had late in life become boatman and guide for marine and island biologists. In his late sixties, Alfredo often taught younger Seri about the days when their people had subsisted on seafood, wild game, and desert plants like cactus fruit and mesquite pods. He was keenly aware of the traditional diet of his own people, and of his neighbors as well. When he sat down with me, I asked him if any of his forefathers happened to be from other tribes. He answered that one of his great-grandmothers was from a Papago-Pima community.

"But *Hant Coáaxoj*," he called me by my Seri nickname, Horned Lizard, "I have a question for you. What does that have to do to my diabetes?"

"Well, I'm not yet sure that it does. But here's why I'm asking. The Pima and Papago suffer from diabetes more than any other tribes. It might be in their blood," I conjectured, groping for a way to explain the concept of *genetic predisposition* to a person whose native language does not contain the exact concept of "genes." "If people have Pima blood in them, maybe they are more prone to diabetes."

"*Hant Coáaxoj*," he said dryly, "sometimes you scientists don't know much history. If diabetes is in their blood—or for that matter, in our blood—why did their grandparents not have it? Why were the old-time Pima and Papago who I knew skinny and healthy? It is a change in the diet, not their blood. They are no longer eating the bighorn sheep, mule deer, desert tortoise, cactus fruit, and mesquite pods. *Pan Bimbo* bread, Coke, sandwiches, and *chicharrones* are the problem!"

The old man—whose sister died within a year of that conversation due to circulatory complications from her own diabetes—was pretty much right on the mark. Or at least that is what Laurie's interpretation of her screening and my genealogical interviews later showed. Diabetes, aggravated by diet change, was clearly on the rise among the Seri, with more than 27 percent of the adults screened by Laurie showing impaired glucose tolerance. But there were also interesting differences between the village where Alfredo lived, Punta Chueca, and the more remote Seri village to the north, Desemboque, where Western foods and other signs of acculturation were much less prominent. While diabetes prevalence in Desemboque had only recently reached 20 percent of the adults in Laurie's sample, it exceeded 40 percent in Punta Chueca.

Other public-health surveys of the Seri suggested why this might be the case. Punta Chueca's residents had easier access to fast-food restaurants and minimarts than did Desemboque dwellers. Roughly 15 percent more of Punta Chueca's residents consumed groceries purchased in nearby Mexican towns on a daily basis, rather than relying more heavily on native foods from the desert and sea. The people of Punta Chueca consumed significantly more store-bought fats (such as lard), alcohol, and cigarettes.

When data from both villages were pooled, Seri individuals with some Papago-Pima ancestry did not show up as suffering from diabetes any more than those with

100 percent Seri ancestry. And yet, comparing the villages, there was one telling difference: those with Papago-Pima ancestry who ate more acculturated, modernized diets in Punta Chueca had the highest probability of the disease. As long as the Desemboque dwellers with Papago-Pima blood remained close to their traditional diet, diabetes among them was held more in check.

This trend held even though traditional Seri individuals in Desemboque appeared to weigh somewhat more than their counterparts in Punta Chueca. This suggests that it may not be the sheer quantity of food metabolized that triggers diabetes as much as the qualities of the foods the Seri now eat—especially the kinds of fats and carbohydrates regularly consumed.

This key distinction has slipped past the U.S. National Institutes of Health Indian Diabetes Project in the Sonoran Desert, which for nearly four decades has spent hundreds of millions of dollars trying to identify the underlying cause of the diabetes epidemic among the Pima and other indigenous communities. Its scientists and educators have all but ignored qualitative differences between Native American diets, preferring to seek a quick genetic fix to everyone's problem at the same time. Several years ago, *New Yorker* writer Malcolm Gladwell called it the "Pima paradox": "All told, the collaboration between the NIH and the Pima is one of the most fruitful relationships in modern medical science—with one fateful exception. After thirty-five years, no one has had any success in helping the Pima lose weight [and control diabetes]. For all the prodding and poking, the hundreds of research papers describing their bodily processes, and the determined efforts of health workers, year after year the tribe grows fatter."

At most, the NIH epidemiologists have quantified how the contemporary Pima and their Indian neighbors eat more fast foods than ever before, especially ones detrimentally high in animal fats and simple sugars. But what the NIH has failed to discuss with Native Americans are the countless studies, including my own collaborations with nutritionists, that demonstrate how traditional diets of desert peoples formerly *protected* them from diabetes and other life-threatening afflictions now known as Syndrome X. This Syndrome X is not some sinister new disease, but rather a cluster of conditions that, when expressed together, may reflect a predisposition to diabetes, hypertension, and heart disease. The term—first coined by members of a Stanford University biomedical team—describes a cluster of symptoms, including high blood pressure, high triglycerides, decreased HDL ("good" cholesterol), and obesity. These symptoms tend to appear together in some individuals, increasing their risk for both diabetes and heart disease. And of course, all of these symptoms are influenced by diet, but what kind of diet most effectively reduces their expression was something that I seemed more interested in than anyone at the NIH or at Stanford.

In the 1980s, I began to collect traditionally prepared desert foods for nutritional analysis by Chuck Weber and Jim Berry in their University of Arizona Nutrition and Food Sciences lab, and for glycemic analysis by Jennie Brand-Miller and her colleagues who had already done similar work analyzing the desert foods traditionally consumed by Australian aborigines. By glycemic analysis, I refer to a simple finger-prick test for blood-sugar and insulin levels done as soon as a particular food is eaten, and every half hour afterwards; the test determines whether the food in question causes blood-sugar levels to rapidly spike after its ingestion, thereby causing pancreatic stress and asynchronies with insulin production.

Jennie Brand-Miller, a good friend as well as colleague, determined with her students that native desert foods—desert legumes, cacti, and acorns in particular— were so slowly digested and absorbed that blood-sugar levels remained in sync with insulin production, without any adverse health effects generated. Jennie called these native edible plants "slow-release foods" to contrast them with spike-inducing fast foods such as potato chips, sponge cakes, ice cream, and fry breads. The fast foods had glycemic values two to four times higher than the native desert foods, whose slow-release qualities Weber and Berry had shown to be derived from the foods' higher content of soluble fiber, tannins, and complex carbohydrates.

Jennie had found the same trend when comparing Western fast foods with the native desert foods that Aussies call "bush tuckers"—the mainstays of aboriginal diets up until a half century ago, before which diabetes was virtually absent in indigenous communities of Australia. As with the desert tribes of North America, once these protective foods were displaced from aboriginal diets, the incidence of diabetes skyrocketed.

Back on an autumn night in 1985, Jennie and I were sipping prickly pear punch, having spent the day comparing the qualities of Australian and American desert foods. I could see that she was brewing over some large question, and she finally teased it out.

"Gary, I've wondered if there might be some explanation [for why desert peoples are vulnerable to diabetes, other than what the NIH promotes], one that you as a desert plant ecologist might help me figure out. I don't know if I'm framing this question precisely enough, but let me give it a try: is there something that helps a number of desert plants adapt to arid conditions which might help control blood-sugar and insulin levels in the humans that consume them?"

"*What?*" I blurted out. "Could you say that again?" Much later, I thought of a famous comment about the heart of science: "Ask an impertinent question and you are on your way to a pertinent answer."

Jennie laughed, aware that she was asking a question far too complex to consider in the midst of the frivolity of a dinner party. "Oh, that's OK," she said quietly. "I just wondered if desert plants from around the world could have evolved the same protective mechanism against drought that somehow. . . ."

"Oh, I think I get it now, some kind of convergent evolution," I said. "If the same drought-adapted chemical substances show up in plants from various deserts that are scattered around the world, perhaps these substances formerly protected the people who consumed them from the risk of diabetes . . ." Then, once diets changed, the desert peoples who once had the best dietary protection from diabetes suddenly had their genetic susceptibility expressed!

Although Jennie posed it in passing, I could not forget her impertinent question, not that night, not that week, and not for a long time. Friends like Gabriel, as well as Alfredo López's sister, Eva, had died of diabetes, but they still inhabited my memory. I mused over Jennie's question whenever I was out studying plants in the desert, and I brought it to the attention of some physiological ecologists who had a far deeper understanding of plant adaptations to drought than I did. They reminded me that desert plants and animals adapted to drought conditions by many different anatomical, physiological, and chemical means, and that there was probably not a single protective substance found in all arid-adapted biota.

In other words, the flora and the fauna from different deserts emphasized distinctive sets of these adaptive strategies. It was simply too much for these ecologists to imagine that a cactus from American deserts and a *wichitty* grub from the Australian outback might all share some dietary chemical that controlled diabetes among the Pima, Papago, and Seri of American deserts as well as among the Warlpiri and Pinkjanjara of Australian deserts.

Still, Jennie's question was rooted in a valid observation: there was an apparent correlation between the extraordinarily high susceptibility of diabetes among desert peoples and the quantity of drought-adapted plants in their diets. If some aboriginal cultures had subsisted on drought-adapted plants and associated wildlife for upwards of 40,000 years, was it not plausible that these people's metabolisms had adapted to the prevailing substances in these foodstuffs? And if, within the last fifty years, the prevalence of these foodstuffs had declined precipitously in their diets, was it not just as plausible that they had suddenly become susceptible to nutrition-related diseases because they had *lost their protection*? The question to pursue, then, was what dietary chemicals—nutrients or even antinutritional factors—might be more common in drought-adapted plants than those occurring in wetter environments?

With the help of ecophysiologist Suzanne Morse, I tried to imagine how water loss from a plant's tissue was slowed by the adaptations developed by a desert-dwelling organism to deal with scant and unpredictable rainfall. At the time, I was involved in a number of field evaluations of drought tolerance in desert legumes, cacti, and century plants. I soon learned that prickly pear cactus pads contain *extracellular mucilage*, that is, gooey globs of soluble fiber that holds onto water longer and stronger than the moisture held within photosynthesizing cells. If a cactus is terribly stressed by drought, it may shut down its photosynthetic apparatus, shut its stomatal pores, and shed most of its root mass, going "dormant" until rain returns. But if stress is not so severe, the cactus will instead gradually shunt the moisture in its extracellular mucilage into photosynthetically active but water-limited cells, thereby slowing the plant's total water loss while keeping active tissues turgid.

In explaining this concept—called "leaf capacitance"—to me, Morse offered me a parallel to slow (sugar) release foods: slow (water) release plant tissues. The very mucilage and pectin that slow down the digestion and absorption of sugars in our guts are produced by prickly pears to slow water loss during times of drought. And prickly pear, it turns out, has been among the most effective slow-release foods in terms of helping diabetes-prone native peoples slow the rise in their blood-glucose levels after a sugar-rich meal. In fact, it was among the first foods native to the Americas demonstrated to lower the blood glucose and cholesterol of indigenous people susceptible to diabetes. As Morse and I followed up on that research, we documented that most of the twenty-two species of cacti traditionally used by the Seri have the same slow-release qualities and are available along the desert coast much of the year.

Soon, Jennie Brand-Miller, in Sydney, and Boyd Swinburn, an endocrinologist from New Zealand, gave me greater insight into how slow-release foods differ from conventional foodstuffs in they way they are digested and absorbed. As I read reports about the "low gastric motility" of slow-release foods, I began to imagine how these foods make a viscous, gooey mass in our bellies. Even when our digestive juices cleave them into simpler sugars, the sugars have a tough time moving through the goo to reach the linings of our guts, to be absorbed and then transported to where they fuel our cells.

Here then, in the prickly pear—one of the food plants in the Americas with the greatest antiquity of use—was the convergence that Jennie had been seeking: the existence of slow-water-release mucilage in cactus pads and fruit explained why desert food plants were likely to produce slow-sugar-release foods. Five years after our conversation over prickly pear punch, I found a potential answer in the very plant Jennie and I had been consuming at the time she asked her impertinent question! The trouble was, prickly pear and other cacti are not native to Australian deserts; I began to investigate if there were plants in other deserts that also contained slow-water-release mucilages.

I soon learned that cacti are not special cases that occur only in the diets of desert-dwelling Native Americans; there are dozens of other plants in both American and Australian deserts that have similar slow-sugar-release/slow-water-loss qualities, albeit with different morphologies and different chemical mechanisms. Given that desert peoples have been exposed to such plants for upwards of 10,000 years—more than 40,000 years in Australia—is there any evidence that these people's metabolisms have adapted over time to the presence of these protective foods?

With regard to the Seri, the only general genetic survey comparing them to neighboring agricultural tribes indicates that the Seri exhibit "several micro-polymorphisms [that] may be important in conferring a biological advantage" in their desert coastal homeland. The study claimed that "these may emphasize the relevance of interactions between genes and environment," for Seri hunter-gatherers express several alleles not found in more agriculture-dependent U.S. and Mexican indigenous peoples (Infante et al. 1999).

But do long-time hunter-gatherers with such polymorphisms respond to certain desert and marine foods differently than other people do? The answer can be found in research that Jennie and colleagues have done contrasting various ethnic populations' responses to foods common to one group's traditional diet, but not the other's. As Jennie and her fellow researcher Anne Thorburn have explained:

> the aim of [our] next series of experiments was to compare the responses of healthy Aboriginal and Caucasian subjects to two foods, one a slow release Aboriginal bush food—bush potato (*Ipomoea costata*)—and the other a fast release Western food—[the domesticated] potato (*Solanum tuberosum*). Both Aborigines and Caucasians were found to produce lower plasma insulin responses to the slow release bush food than to the fast release Western food. But the differences were more marked in Aborigines, with the areas under the glucose and insulin curves being one-third smaller after bush potato than potato (Brand-Miller and Thorburn 1987).

In other words, the Aborigines were protected from diabetic-inducing pancreatic stress by a bush food that their metabolisms had genetically adapted to over 40,000 years. Caucasians, with hardly any exposure to this or similar bush foods since colonizing Australia, did not experience such marked benefits.

When many scientists learn of these differences, they recall the theory of a *thrifty gene* that indigenous hunter-gatherers are presumed to maintain as an adaptation to a feast-or-famine existence, and they attribute the differences in insulin response to that gene. As originally hypothesized by James Neel in 1962, hunter-gatherers were likely to exhibit a thrifty genotype that was a vestigial survival mechanism from eras during which they suffered from irregular food availability. "During the first 99 percent or more of man's life on earth while he existed as a hunter-gatherer," Neel wrote, "it was often feast or famine. Periods of gorging alternated with periods of greatly reduced food intake" (Neel 1962).

Neel persuasively argued that repeated cycles of feast and famine over the course of human evolution had selected for a genotype that promoted excessive weight gain during times of food abundance and gradual weight loss of those "reserves" during times of drought. Neel focused on food quantity—the evenness of calories over time—and not food quality, arguing that when former hunter-gatherers were assured regular food quantities over time, the previously adaptive genetic predisposition to weight gain became maladaptive.

However, the only early NIH attempt to characterize the diets of Pima women with traditional versus acculturated (modern) lifestyles found insignificant differences between the calorie amounts consumed by the two groups, nor was there much difference when both groups' diets were compared to what surrounding Anglo populations ate. In other words, despite Neel's hypothesis, food quantity alone did not account for the rise in diabetes among acculturated Pima Indian women.

Nevertheless, Neel's argument has been cited by hundreds of scientific papers on diabetes and other diseases and has reached millions of other readers through "popular science" magazine essays written by such science-literate writers as the *New Yorker's* Malcolm Gladwell, *Harper's* Greg Cristner, *Outside's* David Quammen, and *Natural History's* Jared Diamond. What's more, Neel's hypothesis essentially drove the first thirty-five years of research at the NIH Indian Diabetes Project in Phoenix, Arizona, whose director and staff set their sights on becoming the first to discover *the* thrifty gene. Hundreds of millions of research dollars later, it is clear that their focus on a single gene and on sheer food *quantity* has blinded researchers to a variety of gene-food-culture interactions that may trigger or prevent diabetes.

Thirty-six years after proposing his famous hypothesis, Neel himself conceded that "the term 'thrifty genotype' has [already] served its purpose, overtaken by the growing complexity of modern genetic medicine," adding that while type 2 diabetes may still be "a multifactorial or oligogenic trait, the enormous range of individual or group socioeconomic circumstances in industrialized nations badly interferes with an estimate of genetic susceptibilities" (Neel 1998).

Neel's colleagues in biomedical research are much more direct in their assertion that there is no single thrifty gene that confers susceptibility to type 2 diabetes among all ethnic populations, or even among all hunter-gatherers. Assessing the recent identification of several genes that heighten or trigger diabetes, geneticist Alan Shuldiner of the University of Maryland School of Medicine told *Science News*, "I expect there would be dozens of diabetes-susceptibility genes [and that] specific combinations of these genes will identify risk" (Seppa 2002).

What these genes actually do is also different from what Neel and other proponents of the thrifty genotype suspected they would do. When the NIH worked to determine whether *the* thrifty gene they had identified in the Pima was actually a gene for insulin resistance—which causes reduced metabolic sensitivity to sugar loads—researchers found this gene's true function to be weight maintenance and *not* weight gain.

As molecular biologist Morris White of the Joslin Diabetes Center recently concluded in the pages of *Science*, "We used to think type 2 diabetes was an insulin receptor problem, and it's not. We used to think it was solely a problem of insulin resistance, and it's not. We used to think that muscle and fat were the primary tissues involved, and they are not. Nearly every feature of this disease that we thought was true 10 years ago turned out to be wrong" (White 2000).

Once again, it was my friend Jennie Brand-Miller who hammered the coffin closed on the thrifty gene hypothesis by refuting its very underpinnings—that famines were more frequent among hunter-gatherers than among agriculturists, leading to the former's extraordinary capacity to accumulate fat reserves. In scanning the historic anthropological literature on periodic famine and starvation among various ethnic groups, Jennie and her colleagues found scant evidence that hunter-gatherers suffered from these stresses anywhere near as frequently as agriculturalists did. In fact, periodic starvation and wide-spread famines increased in frequency less than 10,000 years ago, after various ethnic groups became fully dependent on agricultural yields. In particular, Jennie noted, since Caucasians living in Europe have repeatedly suffered from famines in historic times, they ought to be predisposed to insulin resistance and diabetes if Neel's hypothesis is correct. And yet, Caucasians are one of the few groups that do not exhibit much insulin resistance or heightened susceptibility to type 2 diabetes when they consume modern industrialized agricultural diets.

"The challenge," Jennie and her colleagues argue, "is to explain how Europeans came to have a low prevalence and low susceptibility to adult-onset diabetes ..." (Cordain et al. 2000). Indeed, Europe harbors most of the world's ethnic populations who have *not* suffered dramatic rises in this nutrition-related disease since 1950.

At an international workshop that Jennie and I hosted at Kims Toowoon Bay on the coast of New South Wales in May of 1993, we elucidated four factors that could explain why individuals of European descent appear to be less vulnerable to Syndrome X maladies—including diabetes—than do ethnic populations that have adopted agricultural and industrial economies more recently. With colleagues from four countries, including Australian Aborigines and Native Americans, we identified that the incidence of diabetes rapidly increases under the following four circumstances.

First, when an ethnic population shifts to an agricultural diet and abandons a diverse cornucopia of wild foods, its members lose many secondary plant compounds that formerly protected them from impaired glucose tolerance. This is particularly true for populations that have coevolved with a certain set of wild foods over millennia, ones that are rich in antioxidants.

Second, when the remaining beneficial compounds in traditional crops and free-ranging livestock are selected out of a people's diet through breeding and restricted livestock management practices, their diet is further depleted of protective factors. For instance, modern bean cultivars have been bred to contain less soluble fiber, and livestock raised on cereal grains under feedlot conditions lack omega-3 fatty acids.

Third, the industrial revolution that began in Europe in the seventeenth century changed the quality of carbohydrates in staple foods by milling away most of the fiber in them. High-speed roller mills now grind grains into easily digested and rapidly absorbed cereals and flours, which results in blood-sugar and insulin responses two to three times higher than those reported from whole grains or coarse-milled products like bulgur wheat.

Fourth, the last fifty years of highly industrialized foods has introduced additives such as trans-fatty acids, fiber-depleted gelatinous starches, and sugary syrups, which ensure that most fast foods are truly *fast-release* foods. Jennie estimates that the typical fast-food meal raises blood-sugar and insulin levels three times higher than humans ever experienced during preagricultural periods in our evolution. Combined with the trend toward oversize servings of convenience foods and a more sedentary lifestyle, the dominance of fast foods in modern diets has made contemporary humans less fit than ever.

Although nearly all ethnic populations have come to suffer from fast foods over the last quarter century, the other changes took place in European societies over thousands of years. Whereas the genetic constituency of European peoples may have slowly shifted with these technological and agricultural changes as they emerged, the Seri and Warlpiri have had less than fifty years to accommodate these changes, and their genes are not in sync with them. Significant adaptation through evolutionary processes to new diets rarely occurs over the course of two to three generations.

And yet, most people now living in the world fall somewhere between the French and German farmers on the one hand, and the Seri and Warlpiri hunter-gatherers on the other. The majority of traditional diets have historically been more like the Pima and Papago in the Arizona deserts, where perhaps 60 percent of foods were harvested from domesticated crops in wet years while the rest came from wild and weedy plants and free-ranging game or fish. In dry years, the Papago-Pima diet shifted more toward the reliable harvests of drought-tolerant wild perennials.

While details richly vary around the world—from coastal habitats where fish were once abundant to rain forests where birds and root crops proliferated—most indigenous peoples in developing countries have maintained, until recently, a healthy mix of wild foods and diverse cultivated crops. Today, following dramatic economic shifts that have favored a few cereal grains and livestock production for export over mixed cropping, the bulk of the world's population has been left vulnerable to diabetes. One recent reckoning suggests that upwards of 200 million people are now susceptible to diabetes and the other killers associated with Syndrome X. This is not the exception among the diverse peoples of the world; it is a pathology that has become the norm.

But while fast foods lead to rapid deterioration of healthy carbohydrate metabolism in most people—with or without the existence of a thrifty gene—a return to the traditional foods of one's own ancestry leads to rapid recovery. This is what New Zealand endocrinologist Boyd Swinburn found when he asked me to help him reconstruct a semblance of the nineteenth-century dietary regime for the Pima and Papago. Swinburn wanted to compare the effects of a traditional versus a fast-food diet, both consisting of the same number of calories and percentages of carbohydrate and fats.

When twenty-two Pima Indians in his study were exposed to the fast-food diet, their insulin metabolism deteriorated enough to trigger diabetic stress without the need to conjure up any other explanation to explain it. Yet when the same individuals were placed on the traditional diet rich in soluble fiber and other secondary plant compounds, their insulin sensitivity and glucose tolerance improved. Swinburn and his coworkers concluded that "the influence of Westernization on the prevalence of diabetes may in part be due to changes in dietary composition [as opposed to food quantity]" (Swinburn et al. 1993).

I followed Swinburn's clinical study with a demonstration project at the National Institute for Fitness outside St. George, Utah, where eight Pima, Papago, Hopi, and Southern Paiute friends suffering from diabetes came together for ten days of all-you-can-eat slow-release foods and outdoor exercise. Within ten days, their weight and their blood-sugar levels had been dramatically reduced, and everyone felt healthier. The changes began so immediately that several participants had to seek medical advice to figure out how to reduce the hypoglycemic medications they had been self-administering for years.

In yet another example, in what may be one of the most dramatic gains in health conditions ever witnessed in a short period of time, Kieran O'Dea documented the marked improvement in diabetic Australian aborigines after they reverted for a month to a nomadic foraging lifestyle in western Australia. Even though study subjects "poached" several free-ranging cows as part of their meat consumption, their diet primarily consisted of bush foods that their ancestors had long eaten. The aboriginal participants moved frequently to take advantage of hunting and plant-gathering opportunities, and they lost considerable weight while doing so.

Their consumption of calories from macronutrients was 54 percent protein, about 20 percent plant carbohydrates, and 26 percent fat. These proportions had a dramatic effect on lowering blood-sugar levels and increasing insulin sensitivity. While some critics have conjectured that their insulin sensitivity, glucose tolerance, and cholesterol levels improved merely because of the subjects' weight loss, others have pointed out that the ratio of macronutrients they consumed certainly did not worsen their condition. While not necessarily optimal for all ethnic populations, a diet with this mixture of macronutrients clearly brought health benefits to the Australian desert dwellers that participated.

Inspired by O'Dea's collaboration with indigenous sufferers of diabetes, I organized a similar moveable feast in the spring of 1999, engaging more than twenty Seri, Papago, and Pima individuals who also suffered from diabetes. We walked 220 miles through the Sonoran Desert during a twelve-day pilgrimage, fueled only by native slow-release foods and beverages. Although we did not measure our blood-sugar and insulin levels each day to compare our health status before and after our journey, we took note of something perhaps far more significant: the native foods we ate were considered by all the participants to be nutritious, satisfying, and filling enough to sustain our arduous pilgrimage. These foods enabled us to hike across rugged terrain for ten hours a day, followed by another hour or two of celebratory dancing. Our collective effort made us more deeply aware that our own energy levels could be sustained for hours by slow-release foods. At the same time, we took a good hard look at the health of our neighbors and of the land itself. The pilgrimage allowed us to clearly see for the first time all the damage that had been done to our homeland and its food system, damage that was echoed in our very own bodies.

There was something else going on among my Native American companions during that walk. The Seri, Papago, and Pima pilgrims frequently expressed that their cultural pride, spiritual identity, and sense of curiosity were being renewed. And so, a return to a more traditional diet of their ancestral foods was not merely some trip to fantasy land for nostalgia's sake; it provided them with a deep motivation for improving their own health by blending modern and traditional medical knowledge in a way that made them feel *whole*. They were not eating native slow-release foods merely to benefit a single gene—thrifty or not. Instead, they were communing to keep their entire bodies, their entire communities, and the entire Earth healthy.

Yes, genes matter, but diverse diets and exercise patterns matter just as much. And when the positive interaction among all three of these factors is reinforced by strong cultural traditions, our physical health improves, as does our determination to keep it that way. The Native American folks I walked with on that pilgrimage have re-doubled their commitments to keep their traditional slow-release foods accessible in their communities; they serve them at village feasts and at wakes honoring those who have succumbed to the complications of diabetes for lack of earlier access to these foods. When the persistence of traditional foods is more widely recognized as a source of both cultural pride and as an aid to physical survival and well-being, I doubt that many Native American communities will abandon what many of them feel to be a true gift from their Creator.

References

Brand-Miller, J. C., J. Snow, G. P. Nabhan, and A. S. Truswell (1990) Plasma glucose and insulin responses to traditional Pima Indian meals. *American Journal of Clinical Nutrition* 51:416–420.

Brand-Miller, J. C., and A. W. Thorburn (1987) Traditional foods of Australian aborigines and Pacific Islanders. In *Nutrition and health in the tropics*, ed. C. Rae and J. Green, 262–270. Canberra, Australia: Menzies Symposium.

Brand-Miller, J. C., and S. Colaguiri (1994) The carnivore connection: Dietary carbohydrate and the evolution of NIDDM. *Diabetologica* 37:1280–1286.

—— (1999) Evolutionary aspects of diet and insulin resistance. In *Evolutionary aspects of nutrition and health: diet, exercise, genetics, and chronic disease*. Vol. 84 of *World review of nutrition and diet*. Basel, Switzerland: Karger.

Cordain, L., J. C. Brand-Miller, and N. Mann (2000) Scant evidence of periodic starvation among hunter-gatherers. *Diabetologica* 24 (3): 2400–2408.

Cowen, R. (1991) Desert foods offer protection from diabetes. *Science News* 32:12–14.

Diamond, J. (1992) Sweet death. *Natural History* 2:2–7.

Gladwell, M. (1998) The Pima paradox. *New Yorker*, February 2, 41–53.

Infante, E., A. Olivo, C. Alaez, F. Williams, D. Middleton, G. de la Rosa, M. J. Pujo, C. Duran, J. L. Navarro, and C. Gorodezky (1999) Molecular analysis of HLA class I alleles in Mexican Seri Indians: Implications for their origin. *Tissue Antigens* 54:35–42.

Nabhan, G. P. (2004) *Cross-pollinations: The marriage of science and poetry*. Minneapolis: Milkweed Editions.

Nabhan, G. P., C. W. Weber, and J. Berry (1979) Legumes in the Papago-Pima Indian diet and ecological niche. *Kiva* 44:173–178.

Neel, J. V. (1962) Diabetes mellitus: A "thrifty genotype" rendered detrimental by progress. *American Journal of Human Genetics* 14 (4): 353–362.

—— (1998) The "thrifty genotype" in 1998. *Perspectives in Biology and Medicine* 42:44–74.

O'Dea, K. (1984) Marked improvement in carbohydrate metabolism in diabetic Australian aborigines after temporary reversion to traditional lifestyle. *Diabetes* 33:596–603.

Seppa, N. (2002) Gene tied to heightened diabetes risk. *Science News* 158:212.

Swinburn, B. A., V. L. Boyce, R. N. Bergman, B. V. Howard, and C. Borgardus (1993) Deterioration in carbohydrate metabolism and lipoprotein changes induced by a modern, high fat diet in Pima Indians and Caucasians. *Journal of Clinical Endrocrinology and Metabolism* 73 (1): 156–164.

Villela, G. J., and L. A. Palinkas (2000) Sociocultural change and health status among the Seri Indians of Sonora, Mexico. *Medical Anthropology* 19:147–172.

Weber, C. W., R. B. Arrifin, G. P. Nabhan, A. Idouraine, and E. A. Kohlhepp (1996) Composition of Sonoran Desert foods used by Tohono O'odham and Pima Indians. *Journal of the Ecology of Food and Nutrition* 26:63–66.

White, M. (2000) New insights into type 2 diabetes. *Science* 289:37–39.

24

Slow Food and the Politics of Pork Fat: Italian Food and European Identity

Alison Leitch

This paper explores the emergence of the Slow Food Movement, an international consumer movement dedicated to the protection of "endangered foods." The history of one of these "endangered foods," lardo di Colonnata, provides the ethnographic window through which I examine Slow Food's cultural politics. The paper seeks to understand the politics of "slowness" within current debates over European identity, critiques of neo-liberal models of rationality, and the significant ideological shift towards market-driven politics in advanced capitalist societies.

In April 1998 I returned to Carrara in central Italy where, a decade earlier, I had conducted ethnographic research on the subject of craft identity among marble quarry workers and the history of local labor politics (Leitch 1993). I hoped to renew my associations with local families and update my previous research by revisiting the quarries, reinterviewing marble workers I knew, and tracking any other significant transformations to the local marble industry.

Pulling out my notebook as I arrived in Milan's Malpensa airport, I began to scribble some initial impressions. Perhaps because I had been away for so long, I was struck by the overtly transnational space of the airport itself. With public announcements made in four languages, it was an explicitly modern European frontier. It became more so during the rush hour bus ride towards the city, surrounded by wildly gesticulating drivers all conversing on mobile phones. However, when I noticed the advertisement for McDonald's printed on the back of the bus ticket— "Buy one: get one Free"—I was slightly taken aback. I could not recall much fondness amongst Italians for such a marked category of American fast food culture, yet soon enough we passed a "McDrive." Adding to my initial disorientation were the visual manifestations of other recent changes in Italian national politics, with fading posters promoting Umberto Bossi's Northern League and its call for the formation of a separate regional entity called Padania, as well as those for Silvio Berlusconi's Forza Italia party. My sense that the cultural and political landscape had indeed changed in ten years was later confirmed in conversations with Milanese friends, who admitted to me that they themselves were confused. The old categories of "left" and "right" in their imagination had somehow merged or become indistinct and, among other things, they mentioned the growing popularity of New Age philosophies amongst their friends.

I arrived in Carrara to find the marble industry in crisis, with the price of high-quality stone at its lowest point and unemployment figures at their highest in ten years. However, much to my surprise, I found that a much humbler, and decidedly more proletarian local product had become newly controversial: pork fat, locally known as *lardo di Colonnata*, had apparently been nominated as the key example of a nationally "endangered food" by an organization called Slow Food.

At the time of my original fieldwork, neither *lardo* nor Slow Food had particularly high media profiles. Indeed, my own interest in the subject of pork fat stemmed from local reverence towards such an obviously, elsewhere, despised food, one often associated, for example, with the notion of fat as "poison" in modern American diets (Rozin 1998; Klein 1996).[1] Every summer since the mid-1970s a festival dedicated to this specialty had been held in Colonnata, a tiny village located at the end of a narrow, winding mountain road traversing one of the three marble valleys of Carrara. And during the years I spent in Carrara, *lardo*-tasting visits to Colonnata became one of the ways I entertained foreign visitors who, more often than not, registered the appropriate signs of disgust at the mere mention of feasting on pork fat. However by the late 1990s, Colonnata had become a major destination for international culinary tourism. Venanzio, a restaurant named eponymously after its owner, a local gourmet and *lardo* purveyor, was one attraction, but Colonnata's pork fat was also being promoted with great acclaim by Pecks, Milan's epicurean mecca. Moreover, it had even been nominated as a delicious, albeit exotic, delicacy by writers as far afield as the food columns of *The New York Times* (*La Nazione* 18/2/1997) and *Bon Appétit* (Spender 2000).

Similarly, the organization now called Slow Food had limited public visibility in the mid-1980s. Founded by Carlo Petrini, a well-known food and wine writer associated with specific elite intellectual circles of the 1960s Italian left, it was known then as a loose coalition opposed to the introduction in Italy of American-style fast-food chains. This relatively small group achieved some initial national notoriety in the context of a spirited media campaign waged in 1986 against the installation of a new McDonald's franchise near the Spanish Steps in Rome. And in 1987, taking a snail as its logo, Slow Food emerged with its first public manifesto signed by leading cultural figures of the Italian left outlining its dedication to the politics and pleasures of "slowness" and its opposition to the "fast life."

In the late 1980s, Slow Food began to intervene within the growing circuits of a vigorous national debate concerning the widening application of new uniform European Union food and safety legislation. Theoretically designed as a measure of standardization for the European food industry, this legislation threatened the production of artisanal foods linked to particular localities and cultural traditions. Thus, whereas elsewhere in the world, anxiety about the homogenizing practices of post-industrial capitalism had taken the form of movements around endangered species, environments and people, in Europe there was a growing concern for "endangered foods." In 1989 Petrini launched the International Slow Food Movement at the Opéra Comique in Paris, and during the 1990s, Slow Food gradually developed into a large international organization, now in 83 countries.

From its inception Slow Food mixed business and politics. In Italy, for example, it developed an extremely successful commercial wing publishing books and travel guides on cultural tourism, food and wine. It has also initiated taste education pro-grams in primary schools and has recently proposed a university of gastronomy. An explicit organizational strategy has been the cultivation of an international net-work of journalists and writers. To this end Slow Food sponsors a star-studded annual food award—a food Oscar—that recognizes outstanding contributions to international food diversity. Although its headquarters remain in Bra, a small Piedmont town of about 70,000 people where Petrini grew up, an indication of Slow Food's institutional and economic weight may be glimpsed in its rapid expansion, with additional offices opening in Switzerland, Germany, New York and most recently in Brussels, where it lobbies the European Union on agriculture and trade policy.

What accounts for this current explosion of public interest in European food politics? There is, of course, a rich body of literature on the potency of food as a political symbol particularly in periods of great economic and social change. Indeed one need only recall the unfortunate consequences of Marie Antoinette's remark about eating cake in the context of a ferocious battle over the production of bread in Paris. Similarly, in the late 19th century, when socialism vied with republicanism as feudalism finally gave way to industrial capitalism, attempts to raise the bread tax in Italy provided the impetus for wide-scale revolts against the monarchy. Social historians researching in this arena have been influenced in particular by E.P. Thompson's (1971) path-breaking study of English food rioters in pre-industrial England; he argued that peasants protesting the rising price of bread were responding not just to increased economic hardship but to the aban-donment of a "just price" system which guaranteed prices on certain basic com-modities for the poor in the feudal economy. A common thread in many sub-sequent historical studies is that food protests, disturbances, and other forms of collective action around food are often motivated by ideas of social justice within moral economies, rather than more pragmatic concerns such as hunger or scarcity (Hobsbawm 1959; Gailus 1994; Gilje 1996; C. Tilly 1975; L. Tilly 1983; Taylor 1996; Orlove 1997).

Food and other items of consumption have also been central as cultural symbols in colonial and post-colonial nationalist struggles. In colonial America tea took on a radical symbolic function uniting colonists of different classes and regions, to eventually become a catalyst for boycotts, riots and even revolution (Breen 1988; Bentley 2001). Under British rule, Ghanian elites increasingly turned from European to African foods as an expression of nationalist sentiment (Goody 1982). In Mexico, corn, a product which was associated with the peasantry and denigrated by colonial elites as nutritionally inferior to wheat, later became central to the development of a national cuisine (Pilcher 1998). Similarly in Algeria, French bread is imbued with complex meanings reflecting post-colonial ambivalence (Jansen 2001). Equally numerous are examples of the political appropriation of food as a symbol of col-lective or contested national identity. Familiar recent cases include the wide-scale Indonesian protests in 1998 over IMF demands to remove subsidies on basic food items such as oil and rice; the 1990s protests by the French over American tariffs on *foie gras*, and, of course, McDonalds as a focal protest symbol for anti-globalization activists.

My assumption in this paper is that deepening concerns in Europe over food policy are linked to questions of European identity, indeed with moral economies and with the imagination of Europe's future as well as its past. Nadia Seremetakis, for example, has discussed how the disappearance of specific tastes and local material cultures of production accompanying widening European Union regimes of standardization constitutes a massive "reorganization of public memory" (1994:3), a rationalizing project which potentially limits the capacity of marginalized rural communities to reproduce themselves as active subjects of history. The Slow Food movement, with its emphasis on the protection of threatened foods and the diversity of cultural landscapes is, perhaps, one response.

In the context of transformations to the global economy, these debates are inevitably also caught up in what has been called the politics of risk discourse (Beck 1986). Issues such as the introduction of genetically modified foods and crops, the widespread use of antibiotics and growth hormones in animal fodder, the spread of diseases such as Bovine Spongiform Encephalopathy (BSE, colloquially known as mad cow's disease), the 1999 Belgium chicken dioxin scandal or the more recent foot and mouth scare in Britain, are now central topics of conversation in most European nations. I would suggest that public anxiety over these risks, both real and imagined, is symptomatic of other widespread fears concerning the rapidity of social and economic change since the Maastricht Treaty of 1992. In sum, food and identity are becoming like the "Euro," a single common discursive currency through which to debate Europeanness and the implications of economic globalization at the beginning of the twenty-first century.

Two other sets of questions also frame this inquiry. One concerns the genesis of the Slow Food movement within the broader context of transformations to Italian political life over the past two decades. This is a period associated with a significant decline in the cultural influence of political parties, labor unions and the Catholic Church, institutions that until recently have been the recognized political voice for collective interests in Italian society. It is also a period marked by tremendous economic growth, the expansion of commercially organized leisure and the passage of cultural power into the hands of the economic elite. Coinciding with these trends has been the rapid emergence of an influential independent non-profit sector of the economy and the development of new civil spaces fostering alternate forms of civic associationism. The appearance of the Slow Food movement at this specific historical conjuncture must, therefore, be tracked in relation to these more general transformations to Italian institutional politics and cultural life.

Finally my analysis is positioned alongside recent attempts to understand consumption as a relatively new ethnographic arena for the analysis of the capitalisms of late modernity. If we understand the global economy of the early twenty-first century to be an "economy of signs" (Lash & Urry 1994; Baudrillard 1981), where the symbolic and aesthetic content of commodities has become increasingly important, then potentially new relationships may be created between consumption and the market. There is already ample ethnographic evidence demonstrating the influence of global cultural shifts in consumer taste on the organization of production (Blim 2000b; Mintz 1985; Roseberry 1996; Heyman 1997; Hernandez & Nigh 1998; Schneider 1994). More polemically, Daniel Miller (1995; 1997) has suggested that consumption has displaced production as the new "vanguard" of the late-capitalist motor and that understanding the practices of consumption cross-culturally may reveal new roles for consumers as international political actors.

Theorists of new social movements also argue for the priority of culture and "symbolically defined action spaces" (Eder 1993:9) as the basis for understanding new forms of collective action in post-industrial societies. In a somewhat problematic, though intriguing analysis of new forms of middle-class consciousness, Klaus Eder has argued that middle-class collective action often manifests itself in struggles for an "identitarian" form of social existence, involving the idea of an authentic life-form where people interact as equals (1993:181). These kinds of struggles for identity and expressive social relations can, of course, also occur within the market.[2] The growth of consumerist forms of identity-production in liberal democratic societies thus coincides with the development of new possibilities for consumer politics in which culture has become a favored idiom of political mobilization.[3]

Leaving aside for the moment these larger questions, let me now return to the local ethnographic context. The first section of this essay is grounded in what might be termed a phenomenology of pork fat.[4] In other words, I am interested in exploring the meanings of pork fat for local people and how these meanings may have changed in relation to its later appropriation as a key symbol of an "endangered food" for the Slow Food movement. I focus on this example since it fortuitously coincides with my previous research and also because it provides an ethnographic window onto the promotional politics and origins of the Slow Food movement outlined in the second section.

Eating *Lardo*

At the time of my original research pork fat was not commonly eaten in any of the households I regularly visited for meals. *Lardo* was, however, almost always nominated in the oral histories I collected detailing the conditions of work over past generations. Many households maintained small vegetable gardens, which kept them going during periods of unemployment, and some households, with access to land, kept pigs or cows. One of the by-products of these pigs, *lardo* or cured pork fat, thus constituted a kind of food safe for families in the region and was an essential daily source of calorific energy in the quarry worker's diet. Like sugar and coffee, *lardo* was a "proletarian hunger killer" (Mintz 1979). Eaten with a tomato and a piece of onion on dry bread, it was a taken-for-granted element in the worker's lunch. *Lardo* was thought to quell thirst as well as hunger and was appreciated for its coolness on hot summer days. Given its dietary importance, it is perhaps not surprising that it was also adopted as a cure for any number of health ailments from an upset stomach to a bad back.

Apart from its nutritional or curative value, *lardo* can easily be seen as the perfect culinary analog of a block of marble (Leivick 1999). Firstly, both materials convey parallel ideas of metamorphosis. Elsewhere I have written about the ways in which quarry workers utilize organic metaphors to talk about the tranformative properties of the stone (Leitch 1993, 1996, 1999). *Lardo* embodies similar ideas of meta-morphosis. It is transformed from its natural state as pig fat through the curing process, and this process is also narrated as one that encapsulates ideas of craft and individual skill. Secondly, although recipes for *lardo* vary in their finer details, marble, preferably quarried from near Colonnata at Canalone, is always cited as an essential production ingredient. Its qualities of porosity and coolness are vital, especially because *lardo* makers do not use any kind of preservative apart from salt. Apparently the crystalline structure of Canalone marble allows the pork fat to "breathe," while at the same time containing the curing brine. If at any stage the *lardo* goes bad, it is simply thrown out. Just like marble workers who have often suggested to me that marble dust is actually beneficial to the body because it is "pure calcium," *lardo* makers say that the chemical composition of marble, calcium carbonate, is a purificatory medium which extracts harmful substances from pork fat, including cholesterol.

The curing process begins with the raw fat, cut from the back of select pigs. It is then layered in rectangular, marble troughs resembling small sarcophagi, called *conche*. The *conche* are placed in the cellar, always the coolest part of the house. The majority of these cellars are quite dank and mouldy. Some still contain underground cisterns, which in the past supplied water to households without plumbing. Once placed in the troughs, the pork fat is covered with layers of rock salt and variety of herbs, including pink-jacketed garlic, pepper, rosemary and juniper berries. Finally, a small slab of bacon is placed on top to start the pickling process, and six to nine months later it is ready to eat. Translucent, white, veined with pink, cool and soft to touch, the end product mimics the exact aesthetic qualities prized in high quality marble.

But *lardo* is of course more than just marble's visual culinary analog. For local people *lardo* is deeply reminiscent of a shared past characterized by poverty and food scarcity. In diets distinguished by protein scarcity, *lardo* was an essential calorific food for men who, in the past, labored up to fifteen hours a day cutting and hauling huge blocks of marble. To eat *lardo*, especially in the carnevalesque space of an annual festival, where hundreds of kilos of pork fat are consumed over four hot days in late August, is to remember and celebrate this past as collective history and cor-poreal memory. This is a performance of sensuous display and consumption where the skin, fat, and flesh of *lardo* is counterpoised to that of its consumers. The juxtaposition of two kinds of beautiful bodies and flesh, *lardo* and human, rephrases, or resculpts, two kinds of smooth, sensuous, luxuriousness: *lardo* and marble. And so just as *lardo* tastes of marble, it also mimics it. Through the curing process, *lardo* and marble metaphorically become one and the same. Through its physical incorporation, memories of place and self are actually ingested.

The Politics of Pork Fat

The events that led to Slow Food's declaration of *lardo di Colonnata* as an "endangered food" began two years before my return to Carrara. In March of 1996 the local police force had descended on the Venanzio Restaurant in Colonnata, "the temple of *lardo*" (*La Nazione* 1/4/1996). Protected by the constabulary, local health authority personnel proceeded to remove several samples of Venanzio's *lardo* and subsequently placed all of his *conche* under quarantine. Later, samples were also taken from several other small *lardo* makers in the village, but Venanzio and one other wholesaler, Fausto Guadagni, were singled out for special attention.

This action led to a barrage of media commentary that soon reached the national dailies. At the local level, the main preoccupation was the possible threat to the 1996 *lardo* festival. Nationally, the quarantine and subsequent application of new European hygiene legislation led to debates over the power of the European Union to regulate Italian food production and determine Italian eating habits. The *lardo* quarantine controversy also provided the perfect media opportunity for the political aspirations of the Slow Food Movement. According to Carlo Petrini, coinage of the term "endangered food" dated to the mid-1990s, just before the *lardo* controversy erupted. Up until then, Slow had been perceived by the public as an association of gourmets mostly concerned with the protection of national cuisines. But by the mid-1990s—a period which coincided with a number of high-profile food scares in Europe and public loss of trust in national food regulatory authorities—Slow began to imagine itself as an international organization concerned with the global protection of food tastes.

For Slow, *lardo di Colonnata* became the example par excellence in a long list of "endangered foods" which included, for example, red onions from Tropea in Calabria, an ancient legume from the region of Le Marche called *la cicerchia*, and a plum and apricot hybrid called *il biricoccolo*. Several "endangered foods" were imagined as under threat, from trends towards farming monocultures, from the disintegration of traditional rural foodways, from pollution of waterways, or from the dearth of alternate distribution networks. In the case of *lardo*, salamis and cheese, the threat was standardization and the imposition of new hygiene legislation, which would considerably diminish the economic viability of many of these artisanal producers.

Pork fat was singled out for a number of other reasons apart from timing. Firstly, due to the success of the *lardo* festival over twenty years and the promotional efforts of people like Venanzio, it had already acquired a certain exotic caché, especially among a group of celebrity chefs to whom Venanzio himself was connected. More importantly, however, *lardo* presented an unambiguous test case for new European Union hygiene rules, which insisted on the utilization of non-porous materials in food production. Although there are certainly good techniques for sterilizing the *conche*, marble is porous and its porosity is clearly essential to the curing process as well as to *lardo*'s claims to authenticity. Local *lardo* makers involved in this dispute thus had a vested interest in lobbying for exceptions to the generic rules designed for large food manufacturers. Their interests coincided perfectly with Slow Food's own political agenda, in particular its campaign to widen the debate over food rules to include cultural issues.

Slow Food's appropriation of *lardo di Colonnata* as a key symbol of its "endangered foods" campaign also had great rhetorical value. In the numerous publicity materials that subsequently appeared in the press, Petrini often likened the protection of pork fat made by local people in dank and mouldy cellars to other objects of significant national heritage, including major works of art or buildings of national architectural note. In valorizing the traditional techniques of *lardo* producers, Petrini was rhetorically distancing his organization from accusations of gourmet elitism, while simultaneously challenging normalizing hierarchies of expert scientific knowledge, including those of the European health authorities. In this kind of strategic symbolic reversal, the food artisan is envisaged not as a backward-thinking conservative standing in the way of progress, but rather, as a quintessential modern subject, a holder par excellence of national heritage.

Ironically, the publicity surrounding these events subsequently amplified into yet another threat: copying. Much to the dismay of local *lardo* wholesalers, big butcheries from all over Italy began manufacturing a product, which they also called *lardo di Colonnata*. When I visited them in 1998, the *lardo* makers in Colonnata were lobbying regional politicians to protect the name of *lardo* through its nomination as *Denominazione d'Origine Protetta* (D.O.P.), a label which would demonstrate that *lardo* was entirely produced in the village of Colonnata. Alternately, if this failed, they wanted the Tuscan regional government to approve the title of *Indicazione Geografica Protetta* (I.G.P.), a less onerous label indicating that the raw material used in *lardo* production is derived from a circumscribed area around the village of Colonnata.

When I returned to Carrara the following year, in 1999, Fausto despondently told me that while they had failed to obtain the protection of a collective trademark for *lardo*, eleven individuals, including himself, had managed to acquire the legal copyright to the name *lardo di Colonnata*. More recently again, this group has formed a legal association—*Associazione Tutela Lardo di Colonnata*—specifically for the protection of the name of their product. Though they own the legal title, not all members of this group actually produce pork fat, while others outside this original group are now no longer technically entitled to sell a product with the name *lardo di Colonnata*. Nevertheless, an ongoing legal battle still rages between the original group of eleven and outside butcheries who have formed a rival group. The city of Carrara has now established a working study group to deal with the controversy, funding the publication of further books, articles and scientific reports on the subject of pork fat, while the village of Colonnata has sponsored a sculpting competition and plans to erect a marble statue of a pig in its main square.

Clearly, there is a deeply ironic conclusion to be drawn from this brief account. Partly as a consequence of Slow Food's promotional campaign, a food which was once a common element in local diets and an essential source of calorific energy for impoverished quarry workers, has been reinvented and repackaged as an exotic item for gourmet consumption. A product associated with a distinct social history and corporeal memory is now privately patented by a group of people who may be entitled to sell the recipe. As I have noted elsewhere (Leitch 2000), this is a story not of the "invention of tradition" (Hobsbawm & Ranger 1983) but of its commodification. The story speaks to how memory replaces tradition as we move from modernity into post-modernity, a process which writers on other culture industries, such as art and music, have tracked as the commodification of nostalgia (Feld 1995).

Apart from its obvious success in the niche marketing of "endangered foods," a project Petrini has recently dubbed "eco-gastronomy" (Stille 2001), Slow Food clearly defines itself as an organization devoted to cultural politics. However, the fact that it so closely engaged in promotional activities which have far-reaching commercial consequences for direct producers raises the question of articulating what kind of politics it actually advocates. What accounts for Slow Food's rapid expansion and evolution as an influential consumer organization in Italy? Why has food in particular become such a nodal point in recent debates over national and European identity? In order to address these questions I now turn to a more detailed consideration of the specific cultural context which partly facilitated Slow Food's emergence as a consumer organization devoted to a new politics of food and pleasure.

The Politics of Pleasure: Food and the Italian Left[5]

In Italian *gola* means "throat" as well as the "desire for food." Although it is commonly translated as "gluttonous," implying a negative state of excess or greed, to be *goloso* has a more positive connotation of craving with pleasure a particular food. As Carole Counihan (1999:180) aptly observes, because *gola* implies both "desire" and "voice," it suggests that desire for food is a voice—a central vehicle for self-expression in Italian cultural life. *La Gola* is also the name of a journal dedicated to epicurean philosophy, first published in the early 1980s by a group of Italian intellectuals, including Carlo Petrini.[6] In turn, Slow Food grew out of a previous organization called *Arci Gola*, where Arci was an acronym for Recreative Association of Italian Communists. First formed in 1957 to counter the influence of ENAL, the state recreational organism that supplanted the Fascist OND at the end of the war, Arci circles quickly evolved in the 1960s and 70s, becoming integral to the Italian Communist Party's political agenda on leisure, youth culture and consumer society (Gundle 2000).

Carlo Petrini's own intellectual biography was forged within the milieu of Arci circles and younger leftish critiques of the Italian Communist Party.[7] Born in 1948 in the city of Bra, Petrini grew up in what he describes as a middle-class family, the son of a teacher and an artisan whose own parents also had long attachments to the region. Politically and culturally this area has strong connections to the Italian aristocracy, as well as deeply entrenched working-class traditions, particularly left-Catholicism. Once noted as a centre for the leather industry, the town's main industries are now the production of laminated plastics and agricultural machinery. Made famous in the literature of distinguished literary figures such as Cesare Pavese, the area surrounding the city known as *Le Langhe* is also acclaimed for its fine quality agricultural produce, truffles, and for the production of one of Italy's most prestigious wines, *Barolo*.

Petrini first studied to become a mechanic but later enrolled in sociology at the University of Trento in a department that, not incidentally, was widely noted for training many of the more prominent leftist leaders of the 1970s extra-parliamentarian groups. Upon graduating, he returned to his hometown, where he became active in local cultural politics, founding one of Italy's first radical-left pirate radio stations called *Radio Bra Onde Rosse* or "Red Waves." This was also the period in which Petrini recalls first becoming involved with a group of friends interested in gastronomy, some of whom would eventually pioneer the core-founding groups of Arci Gola in the early 1980s. According to Petrini, he and his friends were motivated by a desire to create a less hierarchical, youthful alternative to the existing gourmet associations, which they viewed as linked to chauvinistic and elitist cultural politics. Later, Petrini developed his professional credentials as a cultural critic, becoming a self-taught food and wine expert writing for the national media, including *L'Espresso*, a widely circulating national current affairs weekly.

Historically, haute cuisine, consumption and the pursuit of pleasure have not been associated with the cultural imagination of the authoritarian Italian left, particularly the Italian Communist Party. As food journalist Fabio Parasecoli (2000) astutely notes, even during the 1960s, a period of enormous economic and cultural upheaval in Italy, Communist Party events were mostly noted for their extreme asceticism and overtones of Catholic morality. He further suggests that in Communist party discourse and practice, *haute cuisine* was regarded with special contempt as a marker of bourgeois decadence and even amongst avant-garde intellectuals and artists of the era, food was often treated with a great deal of suspicion. But by the mid-1980s, a period in which the cultural heritage of the old and new left was significantly undermined, this dominant image of rigid austerity had undergone a marked reversal. Many public intellectuals, most notably journalists, increasingly turned towards the language of consumption as a form of transformative cultural politics.

For example, from the end of 1986 the independent communist daily newspaper, *Il Manifesto*, began publishing an eight-page monthly "lifestyle" supplement entitled *Gambero Rosso*. According to Parasecoli, *Gambero Rosso* literally translates as the "red shrimp/prawn" but has a further double cultural reference to Italy's most noted political morality tale, *Pinocchio*. *Gambero Rosso* refers here to the name of the tavern where the infamous pair of unscrupulous thugs—the cat and the fox—con the unsuspecting Pinocchio out of his gold coins. The obvious mission of the supplement was intended to "protect Pinocchio's real-life counterparts—innocents abroad as well as trusting customers at home—from finding themselves at the mercy of padded bills, lumpy beds, gruff service, watery wine, or mediocre food" (Parasecoli 2000:7). At the same time the name was apparently also an ironic wink by a younger generation of leftist intellectuals to the supposed dangers of communism—the "red menace"—and the cultural politics of the pleasure-allergic left parties, particularly the PCI.[8]

Ideology coincided with profitability and the magazine subsequently became an enormous publishing success.[9] In part this was due to the media's increasing influence in political communication during the 1980s. For the first time in Italian post-war political history, television and the press assumed a more central role in galvanizing public opinion, shaping political conflict and conveying more information to the public than political parties. Independent newspapers, including *Il Manifesto*, *La Repubblica* and *L'Espresso*, even began intervening in the internal and external media circuits of the Italian Communist Party. At the same time, within the PCI there were substantial generational conflicts over the redefinition of what became known as "the ephemeral," those aspects of popular culture such as music, cinema and sport, which were of consuming interest to contemporary Italian youth, but which were seen by older PCI leaders as superfluous to the party's historic revolutionary project (Gundle 2000: 165–193; Grossi 1985).

The attitude of these younger leftist intellectuals towards what might be termed a new politics of pleasure, was certainly also linked to more general transformations within Italian society during these years. This was a period marked by tremendous economic growth, the rapid expansion of commercially organized leisure and the passage of cultural power into the hands of the economic elite (Gundle 2000; Forgacs 1990). As many scholars of contemporary Italy have argued, these years saw the emergence of a new social paradigm and the parallel growth of new privately oriented individualism; this created the conditions for the advance of commodification and the affirmation of post-industrial capitalism. The victory of capital was thus accompanied by a "cultural revolution" at a mass level (Gundle 2000; Ginsburg 1989; Asor Rosa 1988).

In the 1990s and beyond, this trend towards the commodification of culture has amplified and is now the subject of scholarly debate. For example, food and cuisine have become topics of conversation even in elite journals devoted to cultural critique such as *Micromega*. In November of 1998 a giant food fair—the *Salone del Gusto* or the "Hall of Tastes"—organized by Slow Food in Turin, developed into a major media event attracting leading figures of the left intelligentsia, including Nobel Prize-winning playwright, Dario Fo, singer song-writer Francesco Guccini, as well as prominent national politicians, including the former Italian Prime Minister, Massimo D'Alema. Held in the converted ex-Fiat factory exhibition hall at Lingotto, ironically a site that is iconic of post-war narratives of class struggle in Italy, the fair was an enormous commercial and political success, attracting thousands of visitors and garnering popular support for the region of Piedmont's future winter Olympic bid.[10] But perhaps an even more telling shift in leftist political-culinary consciousness became evident in December of 1998 when the left democrat minister for culture, Giovanna Melandri, apparently missed the opening of the opera season at la Scala in Milan, in order to attend instead the annual dinner organized by *Gambero Rosso* food magazine (Parasecoli 2000).

What accounts for the development of this relatively new, but widely circulating discourse on the moral imperatives of pleasure and food in contemporary Italian cultural life? Could it be linked in any way to the generational shift in cultural politics of the Italian left, these days more accurately termed the Left(s) (Blim 2000a)? While often characterized as a party system based on permanent upheaval, over the last decade the Italian political landscape has undergone a veritable revolution leading to media declarations of the end of the First Republic. Beginning in the late 1980s, the corruption scandals, popularly dubbed *Tangentopoli*, "Kickbackopolis" or "Bribesville" initially as an allusion to Milan, eventually resulted in widespread arrests and the discreditation of the entire ruling political class. Along with some of its allies, the powerful Christian Democratic Party collapsed, while the ruling Socialist Party also virtually disappeared, with its leader Bettino Craxi fleeing to Tunisia. Less directly implicated in these scandals, but still rocked by the events of 1989 and the collapse of the Berlin Wall, the Communist Party split in two, becoming the *Partito Democratico della Sinistra* (PDS) and *Rifondazione Communista*. In the wake of these extraordinary events, entirely new political parties and alliances emerged onto the national stage, now organized into two shifting amorphous coalitions: the Center-Right and the Center-Left.

Perhaps not surprisingly, public disillusionment with the political process has been reflected in falling electoral turnout and a significant decline in strong attachments to political parties. As political scientist Simon Parker (1996) suggests, this move from what Max Weber called a "politics of vocation" to what might be termed a "politics as spectacle" after Guy Debord (1967), is symptomatic of many Western democracies over the last thirty years. This entails a period marked by the emergence of new forms of collective action and social movements, by the diminishing importance of class as a major political cleavage, and by the increasing gap between those employed in regular, well-paid occupations and those either without work or in temporary, low-paid employment (Parker 1996:117).

This trend has been especially significant in Italy where, at least up until the 1980s, political parties were not just about representational electoral politics. They were often important sources of cultural identity and vehicles for social cohesion (Kertzer 1980, 1996). Crucially, the final disintegration of the PCI contributed to the waning in importance of explicitly "workerist" politics, including the centrality of the male factory worker in social struggles, now upheld more or less exclusively by the hard-left *Rifondazione Communista*. More generally, the demise of the post-war party system has deepened the search for new forms of cultural, political and civic associationism, as well as the desire for more varied and dynamic forms of individual realization (Parker 1996). Alongside religious fundamentalism, new-age spiritualism, spectacular virtual politics and the retreat into private worlds, other arenas for the imagination of alternate social worlds and collective action have emerged. The Slow Food movement could, I suggest, be interpreted as one of these arenas. Once associated with admittedly sectarian notions of bourgeois elitism, the consumption of food, even haute cuisine, has become a new metaphorical reference point for the reappraisal of individual, local and national identities. At the turn of the twenty-first century, Italian leftists insist upon the "right to pleasure" through the physical incorporation of good food and wine.

Post Revolutionary Gourmets: From *Neo-Forchettoni* to Ecological Gastronomists[11]

In his 1995 novel entitled *Slowness* Milan Kundera makes an impassioned plea for reclaiming Europe as a site of pleasure and civility, for remembering and savoring the past. Set in an eighteenth-century French chateau, somewhere in the countryside near Paris, the novel works as a literary hall of mirrors, of stories within stories, blurring fact and fiction, past and present, reality and imagination, sex and fantasy, narrative and philosophy. At a crucial point in the narrative, in the eighteenth-century writer's novel within the novel entitled *No Tomorrow*, a young noble and his lover purposefully stage a night of slow lovemaking, waiting until dawn for the final act of consummation in order to recall every delicious detail. The writer then reflects:

> There is a secret bond between slowness and memory, between speed and forgetting … In existential mathematics, that experience takes the form of two basic equations: the degree of slowness is directly proportional to the intensity of memory; the degree of speed is directly proportional to the intensity of forgetting (1995:34).

Kundera seems to be suggesting here that nostalgia may be a socially productive force in late modernity. While everyday life is increasingly experienced at a sensory level by the escalating compression of time and space, through acts of slow contemplation or moments of "stillness" (Seremetakis 1994), sensory memories may be recovered and alternate, forgotten or discarded histories brought to the surface of consciousness. But Kundera's novel is not just a philosophical discussion on speed and modernity, or slowness and its relationship to memory. It is also a literary treatise on Europe and the nature of Europeanism. The book plays with a series of European symbols to ask questions about what kind of place Europe might become. A vignette of an impatient car driver rudely passing another on a crowded highway at the beginning of the novel essentially becomes at the end a metaphor of the future, in which Kundera seems to be suggesting that the past, like the speeding car, is literally up one's ass.

Like Kundera, Carlo Petrini is also interested in socially productive excavations of the past. Both are concerned with the erasure of sensorial memory under modernity. Both are seeking to replace the usual discussion of how Europe is constituted as a political and economic entity with how one might think of Europe as a cultural entity. Both are insisting upon the intimate connections between economy and culture, the past and the future, fantasy and reality.

Media representations generally locate the origin of the Slow Food Movement in the context of a mid-1980s national polemic critiquing the establishment of the first McDonald's restaurant in Rome.[12] According to the *Italy Daily*, it was an almost anti-Proustian moment of the smell of French fries that first stirred Petrini into action:

> Walking in Rome one day, he [Petrini] found himself gazing at the splendid Spanish Steps when the overwhelming odor of French fries disturbed his reverie. To his horror he discovered that not twenty meters along the piazza loomed the infamous golden arches of a well-known food chain. "*Basta!*" he cried. And thus begun a project which would take him all over the world in order to promote and protect local culinary traditions. As a symbol for his cause he chose the snail because it was the slowest food he could think of (11/3/1998).[13]

A consummate media manipulator, Petrini actually denies that Slow Food is simply anti-fast food. Rather, he suggests that Slow Food is against the homogenization of taste, which fast food symbolizes. In other words, for Petrini, fast food is a sign of the more negative effects of modern market rationalities on cultural difference—a world in which speed, or "dromocentrism," as well as placelessness, is the essence of the era (Casey 1997:xiii).[14] Slowness in this formulation becomes a metaphor for a politics of place: a philosophy complexly concerned with the defence of local cultural heritage, regional landscapes and idiosyncratic material cultures of production, as well as international biodiversity and cosmopolitanism.

These ideas are well articulated in the movement's first and often cited manifesto, which states that:

> In order to fight against the universal manner of the Fast Life we need to make a concerted effort to defend the pleasure of slowness. We are against those who confuse efficiency with speed. Our movement is in favor of sensual pleasure to be practised and enjoyed slowly. Through Slow Food, which is against the homogenizing effects of fast foods, we are rediscovering the rich variety of tastes and smells of local cuisine. And it is here, in developing an appreciation for these tastes, that we will be able to rediscover the meaning of culture, which will grow through the international exchange of stories, knowledge and other projects.[15]

Implicit in this manifesto is the notion that memory is entangled in the senses and that through the sensory experience of rediscovering taste memories one recuperates and holds onto the past (Sutton 2001). According to Petrini,

> In this century speed has become a drug. For us slowness is not an absolute value. It is more like a homeopathic medicine. A medicine one takes daily to remind ourselves that it is we who decide the rhythm of life we want to lead, rather than having these rhythms imposed on us from outside. Slowness is a metaphor for understanding and enjoyment, of being able to know who you are and what you taste (Interview with Petrini June 1999).

Slowness, in other words, is linked to pleasure, conviviality and corporeal memory. "Slow life" says Petrini with typical sound-bite finesse, "is not just Slow Food" (Petrini 2001:15).

It is important to recognize that these public manifestos advocating "slowness" are neither explicitly anti-capitalist nor anti-corporate. Rather slowness is employed ideologically in order to promote what Petrini has termed a form of "virtuous globalization," in which members of minority cultures, including niche-food producers, are encouraged to network and thrive (Stille 2001). The notion of slowness for Petrini thus represents a discursive field linked to critiques of modernity *and* an arena of practical action: culture and politics. For Petrini, the questions are: Will the new Europe be a "fast" Europe that protects the interests of fast capitalism, corporate control of food production and the indiscriminate introduction of genetically modified crops? Or can the new Europe be a "slow" Europe that protects small artisanal food producers and the cultural landscapes to which they are attached? What kind of political vision of Europe will prevail? Will it be a Europe committed to neo-liberal models of economic rationality? Or will it be a democratic Europe fostering cultural diversity and communities of memory?

Attempting to construct a cultural politics of the future, Slow Food is, however, inevitably also caught up in the interpretive and cultural frames of its main protagonists. In the 1990s, for example, McDonald's was singled out both by grass-roots "anti-globalization" activists and by the media attention given to French farmer/activist, José Bové's acts of sabotage against McDonald's in France. Petrini explicitly disavows these kinds of guerrilla strategies as being against "Slow style." As he puts it "we prefer to concentrate our efforts on what we are losing, rather than trying to stop what we don't like" (2001:28).

Conclusion

This paper began by tracing the recent "social life" (Appadurai 1986) of *lardo* as it has moved from a commodity with a relatively contained set of meanings for local people to its current role as a widely circulating symbol of an endangered food for the Slow Food movement. As a direct result of this campaign, *lardo* has also acquired new meaning as an exotic item of consumption for middle-class and local consumers alike. Despite occasional accusations of culinary luddism or culinary elitism (Lauder 1999), in Italy Slow Food has succeeded in creating the cultural space for the performance of a new kind of consumer politics. Clearly, Slow Food self-consciously resists easy categorization in terms of any familiar political narratives, including class. But while the "identitarian" consciousness (Eder 1993) of Slow Food adherents is not easily ascertained, Slow Food political strategies amply demonstrate the *power* of consumption practices as shaping forces of modern identity (Klein 1999), as well as the *potential* for new forms of transnational consumer alliances (Miller 1995, 1997).

Scholars, particularly of the media, have pointed to the extent to which promotion has now penetrated the heart of the political process in liberal democratic societies, as well as the way in which media and public relations experts have increasingly become crucial to the management of political meanings in the public sphere (McLagan 2000; Wernick 1991; Marshall 1997). New social movement theorists have also emphasized that political struggles in the contemporary era focus on struggles over meanings, as well as over political and economic conditions (Torraine 1988; Fox & Stan 1997; Eder 1993). Slow Food politics can, I suggest, be interpreted as the product of all these trends. As an emergent political form it slips easily between the realms of advertising, commerce and cultural critique. It is as much concerned with the commodification of rural and proletarian nostalgia as with the actual protection of local material cultures of production and the memories to which they are attached. Thus while consumption is revealed as a realm for the potential emergence of new forms of collective political agency, the case of *lardo di Colonnata* also demonstrates the way in which these new spaces for the performance of collective action continue to be fashioned through the dynamic interplay between consumption, production and distribution practices (Mintz 1985).

Filled as much with irony as nostalgia, the cultural politics of the Slow Food movement are not slow. They are fast, concerned as much with the proliferation of images, as with the marketing of memory. Just as international corporations increasingly appeal to particularistic cultural identities in order to capture greater global market share for their commodities, Slow Food political manifestos promote the idea of cultural diversity by urging consumers to buy niche-marketed foods. With this kind of promotional politics, where consumers are envisaged as international political activists by virtue of market choice, there can be no guaranteed ideological outcomes. Demands to protect local culinary traditions and cultural diversity could just as easily risk appropriation by radical regionalist movements with exclusionary political agendas. Anti-corporate rhetoric combined with narratives of cultural loss may fuel a deepening sense of nationalist nostalgia.

As we have seen, the cultural politics of marketing authenticity may also have unexpected consequences for direct producers. One further ironic example emerged during a return visit to Carrara in the summer of 2002. As Fausto joked with me about refusing to play the part of the "peasant dressed in black," it became clear that the pork fat makers had fallen out with their Slow Food promoters precisely over issues of marketing. In particular they objected to the promotion of pork fat manufactured outside the area of Colonnata in the Co-op, a national supermarket chain with political connections to the old and new lefts, including the Slow Food Movement. Eventually fed up with Slow Food's instrumentalization of *lardo* as a logo for the authenticity of their politics, especially in the absence of continuing economic benefit, pork fat makers from Colonnata were threatening an embarrassing protest outside Slow Food's signature public event, the 2002 trade fair in Turin.

Slow Food's emergence, I have argued, is critically framed within a uniquely Italian post-war cultural and political trajectory that has witnessed, among other things, the gradual demise of the post-war party system and the search for new kinds of civic associationism. It is an attempt to devise or reflect upon current constructions of Europe as purely a debate between regions and nations. I have also suggested that the endangered foods campaign is, in part, a well-orchestrated political response to what Nadia Seremetakis has called the "reorganization of public memory," (1994:3) which has accompanied the intensification of market rationalities in European peripheries. Slow Food's preoccupation with the disappearance of distinctive regional tastes may indeed, as I have argued, be linked to the generational political sensibilities of its founders, but it is equally deeply engaged with current debates over the future of European identity, to moral economies.

There is, however, one final irony. Over the last ten years, thinking about food, and tasting it, has become a pressing political issue. While more generally widespread fears of environmental contamination through the uncontrolled introduction of genetically modified foods and crops mirror other fears of cultural contamination as national boundaries disappear, in Italy a fear of cultural homogenization has manifested itself in a politics of taste, based around the protection of "endangered foods." But whereas Slow Food founders once impishly promoted the "right to pleasure" as a critique of Left asceticism, now pleasure has become a political duty and food is, perhaps, no longer simply a private pleasure.

Acknowledgment

I wish to thank Vivienne Kondos for inviting me to present a first version of this paper in the anthropology seminar at the University of Sydney and Franca Tamisari, Souchou Yao, Linda Barwick and Jadran Mimica for their stimulating discussion. The essay has also benefited from the helpful comments and editorial suggestions of a number of other readers including Jennifer Alexander, Steven Feld, Tiziana Ferrero-Regis, Don Kulick, Meg McLagan, Daniel Niles, Dorothy Zinn, members of the Feast and Famine colloquium at New York University, and the three anonymous readers for *Ethnos*. A small section of this essay has previously been published in *The Asia Pacific Journal of Anthropology*. I would like to thank the editors for allowing me to reproduce it here in the context of a wider argument.

Notes

1. Dietary fashions of course change. Many popular American dietary gurus, for example, Atkins, now recommend diets high in "good" fats and protein, eliminating, instead, carbohydrates.
2. See, for example, Klein 1999 on the way in which market-based invasions of public space and individuals' "life-worlds" have become the impetus for anti-corporate activism. See also Edelman 2001 for a cogent discussion of these trends towards new forms of transnational activism.

3. For a discussion of the emergence of a "metacultural" framework of culture and political action specifically in relation to indigenous social movements in Brazil, see Turner 1993. See also Conklin 1995. For a more general discussion of the rise of culturalist social movements or "postnational" social formations, see Appadurai 1996.

4. For a more detailed account of this controversy see Leitch 2000.

5. Post-communist Italy comprises many "Lefts." My focus here is on the Italian Communist Party (PCI) because of its institutional connections to some of Slow Food's main protagonists, as well as its domination of post-war left cultural politics. As David Kertzer puts it: "For millions of Italians, from the end of the Second World War through the 1980s, personal identity was rooted in the Communist Party and its symbolism: *Sono comunista* (I am a Communist) was a statement not only of people's political allegiance but of their core identity. For many, being identified as Communist was more satisfying than being identified as Italian" (1996:64). See also Blim 2000a for a succinct appraisal of more recent splits in Italian Left politics.

6. Published by a Milanese editorial collective, the monthly journal lasted from 1982–1989.

7. Notes on Carlo Petrini's biography are gathered from the media, as well as from an interview I conducted with him at Slow Food headquarters in Bra in June 1999.

8. My analysis of food and the politics of pleasure among the Italian left is greatly enhanced by my conversation with Fabio Parasecoli, food writer and editor of *Gambero Rosso*. See also Parasecoli's forthcoming article in *Gastronomica*.

9. The last issue of *Gambero Rosso* as a supplement of *Il Manifesto* was published in 1991. It is now a magazine owned by its founder Stefano Bonilli, who had previously worked as a political journalist for *Il Manifesto* from 1971–1982. *Gambero Rosso* now has a regular spot on Italian national television, as well as a book series dedicated to food and wine.

10. In 1998, the Salone del Gusto attracted over 120,000 visitors over four days. Its success prompted the organization of an even larger event in 2000, also held at Lingotto.

11. *Neo-forchettoni* roughly translates as "hearty eaters." The phrase *da neo-forchettoni a eco-gastronomi* comes from a subtitle in Carlo Petrini's (2001) own recently published book on the history of Slow Food.

12. For a more detailed account of the development of McDonald's in Italy see Petrini 2001.

13. The choice of the snail as symbol for the Slow movement was quite deliberate. As Petrini remarked to me, a tortoise is also slow. The snail is a food but, more importantly, it can be found everywhere and it carries its home on its back. According to Petrini, the snail captures the ideals of the Slow Food movement as being about food and connection to place, as well as cosmopolitanism.

14. According to Edward Casey, "Dromocentrism amounts to temporocentrism writ large: not just time but speeded-up time (dromos connotes 'running,' 'race,' 'racecourse') is the essence of the era. It is as if the acceleration discovered by Galileo to be inherent in falling bodies has come to pervade the earth (conceived of as a single scene of communication), rendering the planet a 'global village' not in a positive sense but as a placeless place indeed" (1997:xiii).

15. The original Slow Food manifesto was a collaborative effort put together by a group of left public intellectuals, including writers, journalists, singer-songwriters, for example, Valentino Parlato, Gerado Chiaromonte, Dario Fo, Francesco Guccini, Gina Lagorio, Enrico Menduni, Antonio Porta, Renate Realacci, Gianni Sassi and Sergio Staino.

References

Appadurai, Arjun (1986) Introduction: Commodities and the Politics of Value. In *The Social Life of Things: Commodities in Cultural Perspective*, edited by Arjun Appadurai, pp. 1–63. Cambridge: Cambridge University Press.

—— (1996) *Modernity at Large*. Minneapolis: University of Minnesota Press.

Asor Rosa, Alberto (1988) Il Capitale da Lezione ai Partiti di Sinistra. *La Repubblica*, 13 August, pp. 1, 6.

Baudrillard, Jean (1981) *For a Critique of the Economy of Signs*. St Louis, MO: Telos Press.

Beck, Ulrich (1986 [1992]) *Risk Society: Toward a New Modernity*. London: Sage Publications.

Bentley, Amy (2001) Reading Food Riots: Scarcity, Abundance and National Identity. In *Food, Drink and Identity*, edited by Peter Scholliers, pp. 179–193. Oxford/New York: Berg.

Blim, Michael (2000a) What is Still Left for the Left in Italy? Piecing Together a Post-Communist Position on Labor and Employment. *Journal of Modern Italian Studies*, 5(2):169–185.

—— (2000b) Capitalisms in Late Modernity. *Annual Review of Anthropology*, 29:25–38.

Breen, Timothy, H. (1988) "Baubles of Britain": the American and Consumer Revolutions of the Eighteenth Century. *Past and Present*, 119:73–104.

Casey, Edward, S. (1997) *The Fate of Place*. Berkeley: University of California Press.

Conklin, Beth (1995) Body Paint, Feathers, and VCRs: Aesthetics and Authenticity in Amazonian Activism. *American Ethnologist*, 24(4):711–747.

Counihan, Carole, M. (1999) *The Anthropology of Food and Body: Gender, Meaning and Power*. New York/London: Routledge.

Debord, Guy (1967) *The Society of the Spectacle*. Detroit: Black and Red.

Edelman, Marc (2001) Social Movements: Changing Paradigms and Forms of Politics. *Annual Review of Anthropology*, 30:285–317.

Eder, Klaus (1993) *The New Politics of Class: Social Movements and Cultural Dynamics in Advanced Societies*. London/Newbury Park/New Delhi: Sage Publications.

Feld, Steven (1999) From Schizophonia to Schismogenesis: The Discourses and Practices of World Music and World Beat. In *The Traffic in Culture*, edited by George Marcus & Fred Myers, pp. 96–126. Berkeley: University of California Press.

Forgacs, David (1990) *Italian Culture in the Industrial Era, 1880–1980*. Manchester: Manchester University Press.

Fox, Richard & Orin Starn (eds) (1997) *Between Resistance and Revolution: Cultural Politics and Social Protest*. New Brunswick, NJ: Rutgers University Press.

Gailus (1994) Food Riots in Germany in the Late 1840s. *Past and Present*, 145:157–93.

Gilje, P. (1996) *Rioting in America*. Bloomington: University of Indiana Press.

Ginsburg, Paul (1989) *A History of Contemporary Italy: Society and Politics 1943–1988*. Harmondsworth: Penguin.

Goody, Jack (1982) *Cooking, Cuisine and Class: A Study in Comparative Sociology*. Cambridge: Cambridge University Press.

Grossi, Giorgio (1985) *Rappresentanza e Rappresentazione: Percorsi di Analisi dell'Interazione fra Mass Media e Sistema Politica in Italia*. Milan: Franco Angeli.

Gundle, Stephen (2000) *Between Hollywood and Moscow: The Italian Communists and the Challenge of Mass Culture, 1943–1991*. Durham/London: Duke University Press.

Hernandez, Castillo Rosalva Aida & Ronald Nigh (1998) Global Processes and Local Identity Among Mayan Coffee Growers in Chiapas, Mexico. *American Anthropologist*, 100(1):136–147.

Heyman, J. (1997) Imports and Standards of Justice on the Mexico-United States Border. In *The Allure of the Foreign: Post-Colonial Goods in Latin America*, edited by B. Orlove, pp. 151–83. Ann Arbor: University of Michigan Press.

Hobsbawm, Eric (1959) *Primitive Rebels: Studies in Archaic Forms of Social Movement in the Nineteenth and Twentieth Centuries*. New York: Norton.

Hobsbawm, Eric & Terrence Ranger (eds) (1983) *The Invention of Tradition*. Cambridge: Cambridge University Press.

Italy Daily (1998) Slow Food: An Antidote to Modern Times. November 3.

Jansen, Willy (2001) French Bread and Algerian Wine: Conflicting Identities in French Algeria. In *Food, Drink and Identity*, edited by Peter Scholliers, pp. 195–218. Oxford/New York: Berg.

Kertzer, David (1980 [2000]) *Comrades and Christians*. Cambridge: Cambridge University Press.

—— (1996) *Politics and Symbols: The Italian Communist Party and the Fall of Communism*. New Haven: Yale University Press.

Klein, Richard (1996) *Eat Fat*. New York: The Free Press.

Klein, Naomi (1999) *No Logo: Taking Aim at the Brand Bullies*. New York: Picador.

Kundera, Milan (1995) *Slowness*. London/Boston: Faber and Faber.

La Nazione (1997) Bollino USL sul Lardo di Colonnata Piace anche ai Gourmet Americani. 18 February.

—— (1996) Sequestrati 200 Chili di Lardo. Carenti i Documenti Sanitari. 1 April.

Lash, Scott & John Urry. 1994. *Economies of Signs and Spaces*. London: Sage.

Lauder, Rachel (1999) A World of Inauthentic Cuisine. *Proceedings, Cultural and Historical Aspects of Food*, pp. 136–145.

Leitch, Alison (1993) The Killing Mountain: Work, Gender and Politics in an Italian Marble Quarrying Community. Unpublished Ph.D. dissertation. Department of Anthropology, University of Sydney, Australia.

—— (1996) The Life of Marble: The Experience and Meaning of Work in the Marble Quarries of Carrara. *The Australian Journal of Anthropology*, 7(3):235–57.

—— (1999) Afterword. In *Carrara: the Marble Quarries of Tuscany*, edited by Joel Leivick, pp. 63–74. Stanford: Stanford University Press.

—— (2000) The Social Life of Lardo: Slow Food in Fast Times. *The Asia Pacific Journal of Anthropology*, 1(1):103–118.

Leivick, Joel (1999) *Carrara: the Marble Quarries of Tuscany*. Stanford: Stanford University Press.

Marshall, David. P. (1997) *Celebrity and Power: Fame in Contemporary Culture*. Minneapolis: University of Minnesota Press.

McLagan, Margaret (2000) Spectacles of Difference: Buddhism, Media Management, and Contemporary Tibet Activism. *Polygraph*, 12:101–120.

Miller, Daniel (1995) *Acknowledging Consumption: A Review of New Studies*. New York/London: Routledge.

—— (1997) *Capitalism: An Ethnographic Approach*. Oxford: Berg.

Mintz, Sidney. 1979. Time, Sugar and Sweetness. *Marxist Perspectives*, 2:56–73.

—— (1985) *Sweetness and Power: The Place of Sugar in Modern History*. New York: Penguin.

Orlove, B. (1997) Meat and Strength: The Moral Economy of a Chilean Food Riot. *Cultural Anthropology*, 12:234–68.

Parasecoli, Fabio (2000) Post-Revolutionary Chowhounds: Pleasure, Food and the Italian Left. Unpublished Paper presented at the Millennial Stews Conference, June 1–4, New York University.

—— Forthcoming, Food, Globalization and the Italian Left. *Gastronomica*.

Parker, Simon (1996) Political Identities. In *Italian Cultural Studies*, edited by David Forgacs & Robert Lumley, pp. 107–128. Oxford: Oxford University Press.

Petrini, Carlo (2001) *Slow Food: le Ragioni del Gusto*, Roma/Bari: Editori Laterza.

Pilcher, Jeffrey (1998) *Qu'Vivan los Tamales: Food and the Making of Mexican Identity*. Albuquerque: University of New Mexico Press.

Probyn, Elspeth (2000) *Carnal Appetites: Food Sex Identities*. London/New York: Routledge.

Roseberry, W. (1996) The Rise of Yuppie Coffees and the Re-imagination of Class in the United States. *American Anthropologist*, 98(4):762–775.

Rozin, Paul (1998) Food is Fundamental, Fun, Frightening and Far-Reaching. *Social Research*, 66(1):9–30.

Schneider, Jane (1994) In and Out of Polyester: Desire, Disdain and Global Fibre Competitions. *Anthropology Today*, 10:2–10.

Seremetakis, C. Nadia (1994) The Memory of the Senses, pts. 1 & 2. In *The Senses Still: Perception and Memory as Material Culture in Modernity*, edited by C.N. Seremetakis, pp. 1–43. Boulder: Westview Press.

Spender, Mathew (2000) The Politics of Pork Fat. *Bon Appétit*, May, pp. 54–55.

Stille, Alexander (2001) Slow Food. *The Nation*, August 20–27, pp. 11–16.

Sutton, David, E. (2001) *Remembrance of Repasts: An Anthropology of Food and Memory*. Oxford/New York: Berg.

Taylor, L. (1996) Food Riots Revisited. *Journal of Social History*, 30:483–96.

Thompson, E.P. (1971) The Moral Economy of the English Crowd in the Eighteenth Century. *Past and Present*, 50:76–136.

Tilly, Charles (1975) Food Supply and the Public Order in Modern Europe. In *The Formation of Nation States in Western Europe*, edited by Charles Tilly, pp. 380–455. Princeton: Princeton University Press.

Tilly, Louise (1983) Food Entitlement, Famine and Conflict. *Journal of Interdisciplinary History*, 14:333–49.

Torraine, Alain (1988) *Return of the Actor: Social Theory in Postindustrial Society*. Minneapolis: University of Minnesota Press.

Turner, Terence (1993) Anthropology and Multiculturalism: What is Anthropology that Multiculturalists should be Mindful of It? *Cultural Anthropology*, 8(4):411–429.

Wernick, Andrew (1991) *Promotional Culture: Advertising, Ideology and Symbolic Expression*. London/Newbury Park/ New Delhi: Sage Publications.

25

Taco Bell, Maseca, and Slow Food: A Postmodern Apocalypse for Mexico's Peasant Cuisine?

Jeffrey M. Pilcher

Mexico has a distinguished revolutionary tradition, but is the land of Emiliano Zapata ready for the "delicious revolution" of Slow Food? The question may sound a bit facetious at first, but given the movement's origins in the Italian Communist Party, a disquisition on class consciousness in a postmodern era seems appropriate. Of course, class has been virtually banished from postmodern academic discourse, perhaps from the sheer embarrassment of an intellectual vanguard discovering itself to be a petit bourgeoisie on the brink of proletarianization by the forces of global capital. In the script of contemporary revolution, only the villain—global capital—retains its traditional role. Slow Food speaks the lines of the reformist Social Democratic Party (SDP) to José Bové's militant Bolshevism, the international proletariat is now politically suspect for its consumerist tendencies, and the peasantry has become the progressive motor of history. Although Slow Food offers an admirable program for personal life, it will never represent a genuine revolution until it confronts the dilemmas of class that have been complicated but not obviated by increasing globalization. Indeed, the Mexican case reveals the impossibility of drawing a clear dichotomy between slow and fast food in markets where global and local capital compete for the trade of middle-class tourists and equally cosmopolitan "peasants."

The Slow Food snail mascot would no doubt bask contentedly in the shade of Emiliano Zapata's sombrero, for the Mexican agrarian revolution of 1910 likewise sought to preserve traditional agrarian livelihoods. Emerging from the crisis of Italian leftist politics in the late 1980s, Slow Food exalted leisure and pleasure as an antidote to the blind pursuit of efficiency within the United States (Parasecoli 2003). The movement sought to revive artisanal production and to preserve vanishing biodiversity against the homogenizing influence of multinational agribusiness. Nevertheless, tensions remained between the elitism of the official manifesto—"preserve us from the contagion of the multitude" (Parasecoli 2003:xxiii)—and the democratizing ideal of affordable but tasty *osterie* (regional restaurants).

Recent work on the history of Italian cuisine demonstrates that the traditions Slow Food seeks to preserve are largely invented, a point acknowledged by the movement's leaders (Hobsbawn 1983; Petrini 2003). Well into the twentieth century, peasants subsisted on monotonous porridges of maize, chestnuts, broad beans, or rice, depending on the region and the season (Camporesi 1993; Diner 2001; Helstosky 2004). Italy's regional cuisines are not only modern inventions, they may even have been created in the Americas through the industrial production and canning of olive oil, tomato paste, and cheeses to satisfy migrant workers who could afford foods unavailable to peasants at home (Teti 1992; Gabaccia 2004). More research is needed on the nineteenth-century bourgeois diet, but it seems that contemporary Italian cuisine emerged largely from festival foods such as *maccheroni*, which became items of everyday consumption in the United States. The traditional *osteria* that Slow Food seeks to encourage likewise grew out of the *cucina casalinga*, meals served in urban homes to unattached male workers; they only became restaurants with the postwar growth of tourism. The Italian experience in turn provides a useful model for the emergence of Mexico's regional cuisines.

There can be few foods slower than Mexican peasant cooking. For thousands of years, the preparation of the staple maize tortillas required hours of hard, physical labor each day. Women rose before dawn to grind corn (*masa*) on a basalt stone (*metate*), then patted out tortillas by hand and cooked them on an earthenware griddle (*comal*)—all this before men went out to the fields (Redfield 1929; Lewis 1951). Festivals multiplied the workload of women; in addition to preparing special tortillas and masa dumplings (*tamales*), they labored over the metate grinding chile sauces (*moles*) and cacao while men sat around drinking (Stephen 1991). The metate was so intrinsic to Mexican patriarchy that when mechanical mills capable of adequately grinding masa finally reached the countryside in the first half of the twentieth century, one villager described it as a "revolution of the women against the authority of the men" (Lewis 1951:108). As women transferred their labor from domestic reproduction to market production, tamales and moles were gentrified by restaurants catering to national and international tourists (Pilcher 2004).

Mexico has had two competing visions for the modernization of its cuisine. Attempts by local investors and engineers to industrialize tortilla production have culminated in the rise of Grupo Maseca, a multinational producer of *masa harina* (dehydrated tortilla flour), which can be reconstituted with water to save the trouble of making fresh masa, thus making them popular among migrant workers. Although connoisseurs can distinguish fresh from industrial corn tortillas, both are quite different from the wheat flour tortillas most common in the United States. Meanwhile, Taco Bell has led the way in applying North American fast food technology to Mexican cooking. Neither approach proved satisfactory to aficionados, leaving a space for Slow Food to catch on among middle-class Mexicans and foreign tourists. Yet attempts to save traditional Mexican cuisine have been plagued by the same contradictions of elitism, gender bias, and even a measure of imperial arrogance—treating Native Americans like southern Italians—that typified the movement at home.

Maseca

Although the industrialization of Mexico was generally characterized by imported technology and capital, the modernization of the tortilla was a uniquely national enterprise. The complex skills involved in making tortillas were mechanized in three distinct stages at roughly fifty-year intervals around the beginning, middle, and end of the twentieth century. The arduous task of hand grinding maize at the metate was

first replaced by forged steel mills. Next came the technology for automatically pressing out and cooking tortillas, which facilitated the spread of small-scale tortilla
factories throughout the country. Finally, the industrial production of masa harina
allowed the vertical integration of food processing under Grupo Maseca. Tortilla
futurists envisioned these technologies as a complete package, but the three distinct
processes could in fact be combined or separated according to circumstances, and
each was tied into complex culinary, social, and political relationships. Ultimately,
the fate of the tortilla resulted more from questions of political economy than of
consumer choice.

Corn mills arrived in Mexican cities by the late nineteenth century but took
decades to spread through the countryside, in part because of concern among
women about their position within the family. The ability to make tortillas was long
considered essential to a rural woman seeking marriage, and any who neglected the
metate risked unfavorable gossip. Even the few centavos charged by the mills posed a
significant cost in subsistence communities, but poor women often had the greatest
incentive to grind corn mechanically because the time saved could be used to engage
in artisanal crafts or petty trade. Wealthy families, by contrast, were among the last to
give up the metate on a daily basis; this form of slow food offered a status symbol, in
part because the rich could pay others to do the actual work. As the benefits of
milling gained acceptance, women often organized cooperatives to purchase the
machines. By midcentury, corn mills had arrived in virtually every community in
Mexico, transforming social relationships and helping to incorporate rural dwellers
into the monetary economy (Lewis 1951; Bauer 1990).

The mechanization of the tortilla and the development of masa harina posed more
technological than social problems. Conveyer belt cookers were first introduced
around 1900, but only at midcentury could they produce a tortilla that satisfied
Mexican consumers. Once these machines were created, a cottage industry of tortilla
factories quickly spread to all but the most remote rural communities. Although
these shops generally sold tortillas by the kilo made from commercially purchased
maize, they also ground corn for customers who wished to prepare their own tortillas
at home. Both small-scale tortilla factories and their eventual corporate challenger
developed under the aegis of a welfare program intended to subsidize food for poor
urban consumers. The first masa harina factories were established in 1949 by Molinas
Azteca, S.A. (Maseca) and a parastatal corporation, Maíz Industrializado, S.A.
(Minsa). The two firms collaborated on research and development for more than a
decade before arriving at a suitable formulation that could be transformed into
tortilla masa with just the addition of water. By the 1970s, tortilla flour accounted for
5 percent of the maize consumed in Mexico. Sales grew steadily over the next two
decades until Maseca alone held 27 percent of the national corn market. Nevertheless, the subsidy on maize supplied to small-scale tortilla factories slowed the firm's
expansion.

The dismantling of the state food agency in the 1990s assured Maseca's triumph.
President Carlos Salinas de Gortari (1988–1994) first curtailed the corn subsidy while
also selling Minsa to a consortium rivaling Maseca. For the fifty-one thousand
small-scale tortilla factories, subsidized corn had been essential for their commercial
viability. Led by a trade organization, the Association of Proprietors of Tortilla

Factories, and Nixtamal Mills, they launched a vocal political campaign to retain the subsidy, citing scientific studies concluding that traditional tortillas were more nutritive than those made with Maseca, which they dubbed "MAsaSECA" ("dry masa"). They also accused Maseca of manipulating corn markets and attempting to monopolize supplies, thereby tapping popular memories of hunger that remain vivid in many sectors of Mexican society. But the proprietors found themselves on the wrong side of a neoliberal political avalanche, and the subsidy was eliminated in January 1999. The sudden change left tortilla factories unable to establish competitive sources of supply, and they were reduced to mere vendors for the multinational company, as masa harina quickly cornered an estimated 80 percent of the national tortilla market. Maseca, in turn, claimed nearly three quarters of these sales (Pilcher 1998; Ochoa 2000; Cebreros, n.d.).

The results of this change have been mixed. Proponents of modernization considered tortillas to be essentially tasteless, anyway, and found no difference in the new product. Journalist Alma Guillermoprieto (1999:46) took a different view, claiming that "when the privatization program of Mexico's notorious former President Carlos Salinas delivered the future of the tortilla into their hands . . . [the tortilla magnates] served up to the Mexican people the rounds of grilled cardboard that at present constitute the nation's basic foodstuff." Campesinos are quite sensitive to tortilla quality, and many have resisted Maseca (see, for example, González 2001:173). Nevertheless, the exigencies of subsistence often require small farmers to sell their best produce to urban markets. Meanwhile, Mexican food faced a still more uncertain fate abroad.

Taco Bell

Even as the tortilla market evolved under government protection at home, tacos slipped across the border into greater Los Angeles, where they fell into the hands of scientists and industrialists. While migrants labored in corporate agribusiness, their foods underwent a process of Taylorization to become more standardized and efficient. Neatly packaged under the trademark of Taco Bell, these brave new tacos subsequently traveled around the world, ultimately colonizing their native land. But despite their best efforts, the food formulators and advertising executives could not determine the global reception of the cyborgs they had created.

Sociologist George Ritzer (1993) has examined most comprehensively the threat of corporate fast food to local traditions, extending Max Weber's theory of rationalization, the process whereby modern technology has made society more efficient, predictable, and controlled. One would, indeed, be hard pressed to find a better example of those values than the Big Mac. The result of this process has been, first, the standardization of food, replacing the endless variety of local cuisines with the artificial choice of numbered selections from a "value menu." The creation of so-called McJobs has further alienated labor by requiring only minimally skilled workers who respond to the commands of machinery. By stifling the social interactions customarily associated with dining, fast food has supposedly dehumanized the process of eating.

The McDonaldized taco, like the hamburger, had its origins in southern California in the mid-1950s. Glen Bell, a telephone repairman from San Bernardino, was impressed by the original McDonald's restaurant, but rather than compete head-on for the hamburger trade, he applied the new industrial techniques to a separate market segment, the Mexican American taco stand. As he described it: "If you wanted a dozen, you were in for a wait. They stuffed them first, quickly fried them and stuck them together with a toothpick. I thought they were delicious, but something had to be done about the method of preparation" (Taco Bell 2001). That something was to pre-fry the tortillas, anathema to any Mexican, but a blessing to Anglos who preferred to drive off with their food rather than eat at the stand while the other tacos were freshly cooked.

Corporate mythology attributing the North American taco to Glen Bell is questionable, but whatever the origins of the pre-fried taco, precursor to the industrial taco shell, it made possible the globalization of Mexican food by freeing it from its ethnic roots. No longer would restaurateurs or home cooks need a local supply of fresh corn tortillas to make tacos; the shells could henceforth be mass-produced and shipped around the world, albeit with some breakage. The extent to which the taco has been alienated from the Mexican community can be seen in the job description given by a Taco Bell employee in *The New Yorker* (2000): "My job is I, like, basically make the tacos! The meat comes in boxes that have bags inside, and those bags you boil to heat up the meat. That's how you make tacos."

By the mid-1990s, as the Tex-Mex fad spread worldwide, restaurants in Europe and East Asia abandoned even the pretense of serving Mexican food. Nevertheless, critics have questioned the homogenizing effects of the McDonaldization thesis by emphasizing the diverse ways in which people around the world experience restaurants. Although travelers may find the bland uniformity of the Golden Arches disturbing, locals embrace the new choices made possible by the arrival of fast food. The protests of burger-Luddite José Bové notwithstanding, French cuisine has little to fear from the spread of McDonalds, as sociologist Rick Fantasia (1995) has pointed out, because the fast food chains and upscale restaurants cater to quite different markets. The contributors to James Watson's (1997) volume, *Golden Arches East*, found the experience of U.S. culture, rather than actual food, to be the chain's biggest selling point in Asia. Customers were enthralled by the democratizing influence of waiting in line, the freedom from the social demands of competitive banqueting, and the novelty of clean public bathrooms.

The opening of Taco Bell's first Mexican outlet in 1992 illustrates the divergent expectations of corporate executives and local consumers. The Tricon conglomerate already operated a number of fast food restaurants in the country, and it tested the market by offering a selection of tacos at a Kentucky Fried Chicken location in an upscale mall in the suburbs of Mexico City. Skeptics murmured about "coals to Newcastle," but company spokesmen blithely applied the usual justification for fast food in the United States that "the one thing Mexico lacks is somewhere to get a clean, cheap, fast taco" (quoted in Guillermoprieto 1994:248). In fact, countless *taquerías* offered precisely that, although Guillermoprieto conceded that "no Mexican taco stand looks like a NASA food-preparation station." Even the company acknowledged its doubts by obtaining a supply of fresh tortillas and offering the standard taquería fare of the 1990s, pork *carnitas* and shredded beef. But middle-class customers responded with disappointment, for they could get Mexican tacos anywhere. One woman complained: "This doesn't taste like the real thing, does it? What I wanted was those big taco shells stuffed with salad and Kraft cheese and all *kinds* of stuff, like what you get in Texas" (quoted in Guillermoprieto 1994).

Mexico thus replicates the global experience of fast food as a middle-class privilege, with its own versions of authenticity. The initial opening of Taco Bell in Mexico in 1992 failed due to the economic crisis of that year, but by the end of the decade the company had opened a number of successful outlets. Even in the United States, fast food need not entail the complete obliteration of local culture; witness the spread of "fresh Mex" restaurants such as Chipotle, a subsidiary of the McDonald's corporation. Moreover, by introducing customers to an artificial version of Mexican food, chains may well have expanded opportunities for marketing the genuine article.

Slow Food

In its mission to preserve and advertise distinctive regional foods, Slow Food created an "Ark of Taste" along with awards for biodiversity to promote awareness of and support for ecologically beneficial practices around the world. Despite their relatively small size, organizations from developed nations wield disproportionate power in a country such as Mexico, and as a result, they slant local activist movements toward middle-class agendas with little relevance for the needs of common people. Moreover, the fascination with Mexico's folkloric and Native American heritage diverts attention from the *mestizo* (mixed-race) majority, who do not speak an indigenous language but still suffer economic and political marginalization.

Mexico has done very well through the Slow Food movement, although it is unclear how much of this attention is due to its rich culinary heritage and how much to the Zapatista rebellion in Chiapas. The indigenous uprising gained international acclaim almost from the first shots fired on January 1, 1994, the day the North American Free Trade Agreement (NAFTA) went into effect. While brandishing an AK-47, the charismatic Subcomandante Marcos shunned terrorist bombings and instead waged his campaign with anti-globalization *pronunciamientos* fired across the World Wide Web. Stylish, balaclava-clad guerrillas held particular fascination for the Italian left.

Regardless of the inspiration, Mexicans have topped the list of Slow Food's Award for the Defense of Biodiversity. The jury consists of about five hundred food writers and other culinary authorities from around the world, although the United States and Italy provide the largest number. In the first four years, Mexicans received five out of fifty nominations, worth €3,500 each, and three out of twenty jury prizes, providing an additional €7,500. The United States, by comparison, received four nominations but no jury prize winners. The first Mexican jury prize, in 2000, went to Raúl Manuel Antonio, from Rancho Grande, Oaxaca, for establishing an indigenous vanilla-growing cooperative to supplement the incomes of small coffee producers. The following year, Doña Sebastiana Juárez Broca, known in her Tabasco community as Tía Tana, won a prize for reviving traditional cacao production techniques as an alternative to an official marketing organization that had impoverished farmers. Finally, in 2003, the jury honored historian José Iturriaga de la Fuente for organizing a fifty-four-volume series documenting Mexico's regional and indigenous cuisines (Slow Food n.d.).

Just as Slow Food began in Italy with a 1986 protest against the opening of a McDonald's restaurant below the historic Spanish Steps in Rome, Mexico has had its own showdown with McDonald's on the main square in Oaxaca City, the mecca of indigenous gastronomy. The company had already had one restaurant in a middle-class shopping mall on the outskirts of the city, like hundreds of others from Cancún to Tijuana, but when the local franchise holder sought permission for an outlet on the Zócalo in the summer of 2002, the intelligentsia mobilized its opposition. The campaign, led by renowned Oaxacan artist Francisco Toledo, asserted that burgers and fries were simply too different from the indigenous version of fast food, *chapulines*, fried grasshoppers sold from baskets and eaten with tortillas and guacamole. Bowing to the complaints of such high-profile figures, the government withdrew the permit, but the controversy had little importance, either culturally or economically, for ordinary Oaxacans who could not afford a Big Mac (Weiner 2002).

A similar protest had already played out in Mexico City against Maseca. As masa harina became ubiquitous in tortilla factories throughout the capital, the local branch of Greenpeace began issuing warnings about the use of GM-maize. In September 2001, self-styled ecological guerrillas changed the slogan on a giant billboard depicting flute-shaped crispy tacos from "Flautas with Maseca are tastier," to "Flautas with Maseca are genetically modified." Héctor Magallón explained the action by pointing out that in a recent survey, 88 percent of Mexicans indicated that they wanted GM-food to be labeled as such, and as a significant importer of maize from the United States, Maseca made an obvious target (Greenpeace 2001). While the desire for accurate labeling is certainly understandable, the protest movement has traveled a long way from the struggles of family-owned tortilla shops against a multinational giant.

For most poor Mexicans, surviving the shock of neoliberal reforms meant increased labor migration. Oaxacans were relative latecomers to the United States, but by the 1970s, Zapotecs and Mixtecs had become an important source of seasonal labor for California agriculture, and many lived year-round in Los Angeles. Despite difficult working conditions and vulnerability due to their undocumented status, migrants generally succeeded in improving their economic position and their remittances were vital for the survival of their families in Mexico. Over time, patterns of migration changed, from predominantly single males to include women and children as well, creating complicated transnational family networks. Higher incomes and tastes for consumer goods have caused significant cultural changes among migrants, especially those who settle permanently in the United States, but ties to home communities often remain strong nevertheless (Kearney 1996; Cohen 1999).

Migrant remittances form only one part of broader family survival strategies based on precarious subsistence agriculture in rural Oaxaca, as elsewhere. Farming has always been difficult in this region, but population growth and revenue from migrant labor has fostered urbanization, placing even greater pressure on cultivable land (Cohen 1999). Even successful artisans catering to the tourist trade spend much of their time in agriculture and depend on local food production (Chibnik 2003). Although often unlettered, these peasant farmers possess an extraordinary empirical understanding of their land and crops (González 2001). Nevertheless, such local knowledge may count for little if the economic necessities of competing with agribusiness cause a downward spiral of ecologically devastating practices, a sort of Gresham's Law of farming (Raikes 1988:62).

In seeking to encourage sustainable agriculture, Slow Food recognized a crucial problem for poor commodity-producing nations, the failure of branding. Mexico exports high-quality pork and vegetables but gains relatively little value from them because of the lack of global cachet associated with, for example, prosciutto ham or Tuscan olive oil. The case of cacao is symptomatic; once the drink of Maya lords, it is now a commodity subject to cutthroat international competition. Small farmers in Mexico suffered further exploitation, having just two options for marketing their harvest, a corrupt monopoly union or multinational buyers, both of which paid desperately low prices. With the assistance of two Mexican biologists and the Dutch organic certifying organization NOVIB, Sebastiana Juárez Broca parlayed her indigenous origins into a valuable brand name, Tia Tana Chocolate. But the adoption of capitalist advertising techniques conferred de facto property rights to a single individual rather than a village cooperative. As the Slow Food citation reads, Tia Tana "produces chocolate using traditional methods employing village women and encouraging the expansion of biological and economically compatible cultivation" (Slow Food n.d.). Yet the picturesque image of local production for tourists can conceal exploitative class relations within indigenous communities. Néstor García Canclini (1993) has observed that profits from craft sales to foreigners generally end up in the hands of local merchants and intermediaries.

Corporate control of international marketing further complicates the prospects for such ecological initiatives. Warren Belasco (1993) has shown the skill with which the food industry has co-opted counterculture movements and created its own forms of ersatz authenticity. Cookbook publishers likewise shared in the bonanza, offering up picturesque images of peasant cuisine to affluent readers (Pilcher 2004). The complicated recipes for traditional festival dishes may even be prepared occasionally by the leisured elite, but slow food offers little to single parents working overtime to support a family in the collapsing ruins of the U.S. welfare state.

Apocalypse Now?

The United States' campaign to spread GM food represents only the latest version of the "white man's burden" to uplift backward peoples through modern agricultural technology, although in practice the so-called Green Revolution has succeeded in promoting capitalist farming, rural unemployment, and urban shantytowns. Yet the Slow Food program likewise bears more than a passing resemblance to the *mission civilisatrice* of nineteenth-century imperialism. While more benign than military intervention, the missionary approach still conceals uneven power relationships that limit the opportunities available to the Mexican people. Moreover, the movement's goal of reducing agricultural overproduction resonates clearly with the dilemmas of the European Union's Common Agricultural Policy, with its mountains of surplus foods. International commodity markets can have complex implications for developing countries (see Raikes 1988), but the agricultural trade wars between Europe and the United States have caused tremendous harm to poor farmers around the world. One of the latest victims of this imperial struggle has been famine-ridden Zambia, the subject of widespread criticism for rejecting U.S. food aid out of fear that GM grain would contaminate domestic crops and forfeit European sales (Annear 2004). Viewed at this level, Sebastiana Juárez Broca appears as simply a colorful indigenous brand to sell Mexican chocolate to upscale consumers.

From the global struggles of industrial food retailers, Grupo Maseca has emerged as an unlikely champion of authentic Mexican cuisine. Mechanization has been essential to the survival of tortillas as the daily bread of Mexican wage laborers, who cannot afford the luxury of slow food on an everyday basis. In foreign markets, the company has launched ambitious expansion plans, and United States and Europe make up nearly 50 percent of total sales (Vega 2003). Occupying foreign territory has meant using local knowledge of fresh corn tortillas to challenge taco shell stereotypes. Although condemned at home as the antithesis of traditional cooking, bags of masa harina have become essential care packages for migrants and aficionados in exile from Aztlán.

With multinational corporations increasingly determining the availability and even the authenticity of food, class-based issues of market power become ever more crucial. While in many European countries restaurant service is an honorable trade paying a living wage, in Mexico it is often the last resort of the most impoverished people. Self-exploitation ultimately makes possible the Slow Food ideal of "good regional cuisine at moderate prices" (Petrini 2003:15). Moreover, Michael Kearney (1996:107) has offered a cautionary note about the conservative embrace of ecological projects generally: "one need not be cynical to see in official support of sustainable development and appropriate technology a de facto recognition that rural poverty in the Third World is not going to be developed out of existence. All peoples will not be brought up to the comfort level of the affluent classes and must therefore adapt to conditions of persistent poverty in ways that are not ecologically, economically, or politically disruptive." He concludes that "the de facto project of such right romantics is to sustain existing relations of inequality." Historically minded observers will note also that much of the program and even the slogan of "alimentary sovereignty" adopted by José Bové was promoted earlier by the fascist regime of Benito Mussolini (Parasecoli 2003:37; Helstosky n.d.). Slow Food has likewise replaced the developmental ideal of Prometheus with the more defensive symbol of Noah. As Petrini (2003:86) explains: "Faced with the excesses of modernization, we are not trying to change the world anymore, just to save it." Yet Oaxacan migrant workers, who must cross a militarized frontier in order to save their own communities, may ask themselves: will they find a place on the Ark?

This is not to deny that Slow Food offers hope for the survival of peasant cuisines. Indeed, the contemporary situation in Mexico parallels that of Italy a century ago. Just as half of all Italian labor migrants ultimately returned home, with new attitudes and more money but still part of their old communities, so do Mexican sojourners continue to follow their circular routes, notwithstanding hysterical media accounts of "a border out of control." Of course, it would be absurd to think that NAFTA will ever lead to income redistribution programs comparable to those of the European Union. Nevertheless, attempts to improve conditions in Oaxaca, like those in the Mezzogiorno, must adopt a continental vision. The indigenous political revival, with roots in both southern California and southern Mexico, offers a model for revalorizing ethnic communities (Kearney 1996). Those who sympathize can help most by allowing transnational families to flourish in their own neighborhoods rather than by indulging in exotic tourism to distant lands. The true foundation of sustainable agriculture in Mexico is outside labor paying decent wages—as it is for any American family farm.

Note

Acknowledgments. This essay was inspired by Martín González de la Vara and greatly improved by the suggestions of Richard Wilk, Donna Gabaccia, William Beezley, Sterling Evans, Glen Kuecker, and the anthropologists of SEA, who graciously welcomed an imposter in their midst.

References

Annear, Christopher M. (2004) "GM or Death": Food and Choice in Zambia. *Gastronomica* 4(2):16–23.

Bauer, Arnold J. (1990) Millers and Grinders: Technology and Household Economy in Meso-America. *Agricultural History* 64(1):1–17.

Belasco, Warren (1993) *Appetite for Change: How the Counterculture Took on the Food Industry.* Ithaca: Cornell University Press.

Camporesi, Piero (1993 [1989]) *The Magic Harvest: Food, Folklore, and Society,* Joan Krakover Hall, trans. Cambridge, U.K.: Polity Press.

Cebreros, Alfonso, n.d. Grupo MASECA: Un Caso Exitoso de Transnacionalización Agroalimentaria. Electronic document. http://ciat-library.ciat.cgiar.org/Alacea/v_congreso_memorias/V_grupo_maseca.htm, accessed September 6, 2003.

Chibnik, Michael (2003) *Crafting Tradition: The Making and Marketing of Oaxacan Wood Carvings.* Austin: University of Texas Press.

Cohen, Jeffrey A. (1999) *Cooperation and Community: Economy and Society in Oaxaca.* Austin: University of Texas Press.

Diner, Hasia (2001) *Hungering for America: Italian, Irish, and Jewish Foodways in the Age of Migration.* Cambridge, MA: Harvard University Press.

Fantasia, Rick (1995) Fast Food in France. *Theory and Society* 24:201–43.

Gabaccia, Donna (2004) *The Atlantic Origins of Italian Food. Conference on American Popular Culture.* Toronto, Ontario.

García Canclini, Néstor (1993) *Transforming Modernity: Popular Culture in Mexico.* Lidia Lozano, trans. Austin: University of Texas Press.

González, Roberto J. (2001) *Zapotec Science: Farming and Food in the Northern Sierra of Oaxaca.* Austin: University of Texas Press.

Greenpeace (2001) Greenpeace etiqueta anuncio espectacular de Maseca. Boletín 173, September. Electronic document. http://www.greenpeace.org.mx/php/gp.php?target=%2Fphp%2Fboletines.php%3Fc%3Dtrans%26n%3D173, accessed April 22, 2004.

Guillermoprieto, Alma (1994) *The Heart That Bleeds.* New York: Vintage.

—— (1999) In Search of the Tortilla. *The New Yorker,* November 26:46.

Helstosky, Carol (2004) *Garlic and Oil: The Politics of Italian Food.* London: Berg.

—— . n.d. In press. Mussolini's Alimentary Sovereignty.

Hobsbawm, Eric (1983) Introduction: Inventing Traditions. In *The Invention of Tradition*. Eric Hobsbawm and Terence Ranger, eds. Pp. 1–14. Cambridge: Cambridge University Press.

Kearney, Michael (1996) *Reconceptualizing the Peasantry: Anthropology in Global Perspective*. Boulder, CO: Westview Press.

Lewis, Oscar (1951) *Life in a Mexican Village: Tepoztlán Revisited*. Urbana: University of Illinois Press.

The New Yorker (2000) *Day Job: Taco Bell Employee*. April 24–May 1:185.

Ochoa, Enrique C. (2000) *Feeding Mexico: The Politics of Food Since 1910*. Wilmington, DE: Scholarly Resources.

Parasecoli, Fabio (2003) Postrevolutionary Chowhounds: Food, Globalization, and the Italian Left. *Gastronomica* 3(3):29–39.

Petrini, Carlo (2003) *Slow Food: The Case for Taste*. William McCuaig, trans. New York: Columbia University Press.

Pilcher, Jeffrey M. (1998) *¡Que vivan los tamales! Food and the Making of Mexican Identity*. Albuquerque: University of New Mexico Press.

—— (2004) From "Montezuma's Revenge" to "Mexican Truffles": Culinary Tourism across the Rio Grande. In *Culinary Tourism*. Lucy M. Long, ed. Pp. 76–96. Lexington: University Press of Kentucky.

Raikes, Philip (1988) *Modernising Hunger: Famine, Food Surplus and Farm Policy in the EEC and Africa*. London: James Currey.

Redfield, Margaret Park (1929) Notes on the Cookery of Tepoztlan, Morelos. *American Journal of Folklore* 42(164):167–96.

Ritzer, George (1993) *The McDonaldization of Society*. Thousand Oaks, CA: Pine Forge Press.

Slow Food (n.d.) Premio Slow Food. Electronic document. http://www.slowfood.com/eng/sf_premio/sf_premio.lasso, accessed April 22, 2004.

Stephen, Lynn (1991) *Zapotec Women*. Austin: University of Texas Press.

Taco Bell (2001) History. Electronic document. http://www.tacobell.com, accessed March 17, 2004.

Teti, Vito (1992) La cucina calabrese: è un'invenzione americana? *I viaggi di Erodoto* 6(14):58–73.

Vega, Marielena (2003) Grupo Industrial Maseca. *El Economista*, September 3.

Watson, James L. (ed.) (1997) *Golden Arches East: McDonald's in East Asia*. Palo Alto: Stanford University Press.

Weiner, Tim (2002) *Mexicans Resisting McDonald's Fast Food Invasion*. New York Times, August 24.

26

The Raw and the Rotten: Punk Cuisine[1]

Dylan Clark

This article investigates the ideological content of punk cuisine, a subcultural food system with its own grammar, logic, exclusions, and symbolism. As a shared system of praxis, punk cuisine helps to articulate subcultural identity, purpose, and politics. In the case of Seattle punks in the late twentieth century, their cuisine served to critique Whiteness, corporate-capitalism, patriarchy, environmental destruction, and consumerism.

Having been moved 2,000 miles to the north of its original home on the Rio Grande, a steel government sign was placed along the colorful fence of the Black Cat Café in Seattle, and there it retained something of its original meaning. It was a small white sign with black letters which announced, "U.S. Border." On one side, land administered by the United States; on the other, the sign implied, a space beyond the reach of the American state: an autonomous region.

For five years, this zone was a haven for people called punks and their kindred spirits,[2] an assortment of young adults who exercised and debated punk praxis in and through the premises. At the Cat, punks read, talked, smoked, and ate. They chewed ideas and articulated dietary practices, and rehashed their experiences with one another. Being punk is a way of critiquing privileges and challenging social hierarchies. Contemporary punks are generally inspired by anarchism, which they understand to be a way of life in favor of egalitarianism and environmentalism and against sexism, racism, and corporate domination. This ideology shows up in punk routines: in their conversations, their travels, and in their approach to food.

Food practices mark ideological moments: eating is a cauldron for the domination of states, races, genders, ideologies, and the practice through which these discourses are resisted. Indeed, as Weiss (1996: 130) argues, "Certain qualities of food make it *the* most appropriate vehicle for describing alienation." The theory and practice of punk cuisine gain clarity when they are viewed through the work of Claude Lévi-Strauss (1969), who saw the process of cooking food as the quintessential means through which humans differentiate themselves from animals, and through which they make culture and civilization. Lévi-Strauss's tripolar gastronomic system defines raw, cooked, and rotten as categories basic to all human cuisines. This model is useful for analyzing punk cuisine, and thereby punk culture. Yet this article also toys with the model, using it to give voice to the ardent critics of "civilization." Many punks associate the "civilizing" process of producing and transforming food with the human domination of nature and with White, male, corporate supremacy. Punks believe that industrial food fills a person's body with the norms, rationales, and moral pollution of corporate capitalism and imperialism. Punks reject such "poisons" and do not want to be mistaken for being White or part of American mainstream society. A variety of practices, many dietary, provide a powerful critique against the status quo.

A Punk Culinary Triangle

In the punk community, food serves to elaborate and structure ideologies about how the world works. Through a complex system of rules, suggestions, and arguments, punk cuisine is a code like those posited by Lévi-Strauss (1969, 1997). But punk cuisine is best discussed as a cultural mechanism responsible to its own logic, and in dialogue with what punks perceive to be the normative culture. Lévi-Strauss's ideas about food are insightful, especially when placed in a locally defined context (Douglas 1984). His culinary triangle (Figure 26.1) provides a helpful way to think about how the transformations of food can be cognitively mapped. For example, American food geographies have shifted toward processing (or cooking) food. Industrial food products are milled, refined, butchered, baked, packaged, branded, and advertised. They are often composed of ingredients shipped from remote places, only to be processed and sent once more around the globe. From a Lévi-Strauss perspective, then, punks consider industrial food to be extraordinarily cooked. Punks, in turn, preferentially seek food that is more "raw"; i.e., closer to its wild, organic, uncultured state; and punks even enjoy food that has, from an American perspective, become rotten—disposed of or stolen.

For punks, mainstream food is epitomized by corporate-capitalist "junk food." Punks regularly liken mainstream food geographies to colonialism because of their association with the Third World: destruction of rainforests (allegedly cleared for beef production), the creation of cash-cropping (to service World Bank debts), and cancer (in the use of banned pesticides on unprotected workers and water supplies). Furthermore, punks allege that large-scale stock-raising (cattle, chickens, pigs) and

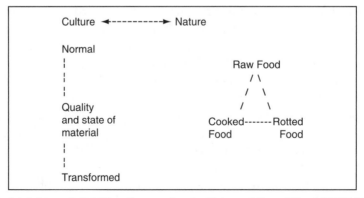

Figure 26.1 Lévi-Strauss' (1969) culinary triangle (Adopted from Wood (1995:11)).

agribusiness destroy whole ecosystems. A representative of this point of view states, "Ultimately this vortex brings about the complete objectification of nature. Every relationship is increasingly instrumentalized and technicized. Mechanization and industrialization have rapidly transformed the planet, exploding ecosystems and human communities with monoculture, industrial degradation, and mass markets" (Watson 1999:164; see also CrimethInc. Workers' Collective 2001:122).

Punk food attempts to break free from the fetishism of food as a commodity. As such, it is ideally purchased in brandless bulk or directly from farmers, self-made or home-grown, and otherwise less commodified, which is to say stolen or reclaimed from a garbage dumpster. By bathing corporate food in a dumpster or by stealing natural foods from an upscale grocery store, punk food is, in a sense, decommodified, stripped of its alienating qualities, and restored to a kind of pure use-value as bodily sustenance. In their organic, unmediated forms, such foods come closer to a "wild" diet, free of commodification and hierarchical relations of production, and closer to Lévi-Strauss' "raw" and "rotten" and further from his "cooked." Comments anarchist Bey (1991:54), "Food, cooked or raw, cannot escape from symbolism. . . . But in the airless vault of our civilization, where nearly every experience is mediated . . . we lose touch with food as nourishment; we begin to construct for ourselves personae based on what we consume, treating *products* as projections of our yearning for the authentic."

The Order of Signs at the Black Cat Café

Hardly a quaint place sweetly nestled in a booming urban landscape, the Black Cat was a boxy structure enclosed by a jagged rampart of fencing and discarded materials. Part of the fence was topped with a tangled line of bicycle frames, reminiscent of a wall of thorns. It enclosed a café yard of scattered benches, tables, and cigarette butts. Against the side of the café a mass of bicycle frames and parts made a tangled mound of metal. To beautify the courtyard, scrap-wood planters held salvaged greenery. The place looked more like a junkyard than a restaurant, for it violated normal aesthetic conventions associated with dining. Unlike other restaurants, the café did not strive to declare sanitation and safety. If the space of modern authority is clean, empty, and clearly marked (Sennett 1990:38), the façade and decor of the Black Cat Café suggested the antithesis. The café was cluttered, soiled, its interior covered with posters, art, and canvas coffee sacks: packed with bulky, dilapidated furniture, it felt cramped.

This ambiance was precisely what drew punks to the Black Cat collective. On a dirty cement floor that would offend mainstream good taste, punks tossed their rucksacks holding all their worldly possessions. Where others were repelled by the body odors of the unwashed, punks recognized kindred spirits. Where others feared to eat food prepared by grimy, garishly pierced cooks, patrons appreciated the ambiance of food lovingly prepared.

Food itself was one of the centrally reversed signs here, perhaps because food was the ostensible *raison d'être* for the restaurant. Black Cat food, like the café itself, was a declaration of autonomy and organic creation, a rejection of commodification.

Meat and dairy products were proudly excluded. Vegetables with peanut sauce, tofu scrambles, and other vegan creations served as entrees. The place and the food rejected strict adherence to conventional conceptions of hygiene, where even the appearance of filth somehow infects the object or the body. Here hygiene was associated with bleached teeth, carcinogenic chemicals, and freshly waxed cars, and operated as a code for sterility, automation, and alienation. Hygiene meant "idiot box" sitcoms and suburban fears of dark bodies. At the café, hygiene was a projection of Whiteness, and rejected.[3]

In rejecting the image of sterility, the Black Cat collective scorned decades of market research, and refuted dominant mantras of modernity. Marketing doctrine in the United States urges restaurants to emphasize scrubbed surfaces, clarity, and predictability. As a rule, the food industry seeks to provide a product so clean and neat that its human creation is not readily apparent. In this sense, the commodity fetishism as a corporate mandate is more apparent to the senses than is the migrant laborer in the field or the minimum-wage dishwasher. For the greater punk community, interchangeability, consistency, and hygienic food represent food that is utterly cooked and gastronomically problematic: "When we accept their definition of 'cleanliness,' we are accepting their economic domination of our lives" (CrimethInc. Workers' Collective 2001:123).

What to make, then, of a restaurant which rarely produces a tahini salad dressing the same way twice or a pile of home fries without a good many charred? What of a restaurant with spotty service, spotty dishes, where the roof leaks, and the bathroom reeks? For five years, the Black Cat found a way to thrive in spite of, or because of, its unorthodox practices. The workers and patrons of the café are a different breed who seek out what is "rotten" in mainstream society. One worker-owner, Ketan, talked about how the marginality of the Black Cat scared away many potential patrons, but noted with a laugh:

> I hope . . . people realize that this not a café. This is not a café. This is not a restaurant. . . . That's not what this place is about. This is a safe space. It is a haven for people who want to live their lives away from the bullshit of corporate oppression. That's what this space is about. It's not about anything else other than that. It's for people who want to believe what they want to believe and not be ridiculed, and be free from control by governments or other forms of systematic, abusive power things.

Food as Gender/Power

As a site of resource allocation, food tends to recapitulate power relations. Around the globe, unequal allocations of food according to a patriarchal system are common. A working-class male comes of age in France when he is able to help himself to large volumes of food (Bourdieu 1984). Men, and sometimes boys, often receive larger amounts of food and have culinary choices catered to their taste (see Narotzky 1997:136–37; Mintz 1985:144–45; Appadurai 1981). Thus, food displays practices through which unequal gender power is acted out, resisted, and reproduced (Counihan 1999). Punks, too, play out gender/power relations in their diets. In recent years, the punk ethic has become more committed to anarchist, egalitarian principles that celebrate and practice an antihierarchical social order, including one that prohibits a hierarchy of gender.

Feminist praxis in punk explicitly critiques food as a site of repression, using the Victorian age as an example of a discourse disciplining female bodies through food (Mennell 1985). This discourse was fostered in part by capitalist food and pharmaceutical industries eager to create new products for dieting and beauty (Bordo 1993; Chapkis 1986). Feminists identify this discourse as a form of control over women that at times leaves them malnourished, anorexic, or bulimic, and fixated on manipulating their body shape and diets. As a gendered and specifically American national project, by the early twentieth century, through women's magazines, newspapers, churches, cookbooks, and civic societies, native-born and immigrant women were educated in "home economics," a correlated set of technologies intended to produce an idealized femininity schooled in Whiteness, to produce the right kind of patriarchy and racial order of the U.S. nation-state. Such ideological uses of food are routinely referenced in punk food discourse, in everyday talk, by bands such as Tribe 8, and in "zines" (the popular broadsheets of punk) such as *Fat Girl*.

Thus, many punks identify the body as a place where hegemony is both made and resisted. Punks are critical of the beauty industry and of the commodification of the body. They argue that food is part of a disciplinary order in which women are taught to diet and manage their bodies so as to publicly communicate in the grammar of patriarchy. "Riot girl" punks, in particular, have produced a large volume of zines, music, conversations, and practices that challenge the sexist politics of food. In the ongoing evolution and critique of punk culture, diet is one of the many places where feminist ideas have been advanced and largely won out.

Indeed, vegetarianism for many punks is partly a feminist practice, but it also reveals ideological fissures within punk culture. Meat, with its prestige, caloric content, and proximity to physical violence, has been widely associated with masculinity (Adams 1990; Rifkin 1992). Yet even within punk culture, which is critical of both sexism and meat-eating (O'Hara 1999), some punks continue to produce an overtly sexist, masculine presence (Nguyen 1999) and one associated with eating meat. Meat for some punks is a way to challenge feminism in punk and to reassert masculine power. Other punk meat-eating falls into the categories of those who are apolitical about food, and those who flaunt meat-eating as a way of challenging punk orthodoxy.

For most punks, however, meat-eating is collaborative with an unjust social order, which punks typically portray as a patriarchy. Opposing social hierarchies, and living in staunchly patriarchal societies, they need to subvert male supremacy in everyday life, and vegetarianism, widely stigmatized as an oriental and feminine practice, helps to differentiate punks from the mainstream.

Punk Veganism

In punk veganism, the daily politics of consumption and the ethical quandaries of everyday life are intensified. In part, the decade-long struggle to make food and animal products overtly political was carried out by bands such as Vegan Reich and in zines. Zines regularly comment on animal rights, industrial food, and veganism. Often drawing upon Rifkin (1992) and Robbins (1987), many zines recount details of cruelty toward animals, contaminated meat, and the unhealthful effects of meat and dairy products on the human body. Other punk writing describes environmental consequences of industrial food production. "Even Punks who do not acknowledge the concept of animal rights and hold strong anthropocentric views have been known to change their diet purely for environmental reasons" (O'Hara 1999:135). In the daily praxis of punk, vegetarianism and veganism are strategies through which many punks combat corporate capitalism, patriarchy, and environmental collapse.

The emphasis on a radical diet was not always a dominant part of punk cultures. But by the 1990s, veganism was a rapidly ascending force within the greater punk landscape in North America. Led by the "straight edge" punk movement, veganism gained credence across the punk spectrum, including those who scorned the drug-abstaining politics of straight edge, as did most Seattle punks in this study.

At the Black Cat Café, punks said that to eat animal-based products was not only unhealthful, it participated in the bondage and murder of animals. Many punks were concerned about the cruel conditions of factory farms, where animals were kept in cramped quarters, pumped with hormones and antibiotics, and "tortured" in sundry ways. Near the middle of its tenure, the Black Cat discontinued its use of milk and eggs. A vegetarian café from its outset, the Cat became more orthodox when its menu was made completely vegan. The transition to a vegan menu marked a turning point for the collective. The original members had dropped out, and a younger, more militant membership had taken control. The café became less tolerant, less com-promising, and more thoroughly punk in its clientele and ambiance. Ketan expressed the urgency that many punks feel about veganism:

> There's this line that occurs with being vegan and being activist: at what point does the freedom of people who believe what they believe cross over to the point where people are being harmed? You know? Like, yeah: people are free to eat meat. But actually, in this day and age, they *can't* eat meat because it's killing animals. Because someone is eating meat, land that could potentially benefit all of us is being destroyed. I have a lot of problems with that line: I don't want to impede people's freedom, but what everyone does affects everyone else. . . . I honestly believe that people have to stop eating meat now. Now! I'm not gonna force anyone to stop eating meat, but they're hurting me, my children's future, my friends, my family— because they're eating meat. And they're hurting the Earth, which is most important of all.

Many punks around the nation were part of the growing politicization of the culture, with veganism at the forefront of the politics. To be vegan in America is to per-petually find oneself in the minority, chastised, excluded, challenged, and reminded of one's difference. In this sense, veganism also served as an incessant critique of the mainstream, a marker of Otherness, and an enactment of punk.

Raw as a Critique of Cooked

In punk cuisine, the degree to which food is processed, sterilized, brand named, and fetishized is the degree to which it is corrupted, distanced from nature, and "cooked." Punks describe a world under the assault of homogenized foods and culture, a world of vast monocropped cornfields and televisions lit with prefabricated corporate "infotainment." Whereas industrial agriculture is associated with genetic engineering, monocropping, pesticides, animal cages, chemical fertilizers, and commodification, "raw" food tends toward wildness and complexity.

Punks perceive in everyday American food an abject modernity, a synthetic destroyer of locality and diversity. The "cooking" of foods, to which punks vociferously object, is an outcome of the industrialization and commercialization of modern food production, which are made visible and critiqued through punk culinary practices. The following trends in modern food manufacture and consumption comprise the increasingly cooked qualities of food against which punks can be said to form their culinary triangle.

From a punk perspective, American food has reached an unprecedented and remarkable state: nearly all the food that Americans eat is received in the form of a commodity, and the fetishism of food goes far beyond the simple erasure of labor. Lears (1994:171) describes the emergence of the industrialization of eating:

> By the 1920s and 1930s, advertisements for food displayed an almost panicky reassertion of culture over nature—an anxious impulse to extirpate all signs of biological life from one's immediate personal environment. That impulse has been spreading widely for decades, as methods of mass production were brought to food processing and distribution.

Such logics, for example, are apparent in the segregated meat products, in which the animal carcass is hidden. The animal's head, feet, and tongue (its recognizable body parts) have disappeared from most American butcher displays.

Through the most sophisticated branding, packaging, and advertising, American food commodities work hard to conceal the labor, spatial divides, and resources that went into making the food. Or, as Weiss (1996:131) shows, "the effects of encompassing transformations in political economy (colonialism, wage labor, commoditization and the like) . . . [have] their greatest and gravest consequences for food." In modern advertising, images of food often divert attention from the industrialized production of food, and draw attention to its consumption (DuPuis 2000). Rather than depict the mechanized dairy factory, ads show celebrities and athletes wearing smiles and milk mustaches. Notes Harvey (1989:300):

> The whole world's cuisine is now assembled in one place. . . . The general implication is that through the experience of everything from food, to culinary habits, music, television, entertainment, and cinema, it is now possible to experience the world's geography vicariously, as a simulacrum. The interweaving of simulacra in daily life brings together different worlds (of commodities) in the same space and time. But it does so in such a way as to conceal almost perfectly any trace of origin, of the labour processes that produced them, or of the social relations implicated in their production.

Perhaps these postmodern geographies, along with relentless commodification, heighten the fetishism of the commodity, hiding as much as possible the making of a product; the alienating conditions of production, cooking in the extreme.

Punks see industrialized food production not as a desired convenience, but as one of the hallmarks of monoculture. The anarchist idea of monoculture plays on the "culture" part of the term, thus expanding it to cover not only modern industrial agriculture, but also mainstream culture. For punks, monoculture encapsulates the idea that societies around the world are being devoured and homogenized by consumerism; it invokes the idea that humans everywhere increasingly eat, dream, work, are gendered, and otherwise live according to a narrow and hegemonic culture sold to them by global capitalism. Across the globe, punks argue, humans are losing their cultural, ecological, temporal, and regional specificity. Among other things, this means that people are often eating foods grown and flavored elsewhere: people everywhere are increasingly alienated from that which keeps them alive.

"Raw" food, which is to say, organic, home-grown, bartered food, was one way punks resisted the spread of monoculture. At the Black Cat Café, customers could trade home-grown organic produce for meal credits, they could trade their dishwashing labor for meals, and they could drink "fair trade" coffee. Moreover, the café strove to subvert profiteering at every step in the food's production. At the Cat, people who might be called punks contrasted the synthetic, processed, and destructive diet of the mainstream with their own, and declared that their bodies and minds were healthier for it, unpolluted by toxic chemicals and capitalist culture.

Stealing Yuppie-Natural Foods

Not far from the Black Cat Café, Seattle hosted a variety of natural-foods retailers, who attracted both the contempt and the palates of punks. Such places offered organically grown foods marketed to an upscale clientele. Indeed, the natural-foods industry in 1990s Seattle was part of a vast reconfiguration of food in America, which witnessed a hitherto unprecedented niche marketing of what punks saw as foods which fed egos more than bodies. The punk narrative critique of the natural-foods movement was extended by stealing, for by this the food was remade.

Punk discourses of food are partly a response to the heightening of identity marketing in foods over the last few decades of the twentieth century. Although locating identity and prestige in food is an ancient practice, it has historically been limited by income, tradition, and spatial divides. But in contemporary America, the bewildering array of food choices challenges the consumer, whose choices are understood to "express" or manufacture him- or herself. Americans have reached the point at which food as essential for survival has been sublimated under the ideology of food as self-gratification and consumer identity.

Such formulas were apparent to punks in the commercial discourse on natural foods. Punks regard these foods, while ostensibly pure and simple, as much commodities as the food products that preceded them, and derisively locate "yuppie," "individualistic," and "White" behavior in an expensive obsession about one's own purity and health (see Bey 1991:53). The natural-foods industry, then, is a target of punk critical practices. In Seattle, the Puget Consumers' Co-op (PCC) bore the brunt of the punk natural-foods critique. Fashionable, expensive, and allegedly catering to a mostly White and upscale clientele, the PCC was scorned by punks.

While commodified natural foods were repulsively overcooked, they were simultaneously closer to the raw forms of food that punks preferred: organic, bulk, and whole grain. So, while the PCC market offered the organic products that punks preferred (as well as a relatively tolerable and tolerant workplace for those who opted for wage labor), the high prices and upscale marketing represented the cooking of foods; the heightened state of gastronomic fetishism from which punks felt alienated. Cleansed of their commodification, these foods would be perfectly suited to the punk culinary system. Thus, many punks, whether as workers or customers, targeted natural-foods supermarkets for theft (c.f. Himelstein and Schweser 1998:18–21, 24). In this manner, the kitchen of the Black Cat Café was routinely stocked with products stolen from chain supermarkets and natural-foods stores. This behavior suggests an axiom of punk culinary geometry: in the act of being stolen, heavily cooked food is transformed into a more nutritive, gustative state. Stolen foods are outlaw foods, contaminated or rotten to the mainstream, but a delicacy in punk cuisine.

The Rotten Logic of Dumpster Diving

Each night American supermarkets and restaurants fill their dumpsters with food, and each night punks arrive to claim some of it. A host of foods become rotten in corporate-capitalist food production: food with an advanced expiration date, cosmetically damaged produce, food in dented packaging, day-old baked goods, and the like. As punks saw it, people were hungry in Seattle, in America, and around the world. To punks it was obscene that businesses were trashing good food (Resist 2003:67).

Unlike raw foods, dumped food tends to be commercialized, nonorganic, and highly processed. Baked goods, donuts, produce, vegetables, pizza, and an array of junk foods are foraged by punks, who otherwise disdain such products. Yet in the process of passing through a dumpster, such foods are cleansed or rotted, as it were, and made nutritious and attractive to the punk being.

It was ironic to punks that people are hassled by security guards, store employees, and police merely for taking things out of a dumpster. Not only did the mainstream waste food, it protected its garbage with armed guards. Commented one punk: "There is the odd paradox—the casualness with which they will throw something into the dumpster, and the lengths they will go to protect it once it's there. How an innocent and harmless act—dumpster diving—will be confronted by greedy shopkeepers, store managers, and employees with scathing words, rage, and violence" (Anonymous 2001:72). Taken in tandem, the waste of food and the protection of waste were seen by punks as the avaricious gluttony of American society. Food in dumpsters is, for most Americans, garbage and repulsive. It goes beyond the pale of Whiteness to eat food classified as garbage: only untouchables, such as the homeless, eat trash. So for those punks who were raised White or middle class, dumpsters and dumped food dirty their bodies and tarnish their affiliation with a White, bourgeois power structure. In this sense, the downward descent into a dumpster is literally an act of downward mobility. Moreover, eating garbage (food deemed rotten) is a forceful condemnation of societal injustices. On an ecologically strained planet home to two billion hungry people, punks see their reclamation of rotten food as a profoundly radical act.

Gastro-Politics in Punk Activism

For its five years of existence, the Black Cat Café was the kitchen of Seattle's punk scene. It was a decidedly anticorporate environment, where mainstream types were not always welcomed, and where there was always room for young wayfarers. As with many cultures, punk food practices helped shape community, symbolize values, and foster group solidarity. The Black Cat was a place where anarcho-punk "dis-organizations" could put up flyers, recruit members, and keep their limited dollars circulating. At the café, feelings of alienation from the mainstream were converted into punk sentiments and channeled into anarchist practices.

Various activist groups were associated with punk culture. One of the foremost was Food Not Bombs, an anarchist dis-organization. It served to collect, prepare, and distribute free food to the homeless and the hungry. The hostility of the Seattle City Council and Seattle police toward Food Not Bombs was received at the Cat as another sign of American class warfare and a coercive attempt to force even the homeless to turn to commodities for their survival (see also Narotzky 1997:114). When Food Not Bombs was cited for giving meals to the poor, this revealed the militancy of the ruling class to punks. Despite—and because of—the hassles from authorities, Food Not Bombs drew many volunteer hours from people who were affiliated with the Black Cat. Ketan mentioned Food Not Bombs as inspiring him to become a punk:

> I think the reason I chose not to [be a part of the mainstream] is . . . empathy . . . empathy and recognition of . . . what we're going through. I myself have been helping out with Food Not Bombs for a year straight, and [so] I've got a pretty good idea of what [poor] people go through. And I myself have [suffered] in the sense that I've not had my own space, and it's drove me crazy—you know, not knowing where I was going to sleep the next night. . . . Certainly I can't say that I know exactly what's going on [with the homeless], but I'm just trying to say that I have some understanding of it, you know? Just knowing that [poverty's] happening. And knowing that that's happening in the midst of that CEO making 109 million dollars [a year] . . . just knowing that makes me not want to be a part of that [wealth]. And that's happened with a lot of people here. I don't want to say what they believe, but—people here try to be as aware as they can of what's going on.

Another member of the scene, Karma, said that the "sense of family" drew her to the Cat.

> I like the fact that it's not run to make money. It's run for people, not profit. There's always some cause happening, some flyer up about something to go to: Books to Prisoners or Food Not Bombs or the Art and Revolution thing. . . . I think [activism] has a lot to do with it—certainly not the majority of why people come here. I think the majority of why people come here is because there's cheap food that's damn good. But because the food is specifically vegan, and that on a level by itself is activism, a lot of activists are vegan so they end up coming here. [Laughs.] And that kind of spurs the whole activism-crowd thing. Because they're all coming here, leaving their flyers, more people are coming, they're seeing the flyers, "Oh yeah, look: this is going on."

By making its political content explicit, food became a primary site of discussion and recruitment. In these movements, punk cuisine took shape and with it punks at the Black Cat concocted a daily life of meaningful situations, anarchist discourse, and resistance to "the System."

Conclusions

Contemporary punks—largely anarchist, antiracist, and feminist—use food as a medium to make themselves, and to theorize and contest the status quo. As an integral part of their daily practice, punks politicize food. For punks, everyday American food choices are not only nutritionally deficient, they are filled with a commodified, homogenous culture, and are based in White-male domination over nature, animals, and people around the world. Punk cuisine is one way punks critique these power relations, and one substance with which to remake themselves outside of those relations. Punk cuisine is a way to make punk ideas knowable, ritualized, and edible; a way to favor the less mediated anarchist food over the capitalist product: the raw over the cooked. From punk vantage points, modern American food is transformed to a cultural extreme; its origins in nature and labor are cooked away, leaving a fetishized byproduct. Punk cuisine aspires toward food that is free of brand names, pesticides, and exploited labor, and toward food that is as raw as possible. In punk poesis, raw is a metaphor for wild, and one of the most important tropes in punk culture. Where mainstream society is said to control, exploit, and homogenize foods and people, punks idealize freedom, autonomy, and diversity.

For five years, the Black Cat Café brought punks together in a cultural space where they critiqued modernity, capitalism, Whiteness, and mainstream America. In their cuisine, punks identify and challenge Fordism, sexism, greed, cruelty, and environmental destruction. They choose to avoid eating American cuisine, for they see the act of eating everyday American food as a complicit endorsement of White-male corporate power. Reared White and middle class, and raised on foods that are seemingly nonideological in American culture, punks come to reject their ethno-class identities and cuisine, for they believe that mainstream American foods recapitulate a violent and unjust society. Mainstream American food, with its labor and natural components cooked beyond recognition, is countered with the raw and rotten foods of punks; foods that are ideally natural, home grown, stolen, discarded, and uncommodified. These foodways define punk cuisine and punks themselves.

Notes

1. My deep gratitude goes to Ratna Saptari, Marcel van der Linden, and to the International Instituut voor Sociale Geschiedenis (International Institute for Social History) of Amsterdam for support during the writing of this paper. Many thanks also to the Bushuis Bibliotheek in Amsterdam.
2. Research for this article stems from my participant-observation in the Black Cat Café in Seattle from 1993 to 1998. The café was owned and operated by people called "punk" in their culture for their anarchist philosophy. Punks are diverse, and though these punks might be called "anarcho-punks" (to distinguish them from gutter punks, straight-edge punks, and other types), all punk ideologies are related. The cuisine of punks is always changing, always being argued over, and always responding to new circumstances and ideologies.
3. The restaurant was never in violation of health codes except for minor offenses (once, for example, an inspector prohibited leaving rice in the rice cooker and the collective grudgingly had to buy a food warmer). Dishes, food, and hands were washed, and no customer ever reported suffering from food poisoning.

References

Adams, C. J. (1990) *The Sexual Politics of Meat: A Feminist-Vegetarian Critical Theory.* New York.

Anonymous (2001) Evasion. Atlanta.

Appadurai, A. (1981) Gastro-Politics in Hindu South-Asia. *American Ethnologist* 8:494–511.

Bey, H. (1991) *T.A.Z.: The Temporary Autonomous Zone, Ontological Anarchy, Poetic Terrorism.* Brooklyn.

Bordo, S. (1993) *Unbearable Weight: Feminism, Western Culture and the Body.* Berkeley.

Bourdieu, P. (1984) *Distinction: A Social Critique of the Judgment of Taste,* transl. R. Nice. Cambridge MA.

Chapkis, W. (1986) *Beauty Secrets: Women and Politics of Appearance.* Boston.

Counihan, C. M. (1999) *The Anthropology of Food and the Body: Gender, Meaning, and Power.* New York.

CrimethInc. Workers' Collective (2001) *Days of War, Nights of Love: Crimethink for Beginners.* Atlanta.

Douglas, M. (ed.) (1984) *Food in the Social Order: Studies of Food and Festivities in Three American Communities.* New York.

DuPuis, E. M. (2000) *The Body and the Country: A Political Ecology of Consumption. New Forms of Consumption: Consumers, Culture, and Commodification* ed. M. Gottdiener, pp. 131–52. New York.

Harvey, D. (1989) *The Condition of Postmodernity: An Enquiry into the Origins of Cultural Change.* Oxford.

Himelstein, A. S., and J. Schweser (1998) *Tales of a Punk Rock Nothing.* New Orleans.

Lears, J. (1994) *Fables of Abundance: A Cultural History of Advertising in America.* New York.

Lévi-Strauss, C. (1969) (1964). *The Raw and the Cooked,* transl. J. Weightman and D. Weightman. New York.

Mennell, S. (1985) *All Manners of Food.* London.

Mintz, S. W. (1985) *Sweetness and Power: The Place of Sugar in Modern History.* New York.

Narotzky, S. (1997) *New Dimensions in Economic Anthropology.* London.

Nguyen, M. (1999) It's (Not) a White World: Looking for Race in Punk. *Punk Planet,* Nov/Dec. Also available at: www.worsethanqueer.com.

O'Hara, C. (1999) *The Philosophy of Punk: More Than Noise.* San Francisco.

Resist, M. (2003) *Resist #44.* Minneapolis.

Rifkin, J. (1992) *Beyond Beef: The Rise and Fall of the Cattle Culture.* New York.

Robbins, J. (1987) *Diet for a New America.* Walpole NH.

Sennett, R. (1990) *The Conscience of the Eye: The Design and Social Life of Cities.* New York.

Watson, D. (1999) (1991) *Civilization in Bulk. Against Civilization: Readings and Reflections,* ed. J. Zerzan, pp. 155–64. Eugene.

Weiss, B. (1996) *The Making and Unmaking of the Haya Lived World: Consumption, Commoditization, and Everyday Practice.* Durham.

Wood, R. C. (1995) *The Sociology of the Meal.* Edinburgh.

27

Salad Days: A Visual Study of Children's Food Culture

Melissa Salazar, Gail Feenstra, and Jeri Ohmart

Please visit the companion website at www.routledge.com/textbooks/ 9780415977777 to see this chapter in vivid color

Introduction

As the food industry is well aware, making food choices is a highly visual activity. We see food before we touch or even taste it, and use our eyes to determine whether or not something "looks good enough" to eat. These visual cues are not all innate; instead we learn what attractive and unattractive foods and meals look like through our individual visual, taste, and smell experiences, as well as through watching what others around us do with food (Birch et al. 1986; Birch et al. 1996).

It makes sense, then, that visually-based research methods are particularly relevant tools for explorations of food behavior. As nutrition researchers have discovered, asking people to describe their quotidian food behaviors can be challenging, and documenting food behavior through photography and video can be more accurate than questionnaires or memory recalls (Williamson et al. 2003, 2004). In addition, image-based researchers have pointed out that the production and examination of photographs and videos of social life can act as a concrete activity around which researchers and participants develop a greater understanding of both individual and cultural norms (Collier 1967; Prosser 1998; Wagner 2004). In this respect, images taken of food in everyday social use can not only help to bring routine eating behaviors to the forefront of subjects' consciousness, but also assist researchers and participants in "visualizing" socio-cultural food norms and values that might be hard to articulate through other means.

A third, and often-overlooked benefit of employing visual recording into food related fieldwork, however, is that visual data provide rich "snapshots" of social life that can be viewed (and re-viewed) as research inquiries develop and change. Images, though fixed in a certain moment in time, can contain a surprising amount of information—often beyond what the photo- or video-grapher set out to capture. Researchers and participants can mine visual data again and again, each time with fresh eyes, moving background activities in the photograph or video to the forefront, and re-analyze photos and video not only for what "truth" they portray but also for what they do not (Becker 1986; Prosser 1998).

For all these reasons, visually-based methodologies can be particularly useful in examinations of food in cultural life. In this photo-essay we will explore such a case—specifically, how taking photographs of children's lunch plates provided our team of nutritional researchers with additional (and unexpected) information about children's food culture. In the next section, we will discuss our case in more detail, first by presenting the purpose of the project and where our photographs came from, and then follow with some specific examples of how our visual data became useful to us in other ways. We'll conclude with a discussion of how the photographs allowed us to better understand the ways children go about assigning meaning to their food choices, and, in a sense helped us to "see" food through child rather than through adult eyes.

"Can we please take a picture of your lunch?"

The photographs presented here were taken as part of a comprehensive evaluation of a salad bar school lunch program[1] operating in several elementary schools in Northern California during the 2004–2005 school year (Ohmart & Feenstra 2004). The school districts had spent considerable time and money in instituting the program and were interested in finding out more about students' use of their new school lunch program. The school lunch salad bars featured bins of fresh fruits and vegetables in a self-serve format, with the intent of encouraging children to select (and consume) a wider variety and greater amount of produce as part of their school lunch. As a result, a main objective of our evaluation was to determine if the produce portion sizes of children's choices within this context were greater than the pre-packaged and pre-selected fruits and vegetables that were served with the ordinary school hot lunch.

The self-serve aspect of the new school lunch salad bar was the driving force behind our initial decision to use photography as an evaluative device: during lunch-time, hundreds of children went through the school lunch line within the course of an hour. Rather than trying to sample random students, or record each student's individual meal with written field notes, we decided that a more efficient and accurate method would be to photograph every student's individual lunch plate as they left the salad bar lunch line. To move as quickly as possible and not miss any of the students, two researchers were stationed with hand-held "point and shoot" digital cameras at the exits of the lunch line. We approached students quietly as they left the line, asking them if we could take a picture of their lunch plate "in order to learn more about what kids like." Since it was important to our evaluation that the children's normal lunch behavior would not be altered by our presence, we tried to be as inconspicuous as possible, and not have students entering the line knowing that we were going to take their picture[2].

We collected approximately 850 digital photographs of individual students' lunches over a period of three months, both on salad bar days and regular hot lunch days where the school served a pre-portioned meal. Because we had intended to use the photos to quantify the students' serving sizes of various food components, we zoomed in quite closely to the meal, and our photos were filled with a great deal of food detail.[3] As we sorted through the hundreds of photos, our eyes glazing over as we meticulously counted croutons and the numbers of carrot coins on each plate, we began to notice several qualitative themes from our photographs—particularly those taken on salad bar days where children had wider freedom in arranging food on their plates.

As a result, we decided to revisit our salad bar photographs with new questions in mind. Instead of simply cataloging the quantities of fruits and vegetables on students' plates, we took a step back and examined each photographed school lunch plate as a *cultural construction*: a visual depiction of a series of conscious actions and decisions made by the children that reflected the cultural meaning behind their food choices. We were particularly interested in the plates of food that we (as adults) thought looked "disgusting," or "weird"—in other words plates that caught our eye because they deviated from our adult norms of what a "proper" lunch salad should look like, and challenged our expectations of what we had thought children would do with the choices we had laid before them at the salad bar. The following pages present some of these photographs, accompanied by text that describes some of our thinking about these images. Some of the photos made us laugh, others perplexed us, but overall we found the following images valuable in raising our awareness of the various disconnects between adults and young children's food rules, norms, and values in their lunchtime agendas, and thinking more broadly about the design of school lunch programs for elementary age children.

Deconstructing a salad

Order in the (Food) Court

These photos show the general set-up of the salad bar program. Children lined up on both sides of the bar, picked up a plate and utensil set, and then slid their tray along the bar. There were several aspects of the program design that encouraged children to copy adult food norms, and hinted at adult's expectations of how children should go about building their salads. For example, the salad bar equipment closely resembled self-serve salad bars found in adult restaurants, complete with germ-fighting plastic sneeze guards, although the height of the salad bar was adjustable to allow the bar to go lower to the ground.

The salad bar was messy. The tongs and spoons were adult rather than kid-sized, and meant that younger children with still-developing motor skills had a harder time scooping food neatly onto their trays without spilling. At times we also observed children using their hands to scrape food off of their serving spoons or simply picking up the food out of the bins with their fingers. Food would often end up on the floor or on the sliding area of the salad bar setup.

The loose bins of food and the resultant lack of "order" in the service area clearly bothered the food service workers. Many of them carried a wet towel with them in order to wipe up the inevitable spills, and constantly reminded students about food safety and proper handling of food. But their exasperation at the students' messiness was due to their concern about the bottom line—cleaning up after the salad bar took much longer than on a regular hot lunch day, causing them to work overtime and raised labor costs (and food service worker resistance) to the program.

What adults planned . . .

Another kind of "adult order" was apparent from the order of the ingredients within the bar. The students encountered the salad bar from left to right, and would encounter lettuce first, then toppings like carrots and cheese, followed by a variety of fresh fruit. Then, students would move to a second table that held bins of croutons, protein options like beans, chicken, and tuna fish, and finally salad dressings and cartons of milk. From this order of ingredients, a child was most likely expected to construct something like this salad made by a teacher going through the lunch line with her class.

... and what kids did

The following photographs depict the "salads" constructed by the elementary age students. At first glance, we felt these salads looked odd, as they challenged our adult images of what a correct salad should look like. Once we completed our nutritional calculations, however, we saw that the majority of children's salads met or exceeded the nutritional requirements for a balanced school lunch.

While both the teacher- and student-constructed salads had similar components, visually they were strikingly different. Children's plates typically showed careful arrangements of foods into piles, some into compositions worthy of Mondrian. It appeared children wanted to eat small collections of hors d'oeuvres plates, rather than a mixed salad typically made by adults. We wondered whether this occurred due to children's lack of visual knowledge of what a "proper" salad should "look like," or if they knew about—but chose to ignore—the expectation of building a mixed salad.

In either case, it was apparent to us from these photographs that what children did was no accident; the clearly defined, discrete homogenous food piles they created indicated that they were thoughtful about their food choices and placements. The use of recyclable brown lunch trays was a major accomplishment for the eco-friendly parent advocacy group in this area, but they were an annoyance to the students. They complained bitterly about the trays for making it difficult to take a larger pile of food without having their foods mix or touch. Many students went so far as to use objects like utensil packets and milk cartons to create makeshift sections in their trays that would keep various components from mixing.

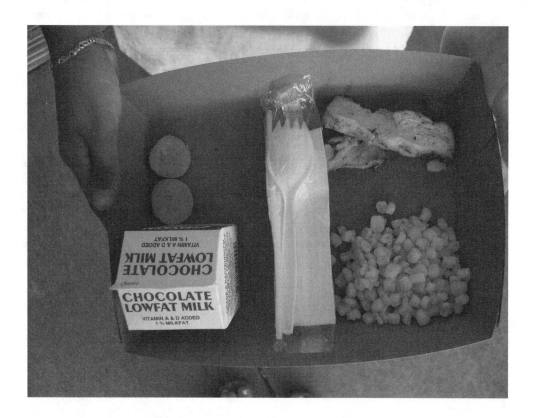

Freedom of Choice?

Creative meal combinations

Our photos also showed us the potential dangers, as well as the pleasures, of allowing children to make their own combinations and amounts of food at the salad bar. On hot lunch days, children had to make very few (if any) of their own decisions—they simply walked through the line and picked up a pre-packaged meal with pre-selected food combinations. On salad bar days, however, children told us they enjoyed the freedom of the salad bar program as they could accommodate their variable appetites and tastes. With the salad bar, they said, they would take only the amounts and types of food that they really "felt like" eating that day.

This led to some unusual visual combinations of food but perhaps meals that were more reflective of the "mini-meal" eating style that this age of children seemed to prefer.

Kiwis, cheese and croutons.

Lopsided plates

There were several institutional tensions, however, that limited students' self-serve freedoms. Adults working at the salad bar did frequent visual checks of student's plates for correct portion sizes and variety. These visual checks were not arbitrarily imposed—under the "offer vs. serve" USDA model of school lunch, students' salad plates were required to contain at least 3 out of the 5 nutritional components (milk, fruit, vegetables, bread, protein). But the larger battle between adults and children was fought over portion size. The economics of the program meant adults had to limit children's servings of more expensive (but highly popular) items like cherries, strawberries, oranges, and kiwis, or else these items would run out by the end of the first lunch period. The small ratio of adults to students, however, meant that some students snuck larger serving sizes than they were allowed.

Making economic matters worse, lunch workers also had to allocate more food to the program than the students would actually take. Apparently this was necessary for psychological reasons—according to the workers, students coming in during the final lunch period were more tempted to buy the salad bar lunch by seeing full "fresh looking" (rather than mauled through) bins of carrots, lettuce, and fruit, rather than an almost empty bin.

Apple and oranges.   *Chicken and cherries.*

Skip the lettuce

Building crouton-based "salads"

Croutons were popular, and were used by the children in creative ways different than adults. Students seemed to treat croutons as adults would treat lettuce in a salad, using it as a crunchy base that would support (and soak up) salad dressing and other toppings. But since adults often place croutons on top of salad, the crouton bin was located at the very end of the salad bar setup. Students were careful planners in this respect, saving room for their croutons on their trays while loading up on fruit and other items in the beginning of the line.

Croutons and dressing, with a side of chicken and oranges.

While crouton abuse was a common complaint from adults watching over the salad bar, our nutrition analyses showed a different story. On average most students took only one or two servings of croutons, and the majority of students' crouton servings were within the USDA guidelines of bread servings. But crouton-heavy plates like these pictured here were such powerful visuals to the adults running the program that eventually a crouton management system was set in place where students would be handed prepackaged and pre-measured serving size packets of croutons instead of being able to serve themselves[4].

Croutons topped with chopped egg, tuna fish and beans.

What our photos helped us to 'see' about children's food culture

Our choice to use images in our research on children's lunch plates provided several benefits to our evaluation process. The photos allowed us a more comprehensive, and holistic view of children's use of the salad bar program. For example, from our photos we could see not only *what* children took from the salad bar, but also *how* they took the materials and arranged them on their plates to construct a different type of "salad." While the nutritional analysis from our photos confirmed that children were indeed taking a greater number and a wider variety of fruits and vegetables, the photos also provided some clues as to why that was so. From the high number of lunch plates that contained diverse (and sometimes abnormal-looking) combinations and arrangements, it appeared that children enjoyed the freedom to feed their highly changeable and individualistic tastes and appetites.

The photos also helped us to gain new insights and understandings into children's food culture. The loose, self-service style of the salad bar, where children were relatively free to choose their own ingredients, combinations, and amounts of food, was an ideal vantage point from which to view children's own food values and material culture as distinct from adults. Rather than building what adults might consider a "salad" (mixed ingredients with certain foods like lettuce on the bottom and others like carrots and croutons on top) children instead crafted carefully separated piles of bite-size pieces of fruits, vegetables and meats—a deconstructed "salad" that appeared to an adult eye to be unevenly balanced and visually monotonous.

Children's distinctly different visual food culture also served as an especially important counter-point to deterministic models of children's food socialization; namely that children will simply copy adult ideas of what looks good to eat when presented with the freedom to conduct their own food experiments. Instead, our photos showed exactly the opposite–children built salads that were strikingly different than adults, and were also exceedingly thoughtful about how best to utilize the food selections and materials presented to them. Sociologist William Corsaro characterizes this as the process of *interpretative reproduction*, a dynamic process by which children actively contribute to their own cultural production while living within an existing adult-imposed social structure (1997). This model supports a new paradigm in child development that treats children as beings that not only are capable of complex thought but also able to actively construct their own lives and experiences (Prout & James 1997).

The photographs also helped us to explore tensions between satisfying children's cultural food preferences and the institutional tensions of operating a school food program. Allowing children to make their own meals might account for happier lunchtime eating[5], but giving children these freedoms to construct their own lunches has other costs. In this case there was considerably more prep and clean up work for the food service workers, which raised costs. In addition, the many choices also slowed children's passage through the lunch line, turning away children who were too impatient to wait.

Without a major overhaul of the financial and time constraints that school lunch programs operate under, however, caring about children's individual tastes and their ability to express variety on their lunch plates will always take a back seat to economic concerns. In the case of this program, due to the high cost of the salad bar in labor and overtime paid for cleaning it up, after a few years the salad bars in almost all of these schools were eventually phased out and replaced with pre-packaged lunch salads. Although we were disappointed in this outcome, our photos still played a role in the design of the new program. The school district food service manager used them to help decide what selections of foods should be in the pre-packaged salads offered to students (although, in the end, we saw no kiwi, cheese and crouton salads being offered).

Overall, children's food actions that were depicted in the photos spoke louder than any words the children had used in interviews to describe their use of the program. As nutrition researchers and reformers, the re-examination of our photos for these cultural behaviors added considerably to our understanding of how young children thought about food choices, and how adults' best intentions might not mesh with children's food agendas. While we knew that kids thought about food differently than adults, and we had caught glimpses of this behavior from our live observations, only from the systematic review of the hundreds of photographs after the event did we realize the *extent* to which children were re-institutionalizing the program to meet their own needs.

In this respect, we contend that those with food reform projects and nutrition interventions in mind would similarly benefit from ethnographic visually-based studies of the program in use as a way to better understand the food culture of the people that the program is intended for. Although photographs and video of cooking, eating, and procurement of food are fixed in a certain time, they are still important discussion catalysts as they can be examined and re-examined for what "truth" they tell about people's use of food in place, and additionally can counteract researchers' own biases and assumptions about what "should be" versus what really "is" happening. Considering that people's food habits of all ages and cultures are becoming increasingly important to watch and understand, image-based fieldwork appears to be, more and more, an ideal methodology for the food researcher.

Notes

1. The salad bar programs were running in each of the schools two days of the week and took the place of the normal "hot lunch" program the schools offered as part of the USDA's National School Lunch Program (NSLP). As such, the salad bar was a nutritionally complete meal and contained all the components necessary to meet USDA school lunch requirements (i.e. protein, bread, fruit, vegetables, and milk).

2. The entire school community (as well as students' parents), however, was notified in advance about the salad bar evaluation and allowed to decline participation in the study by requesting to opt out beforehand. We also gave all students an additional opportunity to decline before we took their photograph. After we examined the pictures, we were also careful to remove any identifying characteristics like hands and clothing that might have appeared in the picture. In the end, only a few students declined to have their plate of food photographed. More often, students were disappointed not to be included in the picture as well.

3. We used a set of "baseline" photographs of pre-measured and therefore known portion sizes of fruits and vegetables. Using these visual markers, we then quantified the serving sizes of each food type (bread, fruit, vegetable, etc.) on individual students' plates. For more details on the results of our nutritional study, contact the authors. Partial results have been published in the UC SAREP Sustainable Agriculture Newsletter currently available online at http://www.sarep.ucdavis.edu/newsltr/v16n3/sa-1.htm

4. This change occurred after our photographic study, so we were unable to see the differences in salads.
5. Although we did conduct a plate waste study as part of our evaluation of the salad bar program, we were not able to conclusively determine if the salads children chose were also more likely to be eaten.

References

Becker, H. (1986) "Do Photographs Tell the Truth?" In *Doing Things Together*. Evanston, Illinois: Northwestern University Press.

Birch, L. L. & Fisher, J. A. (1996) "The Role of Experience in the Development of Children's Eating Behavior." In E. D. Capaldi (ed.), *Why We Eat What We Eat: The Psychology of Eating*. Washington, DC: American Psychological Association.

Birch, L. L. & Marlin, D. W. (1982) "I Don't Like it; I Never Tried it: Effects of Exposure to Food on Two-year-old Children's Food Preferences," *Appetite*, 4, 353–360.

Collier, J. J. (1967) *Visual Anthropology: Photography as a Research Method*. New York: Holt, Reinhart, and Winston.

Corsaro, William (1997) *The Sociology of Childhood*. Thousand Oaks, CA: Pine Forge Press.

Ohmart, J. & Feenstra, G. W. (2004) "'Plate Tectonics': Do Farm-to-school Programs Shift Students' Diets?" *Sustainable Agriculture Newsletter*, 16 (3), Fall 2004.

Prosser, J. (ed.) (1998) *Image Based Research: A Sourcebook for Qualitative Researchers*. Bristol, PA: Taylor & Francis.

Prout, A. & James, A. (1997) "A New Paradigm for the Sociology of Childhood: Provenance, Promise, and Problems", pp. 7–33. In A. James & A. Prout (eds) *Constructing and Reconstructing Childhood: Contemporary Issues in the Sociological Study of Childhood*, 2nd ed. London: Routledge Falmer.

Wagner, J. (2004) "Constructing Credible Images: Documentary Studies, Social Research, and Visual Studies." *American Behavioral Scientist*, 47 (12): 1477–1506.

Williamson, D. A., Allen, H. R., Martin, P. D., Alfonso, A. J. et al. (2003) "Comparison of Digital Photography to Weighed and Visual Estimation of Portion Sizes." *Journal of American Dietary Association*, 103 (9): 1139–1145

Williamson, D. A., Allen, H. R., Martin, P. D., Alfonso, A., et al. (2004) "Digital Photography: A New Method for Estimating Food Intake in Cafeteria Settings." *Eating and Weight Disorders*, 9 (1): 24–28.

Political Economy of Food:
Transformation and Marginalization

28 The Chain Never Stops
   Eric Schlosser

29 Whose "Choice"? "Flexible" Women Workers in the Tomato Food Chain
   Deborah Barndt

30 The Politics of Breastfeeding: An Advocacy Update
   Penny Van Esterik

31 The Political Economy of Obesity: The Fat Pay All
   Alice Julier

32 Of Hamburger and Social Space: Consuming McDonald's in Beijing
   Yunxiang Yan

33 "Plastic-bag Housewives" and Postmodern Restaurants?: Public and Private
   in Bangkok's Foodscape
   Gisèle Yasmeen

34 The Political Economy of Food Aid in an Era of Agricultural Biotechnology
   Jennifer Clapp

35 Street Credit: The Cultural Politics of African Street Children's Hunger
   Karen Coen Flynn

36 Want Amid Plenty: From Hunger to Inequality
   Janet Poppendieck

28

The Chain Never Stops

Eric Schlosser

American slaughterhouses are grinding out meat faster than ever—and the production line keeps moving even when the workers are maimed by the machinery.

> In the beginning he had been fresh and strong, and he had gotten a job the first day; but now he was second-hand, a damaged article, so to speak, and they did not want him. They had worn him out, with their speeding-up and their carelessness, and now they had thrown him away!
>
> Upton Sinclair, *The Jungle* (1906)

Kenny Dobbins was hired by the Monfort Beef Company in 1979. He was 24 years old, and 6 foot 5, and had no fear of the hard work in a slaughterhouse. He seemed invincible. Over the next two decades he suffered injuries working for Monfort that would have crippled or killed lesser men. He was struck by a falling 90-pound box of meat and pinned against the steel lip of a conveyor belt. He blew out a disc and had back surgery. He inhaled too much chlorine while cleaning some blood tanks and spent a month in the hospital, his lungs burned, his body covered in blisters. He damaged the rotator cuff in his left shoulder when a 10,000-pound hammer-mill cover dropped too quickly and pulled his arm straight backward. He broke a leg after stepping into a hole in the slaughterhouse's concrete floor. He got hit by a slow-moving train behind the plant, got bloodied and knocked right out of his boots, spent two weeks in the hospital, then returned to work. He shattered an ankle and had it mended with four steel pins. He got more bruises and cuts, muscle pulls and strains than he could remember.

Despite all the injuries and the pain, the frequent trips to the hospital and the metal brace that now supported one leg, Dobbins felt intensely loyal to Monfort and ConAgra, its parent company. He'd left home at the age of 13 and never learned to read; Monfort had given him a steady job, and he was willing to do whatever the company asked. He moved from Grand Island, Nebraska, to Greeley, Colorado, to help Monfort reopen its slaughterhouse there without a union. He became an outspoken member of a group formed to keep union organizers out. He saved the life of a fellow worker—and was given a framed certificate of appreciation. And then, in December 1995, Dobbins felt a sharp pain in his chest while working in the plant. He thought it was a heart attack. According to Dobbins, the company nurse told him it was a muscle pull and sent him home. It was a heart attack, and Dobbins nearly died.

While awaiting compensation for his injuries, he was fired. The company later agreed to pay him a settlement of $35,000.

Today Kenny Dobbins is disabled, with a bad heart and scarred lungs. He lives entirely off Social Security payments. He has no pension and no health insurance. His recent shoulder surgery—stemming from an old injury at the plant and costing more than $10,000—was paid for by Medicare. He now feels angry beyond words at ConAgra, misused, betrayed. He's embarrassed to be receiving public assistance. "I've never had to ask for help before in my life," Dobbins says. "I've always worked. I've worked since I was 14 years old." In addition to the physical pain, the financial uncertainty, and the stress of finding enough money just to pay the rent each month, he feels humiliated.

What happened to Kenny Dobbins is now being repeated, in various forms, at slaughterhouses throughout the United States. According to the Bureau of Labor Statistics, meatpacking is the nation's most dangerous occupation. In 1999, more than one-quarter of America's nearly 150,000 meatpacking workers suffered a job-related injury or illness. The meatpacking industry not only has the highest injury rate, but also has by far the highest rate of serious injury—more than five times the national average, as measured in lost workdays. If you accept the official figures, about 40,000 meatpacking workers are injured on the job every year. But the actual number is most likely higher. The meatpacking industry has a well-documented history of discouraging injury reports, falsifying injury data, and putting injured workers back on the job quickly to minimize the reporting of lost workdays. Over the past four years, I've met scores of meatpacking workers in Nebraska, Colorado, and Texas who tell stories of being injured and then discarded by their employers. Like Kenny Dobbins, many now rely on public assistance for their food, shelter, and medical care. Each new year throws more injured workers on the dole, forcing taxpayers to subsidize the meatpacking industry's poor safety record. No government statistics can measure the true amount of pain and suffering in the nation's meatpacking communities today.

A list of accident reports filed by the Occupational Safety and Health Administration gives a sense of the dangers that workers now confront in the nation's meatpacking plants. The titles of these OSHA reports sound more like lurid tabloid headlines than the headings of sober government documents: Employee Severely Burned After Fuel From His Saw Is Ignited. Employee Hospitalized for Neck Laceration From Flying Blade. Employee's Finger Amputated in Sausage Extruder. Employee's Finger Amputated in Chitlin Machine. Employee's Eye Injured When Struck by Hanging Hook. Employee's Arm Amputated in Meat Auger. Employee's Arm Amputated When Caught in Meat Tenderizer. Employee Burned in Tallow Fire. Employee Burned by Hot Solution in Tank. One Employee Killed, Eight Injured by Ammonia Spill. Employee Killed When Arm Caught in Meat Grinder. Employee Decapitated by Chain of Hide Puller Machine. Employee Killed When Head Crushed by Conveyor. Employee Killed When Head Crushed in Hide Fleshing Machine. Employee Killed by Stun Gun. Caught and Killed by Gut-Cooker Machine.

The most dangerous plants are the ones where cattle are slaughtered. Poultry slaughterhouses are somewhat safer because they are more highly mechanized; chickens have been bred to reach a uniform size at maturity. Cattle, however, vary enormously in size, shape, and weight when they arrive at a slaughterhouse. As a result, most of the work at a modern beef plant is still performed by hand. In the age of the space station and the microchip, the most important slaughterhouse tool is a well-sharpened knife.

Thirty years ago, meatpacking was one of the highest-paid industrial jobs in the United States, with one of the lowest turnover rates. In the decades that followed the 1906 publication of *The Jungle*, labor unions had slowly gained power in the industry, winning their members good benefits, decent working conditions, and a voice in the workplace. Meatpacking jobs were dangerous and unpleasant, but provided enough income for a solid, middle-class life. There were sometimes waiting lists for these jobs. And then, starting in the early 1960s, a company called Iowa Beef Packers (IBP) began to revolutionize the industry, opening plants in rural areas far from union strongholds, recruiting immigrant workers from Mexico, introducing a new division of labor that eliminated the need for skilled butchers, and ruthlessly battling unions. By the late 1970s, meatpacking companies that wanted to compete with IBP had to adopt its business methods—or go out of business. Wages in the meatpacking industry soon fell by as much as 50 percent. Today meatpacking is one of the nation's lowest-paid industrial jobs, with one of the highest turnover rates. The typical plant now hires an entirely new workforce every year or so. There are no waiting lists at these slaughterhouses today. Staff shortages have become an industrywide problem, making the work even more dangerous. In a relatively brief period of time, the meatpacking industry also became highly centralized and concentrated, giving enormous power to a few large agribusiness firms. In 1970, the top four meatpackers controlled just 21 percent of the beef market. Today the top four—IBP, ConAgra, Excel (a subsidiary of Cargill), and National Beef—control about 85 percent of the market. While the meatpackers have grown more powerful, the unions have grown much weaker. Only half of IBP's workers belong to a union, allowing that company to set the industry standard for low wages and harsh working conditions. Given the industry's high turnover rates, it is a challenge for a union simply to remain in a meatpacking plant, since every year it must gain the allegiance of a whole new set of workers.

In some American slaughterhouses, more than three-quarters of the workers are not native English speakers; many can't read any language, and many are illegal immigrants. A new migrant industrial workforce now circulates through the meatpacking towns of the High Plains. A wage of $9.50 an hour seems incredible to men and women who come from rural areas in Mexico where the wages are $7 a day. These manual laborers, long accustomed to toiling in the fields, are good workers. They're also unlikely to complain or challenge authority, to file lawsuits, organize unions, fight for their legal rights. They tend to be poor, vulnerable, and fearful. From the industry's point of view, they are ideal workers: cheap, largely interchangeable, and disposable.

One of the crucial determinants of a slaughterhouse's profitability is also responsible for many of its greatest dangers: the speed of the production line. Once a plant is fully staffed and running, the more head of cattle slaughtered per hour, the less it costs to process each one. If the production line stops, for any reason, costs go up. Faster means cheaper—and more profitable. The typical line speed in an American slaughterhouse 25 years ago was about 175 cattle per hour. Some line speeds now approach 400 cattle per hour. Technological advances are responsible for part of the increase; the powerlessness of meatpacking workers explains the rest. Faster also means more dangerous. When hundreds of workers stand closely together, down a single line, wielding sharp knives, terrible things can happen when people feel rushed. The most common slaughterhouse injury is a laceration. Workers stab themselves or stab someone nearby. They struggle to keep up with the pace as carcasses rapidly swing toward them, hung on hooks from a moving, overhead chain. All sorts of accidents—involving power tools, saws, knives, conveyor belts, slippery floors, falling carcasses—become more likely when the chain moves too fast. One slaughterhouse nurse told me she could always tell the line speed by the number of people visiting her office.

The golden rule in meatpacking plants is "The Chain Will Not Stop." USDA inspectors can shut down the line to ensure food safety, but the meatpacking firms do everything possible to keep it moving at top speed. Nothing stands in the way of production, not mechanical failures, breakdowns, accidents. Forklifts crash, saws overheat, workers drop knives, workers get cut, workers collapse and lie unconscious on the floor, as dripping carcasses sway past them, and the chain keeps going. "The chain never stops," Rita Beltran, a former IBP worker told me. "I've seen bleeders, and they're gushing because they got hit right in the vein, and I mean they're almost passing out, and here comes the supply guy again, with the bleach, to clean the blood off the floor, but the chain never stops. It never stops."

Albertina Rios was a housewife in Mexico before coming to America nearly 20 years ago and going to work for IBP in Lexington, Nebraska. While bagging intestines, over and over, for eight hours a day, Rios soon injured her right shoulder. She was briefly placed on light duty, but asked to be assigned to a higher-paying position trimming heads, an even more difficult job that required moving heavy baskets of meat all day. When she complained about the pain to her supervisor, she recalls, he accused her of being lazy. Rios eventually underwent surgery on the shoulder, as well as two operations on her hands for carpal tunnel syndrome, a painful and commonplace injury caused by hours of repetitive motion.

Some of the most debilitating injuries in the meatpacking industry are also the least visible. Properly sutured, even a deep laceration will heal. The cumulative trauma injuries that meatpacking workers routinely suffer, however, may cause life-long impairments. The strict regimentation and division of labor in slaughterhouses means that workers must repeat the same motions again and again throughout their shift. Making the same knife cut 10,000 times a day or lifting the same weight every few seconds can cause serious injuries to a person's back, shoulders, or hands. Aside from a 15-minute rest break or two and a brief lunch, the work is unrelenting. Even the repetition of a seemingly harmless task can lead to pain. "If you lightly tap your finger on a desk a few times, it doesn't hurt," an attorney for injured workers told me. "Now try tapping it for eight hours straight, and see how that feels."

The rate of cumulative trauma injuries in meatpacking is the highest of any American industry. It is about 33 times higher than the national average. According to federal statistics, nearly 1 out of every 10 meatpacking workers suffers a cumulative trauma injury every year. In fact, it's very hard to find a meatpacking worker who's not suffering from some kind of recurring pain. For unskilled, unschooled manual laborers, cumulative trauma injuries such as disc problems, tendonitis, and "trigger finger" (a syndrome in which a finger becomes stuck in a curled position) can permanently limit the ability to earn a decent income. Much of this damage will never be healed.

After interviews with many slaughterhouse workers who have cumulative trauma injuries, there's one image that stays with me. It's the sight of pale white scars on dark skin. Ana Ramos came from El Salvador and went to work at the same IBP plant as Albertina Rios, trimming hair from the meat with scissors. Her fingers began to lock up; her hands began to swell; she developed shoulder problems from carrying 30- to 60-pound boxes. She recalls going to see the company doctor and describing the pain, only to be told the problem was in her mind. She would leave the appointments crying. In January 1999, Ramos had three operations on the same day—one on her shoulder, another on her elbow, another on her hand. A week later, the doctor sent her back to work. Dora Sanchez, who worked at a different IBP plant, complained for months about soreness in her hands. She says the company ignored her. Sanchez later had surgery on both hands. She now has a "spinal cord stimulator," an elaborate pain-reduction system implanted in her body, controlled from a small box under the skin on her hip. She will need surgery to replace the batteries every six or seven years.

Cumulative trauma injuries may take months or even years to develop; other slaughterhouse injuries can happen in an instant. Lives are forever changed by a simple error, a wrong move, a faulty machine. Paul Lopez worked as a carpenter in Mexico, making tables, chairs, and headboards, before coming to the United States in 1995 to do construction work in Santa Fe, New Mexico. He was 20 years old at the time, and after laying concrete foundations for two years, he moved to Greeley and got a job at the Monfort Beef plant, where the pay was higher. He trimmed hides after they came up from the kill floor, cutting off the legs and heads, lifting them up with mechanical assistance, and placing the hides on a hook. It was one of the most difficult jobs in the plant. Each hide weighed about 180 pounds, and he lifted more than 300 of them every hour. He was good at his job and became a "floater," used by his supervisor to fill in for absent workers. Lopez's hands and shoulders were sore at the end of the day, but for two years and two months he suffered no injuries.

At about seven in the morning on November 22, 1999, Lopez was substituting for an absent worker, standing on a four-foot-high platform, pulling hides from a tank of water that was washing blood and dirt off them. The hides were suspended on hooks from a moving chain. The room was cold and foggy, and it was difficult to see clearly. There were problems somewhere up ahead on the line, but the chain kept moving, and Lopez felt pressure to keep up. One of his steel-mesh gloves suddenly got snagged in the chain, and it dragged him down the line toward bloody, filthy water that was three feet deep. Lopez grabbed the chain with his free hand and screamed for help. Someone ran to another room and took an extraordinary step: He shut down the line. The arm caught in the chain, Lopez's left one, was partially crushed. He lost more than three pints of blood and almost bled to death. He was rushed to a hospital in Denver, endured the first of many operations, and survived. Five months later, Lopez was still in enormous pain and heavily medicated. Nevertheless, he says, a company doctor ordered him back to work. His previous supervisor no longer worked at the plant. Lopez was told that the man had simply walked off the job and quit one day, feeling upset about the accident.

I recently visited Lopez on a lovely spring afternoon. His modest apartment is just a quarter mile down the road from the slaughterhouse. The living room is meticulously neat and clean, filled with children's toys and a large glass display case of Native American curios. Lopez now works in the nurse's office at the plant, handling files. Every day he sees how injured workers are treated—given some Tylenol and then sent back to the line—and worries that ConAgra is now planning to get rid of him. His left arm hangs shriveled and lifeless in a sling. It is a deadweight that causes severe pain in his neck and back. Lopez wants the company to pay for an experimental operation that might restore some movement to the arm. The alternative could be amputation. ConAgra will say only that it is weighing the various medical options. Lopez is 26 years old and believes his arm will work again. "Every night, I pray for this operation," he says, maintaining a polite and dignified facade. A number of times during our conversation, he suddenly gets up and leaves the room. His wife, Silvia, stays behind, sitting across from me on the couch, holding their one-year-old son in her arms. Their three-year-old daughter happily wanders in and out to the porch. Every time the front door swings open for her, a light breeze from the north brings the smell of death into the room.

The meatpacking companies refuse to comment on the cases of individual employees like Raul Lopez, but insist they have a sincere interest in the well-being of their workers. Health and safety, they maintain, are the primary concerns of every supervisor, foreman, nurse, medical claims examiner, and company-approved doctor. "It is in our best interest to take care of our workers and ensure that they are protected and able to work every day," says Janet M. Riley, a vice president of the American Meat Institute, the industry's trade association. "We are very concerned about improving worker safety. It is absolutely to our benefit."

The validity of such claims is measured best in Texas, where the big meatpackers have the most freedom to do as they please. In many ways, the true heart of the industry lies in Texas. About one-quarter of the cattle slaughtered every year in the United States—roughly 9 million animals—are processed in Texas meatpacking plants. One of the state's U.S. senators, Phil Gramm, is the industry's most powerful ally in Congress. His wife, Wendy Lee, sits on the board of IBP. The state courts and the legislature have also been friendly to the industry. Indeed, many injured meatpacking workers in Texas now face a system that has been devised not only to prevent any independent scrutiny of their medical needs, but also to prevent them from suing for on-the-job injuries.

In the early years of the 20th century, public outrage over the misfortune of industrial workers hurt on the job prompted legislatures throughout the United States to enact workers' compensation laws. Workers' comp was intended to be a form of mandatory, no-fault insurance. In return for surrendering the legal right to sue their employer for damages, injured workers were guaranteed immediate access to medical care, steady income while they recuperated, and disability payments. All 50 states eventually passed workers' comp legislation of one sort or another, creating systems in which employers generally obtained private insurance and any disputes were resolved by publicly appointed officials.

Recent efforts by business groups to "reform" workers' comp have made it more difficult for injured employees to obtain payments. In Colorado, the first "workers' comp reform" bill was sponsored in 1990 by Tom Norton, a conservative state senator from Greeley. His wife, Kay, was a vice president at ConAgra Red Meat at the time. Under Colorado's new law, which places limits on compensation, the maximum payment for losing an arm is $37,738. Losing a digit brings you anywhere from $2,400 to $9,312, depending on whether it's a middle finger, a pinkie, or a thumb.

The meatpacking companies have a vested interest in keeping workers' comp payments as low as possible. IBP, Excel, and ConAgra are all self-insured. Every dime spent on injured workers in such programs is one less dime in profits. Slaughterhouse supervisors and foremen, whose annual bonuses are usually tied to the injury rate of their workers, often discourage people from reporting injuries or seeking first aid. The packinghouse culture encourages keeping quiet and laboring in pain. Assignments to "light duty" frequently punish an injured worker by cutting the hourly wage and forbidding overtime. When an injury is visible and impossible to deny—an amputation, a severe laceration, a chemical burn—companies generally don't contest a worker's claim or try to avoid medical payments. But when injuries are less obvious or workers seem uncooperative, companies often block every attempt to seek benefits. It can take injured workers as long as three years to get their medical bills paid. From a purely financial point of view, the company has a strong incentive to delay every payment in order to encourage a less-expensive settlement. Getting someone to quit is even more profitable—an injured worker who walks away from the job is no longer eligible for any benefits. It is not uncommon to find injured workers assigned to meaningless or unpleasant tasks as a form of retaliation, a clear message to leave. They are forced to sit all day watching an emergency exit or to stare at gauges amid the stench in rendering.

In Texas, meatpacking firms don't have to manipulate the workers' comp system—they don't even have to participate in it. The Texas Workers Compensation Reform Act of 1989 allowed private companies to drop out of the state's workers' comp system. Although the law gave injured workers the right to sue employers that had left the system, that provision was later rendered moot. When a worker is injured at an IBP plant in Texas, for example, he or she is immediately presented with a waiver. It reads: "I have been injured at work and want to apply for the payments offered by IBP to me under its Workplace Injury Settlement Program. To qualify, I must accept the rules of the Program. I have been given a copy of the Program summary. I accept the Program."

Signing the waiver means forever surrendering your right—and the right of your family and heirs—to sue IBP on any grounds. Workers who sign the waiver may receive immediate medical care under IBP's program. Or they may not. Once they sign, ISP and its company-approved doctors have control over the worker's job-related medical treatment—for life. Under the program's terms, seeking treatment from an independent physician can be grounds for losing all medical benefits. If the worker objects to any decision, the dispute can be submitted to an IBP-approved arbitrator. The company has said the waivers are designed "to more effectively ensure quality medical care for employees injured on the job." Workers who refuse to sign the IBP waiver not only risk getting no medical care from the company, but also risk being fired on the spot. In February 1998, the Texas Supreme Court ruled that companies operating outside the state's workers' comp system can fire workers simply because they're injured.

Today, an IBP worker who gets hurt on the job in Texas faces a tough dilemma: Sign the waiver, perhaps receive immediate medical attention, and remain beholden, forever, to IBP. Or refuse to sign, risk losing your job, receive no help with your medical bills, file a lawsuit, and hope to win a big judgment against the company someday. Injured workers almost always sign the waiver. The pressure to do so is immense. An IBP medical case manager will literally bring the waiver to a hospital emergency room in order to obtain an injured worker's signature. Karen Olsson, in a fine investigative piece for the Texas Observer, described the lengths to which Terry Zimmerman, one of IBP's managers, will go to get a signed waiver. When Lonita Leal's right hand was mangled by a hamburger grinder at the IBP plant in Amarillo, Zimmerman talked her into signing the waiver with her left hand, as she waited in the hospital for surgery. When Duane Mullin had both hands crushed in a hammer mill at the same plant, Zimmerman persuaded him to sign the waiver with a pen held in his mouth.

Unlike IBP, Excel does not need to get a signed waiver after an injury in Texas. Its waiver is included in the union contract that many workers unwittingly sign upon being hired. Once they're injured, these workers often feel as much anger toward the union as they do toward their employer. In March, the Texas Supreme Court upheld the legality of such waivers, declaring that the "freedom of contract" gave Americans the ability to sign away their common-law rights. Before the waiver became part of the standard contract, Excel was held accountable, every so often, for its behavior.

Hector Reyes is one of the few who has managed to do something productive with his sense of betrayal. For 25 years, his father was a maintenance worker at the Excel plant in Friona, Texas, a couple of hours southwest of Amarillo. As a teenager, Reyes liked to work in the plant's warehouse, doing inventory. He'd grown up around the slaughterhouse. He later became a Golden Gloves champion boxer and went to work for Excel in 1997, at the age of 25, to earn money while he trained. One day he was asked to clean some grease from the blowers in the trolley room. Reyes did as he was told, climbing a ladder in the loud, steam-filled room and wiping the overhead blowers clean. But one of the blowers lacked a proper cover—and in an instant the blade shredded four of the fingers on Reyes' left hand. He climbed down the ladder and yelled for help, but nobody would come near him, as blood flew from the hand. So Reyes got himself to the nurse's office, where he was immediately asked to provide a urine sample. In shock and in pain, he couldn't understand why they needed his urine so badly. Try as he might, he couldn't produce any. The nurse called an ambulance, but said he wouldn't receive any painkillers until he peed in a cup. Reyes later realized that if he'd failed the urine test, Excel would not have been obligated to pay any of his medical bills. This demand for urine truly added insult to injury: Reyes was in training and never took drugs. He finally managed to urinate and received some medication. The drug test later came back negative.

On his fourth night in a Lubbock hospital, Reyes was awakened around midnight and told to report for work the next morning in Friona, two hours away. His wife would have to drive, but she was three months pregnant. Reyes refused to leave the hospital until the following day. For the next three months, he simply sat in a room at the Excel plant with other injured workers or filed papers for eight hours a day, then drove to Lubbock for an hour of physical therapy and an hour of wound cleaning before heading home. "You've already cost the company too much money," he recalls one supervisor telling him. Reyes desperately wanted to quit but knew he'd lose all his medical benefits if he did. He became suicidal, despondent about the end of his boxing career and his disfigurement. Since the union had not yet included a waiver in its Excel contract, Reyes was able to sue the company for failing to train him properly and for disregarding OSHA safety guidelines. In 1999, he won a rare legal victory: $879,312.25 in actual damages and $1 million in punitive damages. Under the current Excel contract, that sort of victory is impossible to achieve.

Federal safety laws were intended to protect workers from harm, regardless of the vagaries of state laws like those in Texas. OSHA is unlikely, however, to do anything for meatpacking workers in the near future. The agency has fewer than 1,200 inspectors to examine the safety risks at the nation's roughly 7 million workplaces. The maximum OSHA fine for the death of a worker due to an employer's willful negligence is $70,000—an amount that hardly strikes fear in the hearts of agribusiness executives whose companies have annual revenues that are measured in the tens of billions. One of President George W. Bush's first acts in office was to rescind an OSHA ergonomics standard on repetitive-motion injuries that the agency had been developing for nearly a decade. His move was applauded by IBP and the American Meat Institute.

The new chairman of the House Subcommittee on Workforce Protections, which oversees all legislation pertaining to OSHA, is Rep. Charles Norwood, a Republican from Georgia. Norwood was an outspoken supporter of the OSHA Reform Act of 1997—a bill that would have effectively abolished the agency. Norwood, a former dentist, became politically active in the early 1990s out of a sense of outrage that OSHA regulations designed to halt the spread of AIDS were forcing him to wear fresh rubber gloves for each new patient. He has publicly suggested that many workers get repetitive-stress injuries not from their jobs, but from skiing and playing too much tennis.

For Kevin Glasheen, one of the few Texas attorneys willing to battle IBP, the plight of America's meatpacking workers is "a fundamental failure of capitalism." By failing to pay the medical bills of injured workers, he says, large meatpackers are routinely imposing their business costs on the rest of society, much as utilities polluted the air a generation ago without much regard to the consequences for those who breathed it.

Rod Rehm, an attorney who defends many Latino meatpacking workers in Nebraska, believes that two key changes could restore the effectiveness of most workers' comp plans. Allowing every worker to select his or her own physician would liberate medical care from the dictates of the meatpacking companies and the medical staff they control. More important, Rehm argues, these companies should not be permitted to insure themselves. If independent underwriters had to insure the meatpackers, the threat of higher insurance premiums would quickly get the attention of the meatpacking industry—and force it to take safety issues seriously.

Until fundamental changes are made, the same old stories will unfold. Michael Glover is still awaiting payment from IBP, his employer for more than two decades. For 16 of those years, Glover worked as a splitter in the company's Amarillo plant. Every 20 to 30 seconds, a carcass would swing toward him on a chain. He would take "one heavy heavy power saw" and cut upward, slicing the animal in half before it went into the cooler. The job took strength, agility, and good aim. The carcasses had to be sliced exactly through the middle. One after another they came at him, about a thousand pounds each, all through the day.

On the morning of September 30, 1996, after splitting his first carcass, Glover noticed vibrations in the steel platform beneath him. A maintenance man checked the platform and found a bolt missing, but told Glover it was safe to keep working until it was replaced. Moments later, the platform collapsed as Glover was splitting a carcass. He dropped about seven feet and shattered his right knee. While he lay on the ground and workers tried to find help, the chain kept going as two other splitters picked up the slack. Glover was taken in a wheelchair to the nurse's station, where he went into shock in the hallway and fell unconscious. He sat in that hall for almost four hours before being driven to an outpatient clinic. A full seven hours after the accident, Glover was finally admitted to a hospital. He spent the next six days there. His knee was too badly shattered to be repaired; no screws would hold; the bone was broken into too many pieces.

An artificial knee was later inserted. Glover suffered through enormous pain and a series of complications: blood clots, ulcers, phlebitis. Nevertheless, he says, IBP pressured him to return to work in a wheelchair during the middle of winter. On snowy days, several men had to carry him into the plant. Once it became clear that his injury would never fully heal, Glover thinks IBP decided to get rid of him. But he refused to quit and lose all his medical benefits. He was given a series of humiliating jobs. For a month, Glover sat in the men's room at the plant for eight hours a day, ordered by his supervisor to make sure no dirty towels or toilet paper remained on the floor.

"Michael Glover played by IBP'S rules," says his attorney, James H. Woods, a fierce critic of the Texas workers' comp system. The day of his accident, Glover had signed the waiver, surrendering any right to sue the company. Instead, he filed for arbitration under IBP's Workplace Injury Settlement Program. Last year, on November 3, Glover was fired by IBP. Twelve days later, his arbitration hearing convened. The arbitrator, an Amarillo lawyer named Tad Fowler, was selected by IBP. Glover sought money for his pain and suffering, as well as lifetime payments from the company. He'd always been a hardworking and loyal employee. Now he had no medical insurance. His only income was $250 a week in unemployment. He'd fallen behind in his rent and worried that his family would be evicted from their home.

On December 20, Fowler issued his decision. He granted Glover no lifetime payments but awarded him $350,000 for pain and suffering. Glover was elated, briefly. Even though its workplace-injury settlement program clearly states that "the arbitrator's decision is final and binding," the company so far has refused to pay him. IBP claims that by signing the waiver, Glover forfeited any right to compensation for "physical pain, mental anguish, disfigurement, or loss of enjoyment of life." IBP has even refused to pay its own arbitrator for his services, and Fowler is now suing the company to get his fee. He has been informed that IBP will never hire him again for an arbitration.

Glover's case is now in federal court. He is a proud man with a strong philosophical streak. He faces the possibility of another knee replacement or of amputation. "How can this company fire me after 23-and-one-half years of service," he asks, "after an accident due to no fault of my own, and requiring so much radical surgery, months and months of pain and suffering, and nothing to look forward to but more pain and suffering, and refuse to pay me an award accrued through its own program?"

There is no good answer to his question. The simple answer is that IBP can do it because the laws allow them to do it. Michael Glover is just one of thousands of meatpacking workers who've been mistreated and then discarded. There is nothing random or inscrutable about their misery; it is the logical outcome of the industry's practices. A lack of public awareness, a lack of outrage, have allowed these abuses to continue, one after another, with a machine-like efficiency. This chain must be stopped.

29

Whose "Choice"? "Flexible" Women Workers in the Tomato Food Chain

Deborah Barndt

> My whole family [works] at McDonald's: my mother, my sisters, my boyfriend, often at different times. And my dad, a police officer, works from eleven in the night 'til six in the morning. So there's no time we can eat together. We just grab something and put it in the microwave.[1]

This narrative, by Tania, a York University student working at McDonald's, may resonate with many young women in the North. At the Southern end of the NAFTA food chain, Tomasa, a Mexican fieldworker for Santa Anita Packers, one of the biggest domestic producers of tomatoes, describes her daily food preparations during the harvest season: "I get up at three a.m. to make tortillas for our lunch, then the truck comes at six to take us to the fields to start working by seven a.m."[2] An hour away at a Santa Anita greenhouse, Sara, a young tomato packer, tells us that the foreign management of Eco-Cultivos has just eliminated the two-hour lunch break, so workers no longer go home for the traditional noontime meal.[3]

These changes in the eating practices of women workers in the continental food system reflect several dimensions of the global economic restructuring that has reshaped the nature of their labor. Shifts in family eating practices have not been the "choice" of the women whose stories are told here, nor have they "chosen" the work shifts that involve them around the clock in growing and preparing food for other people.

"McDonaldization," initiated in the North and spreading to the South, and "maquilization," initiated in the South and now appearing in the North, are inter-related processes in the new global economy. McDonaldization, as George Ritzer[4] describes it, is the model that the fastfood restaurant has offered as a way to reorganize work in all other sectors. This model is based on efficiency, predictability, calculability or quantifiability, substitution of non-human technology, control and the irrationality of rationality. Central to this model is "flexible" part-time labor.

"Maquilization," originating in the maquila free trade zones of northern Mexico, now refers to a more generalized work process characterized by 1) the feminization of the labor force, 2) extreme segmentation of skill categories, 3) the lowering of real wages and 4) a non-union orientation.[5] In the traditional maquila sectors, such as the garment and electronic industries, there is full-time (though not necessarily stable) employment. However, the trade liberalization epitomized by NAFTA has opened the door for the development of maquilas throughout Mexico. "Agromaquilas," in particular, depend on more temporary, part-time and primarily female labor.

Central to both the McDonaldization of the retail and service sectors and the maquilization of the agro-industrial and manufacturing sectors in the continental food chain are the interrelated processes of the "feminization of poverty"[6] and the "flexibilization of labor."[7] Since the 1960s when export processing zones such as the Mexican maquilas began to employ primarily young women in low-skilled and low-wage jobs,[8] women have been key players in this new global formula.[9] In the reorganization of work by global capital, women workers have also become key players in new flexible labor strategies, building on an already established sexual division of labor and institutionalized sexism and racism in the societies where transnational corporations set up shop. In these sectors of the global food system, women bring their own meaning to flexible labor as they juggle their lives as both producers and consumers of food, as both part-time salaried workers and full-time domestic workers in managing households.

Tomasita Comes North While Big Mac Goes South[10]

In the Tomasita Project, the journey of the tomato from the Mexican field through the United States to the Canadian fast-food restaurant reveals the dynamics of globalization. While food production and consumption takes place in all three countries, deep inequities, upon which NAFTA was based, remain among them.

The basic North-South contradiction of this continental (and increasingly hemispheric) system is that Mexico produces fresh fruit and vegetables (in this case, the tomato) for North American consumers, while Northern retail supermarkets[11] and fast-food restaurants, such as McDonald's, are moving South at record speed to market new foods, work and food practices, particularly as a result of NAFTA's trade liberalization. This contradiction is revealed in retail advertising, such as the Loblaws' billboard below. In its promotion of President's Choice products, Loblaws proclaims "Food Means the World to Us." We are seduced by such images into consuming an increasing "diversity" and seemingly endless array of fresh, "exotic"[12] and non-traditional foods. Meanwhile, there are hidden costs under which these foods were produced—the appropriation of Indigenous lands; the degradation of the environment and the health and dignity of workers; increasing poverty; deepening sexist and racist employment practices[13]—which are kept (carefully and consciously) from our view.

The Tomasita Project aims to uncover these costs, particularly by exposing the living and working conditions of the women workers whose labor (not by choice) brings the "world of food" to us. A deconstruction of the Loblaws' ad would reveal these women workers as the producers behind the food product, and show that they, too, are part of a global system that links agro-export economies (such as Mexico) with the increasing consumer demand in the North for fresh produce all year round.

Figure 29.1 A Loblaws' billboard declares "Food Means the World To Us." Toronto 1998.

Figure 29.2 An "adbusted" version of the Loblaws' billboard reveals the hands and labor behind the "world food" we eat.

Tomasita is a both a material and symbolic "ecofeminist"[14] tomato within globalized food production—from biogenetic engineering to intensive use of agrochemicals, from long journeys in refrigerated trucks to shorter journeys across supermarket counters where their internationally standardized "product look up" numbers are punched in. Its fate is paralleled by the intertwining fates of women workers in the different stages of its production, preparation and consumption. If the tomato is shaped by "just-in-time" production practices, women workers make this supply-on-demand possible through their flexible labor.

The tracing of the tomato chain builds on the tradition of "commodity chain analysis,"[15] which examines three interlocking processes: 1) raw material production, 2) combined processing, packaging and exporting activities, and 3) marketing and consumptive activities.[16] The women workers who make the tomato chain come alive represent four different sectors of the food system—two in Mexico and two in Canada. In Mexico, they are the pickers and packers in Santa Anita Packers, a large export-oriented agribusiness, and the assembly line workers producing ketchup in Del Monte, a well-established multinational food processor. In Canada, the workers are cashiers in Loblaws supermarkets and service workers in McDonald's restaurants. How do these women workers (both as producers and consumers) reflect, respond to and resist the "flexible labor strategy" so central to corporate restructuring? There are, of course, obvious differences between the Mexican Indigenous workers moving from harvest to harvest to pick tomatoes and the Canadian women slicing these tomatoes and stacking them into a hamburger. Yet, since NAFTA, there are increasing similarities in the feminization and flexibilization of the labor force in all four sectors and in all three countries. One of the similarities is the increasing participation of young female workers, who, from the perspective of the companies, are seen as both cheaper and more productive than comparable male labor.[17] Gender ideologies, culturally entrenched and reinforced by managerial practices, strongly shape this socially constructed reality.

Flexibilization: From Above and From Below

Key to global economic restructuring is the notion of flexibility. The term, however, changes meaning depending on whose perspective it represents. The perspective from above, from the vantage point of corporate managers, is different than the perspective from below, from the new global workforce. To some, flexibility implies "choice," but "*whose* choice" rules in a food system built on structural inequalities, which are based on differences of national identity, race, class, gender and age? For large transnational corporations, flexibility has meant greater freedom (provided by NAFTA and increasing support from the Mexican government) to set up businesses in Mexico, where businesses are offered lower trade barriers, property laws that allow greater foreign investment, decreasing subsidies, decentralization of production through subcontracting and so forth. For large Mexican domestic producers such as Santa Anita Packers, trade liberalization has meant entering a globally competitive market with comparative advantages of land, climate and cheap labor. Once producing primarily for national consumption, Santa Anita has become ever more export-driven—it now produces 85 percent of crops for export and, in the case of greenhouse production, 100 percent for export. The fruit and vegetable sector is one of the few winners of NAFTA in Mexico.

The meaning of flexibility changes when set in the context of the new global marketplace, where borders and nation-states are less and less relevant, and where production is increasingly decentralized while decision-making is increasingly centralized. In this context, flexibility also refers to the shift from Fordist to post-Fordist production practices. Fordism was based on scientific management principles and organization of tasks in assembly lines for mass production, with the production of large volumes being the objective. Post-Fordist or "just-in-time" production responds to more diversified and specific demands in terms of quality and quantity.[18] It is ultimately very rationalized, of course, as demonstrated by the processes of workplace McDonaldization in which new technologies allow greater control of inventory and labor, while decentralization of production allows companies to shift many of the risks to subcontractors. In talking about the globalized corporate world, or "globalization from above,"[19] then, flexibility is ultimately about maximizing profits and minimizing obstacles (such as trade tariffs, government regulations, underused labor, trade union organization).

Women Workers' Experiences of Flexibility

What does flexibility mean, though, for the women moving the tomato through this continental food system, from Mexican field to Canadian table? If we first look at the consumption-end of the food chain, the fast-food and supermarket workers in Canada, and then move to the source, where women plant, pick, pack and process tomatoes in Mexico, we can learn how flexibilization has affected these women's daily lives.

McDonald's

"Flexible labor strategies" have been key to the model of production of McDonald's and its competitors. McJobs, whether filled by students, seniors or underemployed women, have always been primarily part-time (up to twenty-four hours a week). Part-time jobs do not require certain benefits and, because they are limited to short three- to four-hour shifts, do not require many breaks. Women student workers might be sent home after an hour or two if sales for the day are not reaching their predetermined quota. Karen, a university student, explains:

> They're supposed to make a certain amount of money an hour, say $1,300 between twelve noon and one p.m., and if they make less than that, for every $50 (under the quota), they cut half an hour of labor. Especially if you're newer, there's pressure to go home. It takes me an hour to get to work by bus, and I could be asked to go home after an hour of work.[20]

Flexibility of this temporary labor force is reinforced by the lack of trade union organization. Strong company-induced loyalty is fed by perks such as team outings,

Figure 29.3 Young part-time women workers in a Toronto McDonald's.

weekly treats and training that inculcate a family orientation. It is meant to dissuade employees from seeking unionization or from complaining about their hours. Nonetheless, there are increasing efforts to organize McDonald's workers and there have been recent union successes in BC and Quebec.

Loblaws

The experience of flexibility for women workers in the larger chains of the retail food sector, such as Loblaws in Canada, are just as precarious. Even though they are unionized, the working conditions of part-time workers have been eroded through recent labor negotiations. In the case of Loblaws, for example, the most recent contract negotiated by the United Food and Commercial Workers Union eliminated almost all of the full-time cashier positions. Part-time cashiers are dependent on seniority for being able to choose their working hours. This particularly affects new cashiers, such as Wanda: "When you are low on the seniority list, you are lucky to get any hours. They might call you in once every two weeks for a four-hour shift."[21] This restriction on available hours also affects the cashiers' earning power. A cashier must complete five hundred hours before being eligible for a raise. At this pace, she could work at the starting wage for over two years. From the company's perspective, this shift to primarily part-time flexible labor is a conscious strategy; it is part of "lean production."

Figure 29.4 A part-time cashier at Loblaws' supermarket in Toronto.

Del Monte

What does flexibility look like in the Del Monte food processing plant in Irapuato, Mexico? The production of ketchup in Del Monte takes place during a four-month period, from February through May. In part, this coincides with the peak period for harvesting tomatoes; thus flexibility in the agromaquilas depends, in part, on the seasonal nature of agricultural production (becoming less pronounced with the increasing phenomenon of year-round greenhouse production).

Another reason that production is limited to one period is to maximize the use of the food processing machinery and the skilled labor force. Del Monte's ketchup production employs a combination of Fordist and post-Fordist processes: it is an assembly line production from the dumping and cooking of tomatoes in big vats to the bottling, capping and labelling on a mechanized line. Because other food processing (such as marmalade) uses the same machinery, the same full-time workers can easily shift from one product to another. Many are, in fact, multiskilled and are moved from one process to another, reflecting post-Fordist practices. Such multitasking is another form of flexibility in the experience of the new global workforce. Part-time women workers are brought on for the peak season only and for less skilled tasks. These women sometimes sit in the waiting room of the plant, hoping for a few hours of work, which are determined day by day. Flexibility reigns in a context where there is an oversupply of cheap labor, so companies can make such decisions on the spot, hiring and dismissing workers on a daily basis. This is another example of lean production, dependent on a disposable supply of female labor.

Santa Anita Packers

Finally we reach the source, Santa Anita Packers—the agribusiness that organizes production of tomatoes, from the importing of seeds to the exporting of waxed and packaged tomatoes in refrigerated trucks. Santa Anita, headquartered in Jalisco, in central Mexico, uses a mixture of production practices and diverse applications of the notion of flexibility. It is important to understand the historical development of the agro-export industry in Mexico[22] in the context of North–South political economic relations, which are based on ever-deepening inequalities, both between and within nations. Since the early part of the century, Mexican agriculture has been led by Northern demand for fresh fruit and vegetables, and by the use of cheap Mexican labor by U.S. agribusinesses on both sides of the border. While the Depression in the 1930s led to American workers taking over farm labor jobs from Mexican workers in the U.S. and also to a spurt of farm labor organizing, the availability of cheap Mexican Indigenous migrant labor fed the post-war development of large agribusinesses in both countries from the 1950s onward. This

Figure 29.5 Women workers packing tomatoes in Jalisco, Mexico.

transnationalization of the economy was built upon institutionalized racism and sexism within Mexico and the U.S., employing Indigenous workers often as family units who were brought by the companies from the poorer states.

The sexual division of labor is seen most strongly in the packing plants, where a gendered ideology is used to justify the employment of women, as echoed by one of the company owners:

> Women "see" better than men, they can better distinguish the colors and they treat the product more gently. In selection, care and handling, women are more delicate. They can put up with more than men in all aspects: the routine, the monotony. Men are more restless, and won't put up with it.[23]

The feminization of the global labor force, and thus the feminization of global poverty, has been based on the marginal social role that women play and on a social consensus that their domestic duties are primary. As Lourdes Beneria argues, "the private sphere of the household is at the root of continuing asymmetries between men and women."[24]

In the case of Mexican agro-industry, women are among the most marginalized workers, along with children, students, the elderly and Indigenous peoples. Sara Lara notes that agribusinesses exploit their common situation of "mixedness," referring to the fact that these workers already play socially marginal roles based on their gender, race or age: "women as housewives, Indigenous peoples as 'poor peasants', children as sons and daughters, young people as students, all as the ad hoc subjects of flexible processes,"[25] It is important to integrate national identity, gender, race, class, age and marital status[26] into any analysis of the new global labor force.

Deepening Inequalities: Flexibility for Whom?

In their restructuring, corporations have adopted a dual employment strategy that deepens the inequalities within the workforce and divides it into two groups: a "nucleus" of skilled workers who are trained in new technologies and post-Fordist production processes (quality circles, multiskilling and multitasking) and who have stable employment, and a "periphery" of unskilled workers whose jobs are very precarious. McDonald's and Loblaws both have a small full-time workforce, mainly male, while women make up the majority of the more predominant part-time work-force.[27] Tomato production in Mexico also mirrors this dualism. Small numbers of permanent workers prepare the seedlings and the land for production, and later pack and process the tomatoes; a large number of temporary part-time workers pick tomatoes during the harvest seasons. Santa Anita, for example, employs Mestizos from the local area for the jobs of cultivating the tomato plants, while hundreds of poor Indigenous workers, brought in by trucks and housed in conditions of squalor in makeshift camps, do much of the picking during the three-to-five month harvest season.[28] In this dual employment strategy, Indigenous workers are again required to be the most flexible, which is yet another form of discrimination and exploitation.

Such flexibility has been integral to labor-intensive and seasonal agricultural production for decades, though the composition of the migrant labor force has shifted over time. It is not uncommon for entire families to work together in the field, when the demand for labor is up. Children of local Mestizo peasant workers join their families on weekends during peak season, while children of Indigenous migrant workers, with neither school nor extended family to care for them, often work alongside their parents.[29] With increasing unemployment in Mexico, however, men are taking on agricultural jobs done previously by women, such as picking, and because the current economic crisis has increased the surplus of labor, companies choose the youngest and heartiest workers above the older ones (the ideal age seems to be fifteen to twenty-four, so workers in their thirties can already be considered less desirable). The flexible labor strategies of Mexican agribusinesses are predicated on race, gender and age. And once again, flexibility is determined by the companies and not the workers.

Technological changes within the production process are integral to the application of flexibilization. Differences among workers (of gender, race and skill) are accentuated with the increasingly sophisticated modes of greenhouse production and packing. Tomatoes in those plants, for example, are now weighed and sorted by color in a computerized process, which at the same time records the inventory and monitors the productivity of the workers. Through these changes, foreign managers and technicians are reorganizing production relations and the workday in ways that are also shifting social relations, both in the workplace and at home.

In a Santa Anita greenhouse, unproductive workers are dismissed daily, as there is always a plentiful pool of surplus labor to choose from. There are echoes here of the McDonald's worker being sent home when quotas are down and the Loblaws cashier not being called for weeks when she's not needed, as well as Mexican women waiting for a few hours of work on Del Monte's ketchup production line. Flexibility serves the companies' need to maximize production and profits; it does not always serve the needs of Mexican or Canadian women in this food chain to survive, to complement their family income or to organize their lives and their double-day responsibilities. And as Sara Lara concludes, "Flexibility is not a choice for women," and "labor force management by companies is at the same time family management, that is, it reinforces particular family power relations."[30]

With NAFTA, the Mexican fruit and vegetable industry has been one of the only sectors to benefit from trade liberalization and has maintained an international competitiveness, Mexico has the advantage over its Northern partners in terms of land, climate and cheap labor. The expansion of the agro-export industry, however, reflects a basic North-South contradiction between a "negotiated flexibility" and a "primitive flexibility."[31] Large domestic companies in Mexico, such as Santa Anita, are becoming increasingly multinational, yet are still in the periphery of production decisions (controlled outside Mexico) and often lag behind in technological development. In the agro-export economy of Mexico, there is a growth of unstable and temporary employment in the still labor-intensive processes of production, sorting, packing and processing. In these jobs, women, children and Indigenous peoples (the most flexible workers in a rural labor market) are managed by "primitive flexibility." Transnational companies, however, are located primarily in the more industrialized North and control production through ownership, subcontracting and advanced technology (biogenetic engineering, sophisticated food processing, production of most of the inputs and machinery of production, and architects of the commercialization and distribution systems). These transnationals employ the "nucleus" of skilled workers, with relatively stable employment, and manage this workforce through "negotiated flexibility."

Comparisons Across Borders: Women Workers as Producers and Consumers

Yet there are also increasing similarities between women workers in Mexican agribusinesses and food processing plants and women working as supermarket cashiers and fast-food service workers in Canada. They play key roles in the implementation of corporate flexible labor strategies. As a result, they experience similar contradictions in their efforts to fulfill their dual roles as salaried workers in the food system and as consumers or providers of food for their families. Wanda, a Canadian cashier, feels some common bonds with Tomasa, a Mexican tomato fieldworker:

> Tomasa used to make her own tortillas but now she has to go and work, so she buys ready-made tortillas. And she's feeling that pull just like the North American women are: Should I stay at home with the kids? Should I go to work? She's feeling the economic thing, because everybody has to survive, everybody has to eat. She's taking care of the family, that's priority in her life; I'd like to think that in my life that's a priority.[32]

Wanda has reached a point in her career, after twenty-three years as a part-time cashier, where she now has seniority and so may choose her hours. She "chooses" to work three eight-hour days instead of six four-hour shifts, for example, because she moved out of town a few years ago and must now commute one hour to work, adding two hours to her workday. That "choice" is framed by the fact that if she transferred to a Loblaws that was closer to her home, she would lose her seniority. She also "chooses" to work on weekends, because, as a single mother, it is the only time her former husband can take care of her children, saving her childcare expenses. Her "choice" of hours allows her to be at home during most weekdays:

> As a single parent, I'm taking my kids to school, doing the piano lessons, the Brownies, that kind of thing. So I know which days I don't want to come down to Toronto to work, because it's quite a ways for me. Or if they have a PD day [professional development day for teachers], I don't go into work that day.[33]

Here is where the flexibility of women's labor comes head to head with other social contradictions of an institutionalized sexist culture. Corporate managers, in fact, often point out that their flexible labor strategy suits women who "choose" to have more time with their families, and therefore don't want to work full-time. And there is certainly some truth to this. Even some feminists argue that flexibilization can be reappropriated by women and men, if it challenges the sexual division of labor in the home and promotes more shared responsibility, while also shortening the workweek. But it usually has little to do with "choice" and is often based on the assumption that women, not men or public childcare, will take care of children and feed their families.

In the Mexican context, there is even less of an illusion of "choice" for Indigenous women who are at the bottom of the hierarchy of workers, both locally and globally. While Santa Anita Packers brings Indigenous families to work during the harvest season, they provide neither adequate housing nor childcare, and it has been a struggle to get the children into the local school. It has been reported that company foremen became angry with Indigenous women workers who brought their children tied to their backs to the fields and who stopped work, periodically, to breastfeed them. Here, in the most basic sense, the primary role that women fulfill in feeding their children is regulated by the company's rules. And though they have little choice but to bring their children to the fields, they also take tremendous risks in doing so. When we visited their camp, one baby was reportedly dying because, as the Indigenous workers explained, pesticide residue on the mother's hand had entered the child's mouth during breastfeeding.

Since NAFTA, and with the deepening impoverishment of the rural population in Mexico, these Indigenous families are forced to migrate from one harvest to another for even longer periods of the year. Whereas previously they may have been able to remain home for a few months and raise some of their own food, they are now permanently moving, by necessity, ready to go to wherever there is work.

The Mestizo workers who live near the Santa Anita plant and only work seasonally experience the insecurity in another way. Due to erratic weather conditions, their work periods have been cut short, and the jobs available for them peter out. Describing the situation, Tomasa said:

> In the end, we were working one or two days a week, and then not at all. They don't even say thanks till the day that they return. Only when they begin to plant again in the next season, they come with their truck to take us back to the fields, no?[34]

This sense of never knowing when you are going to work, and often in the case of Indigenous migrant workers even where, is a permanent condition of agricultural fieldworkers. Canadian cashiers and fast-food workers may know a week or two in advance what their shifts are to be, but the constantly changing hours often affect family routines, interactions and, especially, eating practices. It is not uncommon for a family to have no time when they can all sit down to a meal together.

Whose interests are served by this flexible labor strategy? Flexibilization as it plays out in the continental food system, and particularly in the lives of women workers in this food chain from Mexico to Canada, must be seen as "an ideology propagated by firm owners as a desirable future end state, and supported by conservative pro-business forces and governments in order to assist the private sector in achieving this goal."[35] It is part and parcel of lean production, maximizing efficiency and profits and leaving the most vulnerable and marginalized workers bound to the shifting winds of just-in-time production. In the end, they become just-in-time workers with no time of their own.

And what are the real choices for women in this system? Wanda, the Loblaws cashier, has taken a keen interest in this study and has read the stories of the Mexican workers. She concludes:

> I feel an overwhelming sadness and connection to all the women in the "tomato food chain." We all play a seemingly small part, but the ramifications of our work are enormous . . . We are all entrapped in the corporate workings of flexibilization. However, the dilemma still exists for all of us in the food chain: we're trying to survive.[36]

Notes

I gratefully acknowledge the tremendous efforts of the graduate research assistants who worked from 1995 to 1999 on the Tomasita Project, helping to shape it and carrying out the interviews referred to in this chapter. Special thanks to Emily Levitt, Deborah Moffet, Lauren Baker (Mexican interviews), Ann Eyerman (McDonald's interviews), Stephanie Conway (Loblaws' interviews), Egla Martinez-Salazar (review of Mexican interviews), Karen Serwonka (McDonald's interviews), Anuja Mendiratta, and Melissa Tkachyk (glossary).

This chapter also refers to ideas more fully discussed elsewhere in this volume. See Kirsten Appendini, "From Where Have All the Flowers Come?"; Antonieta Barrón, "Mexican Women on the Move"; Ann Eyerman, "Serving Up Service"; Harriet Friedmann, "Remaking 'Traditions' "; Egla Martinez-Salazar, "The 'Poisoning' of Indigenous Migrant Women Workers and Children"; and Ester Reiter, "Serving the McCustomer."

1. Tania (pseudonym), interview with author, Toronto, Ontario, February 1998.
2. Tomasa (pseudonym), interview with author, Gomez Farias, Mexico, April 1997.
3. Sara (pseudonym), interview with author, San Isidro Mazatepec, Mexico, April 1997.
4. See George Ritzer, *The McDonaldization of Society* (Thousand Oaks, CA: Pine Forge Press, 1993). Ritzer notes that the new model of rationalization in our culture is no longer the bureaucracy, as Max Weber suggested, but the fast-food restaurant. He outlines the characteristics of this work organization based on 1) efficiency (from the factory-farm production of the ingredients to the computer scanners at the counter), 2) predictability (from the ambience and the personnel to the limited menu), 3) calculability or quantity, 4) substitution of non-human technology (the techniques, procedures, routines and machines make it almost impossible for workers to act autonomously), 5) control (the rationalization of food preparation and serving gives control over the employees), and 6) the irrationality of rationality (for example, we see McDonald's as rational despite the reality that the chemicals in the food are harmful and that we can gain weight from the high calories and cholesterol levels).
5. The four dimensions of maquilization, developed by J. Carillo as he observed restructuring in the auto industry, are elaborated by Kathryn Kopinak in *Desert Capitalism: What Are the Maquiladoras?* (Montreal: Black Rose Books, 1997), 13.
6. Gita Sen and Caren Grown, *Development, Crises, and Alternative Visions: Third World Women's Perspectives* (New York: Monthly Review Press, 1987), 25.
7. Kirsten Appendini, "Revisiting Women Wage-Workers in Mexico's Agro-Industry: Changes in Rural Labour Markets," Working Paper 95, no. 2, Centre for Development Research (Copenhagen, 1995), 11–12.
8. The categorizing of so-called "low-skilled" work needs to be problematized, particularly when describing the kinds of tasks allotted to women in food production. Job tasks that correlate with women's domestic labor have almost universally been devalued and their counterparts in paid work have suffered a similar fate. While reigning gender ideologies purport that women are "naturally" more suited to certain tasks, Elson and Pearson argue that the famous nimble fingers are not "an inheritance from their mothers," but rather "the result of training they have received from their mothers and other female kin since early infancy in the tasks socially appropriate to woman's role." See Diane Elson and Ruth Roach Pearson, "The Subordination of Women and the Internationalization of Factory Production," in Nalini Visvanathan et al., eds., *The Women, Gender, and Development Reader* (Halifax: Fernwood Publishing, 1997), 191–203.
9. For a classic analysis of this development in the 1980s, see Swasti Mitter, *Common Fate, Common Bond: Women in the Global Economy* (London: Pluto Press, 1986).
10. For a further elaboration of this North–South contradiction, see Deborah Barndt, "Bio/cultural Diversity and Equity in Post-NAFTA Mexico (or: Tomasita Comes North While Big Mac Goes South)," in Jan Drydyk and Peter Penz, eds., *Global Justice, Global Democracy* (Halifax: Fernwood Publishing, 1997), 55–69.

11. While the presence of North American fast-food restaurants in Mexico is more visible, there has also been an incursion of the retail giants. Few are aware, for example, that the big Mexican supermarket chain, Aurera, is now owned by Wal-Mart, the Texas-based company that has become synonymous with corporate take-over, spelling death for smaller retail chains.

12. The "appropriation" of the "exotic other" is the subject of post-colonial theory and cultural studies examination of how difference is constructed within the politics of consumption to entice us into buying the mythical (and essentialist) look, the purity, the passion, the natural freshness of Southern peoples and lands. For an analysis of how Loblaws, and particularly President's Choice, has led the retail market in packaging difference, see C. Sachetti and T. Dufresne, "President's Choice Through the Looking Glass," *Fuse Magazine* (May-June 1994), 23.

13. Ecological economists contribute to unveiling the "hidden costs" in the production of the food we eat, William Rees, for example, advocates that we measure the "ecological footprint" of the goods we consume, and feminist ecological economist Ellie Perkins reminds us of the unpaid labor of women in managing the household. A more popular version of this analysis can be found in the cartoon story "Tomasita Tells All: True Confessions of Tomasita, the Abused Tomato," an ecofeminist tale told from the perspective of the tomato forced onto this continental conveyor belt. Parts of this story appear in *Tomasita's Trail: Women, Work, and the World in a Tomato* (forthcoming).

14. Ecofeminism offers an analysis that links the historical domination of women with the human domination of non-human nature. Although there are many different schools of ecofeminist thought, I support an analysis that proposes an integrative, historically and culturally contingent analysis of structural oppressions based on gender, race, class, as intertwined with the exploitation of nature as a "resource." I don't ascribe to the stream of ecofeminism that suggests women (as an essentialist category) are inherently (biologically) closer to nature. See Noel Sturgeon, *Ecofeminist Natures: Race, Gender, Feminist Theory and Political Action* (New York: Routledge, 1997).

15. See Gary Gereffi and Miguel Korzeniewicz, eds., *Commodity Chains and Global Capitalism* (Westport, CT: Praeger Publishers, 1994).

16. See Laura Reynolds, "Institutionalizing Flexibility: A Comparative Analysis of Fordist and Post-Fordist Models of Third World Agro-Export Production," in Gereffi and Korzeniewicz, eds., *Commodity Chains and Global Capitalism*, 143–160.

17. Elson and Pearson, "The Subordination of Women and the Internationalization of Factory Production," 192.

18. For a useful discussion of Fordist and post-Fordist production practices, particularly in terms of the model of fast-food restaurants, see Ritzer, *The McDonaldization of Society*, 150–153.

19. Jeremy Brecher, John Childs, and J. Cutler, eds., *Global Visions: Beyond the New World Order* (Montreal: Black Rose Books, 1993).

20. Karen (pseudonym), interview with author, Toronto, Ontario, February 1998.

21. Wanda (pseudonym), interview with author, Toronto, Ontario, May 1997.

22. See Sara Lara, "La Flexibilidad del Mercado de Trabajo Rural," *Revista Mexicana de Sociologia* 54, no.1 (January-February 1994), 29–48.

23. Conrado Lomeli, interview with author, Guadalajara, Mexico, December 1996.

24. Lourdes Beneria, "Capitalism and Socialism: Some Feminist Questions," in Visvanathan et al., eds., *The Women, Gender, and Development Reader*, 330.

25. Lara, "La Flexibilidad del Mercado de Trabajo Rural." 41. Translated from the Spanish by the author.

26. Single women are preferred as packers, for example, because they are moved from one production site to another and housed in company homes in women-centred families. In the case of Mexican farm laborers hired by the FARMS program in Ontario to pick and pack our vegetables during the Canadian growing season, however, widows are preferred, reflecting a machista attitude that they're safer than married or single women in a foreign job (Irena [pseudonym], interview with author, Miacatlán, Mexico, December 1998).

27. According to Statistics Canada, women are more likely to work part-time, by a ratio of 3 to 1, compared with men. *The Globe and Mail*, reporting on the study, states that "part-time employment was most prevalent among sales and service occupations, particularly in the food-service industry and among grocery clerks." The *Globe* quotes Gordon Betcherman of Canadian Policy Research Networks: "many employers want to hire staff to work less than 30 hours a week because they can be more flexible in scheduling around peak demand and because they have to provide fewer benefits" ("Part-time Work Stats Questioned," *The Globe and Mail*, 18 March 1998, 6). A related article notes that the most predominant female occupation is "retail sales clerk," with "waitress" as number seven on the list ("He's a Trucker, She Types," *The Globe and Mail*, 18 March 1998, 1).

28. The harvest season has varied tremendously lately, due to erratic weather conditions which are often blamed on El Niño. Unseasonal freezes have cut short the tomato season, causing companies financial losses and sending workers either on to other harvests or home to their villages where they seek casual labor to carry them through til the next harvest. In Gomez Farias, the workers lost three months of expected fieldwork and were eking out a living making and selling straw mats (Tomasa, interview).

29. With the economic crisis in Mexico, and deepening gaps between the rich and the poor, agricultural workers are part of a "family wage economy," requiring all members to work for the survival of the family. In *Desert Capitalism: What Are the Maquiladoras?*, Kathryn Kopinak shows that while in 1981, 1.8 family members had to work to feed a family of 5, by 1996, the number was 5.4. Though Northern economies are described as "family consumer economies" rather than "family wage economies," it is increasingly the case that working-class families also depend on multiple salaries, which are often from combinations of part-time jobs.

30. Lara, "La Flexibilidad del Mercado de Trabajo Rural," 42. Translated from the Spanish by the author.

31. Ibid., 41.

32. Wanda (pseudonym), interview with author, Toronto, Ontario, October 1997.

33. Ibid.

34. Tomasa interview.

35. Kopinak, *Desert Capitalism*, 116.

36. Wanda (pseudonym), interview with author, Toronto, Ontario, August 1998.

30

The Politics of Breastfeeding: An Advocacy Update

Penny Van Esterik[1]

Introduction

Politics may be defined as the practice of prudent, shrewd, and judicious policy. Politics is about power. How, then, can politics have anything to do with breast-feeding? When health, profits and the empowerment of women are at stake, how could politics not be involved? Extraordinary changes in the way power is allocated in the world would be necessary for breastfeeding to flourish in this world. Many people believe such changes are impossible to make, that we have "advanced" too far into industrial capitalism ever to retreat into natural infant feeding regimes not based on profits. But even state policies influencing infant feeding practices can change, particularly when people begin to ask some very basic questions about child survival.

Advocacy on behalf of breastfeeding is incomplete and probably ineffective unless accompanied by a politically informed analysis of the obstacles to breastfeeding. These obstacles include the marketing practices of infant formula manufacturers, physician dominated medical systems, and the relationship between industry and health professionals. This relationship has resulted in widespread misinformation about breastfeeding, including false claims of the equivalence between breastmilk and artificial substitutes, and the devaluing of women's knowledge about the management of breastfeeding.

The purpose of this paper is to trace the development of infant feeding as a public policy issue over the last few decades, to examine the role of non-governmental groups (NGOs) in influencing public policy, and to place breastfeeding within the advocacy debates on the promotion of commercial breastmilk substitutes, with the modest goal of putting the voices of industry critics more directly into discussions of the politics of breastfeeding. The paper concludes with a call for anthropologists to include advocacy discourses as a valid addition to other modes of understanding and interpretation.

The Development of the Controversy

Women have always had choices about infant feeding methods. Throughout history, women have substituted animal milks or wetnursing for maternal breastfeeding. This is, however, the first time in history when many infants lived through these experiments long enough for others to measure the impacts on their health. This is also the first time that huge industries have promoted certain options for women, and profited from mothers' decisions not to breastfeed or to supplement breastmilk with a commercial product. It is this historical and economic fact that requires us to place breastfeeding in a broad political context.

An early presentation on the problem of bottle feeding may be traced to a Rotary Club address made by Dr Cicely Williams in Singapore in 1939 entitled "Milk and Murder." She argued that the increased morbidity and mortality seen in Singapore infants was directly attributable to the increase in bottle feeding with inappropriate breastmilk substitutes, and the decline of breastfeeding. And she dared to call this murder—not something that happens to poor people over there, but murder. Her words:

> If you are legal purists, you may wish me to change the title of this address to "Milk and Manslaughter." But if your lives were embittered as mine is, by seeing day after day this massacre of the innocents by unsuitable feeding, then I believe you would feel as I do that misguided propaganda on infant feeding should be punished as the most criminal form of sedition, and that these deaths should be regarded as murder. (Williams 1986, 70)

Although conditions in other cities in the developing world may have been similar or worse than in Singapore, the voices of warning and reproach were hesitant, isolated, and easily ignored. Conditions in many inner-city and native communities in North America today may be little improved over the conditions Williams found in Singapore in 1939.

Occasionally, reports from missionaries and health workers would confirm the devastating effects of bottle feeding on infant morbidity and mortality. But these were single voices and never stimulated a social movement. And it was easy to assume that the "problem" was "over there" and thus was irrelevant to promotional practices of infant food manufacturers in developed industrial countries. Only recently has the full extent of the dangers of commercial infant formula been acknowledged or publicized (Cunningham et al. 1991, Palmer 1993: 306–312, Walker 1993).

From the 1930s, the promotion of breastmilk substitutes steadily increased, particularly in developed countries. In North America, competition between American pharmaceutical companies and the depression reduced the number of companies producing infant formula to three large firms—Abbott (Ross), Bristol-Myers (Mead-Johnson) and American Home Products (Wyeth) (cf. Apple 1980). Food companies like Nestlé were already producing baby foods before the turn of the century. Both food- and drug-based companies producing infant formula expanded their markets during the post-World War II baby boom, as breastfeeding halved between 1946 and 1956 in America, dropping to 25% at hospital discharge in 1967 (Minchin 1985: 216). By that time, the birth rate in industrialized countries had dropped, and companies sought new markets in the rapidly modernizing cities of developing countries. As industry magazines reported "Bad News in Babyland" as births declined in the sixties in North America, their sales in developing countries increased, with only isolated and occasional protests from health professionals and consumer groups.

Other points of resistance to the increasing collaboration between infant formula manufacturers and health professions in North America came from mothers who wanted to breastfeed their infants and met with resistance or lack of support from the medical profession. These voices of resistance were not raised against the infant formula industry nor against the medical profession per se. Rather, they took the form of mother-to-mother support groups. The prime example is La Leche League, a group founded in 1956 in Chicago by breastfeeding mothers. The founding of La Leche League represented women's growing dissatisfaction with physician-directed bottle feeding regimes. While mother-to-mother support groups in some countries have lent support to infant food industry critics, it is important to remember that since its inception, La Leche League never directed its energies outward against infant formula companies, but rather inward toward the nursing couple. Only in recent years has the linkage been made between advocacy groups oriented towards consumer protests and mother support groups.

One phrase in a speech in 1968 by Dr Derrick Jelliffe caught the attention of a much wider audience. He labelled the results of the commercial promotion of artificial infant feeding as "commerciogenic malnutrition." Like "Milk and Murder," this phrase grabbed headlines and became the focus for advocacy writing. By the mid-1970s, publications like the *New Internationalist* (1973) were bringing the problem to public attention. Reports such as Muller's *The Baby Killer* (1974) and the version by a Swiss group called *Nestlé Totet Babys* (Nestlé Kills Babies) prompted responses from Nestlé. In 1974, Nestlé filed libel charges in a Swiss court for five million dollars against the Third World Action Group for their publication *Nestlé Kills Babies*, leading to a widely publicized trial. The judge found the members of the group guilty of libel and fined members a nominal sum, but clearly recognized publicly the immoral and unethical conduct of Nestlé in the promotion of their infant feeding products. The libel suit and these popular publications provided focal points around which public opinion gradually developed, strengthening the efforts of advocacy groups in two complementary directions, the organization of a consumer boycott and drafting a code to regulate the promotion of baby foods (including bottles, teats, and all breastmilk substitutes, not just infant formula).

Strategy for Change: Consumer Boycotts

Since the mid-1970s, a broad range of people from all walks of life, in many different parts of the world, have participated in a public debate known as the breast–bottle controversy or the baby food scandal. Changes in infant feeding practices do not occur spontaneously, nor as a result of health promotion campaigns. In North America, one catalyst for the "back to the breast" movement and a resurgence of interest in breastfeeding was a consumer movement organized by grass roots advocacy groups that drew attention to how the existence and advertising of commercial infant formula affected women's perceptions of their breasts, breastmilk and breastfeeding. They demonstrated that there was a direct and specifiable link between changes in infant feeding practices and the promotion of commercial infant formula in developing countries. The participation of ordinary people in North America in this debate was mostly through the direct action of a consumer boycott. Without the social mobilization of the consumer boycott, the work to promote a code for the marketing of breastmilk substitutes would not have been as effective.

Both boycott groups and promoters of a code to regulate the way infant formula was being promoted and marketed argued that the decline in initiation rates and the duration of breastfeeding could be linked to the expanding promotion of breast-milk substitutes, usually by multinational food and drug corporations, and to bottle feeding generally. The boycott against Nestlé's products, and eventually those of other infant formula manufacturers generated the largest support of any grass roots consumer movement in North America, and its impact is still being felt in industry, governments, and citizen's action groups around the world. Women were the primary supporters of the boycott against Nestlé and other manufacturers of infant formula, although the movement in North America was strongly male dominated. Nevertheless, many women gained experience in analyzing the relations between corporate power and public health through their experience of working on the boycott campaign.

The groups that took on the task of challenging the infant formula companies were for the most part small, underfunded and in many cases ran on voluntary labor. While they were not the only people to recognize the problems of bottle feeding, they were the first to effectively mobilize to challenge the industries promoting it. Their success against the forces ranged against them, including powerful governments and multinational corporations, is a study in the power of co-operative networking. The importance of these small, non-governmental groups cannot be overstressed.

IBFAN (the International Baby Food Action Network) is a single-issue network of extraordinarily dedicated people—flexible, non-hierarchical, decentralized and international in organization (Allain 1991). IBFAN works to promote breastfeeding worldwide, eliminate irresponsible marketing of infant foods, bottles and teats, advocate implementation of the WHO/UNICEF International Code of Marketing of Breastmilk Substitutes, and monitor company compliance with the Code.

In North America and Europe, advocacy groups also formed around the issue—most notably the Interfaith Centre for Corporate Responsibility (ICCR), the Infant Formula Action Coalition (INFACT) in Canada and formerly in the United States, the Baby Milk Action Group in Britain, the Geneva Infant Feeding Association (GIFA), and the many groups in developing countries that formed part of the IBFAN network. Throughout the late 1970s and early 1980s, these groups provided evidence of the unethical marketing of infant formula in their communities. This evidence was critically important in convincing delegates to the World Health Assembly (WHA, the meetings of the World Health Organization) that a regulatory code of industry practices was necessary.

The New York based ICCR, formed in 1974, monitored multinational corporations, provided information to church groups on responsible corporate investments, and publicized cases such as the lawsuit filed by the Sisters of the Precious Blood against Bristol-Myers in 1976 for misleading stockholders about their infant formula marketing practices. Although the lawsuit was dismissed, information about the marketing of breastmilk substitutes circulated in church basements among groups interested in Third World development and justice issues, bringing a new constituency into the movement. Public education on the promotion of breastmilk substitutes often featured the 1975 film, *Bottle Babies*, a vivid portrayal of the tragic effects of bottle feeding in Kenya.

In 1977, several action networks began the campaign to boycott Nestlé products in North America. The American INFACT (later called Action for Corporate Accountability) grew out of a student group at the University of Minnesota, while the Canadian INFACT groups developed around justice ministries of the Anglican and United Churches, first in Victoria, British Columbia. These groups were linked together through IBFAN to represent the views of coalition members at international health policy meetings such as the World Health Assembly.

It was through these groups that the general public in North America was made aware of the infant formula controversy (or the breast–bottle controversy) through an increasingly sophisticated campaign involving public debates, newsletters, radio and TV shows, petitions, demonstrations, posters, buttons, and the first consumer boycott of Nestlé's products, which ended in 1984.

The advocacy position as defined by the boycott groups is quite straightforward. It argues that the makers of infant formula should not be promoting infant formula and bottle feeding in developing countries where breastfeeding is prevalent and the technology for adequate use of infant formula is absent. Advocacy groups claim that multinational corporations (like Nestlé), in their search for new markets, launched massive and unethical campaigns directed toward medical personnel and consumers that encouraged mothers in developing countries to abandon breastfeeding for a more expensive, inconvenient, technologically complex, and potentially dangerous method of infant feeding—infant formula from bottles. For poor women who have insufficient cash for infant formula, bottles, sterilization equipment, fuel, or refrigerators; who have no regular access to safe, pure drinking water; and who may be unable to read and comprehend instructions for infant formula preparation, the results are tragic. Misuse of infant formula is a major cause of malnutrition and the cycles of gastroenteritis, diarrhoea and dehydration that lead eventually to death. Advocacy groups place part of the blame for this "commerciogenic malnutrition" on the multinational companies promoting infant formula.

The boycott groups have never advocated a ban on the sale of infant formula, although some have advocated its "demarketing" (Post 1985). Nor were women to be pressured to breastfeed against their will, although critics of breastfeeding advocacy groups represented their aims in this light. "Better to bottle feed with love than breastfeed with reluctance" is a cliché cited by many different people convinced that protecting mothers from feelings of guilt for not breastfeeding is more important than removing obstacles to breastfeeding. The intentions of INFACT and other boycott groups are clearly stated in their demands:

1 An immediate halt to all promotion of infant formula.
2 An end to direct product promotion to the consumer, including mass

media promotion and direct promotion through posters, calendars, baby care literature, baby shows, wrist bands, baby bottles, and other inducements.

3   An end to the use of company "milk nurses."

4   An end to the distribution of free samples and supplies of infant formula to hospitals, clinics, and homes of new mothers.

5   An end to promotion of infant formula to the health professions and through health care institutions.

The infant formula companies responded to the boycott groups by modifying their advertising to the public, but they were slow to meet all the demands and certainly never met the spirit of the demands, namely, to stop promoting their products. They simply promoted new products such as follow-on milks for toddlers, developed new marketing strategies, and hired public relations firms to answer their critics and to improve their corporate image.

Nestlé's efforts were concentrated on trying to improve their tarnished public image by hiring a prestigious public relations firm, sending clergy glossy publications about their contributions to infant health, and generally discrediting their critics as being merely uninformed opponents of the free enterprise system (Chetley 1986: 46, 53). Companies such as Nestlé continue their efforts to buy social respectability by sponsoring events at international medical and nutrition conferences, in addition to funding research on infant feeding.

Food boycotts have been a successful tool for social mobilization. Like all mass action social movements, the rhetoric used by advocacy groups oversimplifies the issue and seldom provides all the statistically significant evidence that both the infant formula industry and medical journals call for (cf. Gerlach 1980). But that is the nature of advocacy communication used by all social mobilization groups. At one level of analysis, the issue is both clear and simple; it is made more complex by the many obstacles ranged against breastfeeding. Nevertheless, the words and sentiments voiced in the original advocacy documents still ring clear today.

Strategy for Change: Code Work

Another parallel stream of activities for advocacy groups concerned lobbying and attending drafting sessions on the development of a code to regulate the marketing of breastmilk substitutes. Health professionals called for establishing policy guidelines on infant feeding through United Nation groups such as the Protein-Calorie Advisory Group. In 1979, WHO and UNICEF hosted an international meeting to develop an international code regulating the marketing of breastmilk substitutes. That meeting enabled nine infant formula companies to form the International Council for Infant Food Industries (ICIFI) (Palmer 1993: 237), and to lobby UN agencies for guidelines least damaging to their profits. The code was drafted with the cooperation and consent of the infant formula industry and is very much a compromise, a minimal standard rather than the ideal.

North American advocacy groups in IBFAN "had to divide their very scarce resources and energy between running a boycott of Nestlé and the expensive periodic visits to Geneva for the Code drafting sessions" (Allain 1991: 10). Work in the United States to document abusive marketing practices of infant formula companies was brought to a head in 1978 by the Congressional Hearings on the Marketing and Promotion of Infant Formula in the Developing Nations chaired by Edward Kennedy. During the hearings, Ballarin, a manager of Nestlé's Brazilian operations, claimed—to the amazement of the hearing—that the boycott and the campaign against the infant formula companies was really an "attack on the free world's economic system," led by "a worldwide church organization with the stated purpose of undermining the free enterprise system" (United State Congress 1978: 127).

In May of 1981, the World Health Assembly adopted a non-binding recommendation in the form of the WHO/UNICEF Code for the Marketing of Breastmilk Substitutes with a vote of 118 for, 3 abstentions, and 1 negative vote. The negative vote was cast by the United States, in spite of the fact that the members of the United States Senate proposed the idea of a Marketing Code and had actively participated in the drafting process. The American delegate to the WHA had been an enthusiast for the Marketing Code until shortly before the vote, when direct orders from his government ordered him to vote against its adoption. The Reagan White House had responded to direct lobbying from the infant formula industry (Chetley 1986). The delegate who was ordered to reverse his nation's stance did so, and then resigned his post.

The Marketing Code is not a code of ethics but a set of rules for industry, health workers and governments to regulate the promotion of baby foods through marketing. It covers bottles, teats and all breastmilk substitutes, not just infant formula.

The code includes these provisions:

- No advertising of any of these products to the public.
- No free samples to mothers.
- No promotion of products in health care facilities, including the distribution of free or low-cost supplies.
- No company sales representatives to advise mothers.
- No gifts or personal samples to health workers.
- No words or pictures idealizing artificial feeding, or pictures of infants on labels of infant milk containers.
- Information to health workers should be scientific and factual.
- All information on artificial infant feeding, including that on labels, should explain the benefits of breastfeeding, and the costs and hazards associated with artificial feeding.
- Unsuitable products, such as sweetened condensed milk, should not be promoted for babies.
- Manufacturers and distributors should comply with the Code's provisions even if countries have not adopted laws or other measures.

Since then, subsequent WHA resolutions extended the ban on distribution of free and low-cost supplies to all parts of the health care system, addressed promotion to the general public, and spelled out the responsibilities of different groups in implementing the Code.

After the Code

Following the establishment of the Code, Nestlé and other infant formula companies publicly released special instructions to its marketing personnel to comply with the Code, and asked the International Boycott Committee, a subgroup of IBFAN groups who were working on the boycott, to call it off. However, the boycott continued until 1984, when some means of monitoring company compliance with the Code could be established, and WHO member countries could draft national codes.

The advocacy groups, in the absence of national machinery, continued their monitoring role, recording and publicizing non-compliance of the Code (IBFAN 1991). WHO and UNICEF have never monitored Code compliance, although occasionally they have taken individual companies to task. UNICEF's executive board extracted a promise that manufacturers would end all free supplies of infant formula to hospitals by the end of 1992. They did not comply. In the Philippines, for example, a law banning free supplies was passed, but was evaded by the company tactic of invoicing for milk supplies and not bothering to collect payment. In the face of this and other flagrant violations, a second boycott against Nestlé and American Home Products in the United States, and Nestlé and Milupa in Germany was launched in 1988 by groups who were part of the IBFAN network. Today, Nestlé remains under boycott because the company continues to violate the Code. However, in this era of Internet communication, the boycott is more a global movement to draw attention to obstacles to breastfeeding than national campaigns.

In 1986, a World Health Resolution was adopted that acknowledged the detrimental effect of free or low cost supplies and clarified the relevant Articles in the Code by banning such supplies. According to the resolution, free or low cost supplies of infant formula were not to be given to hospitals. If supplies were donated to an infant, they were to be continued for as long as the infant required the milk. Hospitals that needed small quantities of infant formula for exceptional cases could buy them through the normal procurement channels. Thus, free supplies could no longer be used as sales inducements. Most of the major companies who were giving free supplies ignored the resolution, arguing that they would only stop distributing free supplies if governments brought in laws against them. However, the Code states that "Independently of other measures" manufacturers and distributors should take steps to ensure their conduct at every level conforms to the principles and aims of the Code.

At the World Health Assembly meeting in May of 1994, advocacy groups' successful lobbying reminded delegates that free and low cost supplies of infant formula are marketing devices pure and simple, and not charity, a point made in 1989 by the Nigerian Minister of Health during the WHA. A few European countries including Ireland and Italy and, most forcefully, the United States delegation tried to defeat the resolution to end free supplies. But their efforts were thwarted by a block of African delegates and a very effective Iranian delegate who made it clear that the American position was the industry position as advocated by the International Association of Infant Food Manufacturers (IFM, the successor to ICIFI). The meeting ended with a consensus to withdraw all amendments and support the original text proposed by WHO's Executive Board to end donations of infant formula to all parts of the health care system worldwide. The question still remains how the Code and WHA resolutions can be implemented and monitored. Advocacy groups have continued to take up the challenge, and ensure that the issue does not quietly disappear from the world's conscience.

Allain refers to the "unholy alliance" (1991: 15) between the medical profession and the baby food companies. Certainly, the medical profession and medical associations followed rather than led the advocacy groups in their criticism of industry. Although there was resistance by some doctors to the promotion of commercial baby foods, only occasional voices of protest were heard from health professionals in the 1950s and 1960s, as infant feeding became more completely medicalized.

In the United States, continuing efforts by health professionals, including the late Derrick and Patrice Jelliffe and Michael Latham, continually brought the issue of breastfeeding and promotional practices of industry to the attention of health organizations. Internationally, the advocacy groups turned a number of physicians into more outspoken public advocates for breastfeeding, stimulating a medical consensus on the value of breastfeeding. But many university and medical school research projects on infant nutrition are funded by industry money. Doctors are beginning to speak out against practices in their own hospitals, but they may be criticized by the medical establishment for doing so. As researchers are increasingly being warned (Margolis 1991), there is no such thing as a free lunch, nor do people bite the hand that feeds them.

For all their rhetoric, and what some have decried as their so-called confrontational tactics, the advocacy groups deserve great credit for bringing about what decades of clinical observations alone failed to accomplish: public awareness and concern about the dangers of breastmilk substitutes. This struggle for corporate accountability is often recounted in development education workshops as well as marketing classes (Post 1985). For the first time, non-governmental organizations like INFACT, IBFAN, and ICCR had a direct role in the deliberations at WHO and UNICEF in 1979, and in subsequent meetings regarding infant feeding policy. Chetley points out that in spite of industry's concerns about the "scientific integrity" of allowing popular organizations, mother's groups and consumer groups to participate, delegates to the international meetings were impressed with the contributions of the non-governmental organizations (1986: 65–69). It is the NGOs that keep alive the underlying concern about corporate responsibility, human rights, and infant feeding as a justice issue.

Breastfeeding Advocacy Since the Nineties

In 1990, a global initiative sponsored by a number of bilateral and multilateral agencies resulted in the adoption of the Innocenti Declaration, which reads in part:

> As a global goal for optimal maternal and child health and nutrition, all women should be enabled to practice exclusive breastfeeding and all infants should be fed exclusively on breast

milk from birth to 4–6 months of age. Thereafter, children should continue to be breastfed, while receiving appropriate and adequate complementary foods, for up to two years of age or beyond. . . . Efforts should be made to increase women's confidence in their ability to breast-feed. Such empowerment involves the removal of constraints and influences that manipulate perceptions and behavior towards breastfeeding, often by subtle and indirect means. This requires sensitivity, continued vigilance, and a responsive and comprehensive communic-ations strategy involving all media and addressed to all levels of society. Furthermore, obstacles to breastfeeding within the health system, the workplace and the community must be eliminated. (Innocenti Declaration 1991: 271–272)[2]

This carefully worded statement is nothing less than a challenge to change the priorities of the modern world. The language stresses the empowerment of women rather than their duty to breastfeed, a change that should bring more advocates for women's health to support breastfeeding.[3]

Later in 1990, UNICEF convened a meeting to review progress on breastfeeding programs and concluded that if the Innocenti Declaration were ever to be imple-mented, work would have to be done by NGOs rather than governments alone. This led to the formation of an umbrella group called the World Alliance for Breastfeeding Action (WABA). WABA is a global network of organizations and individuals who are actively working to eliminate obstacles to breastfeeding and to act to implement the Innocenti Declaration. The groups include those who approach breastfeeding from different perspectives—from consumer advocates to mother support groups and lactation consultants.

As part of their social mobilization efforts to gain public support for implement-ing the Innocenti Declaration, WABA sponsors World Breastfeeding Week (August 1–7) to pull together the efforts of all breastfeeding advocates, governments, and the public. The first campaign, in 1992, focused on hospital practices, and was called the Baby Friendly Hospital Initiative (BFHI). This campaign established steps that hospitals should take to support breastfeeding and to implement the Innocenti Declaration, and was based on the WHO/UNICEF statement, Ten Steps to Successful Breastfeeding. By 2005, over 20,000 maternity facilities world-wide had been approved as baby-friendly (Van Esterik 2006). The second campaign, in 1993, tackled the problem of developing mother-friendly workplaces, where breastfeeding and work could be combined. The complexity of the integration of women's productive and reproductive work, and the relevant cultural and policy issues have been explored elsewhere (Van Esterik 1992). Other WBW campaign themes have explored implementing the Code (1994), empowering women (1995), environmental linkages (1997), economic advantages of breastfeeding (1998), human rights (2000), exclusive breastfeeding (2005), complementary feeding (2006), and breastfeeding in the first hour (2007).

New Millennium: New Challenges

The public appeal of the infant formula controversy was that it was presented as a simple, solvable problem. People in North America were attracted to the campaign because it put many of their unspoken concerns about the power of multinational corporations into a clear, concrete example of exploitative behavior that could be acted upon. For some boycott groups, the solution to the problem of bottle feeding with infant formula was for multinational infant formula manufacturers to stop promoting infant formula in developing countries. When the companies agreed to abide by the conditions of the WHO/UNICEF Code and the boycott was lifted, this marked the end of the campaign, a victory of small grass roots organizations over huge corporations. As with other social movements, it was hard to sustain interest in the issue after a "victory" had been declared. But the advocacy groups and most breastfeeding supporters recognized that infant feeding decisions are not related to marketing abuse alone; rather, the issue was embedded in a set of problems that require rethinking broader questions about the status of women, corporate power over the food supply, poverty and environmental issues, among others.

For example, the implications of bottle feeding have not been explored from an environmental perspective, with the exception of the position paper by Radford (1992). The ecology and environmental justice movements have been slow to recognize breastfeeding as part of sustainable development and breastmilk as a unique under-utilized natural resource. The report of the World Commission on the Environment and Development, *Our Common Future* (1987), made no reference to nurturance or infant feeding, although economy, population, human resources, food security, energy, and industry are all discussed as part of sustainable development.

Sustainability refers to courses of action that continue without damaging the environment and causing their own obsolescence. A sustainable infant feeding policy must consider the impact of decisions a number of years in the future, rather than simply examining conditions at the present. If we compare breastfeeding and bottle feeding as modes of infant feeding, each has very different implications for sustainability; breastmilk is a renewable resource, a living product that increases in supply as demand increases. It reinforces continuity with women's natural reproductive phases and is a highly individualized process, adapting itself to the needs of infant and mother. The infant is actively empowered and "controls" its food supply.

By contrast, the bottle feeding mode—most commonly associated with infant formula even in developing countries—is a prime example of using a non-renewable resource that uses even more non-renewable resources to produce and to prepare. It puts demands on fuel supplies and produces solid wastes (for every 3 million bottle fed babies, 450 million tins are discarded). It is a standardized product that does not take into consideration individual needs (although in practice it is not really standardized; it is commonly adulterated in its industrial production with insect parts, rat hairs, iron filings and accidental excesses of chlorine and aluminum, or adjusted by the preparer to individualize it by the addition of herbs and sugar). The bottle fed infant is passive, controlled by others, and becomes a dependent consumer from birth.

Recently, scientists have identified intrinsic contamination of powdered infant formula itself. "Pathogenic micro-organisms" have been found in unopened tins and packages of powdered infant formula. In 2003, the Codex Alimentarius Commission of the FAO identified the presence of harmful bacteria such as Enterobacter sakazakii in powdered infant formula as a "known public health risk." The risk of potentially fatal infections appears to be highest for neonates in hospital settings.

A sustainable development policy for infant feeding must take careful note of the fact that women's capacity to breastfeed successfully is often a gauge for judging when our capacity to adapt to environmental stresses—air and water pollution, environmental toxins, radiation—has been overstrained. But women are not canaries or cows or machines. Breastfeeding promotion that treats women as mere milk producers is bound to fail, and the issue itself will be rejected by women's groups. For this reason, breastfeeding advocacy groups have been working closely with environmental groups and women's groups to reposition breastfeeding in their agendas (cf. Van Esterik 1994, 2002). For example, by agreeing on common language among different movements, we see fewer phrases like "mother's milk poisons infants", and "contaminated breastmilk" in newspaper headlines.

With our increasing knowledge about HIV/AIDS, comes another challenge for breastfeeding advocacy. HIV can be transmitted during pregnancy, childbirth and breastfeeding by a mother who is HIV positive. Consequently, in many parts of the world women who tested positive for HIV were advised not to breastfeed, but instead to feed their newborns with artificial baby milks. But the risks of not breastfeeding were often greater than the risks of transmitting HIV; the risk that HIV can be passed through breastmilk should not undermine breastfeeding support for other mothers and infants. Combining breastfeeding and artificial baby milks appears to be the worst solution for infants. More recent evidence suggests that women who exclusively breastfed can reduce the risk significantly, increasing HIV-free survival for their children (Iliff et al. 2005). Thus, breastfeeding advocates have to work closely with HIV/AIDS groups to insure that policies around preventing mother to child transmission do not undermine local breastfeeding cultures.

Advocacy and Anthropology

Food advocacy is tiring work, and many anthropologists working in academic settings are not full-time activists or breastfeeding counsellors. Yet many of us have been drawn to research on breastfeeding by our personal experiences of mothering or by witnessing the commercial exploitation of women in different countries. Advocacy lessons are personal lessons because they require each and every one of us to put our values on the line—even occasionally to suspend academic canons of reserve and non-involvement, and respond emotionally to things we feel strongly about. In the study of breastfeeding, there is a convergence of different ways of knowing—a convergence of scientific knowledge, experimental knowledge, and experiential knowledge of generations of women, with moral and emotional values that all support action to support, protect and promote breastfeeding. Few areas of research in anthropology encourage such integration. Further, advocacy lessons are never far from us, as advocacy action permeates different parts of our lives and links diverse causes—from the women's movement to environmental concerns.

In this climate of reflexive anthropology, and the increasing responsibility that the profession as a whole is taking in human rights debates, it is important that we clarify our relation to advocacy discourse and action as professional anthropologists and as citizens. Anthropology has a long history of applied work, but more recent and more problematic is the commitment of individual anthropologists to advocacy work (cf. Harries-Jones 1985). But advocacy anthropology is still suspect to some in the profession. Advocacy refers to the act of interceding for or speaking on behalf of another person or group (Van Esterik 1986), or promoting one course of action over another. This takes us beyond presentations, analyses, and discussion of evidence to recommend particular alternatives. Advocacy work draws some anthropologists into taking action with regard to well-defined goals that may best be implemented outside of academic settings. What has made this position acceptable in anthropology? First, the increasing numbers of anthropologists who have become involved in "causes" such as the rights of indigenous peoples, famine, AIDS, and women's rights, have made such commitment more visible within the profession. At the same time, the increasing involvement of indigenous peoples and special interest groups in advocating on their own behalf has resulted in anthropologists working with or for these groups.

Second, these individual and collective initiatives occurred at the same time as theoretical work, arguing that there is no such thing as "scientific objectivity," and that many past examples of applied anthropology were both paternalistic and supportive of the status quo.

Third, feminist anthropology's epistemological stance on the lack of separation between theory and action justifies and even requires advocacy stances. Feminist methodology calls for explicit statements of the positionality of the author. The feminist axiom "the personal is political" breaks down past opposition between "emotional advocacy action" and "cool, detached scientific reasoning," and accepts experience and emotion as valid guides to moral stands. This is particularly true of food activism, where the line between objective and participatory approaches to food is blurred. But as advocacy groups remind us, it is politics that determine whose truth is heard.

Finally, all branches of anthropology continue to be involved in human rights debates. These efforts have changed the way that advocacy is integrated into anthropology.

Advocacy for breastfeeding is one enormous anthropology lesson. Breastfeeding is simultaneously biologically and culturally constructed, deeply embedded in social relations, and yet cannot be understood without reference to varying levels of analysis, including individual, household, community, institutional and world industrial capitalism. As much a part of self and identity as political economy; as personal as skin and as impersonal as the audit sheets of international multinational corporations, breastfeeding research requires a synthesis of multiple methods and theoretical approaches. At a time when anthropology hovers on the brink of self-reflexive nihilism and fragmentation on the one hand, and greater involvement in studying global change, internationalism, and public policy on the other (cf. Givens and Tucker 1994), breastfeeding provides a challenging focus for holistic, biocultural, interdisciplinary research.

Notes

1. The most comprehensive history of the controversy is Andrew Chetley's, *The Politics of Baby Foods* (1986) and *Baby Milk: Destruction of a World Resource* from the Catholic Institute for International Relations (1993). I also review the history in my book, *Beyond the Breast–Bottle Controversy* (1989), and am using this opportunity to update that discussion. This update has benefited from the views and writings of Gabrielle Palmer and Elizabeth Sterken.
2. "The Innocenti Declaration was produced and adopted by participants at the WHO/UNICEF policy-makers' meeting on "Breastfeeding in the 1990s: A Global Initiative," co-sponsored by the United States Agency for International Development (AID) and the Swedish International Development Authority (SIDA), held at the Spedale degli Innocenti, Florence, Italy, on July 30—August 1, 1990. The Declaration reflects the content of the original background document for the meeting and the views expressed in group and plenary sessions." (Innocenti Declaration 1991).
3. The detailed reports on the infractions of the Marketing Code by infant formula companies used to be available only in "fugitive literature"—letters and brief reports in newsletters in many languages. The violations are most accurately recorded in the "SOCs," red and blue folders published by IBFAN since 1988, documenting the State of the Code, by country and by company. Breastfeeding advocacy groups in individual countries now trace and publicize the violations on line. See for example INFACT Canada (www.infactcanada.ca), IBFAN (www.ibfan.org), and the Baby Milk Action Group in Britain (www.babymilkaction.org). WABA (www.waba.org.my) has links to these and other relevant websites. The back files of the campaign over the last thirty years are treasure troves for studying this social movement, but are not easily made to conform to academic standards of citation.

References

Allain, A. (1991) IBFAN: On the cutting edge. *Development Dialogue* offprint, April: 1–36, Uppsala, Sweden.

Apple, R. (1980) To be used only under the direction of a physician: commercial infant feeding and medical practice, 1870–1940. *Bulletin of the History of Medicine* 54: 402–417.

Baby Friendly Hospital Initiative (1994) Progress Report. See endnote 3.

Catholic Institute for International Relations (1993) *Baby Milk: Destruction of a World Resource.* London: Russell Press.

Chetley, A. (1986) *The Politics of Baby Food.* London: Frances Pinter.

Cunningham, A.S., D.B. Jelliffe, and E.F.P. Jelliffe (1991) Breastfeeding and health in the 1980s: A global epidemiologic review. *Journal of Pediatrics* 118 (5): 659–666.

Fildes, V. (1986) *Breasts, Bottles and Babies.* Edinburgh: Edinburgh University Press.

Gerlach, L.P. (1980) The flea and the elephant: infant formula controversy. *Transaction* 17 (6): 51–57.

Givens, D., and R. Tucker (1994) Sociocultural anthropology: the next 25 years. *Anthropology Newsletter* 35 (4): 1.

Harries-Jones, P. (1985) From cultural translator to advocate: changing circles of interpretation. In *Advocacy and Anthropology*, edited by R. Paine, pp. 224–248. St John's, Newfoundland: Institute of Social and Economic Research.

IBFAN (1991) Breaking the rules. Penang. See endnote 3.

Iliff, P. et al. (2005) Early Exclusive Breastfeeding (EBF) reduces the risk of postnatal HIV-1 transmission and increases HIV-free survival. *AIDS* 19 (7): 699–708.

Innocenti Declaration (1991) Innocenti Declaration: on the protection, promotion and support of breastfeeding. *Ecology of Food and Nutrition* 26: 271–273.

Margolis, L.H. (1991) The ethics of accepting gifts from pharmaceutical companies. *Pediatrics* 88 (6): 1233–1237.

Minchin, M. (1985) *Breastfeeding Matters.* Sydney, Australia: George Allen and Unwin.

Muller, M. (1974) *The Baby Killer.* London: War on Want.

Palmer, G. (1993) *The Politics of Breastfeeding.* London: Pandora Press.

Post, J. (1985) Assessing the Nestlé boycott: corporate accountability and human rights. *California Management Review* 27 (2): 113–131.

Radford, A. (1992) *The Ecological Impact of Bottle Feeding.* WABA Activity Sheet No. 1. Penang, Malaysia.

Sussman, G.D. (1982) *Selling Mothers' Milk: The Wet-Nursing Business in France, 1715–1914.* Urbana, IL: University of Illinois Press.

UNICEF News Release (1994) See endnote 3.

United States Congress (1978) *Marketing and Promotion of Infant Formula in the Developing Nations.* Washington, DC: U.S. Government Printing Office.

Van Esterik, P. (1986) Confronting advocacy confronting anthropology. In *Advocacy and Anthropology*, edited by R. Paine, pp. 59–77. St John's, Newfoundland: Institute for Social and Economic Research.

—— (1989) *Beyond the Breast–Bottle Controversy*. New Brunswick, N.J.: Rutgers University Press.

—— (1992) *Women, Work and Breastfeeding*. Cornell International Nutrition Monograph No. 23. Ithaca, New York.

—— (1994) Breastfeeding and Feminism. In *International Journal of Gynecology and Obstetrics* 47 (Suppl.) S.41–54.

—— (2002) *Risks, Rights and Regulation: Communicating about Risks and Infant Feeding*. Penang: WABA.

—— (ed.) (2006) *Celebrating the Innocenti Declaration on the Protection, Promotion and Support of Breastfeeding: Past Achievements, Present Challenges and Priority Action for Infant and Young Child Feeding. Florence*. UNICEF: Innocenti Research Centre, Florence.

Walker, M. (1993) A fresh look at the risks of artificial infant feeding. *Journal of Human Lactation* 9 (2): 97–107.

Williams, C. (1986) Milk and murder. In *Primary Health Care Pioneer: The Selected Works of Dr Cicely Williams*, edited by N. Baumslag, pp. 66–70. Geneva, Switzerland: World Federation of Public Health Associations.

World Commission on Environment and Development (1987) *Our Common Future*. New York: Oxford Press.

31

The Political Economy of Obesity: The Fat Pay All[1]

Alice Julier

Recently, the Learning Channel (TLC) has entered the realm of Reality TV with a new program called "Honey, We're Killing the Kids." In each episode, a heterosexual family with children is chosen and visited by a nutritionist who does an "assessment" of the BMI and eating patterns of the kids. Using computer imaging, the ultra-thin blonde expert shocks the parents with a picture of their children as fat adults: the premise is, of course, that their current food habits and lifestyle put them on a collision course with obesity. TLC calls the program "a wake up call to parents" who will have a "dramatic reality check" when they see the future face of their children. The promotional ads show a mother in a supermarket checkout line, busy but absentmindedly acquiescing to her 8-year-old son's requests for candy—she looks up as she empties her cart to see him aged into a corpulent balding man. Along the way, the nutritional expert berates the parents for their bad behavior, gives them a prescription for changing their habits, and then checks back in three weeks on how they're managing. The family's reward is to see new computer images of how their children will turn out if they stick with the regime forever. As with many reality shows, there is spectacle, public chastisement, and repentance, but no absolution. The viewer wonders how soon after the surveillance ends that the family will fall off the anti-fat bandwagon. Given the prevailing belief in Westerners as undisciplined and self-indulgent, we suspect that a relapse is inevitable.

The show and its advertisements encapsulate much of the public and political anxiety about food and eating that is currently focused on what's been called "the obesity epidemic." It is a spectacle that, in the press, easily out-maneuvers the genocide in Darfur and the war in Iraq as threats to global stability.[2] Fears of an obese nation have led the last three Surgeon Generals as well as the current President to call for greater personal responsibility towards exercise and weight loss as part of the moral obligation of citizenship. Sociologists refer to this kind of event as a moral panic, where a collective fear of a group or its behavior is fomented by intense media coverage (Hunt 1997).

How We Got Fat

Although body size has engendered social concern and state policies in other historical moments, this particular moral panic in the United States builds on decades of what Kim Chernin calls "a tyranny of thinness," which ties a slender appearance to social virtue.[3] Its rise, however, was aided by the moral entrepreneurship of doctors, pharmaceutical companies, and public health professionals who, in the last two decades, began shifting obesity (a descriptor of a physical state) to a disease entity with the potential for contagion (Jutal 2006). In the mid-1990s, the *Journal of the American Medical Association* and other leading medical journals published key studies linking rising rates of obesity and mortality.[4] Around the same time, the World Health Organization issued a report that used new, lower BMI benchmarks for overweight and obesity.[5] In a critical convergence of science and collective framing, William Dietz and colleagues at the Centers for Disease Control published a PowerPoint presentation of a series of U.S. maps over a ten-year period, showing rates of obesity spreading like an infection across the country, as more states crossed into the "red" zone of high BMI.[6] Calling obesity an epidemic frames it such that in 2005 the CDC even sent a team of specialists into West Virginia to investigate an "infectious outbreak" of obesity.

Like most contemporary social problems, groups of claimsmakers have developed on both sides of the issue: on one side, there are anti-fat scientists and activists who use science and the media to lambaste the food industry, Americans' sedentary lifestyles, and a variety of moral causes; on the other, there are fat acceptance researchers and activists, who question the links between body size and morbidity and advocate for "size acceptance" (Saguy & Riley 2005). Using a variety of critical, feminist, and postmodern approaches, social scientists have attempted to understand why body size, a seemingly personal and physical state, has become such a center point for collective outrage and concern.[7] By questioning the very construction of a moral panic around weight and the national diet, they take apart the categorical boxes that often manifest around the growing, producing, and eating of food. These critics try to dismantle the science behind the construction of obesity as a disease state and the vilification of women, poor people, and people of color, who are often seen to epitomize the problem. Feminist scholars offer a reclamation of the meaning of women's bodies as historically shifting, such that contemporary configurations, the constraints on women's body size are directly related to constraints on women's economic, social, and political power (Wann 1999, Bordo 2003). According to Thompson (1994), for lesbians, working class women, and women of color, the seemingly "disordered" relationship to food and eating is related to the stress of living in a sexist, racist, and classist society.[8] Others point to the continued and indeed growing numbers of people, especially women and children, in affluent societies who are food insecure; such debates about obesity draw attention away from the insufficiency of current food distribution and civil entitlements. Within the field of medical research, scientists debate the research methods and results. Even the Centers for Disease Control has been embedded in its own internal conflict about the links between obesity and mortality (Campos 2004).

Meanwhile, as scholars question the veracity of the scientific arguments about obesity as a public health crisis, the public belief in the correlation between fatness, deviance, and illness becomes more entrenched. The degree of media coverage of the obesity–morbidity relationship continues unabated, with some of the public health focus shifting slightly to an emphasis on physical activity rather than weight loss. At the same time, the stigmatization of fat remains perfectly acceptable in a culture that decries overt racial slurs.[9] As the rhetoric increases, fat activists engage in a difficult public legal battle to free body size from the realm of public health crisis, engaging scientific researchers whose work questions the methods and statistics generated by their anti-fat peers.

But sometimes, social problems are not solved by "more science," but rather by questioning the social uses and structural arrangements that go with the science. As Campos et al. (2005b) point out, "very few of those who have announced the arrival of the obesity epidemic have been satisfied with being mere messengers" (p. 6). While others with greater knowledge have the scientific authority to critique the problem on those grounds, I believe in these times, it might be good for some "soft scientists" to rough up the messenger a bit. We might better understand the persistence of this moral panic by applying a more conservative theoretical framework, illuminating how an obesity epidemic benefits some members of society. A political economy of obesity encourages us to look at the systemic distribution of power and rewards associated with this public health crisis. When fatness is conflated with irresponsible behavior, those who are not fat—who "treat" and construct a public agenda based on controlling the obese—gain status.

A Functional Analysis of Obesity

In the 1950s, functionalism was a dominant theoretical tool in American sociology. The theory's grand argument is that institutions have social functions that fill social and individual needs, integrating them into a larger system that remains relatively stable. Critiqued by a new generation of Marxist, feminist, critical race, and postmodern scholars who decried its adherence to the status quo, its presentation of "society" as a cohesive entity, and its inability to understand the changes wrought by the social movements of the time, functionalism fell out of favor with most scholars who unabashedly hoped to challenge persistent inequality, especially in affluent societies.

However, sociologist Herbert Gans has spent much of his distinguished career questioning the ways in which the poor and marginalized are demonized by the media, sociologists, and public policy experts, even in their choice of terminology. In his 1971 article "The Uses of Poverty: The Poor Pay All," Gans was engaged in a twofold sociological and political act. The article was a continuation of his observation that the U.S. has a long history of casting off the poor and the needy and then blaming them for their own marginalization.[10] Gans' second concern was in making use of functionalism as an answer to conservative critics, arguing that, as scholars and citizens, if we do not deal with the *positive* functions of poverty—the ways in which some people benefit from the continued existence of inequality—then public policies would always be inadequately remedial. This way of thinking about the "functionality" of poverty is in many ways a preview of post-structuralist and constructionist theories that unpack the kinds of privileges and entitlements that accrue to wealth, whiteness, and maleness in unequal social systems. As Gans rightly points out, these entitlements remain entrenched because it is difficult to enact policies that ask people to give up or share benefits they perceive as rights.

We can apply Gans' analysis to the current framing of obesity as a social problem. I write as a public sociologist, using sociological frameworks to question how private issues are transformed into social problems and moral panics. Public sociology applies our skills, concepts, and theoretical tools to problems that are deeply and discursively defined by the media, public perception, and scientific ideologies. I use a functionalist lens in order to highlight why such discourses and frameworks dominate the debate.

In this paper, I will be using the terms "overweight," "obese," and "fat," although we need to question the facticity and utility of those categories, recognizing that they are moral and political designations rather than fixed states. Furthermore, such terms presume that people share experiences because of body size. Borrowing Gans' functional analysis of poverty, I ask what are we really concerned about when we talk about an "obesity epidemic"? Ideas about obesity parallel ideas about the poor and indeed, poverty and obesity are often conflated in much of the rhetoric and the science.[11] Although there are clearly relationships between weight and economic or social status, some researchers are actually surprised to see obesity typified as a problem related to poverty, racism, or sexism at the structural level; the ideologies that promote an individualization of economic mobility extend to physical appearance and health.[12] Others have talked about the general concern for "bodies out of bounds," and the nature of citizenship in consumer culture. While there are social constructionist critiques of the structural and institutional forces that place some people in a problematic relationship with food and consumption, it is extremely difficult even for critical frameworks to address the ideological depth of fat phobia and the medicalization of weight and body size.

Thirteen Economic, Political, and Cultural Functions

In his piece, Gans identifies thirteen economic, political, and cultural "functions" of poverty in the U.S. It is striking how easy it is to insert obesity rather than poverty in the statements and come up with roughly the same functions thirty-five years later. Using Gans' list, I draw upon examples from the copious news and academic studies on weight and social life.[13]

First, *the existence of obesity—of a population of people identified visually and scientifically as being too fat—ensures that society's "dirty work" gets done.* Every society has such work: physically dirty or dangerous, temporary, dead end or underpaid, undignified and menial jobs. A society can fill these jobs by paying higher wages than for "clean work" or it can force people who have no other choice to do the dirty work—and at low wages. Sociologists have amply demonstrated that two critical factors affecting economic mobility and stability are discrimination and social capital. People who are perceived to be overweight find it harder to get jobs, are paid less for the same jobs, and are less likely to be promoted (Roehling 1999). Using data from the National Bureau of Labor, Cawley (2000a) suggests that white women are most apt to experience a loss in wages and job opportunities when they are large. Women as a group already experience a measurable wage disparity as well as greater responsibility for unpaid domestic labor—we can surmise that large women are particularly targeted for the low end of the wage scale (Hebl & Heatherton 1998, Fikkan & Rothblum 2005). Researchers at Yale's Rudd Center for Food Policy have compiled an array of studies suggesting that weight bias exists in populations across the life course, affecting student–teacher relations, college admissions, marital prospects, and job advancement.[14] Such biases have direct and indirect economic costs for citizens: one study suggests that one reason the wages for obese workers are lower than those for non-obese workers is because employers perceive a greater cost of providing health insurance for these workers (Bhattacharya & Bundorf 2005). Employers often pay them less; even when they have health insurance, obese workers pay a "wage penalty," calling into question the claims that the epidemic is costing citizens and employers in public health expenditures. Although often difficult to prove legally, obesity is often a criterion for screening people out of jobs (Brownell et al. 2005). As April Herndon (2002) writes, "the prospect of having one's body read as a text about slovenly behavior, inherent flaws, and abnormality—all narratives associated with fat people—robs them of a significant source of power" (p. 134). Fat discrimination becomes an accepted yardstick for measuring workers.

In both media reports and scientific studies, obesity is disproportionately associated with poor African Americans and Latinos, populations who also suffer discrimination in the job market and are increasingly asked to do the jobs that middle class white Americans don't want to do. Often lacking the type of social capital that whites use for employment advantages, people of color are routinely positioned in less desirable jobs.[15] Both academic and popular media focus on obesity among African American and Latino populations. In particular, the thesis is often that assimilation breeds overeating, causing second and third generation immigrants to gain weight in the United States (Popkin & Urdry 1997).[16] These issues become clearer when we look at the CDC's own stated concerns about the convergence between the growing numbers of overweight people and the "browning" of America.[17] Such conflation of marginal statuses plays out in debates about immigration and the reinforcement of national borders. Mainstream environmental organizations like the Sierra Club have a subset of members who see limits on immigration as an environmental issue, arguing that all newcomers quickly become superconsumers as they adapt to America (Knickerbocker 2004). Despite the rhetoric, it is certainly functional for the American middle and upper class to hire immigrants whose work subsidizes a standard of living made easier by relief from undervalued service and caring labor needed to support such families. Growing economic disparities in the U.S. mean that in general, the poor and working class subsidize economic growth for the wealthy by paying more for taxes, goods, and services, and consequently have a higher cost of living.

Second, *because the obese and overweight are working at lower wages and generally higher costs of living, they subsidize a variety of economic activities that benefit the affluent.* Those who are medicalized as obese support innovations in medical practice as patients in hospitals and as guinea pigs in medical experiments. Treatments to "normalize" the obese, such as surgeries and drugs, parallel other treatments to "cure" the deaf, the disabled, the intersexed, or differently embodied (Herndon 2004). The medical and pharmaceutical measures used to re-shape large bodies provide a testing ground for less risky versions of the same procedures for non-obese patients. In particular, the incredible media visibility and advertising of gastric bypass surgery and liposuction make such measures seem ordinary and accessible to a vast swath of American society (Alt 2001, Salant & Santry 2005). Fatness is perceived as a "social emergency," because fat people literally don't fit. Since institutions are difficult to change, individuals are forced to adapt to the indignities of inappropriately shaped clothing, seats, tools, and other materials of everyday life. This lack of acceptance of a variety of embodiments contributes to a general cultural fascination with re-shaping. Television shows that illustrate "extreme" body makeovers normalize other more voluntaristic body interventions. In one horrific reality show, a woman with a cleft palate was offered corrective surgery and, to complete her transformation, was given liposuction, breast implants, and a face lift. Although we cannot make a direct correlation, it is not surprising that the cosmetic surgery industry reports a 300% increase in cosmetic surgeries since 1997 (Williamson 2004). Rather than through doctor referrals, one of the main avenues for the recruitment of new bariatric surgical patients is through website advertisements (Jutel 2006)

Third, *labeling obesity an epidemic creates jobs for a number of occupations and professions that serve or "service" the diet, exercise, and health industries—and perhaps, "protect" the rest of society from the obese.* In sociological terms, obesity has been medicalized, a historical and social process of framing through medical and physiological criteria that institutionalize and rationalize its meaning and treatment (Sobal 2005). Medical professionals (and their organizations and institutions) act as moral entrepreneurs who gain money and status from typifying obesity as a disease and, more recently, as an epidemic. Within the Centers for Disease Control, different researchers have challenged the methodological validity of early studies that overestimated the population of overweight and obese, yet the official position of the CDC stands that current estimates "do not diminish the urgency of obesity as a public health crisis" (Saguy 2006). Although the pharmaceutical industry and health insurance industries also benefit financially, many segments of the medical profession gain a continued source of income and power from labeling obesity a disease. As a crisis, the obesity epidemic requires research funding, institutes and think tanks to study and solve the problem. *Business Week* recently suggested that biotech, drug, and medical device companies all stand to profit in the coming years of the epidemic (Gogoi 2006). Stanton Glantz recently graphed the way that the Robert Wood Johnson Foundation's funding for tobacco research has plummeted in proportion to the increase in their funding for obesity research (Saguy 2006). The International Obesity Task Force and the American Obesity Organization are both funded by pharmaceutical and weight loss companies (Oliver 2006). The obesity epidemic revitalizes the work of nutritionists, dieticians, and social workers, professional workers who, without a socially mandated function as reformers, have less status than doctors and social policy analysts (Ouellette 2004, Biltekoff 2005). Even progressive projects focused on agriculture and food security, like the Yale Sustainable Food program, provide expert status and gainful employment to academics whose work focuses on eradicating obesity. The Rudd Institute embodies the contradictory messages concerning obesity: on the one hand, they suggest punitive public policies, like Kelly Brownell's now-famous idea of "hitting junk food junkies where it hurts: in their wallets" with a so-called junk food tax (Ball 1998). On the other hand, Brownell and his colleagues have co-authored numerous studies focused on de-stigmatizing weight. The moral "goodness" of these public-minded projects deflects collective questioning of what productive social labors would fill the experts' time if the obesity epidemic suddenly ended. Like many other social reform movements, the functional desire to "fix" social ills by curing dysfunctional members obscures the underlying structures of inequality that foster social problems. It also hides the locus of moral responsibility and social control behind the cloak of scientific neutrality (Zola 1986). In another influential article, Gans (1990) points out that by constructing universal and often denigrating terminology, sociologists and policy researchers often contribute to the social ills they hope to eradicate. Gans argues that using the term "underclass" to characterize the chronically poor lumps together a diverse group of people without regard for the possibility of movement in or out of such a rigid identity. Again, we can see a parallel in the use of "the obese" as a descriptor of an entire population, greatly compounded by the tendency to haphazardly extend weight-related risks to those designated "overweight."

Fourth, *the overweight and obese buy goods that others do not want and thus prolong the economic usefulness of such goods.* Consumer goods—including clothing, food, diet aids, drugs, and household items—are marketed to obese people. *Entrepreneur* magazine claims, "there are no market boundaries to the plus size clothing line," which is, as they put it, "morphing into regular size" (Penttila 2003). This occurs simultaneously with the production of clothing and other goods strategically limited to the very thin and the very wealthy. In particular, haute couture and high-end goods are constructed with the presumption that no one who is rich would ever be fat.

The food industry creates three times as many food products as we as a society need and spends billions of dollars selling endless variety and constructing new needs where none existed before. In American culture, food is ubiquitous—from gas stations to banks to department stores and schools, there is food for purchase everywhere, 24 hours a day. To speak of the economic usefulness of 25 kinds of Oreo cookies is to enter the realm of the absurd and yet the food industry depends upon the continued existence of that market while simultaneously providing the products for dieting. Low fat, and low carb products appear almost instantaneously as diet books and nutrition experts propose anew to answer the problem (Nestle 2002).

Consider how the intense marketing of processed food products to children co-exists in the same media that broadcast commercials exhorting kids to get exercise. On a typical television station, you can routinely see public service commercials for kids to "get out and move," whose animation blends seamlessly into the commercial for Cocoa Puffs that follows. For an even more concrete integration, McDonald's offers us Ronald McDonald on a snowboard during broadcasts of the Winter Olympics. The food industry uses a combination of unrestrained desire and one-shot solutions, which allows them to continue to produce, promote and distribute products that are clearly harmful while pushing the government to continue to subsidize the cost of farming and producing processed foods (Nestlé 2002). The links between poverty and obesity are compounded by such things as: food deserts in poor neighborhoods, the unequal distribution of good quality food for a decent price across socioeconomic areas, and the intense pressure for immigrant racial groups to assimilate by "eating American."

Market research that concludes poor people "prefer" Doritos and soda to organic lettuce allows the food industry to justify providing poor quality food in disadvantaged neighborhoods.[18] Overall, the number of food stores in poor neighborhoods is nearly one-third lower than in wealthier areas, and the quality of these stores, their size and physical condition, the range and nutritional content of their merchandise tends to be poorer. People in low-income areas pay more for nutrition and have less access to preventative health and medical services. The health risks and costs of diabetes for the poor may be greater simply because they lack access to health care.

Fifth, moving on to the social and cultural functions of obesity, we can suggest that one *major function of obesity is that the obese can be identified and punished as alleged or real deviants in order to uphold the legitimacy of conventional norms*. Stigma against fat is one of the most acceptable forms of overt prejudice. Erving Goffman's (1963) groundbreaking work *Stigma* provides an original and useful template for understanding how people confront a discredited condition like obesity. The cornerstone of weight bias is the belief that it is a self-induced state from which a self-disciplined individual can escape by hard work or, failing that, the purchase of the right diet book, foods, exercise equipment, or medical interventions. Cultural messages about status anxiety are reinforced by a consumer culture that offers a way out through the possession of consumer products or a commodified self. Failure to do so is moral lapse and deserves public response. Consider the four-year-old Mexican-American child who was taken away from her family by social service workers who labeled her obesity a form of child endangerment, despite the fact that her parents were actively seeking treatment for her abnormal growth patterns (Belkin 2001, Campos 2004). In what is probably an intentional parallel, in one ad for "Honey We're Killing the Kids," a Latina mother is shown throwing her son snacks the way a dog receives a treat—he shows her his report card or plays his trumpet and she distractedly tosses him a cookie which he catches in his mouth. Given the moral panic frame used by the show's producers, it's easy to see how they are equating Latinos with dogs, bad parenting with the off-handed training of a pet, and good grades with social rewards such as food. Campos et al. (2005b) find that news "articles (about the obesity epidemic) that reported on blacks and Latinos were over eight times more likely than articles that did not blame obesity on bad food choices and over 13 times more likely to blame sedentary life styles."

Another form of vilification comes in the shape of punitive public policies: the rise of various "snack taxes" which add to the cost of high sugar, high sodium snack foods neglects to take into account how these foods are distributed across the socioeconomic landscape. To see food as simply good or bad nutrition also medicalizes it, underplaying all the other reasons people eat. To justify the desirability of hard work, thrift, honesty, frugality, self-discipline, for example, the defenders of these norms must find people who can be accused of being lazy, self-indulgent, dishonest, and gluttonous. The "risky behavior" frame argues that fat bodies are read as evidence of both preventable illness and moral failings (Saguy & Riley 2005). Fat acceptance becomes tantamount to accepting bad behavior that knowingly contributes to ill health. Conservative commentator and anti-obesity writer Michael Fumento uses a number of these examples. He argues that "when somebody shows prejudice to an obese person, they are showing prejudice toward overeating and what used to be called laziness. It's a helpful and healthful prejudice for society to have" (Saguy & Riley 2005). Seeing fat people as lazy and ignorant supports an ideology of choice, increasingly popular in a time when governments are pulling back from collective services and obligations.

Sixth, despite the growing numbers of fat activists, *people stigmatized by body size lack the level of political and social power to correct the stereotypes that other people hold of them* and thus continue to be thought of as willing victims, engaged in bad personal decisions. The most obvious is the use of obese people as examples of moral failure (as in "fat jokes"). By unquestioningly connecting "obesity or overweight" to risky health behaviors, medical professionals and public health spokespersons categorize them as "preventable illnesses," which means that people who are fat are willfully creating the social and physical costs. Like blaming the poor for their lack of upward mobility, blaming the fat also takes our gaze away from the structural causes of food as a social emergency. It also misplaces the urgency: we may need a better food system for everyone, but to use "growing obesity" as the basis for food security and sustainability is problematic. As with all moral panics, once the emergency is gone, the funding, support, and moral obligation to support such systems disappears. Ironically, the same problems have operated in the typification and ensuing public policy related to hunger and food insecurity in the USA. As Jan Poppendieck (1998) has illustrated, while hunger remains a constant concern in North America, its visibility as a moral issue is tied to the construction of emergency aid rather than an examination of the underlying structural conditions.

Seventh, *the overweight and obese offer vicarious participation in indulgence to the rest of the population.* Gluttony may be a sin, but in consumer culture, there is a kind of envy of those who appear to consume without restraint. It parallels a history of mythologizing the "freedoms" that working class or poor people seem to have from the panopticon of self-control we expect from middle class citizens. Consider cultural voyeurism around eating a lot or taking pleasure in food. Although the winners of many eating contests in recent years are often tiny Japanese men and women, they are usually shown alongside corpulent American men wearing t-shirts and baseball caps (Berg 2003). In particular, women are defined by their appearance and subject to intense scrutiny about their bodies. Those who cannot or will not participate in the culture of slimness are more deeply scrutinized with a kind of prurient interest. People who focus aesthetic or positive attention on heavier women are seen as having deviant sexual interests.[19] Actor Leonard Nimoy, who recently published a photo art portfolio of large nude women, commented that he is constantly asked about his personal sexual preferences rather than about the subject within a richer vocabulary of representation.

Eighth, *the obese serve a direct cultural function when, seen as a single group, their "culture" is adopted through icons, stereotypes and heroes.* Consider the fat lady who sings at the end of the opera, the fat contestants on "American Idol", and any number of celebrity chefs on the food network. Author, commentator, and Man of Size, Daniel Pinkwater agrees that fat people are stereotyped as funny and jolly—and he claims, historically many of them are. He says, "There's a kind of joyousness to big, ponderous creatures."[20] However, because the media construction of the obesity epidemic is so intense, the cultural celebration of people of size has virtually disappeared except on food television.

One notable exception that begs deeper analysis is the black male comedian in drag and a fat suit. In the past, certain racial and ethnic groups have connected heaviness with power, status, sexiness, and beauty. In some strands of black culture, music and images celebrate a rounder, heavier aesthetic for women. This has taken an odd turn in a recent spate of films, where actors Martin Lawrence, Eddie Murphy, and Tyler Perry have all portrayed heavy people—usually women. The images ambiguously range from repulsion to adulation. In one film, Murphy seems to celebrate the fat professor as the more authentic and sexually attractive man, although his other characters, the heavyset family (particularly the grandmother and mother), never materialize as more than comic fodder. In a more recent film, Murphy more explicitly vilifies the fat black woman he embodies, making her a central object of ridicule, even as he plays opposite himself as her love interest. Perry's character of Medea, is questionably a positive source of folk wisdom, while her size, anger, and dragged-out femininity remain the locus of humor. As feminist race scholars have demonstrated, there is a problematic and lingering history of the image of a large black woman, "reminiscent of 'mammy' symbolizing comfort and reassurance with which many moviegoers want to connect" (Williams-Forson 2006). The Moynihan Report was only the first of many public and scholarly sources that pathologized black families as dysfunctional for relying on single mothers (stereotyped as large women) as the head of household in black communities. For black comedians to adopt this physical parody is a questionable route to power that needs to be examined more fully in light of the limited number of large black women who have cultural and iconic power, even as entertainers. The historical shift in accolades from Hattie McDaniel to Hallie Berry may be simultaneously liberating and constraining. Feminist race scholars have pointed out that recently, "the hegemony of white popular culture and upward class mobility has resulted in increased pressure on African American women to become slender" (Witt 1999, p. 188). This trend coexists uncertainly with the black male comedians' performance of "Big Mama."[21]

Ninth, *seeing obesity as a master identity (and homogeneous grouping) helps guarantee the status of those who are not fat.* It is a way of distinguishing self from others. When fatness is conflated with bad nutrition, bad health, and sedentary lifestyles, those who are not fat gain status through that association. In hierarchical societies, someone (or, more likely, some group) ends up on the bottom. In American society, where social mobility and shifting status are important goals, people need visible signs as to where they stand. The obese function as a reliable measuring rod for status comparisons. This is true whether people accept a discourse that blames genetics or bad choices as the "cause" of obesity. Status is status, whether based on scientifically sanctioned predestination or the Protestant work ethic. At the same time, even epidemiologists who see genetics and environmental factors as key contributors to people's body size generally present the obese as a group with a singular lack of will to change. For example, websites advertising bariatric surgery often induce potential patients with the claim that obesity is a complex disease that is caused by genetic and environmental factors outside the individual's control. However, in the same sites, surgical failure is blamed on the individual's inability to conform to the maintenance routine necessary for weight stability (Salant & Santry 2005). The tinge of personal responsibility is almost impossible to expunge from medical and public health directives. At the same time, many studies relating mortality to body mass fail to control for other variables related to life chances, environment, or more broadly to lifestyle. Discursively, the fluidity of the body connects directly to the fluidity of social class: both are seen as real possibilities in the U.S., and yet there are incredible structural impediments to both weight loss and upward mobility that get masked by the belief in self-discipline. In this historical moment, thinness is an unequivocal visible sign of "health," an undefined value presumably shared by all. As April Herndon (2004) points out, it's incredibly important to recognize that "while there are some people for whom being large presents serious health and life problems . . . we cannot presume on scientific evidence that this is the case for all people of a certain body size or mass. Despite our comfort with statistical norms, there are no populations for whom material experiences are identical."

Tenth, *the obese also aid in the status mobility of those who are just above them in the hierarchy—those who were heavy and lose weight.* We are all familiar with the visibility of so-called "success stories"—Al Roker with his surgically clamped stomach; the endless diet ads with before and after pictures; Jared, the man who lost 100 pounds eating Subway sandwiches. They all profit from their disassociation with a stigmatized identity. Others also gain status by financing their own success from the change—becoming diet and exercise gurus, selling products and their own success. Kirstie Alley seemed to buck the trend in a short-lived TV series called "Fat Actress," but she's received more publicity—and money—for her weight loss through a national diet chain. The number of self-help and autobiographical books written by "former fat people" is astounding. Interestingly, people who lose weight are granted expert status generated from their own experiences, but those who are fat activists are often challenged on their ability to speak with authority about the lives of people of size. As Charlotte Biltekoff (2005) points out, "the pursuit of thinness may have more to do with its power as a signifier of self-control than with its promise of health."

Eleventh, *the obese keep the thin busy, particularly those who assimilate into thinness via diets and surgery, by becoming trainers, diet experts, and icons of what is right.* The existence of obesity provides the fuel behind much of the entertainment industry, both in terms of producing a steady stream of images dedicated to extreme thinness, but also in endless speculation on weight and body in the media. Oprah Winfrey, one of the most influential and richest people in America receives more attention for her "battle" to stay slim than for her other social and cultural endeavors. Those who assimilate into thinness are lauded like terminal disease patients who survive.

Conservative commentator Michael Fumento calls obesity a "disease that is socially contagious," meaning that those who are not yet labeled overweight need to remain vigilant. For everyone, but for women in particular, the culture of fear keeps them engaged in endless acts of self-disciplining. Comparing disparate but interesting attitudinal surveys, one could surmise that the number of people worried about their weight at any given time is greater than the number of people worried about terrorist attacks.

Twelfth, *being relatively powerless to affect public policy and cultural production, the obese and overweight are made to absorb the costs of changes and growth in American society.* Fitness can signify physical health but also who is or who is not worthy of status within a society. In 2001, instigating a national war on obesity, former Health and Human Services director Tommy Thompson claimed that as their patriotic duty, all Americans should lose ten pounds. Communities vied to see which town could collectively lose the most weight and political candidates of both genders are now routinely evaluated for their "fitness" for the job, in terms of pounds as well as skills. As April Herndon (2004) illustrates, "In order to be a proper American one must meet a certain corporeal standard and take seriously the moral responsibility to be a patriotic citizen" (p. 128). However, the critical question becomes, "What kind of responsibilities are people being asked to take?" In this case, good citizenship means doing more to improve their own health, and presupposes that people have the capacity to do more. Indeed, the responsibility of citizens to lose weight coexists with encouragement to consume more as a form of patriotism, particularly post-9/11, where consumption was presented as a way of preserving "the American way of life."

Constructing the problem as an obesity epidemic functions as a means of chastising all who fall outside the confines of ideal citizenship. This is particularly noticeable when it becomes a means of talking indirectly about poverty, race, and immigration without appearing to be racist or classist. Saguy and Riley (2005) conclude that "the war against obesity targets a specific group of people who are already, in some sense, second-class citizens" (p. 129). New versions of racism and sexism are played out through national discourses and programs aimed at reducing fat rather than poverty. Fear of obesity is yet a new way to vent anxiety about changes in the gender or racial order without fear of reproach. When asked why people fear fat, Daniel Pinkwater said, "It can't be fat. Fat is too trivial for it to be about fat. Just think, people see you as a defenseless person."[22]

It is particularly disconcerting when the obesity epidemic is used as a justification for sustainable food projects, using the premise that it is morally right to secure school gardens and good institutional food by claiming obesity as the "problem" rather than the agro-industrial food system and the lack of government support for children's health. Critics move seamlessly from the suggestion that current industrial agricultural policies would be radically changed if we all consumed less to the idea that fat people are "overconsumers" whose behavior pushes the food industry to produce more food and the government to maintain agricultural subsidies (Cafaro et al. 2006). It seems contradictory to talk about people's "desire" for more food when the motivation for more cannot be separated from a capitalist system that fetishizes the market as an entity requiring endless development and promotion of new products, despite any discernable consumer demand. Movements aimed at changing food and agriculture must be careful to escape their historical roots in paternalistic and class-based reform movements designed to produce citizens according to a universalist and restrictive model.

Thirteenth, *the obese facilitate the American political process.* The "obesity epidemic" is more about social and political norms than about the scientific and medical realities of body size. American political institutions have embraced the rhetoric of personal responsibility undergirding the fears of a fat nation. This serves purposes beyond the care of citizenry or concerns about overall costs to a society. As Pinkwater aptly points out, it is about discrediting or disenfranchising people from visibility in the political process. We can see how this works in public condemnation, which is often focused on women. In the 1980s, Rush Limbaugh (not a thin person himself) famously called feminists "a bunch of fat cows who can't get dates." Conservative commentator and anti-obesity crusader Michael Fumento covers two bases with one insult by lambasting global environmental activist and physicist Vandana Shiva as being a "blubbery bourgeois hero." Anti-obesity activists, especially scientists, are often dismissive of fat activists because the vast majority of them are women. This framing of obesity is useful for those whose ideological commitments are focused on "individual responsibility" rather than universal health insurance. Using a historical analysis of American food reform movements, from domestic science to Alice Waters, Charlotte Biltekoff (2005) reminds us that this process has consistently "produced an equivalence between thinness and good citizenship and made the position of 'fat citizen' a conceptual impossibility." Most importantly, it shifts the focus away from structural and cultural reasons for obesity and allows some groups in society to scapegoat other groups, while avoiding questions about the nature of food distribution, poverty, lack of jobs at a livable wage, and a time bind that encourages people to work longer for less pay. To blame the individual for lack of willpower is to ignore the ways in which work has increased, pay has decreased, and avenues for fulfillment are structurally constrained for women, people in poverty, and racial-ethnic groups. Current entitlement programs focused on food security (such as food stamps) provide as little as $3.00 a day per individual. These programs have been chipped away to the extent that hunger relief is almost entirely governed by non-profit charitable organizations, supplemented by a backhanded and unsteady supply of federal funds (Poppendieck 1998). Such an approach makes good food, leisure, and physical exercise into commodities that are only affordable to those who have disposable income, time, and cultural capital. Despite Bush's recent public health initiatives for fitness, all of which are focused on "personal responsibility," there are major structural impediments that prevent people from gaining access to physical activity. Even a simple walk around the block or unstructured outdoor play is problematic in urban neighborhoods with high crime rates, in suburban areas without sidewalks, in school districts that must use their decreasing budgets to pay for "No Child Left Behind" rather than "specials" like physical education, art, or music. In 2000, the IRS passed regulations that allowed people willing to pay for their own bariatric surgeries to take a reduction in taxes (Herndon 2002, p. 133). Rather than come up with national health insurance or preventative health campaigns (supplementing gym memberships for everyone rather than extreme weight control for a few), the government rewards those who are willing to take on personal medical risks for the sake of some vague future reduction in public health costs.

The concept of responsibility has many meanings. By making public health promotion entirely about individual behavior, we limit people's autonomy regarding the vast number of reasons they choose to eat what they eat or lead their lives the way they do. As the Food Ethics council points out, we lose a great deal by valuing food for little but its nutrients, and, I would add, by valuing our citizens for little but their appearance. In *Deconstructing the Underclass*, Gans (1990) states:

> Those who argue that all people are entirely responsible for what they do sidestep the morally and otherwise crucial issue of determining how much responsibility should be assigned to people who lack resources, who are therefore under unusual stress, and who lack effective choices in many areas of their lives in which even moderate income people can choose relatively freely.

Pointing out the important functions of obesity does not intend to suggest that categorizing and stigmatizing people because of their size is a necessary condition for social life. As fat activists and sociologists have demonstrated, it is possible to suggest functional alternatives. Some things that might be more useful for all members of society might be: a different food system that provides access to good food to everyone, regardless of economic status; a wage system that compensates caregivers, food producers, service workers, and manual laborers in a more equitable fashion; dismantling the structures of racial and gender inequality; a food industry that focuses more on feeding people than profiting from their desires; a health care system that is available to all and offers a range of preventative care options; and a government that insures everyone's right to adequate food. There are certainly historical examples of an acceptance and even glorification of a variety of body sizes and shapes, of government programs that supplement individual responsibility with institutional support, and of corporations working for consumers. (Maybe there aren't any examples of that last point.) However, the "risky behavior" frame and the medicalization of food and eating make it extremely difficult to assert an alternative that would easily resonate with an already terrified population. This is particularly true in a time when there is a huge and growing gap between wealthy and poor citizens, both within the U.S. and across the globe. As Gans suggests, a functional analysis must conclude that the obesity epidemic exists because it fulfills a number of positive functions but also because many of the functional alternatives to the obesity epidemic would be quite dysfunctional for the affluent and privileged members of society, including many contemporary organizations, professionals (including academics and doctors), corporations, and institutions. Phenomena like the obesity epidemic can be eliminated only when it becomes dysfunctional for those who benefit from its construction or when the powerless obtain enough power to change society.

Notes

1. This piece is written with admiration for and deference to Herbert Gans. I am indebted to Tom Henricks for his excitement and quick thinking in applying Gans' article to the obesity debate as well as his encouragement in

having me following the line of thought. The ideas for this paper were generated during a seminar sponsored by the Centers for Disease Control and the Academy for Educational Development in the fall of 2005. I would like to thank Peter Conrad, Jeff Sobal, Lisa Heldke, Tom Henricks and a host of ASFS conference participants for their invaluable feedback on the original talk. My thinking about morality, weight, and citizenship originated from talks with Charlotte Biltekoff, who points out numerous historical instances of moral reform played out through bodies and eating. This article was originally presented as my fourth and final presidential address for the joint conference of the Association for the Study of Food and Society and the Agriculture, Food, and Human Values Society. Boston, Mass. 2006.

2. Campos, P. (2004) *The Obesity Myth*, p. 3. Campos cites: "Obesity is America's Greatest Threat, Surgeon General Says." *Orlando Sentinel*, January 22, 2003.

3. For a history of body size and morality in countries like the U.S., see Stearns, P. (1997) *Fat History: Bodies and Beauty in the Modern West*. NY: NYU Press.

4. See, for example, Kuczmarski, R. et al. (1994) Increasing Prevalence of Overweight among U.S. Adults: The National Health and Nutrition Examination Surveys, 1960–1991, *JAMA*, 272. Campos points to the following four studies as being the most frequently cited as evidence for the link between increased morbidity and obesity: Manson, et al. (1995) Body Weight and Mortality Among Women. *New England Journal of Medicine* 3333; Allison, et al. (1999) Annual Deaths Attributable to Obesity, *JAMA*, 289; Fontaine, et al. (2003) Years of Life Lost to Obesity, *JAMA* 289; Calle et al. (2003) Overweight, Obesity, and Mortality from Cancer in a Prospectively Studied Cohort of U. S. Adults. *New England Journal of Medicine*, 248 (Campos, p. 255).

5. Oliver (2006) points out that the 1995 WHO Report on Overweight and Obesity was based on data from the International Obesity Task Force, a private research and policy organization funded by pharmaceutical companies with major stakes in weight-loss drug development (p. 29).

6. The maps and PowerPoint presentation are available to the public on the Centers for Disease Control and Prevention website, www.cdc.gov

7. For example, Campos et al. (2005) evaluate four of the central claims related to the obesity epidemic to argue that the rhetoric is far greater than the real or potential public health crisis. They find limited scientific evidence for the following claims: "that obesity is an epidemic; that obesity and overweight are major contributors to mortality; that higher than average adiposity is pathological and a primary cause of disease; and that significant long-term weight loss is both medically beneficial and a practical goal." For more in-depth analysis of the same issues, see Gard and Wright (2005).

8. Thompson, B. (1994) *A Hunger So Wide and So Deep: American Women Speak Out on Eating Problems*. Minnesota: Univ. of Minnesota Press. For social constructionist critiques, see Saguy and Riley (2005), Campos et al. (2006). For feminist critiques, see Chernin 1981; Orbach, 1988; Bordo, 2003; and most recently Braziel and LeBesco, 2001. Feminist analyses of the body and gender are wide-ranging, often encompassing the issue of size. See, for example the edited volume: Weitz, R. (2003) *The Politics of Women's Bodies: Sexuality, Appearance, and Behavior*. New York: Oxford Press. According to Gard and Wright (2005), "Central to contemporary feminist understanding of the body is the idea that what it means to be female is shaped by and in historical relations of power" (p. 154). For a blending of personal narrative and sociological analysis see also: Thomas (2005).

9. See for example: Fierstein, H. (2007) Our Prejudices, Ourselves. *New York Times*, April 13, 2007. Commenting on the firing of radio host Don Imus for a racist and sexist remark, Fierstein argues that American culture is selective in its outrage over hate speech, tolerating homophobia and fat prejudice while decrying racism and sexism—ignoring the way material discrimination against gays, people of color, women, and heavy people continues.

10. What Gans proposes is really a "radical functionalism," where social life is less beneficial to a singular entity and more so to privileged groups within. In this and other articles, he suggests that the idea of "society" is easily confused with the stability and well-being of its upper classes. He subtly argues that poverty is functional for the middle and upper groups; and the alternatives to it are dysfunctional for these same groups.

11. For example, Critser, G. (2003) *Fat Land: How Americans Became The Fattest People in the World*. New York: Houghton Mifflin.

12. See Crawford, R. (1980) Healthism and the medicalization of everyday life. *International Journal of Health Services* 10 (3). Crawford coins the term "healthism" to capture how new health movements draw upon existing medical frameworks, which situate problems, prevention, and solutions to health and disease at the level of the individual. Crawford argues that such an approach elevates health to a super-value, which de-politicizes health issues and encourages private, personal approaches to health rather than collective or structural strategies.

13. For an impressive compilation of recent studies of obesity, weight stigma, and the food industry, see the Rudd Center for Food Policy (www.yaleruddcenter.org); see also Sobal and Maurer (1999b) *Weighty Matters*. For the shift from physical condition to disease see Jutel (2006).

14. Brownell K. et al. (eds) (2005) *Weight Bias: Nature, Consequences, and Remedies*. New York: Guilford Publications; for articles on weight bias see www.yaleruddcenter.org

15. For an overview of the relationship between employment advantages, race, and social capital see the following: Smith, R. (2002) Race, Gender, and Authority in the Workplace: Theory and Research, *Annual Review of Sociology*; Mason, P. (2000) Understanding Recent Empirical Evidence on Race and Labor Market Outcomes in the U.S.A. *Review of Social Economics*, 58 (3).

16. See Saguy and Riley (2005) for a quantitative analysis of articles on race and obesity. See also Almeling, R. and Saguy, A. C. (2005, Aug.) Fat Panic! The "Obesity Epidemic" as Moral Panic. Paper presented at the annual meeting of the American Sociological Association, Marriott Hotel, Loews Philadelphia Hotel, Philadelphia, PA.

17. Source: Centers for Disease Control "Overview of the Obesity Epidemic" prepared for expert seminar at the Academy for Educational Development, Fall 2005.

18. Although most of these studies are done for market research, see an academic example in Chrzan, J. (2003) Performing the Good Pregnancy: Teen Mothers and Diet in West Philadelphia. Paper presented for the joint conference of the Association for the Study of Food and Society and the Agriculture, Food, and Human Values Society. Chrzan critiques the food deserts of the inner city but ultimately decries the bad food choices of low income women prior to pregnancy.

19. See, for example, Goode's (2002) interpretation of NAAFA as mainly "a dating service for men who like large women," and the critique articles that follow in the same issue. *Qualitative Sociology*, 25(4).

20. Interview with Marilyn Wann, www.fat!so.com

21. Witt suggests that black women are conspicuously absent in the national discourse on eating disorders while simultaneously being highly visible in the specularization of corpulence in U.S. culture (p. 189).

22. Ibid.

References

Allon, N. (1982) "The Stigma of Overweight in Everyday Life." In B. Wolman, ed., *Psychological Aspects of Obesity: A Handbook*. New York: Van Nostrand Reinhold.

Alt, S. J. (2001) "Bariatric Surgery programs growing quickly nationwide." *Health Care Strategic Management*, 19 (9).

Ball, M. (1998) "Brownell Calls for Food Tax to End 'Epidemic'." *Yale Herald Online*, February 18, 1998.

Belkin, L. (2001) Watch Her Weight. *New York Times*, July 8, 2001.

Berg, J. (2003) "Bringing Back the Belt: Nathan's Hot Dogs and American Nationalism." Unpublished paper presented at the Association for the Study of Food and Society/Agriculture, Food, and Human Values Society joint conference. June 11–13, 2003. Austin, Texas.

Bhattacharya, J. and Bundorf, M. (2005) "The Incidence of the Health Care Costs of Obesity." *National Bureau of Economic Research, Working Paper* No. 11303 May.

Biltekoff, C. (2005) "The Terror Within: Citizenship and Self Control in the Fat Epidemic." In *Hidden Hunger: Eating and Citizenship from Domestic Science to the Fat Epidemic*. Ph.D. thesis for the American Civilization Program. Brown University, Rhode Island.

Bordo, S. (2003) *Unbearable Weight: Feminism, Western Culture, and the Body*. Berkeley: University of California Press.

Braziel, J. and LeBesco, K. (eds) (2001) *Bodies out of Bounds: Fatness and Transgression*. Berkeley: University of California Press.

Brownell, K. D. and Horgan, K. B. (2003) *Food Fight: The Inside Story of the Food Industry, America's Obesity Crisis, and What We Can Do About It*. Chicago: Contemporary Books.

Brownell, K., Puhl, R., Schwartz, M. and Rudd L. (eds) (2005) *Weight Bias: Nature, Consequences, and Remedies*. New York: Guilford Publications.

Cafaro, P., Primack, R. and Zimdahl, R. (2006) "The Fat of the Land: Linking American Food Overconsumption, Obesity, and Biodiversity Loss." *Journal of Agricultural and Environmental Ethics*, 10.

Campos, P. (2004) *The Obesity Myth: Why America's Obsession with Weight is Hazardous to Your Health*. New York: Gotham Books

Campos, P., Saguy, A., Ernsberger, P., Oliver, E. and Gaesser, G. (2005a) "Response: Lifestyle Not Weight Should be the Primary Target." *International Journal of Epidemiology*, 35 (1): 80–81.

Campos, P., Saguy, A., Ernsberger, P., Oliver, E. and Gaesser, G. (2005b) "The epidemiology of overweight and obesity: public health crisis or moral panic?" *International Journal of Epidemiology*, 35 (1): 55–60.

Cawley, J. (2000a) "Body Weight and Women's Labor Market Outcomes." *National Bureau of Economic Research, Working Paper* No. 7841 August.

Cawley, J. (2000b) "An Instrumental Variables Approach to Measuring the Effect of Body Weight on Employment Disability." *Health Services Research*, 35 (5).

Chernin, K. (1981) *The Tyranny of Slenderness*. London: Woman's Press.

Conrad, P. (2005) "The Shifting Engines of Medicalization." *Journal of Health and Social Behavior*, 46.

Cramer, P. and Steinvert, T. (1998) "Thin is good, fat is bad: How early does it begin?" *Journal of Applied Developmental Psychology*, 19 (3): 229–252.

Ferraro, K. and Holland, K. (2002) "Physician evaluation of obesity in health surveys: 'Who are you calling fat?'" *Social Science and Medicine*, 55.

Fikkan, J. and Rothblum, E. (2005) "Weight Bias in Employment." In Brownell, K. et al. (eds) *Weight Bias: Nature, Consequences, and Remedies*. New York: Guilford Publication.

Gans, H. (1971) "The Uses of Poverty: The Poor Pay All." *Social Policy*, 2 (2) (July/August): 20–24.

Gans, H. (1990) "Deconstructing the Underclass: The Term's Danger as a Planning Concept." *Journal of the American Planning Association*.

Gard, M. and Wright, J. (2005) *The Obesity Epidemic: Science, Morality, and Ideology*. New York: Routledge.

Goffman, E. (1968) *Stigma: Notes on the Management of Spoiled Identity*. London: Penguin Books.

Gogoi, P. (2006) "All Eyes on the Obesity Prize." *Business Week Online*, June 6, 2006.

Goode, E. (2002) "Sexual Involvement and Social Research in a Fat Civil Rights Organization." *Qualitative Sociology*, 25 (4).

Herndon, A. (2002) "Disparate but Disabled: Fat Embodiment and Disability Studies." *NWSA Journal*, 14 (3) (Fall).

Herndon, A. (2004) "Collateral Damage from 'friendly fire?' Race, Nation, and Class and The 'War Against Obesity'." *Social Semiotics*, 15 (2) (August).

Hesse-Biber, S. (1996) *Am I Thin Enough Yet?: The Cult of Thinness and the Commercialization of Identity*. New York: Oxford Press.

Hebl, M. and Heatherton, T. (1998) "The stigma of obesity in women: The difference is black and white." *Personality and Social Psychology Bulletin*, 24.

Hunt, A. (1997) "The 'Moral Panic' and Moral Language in the Media." *British Journal of Sociology* (online) 48 (4).

Jutel, A. (2006) "The emergence of overweight as a disease entity: measuring up normality." *Social Science and Medicine*, 63.

Knickerbocker, B. (2004) "A hostile takeover bid at the Sierra Club." *Christian Science Monitor*, January 20, 2004.

Kulick, D. and Meneley, A. (ed.) (2005) *Fat: The Anthropology of an Obsession*. New York: Tarcher/Penguin Publishing.

Lupton, D. (1993) "Risk as Moral Danger: The Social and Political Functions of Risk Discourse in Public Health." *International Journal of Health Services*, 23 (3).

Maurer, D. and Sobal, J. (1999) *Weighty Issues: Fatness and Thinness As Social Problems*. New York: Aldine de Gruyter.

Nestle, M. (2002) *Food Politics: How the Food Industry Influences Nutrition and Health*. Berkeley: University of California Press.

Oliver, J. (2006) *Fat Politics: The Real Story Behind America's Obesity Epidemic*. New York: Oxford Press.

Orbach, S. (1988) *Fat is a Feminist Issue*. London: Arrow Books.

Ouellette, L. (2004) Review of "Honey We're Killing the Kids." www.flowtv.org

Penttila, T. (2003) "Hot Market: Overweight." *Entrepreneur Magazine*, December 2003.

Popkin, B. and Urdry, J. (1997) "Adolescent Obesity Increases Significantly in Second and Third Generation U.S. Immigrants: The National Longitudinal Study of Adolescent Health." *Journal of Nutrition*, 128 (4): 701–706.

Poppendieck, J. (1998) *Sweet Charity: Emergency Food and the End of Entitlement*. New York: Penguin Putnam Books.

Powdermaker, H. (1997) "An Anthropological Approach to the Problem of Obesity" from the *Bulletin of the New York Academy of Medicine*, 36, May 5, 1960. (reprinted in *Food and Culture*, 1st edition, eds C. Counihan and P. Van Estrik.)

Roehling, M. (1999) "Weight-Based Discrimination in Employment: Psychological and Legal Aspects." *Personnel Psychology*, 52.

Saguy, A. (2006) "Are Americans Too Fat? Conversation with Stanton Glantz." *Contexts* 5 (2) Spring.

Saguy, A. and Riley, K. (2005) "Weighing Both Sides: Morality, Mortality, and Framing Contests over Obesity." *Journal of Health Politics, Policy, and Law*, 5,

Salant, T. and Santry, H. (2005) "Internet Marketing of Bariatric Surgery: Contemporary Trends in the Medicalization of Obesity." *Social Science and Medicine*, 62 (14).

Sobal, J. (1995) "The Medicalization and Demedicalization of Obesity." In Maurer, D. and Sobal, J. (eds) *Eating Agendas: Food and Nutrition as Social Problems*. New York: Aldine de Gruyter.

Sobal, J. (2005) "Social Consequences of Weight Bias by Partners, Friends, and Strangers." In Brownell, et al. (eds) *Weight Bias: Nature, Consequences, and Remedies*. New York: Guilford Publication.

Sobal, J. and Maurer, D. (eds) (1999a) *Interpreting Weight: The Social Management of Fatness and Thinness*. New York: Aldine de Gruyter.

Sobal, J. and Maurer, D. (eds) (1999b) *Weighty Matters: Fatness and Thinness as Social Problems*. New York: Aldine de Gruyter.

Thomas, P. (2005) *Taking Up Space: How Eating Well and Exercising Regularly Changed My Life*. Nashville: Pearlsong Press.

Thompson, B. (1994) *A Hunger so Wide and so Deep: American Women Speak Out on Eating Problems.* Minnesota: University of Minnesota Press.

Wann, M. (1999) *Fat! So?: Because You Don't Have to Apologize for Your Size.* Berkeley: Tenspeed Press.

Wann, M. *Daniel Pinkwater and the Afterlife.* Interview. www.fat!so.com.

Williams-Forson, P. (2006) *Building Houses Out of Chicken Legs: Black Women, Food, and Power.* Chapel Hill: University of North Carolina Press.

Williamson, B. (2004) "The Surge In Surgery: Why Cosmetic Enhancements Are So Popular." *Elevate Magazine,* October 20, 2004.

Witt, D. (1999) *Black Hunger: Food and the Politics of U.S. Identity.* New York: Oxford Press.

Zola, I. (1986) "Medicine as an institution of social control." In P. Conrad & R. Kern (eds) *The Sociology of Health and Illness.* New York: St Martin's Press.

32

Of Hamburger and Social Space: Consuming McDonald's in Beijing

Yunxiang Yan

This chapter is based on fieldwork in Beijing, August to October 1994, supported by a grant from the Henry Luce Foundation to the Fairbank Center for East Asian Research, Harvard University, and on further documentary research supported by the 1996 Senate Grant, University of California, Los Angeles. I am grateful to Deborah Davis, Thomas Gold, Jun Jing, Joseph Soares, and other participants at the American Council of Learned Societies conference at Yale University for their valuable comments on earlier drafts of this chapter. I also owe special thanks to Nancy Hearst for editorial assistance.

In a 1996 news report on dietary changes in the cities of Beijing, Tianjin, and Shanghai, fast-food consumption was called the most salient development in the national capital: "The development of a fast-food industry with Chinese characteristics has become a hot topic in Beijing's dietary sector. This is underscored by the slogan 'challenge the Western fast food!' "[1] Indeed, with the instant success of Kentucky Fried Chicken after its grand opening in 1987, followed by the sweeping dominance of McDonald's and the introduction of other fast-food chains in the early 1990s, Western-style fast food has played a leading role in the restaurant boom and in the rapid change in the culinary culture of Beijing. A "war of fried chicken" broke out when local businesses tried to recapture the Beijing market from the Western fast-food chains by introducing Chinese-style fast foods. The "fast-food fever" in Beijing, as it is called by local observers, has given restaurant frequenters a stronger consumer consciousness and has created a Chinese notion of fast food and an associated culture.

From an anthropological perspective, this chapter aims to unpack the rich meanings of fast-food consumption in Beijing by focusing on the fast-food restaurants as a social space. Food and eating have long been a central concern in anthropological studies.[2] While nutritional anthropologists emphasize the practical functions of foods and food ways in cultural settings,[3] social and cultural anthropologists try to explore the links between food (and eating) and other dimensions of a given culture. From Lévi-Strauss's attempt to establish a universal system of meanings in the language of foods to Mary Douglas's effort to decipher the social codes of meals and Marshal Sahlins's analysis of the inner/outer, human/inhuman metaphors of food, there is a tradition of symbolic analysis of dietary cultures, whereby foods are treated as messages and eating as a way of social communication.[4] The great variety of food habits can be understood as human responses to material conditions, or as a way to draw boundaries between "us" and "them" in order to construct group identity and thus to engage in "gastro-politics."[5] According to Pierre Bourdieu, the different attitudes toward foods, different ways of eating, and food taste itself all express and define the structure of class relations in French society.[6] Although in Chinese society ceremonial banqueting is frequently used to display and reinforce the existing social structure, James Watson's analysis of the *sihk puhn* among Hong Kong villagers—a special type of ritualized banquet that requires participants to share foods from the same pot—demonstrates that foods can also be used as a leveling device to blur class boundaries.[7]

As Joseph Gusfield notes, the context of food consumption (the participants and the social settings of eating) is as important as the text (the foods that are to be consumed).[8] Restaurants thus should be regarded as part of a system of social codes; as institutionalized and commercialized venues, restaurants also provide a valuable window through which to explore the social meanings of food consumption. In her recent study of dining out and social manners, Joanne Finkelstein classifies restaurants into three grand categories: (1) "formal spectacular" restaurants, where "dining has been elevated to an event of extraordinary stature"; (2) "amusement" restaurants, which add entertainment to dining; and (3) convenience restaurants such as cafes and fast-food outlets.[9] Although Finkelstein recognizes the importance of restaurants as a public space for socialization, she also emphasizes the antisocial aspect of dining out. She argues that, because interactions in restaurants are conditioned by existing manners and customs, "dining out allows us to act in imitation of others, in accord with images, in responses to fashions, out of habit, without need for thought or self-scrutiny." The result is that the styles of interaction that are encouraged in restaurants produce sociality without much individuality, which is an "uncivilized sociality."[10] Concurring with Finkelstein's classification of restaurants, Allen Shelton proceeds further to analyze how restaurants as a theater can shape customers' thoughts and actions. Shelton argues that the cultural codes of restaurants are just as important as the food codes analyzed by Mary Douglas, Lévi-Strauss, and many others. He concludes that the "restaurant is an organized experience using and transforming the raw objects of space, words, and tastes into a coded experience of social structures."[11] Rick Fantasia's analysis of the fast-food industry in France is also illuminating in this respect. He points out that because McDonald's represents an exotic "Other" its outlets attract many young French customers who want to explore a different kind of social space—an "American place."[12]

In light of the studies of both the text and context of food consumption, I first review the development of Western fast food and the local responses in Beijing during the period 1987 to 1996. Next I examine the cultural symbolism of American fast food, the meanings of objects and physical place in fast-food restaurants, the consumer groups, and the use of public space in fast-food outlets. I then discuss the creation of a new social space in fast-food restaurants. In my opinion, the transformation of fast-food establishments from eating place to social space is the key to understanding the popularity of fast-food consumption in Beijing, and it is the major reason why local competitors have yet to successfully challenge the American fast-food chains. This study is based on both ethnographic evidence collected during my fieldwork in 1994 (August to October) and documentary data published in

Chinese newspapers, popular magazines, and academic journals during the 1987–96 period. Since McDonald's is the ultimate icon of American fast food abroad and the most successful competitor in Beijing's fast-food market, McDonald's restaurants were the primary place and object for my research, although I also consider other fast-food outlets and compare them with McDonald's in certain respects.[13]

Fast-food Fever in Beijing, 1987–1996

Fast food is not indigenous to Chinese society. It first appeared as an exotic phenomenon in novels and movies imported from abroad and then entered the everyday life of ordinary consumers when Western fast-food chains opened restaurants in the Beijing market. *Kuaican*, the Chinese translation for fast food, which literally means "fast meal" or "fast eating," contradicts the ancient principle in Chinese culinary culture that regards slow eating as healthy and elegant. There are a great variety of traditional snack foods called *xiaochi* (small eats), but the term "small eats" implies that they cannot be taken as meals. During the late 1970s, *hefan* (boxed rice) was introduced to solve the serious "dining problems" created by the lack of public dining facilities and the record number of visitors to Beijing. The inexpensive and convenient *hefan*—rice with a small quantity of vegetables or meat in a styrofoam box—quickly became popular in train stations, in commercial areas, and at tourist attractions. However, thus far boxed rice remains a special category of convenience food—it does not fall into the category of *kuaican* (fast food), even though it is consumed much faster than any of the fast foods discussed in the following pages. The intriguing point here is that in Beijing the notion of fast food refers only to Western-style fast food and the new Chinese imitations. More important, as a new cultural construct, the notion of fast food includes nonfood elements such as eating manners, environment, and patterns of social interaction. The popularity of fast food among Beijing consumers has little to do with either the food itself or the speed with which it is consumed.

American fast-food chains began to display interest in the huge market in China in the early 1980s. As early as 1983, McDonald's used apples from China to supply its restaurants in Japan; thereafter it began to build up distribution and processing facilities in northern China.[14] However, Kentucky Fried Chicken took the lead in the Beijing market. On October 6, 1987, KFC opened its first outlet at a commercial center just one block from Tiananmen Square. The three-story building, which seats more than 500 customers, at the time was the largest KFC restaurant. On the day of the grand opening, hundreds of customers stood in line outside the restaurant, waiting to taste the world-famous American food. Although few were really impressed with the food itself, they were all thrilled by the eating experience: the encounter with friendly employees, quick service, spotless floors, climate-controlled and brightly-lit dining areas, and of course, smiling Colonel Sanders standing in front of the main gate. From 1987 to 1991, KFC restaurants in Beijing enjoyed celebrity status, and the flagship outlet scored first for both single-day and annual sales in 1988 among the more than 9,000 KFC outlets throughout the world.

In the restaurant business in Beijing during the early 1980s, architecture and internal decoration had to match the rank of a restaurant in an officially prescribed hierarchy, ranging from star-rated hotel restaurants for foreigners to formal restaurants, mass eateries, and simple street stalls. There were strict codes regarding what a restaurant should provide, at what price, and what kind of customers it should serve in accordance with its position in this hierarchy. Therefore, some authorities in the local dietary sector deemed that the KFC decision to sell only fried chicken in such an elegant environment was absurd.[15] Beijing consumers, however, soon learned that a clean, bright, and comfortable environment was a common feature of all Western-style fast-food restaurants that opened in the Beijing market after KFC. Among them, McDonald's has been the most popular and the most successful.

The first McDonald's restaurant in Beijing was built at the southern end of Wangfujing Street, Beijing's Fifth Avenue. With 700 seats and 29 cash registers, the restaurant served more than 40,000 customers on its grand opening day of April 23, 1992.[16] The Wangfujing McDonald's quickly became an important landmark in Beijing, and its image appeared frequently on national television programs. It also became an attraction for domestic tourists, a place where ordinary people could literally taste a piece of American culture. Although not the first to introduce American fast food to Beijing consumers, the McDonald's chain has been the most aggressive in expanding its business and the most influential in developing the fast-food market. Additional McDonald's restaurants appeared in Beijing one after another: two were opened in 1993, four in 1994, and ten more in 1995. There were 35 by August 1997, and according to the general manager the Beijing market is big enough to support more than a hundred McDonald's restaurants.[17] At the same time, Pizza Hut, Bony Fried Chicken (of Canada), and Dunkin' Donuts all made their way into the Beijing market. The most interesting newcomer is a noodle shop chain called Californian Beef Noodle King. Although the restaurant sells Chinese noodle soup, it has managed to portray itself as an American fast-food eatery and competes with McDonald's and KFC with lower prices and its appeal to Chinese tastes.

The instant success of Western fast-food chains surprised those in the local restaurant industry. Soon thereafter, many articles in newspapers and journals called for the invention of Chinese-style fast food and the development of a local fast-food industry. April 1992 was a particularly difficult month for those involved in this sector: two weeks after the largest McDonald's restaurant opened at the southern end of Wangfujing Street, Wu Fang Zhai, an old, prestigious restaurant at the northern end of Wangfujing Street, went out of business; in its stead opened International Fast Food City, which sold Japanese fast food, American hamburgers, fried chicken, and ice cream. This was seen as an alarming threat to both the local food industry and the national pride of Chinese culinary culture.[18]

Actually, the local response to the "invasion" of Western fast food began in the late 1980s, right after the initial success of KFC. It quickly developed into what some reporters called a "war of fried chickens" in Beijing. Following the model of KFC, nearly a hundred local fast-food shops featuring more than a dozen kinds of fried chicken appeared between 1989 and 1990. One of the earliest such establishments was Lingzhi Roast Chicken, which began business in 1989; this was followed by

Chinese Garden Chicken, Huaxiang Chicken, and Xiangfei Chicken in 1990. The chicken war reached its peak when the Ronghua Fried Chicken company of Shanghai opened its first chain store directly opposite one of the KFC restaurants in Beijing. The manager of Ronghua Chicken proudly announced a challenge to KFC: "Wherever KFC opens a store, we will open another Ronghua Fried Chicken next door."

All of the local fried chicken variations were no more than simple imitations of the KFC food. Their only localizing strategy was to emphasize special Chinese species and sacred recipes that supposedly added an extra medicinal value to their dishes. Thus, consumers were told that the Chinese Garden Chicken might prevent cancer and that Huaxiang Chicken could strengthen the yin-yang balance inside one's body.[19] This strategy did not work well; KFC and McDonald's won out in that first wave of competition. Only a small proportion of the local fried chicken shops managed to survive, while KFC and McDonald's became more and more popular.

Realizing that simply imitating Western fast food was a dead end, the emerging local fast-food industry turned to exploring resources within Chinese cuisine. Among the pioneers, Jinghua Shaomai Restaurant in 1991 tried to transform some traditional Beijing foods into fast foods. This was followed by the entry of a large number of local fast-food restaurants, such as the Beijing Beef Noodle King (not to be confused with the California Beef Noodle King). The Jinghe Kuaican company made the first domestic attempt to develop a fast-food business on a large scale. With the support of the Beijing municipal government, this company built its own farms and processing facilities, but it chose to sell boxed fast foods in mobile vans parked on streets and in residential areas.[20] Thus it fell into the pre-existing category of *hefan* (boxed rice) purveyors. Although the price of boxed fast foods was much lower than that of imported fast food, the boxed fast foods did not meet consumers' expectations of fast food. The Jinghe Kuaican Company disappeared as quickly as it had emerged. In October 1992, nearly a thousand state-owned restaurants united under the flag of the Jingshi Fast Food Company, offering five sets of value meals and more than 50 fast-foods items, all of which were derived from traditional Chinese cuisines. This company was also the first fast-food enterprise to be run by the Beijing municipal government, thus indicating the importance of this growing sector to the government.[21] The Henan Province Red Sorghum Restaurant opened on Wangfujing Street in March 1996, immediately across the street from the McDonald's flagship restaurant. Specializing in country-style lamb noodles, the manager of Red Sorghum announced that twelve more restaurants were to be opened in Beijing by the end of 1996, all of which would be next to a McDonald's outlet. "We want to fight McDonald's," the manager claimed, "we want to take back the fast-food market."[22]

By 1996 the fast-food sector in Beijing consisted of three groups: The main group was made up of McDonald's, KFC, and other Western fast-food chains. Although they no longer attracted the keen attention of the news media, their numbers were still growing. The second group consisted of the local KFC imitations, which managed to survive the 1991 "chicken war." The most successful in this group is the Ronghua Chicken restaurant chain, which in 1995 had eleven stores in several cities and more than 500 employees.[23] The third group included restaurants selling newly created Chinese fast foods, from simple noodle soups to Beijing roast duck meals.

Many believe that the long tradition of a national cuisine will win out over the consumers' temporary curiosity about Western-style fast food.

Thus far, however, Chinese fast food has not been able to compete with Western fast food, even though it is cheaper and more appealing to the tastes of ordinary citizens in Beijing. Red Sorghum was the third business to announce in public the ambitious goal of beating McDonald's and KFC (after the Shanghai Ronghua Chicken and Beijing Xiangfei Chicken), but so far none have come close. By August 1996 it was clear that Red Sorghum's lamb noodle soup could not compete in the hot summer with Big Mac, which was popular year-round.[24]

The lack of competitiveness of Chinese fast food has drawn official attention at high levels, and in 1996 efforts were made to support the development of a local fast-food sector.[25] Concerned experts in the restaurant industry and commentators in the media attribute the bad showing of the Chinese fast-food restaurants to several things. In the mid-1990s, at least: (1) the quality, nutritional values, and preparation of Western fast foods were highly standardized, while Chinese fast foods were still prepared in traditional ways; (2) Chinese fast-food establishments did not offer the friendly, quick service of Western fast-food restaurants; (3) the local establishments were not as clean and comfortable as the Western fast-food restaurants; and (4) most important, unlike McDonald's or KFC, Chinese restaurants did not employ advanced technologies or modern management methods.[26] From a Marxist perspective, Ling Dawei has concluded that the race between imported and local fast foods in Beijing is a race between advanced and backward forces of production; hence the development of the local fast-food industry will rest ultimately on modernization.[27]

There is no doubt that these views have a basis in everyday practice; yet they all regard food consumption as purely economic behavior and fast-food restaurants as mere eating places. A more complete understanding of the fast-food fever in Beijing also requires close scrutiny of the social context of consumption—the participants and social settings, because "The specific nature of the consumed substances surely matters; but it cannot, by itself, explain why such substances may seem irresistible."[28]

The Spatial Context of Fast-food Consumption

As Giddens points out, most social life occurs in the context of the "fading away" of time and the "shading off" of space.[29] This is certainly true for fast-food consumption. Fast-food restaurants, therefore, need to be examined both as eating places and as social spaces where social interactions occur. A physical place accommodates objects and human agents and provides an arena for social interactions, and it follows that the use of space cannot be separated from the objects and the physical environment.[30] However, space functions only as a context, not a determinant, of social interactions, and the space itself in some way is also socially constructed.[31] In the following pages I consider, on the one hand, how spatial context shapes consumers' behavior and social relations, and how, on the other hand, consumers appropriate fast-food restaurants into their own space. Such an inquiry must begin with a brief review of Beijing's restaurant sector in the late 1970s in order to assess the extent to which Western fast-food outlets differ from existing local restaurants.

Socialist Canteens and Restaurants in the 1970s

Eating out used to be a difficult venture for ordinary people in Beijing because few restaurants were designed for mass consumption. As mentioned earlier, the restaurants in Beijing were hierarchically ranked by architecture, function, and the type and quality of foods provided. More important, before the economic reforms almost all restaurants and eateries were state-owned businesses, which meant that a restaurant was first and foremost a work unit, just like any factory, shop, or government agency.[32] Thus a restaurant's position and function were also determined by its administrative status as a work unit.

Generally speaking, the restaurant hierarchy consisted of three layers. At the top were luxury and exclusive restaurants in star-rated hotels, such as the Beijing Hotel, which served only foreigners and privileged domestic guests. At the next level were well-established formal restaurants, many of which specialized in a particular style of cuisine and had been in business for many years, even before the 1949 revolution. Unlike the exclusive hotel restaurants, the formal restaurants were open to the public and served two major functions: (1) as public spaces in which small groups of elites could socialize and hold meetings; and (2) as places for ordinary citizens to hold family banquets on special, ritualized occasions such as weddings. At the bottom of the hierarchy were small eateries that provided common family-style foods; these were hardly restaurants (they were actually called *shitang*, meaning canteens). The small eateries were frequented primarily by visitors from outlying provinces and some Beijing residents who had to eat outside their homes because of special job requirements. The majority of Beijing residents rarely ate out—they normally had their meals at home or in their work-unit canteens.

In the 1950s the development of internal canteens (*neibu shitang*) not only constituted an alternative to conventional restaurants but also had a great impact on the latter. Most work units had (and still have) their own canteens, in order to provide employees with relatively inexpensive food and, more important, to control the time allotted for meals. Because canteens were subsidized by the work units and were considered part of employees' benefits, they were run in a manner similar to a family kitchen, only on an enlarged scale. The central message delivered through the canteen facilities was that the work unit, as the representative of the party-state, provided food to its employees, just as a mother feeds her children (without the affectionate component of real parental care). The relationship between the canteen workers and those who ate at the canteens was thus a patronized relationship between the feeder and the fed, rather than a relationship of service provider and customers. The tasteless foods, unfriendly service, and uncomfortable environment were therefore natural components at such public canteens, which prevailed for more than three decades and still exist in many work units today.

The work-unit mentality of "feeding" instead of "serving" people also made its way into restaurants in Beijing because, after all, the restaurants were also work units and thus had the same core features as all other work units—that is, the dominating influence of the state bureaucracy and the planned economy. Commercial restaurants also shared with the work-unit canteens the poor maintenance of internal space, a limited choice of foods, the requirement that the diner pay in advance, fixed times for meals (most restaurants were open only during the short prescribed lunch and dinner times), and of course, ill-tempered workers who acted as if they were distributing food to hungry beggars instead of paying customers.[33] It is true that the higher one moved up the ladder of the restaurant hierarchy the better dining environment and service one could find. But in the famous traditional restaurants and the star-rated hotel restaurants, formality and ritual were most likely the dominating themes. Still, until the late 1980s it was not easy for ordinary people to enjoy dining out in restaurants.

In contrast, Western fast-food restaurants offered local consumers a new cultural experience symbolized by foreign fast food, enjoyable spatial arrangements of objects and people, and American-style service and social interactions.

The Cultural Symbolism of Fast Food

It is perhaps a truism to note that food is not only good to eat but also good for the mind. The (Western) fast-food fever in Beijing provides another example of how in certain circumstances customers may care less about the food and more about the cultural messages it delivers. During my fieldwork in 1994 I discovered that although children were great fans of the Big Mac and french fries, most adult customers did not particularly like those fast foods. Many people commented that the taste was not good and that the flavor of cheese was strange. The most common complaint from adult customers was *chi bu bao*, meaning that McDonald's hamburgers and fries did not make one feel full: they were more like snacks than like meals.[34] It is also interesting to note that both McDonald's and KFC emphasized the freshness, purity, and nutritional value of their foods (instead of their appealing tastes). According to a high-level manager of Beijing McDonald's, the recipes for McDonald's foods were designed to meet modern scientific specifications and thus differed from the recipes for Chinese foods, which were based on cultural expectations.[35] Through advertisements and news media reports, this idea that fast foods use nutritious ingredients and are prepared using scientific cooking methods has been accepted by the public. This may help to explain why few customers compared the taste of fast foods to that of traditional Chinese cuisine; instead customers focused on something other than the food.

If people do not like the imported fast food, why are they still keen on going to Western fast-food restaurants? Most informants said that they liked the atmosphere, the style of eating, and the experience of being there. According to an early report on KFC, customers did not go to KFC to eat the chicken but to enjoy "eating" (consuming) the culture associated with KFC. Most customers spent hours talking to each other and gazing out the huge glass windows onto busy commercial streets—and feeling more sophisticated than the people who passed by.[36] Some local observers argued that the appeal of Chinese cuisine was the taste of the food itself and that, in contrast, Western food relied on the manner of presentation. Thus consumers would seem to be interested in the spectacle created by this new form of eating.[37] In other words, what Beijing customers find satisfying about Western fast-food restaurants is not the food but the experience.

The cultural symbolism that McDonald's, KFC, and other fast-food chains carry with them certainly plays an important role in constructing this nonedible yet fulfilling experience. Fast food, particularly McDonald's fast food, is considered quintessentially American in many parts of the contemporary world. In France, the most commonly agreed "American thing" among teenagers is McDonald's, followed by Coca-Cola and "military and space technologies."[38] In Moscow, a local journalist described the opening of the first McDonald's restaurant as the arrival of the "ultimate icon of Americana."[39] The same is true in Beijing, although the official news media have emphasized the element of modernity instead of Americana. The high efficiency of the service and management, fresh ingredients, friendly service, and spotless dining environment in Western fast-food restaurants have been repeatedly reported by the Beijing media as concrete examples of modernity.[40]

Ordinary consumers are interested in the stories told in news reports, popular magazines, and movies that the Big Mac and fried chicken are what make Americans American. According to a well-known commentator on popular culture in Beijing, because of the modernity inherent in the McDonald's fast-food chain, many American youths prefer to work first at McDonald's before finding other jobs on the market. The experience of working at McDonald's, he argues, prepares American youth for any kind of job in a modern society.[41] To many Beijing residents, "American" also means "modern," and thus to eat at McDonald's is to experience modernity. During my fieldwork I talked with many parents who appreciated their children's fondness for imported fast food because they believed it was in good taste to be modern. A mother told me that she had made great efforts to adapt to the strange flavor of McDonald's food so that she could take her daughter to McDonald's twice a week. She explained: "I want my daughter to learn more about American culture. She is taking an English typing class now, and I will buy her a computer next year." Apparently, eating a Big Mac and fries, like learning typing and computer skills, is part of the mother's plan to prepare her daughter for a modern society.

Inspired by the success of the cultural symbolism of McDonald's and KFC, many Chinese fast-food restaurants have tried to use traditional Chinese culture to lure customers. As I mentioned in the preceding section, almost all local fried-chicken outlets during 1990–91 emphasized the use of traditional medicinal ingredients and the idea of health-enhancing food.[42] Others used ethnic and local flavors to stress the Chineseness of their fast foods, such as the Red Sorghum's promotion of its lamb noodle soup.[43] And some directly invoked the nationalist feelings of the customers. For instance, Happy Chopsticks, a new fast-food chain in Shenzhen, adopted "Chinese people eat Chinese food" as the leading slogan in its market promotion.[44] The power of cultural symbolism in the fast-food sector also has made an impact on the restaurant industry in general: the cultural position of the restaurant business is regarded as an important issue, and the debate about the differences between Western and Chinese cuisine continues in professional journals.[45]

A Place of Entertainment for Equals

According to older residents, in addition to different cuisine styles, traditional restaurants in pre-1949 Beijing also differed in their interior decorations, seating arrangements, and interactions between restaurant employees and customers. During the Maoist era, such features were considered inappropriate to the needs of working-class people and thus gradually disappeared. Under the brutal attack on traditional culture during the Cultural Revolution period, some famous restaurants even replaced their old names with new, revolutionary names, such as Workers and Peasants Canteen (*Gongnong shitang*). As a result, by the late 1970s most restaurants looked similar both inside and out, which, combined with the canteen mentality in restaurant management and poor service, turned Beijing restaurants into unpleasant eating places.

When KFC and McDonald's opened their outlets in Beijing, what most impressed Beijing consumers was their beautiful appearance. As mentioned earlier, both the first KFC and first McDonald's are located near Tiananmen Square in the heart of Beijing, and both boast that they are the largest outlets of their kind in the world, one with a three-floor, 500-seat building and the other with a two-floor, 700-seat building. The statues of Colonel Sanders and Ronald McDonald in front of the two establishments immediately became national tourist attractions.

Once inside the restaurants, Beijing customers found other surprises. First, both McDonald's and the KFC restaurants were brightly lit and climate-controlled. The seats, tables, and walls were painted in light colors, which, together with the shiny counters, stainless-steel kitchenware, and soft music in the background, created an open and cheerful physical environment—a sharp contrast to traditional Chinese restaurants. Moreover, social interaction at McDonald's or KFC was highly ritualized and dramatized,[46] representing a radical departure from the canteen-like restaurants in Beijing. Employees wore neat, brightly colored uniforms, and they smiled at customers while working conscientiously and efficiently. As one observant informant remarked, even the employee responsible for cleaning the toilets worked in a disciplined manner. In his study of restaurants in Athens, Georgia, Allen Shelton commented: "The spectacle of McDonald's is work: the chutes filling up with hamburgers; the restaurant and the other diners are secondary views."[47] In contrast, both the work and the restaurant itself constituted the spectacle at McDonald's and KFC in Beijing.

One of the things that most impressed new customers of the fast-food outlets was the menu, which is displayed above and behind the counter, with soft backlighting and photographic images of the food. The menu delivers a clear message about the public, affordable eating experience that the establishment offers. This was particularly important for first-timers, who did not know anything about the exotic food. Another feature is the open, clean, kitchen area, which clearly shows the customers how the hamburgers and fried chickens are prepared. To emphasize this feature, Beijing's McDonald's also provides a five-minute tour of the kitchen area on customer request.

The Western fast-food restaurants also gave customers a sense of equality. Both employees and customers remain standing during the ordering process, creating an equal relationship between the two parties. More important, the friendly service and the smiling employees give customers the impression that no matter who you are you will be treated with equal warmth and friendliness. Accordingly, many people patronize McDonald's to experience a moment of equality.[48] The restaurants also seem to convey gender equality and have attracted a large number of female customers (I will return to this point later).

All these details in internal space are important in understanding the success of McDonald's and KFC in Beijing: objects have a voice that originates in those who use them, just as the scenery on a stage shapes the movements of an actor.[49] The impact of spatial context on people's behavior in McDonald's restaurants is well addressed by Peter Stephenson. He observed that some Dutch customers lost their cultural "self" in such a culturally decontextualized place because "there is a kind of instant emigration that occurs the moment one walks through the doors, where Dutch rules rather obviously don't apply."[50] Rick Fantasia observed that French customers undergo similar changes or adjustments in behavior in McDonald's outlets in Paris.[51] Given the sharper and deeper cultural differences between American and Chinese societies, it is natural to expect the cultural decontextualization to be even stronger in Beijing's McDonald's and KFC restaurants.

The interesting point is that, owing to the powerful appeal of modernity and Americana projected by McDonald's and KFC, when experiencing the same "instant emigration," Beijing customers seem to be more willing to observe the rules of American fast-food restaurants than their counterparts in Leiden or Paris. For instance, in 1992 and 1993 customers in Beijing (as in Hong Kong and Taiwan) usually left their rubbish on the table for the restaurant employees to clean up: people regarded McDonald's as a formal establishment at which they had paid for full service. However, during the summer of 1994 I observed that about one-fifth of the customers, many of them fashionably dressed youth, carried their own trays to the waste bins. From subsequent interviews I discovered that most of these people were regular customers, and they had learned to clean up their tables by observing the foreigners' behavior. Several informants told me that when they disposed of their own rubbish they felt more "civilized" (*wenming*) than the other customers because they knew the proper behavior. My random check of customer behavior in McDonald's and in comparably priced and more expensive Chinese restaurants shows that people in McDonald's were, on the whole, more self-restrained and polite toward one another, spoke in lower tones, and were more careful not to throw their trash on the ground. Unfortunately, when they returned to a Chinese context, many also returned to their previous patterns of behavior. As a result, the overall atmosphere in a Western fast-food outlet is always nicer than that in Chinese restaurants of the same or even higher quality.[52]

A Multidimensional Social Space

In part because of the cultural symbolism of Americana and modernity and in part because of the exotic, cheerful, and comfortable physical environment, McDonald's, KFC, and other foreign fast-food restaurants attract customers from all walks of life in Beijing. Unlike in the United States, where the frequenters of fast-food restaurants are generally associated with low income and simple tastes, most frequenters of fast-food restaurants in Beijing are middle-class professionals, trendy yuppies, and well-educated youths. Unfortunately, there has yet to be a systematic social survey of Chinese fast-food customers. Nevertheless, according to my field observations in 1994, a clear distinction can be drawn between those who occasionally partake of the imported fast foods and those who regularly frequent fast-food restaurants.

Occasional adventurers include both Beijing residents and visitors from outlying provinces and cities. It should be noted that a standard one-person meal at McDonald's (including a hamburger, a soft drink, and an order of French fries, which is the equivalent of a value-meal at McDonald's in the United States) cost 17 renminbi (rmb) ($2.10) in 1994 and 21 rmb ($2.60) in 1996.[53] This may not be expensive by American standards, but it is not an insignificant amount of money for ordinary workers in Beijing, who typically made less than 500 rmb ($60) per month in 1994. Thus, many people, especially those with moderate incomes, visited McDonald's restaurants only once or twice, primarily to satisfy their curiosity about American food and culinary culture. A considerable proportion of the customers were tourists from other provinces who had only heard of McDonald's or seen its Golden Arches in the movies. The tasting of American food has recently become an important part of the tourist beat in Beijing; and those who partake of the experience are able to boast about it to their relatives and friends back home.

There are also local customers who frequent foreign fast-food outlets on a regular basis. A survey conducted by Beijing McDonald's management in one of its stores showed that 10.2 percent of the customers frequented the restaurant four times per month in 1992, in 1993 the figure was 38.3 percent.[54] The majority of customers fell into three categories: professionals and white-collar workers; young couples and teenagers; and children accompanied by their parents. Moreover, women of all age groups tended to frequent McDonald's restaurants more than men.

For younger Beijing residents who worked in joint-venture enterprises or foreign firms and had higher incomes, eating at McDonald's, Kentucky Fried Chicken, and Pizza Hut had become an integral part of their new lifestyle, a way for them to be connected to the outside world. As one informant commented: "The Big Mac doesn't taste great; but the experience of eating in this place makes me feel good. Sometimes I even imagine that I am sitting in a restaurant in New York or Paris." Although some emphasized that they only went to save time, none finished their meals within twenty minutes. Like other customers, these young professionals arrived in small groups or accompanied by girl/boy friends to enjoy the restaurant for an hour or more. Eating foreign food, and consuming other foreign goods, had become an important way for Chinese yuppies to define themselves as middle-class professionals. By 1996, however, this group had found other types of activities (such as nightclubs or bars), and gradually they were beginning to visit foreign fast-food restaurants for convenience rather than for status.

Young couples and teenagers from all social strata were also regular frequenters of McDonald's and KFC outlets because the dining environment is considered to be romantic and comfortable. The restaurants are brightly-lit, clean, and feature light Western music; and except during busy periods they are relatively quiet, making them ideal for courtship. In 1994, McDonald's seven Beijing restaurants had all created relatively isolated and private service areas with tables for two. In some, these areas were nicknamed "lovers' corners." Many teenagers also considered that, with only the minimum consumption of a soft drink or an ice cream, fast-food establishments were good places simply to hang out.

As in many other parts of the world, children in Beijing had become loyal fans of Western fast food. They were so fond of it that some parents even suspected that Big Mac or fried chicken contained a special, hidden ingredient. The fast-food restaurants also made special efforts to attract children by offering birthday parties, dispensing souvenirs, and holding essay contests, because young customers usually did not come alone: they were brought to McDonald's and KFC by their parents or grandparents. Once a middle-aged woman told me that she did not like the taste of hamburgers and that her husband simply hated them. But their daughter loved hamburgers and milk-shakes so much that their entire family had to visit McDonald's three to five times a month. It is common among Beijing families for children to choose the restaurant in which the whole family dines out. Fast-food outlets were frequently the first choice of children.

A gender aspect of fast-food consumption is highlighted in He Yupeng's 1996 study of McDonald's popularity among female customers. In conducting a small-scale survey at four restaurants in Beijing—a formal Chinese restaurant, a local fast-food outlet, and two McDonald's outlets—He found that women were more likely than men to enjoy dining at fast-food restaurants. According to his survey, while 66 percent of the customers (N=68) at the formal Chinese restaurant were men, 64 percent of the customers (N=423) at the local fast-food outlet were women. Similar patterns were observed in the two McDonald's restaurants, where women constituted 57 percent of a total of 784 adult customers.[55] The most intriguing finding of this survey was that women chose McDonald's because they enjoyed ordering their own food and participating in the conversation while dining. Many female customers pointed out that in formal Chinese restaurants men usually order the food for their female companions and control the conversation. In contrast, they said, at a McDonald's everyone can make his or her own choices and, because smoking and alcohol are prohibited, men dominate less of the conversation.[56]

Furthermore, the imported fast-food restaurants provide a venue where women feel comfortable alone or with female friends. Formal Chinese restaurants are customarily used by elite groups as places to socialize and by middle-class people as places to hold ritual family events such as wedding banquets. In both circumstances, women must subordinate themselves to rules and manners that are androcentric, either explicitly or implicitly (the men order the dishes; the women do not partake of the liquor). These customs reflect the traditional view that women's place is in the household and that men should take charge of all public events. There is a clear division between the private (inside) and the public (outside) along gender lines.

A woman who eats alone in a formal Chinese restaurant is considered abnormal; such behavior often leads to public suspicion about her morality and her occupation. For instance, a young woman I interviewed in a McDonald's outlet in 1994 recalled having lunch alone in a well-known Chinese restaurant frequented mostly by successful businessmen. "Several men gazed at me with lascivious eyes," she said, "and some others exchanged a few words secretly and laughed among themselves. They must have thought I was a prostitute or at least a loose woman. Knowing their evil thoughts, I felt extremely uncomfortable and left the place as quickly as I could." She also commented that even going to a formal Chinese restaurant with female friends would make her feel somewhat improper about herself, because the "normal" customers were men or men with women. But she said that she felt comfortable visiting a McDonald's alone or with her female friends, because "many people do the same." This young woman's experience is by no means unique, and a number of female customers in McDonald's offered similar explanations for liking the foreign fast-food restaurants. Several elderly women also noted the impropriety of women dining in formal Chinese restaurants, although they were less worried about accusations about their morals.[57]

In his survey, He Yupeng asked his respondents where they would choose to eat if there were only a formal Chinese restaurant and a McDonald's outlet. Almost all the male respondents chose the former, and all the female respondents chose the latter. One of the main reasons for such a sharp gender difference, He argues, is the concern of contemporary women for gender equality.[58] The new table manners allowed in fast-food restaurants, and more important, the newly appropriate gender roles in those public places, seem to have enhanced the image of foreign fast-food restaurants as an open place for equals, thus attracting female customers.

The Appropriation of Social Space

Finally, I would point out that Beijing customers do not passively accept everything offered by the American fast-food chains. The American fast-food restaurants have been localized in many aspects, and what Beijing customers enjoy is actually a Chinese version of American culture and fast foods.[59] One aspect of this localization process is the consumers' appropriation of the social space.

My research confirms the impression that most customers in Beijing claim their tables for longer periods of time than Americans do. The average dining time in Beijing (in autumn 1994) was 25 minutes during busy hours and 51 minutes during slack hours. In Beijing, "fastness" does not seem to be particularly important. The cheerful, comfortable, and climate-controlled environment inside McDonald's and KFC restaurants encourages many customers to linger, a practice that seems to contradict the original purpose of the American fast-food business. During off-peak hours it is common for people to walk into McDonald's for a leisurely drink or snack. Sitting with a milk-shake or an order of fries, such customers often spend 30 minutes to an hour, and sometimes longer, chatting, reading newspapers, or holding business meetings. As indicated earlier, young couples and teenagers are particularly fond of frequenting foreign fast-food outlets because they consider the environment to be romantic. Women in all age groups tend to spend the longest time in these establishments, whether they are alone or with friends. In contrast, unaccompanied men rarely linger after finishing their meals. The main reason for this gender difference, according to my informants, is the absence of alcoholic beverages. An interesting footnote in this connection is that 32 percent of my informants in a survey among college students (N=97) regarded McDonald's as a symbol of leisure and emphasized that they went there to relax.

Beijing consumers have appropriated the restaurants not only as places of leisure but also as public arenas for personal and family ritual events. The most popular such event is of course the child's birthday party, which has been institutionalized in Beijing McDonald's restaurants. Arriving with five or more guests, a child can expect an elaborate ritual performed in a special enclosure called "Children's paradise," free of extra charge. The ritual begins with an announcement over the restaurant's loudspeakers—in both Chinese and English—giving the child's name and age, together with congratulations from Ronald McDonald (who is called Uncle McDonald in Beijing). This is followed by the recorded song "Happy Birthday," again in two languages. A female employee in the role of Aunt McDonald then entertains the children with games and presents each child with a small gift from Uncle McDonald. Although less formalized (and without the restaurant's active promotion), private ceremonies are also held in the restaurants for adult customers, particularly for young women in peer groups (the absence of alcohol makes the site attractive to them). Of the 97 college students in my survey, 32 (including nine men) had attended personal celebrations at McDonald's: birthday parties, farewell parties, celebrations for receiving scholarships to American universities, and end-of-term parties.

The multifunctional use of McDonald's space is due in part to the lack of cafes, tea houses, and ice-cream shops in Beijing; it is also a consequence of the management's efforts to attract as many customers as possible by engendering an inviting environment. Although most McDonald's outlets in the United States are designed specifically to prevent socializing (with less-comfortable seats than formal restaurants, for instance) it is clear that the managers of Beijing's McDonald's have accepted their customers' perceptions of McDonald's as a special place that does not fit into pre-existing categories of public eateries. They have not tried to educate Beijing consumers to accept the American view that "fast food" means that one must eat fast and leave quickly.[60] When I wondered how the management accommodated everyone during busy periods, I was told that the problem often resolved itself. A crowd of customers naturally created pressures on those who had finished their meals, and more important, during busy hours the environment was no longer appropriate for relaxation.

In contrast, managers in Chinese fast-food outlets tend to be less tolerant of customers who linger. During my fieldwork in 1994 I conducted several experimental tests by going to Chinese fast-food outlets and ordering only a soft drink but staying for more than an hour. Three out of four times I was indirectly urged to leave by the restaurant employees; they either took away my empty cup or asked if I needed anything else. Given the fact that I was in a fast-food outlet and did all the service for myself, the disturbing "service" in the middle of my stay was clearly a message to urge lingering customers to leave. I once discussed this issue with the manager of a Chinese fast-food restaurant. He openly admitted that he did not like customers claiming a table for long periods of time and certainly did not encourage attempts to turn the fast-food outlet into a coffee shop. As he explained: "If you want to enjoy nice coffee and music then you should go to a fancy hotel cafe, not here."

Concluding Remarks: Dining Place, Social Space, and Mass Consumption

In the United States, fast-food outlets are regarded as "fuel stations" for hungry yet busy people and as family restaurants for low-income groups. Therefore, efficiency (speed) and economic value (low prices) are the two most important reasons why fast foods emerged as a kind of "industrial food" and remain successful in American society today. These features, however, do not apply in Beijing. A Beijing worker who loads the whole family into a taxi to go to McDonald's may spend one-sixth of his monthly income; efficiency and economy are perhaps the least of his concerns. When consumers stay in McDonald's or KFC restaurants for hours, relaxing, chatting, reading, enjoying the music, or celebrating birthdays, they take the "fastness" out of fast food. In Beijing, the fastness of American fast food is reflected mainly in the service provided; for consumers, the dining experience is too meaningful to be short-ened. As a result, the American fast-food outlets in China are fashionable, middle-class establishments—a new kind of social space where people can enjoy their leisure time and experience a Chinese version of American culture.

As I emphasize repeatedly throughout this chapter, eating at a foreign fast-food restaurant is an important social event, although it means different things to different people. McDonald's, KFC, and other fast-food restaurants in Beijing carry the sym-bolism of Americana and modernity, which makes them unsurpassable by existing standards of the social hierarchy in Chinese culture. They represent an emerging tradition where new values, behavior patterns, and social relationships are still being created. People from different social backgrounds may enter the same eating place/ social space without worrying about losing face; on the contrary, they may find new ways to better define their positions. For instance, white-collar professionals may display their new class status, youngsters may show their special taste for leisure, and parents may want to "modernize" their children. Women of all ages are able to enjoy their independence when they choose to eat alone; and when they eat with male companions, they enjoy a sexual equality that is absent in formal Chinese restaurants. The fast-food restaurants, therefore, constitute a multivocal, multidimensional, and open social space. This kind of all-inclusive social space met a particular need in the 1990s, when Beijing residents had to work harder than ever to define their positions in a rapidly changing society.[61]

By contrast, almost all local competitors in the fast-food sector tend to regard fast-food restaurants merely as eating places, and accordingly, they try to compete with the foreign fast-food restaurants by offering lower prices and local flavors or by appealing to nationalist sentiments. Although they also realize the importance of hygiene, food quality, friendly service, and a pleasant physical environment, they regard these features as isolated technical factors. A local observer pointed out that it is easy to build the "hardware" of a fast-food industry (the restaurants) but that the "software" (service and management) cannot be adopted overnight.[62] To borrow from this metaphor, I would argue that an understanding of fast-food outlets not only as eating places but also as social space is one of the "software problems" waiting to be resolved by the local competitors in the fast-food business.

Why is the issue of social space so important for fast-food development in Beijing? It would take another essay to answer this question completely; here I want to highlight three major factors that contribute to fast-food fever and are closely related to consumers' demands for a new kind of social space.

First, the trend of mass consumption that arose in the second half of the 1980s created new demands for dining out as well as new expectations of the restaurant industry. According to 1994 statistics released by the China Consumer Society, the average expenditure per capita has increased 4.1 times since 1984. The ratio of "hard consumption" (on food, clothes, and other necessities of daily life) to "soft consumption" (entertainment, tourism, fashion, and socializing) went from 3:1 in 1984 to 1:1.2 in 1994.[63] In 1990, consumers began spending money as never before on such goods and services as interior decoration, private telephones and pagers, air conditioners, body-building machines, and tourism.[64] As part of this trend toward consumerism, dining out has become a popular form of entertainment among virtually all social groups, and people are particularly interested in experimenting with different cuisines.[65] In response to a survey conducted by the Beijing Statistics Bureau in early 1993, nearly half of the respondents said they had eaten at Western-style restaurants (including fast-food outlets) at least once.[66] A central feature of this development in culinary culture is that people want to dine out as active consumers, and they want the dining experience to be relaxed, fun, and healthful.

In response to increasing consumer demands, thousands of restaurants and eateries have appeared in recent years. By early 1993 there were more than 19,000 eating establishments in Beijing, ranging from elegant five-star hotel restaurants to simple street eateries. Of these, about 5,000 were state-owned, 55 were joint ventures or foreign-owned, and the remaining 14,000 or so were owned by private entrepreneurs or independent vendors (*getihu*).[67]

These figures show that the private sector has played an increasingly important role in the restaurant business. Unlike the state-owned restaurants, some private restaurants have used creativity to meet consumers' demands for a new kind of dining experience. The best example is the emergence of country-style, nostalgic restaurants set up by and for the former sent-down urban youths. In these places customers retaste their experience of youth in the countryside: customers choose from country-style foods in rooms and among objects that remind them of the past. Like customers in McDonald's or KFC, they are also consuming part of the subculture and redefining themselves in a purchased social space. The difference is that the nostalgic restaurants appeal only to a particular social group, while the American fast-food outlets are multivocal and multidimensional and thus attract people from many different social strata.

The rise of new consumer groups is the second major factor that has made the issue of social space so important to understanding fast-food fever in Beijing. Urban youth, children, and women of all ages constitute the majority of the regular frequenters of American fast-food restaurants. It is not by accident that these people are all newcomers as restaurant customers—there was no proper place for them in the pre-existing restaurant system, and the only social role that women, youth, and children could play in a formal Chinese restaurant was as the dependents of men. Women's effort to gain an equal place in restaurant subculture was discussed earlier, so here I briefly examine the place of youth and children.

Young professionals emerged along with the development of the market economy, especially with the expansion of joint-venture and foreign-owned business in Beijing in the 1990s. To prove and further secure their newly obtained social status and prestige, the young elite have taken the construction of a different lifestyle seriously, and they often lead the trend of contemporary consumerism in Chinese cities. Urban youth may be less well off than young professionals, but they are equally eager to embrace a new way of personal life. According to a 1994 survey, the purchasing power of Beijing youth increased dramatically over the previous decade, and nearly half of the 1,000 respondents in the survey had more than 500 rmb per month to spend on discretionary items.[68] With more freedom to determine their lifestyles and more economic independence, these youngsters were eager to establish their own social space in many aspects of life, including dining out.[69] A good example in this connection is the astonishing popularity among young people in mainland China of pop music, films, and romance novels from Hong Kong and Taiwan.[70]

The importance of teenagers and children in effecting social change also emerged in the late twentieth century, along with the growth of the national economy, the increase in family wealth, and the decline of the birth rate. The single-child policy—which is most strictly implemented in the big cities—has created a generation of little emperors and empresses, each demanding the attention and economic support of his or her parents and grandparents. Parental indulgence of children has become a national obsession, making children and teenagers one of the most active groups of consumers. Beijing is by no means exceptional in this respect. According to Deborah Davis and Julia Sensenbrenner, ordinary working-class parents in Shanghai normally spend one-third of their monthly wages to provide their children with a lifestyle that is distinctly upper middle class in its patterns of consumption. For many parents, toys, trips, fashionable clothes, music lessons, and restaurant meals have become necessities in raising their children. This suggests a significant change in patterns of household expenditure, and accordingly there is an urgent need to meet the market demands and special tastes of this important group of consumers.

The emerging importance of women, youth, and children as consumers results from a significant transformation of the family institution in contemporary Chinese society, which is characterized by the nuclearization of the household, the centrality of conjugality in family relations, the rising awareness of independence and sexual equality among women, the waning of the patriarchy, and the rediscovery of the value of children.[71] As far as fast-food consumption is concerned, the link between new groups of independent consumers and shifts in family values is found in other East Asian societies as well. After analyzing the relationship between the McDonald's "takeoff" in five cities (Tokyo, Hong Kong, Taipei, Seoul, and Beijing) and the changes in family values (especially the rising status of teenagers and children), Watson concluded: "More than any other factor ... McDonald's success is attributable to the revolution in family values that has transformed East Asia."[72]

A third important factor in the success of Western fast-food enterprises is the new form of sociality that has been developing in market-controlled public places such as restaurants. A significant change in public life during the post-Mao era has been the disappearance of frequent mass rallies, voluntary work, collective parties, and other forms of "organized sociality" in which the state (through its agents) played the central role. In its place are new forms of private gatherings in public venues. Whereas "organized sociality" emphasized the centrality of the state, the official ideology, and the submission of individuals to an officially endorsed collectivity, the new sociality celebrates individuality and private desires in unofficial social and spatial contexts. The center of public life and socializing, accordingly, has shifted from large state-controlled public spaces (such as city squares, auditoriums, and workers' clubs) to smaller, commercialized arenas such as dancing halls, bowling alleys, and even imaginary spaces provided by radio call-in shows. The new sociality has even emerged in conventionally state-controlled public spaces, such as parks, and has thus transformed them into multidimensional spaces in which the state, the public, and the private may coexist (see Richard Kraus's chapter in this volume).

Restaurants similarly meet the demand for a new kind of sociality outside state control—that is, the public celebration of individual desires, life aspirations, and personal communications in a social context. As indicated above, in earlier decades the socialist state did not encourage the use of restaurants as a social space in which to celebrate private desires or perform family rituals. Rather, by institutionalizing public canteens in the workplace, the state tried to control meal time and also change the meaning of social dining itself. This is particularly true in Beijing, which has been the center of national politics and socialist transformation since 1949. Any new form of social dining was unlikely to develop from the previous restaurant sector in Beijing, which consisted primarily of socialist canteens. It is thus not accidental that by 1993 nearly three-quarters of the more than 19,000 eating establishments in Beijing were owned by private entrepreneurs (local and foreign) or were operating as joint ventures.[73] McDonald's and other foreign fast-food restaurants have been appropriated by Beijing consumers as especially attractive social spaces for a new kind of socializing and for the celebration of individuality in public. Moreover, consuming at McDonald's and other foreign fast-food outlets is also a way of embracing modernity and foreign culture in public.

To sum up, there is a close link between the development of fast-food consumption and changes in social structure, especially the emergence of new social groups.[74] The new groups of agents demand the creation of new space for socialization in every aspect of public life, including dining out. Fast-food restaurants provide just such a space for a number of social groups. The new kind of sociality facilitated by fast-food restaurants in turn further stimulates consumers' demands for both the food and the space. Hence the fast-food fever in Beijing during the 1990s.

Notes

1. Liu Fen and Long Zaizu 1996.
2. For a general review, see Messer 1984.
3. See, e.g., Jerome 1980.
4. See Douglas 1975; Lévi-Strauss 1983; and Sahlins 1976.
5. See Harris 1985; Murphy 1986; and Appadurai 1981.
6. Bourdieu 1984, pp. 175–200.

7. Watson 1987. For more systematic studies of food in China, see Chang 1977 and E. Anderson 1988.

8. See Gusfield 1992, p. 80.

9. Finkelstein 1989, pp. 68–71.

10. Ibid. p. 5.

11. Shelton 1990, p. 525.

12. See Fantasia 1995, pp. 213–15.

13. For an anthropological study of sociocultural encounters at McDonald's in Hong Kong, Taipei, Beijing, Seoul, and Tokyo, see chapters in Watson, ed., 1997.

14. See Love 1986, p. 448.

15. See Zhang Yubin 1992.

16. See New York Times, April 24, 1992. For a detailed account, see Yan 1997a.

17. See China Daily, September 12, 1994; and Service Bridge, August 12, 1994.

18. See Liu Ming 1992; Mian Zhi 1993.

19. Duan Gang 1991.

20. Zhang Zhaonan 1992a.

21. See Zhang Zhaonan 1992b; You Zi 1994; and Zhang Guoyun 1995.

22. Yu Bin 1996; "Honggaoliang yuyan zhongshi kuaican da qushi" 1996.

23. Yu Weize 1995.

24. See Liu Fen and Long Zaizu 1996.

25. The development of Chinese fast food is incorporated into the eighth national five-year plan for scientific research. See Bi Yuhua 1994; see also Ling Dawei 1996.

26. For representative views on this issue, see Guo Jianying 1995; Huang Shengbing 1995; Jian Feng 1992; Xiao Hua 1993; Ye Xianning 1993; Yan Zhenguo and Liu Yinsheng 1992a; and Zhong Zhe 1993.

27. Ling Dawei 1995.

28. Mintz 1993, p. 271.

29. Giddens 1984, p. 132.

30. See Sayer 1985, pp. 30–31.

31. See Lechner 1991; Urry 1985.

32. For a comprehensive study of the work-unit system, see Walder 1986.

33. In prereform Beijing even the hotel restaurants and guesthouse canteens were open only during "proper" meal times. So if a visitor missed the meal time, the only alternative was to buy bread and soft drinks from a grocery store.

34. For more details on the results of the survey, see Yan 1997a.

35. See discussions in Xu Chengbei 1994.

36. Zhongguo shipinbao (Chinese food newspaper), November 6, 1991.

37. Jingji ribao, September 15, 1991.

38. Fantasia 1995, p. 219.

39. Ritzer 1993, pp. 4–5.

40. Every time McDonald's opened a new restaurant in the early 1990s, it was featured in the Chinese media. See e.g., Tianjin qingnianbao (Tianjin youth news), June 8, 1994; Shanghai jingji ribao (Shanghai economic news), July 22, 1994; Wenhui bao (Wenhui daily), July 22, 1994. See also Han Shu 1994; Xu Chengbei 1993, p. 3.

41. Xu Chengbei 1992. In fact, I applied to work in a McDonald's outlet in Beijing but was turned down. The manager told me that the recruitment of employees in McDonald's involves a long and strict review process, in order to make sure that the applicants' qualifications are competitive.

42. The relationship between medicine and food has long been an important concern in Chinese culinary culture. See E. Anderson 1988, pp. 53–56.

43. See Yu Bin 1996; and "Honggaoliang yuyan" 1996.

44. Liu Guoyue 1996.

45. See Zhao Huanyan 1995; Xu Wangsheng 1995; Xie Dingyuan 1996; and Tao Wentai 1996.

46. For an excellent account, see Kottak 1978.

47. Shelton 1990, p. 520.

48. Gaige Daobao (Reform herald), no. 1 (1994), p. 34.

49. See Douglas and Isherwood 1979.

50. Stephenson 1989, p. 237.

51. Fantasia 1995, pp. 221–22.

52. For an interesting study of eating etiquette in southern China, see Cooper 1986, pp. 179–84. As mentioned near the beginning of this chapter, Finkelstein offers an interesting and radically different view of existing manners and custom in restaurants. Since manners and behavior patterns are socially constructed and imposed on customers, they make the "restaurant a diorama that emphasizes the aspects of sociality assumed to be the

most valued and attractive" (Finkelstein 1989, p. 52). Accordingly, customers give up their individuality and spontaneity and thus cannot explore their real inner world in this kind of socially constructed spatial context (ibid., pp. 4–17).

53. The 1994 figure comes from my fieldwork; the 1996 figure is taken from Beijing Dashiye Jingji Diaocha Gongsi (Beijing big perspective economic survey company), quoted in "Kuaican zoujin gongxin jieceng" (Fast food is coming closer to salaried groups), *Zhongguo jingyingbao*, June 21, 1996.
54. Interview with General Manager Tim Lai, September 28, 1994.
55. He Yupeng 1996.
56. Ibid. p. 8.
57. See Yan 1997a.
58. He Yupeng 1996, pp. 8–9.
59. See Yan 1997a.
60. According to John Love, when Den Fujita, the founder and owner of McDonald's chain stores in Japan, began introducing McDonald's foods to Japanese customers, particularly the youngsters, he bent the rules by allowing his McDonald's outlets to be a hangout place for teenagers. He decorated one of the early stores with poster-sized pictures of leather-jacketed members of a motorcycle gang "one shade removed" from Hell's Angels. Fujita's experiment horrified the McDonald's chairman when he visited the company's new branches in Japan. See Love 1986, p. 429.
61. Elsewhere I have argued that Chinese society in the 1990s underwent a process of restructuring. The entire Chinese population—not only the peasants—was on the move: some physically, some socially, and some in both ways. An interesting indicator of the increased social mobility and changing patterns of social stratification was the booming business of name-card printing, because so many people changed jobs and titles frequently and quickly. Thus consumption and lifestyle decisions became more important than ever as ways for individuals to define their positions. For more details, see Yan 1994.
62. Yan Zhenguo and Liu Yinshing 1992b.
63. See Xiao Yan 1994.
64. See, e.g., Gao Changli 1992, p. 6; Dong Fang 1994, p. 22.
65. Gu Bingshu 1994.
66. *Beijing wanbao*, January 27, 1993.
67. *Beijing qingnianbao* (Beijing youth daily), December 18, 1993.
68. Pian Ming 1994.
69. For a review of changes in consumption and lifestyles among Chinese youth, see Huang Zhijian 1994.
70. See Gold 1993.
71. On changing family values and household structure, see chapters in Davis and Harrell 1993. For a detailed study of the rising importance of conjugality in rural family life, see Yan 1997b.
72. Watson 1997, p. 19.
73. See *Beijing qingnianbao*, December 18, 1993.
74. See especially Mintz 1994; see also sources cited in notes 2 to 13.

References

Anderson, Eugene. 1988. *The Food of China*. New Haven, Conn.: Yale University Press.

Appadurai, Arjun. 1981. "Gastro-Politics in Hindu South Asia." *American Ethnologist* 8, no. 3: 494–511.

Bi Yuhua. 1994. "Kuaicanye zhengshi chengwei xin de redianhangye" (Fast food officially becoming a new hot sector). *Shichang bao* (Market news), September 19.

Bourdieu, Pierre. 1984. *Distinction: A Social Critique of the Judgement of Taste*. Cambridge, Mass.: Harvard University Press.

Chang, Kwang-chih, ed. 1977. *Food in Chinese Culture: Anthropological and Historical Perspectives*. New Haven, Conn.: Yale University Press.

Cooper, Eugene. 1986. "Chinese Table Manners: You Are How You Eat." *Human Organization* 45: 179–84.

Davis, Deborah, and Stevan Harrell, eds. 1993. *Chinese Families in the Post Mao Era*. Berkeley: University of California Press.

Dong Fang. 1994. "Zhongguo chengshi xiaofei wuda redian" (The five hot points in Chinese urban consumption). *Jingji shijie*, no. 1.

Douglas, Mary. 1975. "Deciphering a Meal." In *Myth, Symbol, and Culture*, edited by Clifford Geertz. New York: W. W. Norton.

Douglas, Mary, and Baron Isherwood. 1979. *The World of Goods*. New York: Basic Books.

Duan Gang. 1991. "Kuaican quanji hanzhan jingcheng" (Fast food chickens are fighting with each other in Beijing). *Beijing Youth Daily*, April 2.

Fantasia, Rick. 1995. "Fast Food in France." *Theory and Society* 24: 201–43.

Finkelstein, Joanne. 1989. *Dining Out: A Sociology of Modern Manners*. New York: New York University Press.

Gaige daobao (Reform herald), no. 1 (1994), p. 34.

Gao Changli. 1992. "Woguo jiushi niandai chengxian duoyuanhua xiaofei qushi". (Consumption trends are diversified in China during the 1990s). *Shangpin pingjie* (Review of commodities), no. 10.

Giddens, Anthony. 1984. *The Constitution of Society: Outline of the Theory of Structuration*. Berkeley: University of California Press.

Gold, Thomas B. 1993. "Go with Your Feelings: Hong Kong and Taiwan Popular Culture in Greater China." *China Quarterly* 136 (December): 907–25.

Gu Bingshu. 1994. "Waican: dushi xin shishang" (Eating out: a new fashion in cities), *Xiaofeizhe*, no. 3: 14–15.

Guo Jianying. 1995. "Tantan kuaican de fuwu" (On service in the fast food sector) *Fuwu jingji* (Service economy), no. 2: 27–8.

Gusfield, Joseph. 1992. "Nature's Body and the Metaphors of Food." In *Cultivating Differences: Symbolic Boundaries and the Making of Inequality*, edited by Michele Lamont and Marcel Fournier. Chicago: University of Chicago Press.

Han Shu. 1994. "M: changsheng jiangjun" (M [McDonald's]: the undefeated general). *Xiaofei zhinan* (Consumption guide), no. 2: 10–11.

Harris, Marvin. 1985. *Good to Eat: Riddles of Food and Culture*. New York: Simon and Schuster.

He Yupeng. 1996. "Zuowei nuxin ripin de Beijing maidanglao" (McDonald's as feminine food in Beijing). Paper presented at the conference "Changing Diet and Foodways in Chinese Culture," Chinese University of Hong Kong, Hong Kong, June 12–14.

"Honggaoliang yuyan zhongshi kuaican da qushi" (The red sorghum predicts the trend of Chinese fast food). 1996. *Zhongguo jingyingbao* (Chinese business). June 11.

Huang Shengbing. 1995. "Kuaican xiaofie xingwei de bijiao yanjiu" (Comparative study of fast food consumption behavior). *Xiaofei jingji*, no. 5: 33–4.

Huang Zhijian. 1994. "Yi ge juda de qingnian xiaofei shichang" (A huge market of youth consumers). *Zhongguo qingnina yanjiu* (China youth studies), no. 2: 12–16.

Jerome, N. W., ed. 1980. *Nutritional Anthropology: Contemporary Approaches to Diet and Culture*. New York: Redgrave.

Jian Feng. 1992. "Zhongshi kuaican, na chu ni de mingpai" (Chinese fast food, show your best brand). *Shichang bao*, November 10.

Kottak, Conrad. 1978. "Rituals at McDonald's." *Journal of American Culture* 1: 370–86.

Lechner, Frank. 1991. "Simmel on Social Space." *Theory, Culture and Society* 8: 195–201.

Lévi-Strauss, Claude. 1983. *The Raw and the Cooked*. Chicago: University of Chicago Press.

Ling Dawei. 1996. "Nuli ba zhongshi kuaican gao shang qu" (Endeavor to develop Chinese fast food). *Zhongguo pengren*, no. 6: 4–5.

——. 1995. "Zhongxi kuaican jingzheng zhi wo jian" (My views on the competition between Chinese and Western fast foods). *Xinshiji zhoukan* (New century weekly), November.

Liu Fen and Long Zaizu. 1986. "Jing, jin, hu chi shenmo?" (What are people eating in Beijing, Tianjin, and Shanghai?). *People's Daily* (overseas edition), August 9.

Liu Guoyue. 1996. "Shenzhen kuaican shichang jiqi fazhan" (The fast food market in Shenzhen and its development). *Zhongguo pengren*, no. 8: 20–2.

Liu Ming. 1992. "Guoji kuaicancheng de meili" (The charming international fast food city). *Zhongguo shipinbao* (Chinese food newspaper), July 13.

Love, John. 1986. *McDonald's: Behind the Arches*. New York: Bantam Books.

Messer, Ellen. 1984. "Anthropological Perspectives on Diet." *Annual Review of Anthropology* 13: 205–49.

Mian Zhi. 1993. "Xishi kuaican fengmi jingcheng; zhongshi kuaican zemmaban?" (Western fast food is the fashion; what about Chinese fast food?). *Lianai, hunyin, jiating* (Love, marriage, and family), no. 6: 10–11.

Mintz, Sidney. 1994. "The Changing Role of Food in the Study of Consumption." In *Consumption and the World of Goods*, edited by John Brewer and Roy Porter. London: Routledge.

Murphy, Christopher. 1986. "Piety and Honor: The Meaning of Muslim Feasts in Old Delhi." In *Food, Society, and Culture*, edited by R. S. Khare and M. S. A. Rao. Durham, NC: Carolina Academic Press.

Pian Ming. 1994. "Jingcheng qingnian qingxin gaojia shangpin" (Beijing youth are keen on expensive commodities). *Zhonghua gongshang shibao* (China industrial and commercial times), July 16.

Ritzer, George. 1993. *The McDonaldization of Society*. Newbury Park, Calif.: Pine Forge.

Sahlins, Marshall. 1976. *Culture and Practical Reason*. Chicago: University of Chicago Press.

Sayer, Andrew. 1985. "The Difference That Space Makes." In *Social Relations and Spatial Structures*, edited by Derrek Gregory and John Urry. London: Macmillan.

Shelton, Allen. 1990. "A Theater for Eating, Looking, and Thinking: The Restaurant as Symbolic Space." *Sociological Spectrum* 10: 507–26.

Stephenson, Peter. 1989. "Going to McDonald's in Leiden: Reflections on the Concept of Self and Society in the Netherlands." *Ethos* 17, no. 2: 226–47.

Tao Wentai. 1996. "Guanyu zhongwai yinshi wenhua bijiao de ji ge wenti" (Issues in comparing Chinese and foreign culinary cultures). *Zhongguo pengren*, no. 8: 26–8.

Urry, John. 1985. "Social Relations, Space and Time." In *Social Relations and Spatial Structures*, edited by Derek Gregory and John Urry. London: Macmillan.

Walder, Andrew G. 1986. *Communist Neo-Traditionalism: Work and Authority in Chinese Industry*. Berkeley: University of California Press.

Watson, James. 1997. "Introduction: Transnationalism, Localization, and Fast Foods in East Asia." In *Golden Arches East: McDonald's in East Asia*, edited by James Watson. Stanford, Calif.: Stanford University Press.

——. 1987. "From the Common Pot: Feasting with Equals in Chinese Society." *Anthropos* 82: 389–401.

——, ed. 1997. *Golden Arches East: McDonald's in East Asia*. Stanford University Press.

Xiao Hua. 1993. "Da ru Zhongguo de yangkuaican" (The invasion of Western fast food), *Jianting shenghuo zhinan* (Guidance of family life), no. 5.

Xiao Yan. 1994. "Xiaofei guannian xin qingxie" (New orientations of consumption perception). *Zhongguo xiaofeizhe bao* (China consumer news), September 12.

Xie Dingyuan. 1996. "Pengren wangguo mianlin tiaozhan" (The kingdom of cuisine is facing a challenge). *Zhongguo pengren*, no. 2: 27–9.

Xu Chengbei. 1994. "Kuaican, dacai, yu xinlao zihao" (Fast food, formal dishes, and the new and old restaurants). *Jingji ribao* (Economic daily), September 17.

——. 1993. "Cong Maidanglao kan shijie" (Seeing the world from McDonald's). *Zhongguo pengren*, no. 8.

——. 1992. "Maidanglao de faluu" (McDonald's law). *Fazhi ribao* (Legal system daily), September 9.

Xu Wangsheng. 1995. "Zhongxi yinshi wenhua de qubie" (The differences between Chinese and Western culinary cultures). *Zhongguo pengren*, no. 8: 28–30.

Yan, Yunxiang. 1997a. "McDonald's in Beijing: The Localization of Americana." In *Golden Arches East: McDonald's in East Asia*, edited by James Watson. Stanford, Calif.: Stanford University Press.

——. 1997b. "The Triumph of Conjugality: Structural Transformation of Family Relations in a Chinese Village." *Ethnology* 36, no. 3: 191–212.

——. 1994. "Dislocation, Reposition and Restratification: Structural Changes in Chinese Society." In *China Review 1994*, edited by Maurice Brosseau and Lo Chi Kin. Hong Kong: Chinese University Press.

Yan Zhenguo and Liu Yinsheng. 1992a. "Yangkuaican chongjipo hou de chensi" (Pondering thoughts after the shock wave of Western fast food). *Shoudu jingji xinxibao* (Capital economic information news), December 3.

——. 1992b. "Zhongguo kuaican shichang shu zhu chenfu?" (Who will control the fast food market in China?). *Shoudu jingji xinxibao*, December 8.

Ye Xianning. 1993. "Jingcheng kuaican yi pie" (An overview of fast food in Beijing), *Fuwu jingji*, no. 4.

You Zi. 1994. "Jingcheng zhongshi kuaican re qi lai le" (Chinese fast food has become hot in Beijing). *Jingji shijie* (Economic world), no. 6: 60–1.

Yu Bin. 1996. "Zhongwai kuaican zai jingcheng" (Chinese and foreign fast foods in Beijing). *Zhongguo shangbao* (Chinese commercial news), June 20.

Yu Weize. 1995. "Shanghai Ronghuaji kuaican liansuo de jingying zouxiang" (The management directions of the Shanghai Ronghua Chicken fast food chain). *Zhongguo pengren* (Chinese culinary art), no. 9: 19–20.

Zhang Guoyun. 1995. "Zhongshi kuaican, dengni dao chuntian" (Chinese fast food, waiting for you in the spring). *Xiaofei jingji* (Consumer economy), no. 3: 54–6.

Zhang Yubin. 1992. "Xishi kuaican jishi lu" (Inspirations from Western fast food) *Zhongguo shipinbao* (Chinese food newspaper), September 4.

Zhang Zhaonan. 1992a. "Gan yu yangfan bi gaodi" (Dare to challenge the foreign fast food). *Beijing wanbao* (Beijing evening news), September 13.

——. 1992b. "Kuaicanye kai jin Zhongguo budui" (The "Chinese troops" entering the fast food sector). *Beijing wanbao*, October 9.

Zhao Huanyan. 1995. "Shilun fandian yingxiao zhong de wenhua dingwei" (On the cultural position of the restaurant business). *Fuwu jingji* (Service economy), no. 8: 10–11.

Zhong Zhe. 1993. "Meishi kuaican – gongfu zai shi wai" (American fast food – something beyond foods). *Xiaofeizhe* (Consumers), no. 2.

33

"Plastic-bag Housewives" and Postmodern Restaurants?: Public and Private in Bangkok's Foodscape [1]

Gisèle Yasmeen

Abstract: This paper provides evidence of alternative urban food strategies challenging ethnocentric public/private models of urban space that have their roots in a particular historical experience of the Western city. This will be achieved by painting a brief "foodscape" of the sprawling, dynamic metropolis of Bangkok. Specific attention will be devoted to the nature of gender relations in this urban food system, particularly with respect to the culture of public eating. I will argue that, traditionally, women and men in Bangkok occupy different sociospatial positions in the Bangkok food system and that, more recently, femininity and masculinity are being spatially re-orchestrated in light of "postmodern" developments. The first example will detail the arrival of the "plastic-bag housewife," who is instrumental in blurring public and private space in Bangkok. Quotidian, mass-based food strategies will be compared and contrasted with élite food establishments of recent vintage in the city. Many of these have a postmodern quality, blending nostalgic architectural and design *pastiche* with the contemporary restaurant. "Food gardens" and restaurants in an ostensibly "royal style" represent and promote specific and new forms of gender relations. The place of women within the constellation of these relations does not reflect the traditional economic and social prominence of women in the community. Instead, élite establishments either display women publicly as aesthetic objects or capitalize on their talents in the "back" areas of the kitchen. Thai men, who are a minority of small foodshop owners, are involved in key roles in the larger, more profitable food establishments. The nature of "public" and "private" in Bangkok civil society, as well as the postmodern blurring of points along the continuum, cannot be separated from the locally specific, dynamic Thai constructions and practices of gender.

Introduction

My general objective in this paper is to explore urban eating habits from a feminist perspective, which, by definition, focuses on gendered patterns of activity and how these reflect distributions of power in society. The specific purpose of this paper is to characterize and interpret the "restaurant culture" and related "public" eating habits of Bangkok residents in relation to larger patterns of social change and concomitant gendered identities and practices. The geography of Thai eating habits, or foodways,[2] illustrates the culturally specific construction and practice of "public" and "private" and the postmodern blurring of these binary categories both empirically and analytically. This can be illustrated by profiling ubiquitous and inexpensive "foodshops," or small restaurants and stalls, as well as idiosyncratic élite restaurants.

In Thailand and much of Southeast Asia, women have access to the streets and lanes of the city as consumers and producers of prepared food. This "access" is relative to that of their counterparts in South and East Asia, where women are customarily confined to the home, particularly in urban areas (Sheridan and Salaff, 1984; Papanek and Minault, 1982). Based on ethnographic, survey, and statistical research, this paper will examine notions of public and private in Bangkok through the lens of eating habits. The daily food strategies in operation in Bangkok blur the boundaries between private and public space and the gendered identities associated with those spaces. The appearance of restaurants with pastiche styles creating *simulacra* (Baudrillard, 1981) that help forge new urban identities will illustrate the non-Western experience of postmodernity. The élite *foodscape* of Bangkok illustrates a range of contentious issues of power and gender in relation to postmodern forms of consumption in a Southeast Asian context.

Unique Thai styles of production and consumption of prepared food are imbricated in wider socio-spatial processes evident at regional and international scales. In the case of convenience foods, the Bangkok strategy remains primarily labor intensive and involves large numbers of female micro-entrepreneurs catering to the proverbial "plastic-bag housewife." This nameless and faceless woman is a newcomer to the Thai urban scene; she is seen frequenting small "foodshops"[3] as a takeout customer who brings home the family meal in a small plastic bag. At the other end of the spectrum are the large postmodern garden restaurants (*suan ahaan*), which are helping redefine Thai middle-class identity, whereas the most distinctive establishments are very expensive restaurants ostensibly in the "royal style."

Several years ago, one might have asserted that little had been written on issues related to the food habits of city dwellers (Konvitz, 1987). Recent studies have begun to explore the significance of food for geographical inquiry at regional, national, and international scales (Cook, 1994, 1995; Wagner, 1994). Traditionally, Simoons is perhaps the best-known geographer of food (1961, 1991). Other geographers have studied urban food distribution within the framework of food-systems analysis (Watts, 1983; Armstrong and McGee, 1985; Drakakis-Smith, 1990). The preparation and consumption of food are spatially defined and defining activities that should have greater appeal to geographers in general and urbanists in particular.

Likewise, feminist geographers and urbanists have tended to disregard issues related to food (see Bowlby, 1988) despite the fact that, in the words of Sarah Sargeant, "food is an agonizingly feminist issue" (Sargeant, 1985, p. 150). It is unnecessary to go into detail here about the gender division of labor with regard to food preparation and related tasks, such as shopping and cleaning, as this has been addressed in detail elsewhere (Giard and Mayol, 1980; Charles and Kerr, 1988). Trans-historically and cross-culturally, it appears that the provisioning and preparing of food for household consumption traditionally has been defined as women's work and has been associated with socially constructed femininity. However, in much of Southeast Asia, even if women still tend to dominate the activity, food-related work is more flexibly the domain of both sexes (Hauck, Saovanee, and Hanks, 1958; Klopfer, 1993; McKay, 1994).

This one important aspect of so-called "reproductive work" used to be characterized as undervalued, usually unpaid, and typically located in the "private sphere" (Rosaldo and Lamphere, 1974; Elshtain, 1981). Feminist geographers' conceptualizations of public and private often are drawn from the more general social theorists who have dealt with the subject, particularly Habermas, as advanced in *The Structural Transformation of the Public Sphere* (1989). However, as a recent critic remarked, Habermas's mediatized view of public and private may not be as useful as Arendt's more geographical depiction of private and public spaces (Benhabib, 1992; Howell, 1993).

The public and private model of gendered urban space has been the subject of considerable debate, with feminist revisionist historians offering plenty of examples of women's involvement in activities outside the home (Hayden, 1981; Ryan, 1990; Davidoff, 1995). Other critiques emanating from a post-colonialist perspective assert that notions of public and private are based largely on the historical experiences of white, Western, middle-class society and the ideals that conditioned this experience (hooks, 1984, 1990; Mohanty et al., 1991). We need to distinguish between activities classified as either "public" or "private" and the spaces uncritically associated with these activities. Political organizing often has taken place in "private" kitchens, and nurturing activities, such as child-rearing, often are conducted in what are thought of as very public places.

Bondi (1992) suggests that we refer to the private and public spheres as ideological constructs rather than empirical categories. Moore-Milroy and Wismer (1994) take a similar position by critiquing the essentialism of sphere models, but maintain that the public/private framework has allegorical utility. "Postmodern" and "poststructuralist" positions question the validity of all binary models and suggest that we should find conceptual replacements for these and other enlightenment categorizations, which is a laudable and challenging goal (see Nicholson, 1990; Davidoff, 1995). It is hoped that this paper will contribute to the realization of this goal.

Gender and Foodways

By now, it should be quite clear to geographers and those in the environmental disciplines that work patterns that occur through space and indeed places themselves are gendered. Gender is a social construction and a dynamic practice, a performance, rather than a simple biological category (cf. Butler, 1990; Spivak, 1992; Rose, 1993; England, 1994).

Food, and the institutions and people that govern its distribution and consumption, is a medium through which social power relations are expressed (cf. Douglas, 1984; Walker, 1991; Van Esterik, 1992b). The issue at stake is to explore not so much what food is being eaten but how and *where* it is consumed and in what political and economic circumstances. This is the contribution of a geographical approach to "foodways" or the complex set of rules and activities guiding the quest for human nourishment. I have proposed the term "foodscape" to emphasize the spatialization of foodways and the interconnections between people, food, and places. "Foodscape," drawn from "landscape," is a term used to describe a process of viewing place in which food is used as a lens to bring into focus selected human relations.

The Case of Bangkok

A city like Bangkok presents the researcher with a challenging environment in which to study issues of gender, food, and place. A stereotypical, outdated, but touristically important image of the city is of market women selling produce early in the morning on the canals of the city, creating a "floating market." Otherwise, women and men are immediately noticed on the sidewalks, lanes, and markets of the metropolis selling prepared food around the clock. What immediately confronts the foreign observer is the public presence of women as agents in small-scale commerce, as in West Africa, the Caribbean, and parts of Central America. Historically, outsiders sojourning in the region have commented on the contrast this situation presents to those familiar with Indian, Chinese, and traditional European societies (Crawfurd, 1820; also see Manderson, 1983). Travellers were astounded to find women physically present and active in nearly every sphere of existence rather than cloistered within the home.

> The social position of the Thai peasant woman is powerful: she has long had a voice in village governmental affairs; she often represents her household at village meetings when her husband cannot attend; she almost always does the buying and selling in the local markets. (It is so unusual for a Thai male to do this that it elicits comment if he does.) Through their marketing activities Thai farm women produce a sizeable portion of the family cash income, and they not only handle the household money, but usually act as the family treasurer and hold the purse strings (de Young, 1955, p. 24).

Women's control of local commerce is based on their reputation for being more financially and socially responsible than men, at least at the household level.

The Status of Women in Southeast Asia

Traditionally, then, women in Thailand and most of Southeast Asia in general have been associated with economic and commercial involvement, whereas men have dominated the political-bureaucratic and religious spheres (Kirsch, 1982, Van Esterik, 1982). This state of affairs, combined with patterns of matrilinearity and bilateral inheritance, has resulted in considerable debate surrounding the ostensibly "high" status of women is Southeast Asia (Khin, 1980; Kirsch, 1982; Keyes, 1984; Ong, 1991; Van Esterik, 1992a). This paper will not resolve these debates. My position is that women occupy a *contradictory* position in Thailand and Southeast Asia, as they often are financially independent and mobile, but the avenues to positions of power in spheres such as religion, politics, big business, and the military are blocked to them. Southeast Asian languages, childrearing practices, and the general sexual division of labor all point to a high degree of flexibility. The Southeast Asian sex-gender system is less physically and socially confining to women than are traditional ideologies and practices in other regions of the world.

One of the most widespread commercial activities of women in the region today is the selling of prepared food on a small scale. Indeed, over 80% of food vendors and employees of small foodshops are women. These micro-enterprises have been interpreted as extensions of the household economy. Most Southeast Asian women are expected to contribute substantially to household income, which is a partial explanation of their entrepreneurial activities. Thai femininity, like that of the Filipinas, includes acting as economic provider. Thai women are then "private" housekeepers and childrearers as well as "public" income earners. Bangkok's food-scape, where cooking and eating often take place outside the home, is a reminder of the haphazard way in which ostensibly public and private activities and spaces are identified together.

Female labor force participation rates in Thailand are the highest in Southeast Asia, which is a region already characterized as one where large numbers of women engage in remunerated work (Figure 33.1). Women's employment has stimulated the demand for convenient, inexpensive alternatives to home-cooked meals. The wide-spread sale of prepared food also is linked to the prevalence of kitchenless apartments built over the past 15 years (Wathana, 1990) and the labor-intensive nature of much Thai cooking, particularly curries and fermented dishes.

While Western industrialized nations and Japan have turned overwhelmingly toward the development of a capital-intensive food system involving the massive use of labor-saving household devices and convenience foods, the Bangkok food system still is primarily labor intensive, although increased commercialization certainly is apparent. This takes the form of packaged noodles and the use of refrigerators and microwaves by the middle classes (*The Economist*, December 4, 1993, special insert). A conversation between "Uncle," the 60-year-old owner of a small shop selling duck

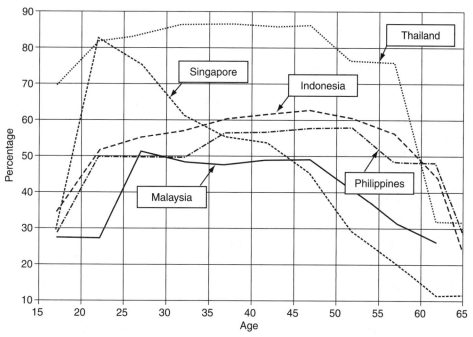

Figure 33.1 Female labor force participation rates in percent, Southeast Asia, 1990s (after ILO, 1993, *Labor Force Statistics*).

noodles, and one of his customers, recorded during my field work, illustrates one perception of the phenomenon.[4]

> Customer: Now there are microwaves, people with money can buy them.
> Uncle: But microwaves can only re-heat the food, isn't that right?
> Customer: No, you can cook in that machine. When you put the food inside, it cooks!
> Ordinary people can't buy it because it's expensive. Only the rich can have it.

The very definition of "prepared food" is becoming rather tenuous. Recently, many supermarkets have begun selling semi-prepared food, ready to be cooked in the microwave. This trend is said to be very popular among members of the middle classes.

Mass-Based Food Strategies

Today, as in most Southeast Asian cities, Bangkok residents spend 50% of their monthly food budget on dishes prepared outside the home, most of this being consumed away from home rather than "taken out" (Table 33.1). Indeed, the 1990 *Food Consumption Survey* (FCS) indicates that 11% of Bangkok residents *never* cook at home (Walker, 1990). The trend of purchasing prepared food away from home began in the post-World War II period and has grown significantly in the last 20 years, with large numbers of women entering the paid urban workforce (Van Esterik, 1992a).

A "traditional" strategy common throughout Southeast Asia is subscription to neighborhood catering networks in which food, normally one soup, one vegetable, and one meat dish (often a curry), is delivered at a regular time every day in a "tiffin," or tiered lunch kit (Thai—*phinto*). The tiffin-network strategy is being eclipsed by the small foodshop sector, where food is available anywhere, anytime—an important attribute in a city where traffic is gridlocked during rush hours. Women can be seen stopping at a foodshop in the evenings on their way home from work to pick up dinner for themselves and/or the family. Main courses are placed in small plastic bags, with rice being prepared easily at home in a rice cooker. Bangkok residents hence refer to *mae baan tung plastic*, or "plastic-bag housewives" (Van Esterik, 1992a). Obviously, however, these women are not solely "housewives," as they are

Table 33.1 Average monthly expenditures per household by type of food consumed (in Thai *baht*)[a].

Type of food consumed	Entire Kingdom	Greater Bangkok
Food prepared at home	1,494 (76.0%)	1,616 (52.3%)
Prepared food taken home	173 (8.8%)	457 (14.8%)
Food eaten away from home[b]	300 (15.2%)	1,014 (32.9%)

[a] One U.S. dollar equals approximately 20 baht; percentages of total in category given in parentheses.
[b] Excludes alcoholic drinks away from home.
Source: NSO, 1991.

engaged in remunerative employment. The following vignette illustrates the case of one informant I met in Bangkok.

Ajaan (Professor) Prinyathip teaches at a university in central Bangkok and is married to an engineer. She has no children and lives in a housing estate in a suburban area of the city. At lunch time, if she doesn't have time to go to the faculty cafeteria, she asks the janitor to bring lunch back to her office. Every day, on her way home from work, she stops at a small roadside curry shop and picks up supper—usually a curry and a vegetable dish or a soup. Since she drives a car, she often frequents one of the shopping centers on the way home where she can park her car and purchase from a large selection of take-home food in small plastic bags on the "food floor." When asked whether she ever sits down to eat in roadside foodshops or stalls, she answers, "Never, not with the heat, dust, and noise . . . it's so unpleasant." If she's going to spend time eating out, Ajaan Prinyathip would rather go to a nice restaurant with air conditioning and beautiful surroundings.[5]

Eating in stalls is associated with unpleasant environmental conditions that can be avoided in a quiet, air-conditioned, middle-class home. Thai professionals, such as Ajaan Prinyathip, tend to be very well dressed, in high-heels and tailored suits, and would rather not deal with the discomfort of eating in a foodstall. Furthermore, pollution and the noise of traffic makes the foodshop experience less aesthetically pleasing and a health hazard. As Walker's FCS (1990) clearly demonstrated, the ideal locus of everyday family commensality is the home. Special occasions, however, merit an outing to a "special" restaurant, funds permitting.

The urban masses are relegated to purchasing food on the streets and *soi* (side-streets or lanes) from vendors both mobile and stationary, and from small food-shops specializing in noodles, curried dishes, or other fare (cf. Yasmeen, 1992).[6] The FCS confirms this observation by indicating that noodle shops and small Thai "restaurants" are the most highly patronized eating establishments in Bangkok, with 91% and 84%, respectively, of the 4,198 respondents reporting that they frequent these places (Walker, 1990, p. 6) (see Table 33.2).

Small restaurants actually are quite diverse in terms of the food that is sold, their access to clientèle, and the types of functions that they perform in the city, and they cannot, therefore, be classified as in many studies that focus on the so-called

Table 33.2 Most highly patronized eating establishments in Bangkok.

Type of establishment	Respondents (%)[a]
Noodle shops	91
Thai restaurants	84
Garden restaurants	65
Regional Thai restaurants	63
Chinese restaurants	47
Western fast food	44
Western restaurants	27
Others	3

[a] Indicates percentage of all those responding that reported patronizing the particular type of establishment.
Source: Walker, 1990.

"informal sector" or "street foods" (McGee, 1971; Napat and Szanton, 1986; Tinker, 1987; FAO, 1988; Amin, 1991). Small foodshops are finding niches in shopping centers, educational institutions, and office complexes, where they are bridging the divide between kinship-based informal enterprise and other types of commerce (Yasmeen, 1995). The next section will profile how these types of arrangements also are blurring boundaries between what are customarily distinguished as public and private spheres.

Blurring Boundaries: Everyday Public Eating

In Bangkok, women, and some men, *contract out* to mostly women food producers by subscribing to neighborhood tiffin networks or, more commonly, by purchasing prepared food from curry shops for the evening meal. Lunch typically is purchased at or near work, often in a Chinese noodle shop or other small, inexpensive eating establishment. Even the early-morning offering of food for monks often is purchased from a neighborhood vendor or shop by urban working women. If food associated with sacredness is being commercialized, there are plenty of illustrations of profanity in the foodscape. Late-night food stalls cluster near the ubiquitous and infamous massage parlors and bars, thereby establishing an "intervening opportunity" for business. Likewise, many restaurants are fronts for the commercialization of sexual services. Any study of the food system or description of the foodscape cannot therefore be separated from other aspects of social relations in the city.

Bangkok therefore houses ubiquitous small food shops that act as a life-support system for many urbanites. Small restaurants serve a number of latent social functions in addition to providing meals. For example, children often are cared for in these environments, and young people help cook and clean, thereby learning skills and meeting others. Information on local affairs can be gleaned in these spaces. These small eating establishments can be interpreted as realms of femininity, where women are employed in and, to a great degree, control their micro-enterprises. Small food shops are "everyday" spaces for the majority of urbanites, whereas larger establishments cater to the "occasion" (Walker, 1991).

The smallest food establishments can be classified as semi-public/private spaces. Here, behavior associated with the home is relocated to commercial spaces. Local residents are sometimes seen sitting in foodshops near their homes in their pajamas having breakfast. This resembles the observations of de Certeau's students in Lyon (Giard and Mayol, 1980). Housewives in the *Croix-Rousses* quarter were seen emerging into the neighborhood bakeries in the early morning clad in housecoats, slippers, and curling rods. The bakery was seen as an extension of the home and the private sphere, and gendered patterns of activity were commensurate with this view. In many foodshops, regular customers sometimes prepare food themselves, wash their own dishes, and even help serve the clientèle. Regular customers sometimes are originally from the same village as the shopowner, or may be kin. The foodshop, then, is both a private homespace and a public commercial place. These neighborhood eateries are examples of the blurring of the socio-spatial boundaries of public and private, extensions of the home sphere, and the domestication of public space.

Traditional cooking activities and kinship networks are re-created within a commercial establishment. Bangkok's small foodshops are instrumental in establishing contemporary Thai public life, similar to the roles played by pubs and coffee houses in industrializing Europe.

> Habermas argues that . . . it was the growth of an *urban* culture—of meeting houses, concert halls, opera houses, press and publishing ventures, coffee houses, taverns and clubs, and the like—. . . which represents the expansion of the public sphere (Howell, 1993, p. 310).

Small foodshops are products of urbanization and industrialization and concomitant social change but, at the same time, reproduce traditional social relations. As such, they represent the simultaneous modernization and postmodernization of Thai urban society.

Similarly, foodshops that locate in "modern" environments such as shopping center foodcourts, office buildings, and educational institutions are instrumental in reorchestrating spatial relations. Again, mostly women micro-entrepreneurs and small restaurant owners enter into a contract with larger-scale commerce, but often continue to rely on unpaid family labor and cook the same types of food. Differences revolve around price, payment arrangements (which sometimes involve the use of a coupon system), and more formalized relations with customers who are not "regulars" making themselves "at home," as is the case in many *soi*-based foodshops. Eating in a food center is a more expensive proposition than eating on the street and takes place in a highly controlled environment replete with security, cameras, air conditioning, and music. Whereas street- and *soi*-based foodshop owners are constrained by property owners and the police, who condition access to space, those who locate in privately owned indoor environments are subject to the rules and regulations of the authorities who own the property. Both are operating in a competitive environment where the market's whims play a decisive role. Micro-entrepreneurs and small restaurant owners, much like their counterparts elsewhere, have relatively little power compared to the agency exercised by state officials, property owners, and large-scale commerce (Jellinek, 1991). This is certainly not to imply that they are without power, but that their agency is limited by the lack of economic and political resources.

Foodshop outlets in shopping centers represent a transitional space between the "mass-based" strategy associated with the small neighbourhood stall and larger formal restaurants that cater to the wealthier middle classes. The next part of this paper will profile élite eating establishments and contrast them with female-dominated small foodshops.

Bangkok's Elite Foodscape

Thai economist Pasuk Pongpaichit recalled growing up in Bangkok in the late 1960s when there were only a handful of "proper" restaurants where one would go as a family on a special occasion (Pasuk Pongpaichit, pers. comm., 1993). Most of these establishments were Chinese and some, such as *Somboon Pattakarn* and *See Faa* remain successful today. *Somboon Pattakarn* is a famous Chinese restaurant near the boxing stadium. *Pattakarn* is a formal Thai word for "restaurant," traditionally used to refer to a grand Chinese establishment where banquet fare is served, although the use of the term has expanded to include any formal eating establishment (see note 3). *See Faa* is now a chain with locations all over the city. It specializes in the serving of *bami*—a thick Chinese egg noodle (Van Esterik, 1992a). Both establishments are popular with university faculty. Since the 1970s, formal establishments modeled on the Chinese and Western traditions of restaurants catering to the special occasion have emerged and specialize in Thai cuisine as well as other specialties.

As documented by Van Esterik and Walker, there is a fairly recent development of hundreds of open-air mega-restaurants located in the suburbs, referred to generically as "food gardens" (*suan ahaan*). These are often designed in a *sala thai*, or open pavilion style. The use of this style is reminiscent of traditional Thai architecture, yet, as a recent invention, they enclose "untraditional" activities. The world's largest restaurant, *Tum Nuk Thai*, which has a staff of 1,000, many on roller skates, is the most extreme example of this phenomenon of Thai restaurant giantism. The restaurant grounds include *khlong* or canals, replete with fish that can be fed by customers who purchase bread from a vendor. Thai classical and folk dances are staged at regular intervals to taped music and there are souvenir shops located at the exit. Most food gardens are slightly less spectacular and are establishments catering to Thai middle-class families and groups rather than tourists. They generally are decorated with bright lights and many plants or trees. Fresh fish and seafood are on display in cases at the entrance for customers to examine and choose.

Discussions of postmodern architecture and design cite the use of vaguely traditional architectural and lifestyle references as a key element, often resulting in the production of "depthless images," or *simulacra* (Baudrillard, 1981; Jameson, 1983; Dear, 1993). Simulacra are fanciful "re-creations" of a past that never *really* existed, much like "Main Street USA" in Disneyland or "Europa Boulevard" in the West Edmonton Mall (Hopkins, 1990; Warren, 1993). The postmodern experience also is framed by the use of *pastiche*, or exaggerated representations that mock the original form being alluded to.

> Postmodern design aims explicitly at expressing specific localities and their history and tradition.... This use of past styles, which is simultaneous with a tendency to erase style as a consistent and distinctive set of features, incorporates a certain nostalgia and leads to a kind of pastiche, revealing, maybe, innovative and unexpected combinations. ... It goes without saying that all these aspects of postmodern design point to some different conception and experience of space on the part of its producers, while also calling its consumers to participate in this new experience (Lagopoulos, 1993, p. 260).

Patrons of suburban garden restaurants are participating in a new experience of place inspired by traditional Thai spatial design. This phenomenon is reproduced elsewhere as well. For example, a well-known housing development in central Bangkok, *Suan Parichat*, is designed in a "traditional" Thai style, with all the modern conveniences of air conditioning and indoor plumbing. The houses have sloping roofs and have floors made of polished teakwood. Mini-canals are found throughout the walled and guarded compound for the aesthetic enjoyment of residents and guests, but are not used for transportation, fishing, or dumping waste, which are the practical uses of canals in central Thailand. Traditional images, void of function and content, therefore are used to add symbolic capital to both restaurants and other new spaces in Thai society. These spaces of food consumption and leisure are instrumental in the creation of a new Thai middle-class aesthetic, lifestyle, and identity. This identity is one that revolves around the accumulation of wealth and status, worldliness, and a reconstruction of Thainess that emphasizes the traditional arts such as cooking, classical dancing, and certain religious images—for example, spirit houses (cf. Askew, 1994; Walker, 1991).

An Emerging Gender Division of Labor

Department of Labour statistics indicate that as an eating establishment, or restaurant, becomes larger and more lucrative, men are involved in greater proportions (Table 33.3). A closer examination reveals that a gender division of labor emerges in the larger establishments, whereby men or boys (and sometimes girls) will work as waiters, whereas women work as cooks or chefs. Some of the most well-known Thai restaurants in the city hire only women cooks. In the "front," however, it is women and girls who act as hostesses and entertainers in the larger, more expensive establishments.

One rather stylish example is *Bussaracum. Bussaracum* markets itself as a venue serving *authentic* Thai food prepared in the royal style. Walker (1991) outlines the debates surrounding this claim. She concludes that the restaurant is geared primarily toward a Western clientèle because the menu breaks with the conventions of Thai gastronomy. The decorations are lavish and reminiscent of fine palace architecture and ornamentation. Women, especially, are a crucial part of the decor, costumed in silk sarongs and sashes. Thais like to invite foreign guests to one of the locations of *Bussaracum* in order to give their visitors a distinctly Thai "experience."

Table 33.3 Number of employees in eating/drinking establishments by sex and size of establishment, metropolitan Bangkok[a].

Size of establishment	Female employees	Male employees
Total	30,321 (61)	19,239 (39)
1–4 employees	7,319 (83)	1,542 (17)
5–9 employees	3,858 (64)	2,131 (36)
10–19 employees	4,741 (57)	3,552 (43)
20–49 employees	7,016 (58)	5,159 (42)
50–99 employees	3,676 (52)	3,353 (48)
100–299 employees	2,975 (51)	2,916 (49)
300–499 employees	736 (56)	586 (44)

[a] Percentage shares of total employees indicated.
Source: Thailand Ministry of Labour, 1991.

Women in the larger establishments, such as *Bussaracum* and *Tum Nuk Thai*, are being more firmly recast as nurturers and pleasure-giving objects. Their historical capacity as entrepreneurs is being overlooked. An extreme example of the specialization of women's roles in food establishments are the "no hands" restaurants, where young women employees spoon- and hand-feed adult male customers. One might write a fascinating exposé of these sorts of food places from a feminist psychoanalytical perspective.

Likewise, Thai men's greater involvement in the restaurant sector, particularly large, lucrative eating establishments, initially appears to be a shift from the ascetic ideals of masculinity outlined by such anthropologists as Keyes (1984, 1986) and Kirsch (1985). Thai, as opposed to Chinese or Sino-Thai, men have not, until recently, played important roles in commerce, presumably because of the low status of such worldly occupations in the Theravada Buddhist value system (cf. Khin, 1980; Kirsch, 1982). Further scrutiny suggests that the idea of asceticism may be spurious, since the law of Karma (Thai—*kamma*) stipulates that one is born into a higher station in subsequent lives following the accumulation of merit. The wealthy and powerful are by association viewed as spiritually superior. Walker (1991) indicates quite clearly that the desire of middle-class Thai to acquire the trappings of aristocracy could, in fact, be related to the Buddhist influence rather than solely materialism and Westernization. Materialism and asceticism occupy competing positions subject to changing interpretations of Buddhism, emerging relations of production (i.e., the introduction of industrial capitalism), and foreign cultural influences.

Conclusion

It is clear from the preceding discussion that Bangkok's foodscape, specifically the culture of public eating, presents us with a unique portrait of urban food consumption. Thai restaurant culture, especially at the small scale, involves large amounts of self-employed female labor and a myriad of different strategies for obtaining prepared meals every day. This ranges from the tiffin network through the classic curry shop to cafeterias and food centers that cater to workers, students, and shoppers.

On the one hand, it appears that Bangkok's micro-entrepreneurs in the prepared-food sector have designed innovative strategies to deal with the work of daily household food provision. Women (and men) are contracting out to other women (and some men); the work is remunerated and visible, and it creates public urban spaces where some of society's physical and social needs can be met. This is a conveniently designed and flexible system which, at first glance, is in the hands of women. Closer scrutiny suggests, however, that food micro-entrepreneurs are not acting in economic circumstances under their control or favorable to them. Most of them barely eke out an existence, as a result of rising rents, food costs, and remittances to relatives in impoverished rural areas.

It is clear that as Bangkok continues its rapid growth and transformation, new types of eating spaces are emerging. Some of these incorporate traditional foodshops within larger institutions controlled by the state or large capitalists. A conventional interpretation, certainly not without its merits, might view this development as an example of micro-entrepreneurs finding niches in the new economy. I take a more critical view and see this integration between the local and global as a potentially hazardous alliance between unequal actors who have disparate access to resources, leading to thorny issues of power and control. In sum, the situation of Thai women in the restaurant sector is fraught with the contradictions and ambiguity that parallel their "status" within Southeast Asia in general.

No matter how one interprets Bangkok's foodscape, it appears as though there is a direct relationship between the size and profitability of a food establishment and the degree to which men become involved. The preceding discussion has argued that definitions of femininity are substantially different in the small establishments, where women tend to manage all the affairs of the business, and larger restaurants, where a less favorable gender division of labor emerges. The larger establishments appear to be limiting women's involvement to cooking, hostessing, entertaining, and, occasionally, waiting. This neglects Southeast Asian women's traditional importance as money managers and entrepreneurs. At the same time, masculinity in Bangkok appears to be caught between competing ideals of conservative asceticism versus materialistic emphasis on conspicuous consumption and lavish displays of wealth. Even some monks in the Kingdom now have credit cards, since they are prohibited from handling cash, and the abbot of at least one *wat* (temple) on the outskirts of Bangkok is known to have a Mercedes (C. Greenberg, 1993, pers. comm.)! This challenges the traditional Thai gender roles, as men now clearly have an interest in controlling economic affairs that are no longer deemed "worldly" and therefore of lower status.

The case of Bangkok presents us with an interesting refutation of conventional "public/private" models of the city that propose that women tend(ed) to be associated with home-based activities and that men dominate(d) the world of paid employment. Thai women have long occupied both the domestic spheres and the world of small business. Indeed, the contemporary Thai urban food system points to the socially constructed definition and practice of "domesticity," since cooking food is performed in public space and is remunerated. The emergence of a Thai middle class and its spaces of food consumption exemplify a postmodern Southeast Asian urban experience. Middle-class and élite Thai restaurants often employ the modified use of traditional architectural styles that, in their more gargantuan forms, such as in *Tum Nuk Thai*, result in the creation of a pastiche. These *simulacra* recursively create and are created by new Thai middle-class urban identities. Femininity and masculinity are perceived, organized, and performed in new ways in these spaces, typically leading to a more rigid sexual division of labor and aestheticized coding of sexual "difference." Indeed, traditional Thai and other Southeast Asian societies have been characterized as ascribing to gender flexibility, at least for the masses. These recent changes appear to be more confining and oppressive, particularly for women.

Bangkok's restaurant culture is indeed more complex than the foodscape this paper has described and interpreted. A discussion of Bangkok's ethnogastronomy helps us explore issues of gender, food, and place in relation to ethnic, class, and gendered patterns of interaction that constitute place and are in turn reproduced in them. Further research could concentrate on comparative research between Thailand's "foodshop and restaurant culture" and other societies, for example, Bombay's *biriyani*-shop culture. The possibilities for in-depth historical research on related topics also would add to our current interpretations. Finally, assessment of the impact of Westernization in Bangkok's foodscape, despite the pride and cultural resilience of Thais, would help us predict the future of foodways in this and similar societies.

Notes

1. Information for this paper is based on the author's ethnographic field work in Thailand in 1992/1993 and the latter half of 1994. Additional data are derived from government statistical sources and from Marilyn Walker's *Food Consumption Survey* (FCS) conducted throughout Thailand using a weighted sample of 4,198 respondents in 1990. Additional materials of empirical importance are listed in the bibliography. Some of the ideas advanced in this paper were generated in Yasmeen (1995). I would like to thank David Ley, Deirdre McKay, Marilyn Walker, and one anonymous reviewer for helpful comments and edits. The financial support provided by the Social Sciences and Humanities Research Council, the International Development Research Centre, and the Canada-ASEAN Centre is gratefully acknowledged, as well as the institutional support provided by the Asian Institute of Technology and the Chulalongkorn University Social Research Institute. My supervisor at UBC (Terry McGee) and advisors in Thailand (Drs. Nurul Amin and Amara Pongsapich), as well as Ms. Napat Sirisambhand, greatly facilitated field work.
2. "Foodways" is defined as the gamut of activities, ranging from the cultural and symbolic to the economic and political, that are part of and condition all aspects of human nourishment.
3. "Foodshop" is a direct translation of the Thai term *raan ahaan*, which means any eating establishment in general. In this paper it is used to refer to small informal street restaurants as well as those located in buildings (such as shophouses). "Foodshop" will not be used with reference to expensive, highly capitalized restaurants. These, instead, will be termed *pattakarn*—a formal Thai expression—or simply "restaurants."
4. Referring to an older person as "uncle" or "auntie" is a form of respect in Thailand and among Asians in general. This is particularly the case for those in the service industries, such as vendors and small foodshop owners. It would not be appropriate to use the term to address a person of obviously "high" status, such as a professor or government official.
5. Based on the author's field work, 1994.
6. It is difficult to ascertain how economically viable this system of "contracting out" is at the household budget level. Certainly, it is clear that individuals are trading potential monetary savings for convenience and time (which presumably can be used to earn extra income). A related concern is the possible sacrifice of nutrition for convenience.

References

Amin, N. (1991) A Policy Agenda for the Informal Sector in Thailand. Report for Submission to the National Economic and Social Development Board, Royal Government of Thailand and Asian Employment Programme, International Labour Organization, New Delhi, India.

Armstong, W.R. and McGee, T.G. (1985) *Theatres of Accumulation*. London: Methuen.

Askew, M. (1994) *Interpreting Bangkok: The Urban Question in Thai Studies*. Bangkok: Chulalongkorn University Press.

Baudrillard, J. (1981) *Simulacres et simulation*. Paris: Editions Galilée.

Benhabib, S. (1992) Models of public space: Hannah Arendt, the liberal tradition and Jürgen Habermas. In C. Calhoun, editor, *Habermas and the Public Sphere*. Cambridge, MA: MIT Press.

Bondi, L. (1992) Gender and dichotomy. *Progress in Human Geography*, Vol. 16, 98–104.

Bowlby, S. (1988) From corner shop to hypermarket: Women and food retailing. In J. Little, L. Peake, and P. Richardson, editors, *Women in Cities: Gender and the Urban Environment*. New York: New York University Press, 61–83.

Butler, J. (1990) *Gender Trouble: Feminism and the Subversion of Identity*. New York: Routledge.

Charles, N. and Kerr, M. (1988) *Women, Food and Families.* Manchester: Manchester University Press.

Cook, I. (1994) New fruits and vanity: The role of symbolic production in the global food economy. In A. Bonanno et al., editors, *From Colombus to ConAgra: The Globalization of Agriculture and Food.* Lawrence, KS: University Press of Kansas, 232–248.

—— (1995) Constructing the exotic: The case of tropical fruit. In J. Allen and D. Massey, editors, *The Shape of the World.* Oxford: Oxford University Press, 137–142.

Crawfurd, J. (1820)(1987) *Journal of an Embassy to the Courts of Siam and Cochin China.* Singapore: Oxford University Press.

Davidoff, L. (1995) *Worlds Between: Historical Perspectives on Gender and Class.* London: Routledge.

de Young, J. (1955) *Village Life in Modern Thailand.* Berkeley, CA: University of California Press.

Dear, M. (1994) Postmodern human geography: A preliminary assessment. *Erdkunde,* Vol. 48(1), 2–13.

Douglas, M., editor (1984) *Food in the Social Order: Studies of Food and Festivities in Three American Communities.* New York: Russell Sage Foundation.

Drakakis-Smith, D. (1990) Food for thought or thought about food: Urban food distribution systems in the Third World. In R. Potter and A. Salau, editors, *Cities and Development in the Third World.* London: Mansell Publishing.

Elshtain, J. (1981) *Public Man, Private Woman: Women in Social and Political Thought.* Princeton, NJ: Princeton University Press.

England, K. (1994) Getting personal: Reflexivity, positionality, and feminist research. *Professional Geographer,* Vol. 46, 80–89.

FAO (1989) *Street Foods: Report of an FAO Expert Consultation, Jogjakarta, Indonesia, 5–9 December, 1988.* Rome: United Nations Food and Agriculture Organization.

Giard, L. and Mayol, P. (1980) *L'Invention du Quotidien. Tome 2. Habiter, Cuisiner.* Paris: Union Géneral des Editions.

Habermas, J. (1989) *The Structural Transformations of the Public Sphere.* Cambridge, MA: MIT Press.

Hauck, H.M., Saovanee Sudsaneh and Hanks, J.R., *Food Habits and Nutrient Intakes in a Siamese Rice Village: Studies in Bang Chan, 1952–54.* Ithaca, NY: Cornell University Southeast Asia Program, Data Paper No. 29 (1958).

Hayden, D. (1981) *The Grand Domestic Revolution: A History of Feminist Designs for American Homes, Neighborhoods and Cities.* Cambridge, MA: MIT Press.

hooks, b. (1984) *Feminist Theory: From Margin to Center.* Boston: South End Press.

—— (1990) *Yearning: Race, Gender and Cultural Politics.* Boston: South End Press.

Hopkins, J. (1990) West Edmonton Mall: Landscape of myths and elsewhereness. *The Canadian Geographer,* Vol. 34, 2–17.

Howell, P. (1993) Public space and the public sphere: Political theory and the historical geography of modernity. *Environment and Planning D: Society and Space,* Vol. 11, 303–322.

ILO (International Labor Organization) (1993) *Labor Force Statistics.* Geneva: International Labor Organization.

Jameson, F. (1983) Postmodernism and consumer society. In H. Foster, editor, *The Anti-Aesthetic: Essays on Postmodern Culture.* Port Townsend, WA: Bay Press.

Jellinek, L. (1991) *The Wheel of Fortune: The History of a Poor Community in Jakarta.* Sydney: Allen and Unwin.

Keyes, C.F. (1986) Ambiguous gender: Male initiation in a northern Thai Buddhist society. In C.W. Bynum, S. Harrell, and P. Richman, editors, *Gender and Religion: On the Complexity of Symbols.* Boston: Beacon Press, 66–96.

Keyes, C.F. (1984) Mother or mistress but never a monk: Buddhist notions of female gender in rural Thailand. *American Ethnologist,* Vol. 11, 223–241.

Khin Thitsa (1980) *Providence and Prostitution: Women in Thailand.* London: Change International Reports.

Kirsch, A.T. (1982) Buddhism, sex-roles and Thai society. In P. Van Esterik, editor, *Women of Southeast Asia.* De Kalb, IL: Northern Illinois University Series on Southeast Asia, Occasional Paper No. 9, 16–41.

—— (1985) Text and context: Buddhist sex-roles/culture of gender revisited. *American Ethnologist,* Vol. 12, 302–320.

Klopfer, L. (1993) Padang restaurants: Creating "ethnic" cuisine in Indonesia. *Food and Foodways,* Vol. 5, 293–304.

Konvitz, J. (1987) Gastronomy and urbanization. *South Atlantic Quarterly,* Vol. 86, 44–56.

Lagopoulos, A.P. (1993) Postmodernism, geography and the social semiotics of space. *Environment and Planning D: Society and Space,* Vol. 11, 255–278.

Manderson, L., editor (1983) *Women's Work and Women's Roles: Economics and Everyday Life in Indonesia, Malaysia and Singapore.* Canberra: The Australian National University, Development Studies Centre Monograph No. 32.

McGee, T.G. (1971) *The Urbanization Process in the Third World: Explanations in Search of a Theory.* London: G. Bell and Sons.

—— and Yeung, Y.M. (1977) *Hawkers in Southeast Asian Cities: Planning for the Bazaar Economy.* Ottawa: International Development Research Centre.

McKay, D. (1994) Engendering ecologies: Interpretations from the uplands of the Philippines. Paper presented to the annual meeting of the Canadian Asian Studies Association, Calgary.

Mohanty, C., Russo, A., and Torres, L., editors, *Third World Women and the Politics of Feminism.* Bloomington, IN: Indiana University Press.

Moore-Milroy, B. and Wismer, S. (1994) Communities, work and public/private sphere models. *Gender, Place and Culture*, Vol. 1, 71–90.

Napat Sirisambhand and Szanton, C. (1986) *Thailand's Street Food Vending: The Sellers and Consumers of "Traditional Fast Foods."* Bangkok: Chulalongkorn University Social Research Institute, Publication No. 5/1990.

Nicholson, L., editor (1990) *Feminism/Postmodernism*. New York and London: Routledge.

NSO, Office of the Prime Minister (1991) *Preliminary Report of the 1990 Household Socio-Economic Survey*. Bangkok: Royal Thai Government.

Ong, A. (1989) Center, periphery and hierarchy: Gender in Southeast Asia. In S. Morgan, editor, *Gender and Anthropology*. Washington, DC: American Anthropological Association, 294–312.

Papanek, H. and Minault, G. (1982) *Separate Worlds: Studies of Purdah in South Asia*. Delhi: Chanakya Publications.

Rosaldo, M., and Lamphère, L., editors (1974) *Woman, Culture and Society*. Palo Alto, CA: Stanford University Press.

Rose, G. (1993) *Feminism and Geography: The Limits of Geographical Knowledge*. Cambridge, UK: Polity Press.

Ryan, M.P. (1990) *Women in Public: Between Banners and Ballots, 1825–1880*. Baltimore: Johns Hopkins University Press.

Sargeant, S. (1985) *The Foodmakers*. Sydney: Penguin Books.

Sheridan, M. and Salaff, J., editors, *Chinese Working Women*. Bloomington, IN: Indiana University Press.

Simoons, F.J. (1961) *Eat Not This Flesh: Food Avoidances in the Old World*. Madison, WI: University of Wisconsin Press.

—— (1991) *Food in China: A Cultural and Historical Inquiry*. Boca Raton, FL: CRC Press.

Spivak, G. (1992) Acting bits/identity talk. *Critical Inquiry*, Summer, 770–803.

Thailand Ministry of Labour, 1991, *Labour Force Survey 1990*. Bangkok: Thailand Ministry of Labour.

Tinker, I. (1987) Street foods: Testing assumptions about informal sector activity by women and men. *Current Sociology*, Vol. 35, entire issue.

Van Esterik, P., editor (1982) *Women of Southeast Asia*. De Kalb, IL: Northern Illinois University Series on Southeast Asia, Occasional Paper No. 9.

—— (1992a) From Marco Polo to McDonald's: Thai cuisine in transition. *Food and Foodways*, Vol. 5, 177–193.

—— (1992b) *Nurturance and Reciprocity in Thai Studies: A Tribute to Lucien and Jane Hanks*. Downsview, Ont.: York University, Paper Number 8, Working Paper Series, Thai Studies Project, Women in Development Consortium in Thailand.

Wagner, P.L. (1994) Foodways: Culture and geography. *Western Geography*, Vol. 4, 84–93.

Walker, M. (1990) *Food Consumption Survey* (Thailand). Conducted by Frank Small and Associates (unpublished).

—— (1991) Thai Elites and the Construction of Socio-Cultural Identity. Unpublished Ph.D. Dissertation, York University.

Warren, S. (1993) "This heaven gives me migraines": The problems and promises of landscapes of leisure. In J. Duncan and D. Ley, editors, *Place/Culture/Representation*. London and New York: Routledge, 173–187.

Wathana W. (1990) Tenants in rental apartments in Bangkok. *Journal of Social Research* (Chulalongkorn University, Bangkok), Vol. 13, 58–61.

Watts, M. (1983) *Silent Violence: Food, Famine and Peasantry in Northern Nigeria*. Berkeley, CA: University of California Press.

Yasmeen, G. (1992) Bangkok's restaurant sector: Gender, employment and consumption. *Journal of Social Research* (Chulalongkorn University, Bangkok), Vol. 15, 69–81.

—— (1995) Exploring a foodscape: The case of Bangkok. *Malaysian Journal of Tropical Geography*, Vol. 26, 1–11.

34

The Political Economy of Food Aid in an Era of Agricultural Biotechnology

Jennifer Clapp

Abstract:

Recent years have seen numerous rejections of food aid containing genetically modified organisms (GMOs). The U.S., as the principal donor of this aid, went on the defensive, and blamed the European Union for hunger in developing countries. Rarely is food aid rejected. And rarely do food aid donors act so strongly to blame other donors. The reaction of both donors and recipients is also puzzling because it contradicts much of the literature from the 1990s which argued that the international food aid regime had become largely "depoliticized" following reforms to food aid policies in the 1980s. The current literature on food aid has not adequately addressed the ways in which the advent of GMOs has affected the food aid regime. I argue that scientific debates over the safety of GMOs, and economic factors tied to the market for genetically modified crops—both highly political issues—are extremely relevant to current debates on food aid.

In 2002 the U.S. sent significant quantities of food aid, in the form of whole kernel maize, to southern Africa in response to the looming famine in the drought stricken region. It soon became apparent that the aid contained genetically modified organisms (GMOs), though the recipients had not been notified prior to the shipments being sent. Many southern African countries initially refused to accept the GM food aid, partly as a health precaution, and partly on the grounds that it could contaminate their own crops, thus hurting potential future exports to Europe. A number of the countries eventually accepted the food aid provided it was milled first, but Zambia continued to refuse even the milled maize. The U.S. argued that it could not supply non-genetically modified (GM) food aid, and it refused to pay for the milling. The U.S. then blamed Europe's moratorium on imports of GM foods and seed for contributing to hunger in southern Africa.

This incident highlights a new aspect of the recent global predicament over how the international movement of GMOs should be governed. While there has been much analysis of this question with respect to commercial trade in recent years, particularly regarding the adoption of the biosafety protocol[1], the literature on food aid has not kept up with these new developments. Recent academic analyses of food aid have paid little attention to the question of GMOs. The literature on food aid has focused mainly on the motivations for donating food aid, and its potential as a development tool. Some have argued that while economic and political considerations are present to some degree in the motivation for giving food aid, today it is mainly given as part of a development regime that aims primarily to promote food security and rural development rather than as a means to serve donor countries' domestic economic and political interests.[2]

In light of the changes in global agriculture over the past decade, especially the rise of the U.S. as a major producer of GM crops, it is important to re-examine the political economy of food aid. There appear to be strong economic motivations for the U.S. to pursue the food aid policy described above, as well as scientific motivations, not addressed by the earlier food aid literature. Both of these motivations are highly politicized. Europe has not followed the lead of the U.S. on GM food aid policy. The divergence of the policies of the EU and the U.S. on this issue may well lead to interesting politics in the years to come in the international battle over GMOs. Only this time, the debate looks set to be played out globally, with some of the world's poorest countries as unwitting participants in the conflict.

Why Revisit Food Aid Politics?

The modern era of food aid was instituted in the U.S. in 1954, with the passage of U.S. Public Law (PL) 480. Since that time food aid has been an important feature of U.S. assistance to developing countries, though its role has changed over time. In the 1950s food aid accounted for nearly a third of U.S. agricultural exports, whereas in the mid 1990s it was closer to 6 percent.[3] Food aid under PL 480 is given under different categories of assistance. Title I is government to government aid in the form of concessional sales with the express aim of opening new markets for U.S. grain. Title II is grant food aid distributed in emergencies. Such aid can be distributed via NGOs and the World Food Program. Title III is government to government grants of food aid for development activities.[4]

From its origins, U.S. food aid was largely seen as a multi-purpose tool. On the surface, the idea of the PL 480 was to provide the world's food deficit countries with food from the U.S., as part of a broader humanitarian effort. It was also clearly a mechanism for surplus disposal and export promotion in the U.S. It created a market for surplus food and as such it had the effect of raising U.S. domestic prices for grain. Further, shipping free or concessionally-priced U.S. grain to poor countries, it was hoped, would create new markets in the future for commercially traded U.S. grain.[5] U.S. food aid was, however, soon seen as a political tool as well. The U.S. had even gone so far as to amend PL 480 in the 1960s to explicitly tie the donation of food aid to political goals, in particular to favor non-communist countries.[6] Other countries followed the U.S. in giving food aid, including Canada, which began its program in the 1950s, and the European Economic Community, which began to give food aid in the late 1960s.[7] Canada and the European donors have been less overtly political and economic in their rationale for food aid, though some surplus disposal mechanism has been part of their food aid donations at various times.

The World Food Program (WFP) of the United Nations was set up in 1963 as a multilateral channel for food aid from donor countries. In its early years the WFP distributed around 10 percent of food aid, while today nearly 50 percent of all food aid is channeled through the WFP.[8] The Food Aid Convention (FAC) was first established in 1968 as part of the Kennedy Round of GATT negotiations, and was attached to the International Wheat Trade Convention.[9] The FAC members are the donor countries, and the agreement sets out minimum amounts of food aid per year to be given by each donor (denominated in tonnes of wheat). The FAC, re-negotiated periodically, now stipulates that food aid can be given either in-kind or in cash equivalent, and other commodities apart from wheat can be given (but they are still measured in wheat equivalents). Today the donor members of the FAC include Argentina, Australia, Canada, the European Community and its member states, Japan, Norway, Switzerland, and the United States.[10] From the early days of food aid to the present, the U.S. has remained the principal donor country, and gives its aid primarily in-kind. Other donors have over the past decade increasingly given their food aid in the form of cash, which is directed toward third country purchases or local purchases in the recipient region.

Since the 1980s food aid policies in the U.S. have been reformed significantly, with the overt political goals removed from the PL 480. And in the mid-1990s the surplus of grain in the U.S. diminished, making the surplus disposal element of food aid appear to be less significant than it was in the past.[11] In the European Union, food aid policies since the 1990s have focused on giving aid in the form of cash to finance food distribution programs with local or third party sources of the food, rather than in-kind. This policy has been reinforced by the EU's regulation requiring the shift toward cash-based food aid in 1996.[12] This policy on the part of the EU was largely in response to studies which showed that cash spent on local purchases of food in aid recipient regions boosts the local economies and allows for much more flexibility in terms of sourcing culturally appropriate foods.[13]

Because of these policy shifts, some have argued that food aid by the late 1980s and early 1990s had in fact become largely a development tool, with the motivations for donating food governed more by the existence of an international regime (and desire to cooperate) than by donor economic and political considerations, which is in line with a liberal institutionalist perspective on international relations.[14] In other words, a "depoliticized" food aid regime was seen to have emerged, which was not merely serving the interests of the donors, but rather was promoting international development. This was especially the case for European donors. Uvin argued that by the early 1990s most EU food aid and about 60 percent of food aid from the U.S. was clearly not driven by economic or political motivations.[15] Other, more recent studies have made similar arguments. Eric Neumayer, for example, argues that in the 1990s while U.S. economic and military-strategic interests had some influence on food aid donations for program, or longer term food aid, it is an insignificant influence on emergency relief.[16]

In light of the development of agricultural biotechnology in recent years, I argue that it is important to re-examine the political economy of food aid. Important factors influencing donor motivations in an age of agricultural biotechnology are not adequately considered in the food aid literature. Economic factors may once again be key motivating factors for food aid policy. These factors are especially important to consider given the growing corporate concentration in the agricultural biotechnology sector and its close ties with U.S. government agencies, as well as the U.S.

dispute with the EU over its 1998–2004 moratorium on GMOs. Moreover, new factors that may influence food aid policy must also be considered, such as the scientific debates over the safety of genetically modified food. These factors appear to be influencing food aid policies on the part of both donors and recipients, and they are highly political.

GMOs in Food Aid

GMOs have been present in food aid since GM soy and maize were initially approved for production in the U.S. in the mid-1990s. Its presence in food aid was inevitable for a number of reasons. U.S. food aid is predominantly given in-kind and is made up of food (mainly wheat, corn and soy) grown in the U.S. The U.S. is by far the largest producer of GM crops, accounting for over 60 percent of the global acreage planted with GM varieties. Between 1996 and 2003 the global area planted with GM crops increased by 40 fold, in 2003 covering some 67.7 million hectares.[17] In 2002 some three-quarters of the soy and over a third of maize grown in the U.S. were GM varieties.[18] Moreover, there is not a segregated system for GM and non-GM crops in the U.S., and cross contamination has been widespread. This is important, as the U.S. accounts for 60 percent of all food aid donations.

Though negotiations began in 1995 on a protocol on biosafety under the Convention on Biodiversity to address the safety issues related to trade in GMOs, little attention was paid to their presence in food aid transactions at that time. It is not surprising, then, that when it was discovered that some food aid donations contained GMOs in 2000, many were caught unaware. Both USAID and the WFP had sent shipments of food aid containing GMOs, amounting to some 3.5 million tonnes per year.[19] Such shipments were often in contravention of the national regulations in the recipient country.

Ecuador was the first known developing country to receive food aid containing GMOs in a shipment of soy from the U.S. and channeled through the WFP. The product was eventually destroyed following complaints by Ecuador.[20] There were also cases of GMOs being sent in food aid shipments to Sudan and India in 2000. In 2001 GMO soy was found in food aid shipments sent to Columbia and Uganda. Food aid maize from the U.S. containing GMOs was also reportedly sent to Bolivia in 2002, despite the fact that the country had a moratorium in place on the import of GMO crops. The GMOs found in the Bolivian aid contained StarLink corn, which is a modified form of corn that was not approved in the U.S. for human consumption (it was approved as animal feed), but which nonetheless managed to find its way into the human food system in the U.S. in the fall of 2000. NGOs claim that despite the fact that when StarLink was found in the U.S. food supply it was immediately removed from the market, the U.S. did little to remove the maize from Bolivia. In 2002 Nicaragua and Guatemala were also sent GM corn seed as food aid from the WFP. This caused a stir in Nicaragua in particular, as that country is a centre of origin for corn.[21]

By mid-2002, there were enough incidents of GMO food aid to have made the donors fully aware of the issue. Recipient countries expressed concern about the potential health and environmental impacts of GMOs, including allergenicity, and out-crossing of GMOs with wild relatives (if the grains are planted rather than eaten) which could reduce biodiversity by contaminating and driving out local varieties. Once GMOs are released into an environment, they are difficult, if not impossible, to remove. Food aid, when given in whole grain form, is often planted by local farmers, who may have exhausted their seed supply as food in times of crisis.[22] The fact that GM crops have not been approved in many countries, including in the European Union which had placed a moratorium on their imports and new approvals of GM crops in 1998, fueled many of these concerns, especially for those countries which have export markets in the EU.

Until mid-2002 the food aid shipments identified as containing GMOs were mainly to areas, which, while in food deficit, were not facing an acute food shortage. This changed in mid-2002 with the looming famine in southern Africa. Some 14 million people in 6 countries faced imminent food shortages and famine at that time. The situation was seen to have been precipitated by a number of factors. Drought and floods were identified as one of the immediate causes. However, underlying factors were just as important. These include the high prevalence of HIV/AIDs in the region, as well as conflict in Angola (and refugees from Angola in neighbouring countries), domestic agricultural policies, as well as the impacts of trade liberalization under structural adjustment in some countries in the region.[23] It was the worst food shortage faced by the region in 50 years.

In response to this crisis the U.S. sent 500,000 tonnes of maize in whole kernel form to the region in the summer and fall of 2002 as food aid. It was estimated by the WFP that around 75 percent of food aid to the region at that time contained GMOs.[24] The countries that received the shipments were Zambia, Zimbabwe, Malawi, Swaziland, Mozambique and Lesotho. The aid was channeled through the WFP as well as NGOs. When the countries discovered that the aid contained GMOs, they were forced to consider whether they wanted to accept the aid. Zimbabwe and Zambia said they would not accept the food aid at all, while Mozambique, Swaziland and Lesotho said they would accept it if it was milled first. Malawi accepted it with strict monitoring to ensure that its farmers do not plant it. Zimbabwe eventually said it would accept it if it was milled and labelled first.[25]

Zambia stood firm in not accepting it for its own people. It did eventually accept it in milled form but only for the 130,000 Angolan refugees in camps within its borders, but not for the general population.[26] Zambia expressed its concern that any health problems that might arise from eating GMOs would be too costly for the country to address. Since the Zambian diet consists of far more maize than the diets of North American consumers, such health problems may not be foreseen. Moreover, Zambia does have some maize exports to Europe, and contamination of its maize with GMOs could affect its exports if the EU moratorium continues.[27] The WFP scrambled to find non-GMO aid for Zambia, which had some 3 million people at risk of starvation. In the end, the WFP was only able to source about one-half of the necessary 21,000 tonnes of maize needed for Zambia.[28]

The WFP responded by saying that it was impossible to mobilize non-GMO food aid fast enough. The WFP made it clear that it respected the right of the countries to refuse to accept the aid, and it did what it could to organize the milling of the maize for those countries that would accept it in that form, and to source non-GMO aid for Zambia. The WFP had to quickly arrange local milling, and in the case of Zambia, it had to remove shipments which had already been delivered. The milling did provide the WFP with the ability to fortify the grain to raise its micronutrient content, however, which was seen as an unexpected benefit. Further, the WFP did manage to solicit donations from non-traditional donors of aid for food, including a number of developing countries.[29]

The response of the U.S. was much more defensive. The U.S. refused to mill the maize before sending it to the region, claiming that it was too expensive to do so, raising the cost of the food by as much as 25 percent, and reducing its shelf life. It also initially refused to send non-GM varieties of corn to the region, claiming that it was impossible to source non-GM crops from the U.S. It refused to give cash instead of in-kind aid, on the grounds that it has traditionally given in-kind aid, and has done so for 50 years. The U.S. did, however, stress that it would respect the wishes of the countries that did not want GMO food aid sent to them. The U.S. position was clearly spelled out in a USAID website "questions and answers" on the GMO food aid crisis.[30] The U.S. did eventually give Zambia a donation of GM-free maize of some 30,000 tonnes, however, after heavy international pressure to do so.[31]

The U.S. also blamed Europe for the crisis, saying that its moratorium on approval of seeds and foods containing GMOs was stalling efforts to promote food security.[32] In the midst of the African crisis the U.S. began to seriously consider launching a formal complaint at the WTO over the EU's moratorium on GMOs, claiming that it was in contravention of WTO rules. Though the WTO rules do allow countries to ban imports of a product on food safety concerns while the country seeks further scientific evidence, the U.S. argued that five years was plenty of time and that no such evidence had been gathered. The U.S. was concerned that this position of the EU was influencing too many countries, including those in Africa.[33]

Throughout the fall of 2002 and early in 2003 the U.S. put pressure on Europe to remove its moratorium. The U.S. finally launched the formal complaint against the EU at the WTO in May 2003. Argentina and Canada joined the formal challenge.[34] At the time U.S. President Bush stated: "European governments should join—not hinder—the great cause of ending hunger in Africa."[35] Egypt was initially listed as a co-complainant, but it pulled out. Though Egypt does have an active agricultural biotechnology research program, it withdrew from the dispute because Europe is a very important market for its exports of fresh fruits and vegetables. The U.S. had hoped that having Egypt on board would help it to drive the point home that GM crops are beneficial to Africa. The U.S. retaliated against Egypt for its withdrawal by pulling out of talks on a free trade agreement with that country.[36]

When the U.S. launched the dispute, the European Union issued a press release stating its regret over the U.S. decision to take action on this case. It criticized the U.S. for using the African countries' refusal of GM food aid to pressure the EU:

> The European Commission finds it unacceptable that such legitimate concerns are used by the U.S. against the EU policy on GMOs . . . Food aid to starving populations should be about meeting the urgent humanitarian needs of those who are in need. It should not be about trying to advance the case for GM food abroad.[37]

In the midst of the southern African crisis the European Commission specifically requested the WFP to only purchase non-GM maize as food aid with the money the EU donates for such assistance.[38]

In mid-2003, another dispute over GM food aid emerged. Sudan had been pressured by the U.S. to accept GM food aid, despite its recently passed legislation that requires food aid to be certified GM-free. In response to U.S. pressure, the Sudan issued a temporary six month waiver to this legislation in order to give the U.S. more time to source GM-free food aid. In March 2004, however, the U.S. threatened to cut the Sudan off from food aid completely.[39] This prompted the Sudan to extend the waiver to early 2005.

Unclear Rules on International Trade in GMOs

How is it that such massive shipments of GM food aid could have been sent without the recipients knowing about it before it was sent? The rules regarding trade in GMOs were unclear at both the global and local levels between the mid 1990s and 2003. This was the very period when much of the controversy over GMO food aid was at its highest.

As of mid-2002 when the southern African crisis erupted, only a few developing countries had any domestic legislation dealing with imports of GMOs, let alone GMO food aid. The only sub-Saharan African countries with biosafety laws in place at that time were South Africa and Zimbabwe, though a number have since begun to develop policies dealing with the import of genetically modified organisms. Zimbabwe had a Biosafety board to advise on GMOs, but it has not approved any GM crops for commercial release. South Africa is the only sub-Saharan African country which has approved the commercial planting of genetically modified crops. In July 2001 the Organization of African Unity (OAU, now the African Union) endorsed a Model Law on Safety in Biotechnology which takes a precautionary approach to biotechnology and calls for clear labeling and identification of imports of GMOs. This model legislation was designed as guidance for countries in formulating their own national laws on biosafety as well as a way to develop an Africa-wide system for biosafety. As of 2003, no countries in Africa had yet adopted the model law into its legislation.[40]

In response to the southern African crisis in 2002, the Southern African Development Community (SADC) established an advisory committee to set out guidelines for policy on GMOs in the region. These guidelines stipulate that "food aid that contains or may contain GMOs has to be delivered with the prior informed consent of the recipient country and that shipments must be labeled."[41] But such guidelines were not available at the time of the crisis. Other regional responses include efforts by the Common Market for Eastern and Southern Africa (COMESA) to develop a regional policy on GMOs, including food aid. And the New Partnership for Africa's Development (NEPAD) decided in mid-2003 to establish a panel to advise African countries on biosafety and biotechnology as a means to try to harmonize regulations on these issues across Africa.[42]

At the international level, rules on biosafety and trade in GMOs were also not all that clear prior to 2003. The Cartagena Protocol on Biosafety, which governs trade in GMOs, was negotiated between 1995 and 2000 (when it was adopted), but was not in legal force until September 2003.[43] The Protocol's rules state that GMOs (living modified organisms) intended for release into the environment (seeds) in the importing country, are subject to a formal Advanced Informed Agreement (AIA) procedure for the first international transboundary movement to a country. Importing countries can reject these if they wish, based on risk assessment. Genetically modified commodities (living modified organisms intended for food, feed or processing) are exempted from the formal AIA procedure, and instead are subject to a separate form of notification, in the form of a Biosafety Clearing House, an internet-based database where exporters are required to note whether shipments of such commodities "may contain GMOs." Importers can also reject such shipments, based on risk assessment. In both cases, parties are given the right to make decisions on imports on the basis of precaution in cases where full scientific certainty is lacking.[44] The food aid donations shipped prior to the Protocol's entry into force were not covered by these rules. And now that it is in force, these rules only apply to those countries that have signed and ratified the agreement. The U.S. and Canada, two of the major food aid donors which grow significant quantities of GMOs, have not yet ratified the Protocol, and thus are not bound by its rules.

The Codex Alimentarius Commission, which sets voluntary international guidelines on food standards, was from the late 1990s trying to address questions of biotechnology and food safety. In 1999 the Codex established a special Task Force on Biotechnology to address the wider concerns expressed about biotechnology and food safety, especially those related to risk analysis. The Task Force only released its guidelines in mid-2003. Though they are voluntary, the standards are considered a benchmark for international trade under the WTO. The biotechnology guidelines adopted include safety evaluations prior to marketing of GM products, and measures to ensure traceability in case a GM product needs to be recalled.[45] But because these guidelines were not in place at the time of the food crisis in southern Africa, nothing was done to ensure these guidelines were followed for food aid.

The WFP did not set an explicit policy on how to deal with GMOs until mid-2003. Its policy has long been to give food aid to countries in food deficit if the food met requirements for food safety by both the donor and the recipient. But if neither the recipient nor the donor had a policy of notification, it was difficult for the WFP to keep track of them. It defended its lack of a GMO policy prior to that date by stating that "none of the international bodies charged with dealing with foods derived from biotechnology had ever requested that the Programme handle GM/biotech commodities in any special manner for either health or environmental reasons."[46] Because of the media attention to the issue, and claims that the WFP was negligent, the WFP decided to establish a formal policy for dealing with GMOs in food aid in 2002, which was finalized in 2003. The new policy asks recipient country offices of the WFP to be aware of and comply with national regulations regarding GM food imports. It also maintains its original policy that it will only provide food as aid which is approved as safe in both donor and recipient nations. Countries that clearly state that they do not wish to receive GM food aid will have their wishes respected. The WFP stated that it will still accept GM food aid from donors, but will also respect the wishes of donors who give cash in lieu of in-kind aid if they request that the money not be spent on GM food.[47]

Unpacking Motivations for GM Food Aid Policy

What explains the widely divergent positions on GM food aid on the part of the donors, specifically the U.S. and the EU, and the rejections by the recipient countries? As mentioned above, much of the food aid literature sees the current donor motivation for giving food aid as being driven not so much by economic and political goals as had been the case in the past, especially in the case of emergency aid. In this section I argue that we need to revisit this issue. In an age of agricultural biotechnology, new issues must be considered as having an influence on food aid policy, primarily the scientific debate over the safety of GM food. Further, economic and political incentives, inextricably tied to corporate interests in agricultural biotechnology, appear once again to be important factors behind the U.S. position on GM food aid in particular.

Debates over the Science of GMOs: Differing Interpretations of Risk and Precaution

The southern African crisis fuelled an already existing scientific debate over the safety of GMOs and their role in promoting food security. The debate has largely been over whether there is sufficient risk associated with the planting and consumption of GM crops and foods to warrant precaution with respect to their adoption. In the media accounts of the GM food aid incidents in southern Africa, this scientific dimension has tended to dominate the explanations for the policies pursued by the donors and the recipients.

The North American position on the safety of GM foods and crops is that there is minimal risk attached to them, and that because of this a precautionary approach in their adoption is not warranted. In both the U.S. and in Canada, regulatory procedures for GMOs are built on the notion that if the developer of a genetically-modified crop or food can demonstrate that it is "substantially equivalent" to a conventional counterpart, the GM crop or food does not require an extensive risk assessment prior to its approval.[48] Ongoing scientific uncertainty with respect to the risks of GMOs does not automatically invoke a precautionary approach in these countries. In other words, agricultural biotechnology products are assumed to be innocent until proven guilty. It is further argued that the benefits of GM crops, in terms of higher yields and easier management of weeds, far outweigh the (known) risks associated with them.[49] The U.S. and Canada view their approach to the regulation of agricultural biotechnology as being firmly grounded in "sound science."

The approach to regulating agricultural biotechnology products is very different in the EU and in many developing countries. The EU's interpretation of the potential risks associated with GMOs is much more precautionary. It views the existence of scientific uncertainty with respect to the safety of agricultural biotechnology as enough reason to take more time to evaluate the full range of potential risks associated with these products prior to their approval. Before such products can be approved, they must be subject to a rigorous scientific risk assessment.[50] In this sense, agricultural biotechnology products are assumed to be guilty until proven innocent. Many developing countries lack a regulatory structure for approval of agricultural biotechnology products, and for this reason they have tended to favor the EU approach which applies precaution in the face of scientific uncertainty. Further, there is widespread sentiment in Europe and in many developing countries that the potential risks, such as the potential for out-crossing with wild relatives and creating new allergens, do not outweigh the possible benefits, which reinforces the precautionary mood in those countries.

The different interpretations of the science and risks of GMOs go some way to explaining the widely divergent positions with respect to GM food aid amongst the U.S., the EU, and the recipient countries. The hostility on the part of the U.S. toward those countries that rejected the GM food aid, and placing of the blame on Europe, are partly products of these different viewpoints. In particular, the U.S. sees the EU's regulatory system, which is much more precautionary, as being too "emotional" and not scientifically based. The U.S. would much rather see its own regulatory style, rather than the EU approach, adopted in developing countries that currently lack a regulatory framework. This attitude on the part of the U.S. can be seen in the comments made by U.S. Senator Chuck Grassley, at a speech to the Congressional Leadership Institute in March 2003, just prior to the launch of the trade dispute against the EU:

> By refusing to adopt scientifically based laws regarding biotechnology, the EU has fed the myth that biotech crops are somehow dangerous . . . The European Union's lack of science based biotech laws is unacceptable, and is threatening the health of millions of Africans.[51]

The refusal of the GM food aid on the part of the southern African countries can also be seen as a reflection of their position in the scientific debate, as many of the comments made by African leaders when rejecting the aid made this specific link. For example, Zambian president Levy Mwanawasa expressed his concern that GM food aid was "poison", stating "If it is safe, then we will give it to our people. But if it is not, then we would rather starve than get something toxic."[52] The Zambian government did authorize a scientific delegation to study the issue, which was sponsored by the U.S. government and several European countries. This delegation traveled to South Africa, a number of European countries, and the U.S. The eventual report from the delegation, which came in the fall of 2002, cautioned against the acceptance of GMOs in Zambia, much to the disappointment of the U.S.[53]

Economic Motivations

While the different interpretations of risk and precaution are clearly relevant in explaining motivations for GM food aid policy, they are not the only important factors. In an age of agricultural biotechnology, it appears that economic considerations are re-emerging as explanatory factors for food aid policy, at least on the part of the U.S.

Throughout the history of food aid, surplus disposal has remained important for the U.S.[54] Because stocks have declined over the past 50 years, and over the past decade in particular, however, some say that surplus disposal is no longer as important as it once was.[55] But the advent of GM food aid may be reviving and reinforcing the surplus disposal aspect of U.S. food aid. The European moratorium on imports of GMOs has meant a significant loss of markets for U.S. grain. The U.S. has lost around U.S.\$300 million *per year* in sales of maize to Europe, for example, since 1998.[56] Some 35 countries, comprising half of the world's population, have rejected GM technology, and this is also closing the market opportunities for GMO-producing countries to export their products. In addition to the European Union, Australia, Japan, China, Indonesia and Saudi Arabia, also refuse to approve most agricultural biotechnology for domestic use and import.[57] Because of the loss of these markets, the U.S. may well be looking for other outlets for its GM maize.

The inability to find export markets for its GM grain may well be a principal reason why the U.S. continues to insist on giving its food aid in-kind, rather than in the form of cash. Both the FAC and the WFP encourage food aid donations in cash rather than in-kind, and the EU has been pushing to have cash-only donations of food aid incorporated into WTO rules. In the case of southern Africa, the U.S. was the only donor that gave food aid in kind rather than as cash.[58] This may be in part due to the preferences of the strong grain lobby in the U.S. In a letter to the U.S. trade representative on this issue, the National Wheat Growers Association stated: "We wish to assure you that producers across the nation are strong supporters of humanitarian programs, but will not be willing to support cash-only programs."[59]

A second potential economic motivation for the U.S. in giving GM food aid, not unrelated to surplus disposal, is to subsidize the production and sale of GM crops, as well as the agricultural biotechnology sector more broadly, which is dominated by U.S. transnational corporations (TNCs). Some 80 percent of funds for the PL 480 program are in actual fact spent in the U.S.[60] In 2000 it was reported that Archer Daniels Midland and Cargill, two of the largest grain trading corporations, were granted a third of all food aid contracts in the U.S. in 1999, worth some U.S.\$140 million.[61] The U.S. Department of Agriculture (USDA), which is responsible for regulating biotechnology in the U.S. and which also oversees the Title I food aid, works in close cooperation with the agricultural biotechnology industry.[62] One example of this is the 2002 U.S. Farm Bill which provided funding for the USDA to set up a biotechnology and agricultural trade program with the aim "to remove, resolve or mitigate significant regulatory non-tariff barriers to the export of United States agricultural commodities."[63] USAID, which is responsible for Title II and Title III food aid, also actively promotes the adoption of agricultural biotechnology in the developing world through educational programs, giving some U.S.\$100 million for that purpose in recent years.[64] This includes USAID funding for private-public partnerships such as the African Agricultural Technology Foundation[65] and the Agricultural Biotechnology Support Project[66] both of which have heavy participation from TNCs in the agricultural biotechnology industry. These initiatives seek to promote the use of agricultural biotechnology in the developing world through research, education and training, and they also acknowledge that they hope such efforts will open new markets in the future.[67]

Critics see such efforts as a means by which the U.S. is trying to pave the way for the introduction of pro-GM legislation to facilitate the export of GM crops and seeds around the world.[68] For many the position of the U.S. in the southern African crisis, especially its refusal to mill the GM grain and its attack on Europe's regulatory structure, was seen as a deliberate strategy to spread GMOs as far and as wide as possible, in order to break the remaining resistance to the technology.[69]

Economic considerations in the EU must also be taken into account in unpacking donor motivations. The EU's position on GM food aid is very much tied up in the WTO trade dispute over GMOs more broadly. The U.S. had been pressuring the EU to lift its moratorium on approvals of GM crops and foods before the crisis hit in southern Africa, and so it is not surprising that the EU position was in opposition to that of the U.S. Tied up in this broader dispute is the question of export markets for the EU as well. It may be that the EU is seeking to solidify trade relations with developing countries by creating a non-GM market which would exclude the U.S. The EU has also been pushing for several years now for cash-only food aid to be written into WTO rules as part of the ongoing talks on the revision of the WTO's Agreement on Agriculture. The EU sees the U.S. in-kind food aid, and Title I sales of food aid, as unfair subsidies to the U.S. agricultural industry, and wishes to see these removed in exchange for its own subsidy reductions. This helps to explain the EU's criticism of in-kind food aid.[70]

On the recipient side, economic considerations are also important in helping to explain their acceptance or rejection of food aid. The southern African countries were concerned about their export prospects with the EU if they accepted GM food aid in whole grain form. If GM food aid were planted and crossed with local varieties, this could affect exports of maize. Zambia, for example, exports some maize to European countries, and Zambia and other countries in the region did not want to close the door to potential future markets in Europe for GM free maize exports.[71]

Conclusion

It is unfortunate that the debate over biotechnology has been played out in the developing world through the politics of food aid. It has profoundly affected the recipient countries, and their environments and future trade prospects may suffer from it. The literature on food aid has to date paid insufficient attention to the question of GMOs and the impact they have on the food aid regime.[72] I argue that it is time to insert the question of agricultural biotechnology squarely into the debates on food aid. The food aid regime is being influenced by a number of factors that are unique to an age of agricultural biotechnology. These include the scientific debate over the safety of GMOs, as well as economic considerations linked to markets for GM crops. Both of these factors appear to have had an important influence on the policies on GM food aid pursued by both donors and recipients. In many ways, these factors are hard to separate from one another, and both are highly political. The notion put forward in the early 1990s that the food aid regime had become largely "depoliticized" must today be questioned. It is clear that the advent of agricultural biotechnology has fundamentally changed the nature of the regime.

Jennifer Clapp is Chair of the International Development Studies Program and Associate Professor of Environmental and Resource Studies at Trent University in Canada. She would like to thank Peter Andrée, Derek Hall, Brewster Kneen, Marc Williams and three anonymous reviewers for useful comments. She would also like to thank Marcelina Salazar, Sam Grey, Chris Rompré and Kate Turner for research assistance, and the Social Science and Humanities Research Council of Canada for research support.

Notes

1. See, for example, Robert Falkner, "Regulation Biotech Trade: The Cartagena Protocol on Biosafety," *International Affairs*, 76, no.2, (2000); Peter Newell and Ruth Mackenzie, "The 2000 Cartagena Protocol on Biosafety: Legal and Political Dimensions," *Global Environmental Change*, 10, (2000); Cristoph Bail, Robert Falkner and Helen Marquard (eds), *The Cartagena Protocol on Biosafety: Reconciling Trade in Biotechnology with Environment and Development?* (London: Earthscan, 2002).

2. Peter Uvin, "Regime, Surplus and Self-Interest: The International Politics of Food Aid," *International Studies Quarterly*, 36, (1992), pp. 293–312; Raymond Hopkins, "Reform in the International Food Aid Regime: the Role of Consensual Knowledge," *International Organization*, 46, no. 1, (1992), pp. 225–264; Raymond Hopkins, "The Evolution of Food Aid: Towards a Development-First Regime", in Vernan Ruttan (ed), *Why Food Aid?* (Baltimore: Johns Hopkins, 1993); Edward Clay and Olav Stokke (eds), *Food Aid and Human Security* (London: Frank Cass, 2000).

3. Cheryl Christiansen, "The New Policy Environment for Food Aid: The Challenge of Sub-Saharan Africa," *Food Policy*, 25, (2000), p. 256.

4. USAID, *Food Aid and Food Security Policy Paper* (PN-ABU-219) (Washington, D.C.: USAID, 1995).

5. Harriet Friedmann, "The Political Economy of Food: The Rise and Fall of the Postwar International Food Order", in M. Burawoy and T. Skocpol (eds), *Marxist Inquiries: Studies of Labour, Class and States*, supplement to *American Journal of Sociology*, 88, (1982) S248–86.; Vernan Ruttan, "The Politics of U.S. Food Aid Policy: A Historical Review," in Vernan Ruttan (ed), *Why Food Aid?* (Baltimore: Johns Hopkins, 1993).

6. Mitchel Wallerstein, *Food for War—Food for Peace* (Cambridge, MA: MIT Press, 1980); Shlomo Reutlinger, "From 'Food Aid' to 'Aid for Food': Into the 21st Century," *Food Policy*, 24, (1999), p. 9.

7. See (on Europe), John Cathie, *European Food Aid Policy* (Aldershot: Ashgate, 1997); and (on Canada), Mark Charlton, *The Making of Canadian Food Aid Policy* (Montreal: McGill-Queens, 1992).

8. International Grains Council (IGC), *IGC Annual Report* (London: IGC, 2004).

9. International Wheat Council, *The Food Aid Convention of the International Wheat Agreement* (London: International Wheat Council, 1991).

10. Food Aid Convention, *Food Aid Convention* (London: International Grains Council, 1999).

11. See Christiansen, "The New Policy Environment for Food Aid: The Challenge of Sub-Saharan Africa".

12. See Christopher Barrett and D. Maxwell, *Food Aid After Fifty Years: Recasting Its Role* (London: Routledge, 2005); EU Council Regulation (ED) No. 1292/96.

13. See, for example, Edward Clay and Olav Stokke (eds), *Food Aid Reconsidered: Assessing the Impact on Third World Countries* (London: Frank Cass, 1991).

14. Uvin, "Regime, Surplus and Self-Interest: The International Politics of Food Aid"; Hopkins "Reform in the International Food Aid Regime: the Role of Consensual Knowledge," "The Evolution of Food Aid: Towards a Development-First Regime".

15. Uvin, "Regime, Surplus and Self-Interest: The International Politics of Food Aid," pp. 307–308.

16. Eric Neumayer, "Is the Allocation of Food Aid Free from Donor Interest Bias?" *Journal of Development Studies*, (forthcoming: 2004).

17. International Service for the Acquisition of Agri-Biotech Applications, *Preview: Global Status of Commercialized Transgenic Crops: 2003: Executive Summary, No.30*, pp. 3–4. Internet address: http://www.isaaa.org

18. See USAID, "United States and Food Assistance," *Africa Humanitarian Crisis* (3 July 2003). Internet address: http://www.usaid.gov/about/africafoodcrisis/bio_answers.html#8.

19. Geoffrey Lean, "Rejected GM Food Dumped on the Poor," *The Independent* (London), (18 June 2000); Fred Pearce, "UN is slipping modified food into aid," *New Scientist*, 175, no. 2361, (2003), p. 5.

20. Friends of the Earth International (FOEI), *Playing with Hunger: The Reality Behind the Shipment of GMOs as Food Aid* (Amsterdam: FOEI. April 2003), p. 5.

21. FOEI, *Playing with Hunger: The Reality Behind the Shipment of GMOs as Food Aid*, pp. 6–7; ACDI/VOCA. *Genetically Modified Food: Implications for U.S. Food Aid Programs* (2nd revision, April 2003), pp. 6–10. Internet address: http://www.acdivoca.org/acdivoca/acdiweb2.nsf/news/gmfoodsarticle

22. Genetic Resources Action International (GRAIN) "Better Dead than GM Fed?", *Seedling* (October 2002).

23. For a more detailed explanation of each of these factors, see Oxfam International, "Crisis in Southern Africa," *Oxfam Briefing Paper 23*, (2002), p. 6.

24. World Food Program (WFP), *Policy on donations of foods derived from biotechnology (GM/Biotech foods)*, WFP/EB.3/2002/4-C, (14 October 2002), pp. 4–5.

25. Institute for the Study of International Migration, *Genetically Modified Food in the Southern Africa Food Crisis of 2002–2003* (Georgetown: University School of Foreign Service, 2004), pp. 16–17.

26. Jon Bennett, "Food Aid Logistics and the Southern Africa Emergency," *Forced Migration Review*, 18, no. 5, (2003), p. 29.

27. The value of Zambia's exports of maize in 2002 was U.S.$2.23 million, according to the FAO. FAO, *FAO Statistical Database*. Internet address: http://faostat.fao.org.

28. Rob Crilly, "Children go Hungry as GM food rejected," *The Herald* (Glasgow), (30 October 2002), p. 12.

29. Bennett, "Food Aid Logistics and the Southern Africa Emergency," p. 29.

30. ACDI/VOCA, *Genetically Modified Food: Implications for U.S. Food Aid Programs*, p. 9; USAID "United States and Food Assistance."

31. Matt Mellen, "Who is Getting Fed?", *Seedling*, (April 2003). Internet address: http://www.grain.org/seedline/seed-03-04-3-en.cfm.

32. Chuck Grassley, "Salvation of Starvation? GMO food aid to Africa," *Remarks of Senator Chuck Grassley to the Congressional Economic Leadership Institute*, (5 March 2003).

33. Norman Borlaug, "Science vs. Hysteria," *Wall Street Journal*, (22 January 2003).

34. For an overview of the technical issues involved in the dispute, see Duncan Brack, Robert Falkner and Judith Goll, "The Next Trade War? GM Products, the Cartagena Protocol and the WTO," *Royal Institute of International Affairs Briefing Paper No.8*, (September 2003).

35. Quoted in Charlotte Denny and Larry Elliott, "French Plan to Aid Africa Could be Sunk by Bush," *The Guardian*, (23 May 2003).

36. Edward Alden, "U.S. Beats Egypt with Trade Stick," *Financial Times* (29 June 2002).

37. EU, *Press Release IP/03/681*, (13 May 2003).

38. EU, *Press Release IP/03/681*.

39. Africa Center for Biosafety, Earthlife Africa, Environmental Rights Action—Friends of the Earth Nigeria, Grain and SafeAge, *GE Food Aid: Africa Denied Choice Once Again?*, (4 May 2004).

40. Heike Baumüller, "Domestic Import Regulations for Genetically Modified Organisms and their Compatibility with WTO Rules," *Trade Knowledge Network*, pp. 13–15. Internet address: http://www.tradeknowledgenetwork.org/pdf/tkn_domestic_regs.pdf

41. FOEI, *Playing with Hunger: The Reality Behind the Shipment of GMOs as Food Aid*, p. 9.

42. Baumüller, "Domestic Import Regulations for Genetically Modified Organisms and their Compatibility with WTO Rules," p. 14.

43. For a history and analysis of these negotiations, see Bail, Falkner and Marquard (eds), *The Cartagena Protocol on Biosafety: Reconciling Trade in Biotechnology with Environment and Development?*; Falkner, "Regulation Biotech Trade: The Cartagena Protocol on Biosafety"; Newell and MacKenzie, "The 2000 Cartagena Protocol on Biosafety: Legal and Political Dimensions."

44. Text of the Cartagena Protocol on Biosafety.

45. Codex Alimentarius Commission, *Report of the Fourth Session of the Ad Hoc Intergovernmental Task Force of Foods Derived from Biotechnology ALINORM 02/34A* (Rome: FAO and WHO, 2003).

46. WFP, *Policy on donations of foods derived from biotechnology (GM/Biotech foods)*, p. 5.

47. World Food Program (WFP), *Policy on donations of foods derived from biotechnology*, WFP/EB.A/2003/5-0B/Rev.1, (2003).

48. Aseem Prakash and Kelly Kollman, "Biopolitics in the EU and the U.S.: A Race to the Bottom or Convergence to the Top?", *International Studies Quarterly*, 46, (2003), p. 625.

49. Robert Paarlberg, "The Global Food Fight," *Foreign Affairs*, 79, no.3, (2000), pp. 24–38.

50. Grant Isaac and William Kerr, "Genetically Modified Organisms at the World Trade Organization: A Harvest of Trouble," *Journal of World Trade*, 37, no. 6, pp. 1086–1090; Prakash and Kollman, "Biopolitics in the EU and the U.S.: A Race to the Bottom or Convergence to the Top?", p. 626.

51. Grassley, "Salvation of Starvation? GMO food aid to Africa."

52. Michael Dynes, "Africa Torn between GM aid and starvation," *The Times* (London), 6 August 2002), p. 12.

53. Rory Carroll, "Zambia Slams Door Shut on GM Food Relief," *The Guardian* (London), (30 October 2002); Institute for the Study of International Migration, *Genetically Modified Food in the Southern Africa Food Crisis of 2002–2003*, p. 20.

54. Polly Diven, "The Domestic Determinants of U.S. Food Aid Policy," *Food Policy*, 26, (2001), p. 471.

55. Christiansen, "The New Policy Environment for Food Aid: The Challenge of Sub-Saharan Africa," p. 257.

56. Brack, Faulkner and Goll, "The Next Trade War? GM Products, the Cartagena Protocol and the WTO," p. 3.

57. Katrin Dauenhauer, "Health: Africans Challenge Bush Claim that GM food is Good for Them," *SUNS: South-North Monitor*, #5368, (23 June 2003).

58. WFP official Richard Lee, Quoted in Greenpeace UK, *Statements on the Southern African Food Crisis* (November 2002).

59. National Association of Wheat Growers (NAWG), "Letter to Robert Zoellick," portions reprinted in the *NAWG Weekly Newsletter*, (13 February 2004). Internet address: http://www.wheatworld.org.

60. Oxfam America, *U.S. Export Credits: Denials and Double Standards* (Washington, D.C.: Oxfam, 2003), p.8.

61. Declan Walsh, "America Finds Ready market for GM food—The hungry," *The Independent* (London), (30 March, 2000), p. 18.

62. Katherine Stapp, "Biotech Boom Linked to Development Dollars, say Critics," *SUNS*, (3 December 2003).

63. Section 1543A of the Farm Bill, cited in ACDI/VOCA, *Genetically Modified Food: Implications for U.S. Food Aid Programs*, p. 5.

64. *The Ecologist*, 33, no.2, (March 2003), p. 46.

65. African Agricultural Technology Foundation (AATF), "Rationale and design of the AATF," *AATF Website*, (June 2002). Internet address: http://www.aftechfound.org/rationale.php

66. Agricultural Biotechnology Support Project II (ABSPII), "Scope and Activities," *ABSPII Website*, (2004). Internet address: http://www.absp2.cornell.edu/whatisabsp2/

67. Mellen, "Who is Getting Fed?"

68. Devlin Kuyek, "Past Predicts the Future—GM Crops and African Farmers," *Seedling*, (October 2002). Internet address: http://www.grain.org/seedling/seed-02–10–3–en.cfm; Mellen, "Who is Getting Fed?"

69. Dominic Glover, "GMOs and the Politics of International Trade," *Democratising Biotechnology: Genetically Modified Crops in Developing Countries Briefing Series*, Briefing 5, (Brighton, UK: Institute of Development Studies, 2003); Brewster Kneen, *Farmageddon* (Gabriola, B.C.: New Society, 1999); Kuyek, "Past Predicts the Future—GM Crops and African Farmers"; Mellen, "Who is Getting Fed?"

70. Jennifer Clapp, "WTO Agricultural Trade Battles and Food Aid," *Third World Quarterly*, (forthcoming).

71. Institute for the Study of International Migration, *Genetically Modified Food in the Southern Africa Food Crisis of 2002–2003*, pp. 18–19.

72. An exception to this is Barrett and Maxwell, *Food Aid After Fifty Years: Recasting Its Role*, who do incorporate some discussion of GMOs.

35

Street Credit: The Cultural Politics of African Street Children's Hunger

Karen Coen Flynn

A fleeting, shadowy segment of documentary footage depicts someone off-camera handing a group of hungry African street boys a pot of cooked rice. Immediately joined by others, a melée ensues as they start shoving and punching one another to get a fistful. This bleak scene in *Darwin's Nightmare* (Sauper 2004) provides a real-life glimpse into one aspect of the cultural politics of street children's hunger in the city of Mwanza, Tanzania.[1] And make no mistake, the causes behind these children's hunger are *cultural* and their power struggles *political*; there is nothing *natural* about chronic hunger in our world.

Some observers may interpret the boys' behavior as typical of street children, and certainly destitute girls and boys often fight over these types of food handouts, as well as prime begging areas, and relationships with particular street adults who provide food. Others might also view the boys' struggle as one over *street credit*, commonly portrayed by pop singers, graffiti artists, and others employing urban/street language cross-culturally as the credibility or status one gains by performing bold, illegal and/or violent acts. Certainly Mwanza's street children occasionally obtained food in such ways by daring to rob child vendors selling penny candy, stealing fruit from market vendors or pick-pocketing other street children.

Yet there was more to the food-related street-credit economy in Mwanza than simply the status gained by street children—in the eyes of other street children—for engaging in bold or unlawful acts. For instance, these youth also sought opportunities to acquire street credit in the eyes of the general public, as in the case where a group of street boys, encouraged by an assembly of adult male bystanders, beat up three street girls who inadvertently crossed into the boys' food-begging turf. A street girl also could acquire street credit among her peers in positive ways, such as by sharing a snack she earned by working at a roadside café. In addition, even private citizens sought street credit among their friends and acquaintances by offering food charity from their kitchens to disparagingly stereotyped "dirty" and "dangerous" street children. So while many of the youth's food-acquisition activities were directly linked to the local street-credit economy, the children were neither the sole participants seeking status therein nor were their efforts to obtain food restricted to audacious, criminal and/or cruel acts.

I recognize that in any situation a child's intent to gain street credit or obtain enough food to survive the moment might overlap. I also understand that the means by which street credit was calculated and disbursed were not only dependent on situational specifics but also the given participants, including the audience at hand. Given this complexity I can only offer very general descriptions of the ways in which these children obtained food and street credit. Yet one of my fundamental points is that street children partook—sometimes willingly, often very reluctantly—in food-acquisition activities that largely shaped, and were shaped by, local meanings associated with the concepts of childhood, work, and gender. Moreover, the street credit associated with these activities further illuminated power/prestige differentials among the various participants engaged with Mwanza's politics of hunger.

Another point that I wish to emphasize is that in spite of the diverse ways in which these children fed themselves, their acquisition of food in the form of private charity appeared to be strikingly significant because in these situations the children were not the only ones seeking to gain an advantage. While a street child sought to benefit by getting something to eat, an alms-giver concurrently sought street credit in the form of prestige in the community or personal favor with God. What was key about this relatively innocuous aspect of the local street-credit economy was that by successfully establishing these types of exchange relationships with alms-givers, "marginalized" street children actively positioned themselves more closely to the "mainstream" than is widely recognized. These actions not only had practical on-the-ground consequences for understanding certain street-youth's survival strategies in Mwanza, but also have broader policy implications concerning the relationship of private charity to public-entitlement programs as well as hunger relief (see Chapter 30).

My interest in the cultural politics of hunger stems from my observation that various people in Mwanza had vastly different opportunities to access food despite an ample year-round supply. In addition, Mwanza was populated primarily by self-described "Africans," with self-identifying "Asians" and "Arabs" comprising a small minority.[2] Despite this diversity, all of the street people and the very poor were Africans, which implied that certain social inequalities distinguished Asians and Arabs from Africans. These inequalities exist, on the one hand, because for several generations Asians—and to a lesser degree Arabs—generally enjoyed greater access to food and wealth as urban-based merchant capitalists who had correspondingly superior connections to transport, financial, and communication services, as well as schooling opportunities and kin-support systems that underpinned successful business networks spanning large regions of the world. On the other hand, Mwanza's Africans had generational roots in subsistence/peasant farming and migrant labor, and many were recent arrivals in the city whose connections with kin living elsewhere were weakened by high transport costs. Because of their relative poverty and diminishing support networks, most Africans had fewer opportunities to exchange goods or services for food and, likewise, were at greater risk of experiencing chronic hunger.

Recognizing the social basis of this disparity, I turned to economist Amartya Sen's work because he is known for his simple, widely applicable "entitlement approach," which asks how social relations determine who goes hungry (Sen 1977, 1981; Dreze & Sen 1989).[3] Sen's entitlement approach is built on the idea that in market economies one's success in obtaining food is directly related to one's lawful ownership—via production, exchange or transfer—of a tangible or intangible commodity that can be exchanged for food. Simply stated, one's exchange possibilities and hunger are inversely related—the risk of experiencing hunger increases as one's opportunities to exchange one's resources decreases. Sen asserts that this is especially true if one lacks opportunities to acquire food from what he refers to as "non-entitlement transfers" such as charity (Sen 1981, p. 3).

At the time of my study, approximately 400 extremely poor, largely independent girls and boys survived on Mwanza's streets. This was a largely contemporary phenomenon. Other than during famine, child poverty in Africa was rare from the late pre-colonial period to the mid twentieth century, and references to street children in Tanzania, specifically, were scarce (Hake 1977; Iliffe 1987; Lugalla & Kibasa 2002). A product of region-wide socio-economic, political, and public health-related shocks that first occurred in the 1970s and 1980s, large numbers of children now live on the streets in many African countries. The population of Tanzanian street children, in particular, also originated in this period and is estimated to have doubled from 5,000 to 10,000 between 1989 and 1993 alone (Rajani 1993; Lugalla 1995).

Living in cities throughout the world, such "homeless" or "street" children are not only the subject of an extensive literature but widely recognized as comprising a vastly heterogeneous, ill-defined group because these children survive in different cultural and environmental contexts, through wide-ranging activities, on fluctuating schedules, and under the threat of varying degrees of violence.[4] Some originate in cities while others migrate from the countryside. Some girls and boys are "throw aways" who were involuntarily evicted from home by their "caregivers"; others are "voluntary" runaways often pushed out by dire poverty and hunger, a lack of schooling opportunities and/or abuse.

Nine girls (see Table 35.1) between the ages of 11 and 15 years and 19 boys (see Table 35.2) ranging from 5 to 16 years participated in my research.[5] I identified "street children" as those younger than 18 years of age who lived, slept, and foraged for food unaccompanied by any adult kin.[6] Try as I might to refine it, my definition proved problematic because most of the children did not know their age, some had contact with relatives, and at least one girl described herself as being "married" to a

Table 35.1 Street girls.

Name	Age	Ethnic affiliation	Region of origin	Kin on streets	Reason on streets	Years in town
Adelina	11	Sukuma	Mwanza		abandoned	0.5
Dotti	13	Sukuma	Mwanza		orphaned	2
Jane	13	Haya	Kagera		orphaned	
Josie	15	Haya	Kagera		ran away	0.5
Leah	16	Sukuma	Mwanza		abusive stepmother	5
Lucy	14	Haya	Mwanza		abusive parents	1
Mariam	13	Sukuma	Mwanza	brother	abusive father	3
Pina	13	Haya	Kagera		abusive stepfather	0.5
Regina	15	Sukuma	Shinyanga		abused as housegirl	1

Table 35.2 Street boys.

Name	Age	Ethnic affiliation	Region of origin	Kin on streets	Reason on streets	Years in town
Ciano	16	Sukuma	Mwanza		abused	1.5
Emma	10	Sukuma	Tabora		abusive stepfather	1
Ibrahim	11	Kerewe	Mwanza			
Iddi	10	Sukuma	Mwanza			3
John	12	Kuria	Mara		abusive father	1
Joseph	14	Sukuma	Kagera		came to town seeking work	
Juma	8	Sukuma	Mwanza	brother	orphaned	
Kato	6	Sukuma	Mwanza		abusive father	1
Kessy	5	Sukuma	Mwanza		orphaned	5
Lucas	14	Haya	Kagera		orphaned	1
Nuru	16	Swahili	Coast			1
Nyakato	14	Jita	Mara		ran away	2
Ramadhani	10	Sukuma	Mwanza		abandoned	
Robert	12	Rangi	Mwanza		orphaned	0.1
Salum	16	Sukuma	Mwanza			
Samson	14	Sukuma	Mwanza			1
Steven	12	Sukuma	Mwanza			
Sweddy	15	Sukuma	Mwanza			1
Theo	11	Nyamwezi	Tabora		abusive stepmother	0.5

street man. Moreover some children slept on the "streets" (at the central bus terminal, under bridges or against shop doorways), while others sought refuge at "shelters" (on the ground at the informal roadside cafes where they worked, in alleys near apartments in which they washed floors, or on the cement veranda at the local street children's center).

Further complicating my definitional challenges was that none of the 28 children in this study self-identified as a "street" child. Instead they asserted that this was simply a temporary event in their lives and that "genuine" street children were always much worse off, more completely alone and/or more permanently homeless. In spite of these definitional challenges, all of the children in my Mwanza sample were severely impoverished and responsible for their own survival, and should not be confused with either those who sold cigarettes, candy or snacks on the streets to help bolster their families' incomes or those who played in the alleys, open sewers, and garbage heaps because their families could not afford to send them to school or to provide them with alternative amusement.

I also encountered methodological problems at the outset of my study because the children initially resisted interacting with me. Much of this fear stemmed from police, *sungusungu* (neighborhood vigilante groups), and military "round-ups"— intermittent sweeps or "Back to the Land" campaigns that since 1920 were the official answer to urban poverty throughout the country. These rural repatriation drives, legitimized by law[7] and sometimes undertaken violently, involved herding Mwanza's visibly destitute into trucks. These people, often comprised of street adults suffering from leprosy, epilepsy, or blindness, were then released in the countryside on the presupposition that they would take up farming and contribute to the nation's food supply.[8]

Prior to beginning my research, my encounters with these children, recognizable by their tattered clothing and/or disease- or parasite-ravaged bodies, as well as their constant entreaties for food or money, were uncomfortable and brief. Yet I soon learned that in spite of the children's desperate situations one thing that they yearned for was respect, which at the earliest stages of any social interaction is most explicitly expressed among the Kiswahili speakers of East Africa through a vast repertoire of greetings. By way of greeting-based inquiries into their health and goings-on around town, I developed a personal rapport with several boys. In addition, my daily visits to a local street children's support center,[9] which became a central meeting place for me and my research assistants,[10] also facilitated my contacts and acquaintances with these children.[11]

There were many reasons why these children were on the streets. Locals often blamed the attraction posed by the city's schools and job opportunities, but the children themselves described the reasons they had moved onto the streets as related to changing social obligations among family members, AIDS, and hunger. For instance, familial commitments had been changing as Tanzanians, like people in many other parts of the world, began to engage in sexual relations with more partners over time and consequently had more children with different mates. Personal choice, job-related migration, and/or poverty led some people to view both their sexual and parenting partnerships as temporary, resulting in child-rearing responsibilities that had once been shared to varying degrees by both men and women, to become largely the obligation of women. Women's inferior status often left them with less wealth and education and in poorer health than men—all of which presented women with greater obstacles to properly caring and advocating for their children (Nelson 1978–1979; Abrahams 1981; Lovett 1989; Kilbride & Kilbride 1990; White 1990; TGNP 1993; Karanja 1994; Potash 1995). It was within this familial context that some children described being forced out of their homes. Some were singled-out from among their siblings and chased away because they were disliked. Others decided to leave on their own because destitution had stretched the children's caregiver(s) beyond their capabilities to nurture them. Eleven of the twenty-eight children said that they were victims of sexual, physical and/or emotional abuse, and a few had barely escaped brutal machete attacks or beatings by a drunken father, an abusive step-parent or a parent's spiteful lover (Varkevisser 1973; Bledsoe 1980; Kilbride & Kilbride 1990; Howard & Millard 1997).

The HIV/AIDS pandemic that has hit sub-Saharan Africa so hard also took its toll on many of these children as attention, food, and other scant resources were diverted from them to family members who were ill, forcing some children to seek their survival elsewhere. AIDS-related deaths were on the rise in Tanzania especially in the neighboring Kagera region, the home to an estimated 72,000 orphans by 1994—and the area from which many of Mwanza's street children originated (*The Daily News*, March 26, 1994). Once orphaned, children were sometimes kicked out of their surviving or new caregivers' homes because these adults found it too difficult to cope with the additional childcare and/or financial demands. Some orphans ran away because they were considered less entitled to household food and other resources than their caregivers' biological children (Barnett & Blaikie 1992; Whyte 1995; Hunter 2003).

Hunger also was often claimed by the girls and boys as a reason they took to Mwanza's streets. Some children explained that they went hungry because their single or divorced mothers had little money to buy food, especially any food that tasted good, was nutritious or of a wide variety. Eight-year-old Juma explained that he preferred rice over *ugali* (maize-meal porridge, and a Tanzanian main staple) because "at home . . . I ate *ugali* and *dagaa* (tiny lake sardines) every day until I was sick of it." In other cases the children's caregivers were too poor to provide them with food. Robert, 12 years old, recounted that "after my parents died I was taken to live with my aunt. She didn't have work and raised me only by luck. She would beg for food from her neighbors and friends, but it was never enough." Not surprisingly, more than half of the 122 street children interviewed in another street children's study in Mwanza also cited "hunger" as a primary reason they left home (see Rajani & Kudrati 1994, p. 3).

The anthropological literature on children clearly shows that throughout the world there are a "plurality of childhoods . . . stratified by class, age, gender and ethnicity, by urban or rural locations and by disability and health" (Jenks 1996, pp. 121–2; see also Aries 1962; Prout & James 1990; Reynolds 1991; Vreeman 1992; Blanc 1994; and Seabrook 2001). Cross-cultural models of childhood differ from many present-day, North American or Western European middle-class ones. For instance, many African children often begin to work at an earlier age and commonly spend lengthier periods of time at a greater distance from their natal homes (see Varkevisser 1973; Reynolds 1991; Harkness & Super 1991; Rwezaura 1998; Panter-Brick 2000). In 1993–94 many people in Mwanza appeared to hold this same view of childhood. I often saw four- or five-year-old children caring for younger siblings outside their houses or apartment flats. Very young girls helped their mothers pick impurities out of rice and rinse laundry; boys corralled wayward chickens and tended to goats or cows staked in their yards. Explained 42-year-old Rose, a mother of ten, "Even before the children are old enough to go to school they learn their responsibilities." According to her, when they were young all of her children had carried groceries, swept the compound, washed cooking pots, and collected water because "even when their hands are little they can do many things."

Yet at the same time, argues law scholar Bart Rwezaura, this image of child-as-laborer has changed and expanded so that in diverse situations throughout contemporary Africa there are many "competing images" of childhood (1998, p. 253). I witnessed this in Mwanza. At the time of my study there were upper-income families who proudly sent their children to boarding schools in other parts of Tanzania, neighboring Kenya or India with the expectation that they would work hard for good grades, get well-paying jobs, and eventually support their parents in their old age. Middle-income families routinely sent their children to local schools, as well as expected them to labor to meet the daily domestic demands of the household. Many poor caregivers could not afford to send their children to school at all and instead sent them to wealthier families in Mwanza or elsewhere to work as household, farm, and business laborers. Rwezaura emphasizes that "the decision to give away children as labour is not always self-evident or taken without a heavy heart . . . [some believe] this to be a traditional practice, others do so out of economic necessity" (1998, p. 258).[12]

During my research in Mwanza I learned that the association of children with work was very strong and some of the city's residents viewed the street children disdainfully as lazy escapees of rural village life. A middle-aged Asian pharmacist sputtered that the children "believe that life here [in town] is easy [and] they come here to escape their chores." I learned that in the eyes of the general public, most of the street children were viewed as old enough and apparently healthy enough to work to support themselves—as opposed to requiring charity to survive. Explained an approximately fifty-year-old Asian woman, who managed a small dry-goods shop,

> I see so many of the children washing taxis. I know they can work. Very few are lepers. If they want to sweep the doorway [to the shop] I will give them a few shillings. But when they beg I just laugh . . . If they want to work, there is work.

Mwanza's politics of hunger were played out in this cultural context wherein street children's food-acquisition activities shaped and were shaped by the meanings people associated with childhood and work. For instance, casual employment in a relatively wide variety of jobs was the primary means through which many of the older boys, ranging in age from 10 to 16, tried to survive. But finding work was difficult, which lent such an accomplishment a correspondingly high level of street credit. While colonial-era laws that were still on the books restricted the employment of children (to protect them from harm and abuse, to discourage children from coming to town, and to emphasize the unemployability of independent children), it was the widely held stereotype of street boys as thieves that left open to them only a few very low-paying, casual, one-time jobs in which they could be closely supervised and did not have any responsibility over objects of any value (Coppock 1951; Lugalla 1995). Boys washed taxis and cars, swept storefronts, and scrubbed laundry or floors in local homes. The boys also labored in the food trade scouring cooking pots and hauling water for food-stand operators. Some entrepreneurial boys risked cuts and tetanus as they wandered barefoot through roadside trash heaps collecting recyclable valuables to sell, such as unburned pieces of charcoal and tobacco from discarded cigarette butts. Others scavenged to gather insects, small fish, and food waste to sell as bait to fishermen. Some boys fished from shore and then sold their catch to roadside vendors to fry and resell. Still others roamed the streets selling penny candy, gum, peanuts, and cigarettes or worked in roadside stands selling soda and sundries.

The children's work opportunities also were governed by local meanings associated with gender. Even fewer jobs were deemed as suitable for the girls and this situation held important implications not only for the girls' street-credit status, but their visibility in public as well. My first impressions in Mwanza were that street boys drastically outnumbered girls, but the ratio of boys to girls was narrower than appearances implied because many of the city's street girls, like elsewhere in Tanzania, cooked, cleaned, and cared for children in more hidden settings as house-girls (Lugalla & Kibasa 2003). Five of the nine girls in my study mentioned previously working as housegirls. While all housegirls were vulnerable to their employer's exploitation, street girls were particularly susceptible because they had neither supportive kin to return home to nor any other caring authority to whom they could voice complaints.[13] In addition to suffering from burns, breathing problems, and upper-respiratory infections (from tending smokey fires and sleeping on cold, bare floors), chronic malaria (from sleeping without mosquito nets), and malnourishment (from eating only leftovers), some girls reported being pressured to have sex with their male employer or his "friends" (Onyango 1983; Stichter 1985; Sheikh-Hashim 1990; Rajani & Kudrati 1994).

Yet while they often were exploited, the street girls in my study were not powerless victims. They cited intentionally quitting their jobs without notice or manipulating employer's sexual demands to acquire money or other gifts—or just to spite the "mother of the house" (TGNP 1993, p. 69). In short, the former housegirls included in my sample found life on the streets preferable to working under such dire conditions, and those who left such jobs to move onto the streets were often welcomed by small groups of existing street girls and accorded the street-credit status commensurate with making such a bold move.

In addition, there is a long history of independent females in African urban areas being viewed as prostitutes and this stereotype certainly hurt the street girls' job prospects (White 1990; Robertson 1995). The girls also feared pursuing the same income-generating activities undertaken by the boys, such as washing cars and selling peanuts, candy, and cigarettes because they believed that the boys would harass them and generally "sabotage" their efforts (Plummer 1994, p. 86). Working at local food stands was just about the only place outside of domestic service where they could get jobs. "I wash cooking pots and utensils and fetch water," recounted eleven-year-old Adelina. "[Some of] the money I get from Mama Nitilie is for helping her carry the pots and utensils from her stand to her home on Bugando hill."

Because of the severe limitations getting work, the girls were often forced to survive via sexual exchanges. While none of the boys in this study acknowledged exchanging sex for food, some of the youngest street boys in Mwanza and elsewhere in Tanzania were known to have done so. This is not to say that older boys never had sex or did not exchange sex for food or money. Boys in Mwanza had consensual sex with and raped the street girls, in addition to practicing *kunyenga* (slang for non-consensual, anal-penetrative sex) among themselves as a group-member initiation rite (Rajani & Kudrati 1996; Lugalla & Mbwambo 2002; Lockhart 2002; Lugalla & Kibasa 2003). But sex-for-food practices did not appear to be a regular occurrence among the self-provisioning practices of older boys in Mwanza or other places in Tanzania (Muhimbo Mdoe and Fred Nyiti[14], personal communication; Lugalla 1995).

Girls more commonly relied on a complex variety of sexual relationships ranging from quick one-time encounters with strangers to long-term "boyfriends."[15] Fifteen-year-old Josie described how "[some]times I go directly to have sex with my boyfriend, John, and he gives me money to buy food, soda and juice." Girls like Josie found sex a viable survival strategy because it required neither a formal education nor a high input of capital. Yet their "survival sex" activities could take different forms, thereby blurring the conceptual boundaries between quick commercial/exploitative sex and other types of friendships and nurturing partnerships.

Most of these girls had what they called "*rafiki* or *marafiki*" (literally "friend or friends" but also in this case "boyfriend or boyfriends")[16] with whom they exchanged affection, protection, sex, and food. These males usually were young (under 18) and poor and, even though they were not themselves impoverished or homeless, they were in no financial position to permanently house, feed and/or clothe the girls.

In other instances girls exchanged sex with street boys as well as adult males, some of whom also lived on the streets. But while these partnerships could be long-term they also appeared to be based largely on food- or money-for-sex transactions. The other adult men with whom these girls had sexual encounters were often complete strangers who accosted and frightened the girls into having sex with them. Given the girls' lack of bargaining power and the limited financial resources of their partners, one sexual encounter usually generated only enough food to sustain the girls for the moment—if the girls were given anything at all. For all of these reasons the girls' sex-for-food relationships could involve an intricate mixture of dependency, support, affection, threats and exploitation.[17] They also involved a continuum of street credit, social status and power. On the one hand, the girls' relative assertiveness and status were apparent in their ability to command money or food in exchange for sex, whether from a "boyfriend" or one-time partner. On the other hand, their power-lessness, worthlessness, and hopelessness were apparent in that they often went uncompensated for consensual sex or were forcibly raped (see also Bamurange 1998).[18] In sum, the public stereotype of street girls as prostitutes was easily self-replicating; the public restricted their survival strategies to such an extent that the girls had to engage in casual sex work to survive.

Besides casual employment and sex work, Mwanza's street boys and girls also engaged in theft.[19] While theft falls outside the analytical frame of Sen's entitlement approach because it is illegal, I mention it here because some of the children were proficient thieves, and accorded relatively high amounts of street credit for accomplishing these brazen acts, even though they stole things to a much lesser extent than their widespread stereotype might imply. The girls often described stealing food and clothing. The boys often told of, or were seen, stealing oranges, carrots, or tomatoes from market vendors, stealthily pick-pocketing rival street boys while they slept, violently beating and/or robbing one another, or grabbing car parts or valuables from vehicles belonging to careless drivers. Yet most of the children described theft as a safety net used only when they were experiencing dire hunger; they knew that if they were caught they risked a brutal—even fatal—beating in the name of mob "justice" or arrest and then months languishing in jail.[20]

Local meanings associated with childhood, work, and street "girl" as opposed to "boy" created clear gradients of power that influenced these children's entitlement to food and their corresponding street credit, both among one another and within the wider community. Twelve-year-old John was given special street credit among his peers for his thieving exploits, because as one younger boy explained, "He always gets what he wants, nobody hits him . . . and sometimes he shares with us." Certain older girls were held in particularly high esteem, especially among younger ones, because of their cooperation in helping the latter to secure nourishment. Older girls in Mwanza, like those elsewhere in Tanzania, sometimes offered their services if only one had to have sex to acquire enough food or money to provision the group (Bamurange 1998). The relatively high amounts of street credit given these girls was usually fleeting at best, however, because the girls lacked the emotional maturity needed to sustain long-term group membership without fighting over their meager resources and connections to others.

The street credit accorded the boys by the general public also varied and was often short-lived. For instance, while some of the boys were able to find repeat work from sympathetic patrons, the majority of the boys' jobs changed from day to day. "To get food myself is very difficult because I do not have important work," explained Theo, 11 years old. "Some days I remain unemployed the entire day and just drink water and sleep." Still, a few boys and girls were successful in acquiring more durable street credit and status in the community. For example, Leah, 16, had diverse employment opportunities, had lived in Mwanza for many years, and knew many community contacts. She acquired food and money through her sexual relationships with several regular partners and ate at the food stands in return for carrying water and cleaning cooking pots. "Getting food is easy," she said. Yet Leah's situation was unique. She had lived in Mwanza for five years, which was longer than any of the other children included in this study.

As a whole, both street boys and girls suffered at the hands of the local community, but the girls had even fewer opportunities to gain entitlement to food. Almost all of the children depended on food-stand operators and roadside-snack vendors for their "meals," although many of the boys and girls in my sample ate a daily snack (usually a banana or some biscuits/crackers) distributed by the street children's support center. Yet for all of their efforts the girls, in particular, usually acquired little by way of quantity, quality or variety of food. Their diets were comprised largely of *ugali*,[21] *uji* (watery gruel made from millet, wheat, maize, or cassava flour) or rice with a side relish of kidney beans and/or spinach (or other cooked vegetable leaves). As a group the boys ate a slightly better diet because they could afford a little more animal protein such as fish, chicken, and goat meat. Six of the boys who attended school ate school lunches. They also were given a simple evening meal of beans and rice cooked at the local street children's center because their schooling left them little time to earn extra money. In addition, several boys ate at the houses or apartments in which they washed clothes and scrubbed floors. If the girls scavenged raw pieces of dried cassava root or caught grasshoppers they roasted them over bits of charcoal found in waste heaps, but none of them regularly cooked food. Although the quantity and the quality of the food varied, the girls and boys employed at the food-stands commonly were successful in eating at least two meals a day, which was a relative bonanza for some of the children. As 15-year-old Regina stated, "I get more food from the stand than I did at home in Sengerema."

Still, there was a particular way in which some of Mwanza's street youth obtained food—via *private charity*—that was key in delineating the street-credit economy and power gradients therein. The children's experiences with private charity also hold important implications for expanding the utility of Sen's entitlement approach. For example, begging for alms was the mainstay of the youngest street boys. Five-year-old Kessy made daily visits to his benefactors at a gas station, a roadside café, and a large newspaper stand—all in hope of collecting a ten-shilling piece or two (U.S.$0.02 or

0.04). Due to their ages Kessy and his best friend and nearly constant companion, six-year-old Kato, were both provided a free daily meal of rice and beans at the expense of the local street children's center. But they regularly supplemented their diet with peanuts and penny candies that they bought with money collected along their donor circuits, as well as with donations that they acquired from a few roadside snack vendors.

Many of the older boys also begged for alms from people they met on the streets, especially any visiting expatriate tourists, researchers, or staff overseeing development projects in greater Mwanza region. Some of the boys were very accomplished at pulling at the heartstrings of these visitors, many of whom had seen aid-agency advertisements portraying "typical" dirty, hungry, "Third World" children. For this audience the boys made pitifully sad facial expressions and used body language conveying hunger. One to two hours of begging outside the town's largest western-style hotel or the popular ice cream shop could earn a boy anywhere from 100 to 1,000 shillings (U.S. $0.20–2) and it could take all day to earn that by selling candy on the streets. "Some boys don't like to work hard," explained 11-year-old Ibrahim, "so they beg instead." Two boys also ate meals and snacks offered in some of Mwanza's private kitchens as alms on Fridays, the Muslim holy day. The few boys who were regularly fed by charitable households were commonly offered pilau (spiced rice), chicken, and beef at Arab or African houses and bean-based dishes with rice, flat breads, and yogurt at Asian houses.

While the street girls tried their hand at begging, they were less likely to do so at the central expatriate hangouts in town because the boys regularly controlled this turf and the girls risked severe beatings if they attempted to beg there. Instead, the girls commonly approached taxi drivers, roadside food-stand customers, and produce vendors with their pleas for alms. Yet again, the street girls were less successful than the boys in gaining access to charity because females are so widely associated with food in Mwanza, as well as throughout much of sub-Saharan Africa. Females grow, prepare, and distribute the vast majority of the food on the continent and street girls were commonly viewed as idle-but-capable food provisioners—and certainly not suitable for charity (Raikes 1988). Mama Gina, approximately thirty-five and a banana vendor in Soko Kuu, underscored this point for me as three street girls passed by her table, "They [the girls] come here to have sex with the market laborers. Then they get sick. It is easy for girls to get started in the food trade. They can sell bananas, mangoes, tomatoes. I did. They just eat their capital."

In thinking over the boys' and girls' various experiences with begging, I wondered how their dependence on private charity meshed with Sen's entitlement approach. Sen writes in *Poverty and Famines* that once people lose entitlement to food—that is lose their ability to exchange their resources for food—starvation is inevitable unless they can acquire food from *"non-entitlement transfers (e.g. charity)"* (1981, p. 3, italics mine). Why does Sen consider private (non-state-sponsored) charity as non-entitlement transfers? Two reasons come to mind: (1) with his emphasis on entitlement as based on an *exchange* of resources between parties, and with his use of the term "transfer" in association with charity, perhaps Sen views private charity as involving only a one-way conveyance between parties as opposed to a two-way exchange; and/or (2) given the entitlement approach's emphasis on the legal owner-ship of one's resources, perhaps Sen does not view charity as legal.

But is private charity only a one-way transfer as Sen seems to imply? To answer this question one needs to delve deeper into the meanings alms-givers assigned to their activities. While it was often impossible to authenticate donors' goals in offering charity to others, I offer an example as to what compels alms-givers to extend charity to street children. An Arab family, comprised of Mama Mahmedi and her grown children, sought religious favor through their charitable deeds in which they had shared their limited resources over the years with nearly 20 street boys, whom they circumcised, converted to Islam, and sent to school so that the family would "prosper in the eyes of God." I believe that the blessings/salvation these alms-givers sought are another form of street credit that formed a key component of the *two-way exchanges* taking place between alms-givers and street children. Alms-givers offered primarily tangible commodities (food, money, clothing, shelter and/or education) and street child recipients provided largely intangible ones (opportunities to acquire street-credit for helping the needy, or religious favor in this life or salvation in the after-life, and/or perhaps even a link to inexpensive labor). If Sen views charity as falling outside the entitlement approach because it appears to be only a one-way transfer that is provided voluntarily and spontaneously by those who are genuinely con-cerned, the experiences of some of Mwanza's street children demonstrate otherwise. These incidents also support what French sociologist Marcel Mauss asserted in his classic text *The Gift:* "Prestations which are in theory voluntary, disinterested and spontaneous are in fact obligatory and interested" (1967, p. 1).

The other reason that Sen might avoid including charity as an exchange entitle-ment is because of charity's other-than-legal status. Development economist Siddiq Osmani argues that in Sen's entitlement approach:

> The transfer component ... includes only those transfers to which a person is legally entitled—for example, social security provisions of the state ... [a]lthough there is nothing illegal about receiving charity, it is not counted as part of [one's possibilities of exchange] for the simple reason that one is not legally entitled to charity, whatever may be one's view about the poor's moral entitlement to it (1995, p. 255).

Yet while Sen and Osmani make evident their assessments of a legal and non-legal entitlement, what remains confusingly vague are their assumptions about private charity that make it irrelevant to the entitlement approach. I believe that this may stem from their more quantitative economic approach, which may inadvertently contribute to their glossing over of vital considerations of both *time* and *social process* in their characterizations of charitable goods and exchanges.

Anthropologist Arjun Appadurai's (1986) view of the "social life of things" is particularly useful in revealing the temporal and processual aspects underlying the direct relationship between private charity and many government-sanctioned entitlements. As Appadurai argues, it is typically the "phase" or historical context of an item's culture-based "career" that defines its exchangeability. In other words, removing a charitable item from its given social and historical context may cause problems in determining its real value. I believe, in the same vein, that removing charitable *exchanges* from their "career" trajectories also can unintentionally conceal the close relationship between private charity and legal/public/governmental entitlements (such as the social security provisions to which Osmani referred above) because doing so erases their cultural history and development. As demonstrated by economist Robert William Fogel (1999, 2000), legal entitlements in the United States and in other industrial countries very often were born out of wide-ranging, non-legal, private charitable exchanges employed and promoted by religious or other morally motivated organizations remonstrating and advocating for political change. Key to my extension of Sen's entitlement approach, therefore, is the line of reasoning that in the instant a particular non-legal form of charity (such as faith-based organizations' food hand-outs to the un- or under-employed, for example) becomes legal (as in the form of government-subsidized food stamps) vast numbers of people may already depend on it for food and their survival (see Fogel 1999, 2000; Flynn 2005).

This research opens up many new questions. What exactly is the nature of charitable exchange? If the acquisition of certain kinds of street credit is a motivation to alms-donors, how strong is this non-monetary incentive and how can it be further encouraged? To what extent are cast-off, derided and "marginalized" children (and adults) actually integral to the functioning of the "mainstream" status- and salvation-oriented economies in other places in the way that they appear to be in Mwanza? Do the same processes that fueled the transformation of charity-based protest movements into governmental realignments and legally sanctioned entitlements in many industrial countries work in the same way in the world's poorest countries that are undergoing vast political-economic change? Through an understanding of the full range of meanings that people in Mwanza and elsewhere assign to their alms-giving activities, I believe that advocates and policymakers worldwide can more effectively facilitate destitute persons' fulfillment of their food and other basic needs through increased access to private charity. Having said this, I do find it very troubling that we need to develop opportunities for the hungry to gain entitlement through private charity rather than through state-guaranteed entitlement programs. It is this emphasis on gifts instead of rights that is the problem (Poppendieck 1998). But given the trend toward a widening gap between the haves and have-nots both in cities such as Mwanza (and in other parts of the world), coupled with ongoing reductions in public-service offerings, the reality of the situation suggests that more rather than fewer people will be seeking charity in attempts to relieve their hunger, at least in the near future, and access to private charity *must* be part of advocates' and policymakers' assessments of hunger relief, food entitlement, and the local politics of hunger.

Notes

1. This study was set in Mwanza, Tanzania, during ten months of anthropological research in 1993–94 and several weeks in 2000. In 1993 Mwanza had a rapidly increasing population estimated at nearly 277,000 that was predicted to exceed 1.3 million by 2011 (Mwanza Master Plan (1992–2012) 1992). The city is situated on the southeastern shore of equatorial Lake Victoria, and is an industrial center of fish, textile and leather processing, soda bottling, and furniture building. It is also a governmental seat and transportation hub of the greater Lake Zone.

2. The umbrella term "Asian" is commonly used by the people of East Africa to refer to people of Indian or Pakistani ancestry, including Hindus, Muslims, Christians, Sikhs, and those belonging to related sects. There were significant differences among Mwanza's Asians in regard to Hindu/Muslim splits, allegiance to the Aga Khan, and issues pertaining to the historical partition of India. Yet the strong bonds of ethnicity, place of origin, and a shared history of "otherness" in relation to the Africans and Europeans in East Africa had contributed to the formation of a common Asian identity in Mwanza (Mangat 1969; Bienen 1974).

3. Of course there are important differences between "famine," which Sen's early work addresses specifically, and the "chronic hunger" that he takes up later (see Sen 1987)—and that I witnessed among Mwanza's street children (see Flynn 2005). Yet the approach's broad applicability rests on the way it complements the "food shortage" and "political crisis" schools of thought (Devereux 2001, p. 248; Messer & Shipton 2002, p. 229).

4. See for example Dallape 1987; Iliffe 1987; Hardoy & Satterwhite 1989; Swart 1990; Reynolds 1991; United Nations 1991; Gunther 1992; Williams 1993; Blunt 1994; Campos, Raffaelli, Ude et al. 1994; Ennew 1994a, 1994b; Rajani & Kudrati 1994, 1996; Scheper-Hughes & Hoffman 1994; Lugalla 1995; Mulders 1995; Save The Children Fund (UK) 1997; Hecht 1998; Marquez 1999; Lugalla & Kibasa 2002, 2003.

5. All of these children were African (as opposed to Asian or Arab) and 17 claimed affiliation (through either one or both parents) to the regionally predominant Sukuma. The others associated themselves with Haya, Jita, Rangi, Kuria, Nyamwezi or Kerewe groups. All of these children came from areas outside of the city, and some had originated from as far away as Kenya and Burundi. In the past decade the situation has changed and more children from Mwanza's slums are moving onto the streets.

6. Laws defining the age of majority in Tanzania are inconsistent. I chose this age based on the outreach practices of a local street children center, the *kuleana* Center for Children's Rights.

7. This included laws such as the Townships (Removal of Undesirable Persons) Ordinance of 1944, the *Kilimo cha Kufa na Kupona* (Agriculture for Life or Death) campaign of 1974–75, and those associated with the *Nguvu Kazi* (Hard Work) Act of 1983. Under the law destitute people could be identified as "loiterers" and "criminals," and trucked out into the countryside under the premise that they would work growing crops. Despite their long history these campaigns usually failed to accomplish anything more than temporarily clearing the streets. As soon as they were dropped off in the rural areas, and given any meager food rations and/or cash to begin their new lives, many of the "repatriated" began to make their way back to town (Heijnen 1968; Bryceson 1990; Lugalla 1995).

8. For more on Mwanza's street adults and their food-provisioning practices see Flynn 1999, 2005.

9. The *kuleana* Center for Children's Rights was a non-governmental agency offering medical, legal, educational, and nutritional support to approximately 120 of Mwanza's street children.

10. Several secondary-school graduates assisted me at various times with my larger research, of which our work with street children was only a part. These were Philbert Bugeke, Agripina Cosmos, Gilbert Maganga and Emma John.

11. The 28 children included in this study were formally interviewed once to learn of the children's life histories and food-acquisition activities either by me or one of my assistants. Yet because I was in regular contact with many of them, I was able to reconfirm and collect new information over the course of weeks and even months, depending on the amount of time these very mobile boys and girls spent in town. I acquired much of the information about the boys, in particular, during countless informal discussions with them while walking either individually or in groups through the streets or marketplaces, while sitting near the bus stand, ferry ports or post office, or while hanging around outside of my house or the street children's center. My assistants' interviews and my long conversations with some of the center's staff members also helped me acquire information from the shyest boys and many of the girls.

12. Contemporary child labor is complicated further because children are socialized to respect authority in ways that made voicing complaints both inappropriate and an invitation for corporal punishment, which is practiced widely in Tanzanian homes and schools. In addition, Tanzanian parents, like others, may be powerless to voice grievances against their children's employers not only because of local customs concerning disciplinary relationships between child laborers and their employers, but because of the parental need for income (Thompson 1971; Bledsoe 1990; Blanchet 1996; Rwezaura 1998).

13. Employing girls under the age of 12 is illegal under the Tanzanian Employment Ordinance (Shaidi 1991; TGNP 1993). Yet there is a very fine line separating paid employment from the regular domestic responsibilities of girls. This situation makes it difficult for labor officers to prove that an arrangement is illegal, especially if both the housegirl and the employing family can verify even distant kinship connections between them (Sheikh-Hashim 1990). Anecdotal evidence suggests that some of the authorities appointed to enforce these child-labor laws employ housegirls illegally themselves.

14. Muhimbo Mdoe and Fred Nyiti were senior educators with Street Kids International in Dar es Salaam.

15. Acquiring accurate information from the girls about their sexuality was difficult. Four of the nine street girls described exchanging sex for food or money, although it was very likely that all of them did (Mary Plummer,

personal communication, February 28, 1994). Some might have felt uncomfortable discussing it given the stigma prostitution carried, others might not have identified their sexual encounters as a direct means through which they acquired food, and a few might not have exchanged sex for food on a regular basis. Moreover, the girls may have lacked the language with which to analyze what was happening to them in their sexual relationships. On the whole, however, sex was an integral part of their survival.

16. In Mwanza the plural form of *rafiki* was commonly pronounced without the prefix "ma" that distinguished the plural form *marafiki* (friends, boy/girlfriends) from the singular form of the word *rafiki* (friend, boy/girlfriend) among other Swahili speakers in Tanzania.

17. The girls' sex-for-food exchanges appeared to be a bitter twist on the exchanges of money or gifts that are integral to many adult marriage and sexual relationships in Africa and elsewhere around the world, except these girls were forced into these exploitative relationships at a much younger age, and neither gained economic advantage from these exchanges nor fulfilled their culturally sanctioned roles as provisioners of food.

18. The girls' dependency on sexual exchanges to acquire food also left them vulnerable to the physical and emotional difficulties of unplanned pregnancies, botched or back-alley abortions, miscarriages, premature deliveries and poorly spaced births. They also suffered repeatedly from beatings and rape. Not only were the girls' physically immature bodies more susceptible to sex-related injuries, but these injuries made them more vulnerable to HIV infection. All unprotected sexual activities made them susceptible to sexually transmitted diseases and as many as 20% of the young people living in Mwanza were infected with HIV (Rajani & Kudrati 1994). These girls and their partners also had little access to protective condoms, and when they did the girls lacked the negotiating power to encourage males to wear them.

19. At the same time what constituted "theft" was hardly clear-cut among Mwanza's residents. Street children could be seen scavenging spilled or discarded foods at marketplace peripheries. At one of the local markets street girls competed with the poor elderly women employed to collect *dagaa*, maize and bits of dried cassava that fell to the ground from their deteriorating burlap sacks during the rearranging of inventory. The women angrily branded the scavenging girls as "thieves" because the children "stole" the goods the women collected for pay. Yet the girls were tolerated by some vendors because the foragers cleaned the area at no cost. The girls often gave their uncooked booty to sympathetic "*Mama Nitilies*," who either cooked it for the girls or gave them some other food in return. Some people explained that out of sympathy they ignored thefts-in-progress, especially by the very youngest street children, or else took the opportunity to educate the children on the proper way to beg.

20. Street boys also were regularly used by adult criminals to commit crimes. If caught by the police, the children suffered because the court and prison systems were poorly equipped to deal with the children in an appropriate and timely manner. In 1993 there was a months' long backlog in Mwanza's courts of the hearing of children's cases. This left children languishing in the filthy, overcrowded municipal Butimba jail. Once their cases made it to court, youths were often poorly represented in court hearings and given adult sentences involving corporal punishment. Long-term confinement in the same jail cells as adults was common and resulted in children suffering unbearable trauma from physical and sexual abuse at the hands of their adult cell mates (Shaidi 1991).

21. In 1993–94 Tanzanians lacked any broad, systematic, class-specific patterns of food consumption or in other words what anthropologist Jack Goody (1982) refers to as "high" and "low" cuisines (see also Bourdieu 1984). In spite of some regional variation (cooked bananas in the far northwest and rice along the Indian Ocean coast) *ugali* is what made a Tanzanian meal a meal. Yet there were several grades of maize flour that could form the basis of the *ugali* and the more processed maize flours were viewed by locals as "smoother," "sweeter," "whiter" and more desirable. They were also more expensive. So the children often ate darker grades of *ugali* or more commonly even darker millet-based *uji*, which was widely recognized as "less pure," "the color of soil" and "poor people's food."

References

Abrahams, Raymond (1981) *The Nyamwezi Today: A Tanzanian People in the 1970s*. Cambridge: Cambridge University Press.

Appadurai, Arjun (ed.) (1986) *The Social Life of Things: Commodities in Cultural Perspective*. Cambridge: Cambridge University Press.

Aries, Phillipe (1962) *Centuries of Childhood*. New York: Vintage.

Bamurange, V. (1998) "Relationships for Survival—Young Mothers and Street Youth." In Magdalena Rwebangira and Rita Liliejestrom (eds) *Haraka, Haraka . . . Look Before You Leap: Youth at the Crossroads of Custom and Modernity*, pp. 221–47. Stockholm: Nordiska Afrikainstitutet.

Barnett, Tony and Piers Blakie (1992) *AIDS in Africa: Its Present and Future Impact*. New York: Guilford Press.

Bienen, Henry (1974) *Kenya: The Politics of Participation and Control*. Princeton: Princeton University Press.

Blanc, Cristina Szanton (1994) "Introduction." In Cristina Szanton Blanc, (ed.) *Urban Children in Distress: Global Predicaments and Innovative Strategies*. Yverdon, Switzerland: Gordon and Breach Science Publishers for UNICEF.

Blanchet, Therese (1996) *Lost Innocence, Stolen Childhoods*. Dhaka: University Press.

Bledsoe, Caroline, H. (1980) *Women and Marriage in Kpelle Society*. Stanford: Stanford University Press.

—— (1990) No Success Without Struggle: Social Mobility and Hardship for Foster Children in Sierre Leone. *Man* 25.

Blunt, Adrian (1994) "Street Children and Their Education: A Challenge for Urban Educators." In Nelly P. Stromquist (ed.) *Education in Urban Areas: Cross-National Dimensions*. Westport: Praeger.

Bourdieu, Pierre (1984) *Distinction: A Social Critique of the Judgment of Taste*. London: Routledge and Kegan Paul.

Bryceson, Deborah Fahy (1990) *Food Insecurity and the Division of Labor in Tanzania 1919–1985*. London: MacMillan.

Campos, Regina, Marcela Raffaelli, Walter Ude M. Greco et al. (1994) Social Networks and Daily Activities of Street Youth in Belo Horizonte, Brazil. *Child Development* 65.

Coppock, J. R. (1951) Report to the Provincial Council in 1950 Subsequent to the Month of September. In Proceedings of the Seventh Meeting of the Lake Province Council February 27–March 2, 1950, vol. 1.

Dallape, Fabio (1987) *You Are a Thief: An Experience with Street Children*. Nairobi: Undugu Society of Kenya.

Department of Urban Development (1992) *Mwanza Master Plan (1992–2012)*. Ministry of Lands, Housing and Urban Development. Dar es Salaam: Department of Urban Development.

Devereux, Stephen (2001) Sen's Entitlement Approach: Critiques and Counter-Critiques. *Oxford Development Studies* 29 (3): 245–63.

Dreze, Jean and Amartya Sen (1989) *Hunger and Public Action*. Oxford: Clarendon Press.

Ennew, Judith (1994a) Less Bitter Than Expected: Street Youth in Latin America. *Anthropology in Action* 1 (1): 7–10.

—— (1994b) "Parentless Friends: A Cross-Cultural Examination of Networks among Street Children and Street Youth." In Frank Nestmann and Klaus Hurrelmann (eds) *Social Networks and Social Support in Childhood and Adolescence*. Berlin: Walter de Gruyter.

Flynn, Karen Coen (1999) Food, Gender and Survival Among Street Adults in Mwanza, Tanzania. *Food and Foodways* 8 (3).

—— (2005) *Food, Culture, and Survival in an African City*. New York: Palgrave.

Fogel, Robert William (1999) Catching Up with Economy. *The American Economic Review* 89 (1): 1–21.

—— (2000) *The Fourth Great Awakening and the Future of Egalitarianism*. Chicago: University of Chicago Press.

Goody, Jack (1982) *Cooking, Cuisine, and Class: A Study in Comparative Sociology*. New York: Cambridge University.

Gunther, Hartmut (1992) "Interviewing Street Children in a Brazilian City." *The Journal of Social Psychology* 132 (3).

Hake, Andrew (1977) *African Metropolis*. New York: St. Martin's Press.

Hardoy, Jorge and David Satterwhite (1989) *Squatter Children: Life in the Urban Third World*. London: Earthscan Publications.

Harkness, Sara and Charles M. Super (1991) "East Africa." In Joseph M. Hawes and N. Ray Hiner (eds) *Children in Historical and Comparative Perspective*. New York: Greenwood.

Hecht, Tobias (1998) *At Home on the Street*. Cambridge: Cambridge University.

Heijnen, J. D. (1968) *Development and Education in Mwanza District (Tanzania): A Case Study of Migration and Peasant Farming*. Rotterdam: Bronder-Offset.

Howard, Mary and Ann V. Millard (1997) *Hunger and Shame: Child Malnutrition and Poverty on Mount Kilimanjaro*. New York: Routledge.

Hunter, Susan (2003) *Black Death: AIDS in Africa*. New York: Palgrave Macmillan.

Iliffe, John (1987) *The African Poor: A History*. Cambridge: Cambridge University Press.

Jenks, Chris (1996) *Childhood*. New York: Routledge.

Karanja, Wambui Wa (1994) "The Phenomenon of 'Outside Wives': Some Reflections on its Possible Influence on Fertility." In Caroline Bledsoe and Gilles Pison (eds) *Nuptiality in Sub-Saharan Africa: Contemporary Anthropological and Demographic Perspectives*. Oxford: Clarendon Press.

Kilbride, Philip Leroy and Janet Capriotti Kilbride (1990) *Changing Family Life in East Africa: Women and Children at Risk*. University Park, Pa.: The Pennsylvania State University Press.

Lockhart, Chris (2002) *Kunyenga*, "Real Sex," and Survival: Assessing the Risk of HIV Infection Among Urban Street Boys in Tanzania. *Medical Anthropology* 16 (3).

Lovett, Margot (1989) "Gender Relations, Class Formation, and the Colonial State in Africa." In Jane L. Parpart and Kathleen A. Staudt (eds) *Women and the State in Africa*. Boulder: Lynne Reinner.

Lugalla, Joe (1995) *Crisis, Urbanization and Urban Poverty in Tanzania: A Study of Urban Poverty and Survival Politics*. Lanham, Md.: University Press of America.

Lugalla, Joe and Coletta G. Kibasa (eds) (2002) *Poverty, AIDS, and Street Children in Africa*. Lewiston, N.Y.: Edwin Mellen.

—— (2003) *Urban Life and Street Children's Health: Children's Accounts of Urban Hardship and Violence in Tanzania*. Piscataway, N.J.: Transaction.

Lugalla, Joe and Jessie Kazeni Mbwambo (2002) "Street Children and Street Life in Urban Tanzania: The Culture of Surviving and Its Implications for Children's Health." In Joe Lugalla and Colleta G. Kibassa (eds) *Poverty, AIDS, and Street Children in Africa*. Lewiston, N.Y.: Edwin Mellen.

Mangat, J. S. (1969) *A History of the Asians in East Africa: 1886–1945*. Oxford: Clarendon Press.

Marquez, Patricia C. (1999) *The Street Is My Home*. Stanford: Stanford University Press.

Mauss, Marcel (1967) *The Gift: Forms and Functions of Exchange in Archaic Societies*. New York: W. W. Norton and Company.

Messer, Ellen and Parker Shipton (2002) "Hunger in Africa: Untangling Its Human Roots." In Jeremy MacClancy (ed.) *Exotic No More: Anthropology on the Front Lines*. Chicago: University of Chicago Press.

Mulders, Thea (1995) *Children En Route: A Situation Analysis of Street Children and Street Children's Projects in Dar es Salaam*. UNICEF Tanzania.

Nelson, Nici (1978–79) Female-centered Families: Changing Patterns of Marriage and Family among Buzaa Brewers of Mathare Valley. *African Urban Studies* 3.

—— (1979) "How Women and Men Get By: The Sexual Division of Labour in the Informal Sector of a Nairobi Squatter Settlement." In Ray Bromley and Chris Gerry (eds) *Casual Work and Poverty in Third World Cities*. Chichester, England: John Wiley and Sons.

Onyango, Philista P. M. (1983) The Working Mother and the Housemaid as a Substitute: Its Complications on the Children. *Journal of Eastern African Research and Development* 13: 24–31.

Osmani, Siddiq (1995) "The Entitlement Approach to Famine: An Assessment." In K. Basu, P. Pattanaik and K. Suzumura (eds) *Choice, Welfare and Development: A Festchrift in Honour of Amartya K. Sen*. pp. 253–94. Oxford: Clarendon.

Panter-Brick, Catherine (2000) "Nobody's Children? A Reconsideration of Child Abandonment." In Catherine Panter-Brick and Malcolm T. Smith (eds) *Abandoned Children*. Cambridge: Cambridge University.

Plummer, Mary (1994) "*kuleana* Consultancy Report." Unpublished report, *kuleana* Center for Children's Rights, Mwanza, Tanzania.

Poppendieck, Janet (1998) *Sweet Charity: Emergency Food and the End of Entitlement*. New York: Viking.

Potash, Betty (1995) "Women in the Changing African Family." In Margaret Jean Hay and Sharon Stichter (eds) *African Women South of the Sahara*. New York: Longman.

Prout, Alan and Allison James (1990) "A New Paradigm for the Sociology of Childhood? Provenance, Promise, and Problems." In Allison James and Alan Prout (eds) *Constructing and Reconstructing Childhood*. London: Falmer Press.

Raikes, Philip (1988) *Modernising Hunger: Famine, Food Surplus and Farm Policy in the EEC and Africa*. London: James Currey.

Rajani, Rakesh (ed.) (1993) *The Proceedings of Working with Street Children: A Workshop on Community Awareness and Practical Action. Nyegezi (Mwanza), Tanzania, October 18–20, 1993*. Compiled by Karen Coen Flynn and Mustafa Kudrati. Distributed by UNICEF Tanzania.

Rajani, Rakesh and Mustafa Kudrati (1994) *Street Children of Mwanza: A Situational Analysis*. Mwanza, Tanzania: *kuleana* Center for Children's Rights.

—— (1996) "The Varieties of Sexual Experience of the Street Children of Mwanza, Tanzania." In S. Zeidenstein and K. Moore (eds) *Learning about Sexuality: A Practical Beginning*. New York: Population Council, International Women's Coalition.

Reynolds, Pamela (1991) *Dance Civet Cat*. London: Zed.

Robertson, Claire (1995) "Women in the Urban Economy." In Margaret Jean Hay and Sharon Stichter (eds) *African Women South of the Sahara*.

Robertson, M. (1989) *Homeless Youth in Hollywood: Patterns of Alcohol Use: A Report to the National Institute on Alcohol Abuse and Alcoholism*. Berkeley, CA: Alcohol Research Group, New York: Longman.

Rwezaura, Bart (1998) Competing "Images" of Childhood in the Social and Legal Systems of Contemporary Sub-Saharan Africa. *International Journal of Law, Policy, and the Family* (12): 253–278.

Sauper, Hubert (director) (2004) *Darwin's Nightmare*. Mille et Une Productions: Paris.

Save The Children Fund (U.K.) (1997) *Poor Urban Children at Risk in Dar es Salaam: A Participatory Research Report*. Compiled by Mahimbo Mdoe. Distributed by Save The Children Fund (U.K.).

Scheper-Hughes, Nancy and Daniel Hoffman (1994) Kids Out of Place. *NACLA Report on the Americas* 27 (6).

Seabrook, Jeremy (2001) *Children of Other Worlds*. New Haven, Conn.: Yale University Press.

Sen, Amartya (1977) Starvation and Exchange Entitlements: A General Approach and its Application to the Great Bengal Famine. *Cambridge Journal of Economics* 1.

—— (1981) *Poverty and Famines: An Essay on Entitlement and Deprivation*. Oxford: Clarendon Press.

—— (1987) *Hunger and Entitlement*. Helsinki: WIDER.

Shaidi, Leonard P. (1991) The Rights of Street and Abandoned Children. *University of Dar es Salaam Law Journal* 8: 20–23.

Sheikh-Hashim, Leila (1990) *Occupational Hazards of Working Women: Housegirls (A Survey Conducted in Dar es Salaam)*. Dar es Salaam: Tanzania Media Women's Association.

Stichter, Sharon (1985) The Middle Class Family in Kenya: Changes in Gender Relations. *African Urban Studies* 21 (Spring).

Swart, J. (1990) *Malunde: The Street Children of Hillbrow*. Johannesburg: Witwatersrand University Press.

TGNP (Tanzania Gender Networking Program) (1993) *Gender Profile of Tanzania*. Dar es Salaam: Tanzania Gender Networking Program.

The Daily News: Tanzanian newspaper

Thompson, Edward P. (1971) The Moral Economy of the English Crowd in the Eighteenth Century. *Past and Present* 50.

United Nations (1991) *World Population Prospects*. New York: United Nations.

Varkevisser, Corlien M. (1973) *Socialization in a Changing Society: Sukuma Childhood in Rural and Urban Mwanza, Tanzania*. The Hague: Centre for the Study of Education in Changing Societies.

Vreeman, P. (1992) *The Rights of the Child and the Changing Images of Childhood*. Dordrecht, The Netherlands: Martinus Nijhoff Publishers.

White, Luise (1990) *The Comforts of Home: Prostitution in Colonial Nairobi*. Chicago: Chicago University Press.

Whyte, Susan Reynolds (1995) "Constructing Epilepsy: Images and Contexts in East Africa." In Susan Reynolds Whyte (ed.) *Disability and Culture*. Berkeley: University of California Press.

Williams, Christopher (1993) Who are "Street Children?" A Hierarchy of Street Use and Appropriate Responses. *Child Abuse and Neglect* 17.

Want Amid Plenty: From Hunger to Inequality

Janet Poppendieck

"Scouting has some unacceptables," the Executive Director of the Jersey Shore Council of the Boy Scouts of America told me, "and one of them is hunger."[1] We were talking in the entrance to the Ciba Geigy company cafeteria in Toms River, New Jersey, where several hundred Boy Scouts, their parents, grandparents, siblings, and neighbors were sorting and packing the 280,000 pounds of canned goods that the scouts of this Council had netted in their 1994 Scouting For Food drive. The food would be stored on the Ciba Geigy corporate campus, where downsizing had left a number of buildings empty, and redistributed to local food pantries to be passed along to the hungry. The scouting executive was one of several hundred people I interviewed as part of a study of charitable food programs—so called "emergency food" in the United States. In the years since the early 1980s, literally millions of Americans have been drawn into such projects: soup kitchens and food pantries on the front lines, and canned goods drives, food banks, and "food rescue" projects that supply them.

Hunger Has a "Cure"

What makes hunger in America unacceptable, to Boy Scouts and to the rest of us, is the extraordinary abundance produced by American agriculture. There is no short-age of food here, and everybody knows it. In fact, for much of this century, national agricultural policy has been preoccupied with surplus, and individual Americans have been preoccupied with avoiding, losing, or hiding the corporeal effects of over-eating. Collectively, and for the most part individually, we have too much food, not too little. To make matters worse, we waste food in spectacular quantities. A study recently released by USDA estimates that between production and end use, more than a quarter of the food produced in the United States goes to waste, from fields planted but not harvested to the bread molding on top of my refrigerator or the lettuce wilting at the back of the vegetable bin. Farm waste, transport waste, processor waste, wholesaler waste, supermarket waste, institutional waste, household waste, plate waste; together in 1995 they totaled a startling 96 billion pounds, or 365 pounds—a pound a day—for every person in the nation.[2]

The connection between abundant production and food waste on the one hand, and hunger on the other, is not merely abstract and philosophical. Both public and private food assistance efforts in this country have been shaped by efforts to find acceptable outlets for food that would otherwise go to waste. These include the wheat surpluses stockpiled by Herbert Hoover's Federal Farm Board and belatedly given to the Red Cross for distribution to the unemployed, the martyred piglets of the New Deal agricultural adjustment (which led to the establishment of federal surplus commodity distribution), and the cheese that Ronald Reagan finally donated to the needy to quell the criticism of mounting storage costs. Accumulation of large supplies of food in public hands, especially in times of economic distress and privation, has repeatedly resulted in the creation of public programs to distribute the surplus to the hungry. And in the private sphere as well, a great deal of the food that supplies today's soup kitchens and food pantries is food that would otherwise end up as waste: corporate over-production or labeling errors donated to the food bank, farm and orchard extras gleaned by volunteers after the commercial harvest, and the vast quantities of leftovers generated by hospital, school, government and corporate cafeterias, and caterers and restaurants. All of this is food that is now rescued and recycled through the type of food recovery programs urged by Vice President Al Gore and Agriculture Secretary Dan Glickman at their 1997 National Summit on Food Recovery and Gleaning. "There is simply no excuse for hunger in the most agriculturally abundant country in the world," said Glickman, who urged a 33 percent increase in food recovery by the year 2000 that would enable social service agencies to feed an additional 450,000 Americans each day.[3] For Americans reared as members of the "clean plate club" and socialized to associate our own uneaten food with hunger in faraway places, such programs have enormous appeal. They provide a sort of moral relief from the discomfort that ensues when we are confronted with images of hunger in our midst, or when we are reminded of the excesses of consumption that characterize our culture. They offer what appear to be old-fashioned moral absolutes in a sea of shifting values and ethical uncertainties. Many of the volunteers I interviewed for my study told me that they felt that their work at the soup kitchen or food pantry was the one unequivocally good thing in their lives, the one point in the week in which they felt sure they were on the side of the angels. Furthermore, they perceive hunger as one problem that is solvable—precisely because of the abundant production—one problem about which they can do something concrete and meaningful. "Hunger has a cure," is the new slogan developed by the Ad Council for Second Harvest, the National Network of Foodbanks. It is not surprising, then, that hunger in America has demonstrated an enormous capacity to mobilize both public and private action. There are fourteen separate federal food assistance programs, numerous state and local programs, and thousands upon thousands of local, private charitable feeding projects which elicit millions of hours of volunteer time as well as enormous quantities of donated funds and food. In one random survey in the early 1990s, nearly four-fifths of respondents indicated that they, personally, had done something to alleviate hunger in their communities in the previous year.[4]

The Seductions of Hunger

Progressives have not been immune to the lure of hunger-as-the-problem. We have been drawn into the anti-hunger crusade for several reasons. First, hunger in America shows with great clarity the absurdity of our distribution system, of capitalism's approach to meeting basic human needs. Poor people routinely suffer for want of things that are produced in abundance in this country, things that gather dust in warehouses and inventories, but the bicycles and personal computers that people desire and could use are not perishable and hence are not rotting in front of their eyes in defiance of their bellies. The Great Depression of the 1930s, with its startling contrasts of agricultural surpluses and widespread hunger, made this terrible irony excruciatingly clear, and many people were able to perceive the underlying economic madness: "A breadline knee-deep in wheat," observed commentator James Crowther, "is surely the handiwork of foolish men."[5] Progressives are attracted to hunger as an issue because it reveals in so powerful a way the fundamental shortcomings of unbridled reliance on markets.

Second, progressives are drawn to hunger as a cause by its emotional salience, its capacity to arouse sympathy and mobilize action. Hunger is, as George McGovern once pointed out, "the cutting edge of poverty," the form of privation that is at once the easiest to imagine, the most immediately painful, and the most far-reaching in its damaging consequences.[6] McGovern was writing in the aftermath of the dramatic rediscovery of hunger in America that occurred in the late 1960s when a Senate subcommittee, holding hearings on anti-poverty programs in Mississippi, encountered the harsh realities of economic and political deprivation in the form of empty cupboards and malnourished children in the Mississippi Delta. Hunger was in the news, and journalist Nick Kotz reports that a coalition of civil rights and anti-poverty activists made a conscious decision to keep it there. They perceived in hunger "the one problem to which the public might respond. They reasoned that 'hunger' made a higher moral claim than any of the other problems of poverty."[7] The anti-hunger movement—or "hunger lobby" that they initiated—was successful in enlisting Congressional support for a major expansion of food assistance and the gradual creation of a food entitlement through food stamps, the closest thing to a guaranteed income that we have ever had in this country.

The broad appeal of the hunger issue and its ability to evoke action are also visible in the more recent proliferation of emergency food programs. "I think the reason . . . that you get the whole spectrum of people involved in this is because it's something that is real basic for people to relate to. You know, you're busy, you skip lunch, you feel hungry. On certain levels, everyone has experienced feeling hungry at some point in the day or the year," explained Ellen Teller, an attorney with the Food Research and Action Center whose work brings her into frequent contact with both emergency food providers and anti-hunger policy advocates. The food program staff and volunteers I interviewed recognized the difference between their own, essentially voluntary and temporary hunger and hunger that is externally imposed and of unpredictable duration, but the reservoir of common human experience is there. Hunger is not exotic and hard to imagine; it stems from the failure to meet a basic and incontrovertible need that we all share.

Furthermore, the failure to eliminate hunger has enormous consequences. As the research on the link between nutrition and cognition mounts, the social costs of failing to ensure adequate nutrition for pregnant women and young children become starkly obvious. And this, too, contributes to the broad spectrum that Ellen Teller mentioned. There is something for everyone here—a prudent investment in human capital for those concerned about the productivity of the labor force of tomorrow, a prevention of suffering for the tender hearted, a unifying concern for would-be organizers, a blatant injustice for critics of our social structure. Many anti-hunger organizations with relatively sophisticated critiques of the structural roots of hunger in America have engaged with the "feeding movement," the soup kitchens and the food pantries, in the belief that, as the Bread for the World Institute once put it, "Hunger can be the 'door' through which people enter an introduction to larger problems of poverty, powerlessness, and distorted public values."[8] For those progressives seeking common ground with a wider range of American opinion, hunger is an attractive issue precisely because of the breadth of the political spectrum of people who are moved by it.

Third, progressive have been drawn into the hunger lobby by the utility of hunger as a means of resisting, or at least documenting the effects of, government cuts in entitlements. In the early 1980s, especially, when Ronald Reagan began his presidential assault on the nation's meager safety net of entitlement programs for the poor, progressives of all sorts pointed to the lengthening soup kitchen lines as evidence that the cuts in income supports, housing subsidies, food assistance, and a host of other public programs were cuts that neither the poor nor the society could afford. While Reagan and his team claimed that they were simply stripping away waste and fat from bloated programs, critics on the left kept track of mounting use of emergency food programs as a means of documenting the suffering caused by the erosion of the welfare state. The scenario is being replayed, this time amid an expanding economy, as soup kitchens and food pantries register the effects of "the end of welfare as we know it."

Finally, of course, progressive are drawn to the hunger issue by a sense of solidarity with those in need. Most of us became progressives in the first place because we cared about people and wanted a fairer society that would produce less suffering. Few of us can stomach an argument that says that we should leave the hungry to suffer without aid while we work for a more just future. "People don't eat in the long run," Franklin Roosevelt's relief czar Harry Hopkins is reported to have said; "they eat every day."[9] Many of the more activist and progressive people I interviewed in the course of my emergency food study articulated similar sentiments. A woman who worked in the early eighties helping churches and community groups in southern California set up soup kitchens and food pantries to cope with the fallout from the budget cuts in Washington recalled the dilemma as she had experienced it. "As far as I was concerned, the people in Washington had blood on their hands . . . but I wasn't going to stand by and watch people suffer just to make a political point." As one long-time left activist in Santa Cruz put it when questioned about her work as a member of the local food bank board, "There are numbers of people who are very compatible with my radical philosophy who also feel that foodbanking is very important, because the reality is that there are ever increasing homeless and poor, including working poor,

who need to be fed . . . the need for food has increased and the resources for providing it haven't. And if there weren't foodbanks, I think a lot of people would
starve."

It is easy to see why progressive people have been drawn into anti-hunger activity
in large numbers, and why they have been attracted to the soup kitchens, food
pantries, and food banks, despite misgivings about these private charitable projects.
I, personally, have counted myself an anti-hunger activist since the nation
rediscovered hunger in the late 1960s. Nevertheless, after three decades in the
"hunger lobby," and nearly a decade of observing and interviewing in soup kitchens,
food pantries, food banks, and food recovery projects, I would like to offer a caution
about defining hunger as the central issue.

The Case Against Hunger

The very emotional response that makes hunger a good organizing issue, and the felt
absurdity of such want amid massive waste, makes our society vulnerable to token
solutions—solutions that simply link together complementary symptoms without
disturbing the underlying structural problems. The New Deal surplus commodity
distribution program, which laid the political and administrative groundwork for
most subsequent federal food programs, purchased surplus agricultural commodities
from impoverished farmers in danger of going on relief and distributed them to the
unemployed already receiving public help. It responded to what Walter Lippmann
once called the "sensational and the intolerable paradox of want in the midst of
abundance," by using a portion of the surplus to help some of the needy, without
fundamentally changing the basis for access to food.[10] As Norman Thomas put it in
1936, "We have not had a reorganization of production and a redistribution of
income to end near starvation in the midst of potential plenty. If we do not have such
obvious 'breadlines knee deep in wheat' as under the Hoover administration, it is
because we have done more to reduce the wheat and systematize the giving of crusts
than to end hunger."[11]

For the general public, however, the surplus commodity programs were common
sense, and they made well-fed people feel better. Few asked how much of the surplus
was being transferred to the hungry, or how much of their hunger was thus relieved.
As the *New York Times* predicted in an editorial welcoming the program: "It will
relieve our minds of the distressing paradox."[12] And with the moral pressure relieved,
with consciences eased, the opportunity for more fundamental action evaporated.
Thus the token program served to preserve the underlying status quo.

Something very similar appears to be happening with the private food rescue,
gleaning, and other surplus transfer programs that have expanded and proliferated to
supply emergency food programs since the early 1980s. The constant fund-raising
and food drives that characterize such programs keep them in the public eye, and few
people ask whether the scale of the effort is proportional to the scale of the need.
With the Boy Scouts collecting in the fall and the letter carriers in the spring, with
the convenient barrel at the grocery store door and the opportunity to "check out
hunger" at the checkout counter, with the Taste of the Nation and the enormous

array of other hunger-related fundraisers, with the Vice President and the Secretary of Agriculture assuring us that we can simultaneously feed more people and reduce waste through food recovery, with all this highly visible activity, it is easy to assume that the problem is under control. The double whammy, the moral bargain of feeding the hungry and preventing waste, makes us feel better, thus reducing the discomfort that might motivate more fundamental action. The same emotional salience that makes hunger so popular a cause in the first place makes us quick to relieve our own discomfort by settling for token solutions.

In the contemporary situation, the danger of such tokenism is even more acute. There is more at stake than the radicalizing potential of the contradictions of waste amid want. The whole fragile commitment to public income supports and entitlements is in jeopardy. Food programs not only make the well fed feel better, they reassure us that no one will starve, even if the nation ends welfare and cuts gaping holes in the food stamp safety net. By creating an image of vast, decentralized, kind-hearted effort, an image that is fueled by every fund-raising letter or event, every canned goods drive, every hunger walk, run, bike, swim, or golf-a-thon, every concert or screening or play where a can of food reduces the price of admission, we allow the right wing to destroy the meager protections of the welfare state and undo the New Deal. Ironically, these public appeals have the effect of creating such comforting assurances even for those who do not contribute.

Promoting hunger as a public issue, of course, does not necessarily imply support for the private, voluntary approach. There are undoubtedly social democrats and other progressives who support expanded food entitlements without endorsing the emergency food phenomenon. Unfortunately, however, much of the public makes little distinction. If we raise the issue of hunger, we have no control over just how people will choose to respond. As the network of food banks, food rescue organizations, food pantries, and soup kitchens has grown, so have the chances that people confronted with evidence of hunger in their midst will turn to such programs in an effort to help.

Many private food charities make a point of asserting that they are not a substitute for public food assistance programs and entitlements. Nearly every food banker and food pantry director I interviewed made some such assertion, and the national organizations that coordinate such projects, Second Harvest, Food Chain, Catholic Charities, even the Salvation Army, are on record opposing cuts in public food assistance and specifying their own role as supplementary. When it is time to raise funds, however, such organizations, from the lowliest food pantry in the church basement to national organizations with high-powered fund raising consultants or departments, tend to compare themselves with public programs in ways that reinforce the ideology of privatization. You simply cannot stress the low overhead, efficiency, and cost effectiveness of using donated time to distribute donated food without feeding into the right-wing critique of public programs in general and entitlements in particular. The same fund-raising appeals that reassure the public that no one will starve, even if public assistance is destroyed, convince many that substitution of charitable food programs for public entitlements might be a good idea.

Furthermore, as the programs themselves have invested in infrastructure—in walk-in freezers and refrigerated trucks, in institutional stoves and office equipment, in pension plans and health insurance—their stake in the continuation of their efforts has grown as well, and with it, their need for continuous fund raising, and thus for the perpetuation of hunger as an issue. While many food bankers and food recovery staff argue that there would be a role for their organizations even if this society succeeded in eliminating hunger, that their products also go to improve the meal quality at senior citizen centers or lower the cost of daycare and rehabilitation programs, they clearly realize that they need hunger as an issue in order to raise their funds. Cost effectiveness and efficient service delivery, even the prevention of waste, simply do not have the same ability to elicit contributions. Hunger is, in effect, their bread and butter. The result is a degree of hoopla, of attention getting activity, that I sometimes think of as the commodification of hunger. As Laura DeLind pointed out in her insightful article "Celebrating Hunger in Michigan," the hunger industry has become extraordinarily useful to major corporate interests, but even without such public relations and other benefits to corporate food and financial donors, hunger has become a "product" that enables its purveyors to compete successfully for funds in a sort of social issues marketplace.[13] It does not require identification with despised groups—as does AIDS, for example. Its remedy is not far off, obscure, or difficult to imagine—like the cure for cancer. The emotional salience discussed above, and the broad spectrum of people who have been recruited to this cause in one way or another, make hunger—especially the soup kitchen, food pantry, food recycling version of hunger—a prime commodity in the fund-raising industry, and a handy, inoffensive outlet for the do-gooding efforts of high school community service programs and corporate public relations offices, of synagogues and churches, of the Boy Scouts and the Letter Carriers, of the Rotarians and the Junior League: the taming of hunger.

As we institutionalize and expand the response, of course, we also institutionalize and reinforce the problem definition that underlies it. Sociologists have long argued that the definitional stage is the crucial period in the career of a social problem. Competing definitions vie for attention, and the winners shape the solutions and garner the resources. It is important, therefore, to understand the competing definitions of the situation that "hunger" crowds out. What is lost from public view, from our operant consciousness, as we work to end hunger? In short, defining the problem as hunger contributes to the obfuscation of the underlying problems of poverty and inequality. Many poor people are indeed hungry, but hunger, like home-lessness and a host of other problems, is a symptom, not a cause, of poverty. And poverty, in turn, in an affluent society like our own, is fundamentally a product of inequality.

Defining the problem as hunger ignores a whole host of other needs. Poor people need food, but they also need housing, transportation, clothing, medical care, meaningful work, opportunities for civic and political participation, and recreation. By focusing on hunger, we imply that the food portion of this complex web of human needs can be met independently of the rest, can be exempted or protected from the overall household budget deficit. As anyone who has ever tried to get by on a tight budget can tell you, however, life is not so compartmentalized. Poor people are generally engaged in a daily struggle to stretch inadequate resources over a range of competing demands. The "heat-or-eat" dilemma that arises in the winter months, or the situation reported by many elderly citizens of a constant necessity to choose between food and medications are common manifestations of this reality.

In this situation, if we make food assistance easier to obtain than other forms of aid—help with the rent, for example, or the heating bill—then people will devise a variety of strategies to use food assistance to meet other needs. It is not really difficult to convert food stamps to cash: pick up a few items at the store for a neighbor, pay with your stamps, collect from her in cash. Some landlords will accept them, at a discounted rate of course, then convert them through a friend or relative who owns a grocery store. Drug dealers will also accept them, again at lower than face value, and you can resell the drugs for cash. The list goes on and on. Converting soup kitchen meals is almost impossible, but there are items in many pantry bags that can be resold. In either case, eating at the soup kitchen or collecting a bag from the food pantry frees up cash for other needs, not only the rent, but also a birthday present for a child or a new pair of shoes. By offering help with food, but refusing help with other urgent needs, we are setting up a situation in which poor people are almost required to take steps to convert food assistance to cash.

Conservative critics of entitlements will then seize on these behaviors to argue that poor people are "not really hungry." If they were really hungry, the argument goes, they would not resell items from the pantry bag or convert their food stamps. Such behavioral evidence fits into a whole ideologically driven perception that programs for poor people are bloated, too generous, and full of fraud and abuse; it allows conservatives to cut programs while asserting that they are preserving a safety net for the "truly needy." Progressives meanwhile are forced into a defensive position in which we argue that people are indeed "really hungry," thereby giving tacit assent to the idea that the elimination of hunger is the appropriate goal. In a society as wealthy as ours, however, aiming simply to eliminate hunger is aiming too low. We not only want a society in which no one suffers acute hunger or fails to take full advantage of educational and work opportunities due to inadequate nutrition. We want a society in which no one is excluded, by virtue of poverty, from full participation, in which no one is too poor to provide a decent life for his or her children, no one is too poor to pursue happiness. By defining the problem as "hunger," we set too low a standard for ourselves.

Where to?

The question of where we should direct our organizational efforts is inextricably tied up with the underlying issue of inequality. Above some absolute level of food and shelter, need is a thoroughly relative phenomenon. In an affluent society, the quality of life available at a given level of income has everything to do with how far from the mainstream that level is, with the extent to which any given income can provide a life that looks and feels "normal" to its occupants. In many warm parts of the world, children routinely go barefoot, and no mother would feel driven to convert food resources into cash to buy a pair of shoes, or to demean herself by seeking a charity handout to provide them. In the United States, where children are bombarded with hours of television advertising daily, and where apparel manufacturers trade on "coolness," a mother may well make the rounds of local food pantries, swallowing her pride and subsisting on handouts, to buy not just a pair of shoes, but a particular name brand that her child has been convinced is essential for social acceptance at the junior high school.

In this context, the issue is not whether people have enough to survive, but how far they are from the median and the mainstream, and that is a matter of how unequal our society has become. By every measure, inequality has increased in the United States, dramatically, since the early 1970s, with a small group at the top garnering an ever increasing share of net marketable worth, and the bottom doing less and less well. And it is this growing inequality which explains the crying need for soup kitchens and food banks today, even at a relatively high level of employment that reflects the current peak in the business cycle. Unfortunately, however, a concept like hunger is far easier to understand, despite its ambiguities of definition, than an abstraction like inequality. Furthermore, Americans have not generally been trained to understand the language of inequality nor the tools with which it is measured. Just what is net marketable worth, and do I have any? As the statistics roll off the press, eyes glaze over, and the kindhearted turn to doing something concrete, to addressing a problem they know they can do something about: hunger. Once they begin, and get caught up in the engrossing practical challenges of transferring food to the hungry and the substantial emotional gratifications of doing so, they lose sight of the larger issue of inequality. The gratifications inherent in "feeding the hungry" give people a stake in maintaining the definition of the problem as hunger; the problem definition comes to be driven by the available and visible response in a sort of double helix.

Meanwhile, with anti-hunger activists diverted by the demands of ever larger emergency food systems, the ascendant conservatives are freer than ever to dismantle the fragile income protections that remain and to adjust the tax system to concentrate ever greater resources at the top. The people who want more inequality are getting it, and well-meaning people are responding to the resulting deprivation by handing out more and more pantry bags, and dishing up more and more soup. It is time to find ways to shift the discourse from undernutrition to unfairness, from hunger to inequality.

Notes

1. All quotations not otherwise attributed come from the transcripts of interviews I conducted in conjunction with my study of emergency food. For a more extensive treatment, see Janet Poppendieck, *Sweet Charity? Emergency Food and the End of Entitlement* (New York: Viking, 1998).
2. Foodchain, the National Food Rescue Network, *Feedback* (Fall, 1997), 2–3.
3. Ibid.
4. Vincent Breglio, *Hunger in America: The Voter's Perspective.* (Lanham, MD: Research /Strategy/Management Inc., 1992), 14–16.
5. For a discussion of the so called paradox of want amid plenty in the great depression, see Janet Poppendieck, *Breadlines Knee Deep in Wheat: Food Assistance in the Great Depression.* (New Brunswick, NJ: Rutgers University Press, 1986).
6. George McGovern, "Foreword," in Nick Kotz, *Let Them Eat Promises: The Politics of Hunger in America.* (Englewood Cliffs, NJ: Prentice-Hall, 1969) viii.
7. Nick Kotz, "The Politics of Hunger," *The New Republic* (April 30, 1984), 22.
8. Bread for the World Institute, *Hunger 1994: Transforming the Politics of Hunger.* Fourth Annual Report on the State of World Hunger (Silver Spring, MD, 1993), 19.

9. Quoted in Edward Robb Ellis, *A Nation in Torment: The Great American Depression, 1929–1939*. (New York: Capricorn Books, 1971), 506.
10. Walter Lippmann, "Poverty and Plenty," Proceedings of the National Conference of Social Work, 59th Session, 1932 (Chicago: University of Chicago Press, 1932), 234–35.
11. Norman Thomas, *After the New Deal, What?* (New York: Macmillan, 1936), 33.
12. "Plenty and Want," editorial, *New York Times*, September 23, 1933.
13. Laura B. DeLind, "Celebrating Hunger in Michigan: A Critique of an Emergency Food Program and an Alternative for the Future," *Agriculture and Human Values* (Fall, 1994), 58–68.

Contributors

Anne Allison is Robert O. Keohane Professor and Chair of Cultural Anthropology at Duke University. An anthropologist who specializes in contemporary Japan and globalized mass culture, she is the author of three books: *Nightwork: Pleasure, Sexuality, and Corporate Masculinity in a Tokyo Hostess Club* (1994); *Permitted and Prohibited Desires: Mothers, Comics, and Censorship in Japan* (1996, 2000); and *Millennial Monsters: Japanese Toys and the Global Imagination* (2006).

Arjun Appadurai serves as Senior Advisor for Global Initiatives at The New School in New York City, where he also holds a Distinguished Professorship as the John Dewey Professor in the Social Sciences. Appadurai is the founder and now the President of PUKAR (Partners for Urban Knowledge Action and Research), a non-profit organization based in and oriented to the city of Mumbai (India). He has authored numerous books and scholarly articles including *Fear of Small Numbers: An Essay on the Geography of Anger* (2006); and *Modernity at Large: Cultural Dimensions of Globalization* (1996).

Deborah Barndt is a mother, popular educator, photographer and teacher in the Faculty of Environmental Studies at York University in Toronto, Canada. She has brought her academic, artistic, and activist interests together around food issues in two books, an edited volume, *Women Working the NAFTA Food Chain* (1999); and *Tangled Routes: Women, Work and Globalization on the Tomato Trail* (2008).

In 1976, **Roland Barthes**, teacher and writer, was appointed as the first person to hold the Chair of Literary Semiology at the Collège de France. He pioneered the semiologic interpretation of foodways in articles including the preface of Brillat-Savarin's *Physiology of Taste*. He also authored the anti-autobiography *Roland Barthes by Roland Barthes*.

Susan Bordo holds the Otis A. Singletary Chair in the Humanities and is a professor of English and Gender Studies at the University of Kentucky. She is the author of the Pulitzer Prize nominated *Unbearable Weight: Feminism, Western Culture and the Body; The Male Body: A New Look at Men in Public and in Private*, and other influential books and articles on gender, culture, and the body.

One of the world's pioneering authorities on the diagnosis and treatment of eating disorders, **Hilde Bruch** received an M.D. from the University of Freiburg in 1929 and was Professor of Psychiatry at Baylor College of Medicine in Houston. She has authored scores of journal articles and several books, including *Eating Disorders: Obesity, Anorexia Nervosa, and the Person Within; The Golden Cage: The Enigma of Anorexia Nervosa; and the posthumously published *Conversations with Anorexics*.

Joan Jacobs Brumberg is a social historian and professor emerita at Cornell University. She is the author of *Fasting Girls* (1988); *The Body Project* (1997); and *Kansas Charley* (2004).

Caroline Walker Bynum is Professor of Medieval European history at the Institute for Advanced Study in Princeton. She has written on food imagery in medieval piety and theology, on women's religious movements, and on medieval ideas of personal identity. Her most recent book, *Wonderful Blood* (University of Pennsylvania Press, 2007), is a study of relics and pilgrimage in fifteenth-century northern Europe that places them in the context of a theology of sacrifice.

Christopher Carrington is an Associate Professor of Sociology and Graduate Faculty in the Human Sexuality Studies Department at San Francisco State University. He earned his Ph.D. at the University of Massachusetts, Amherst. His publications include: *No Place Like Home: Relationships and Family Life among Lesbians and Gay Men*, 1999; "Work/Family Patterns among Lesbians and Gay Men," in *Families and Work: New Century Patterns*, 2001. Dr Carrington is currently finishing an ethnography entitled: *Circuit Boys: Into the World of the Gay Dance and Circuit Party Culture*. The book is the result of a ten-year long ethnographic study of the gay dance and circuit-party scene in the U.S., Canada, Australia and Mexico.

Michel de Certeau was a French Jesuit and scholar whose work involved psychoanalysis, philosophy, and social science. He was greatly influenced by Sigmund Freud and founded, along with Jacques Lacan, L'École Freudienne. He taught at universities in Paris, Geneva, and San Diego. He was author of numerous books including *The Practice of Everyday Life*, volume 1 (1984) and volume 2 (with Luce Giard and Pierre Mayol, 1998); *The Capture of Speech and Other Political Writings* (1998); and *The Writing of History* (1988).

Jennifer Clapp is CIGI Chair in International Governance and Professor in the Environment and Resource Studies Department at the University of Waterloo. Her current food-related research projects focus on the global political economy of genetically modified food trade, the politics of food aid reform, the role of transnational corporations in global food governance, and agricultural trade negotiations in the context of the World Trade Organization.

Dylan Clark is an assistant professor of Anthropology at University of Toronto at Mississauga. His areas of expertise include gifts/commodities, anarchism, the youth subculture, Whiteness, punk, anti-globalization, and hegemony. Clark has had his work published in journals such as *Ethnology; Peace Review;* and the *Journal of Thought*.

Carole Counihan is Professor of Anthropology at Millersville University in Pennsylvania. Her books include: *Around the Tuscan Table: Food, Family and Gender in*

Twentieth Century Florence (2004); *The Anthropology of Food and Body: Gender, Meaning, and Power* (1999); and *Food in the USA: A Reader* (2002). She is co-editor of the scholarly journal *Food and Foodways*. She is currently completing *Women's Stories of Food, Identity and Land in the San Luis Valley of Colorado* for the University of Texas Press.

Marjorie L. DeVault is Professor of Sociology and affiliated with the Women's Studies Program at Syracuse University. Her research has explored the "invisible work" in women's household and family lives and in the historically female field of dietetics and nutrition education. She has also written on qualitative and feminist methodologies. She is the author of *Feeding the Family: The Social Organization of Caring as Gendered Work* (1991); and *Liberating Method: Feminism and Social Research* (1999).

Mary Douglas was an anthropology professor at Cambridge University and held a number of other academic appointments, including the Avalon Foundation Chair at Northwestern University. She was a trailblazer in the anthropological study of food, and her books include *Constructive Drinking; Purity and Danger;* and *Food in the Social Order.*

Gail Feenstra is the Food System Analyst at the University of California Sustainable Agriculture Research and Education Program (SAREP), based in Davis, California. Her research and education efforts include direct marketing, farm-to-school program evaluation, regional food system distribution models, food system indicators, urban agriculture, food security, food policy, and food system assessments. Feenstra has a doctorate in nutrition education from Teachers College, Columbia University with an emphasis in public health. Her career has been dedicated to integrating human, environmental and community health through sustainable food systems.

M. F. K. Fisher was a prolific essayist and memoirist whose writings centered on the pleasures of cooking and eating, principally in California and Southern France. She wrote about food with passion and brilliance, encapsulating the complex human relations centering on food. She was author of over sixteen volumes, including *The Art of Eating* and an acclaimed translation of Brillat-Savarin's *The Physiology of Taste*. She spent the last years of her life in Glen Ellen, California, and died in 1992.

Karen Coen Flynn studied cultural anthropology at the University of Cambridge and Harvard University and is an associate professor of anthropology at The University of Akron. Much of her teaching and research addresses food as it intersects with poverty, hunger, and homelessness among both children and adults in Tanzania and the United States. She recently published her research on contemporary foodways in Mwanza, Tanzania, in her 2005 monograph *Food, Culture, and Survival in an African City*.

Luce Giard was a long-term collaborator of Michel De Certeau and author, co-author, and editor of several books including *Michel de Certeau* (1987); *Le voyage mystique: Michel de Certeau* (1988); *Histoire, Mystique et Politique: Michel de Certeau* (1991); and *The Practice of Everyday Life*, volume 2 (with Michel De Certeau and Pierre Mayol, 1998).

Jack Goody, formerly on the faculty of Anthropology at St John's College, Cambridge University, is the author of *Cooking, Cuisine and Class: A Study in Comparative Sociology; Production and Reproduction*; and *The Culture of Flowers*.

Marvin Harris pioneered the cultural materialist interpretation of seemingly quirky human food habits in *Good to Eat: Riddles of Food and Culture* and co-edited *Food in Evolution* with Eric Ross. He is the author of many other books, including *Cultural Materialism; Cows, Pigs, Wars, and Witches*; and *The Rise of Anthropological Theory*.

Lisa Heldke is the author of *Exotic Appetites: Ruminations of a Food Adventurer*; and co-editor of *Cooking, Eating, Thinking: Transformative Philosophies of Food*; and *The Atkins Diet and Philosophy*. She is a cofounder of Convivium: The Philosophy and Food Roundtable. She teaches philosophy at Gustavus Adolphus College, where she has developed several courses on the philosophy of food, including Good Food, an exploration of the philosophical concept of the good, using food as its vehicle.

T. J. M. Holden is currently Professor of Mediated Sociology in the Department of Multicultural Studies at Tohoku University in Sendai, Japan. Most recently, Holden's research has been focused on advertising theory, semiotics, grounded studies of communication, and comparative sociology. He is currently working on a book on Japanese advertising.

Alice Julier has written about shared meals, etiquette, hospitality, and inequality in American households. Other essays on food include: "Hiding Race, Class, Gender in the Discourse of Commercial Food"; "Hospitality and Its Discontents: Beyond Bowling Alone"; and "Mapping Men onto the Menu: Food and Masculinities". She taught courses on food while at Smith College and currently teaches at the University of Pittsburgh. She served as president of the Association for the Study of Food and Society from 2003–2006.

Alison Leitch is a broadly trained social anthropologist, currently teaching in the cultural sociology program at Macquarie University in Sydney. She has carried out a number of ethnographic projects in Italy over the last two decades. Her research on the Slow Food movement developed out of a broad interest in the politics of food and her previous research on work in the marble quarries of Carrara. She is currently engaged in a comparative study looking at transnational food alliances emerging in response to questions surrounding the protection of cultural landscapes and agricultural biodiversity.

Claude Lévi-Strauss is a French cultural anthropologist best known for applying structuralism to the study of symbolism in mythology. He held the Chair of Social Anthropology at the Collège de France from 1959 until his retirement. His books include *Totemism; The Raw and the Cooked; The Origin of Table Manners*; and *The Story of the Lynx*.

Margaret Mead was a cultural anthropologist and curator of ethnology at the American Museum of Natural History in New York. She carried out extensive fieldwork in Oceania, and during World War II, she was Executive Secretary of the Committee on Food Habits of the National Research Council. Some of her many books are *Coming of Age in Samoa; Growing Up in New Guinea; Sex and Temperament in Three Societies;* and *Male and Female.*

Research Professor **Sidney W. Mintz** (Johns Hopkins University) has written on the roles of sugar and soybeans in a global food system. His books include *Sweetness and Power: The Place of Sugar in Modern History* (1985); and *Tasting Food, Tasting Freedom: Excursions into Eating, Culture, and the Past* (1996). His essay "Food, Energy and Culture" introduces a new volume, *Food and Globalization,* edited by Nuetzenadel and Trentmann. A collection of essays, *The World of Soy,* co-edited by Mintz, is scheduled to appear in 2008.

Gary Paul Nabhan co-authored one of the first technical articles relating the composition of slow release desert foods to the rise in Native American diabetes. He is author or editor of twenty books and founder of the Renewing America's Food Traditions campaign. For more of his work, see www.garynabhan.com

Jeri Ohmart has worked in the area of sustainable agriculture since 1999. During this time, the primary focus of her work has been to evaluate Farm to School and Farm to Institution programs as well as to implement Garden Based Learning and Nutrition Education programs. Ms Ohmart is passionate about supporting local farmers, improving children's health and educating people about sustainable agricultural and food systems.

Fabio Parasecoli lives in Rome and New York City. He is the U.S. correspondent for the Italian food and wine magazine *Gambero Rosso.* He concentrates his research on the intersection of food, culture, and politics. He teaches at the Città del Gusto School in Rome and at New York University. Parasecoli published *Food Culture in Italy* in 2004, and contributed with the introduction to *Culinary Cultures in Europe,* published by the Council of Europe in 2005.

Jeffrey M. Pilcher is professor of history at the University of Minnesota. His books include the prize-winning *¡Que vivan los tamales! Food and the Making of Mexican Identity* (1998); *Food in World History* (2006); and *The Sausage Rebellion: Public Health, Private Enterprise, and Meat in Mexico City, 1890–1917* (2006). The essay in this volume is part of his current research project on the globalization of Mexican cuisine.

Janet Poppendieck has taught Sociology at Hunter College, City University of New York, since 1976. Her primary concerns, both as a scholar and as an activist, have been poverty, hunger, and food assistance in the United States. She is the author of *Breadlines Knee Deep in Wheat: Food Assistance in the Great Depression* (1986); *Sweet Charity? Emergency Food and the End of Entitlement* (1998); and articles on hunger, food assistance and public policy. She is currently at work on a book on school meals under contract to the University of California Press.

Melissa Salazar is a Ph.D. candidate in the School of Education at the University of California, Davis. She has conducted research, taught, and written about a wide array of child nutrition and food issues, particularly those related to school food and multicultural student populations. Her dissertation focuses on the food transitions of immigrant children and families in California through ethnographic photo-storytelling.

Eric Schlosser is an award winning journalist, correspondent for *Atlantic Monthly*, and author of three national best-sellers including *Fast Food Nation* (2001), which analyzes how the fast food industry has changed American society. Schlosser has addressed the U.S. Senate about the dangers to the food supply from bioterrorism, as well as lectured at many universities around the country.

Penny Van Esterik is Professor of Anthropology at York University, Toronto. She teaches nutritional anthropology, advocacy anthropology and feminist theory, and works primarily in Southeast Asia (Thailand and Lao PDR). Past books include *Beyond the Breast-Bottle Controversy* (on infant feeding in developing countries); *Materializing Thailand* (on cultural interpretations of gender in Thailand); and *Taking Refuge: Lao Buddhists in North America* (on the reintroduction of Buddhism by Lao refugees to North America). She is a founding member of WABA (World Alliance for Breastfeeding Action) and has been active in developing articles and advocacy materials on breastfeeding and women's work, breastfeeding and feminism, and contemporary challenges to infant feeding such as environmental contaminants and HIV/AIDS.

Richard Wilk is professor of Anthropology and Gender Studies at Indiana University. He has conducted archaeological, ethnohistoric and ethnographic research in Belize for over thirty-five years, and has also done fieldwork in West Africa, Indiana, and the wilds of suburban California. His most recent books is *Home Cooking in the Global Village*, about the history of food and globalization in the Caribbean, and he edited *Fast Food/Slow Food*. His interest in food grew out of his research on consumption and sustainability.

Psyche Williams-Forson is an Assistant Professor in the Department of American Studies and an affiliate of the Women's Studies and African American Studies Departments at the University of Maryland College Park. Her research and teaching interests include cultural studies, material culture, food, women's studies, social and cultural history of the U.S. in the late nineteenth and twentieth centuries. Her recent work, *Building Houses Out of Chicken Legs: Black Women, Food, and Power*, examines the complexity of black women's legacies using food as a form of cultural work. She is the author of several articles and book chapters and the recipient of numerous fellowships including a Ford Foundation Postdoctoral Fellowship and the Winterthur Museum and Library Fellowship.

Yunxiang Yan is a professor of Anthropology at UCLA and the author of two books on social change in China. He also published several articles on the social-cultural impacts of American popular culture brought in by McDonald's and other fast food chains.

Dr. **Gisèle Yasmeen** is vice-president of partnerships of the Social Sciences and Humanities Research Council of Canada. She was previously senior director of the

Outreach, Communications and Research Directorate with Elections Canada and regional director (British Columbia/Yukon) of the Centre for Research and Information on Canada. She was also a French and English-language radio columnist with the CBC in Vancouver, and a consultant to the United Nations Food and Agriculture Organization, the Canadian International Development Agency and the International Development Research Centre.

Credit Lines

M. F. K. Fisher

M. F. K. Fisher, "Foreword" from *The Gastronomical Me*, p. 353. Reprinted with permission of John Wiley & Sons, Inc. from *The Art of Eating* by M. F. K. Fisher. Copyright 1943 by M. F. K. Fisher. Copyright renewed © 1971 by M. F. K. Fisher.

Margaret Mead

Margaret Mead, "The Problem of Changing Food Habits." Reprinted with permission from the National Academies Press, Copyright 1943, National Academy of Sciences.

Roland Barthes

Roland Barthes, "Toward a Psychosociology of Contemporary Food Consumption." Reprinted with permission of Elborg and Robert Forster, from *European Diet from Pre-Industrial to Modern Times*, eds Elborg Foster and Robert Foster. New York: Harper and Row, pp. 47–59.

Claude Lévi-Strauss

Claude Lévi-Strauss, "The Culinary Triangle" from *Partisan Review* 33, 4:586–95, 1966. Reprinted with permission.

Mary Douglas

From "Deciphering a Meal." Reprinted from *Implicit Meaning*, second edition. Mary Douglas, © 1999, originally published by Routledge. Reproduced by permission of Taylor and Francis Books UK.

Marvin Harris

Marvin Harris, "The Abominable Pig." Reprinted by permission of Waveland Press, Inc. from Marvin Harris, *Good to Eat: Riddles of Food and Culture.* (Long Grove, IL; Waveland Press, Inc., 1985 [reissued 1998]) All rights reserved.

Michel de Certeau, Luce Giard and Pierre Mayol

Copyright 1998 by the Regents of the University of Minnesota. Originally published as *L'invention du quotidian, II, habiter, cuisiner*, copyright 1994 Éditions Gallimard.

Christopher Carrington

Christopher Carrington (1999) Chapter 1, "Feeding Lesbigay Families" from *No Place Like Home: Relationships and Family Life among Lesbians and Gay Men*. University of Chicago Press, pp. 29–65. © 1999 by the University of Chicago Press.

Arjun Appadurai

Arjun Appadurai, "How to Make a National Cuisine: Cookbooks in Contemporary India." *Comparative Studies in Society and History*, 30, 1: 3–24, 1988. Reprinted with permission of the author and Cambridge University Press.

Richard Wilk

Richard Wilk, " 'Real Belizean Food': Building Local Identity in the Transnational Caribbean", *American Anthropologist*, 101(2): 244–255. © 1999, American Anthropological Association. Used by permission. All rights reserved.

Lisa Heldke

Lisa Heldke, "Let's Cook Thai. Recipes for Colonialism." Reprinted by permission from *Pilaf, Pozole and Pad Thai: American Women and Ethnic Food*, edited by Sherrie A. Inness, copyright 2001 by the University of Massachusetts Press.

Psyche Williams-Forson

Psyche Williams-Forson. "More than Just the 'Big Piece of Chicken': The Power of Race, Class, and Food in American Consciousness." Reprinted with permission of the author.

Carole Counihan

Carole Counihan. "*Mexicanas*' Food Voice and Differential Consciousness in the San Luis Valley of Colorado." Reprinted with permission of the author.

Gary Paul Nabhan

Gary Paul Nabhan, "Rooting Out the Causes of Disease: Why Diabetes is So Common Among Desert Dwellers." Chapter 7 from *Why Some Like It Hot*, Island Press, 2004. Reprinted with the permission of the author.

Alison Leitch

Leitch, Alison (2003) "Slow Food and the Politics of Pork Fat: Italian Food and European Identity." *Ethnos*, 68, (4): 437–462. Reprinted with permission of the author.

Jeffrey M. Pilcher

Jeffrey M. Pilcher, "Taco Bell Maseca, and Slow Food: A Postmodern Apocalypse for Mexico's Peasant Cuisine?" Reprinted from *Fast Food/Slow Food: The Cultural Economy of the Global Food System*, ed. Richard Wilk. Reprinted with permission of Altamira Press.

Dylan Clark

Dylan Clark, "The Raw and the Rotten: Punk Cuisine." Reprinted with permission from *Ethnology*, 43(1) (Winter 2004), pp. 19–31. Copyright 2004 The University of Pittsburgh.

INDEX

Abbott 468
Abdullah, U. 301
abject, the 199
abominable creatures 45–7, 50, 52
abstinence 73, 106; anorexia nervosa *see* anorexia nervosa; medieval women 4, 121–40
abundance: abundant production and food waste 572–3; illnesses of 74–5
Achler, E. 130
activism 383, 394; consumer boycotts 469–72, 474; punk 420; Slow Food *see* Slow Food Movement
Adonis complex 188
Advanced Informed Agreement (AIA) procedure 546
advertising 180, 417; men's fitness magazines 189–92; and obesity 488–9
advocacy 300–1; anthropology and 478–9; breastfeeding 7, 467–81
Africa: food aid 539, 543–5, 548, 550; street children 7, 554–71; *see also under individual countries*
African Agricultural Technology Foundation 549–50
African Americans: foodways and representations of 5, 342–53; obesity 486, 489, 491
afterbirth 51
agency, masculinity as 213
Agricultural Biotechnology Support Project 549–50
agriculture 407–8, 493–4; abundant production and food waste 572–3; GMOs *see* genetically modified organisms; Mexico 406–7
Ai no Epuron 3 211, 212–13
AIDS/HIV *see* HIV/AIDS
air: creatures of the 46, 47, 52; culinary triangle 40–2
Akerman, C. 70
Albania 66
alcoholic drinks 96
Alembert, J. d' 39
Algeria 383

aligot, l' 87
alimentary sovereignty 408
Alley, K. 492
Allbutt, C. 149, 156
alms-givers 565
Alpaïs of Cudot 130, 134
altar, fitness for the 45–7
alternative masculinities 213–15
Althusser, L. 222–3, 226
American Home Products 468, 474
American Meat Institute 449
American Media 188
American Obesity Organization 487
Angela of Foligno 132
Angolan refugees 543
animals, dietary laws and classifying 3, 44–53
anomalous creatures 47–8
anorexia nervosa: and its differential diagnosis 4, 104–17; history of 4, 141–61; psychopathology 4, 162–86
anthrax 56
anthropological situation 32
anthropology 91–2; advocacy and 478–9; feminist anthropology 355
anticolonialism 338–9
Antonio, R.M. 405
Antonito, Colorado 5, 354–68
Appadurai, A. 565
appearance 223–4
appetite 167–8; awareness of 110; denial of by Victorian anorectics 4, 141–61; irrational 145–50
appropriation of social space 513–14
Archer Daniels Midland 549
Arci Gola 389, 390
Argentina 544
arginine 196
Aristotle 37, 38
asceticism 122, 154–5, 534
Asia: changes in family values in East Asia 517; status of women in Southeast Asia 526–8; *see also under individual countries*

Assiniboin 39
Association of Proprietors of Tortilla Factories
  402–3
Associazione Tutela Lardo di Colonnata
  388
attitudes 23
Augustine, St 166, 167, 168
Austen, J. 152
Austin, T. 93
Australian aborigines 372, 374–5, 378
authority 207–11
axes of continuity 165, 166–80

Baby Friendly Hospital Initiative (BFHI)
  476
Baby Milk Action Group 470
Babylon 65
'Back to the Land' campaigns 557–8
backstage work 279–80
balanced meals 21–2
Balzac, H. de 72
Bangkok 7, 523–38
banquets 129, 501
bariatric surgery 487, 492, 494–5
Barr, A. 334–5
barter 344–5, 418
Barthes, R. 224–5
Beard, G. 154
beauty ideals 179–80
Bedouin 62, 65
begging for alms 563–4
Beijing 7, 500–22
Beijing Beef Noodle King 504
Belize 5, 308–26; national culture 312–15
Bell, G. 404
Berber pork-tolerant village, Morocco 65
Berk, S.F. 240, 275
Berry, J. 371, 372
Bey, H. 413
Biltekoff, C. 492, 494
binge eating 110, 169
Binswanger, L. 163
Biosafety Clearing House 546
birds 46, 50, 63
birthday parties 514
Bistro SMAP 205, 207
black Americans *see* African Americans
Black Cat Café, Seattle 6, 411, 416, 418, 419, 420,
  421; order of signs 413–14
blame, obesity and 490
Blanco, A.L. 369, 370
Blannbekin, A. 131
Bliss, E.L. 104
block plan 24

blood, separation from meat 45, 50–1
board of graces 82
body: body/mind dualism 165, 166–9; control
  of 129–30; eating disorders, culture and
  163–6; failure to interpret bodily signals
  110–11
body-building 171–2, 189–91
body image: disturbance in 109–10; men's fitness
  magazines 4–5, 187–201
boiling 37–42
Bolivia 542
Bonaventura of Siena 130–1
Bonte, P. 72–3
Bony Fried Chicken 503
Bordo, S. 199
borrowing of recipes 334–7
Bottle Babies 471
Bourdieu, P. 282, 321, 323, 501
Bové, J. 394, 408
boxed rice (*hefan*) 502
Boy Scouts of America 572
boycotts, consumer 469–72, 474
braising 43
Branch, C.104
Brand-Miller, J. 371–2, 374, 376
branding 407
bread 30, 99–100
breast-bottle controversy 468–72
breastfeeding: advocacy 7, 467–81; medieval
  women and lactation 125, 133
Brennan, J. 330, 332–3, 338–9
Brillat-Savarin, A. 39, 40
Bristol-Myers 468, 470–1
Britain 93, 96, 97, 99–100
Brownell, K. 488
brucellosis 56
Bruch, H. 23, 163, 168, 170, 174, 175
Brugman, J. 125–6
buccan 40–1, 42
Buddhism 66, 534
budget cuts 575
bulimia 110, 163, 168
Bulmer, R. 44, 51
Burton, R. 178
Bush, G.W. 449, 494, 544
business lunches 33–4
Bussaracum 533, 534
Butler, Josephine 154
Butler, Judith 210
Byron, Lord 152, 153–4

cacti 373–4
Caingang of Brazil 38
Californian Beef Noodle King 503

Californian school salad bars 6, 423–37
camels 60, 61–2
Campbell, R.H. 100
Campos, P. 484, 489
Canada 540, 544, 547; tomato food chain 452–66
cancer 75
cannibalism 39
canteens 506–7, 518
carbohydrates 199
cardiovascular disease 75
Cargill 443, 549
Caribbean: Belize 3, 308–26; sugar and sweetness 3, 91–103
Carrara 381, 382, 388; see also lardo di Colonnata
Cartagena Protocol on Biosafety 546
cash: converting food assistance to 579; food aid in 541, 549
casual employment 560–1
Catherine of Genoa 127, 130, 134
Catherine of Siena 127–8, 131, 132, 134, 154
Catholic Charities 577
cellulose 57
Centers for Disease Control (CDC) 483, 486, 487
chakalis 304
changing food habits, problem of 3, 17–27
character development 23, 114
Charcot, J.-M. 142
charity: and food insecurity in the US 7, 494, 572–81; medieval women 125; private charity and street children 563–6
Charleton, W. 178
chastity 121
cheese 361–2
chef role 208–9
Chernin, K. 162, 167–8
cherogrillus 60–1
Chetley, A. 475
Chiapas rebellion 405
chicken: African Americans 5, 342, 347–51; fried 348, 503–4, 508; see also Kentucky Fried Chicken
chickens 345
child development 114, 435
child initiated clues 114
childhood 559–60
children 6; African street children 7, 554–71; breastfeeding advocacy 7, 467–81; fast food in Beijing 512, 514, 516–17; inculcation of food habits 19; Kleinian theory of paranoid-schizoid position in infants 198–9; lunch-boxes for nursery school children 5, 221–39; salads 6, 423–37; women in tomato food chain and 461, 462–3
China 65, 290; fast foods 7, 500–22

Chinese fast foods 504–5
Chinese Garden Chicken 504
Chipotle 405
chlorosis 146, 147
chocolate 407
choice: children's salads 431–2, 435–6; household work 253–6; where to shop 267–9; women workers and flexibility in tomato food chain 7, 452–66
Chombart de Lauve, P.H. 29
Christian Democratic Party 391
Christianity, medieval 4, 121–40
Christina of Stommeln 129
Chubo Desu Yo 208, 212
Ciba Geigy 572
citizenship 493–4
Clare of Montefalco 129
class: and cultural capital in Belize 323–4; and deciding where to shop 267–8; drug-foods and 99, 101; feeding work and production of class identity 281–3; new middle class in India 291–5; race, food and in American consciousness 5, 342–53; Slow Food and Mexico's peasant cuisine 6, 400–10; taste and 30–1; taste and hierarchy in Belize 316–18, 321–2; Victorian anorectics 155–6; see also status
cleaning up: customer behavior in fast- food outlets 510; lesbigay families 273–4; school salad bar 426–7, 435–6
climate 58
clinical examination 143–5
Codex Alimentarius Commission 546
coffee 34, 92, 93–4, 99, 100–1, 193
Colette of Corbie 128, 134
collagen 190
Collins, P.H. 343–4
Collins, R. 279–80, 282
colonialism 298, 383; Belize 316–19, 321; cultural food colonialism 5, 327–41
Colonnata 382; see also lardo di Colonnata
Colorado 447; Mexican-American women 5, 354–68
Coltrane, S. 275
Columba of Rieti 130, 131
Columbia 542
combinations, meal 431
commerciogenic malnutrition 469, 471
Committee on Food Habits 17, 20, 23, 24
commodification of culture 391
commodity chain analysis 455; see also tomato food chain
Common Market for Eastern and Southern Africa (COMESA) 545

communication: food as a means of 8;
    psychosociology of food consumption 29–32
Communist Party, Italian 389, 390–1, 391–2, 400
communities, food habits in 18–19
community fund-raising cookbooks 339
community organization 24–5
compensation, workers' 446–51
competition, masculinity as 206–7
ConAgra 441–2, 443, 446, 447
conditioning 33
conflict 240–58
Congressional Hearings on the Marketing and
    Promotion of Infant Formula in the
    Developing Nations 473
Connell, R.W. 188
consonant triangle 36
constipation 150
consumer boycotts 469–72, 474
consumer goods 29; obesity and 488–9
consumer groups 510–13, 516–17
consumption: colonial regime in Belize 316–18;
    food–centered life histories 363–4; late
    capitalism, globalization and 384–5;
    psychosociology of food consumption 3,
    28–35; spatial context of fast-food
    consumption 505–14; women workers in
    tomato food chain 462–4
continuity, axes of 165, 166–80
control: anorexia and 107, 111–12, 114, 165, 167,
    169–73; medieval women and 129–32; men's
    fitness magazines 198–200
control axis 165, 169–73
Convention on Biodiversity 542
cony 60–1
cookbooks 5, 68, 86, 88–9, 407; ethnic 5,
    299–302, 327–41; India 5, 289–307, 311
cooked foods: culinary triangle 36–43, 412; raw as
    a critique of cooked 416–18
cooking: Japanese television cooking shows 5,
    202–20; meal preparation and role of women
    4, 67–77; men who cook in heterosexual
    families 247–50; men's fitness magazines and
    191, 194–6; modes of and culinary triangle
    37–43; rejecting cooking but respecting women
    who cook 358–60; thorough cooking and pork
    55–6; see also food preparation
Coon, C. 59, 65
Coon Chicken Inn restaurants 347
Co-op 396
copyright 335, 388
core and periphery workforce 460
corn mills 401, 402
corruption scandals 391
Corsaro, W. 435

corset 165–6, 178–9
cosmetic surgery 487
country-style nostalgic restaurants 516
couples, young 511–12
coupons 271
Covenant 48, 49
cows 57, 60, 66; slaughterhouses 442–6
crana 319–20
cravings 148
Craxi, B. 391
creation 212–13
creativity 72; gender, recipes and 337
Creavalle, L. 195
Cree Indians 40
cricket 80–1
crispness 31
croutons 433–4
cud-chewing, split-hoof formula 46, 56–7, 60–3
culinary practices, ordinary 4, 67–77
culinary triangle 3, 36–43, 411, 412
cultural capital 282, 318, 322–4
cultural food colonialism 5, 327–41
cultural functions 489–94
cultural myth 223–5
cultural practices 165
Cultural Revolution 509
cultural rules 76
cultural substitutes 21
cultural voyeurism 490–1
culture/nature opposition 37–8, 41–2
cumulative trauma injuries 444–5
Cussy, Marquis de 40
customer behavior, in fast-food outlets 510
Cytodine 190

D'Alema, M. 391
Day, H. 304
de Young, J. 526
death 76
deference 240–58
deficiency illnesses 74
deforestation 59, 66
Del Monte 455, 458
DeLind, L. 578
denial of fatigue 110–11
Denominazione d'Origine Protetta (D.O.P.)
    388
depression 115
der Haroutunian, A. 335
Desai, V. 300
Descartes, R. 166–7
desert dwellers 5–6, 369–80
Designer Whey 192
desire 198

DeVault, M. 259, 260, 265, 269, 281
developing countries: food aid and GMOs 7, 539–53; promotion of infant formula in 471–2
diabetes 5–6, 369–80
diagnosis of anorexia nervosa 4, 104–17
Diderot, D. 39
diet: change and diabetes in desert dwellers 5–6, 369–80; weight reducing diets and men's fitness magazines 193–4
Dietz, W. 483
differential consciousness 5, 354–68; food and 356–7
digestive system 148
dining-list 79–81
dinner parties 279–80, 281–2
disease: desert dwellers 5–6, 369–80; food deficiencies or abundance 74–5
distribution problem 574
Dobbins, K. 441–2
Dochi no Ryari Syou 205, 206, 207, 209–10, 214
doctors 475; mothers, anorectic patients and 143–5; *see also* health professionals
dogs 55
domestic violence 244
dominance 256–8
Dorothy of Montau 131
Douglas, M. 57, 199
Dove, F. 19
Downing, G. 198
drama metaphor 315–16
Driver, R.S. 50
drought-adapted plants 372–3
drug-food complex 94–102, 385
dual employment strategy 460
dualist axis 165, 166–9
Dumézil, G. 40
dumpster diving 419
Dunkin' Donuts 503
dyspepsia 146–7

eating, Victorian femininity and 150–7
eating disorders: anorexia nervosa *see* anorexia nervosa; bulimia 110, 163, 168; medieval women and 128–9
Ebers Papyrus 89
economic functions 485–9
economic motivations 548–50
Ecuador 542
Eder, K. 385
edibility 75–6
editorial features 192–3
education 25
Edwin Smith Papyrus 89

Egypt 89, 544; pig taboo 64
elaborated/unelaborated opposition 37, 38–9
Eliot, G. 151–2
Elisabeth of Hungary 128, 131, 134
Elisabeth of Schönau 132
Elizabeth II 320
Ellis, R. 244
emergency food 7, 494, 572–81
emotion: anorexia and recognizing emotional states 111; response to hunger 574–5
emotion management 261–2
employment/work 33–4, 243; children as labor in Africa 559–60; created by medicalization of obesity 487–8; dual employment strategy 460; fast-food outlets 508, 509; female labor force participation rates in Southeast Asia 527; flexibilization of labor 7, 452–66; gender and Bangkok eating places 533–4; obesity and discrimination 485–6; street children 560–1
'endangered foods' 387; *see also lardo di Colonnata*
endo-cuisine 38–9
Engels, F. 356
entertainment: entertainment industry and obesity 492; fast-food outlets 509–10
entitlement approach 555–6, 564–5
entrepreneurship: African Americans 344–5; Mexican-Americans 361–2; micro-entrepreneurs in Thailand 526–7, 528–31, 534–5
environment: breastfeeding vs infant formula and the natural environment 477–8; medieval women and control of 130–2
equality 509–10
ergotic poisoning 74
ethnic cookbooks: cultural food colonialism 5, 327–41: India 299–302
ethnicity/race: and deciding where to shop 268; feeding work and production of ethnic identity 283–4; race, class and food in American consciousness 5, 342–53
eucharist 122, 123, 131–2, 134; Lidwina 126–7
Europe: Italian food and European identity 6, 381–99; sugar and other dessert foods 3, 91–103; *see also under individual countries*
European Economic Community 540
European Union (EU): food aid 541, 544–5, 550; food and safety legislation 382, 387; GMOs 543, 544, 547–8, 549, 550
Everett, G. 151
Everett, S. 151
everyday culinary practices 4, 67–77

Excel 443, 447, 448–9
excess, illnesses of 74–5
exchange: charitable 564–6; entitlement
    approach 555–6; of recipes 292; sexual
    561–2
executive function, masculinity as 207–9
exo-cuisine 38–9
exotic: as familiar 332–3; quest for the 330–1
expert knowledge 209–11
external stimuli, response to 114
extracellular mucilage 373–4

factory system 98–100, 175
familiarity 331, 332–3
family: and anorexia nervosa patients 107–8,
    111–12, 113; changes in family values in East
    Asia 517; changing social obligations in
    Tanzania 558; feeding work in heterosexual
    families 5, 240–58; feeding work in lesbigay
    families 5, 259–86; feeding work and
    production of 284–5; food habits and 21;
    medieval women and manipulation of 130–1;
    see also fathers, mothers
famine 121, 376
Fantasia, R. 501, 510
fast foods 7, 393; China 7, 500–22; cultural
    symbolism of 507–8; Mexico 6, 401, 403–5;
    spatial context of fast-food consumption
    505–14; and Syndrome X 377–8; see also
    Kentucky Fried Chicken, McDonald's
fasting 4, 121–40
fat 199
fathers 113; African American and 'big piece of
    chicken' 342, 348–51
feasting 122
feeding work: character of 260–3; conflict and
    deference in heterosexual families 5, 240–58;
    lesbigay families 5, 259–86; nutritional
    concerns as 264–5
femininity, Victorian 150–7
feminism 1–2, 175–6, 329; punk culture
    415
feminist anthropology 355
feminization of poverty 453, 460
Festinger, L. 22
finances 269–72
Finkelstein, J. 501
first born 48, 49
fish 195
fitness 170–2
fitness magazines, men's 4–5, 187–201
flappers 180
flexibilization of labour 7, 452–66; from above
    and from below 455–6; deepening inequalities

460–2; women workers' experiences of
    flexibility 456–60
Fliess, W. 178
Fo, D. 391
Fogel, R.W. 566
food aid 7, 408, 539–53; GMOs in 542–5;
    motivations for GM food aid policy 547–50;
    politics of 540–2; wartime 20–1
Food Aid Convention (FAC) 541
food assistance programs 7, 494, 572–81
food-centered life histories 5, 354–6, 358–64;
    interview themes 363–4
Food Chain 577
food discourse 204–16
food expenditures 271–2
food gardens 532
food habits: problem of changing 3, 17–27;
    urban 7, 500–38
food intake 73–4
Food Not Bombs 420
food policy 3, 17–27
food preparation: food-centered life histories
    364; learning about 68–9, 262–3; lesbigay
    families 272–3; men's fitness magazines
    and 191, 194–6; and nutrition 21; role of
    women in cooking and 4, 67–77;
    Victorian femininity and 150–1;
    see also cooking
food recovery programs 573
food refusal see abstinence
food riots 383
food scandals 384
food stamps 574; converting to cash 579
food stands 563
food taboos 3, 295, 296; pigs 3, 51–2,
    54–66
food vendors 526–7, 564
food voice 5, 354–68
foodscape, Bangkok 7, 523–38
foodshops, small (Bangkok) 524, 528–31, 536
Fordism 456
Fore tribe, New Guinea 75
formal restaurants 501; Bangkok 531–2; Beijing
    506, 512–13
Foucault, M. 163, 165, 166
Fowler, T. 450–1
France 290, 318, 383, 501, 508, 530;
    psychosociology of food consumption 28–35;
    women and everyday culinary practices 4,
    67–77
Francis of Assisi 128
free supplies of infant formula 474–5
freedom 213; illusion of 214–15
Freud, S. 176, 178

fried chicken 348; local outlets in Beijing 503–4, 508
friendship relationships 24
frontstage work 279–80
fruits, preserving with sugar 96, 99
Fuller, R. 52
Fumento, M. 490, 493, 494
functionalism 484–95; analysis of obesity 485–95
fund-raising 577–8

Gadarene swine 59, 69
Gambero Rosso 390, 391
Gandhi, M. 106
Gans, H. 484–5, 488, 495
garbage 419
garden restaurants 532
Gaskell, E. 152
Gay, P. 179
gay culture 188–9; *see also* lesbigay families
gender: blending gender roles 360–1; conflict, deference and household work 5, 240–58; culinary triangle 39; equality and fast food consumption in Beijing 510, 512–13; feeding work and production of gender identity 274–80; food as gender/power 414–15; and foodways 525; gender division of labor in Bangkok 533–4; gender order in Japanese TV cooking shows 208; gender relations and foodscape in Bangkok 7, 523–38; multiple genders 214; street children and access to charity 564; street children and employment 560–1
gender/power axis 165, 173–80
genetically modified organisms (GMOs) 406, 407; and food aid 7, 408, 539–53; unclear rules on international trade 545–7
Geneva Infant Feeding Association (GIFA) 470
Genovese, E. 346
Gentles, the 309–10, 312–13
Gerard of Cologne 125
Ghana 383
gibnut (royal rat) 320
Gill, T. 213
Gilman, C. Perkins 176
Glasheen, K. 449
Glickman, D. 573
globalization: Belizean national cuisine 5, 308–26; consumption, late capitalism and 384–5; virtuous 394
Glover, M. 450–1
gluttony 151, 490
glycemic analysis 371–2

goats 57, 66
Goffman, E. 207, 489
Goings, K. 346, 347
Gola, La 389
Goldsmith, O. 79, 80
Gore, A. 573
Gotchi Battaru 205
Graham, M.L. 347
Gramm, P. 446
Grandy, C. 346
Grassley, C. 548
Great Depression 574
Greenpeace 406
Griffin, S. 332
Guadagni, F. 387, 388
Guatemala 542
Guayaki of Paraguay 38
Guccini, F. 391
Guiana 40, 41
Guillermoprieto, A. 403, 404
Gusfield, J. 501

Habermas, J. 525
habits, food *see* food habits
Hadewijch 128
Hale, S.J. 151
Hall, S. 203
Hammond, W. 121
Happy Chopsticks 508
hares 60, 62–3
Harland, M. 146, 148, 156
Hartmann, H. 240
Harvey, D. 417
He Yupeng 512, 513
health 32–3, 156; danger and injuries in meatpacking industry 6–7, 441–51; invalidism 146, 175–6; *see also* anorexia nervosa, obesity
health professionals: breastfeeding advocacy and infant formula industry 475; job creation and obesity 487–8
Hebrew dietary laws *see* Jewish dietary laws
hefan (boxed rice) 502
hegemony 315–16
Henry III 93
Herndon, A. 492, 493
Herodotus 64
Hertz, R. 282–3
hierarchy: and taste in Belize 316–18, 321–2; written lists and 79–81
high cuisine 290
Hildegard of Bingen 132
Hinduism 60, 66, 295–7
Hispanic women 5, 354–68
historical theme 32

HIV/AIDS: breastfeeding and 478; street
    children 558–9
Hobsbawm, E. 97–8
Hochschild, A. 261
holiness, degrees of 45
home economics 415
home economists 24
homelessness 420; African street children 7,
    554–71
honey 92, 95, 96
'Honey, We're Killing the Kids' 482, 489
hooks, b. 338, 339
Hoppin, A. 146
host role 207–8
housegirls 560–1
household work: conflict, deference and 5,
    240–58; Mexican-American women 357–62;
    see also feeding work, food preparation
Howell, P. 531
Huaxiang Chicken 504
humanity, woman as symbol of 132–3
hunger: African street children 7, 554–71;
    anorectics and 110, 168, 177–8; charitable
    food programs and hunger as the issue 7,
    572–81
hunger-killers 94–102, 385
hunger lobby 574
hunger strikes 106
hunter-gatherers 5–6, 369–80
Hurt, W. 173
husbands 243–7; involving in cooking
    360–1
hygiene 414
hysteria 108, 144–5, 175–6

icons 491
Ida of Léau 131
Ida of Louvain 130, 131
ideological state apparatus 222–3, 226;
    ideological appropriation of lunch-boxes
    227–30; mothering as gendered ideological
    state apparatus 230–6
images: African Americans 346–7; children's
    salads 6, 423–37; obesity 491
immortality 172–3
import substitution 319–20, 320–1
imports 316–18, 319
in-kind food aid 544, 549, 550
India 106, 542; cattle 60, 66; cookbooks and
    national cuisine 5, 289–307, 311; culinary texts
    and standards in Indian history 295–8; social
    world of the New Indian cuisine 291–5
Indicazione Geografica Protetta (I.G.P.) 388
Indonesia 383, 525

industrialization 98–100, 175, 377, 412–13,
    417–18
ineffectiveness, sense of 111
inequality 579–80; flexibility and the tomato food
    chain 460–2
infant formula 7, 468–75, 476–7; breast-bottle
    controversy 468–72; Marketing Code 472–5
Infant Formula Action Coalition (INFACT) 470,
    471
injuries, meatpacking 6–7, 441–51
Innocenti Declaration 475–6
institutional contexts 203–4
institutional feeding 21
Interfaith Centre for Corporate Responsibility
    (ICCR) 470–1
intermittent shopping trips 266
International Association of Infant Food
    Manufacturers (IFM) 474
International Baby Food Action Network
    (IBFAN) 470, 471, 472–3, 474
International Council for Infant Food Industries
    (ICIFI) 472
International Fast Food City 503
International Obesity Task Force 487
International Wheat Trade Convention 541
intestinal disorders 75
invalidism 146, 175–6
invulnerability 172–3
Iowa Beef Packers (IBP) 443, 444, 445, 447,
    447–8, 449, 450–1
Iron Chef 203, 206–7, 210
Iron Shufu 210–11, 212–13
Islam 54–5, 62, 65–6
Italian Communist Party 389, 390–1, 391–2,
    400
Italy 290; food and the Italian Left 389–92;
    history of Italian cuisine 400–1; Slow Food and
    lardo di Colonnata 6, 381–99
Ito, K. 203
Iturriaga de la Fuente, J. 406

Jaffrey, M. 304, 329
James, W. 154
Japan: masculinities in television cooking shows
    5, 202–20; mothers and lunch-boxes 5,
    221–39
Jeanne Dielman 70
Jelliffe, D. 469, 475
Jelliffe, P. 475
Jewish dietary laws: classifying animals 3, 44–53;
    pork taboo 51–2, 54, 55–6, 56–7, 58–9, 60–4
Jinghe Kuaican Company 504
Jinghua Shaomai Restaurant 504
Jingshi Fast Food Company 504

Johnson, E.P. 349
Johnson, S. 79, 80
Jones, E. Lloyd 149
Juárez Broca, S. 405–6, 407
Julian of Norwich 133

Karam 44–5
Kearney, M. 408
Kempe, M. 130
Kentucky Fried Chicken (KFC) 404; China 500, 502–3, 504, 507, 509
Kimura, I. 208
King Kamali diet 193–4
Klein, M. 198–9
Koos, E.L. 24
Kotz, N. 574
Kristeva, J. 199
Kroeber, A. 102
Kundera, M. 392–3
kuru 75

Labrada Nutrition 192
Lambertini, I. 131
Lambeys, the 310–11, 312–13
land, creatures of the 46–7, 52
Lara, S. 460, 461
lardo di Colonnata 6, 381–99
Lasch, C. 179
Last Supper 49
Latham, M. 475
Lawrence, D.H. 177–8
Lawrence, M. 491
Lazarsfeld, P.F. 30
Leach, E.R. 45, 50, 51
Leacock, E. 356
leaf capacitance 373
Leal, L. 448
learning: about food and food preparation 68–9, 262–3; written recipes and 88
Learning Channel, the 482
Lears, J. 417
Leche League, La 469
Lee, W. 446
Left, Italian political 389–92
leftovers 293
Legion 59
leisure 513–14
lesbigay families 5, 259–86
Lesotho 543
Lévi-Strauss, C. 31, 92, 95, 150, 411, 412
Levites 48
Levy, P. 334–5
Lewin, K. 19, 24
Lidwina of Schiedam 124–7, 134

life histories, food-centered 5, 354–6, 358–64
Limbaugh, R. 494
linguistic triangles 36, 37
Lingzhi Roast Chicken 503
Lippmann, W. 576
lists 4, 78–90
Liu, A. 169, 170, 175, 177
Loblaws 453, 454, 455, 457, 458, 460, 462
lobster 317
local fried chicken fast-food outlets 503–4, 508
local identity, Belize 5, 308–26
Lomax, B. 98
Lopez, R. 445–6
Lukardis of Oberweimar 134
lunch-boxes (obentos) 5, 221–39
Lutgard of Aywières 129, 134
Lyon 538

Macbeth 78
MacKenzie, S. 142
magazines: men's fitness magazines 4–5, 187–201; obentos 223, 230–1, 234
Magallon, H. 406
Maimonides, M. 54–5
Maíz Industrializado (Minsa) 402
Malawi 543
Malaysia 527
malnutrition 74; commerciogenic 469, 471
Manifesto, Il 390
manipulation 224–5, 235
manners, table 152
Maori 39
maquilization 452–3
marble 381, 382, 385–6, 387
Marcos, Subcomandante 405
Margaret of Cortona 131
Marguerite of Oingt 133
Mari royal palace 85–6
marketing 395–6; see also advertising
marmalade 99
Marx, K. 94, 232
Mary, St 132–3
Mary of Oignies 128, 130
masa harina (dehydrated tortilla flour) 401, 402, 403, 408
masala jalebis 304
masculinity: alternative masculinities 213–15; Japanese masculinity in academic literature 203–4; Japanese TV cooking shows 5, 202–20; men's fitness magazines 4–5, 187–201; plural masculinities 187–9
Maseca, Grupo 401–3, 406, 408
mass-based food strategies 528–30

mass consumption 515–18
masturbation 178
materialism 534, 535
Mauss, M. 565
Maxxon 192
McDonaldization 403–4, 452–3
McDonald's 7, 383, 394, 404; China 500–20;
    Italy 381, 382, 393, 406; Mexico 406; Russia
    508; tomato food chain 455, 456–7, 460
McDonnell, L. 345
McGovern, G. 574
meal patterns 21
meal preparation see food preparation
meaning 101–2
meat 195; avoidance by Victorian anorectics
    149–50; cooking modes 37–43; Jewish dietary
    laws and classifying animals 3, 44–53; punk
    culture 415–16; spices, sugar and 96–7;
    see also chicken, pigs
meatpacking workers 6–7, 441–51
mechanization 401–2, 408, 412–13
Mechtild of Hackeborn 132
Mechtild of Magdeburg 128, 133
medical recipes 84–5, 89–90
medieval women 4, 121–40
Meenakshi Ammal, S. 298
Melandri, G. 391
memory: multiple and culinary production 72;
    slowness and 392–4
men: gender relations in heterosexual families 5,
    240–58; involving husbands in cooking 360–1;
    men's fitness magazines 4–5, 187–201; who
    cook 247–50
Men's Fitness 187–8, 191, 192, 193
Men's Health 187, 191, 192, 193, 195, 196–7
Men's Health Guide to Women 196
menstruation 130, 174–5
menus 81–2, 83; fast-food outlets 509; Indian
    cookbooks and 303
Mesopotamia 65, 90
metate 401, 402
methodology 8
Métraux, R. 23
Mexican-American women 5, 354–68
Mexico 383: desert dwellers and diabetes 5–6,
    369–71, 374, 378–9; migrant meatpackers 443;
    Slow Food and peasant cuisine 6, 400–10;
    tomato production chain 452–66
Meyer, B.C.
    115
micro-entrepreneurs 526–7, 528–31, 534–5
middle class, Indian 291–5
Middle East: pig husbandry 58–60; pig taboo
    64–5

migration 406, 443; obesity and employment
    discrimination 486; women in tomato food
    chain 463
milk: breast milk see breastfeeding; Jewish dietary
    laws 45, 51
Mill, J.S. 175, 176
Miller, G. 83
milling 377, 541–2; corn mills 401, 402
Milupa 474
mind/body dualism 165, 166–9
minerals 74
Ministry of Education, Japan (Monbusho) 226,
    227
miracles 123, 125
Mississippi Delta 574
Mitchell, S. Weir 176
Miwa, A. 214
Model Law on Safety in Biotechnology 545
modernity 508
molasses 96, 97
Monford Beef Company 441–2, 445
monitoring supplies, schedules and finances
    269–72
monoculture 417–18
Monroe, M. 164
moral panics 482, 483–4; see also obesity
moral tales 123
Morgan, P. 344
Morgesons, The (Stoddard) 155, 158
Morse, S. 373
mothers 113, 175; African American 350–1; and
    daughter's appetite 146, 148–9; doctors,
    anorectic patients and 143–5; Japanese
    mothers and lunch-boxes 5, 221–39;
    mothering as gendered ideological state
    apparatus 230–6
motivations for food aid policy 540, 541–2,
    547–50
Moynihan Report 491
Mozambique 543
mucilage, extracellular 373–4
Mughlai cuisine 297
Mullin, D. 448
multidimensional social space 510–13, 515, 518
multiple genders 214
multitasking 458
Murphy, E. 491
Muscles and Fitness 187–8, 189–91, 193, 193–4,
    195, 198
muscular body 188–91, 193–4
mutual recognition of racism 338–9
Mwanawasa, L. 548
Mwanza, Tanzania 7, 554–71
myth, cultural 223–5

National Beef 443

national cuisine: Belize 5, 308–26; India 5, 289–307, 311; components of 302–5; significations of food 32

national identity 225

National Institutes of Health (NIH) Indian Diabetes Project 371, 375

National Summit on Food Recovery and Gleaning 573

National Wheat Growers Association 549

Native Americans: desert dwellers and diabetes 5–6, 369–80; modes of cooking and culinary triangle 38, 40–1

natural foods 418–19

nature/culture opposition 37–8, 41–2

neediness 177–8

Neel, J. 375

negativism 111

negotiated flexibility 461–2

neighbors 24–5

Nelson, R. 335

Nestlé 468, 469, 470, 471, 472, 474

Neumayer, E. 541

New Caledonia 37

New Deal surplus commodity distribution program 573, 576

New Guinea 44–5, 75

New Partnership for Africa's Development (NEPAD) 545

new plague of the Crusades 74

new social movements 385; see also Slow Food Movement

Nicaragua 542

Nicholas of Flüe 123

Nimoy, L. 490–1

Nixtamal Mills 403

Nolte, N. 173

nominal lists 78–81

nominal rolls 78–9

non-governmental organizations (NGOs) 475, 540

North American Free Trade Agreement (NAFTA) 405, 409, 452–3, 455–6, 461

Norton, K. 447

Norton, T. 447

Norwood, C. 449

nostalgic country-style restaurants 516

novelty 330–1

nurses 144–5

nursery schools: children's lunch-boxes (obentos) 5, 221–39; ideological appropriation of the lunch-boxes 227–30

nutrition 25; fast foods and 507; food intake and 74; food service and 21–2; nutritional concerns

as feeding work 264–5; nutritional consciousness 33

nutritional change program 24

nutritional supplements 190, 191–2

oatmeal 99–100

Oaxaca City 406

obentos (lunch-boxes) 5, 221–39

obesity 7, 482–99

Occupational Safety and Health Administration (OSHA) 442, 449

O'Dea, K. 378

Olney, R. 335

Olsson, K. 448

Olwell, R. 345

O'Neill, C.B. 169, 174

oral cultures 88–9

order: nursery school lunch-boxes and 224–5, 230, 235; school salad bar 42–30

ordinary culinary practices 4, 67–77

organic foods 418–19

Organization of African Unity (OAU) 545

organized sociality 517–18

originality of recipes 334–7

Ortner, S. 212

Osmani, S. 565

Other: culinary 298–302; as resource 333

overactivity 110–11

ownership, masculinity as 211–12

Papago Indians 370, 377, 378–9

paranoid-schizoid position 198–9

Parasecoli, F. 390

parents: and anorectics 107–8, 111–12, 113; see also fathers, mothers

Parker, D. 177

Parker, S. 392

Partito della Sinistra (PDS) 391

pastiche 532

pastoralism 58–9

Pearson, D. 173

peasant cuisine 87; Mexico 6, 400–10

Pecks of Milan 382

peeling vegetables 73

Pendleton, H. 156

Peoples Unity Party (PUP) 319

performance, blackness as 347, 348–51

Perrot, M. 29

Perry, T. 491

personal responsibility 494–5

personality development 23, 114

Petrini, C. 382, 387–8, 389–90, 393, 394, 408

pharmacopoiea, first written 84

Philip, T.E. 304

Philippines 527
photographs: African Americans 346–7;
   children's salads 6, 423–37
physical activity 170–2, 494–5
Piganid, M. 40
pigs: food taboo 3, 51–2, 54–66; husbandry of
   58–60, 65–6
Pima Indians 370, 371, 375, 377, 378, 378–9
Pinkwater, D. 491, 493, 494
Pizza Hut 503
plagiarism 334–7
planning 241–2, 254, 260–2
plans 83; see also menus, recipes
plantations 94–5
'plastic-bag housewives' 7, 524, 528–31
plate service 21–2
Plato 166, 167, 168
pleasure, politics of 389–92
Pliny 64
Poconachi of Mexico 38
political activism see activism
political parties, Italy 381, 391–2
political process 494–5
politicization of food 2
pork fat 6, 381–99; politics of 386–9;
   see also pigs
portion size 432
possession 211–12
post-Fordism 456
postmodernism 532
postwar era 177
poverty 121, 408; feminization of 453, 460;
   functionalist analysis 484–5; hunger as only
   one aspect of 578–9; and obesity 489
power 8–9; African Americans and intersections
   of food, gender, race, class and power 5,
   342–53; food as gender/power 414–15;
   Japanese masculinity as competition 206–7;
   sources in Belize 323; state ideology 222–3
power/gender axis 165, 173–80
Pratt, M.L. 333
precaution 547–8
pre-packaged salads 436
prescriptions, medical 84–5
presentation 223–4
preservative, sugar as a 96, 99
Price, G. 319–20
prickly pear 373–4
priests 126–7, 131–2, 134
primitive flexibility 461
Primlani, K. 298
printing press 297
private charity: food assistance in US 7, 572–81;
   street children in Africa 563–6

private and public 524–5, 535; blurring
   boundaries in Bangkok 530–1
production: abundant 572–3; and consumption
   by women workers in tomato food chain
   462–4; converting reproductive labor to
   productive work 361–2; masculinity as 212–13
production line speeds 443–4
profession, masculinity in 209–11
professionals, young 511, 516–17
programming mind 72
proletarian hunger-killers 94–102, 385
proteins 74, 199
provisioning 265–6
psychoanalysis 115–16
psychoneurosis 107–8
psychopathology 4, 162–86
psychosociology of food consumption 3, 28–35
psychosomatic medicine 23
puberty 174
public and private 524–5, 535; blurring
   boundaries in Bangkok 530–1
public sector food assistance programs 494, 573,
   574, 575, 576, 577, 579
public space: India 294; new forms of sociality in
   China 517–18
Puget Consumers' Co-op (PCC) 418–19
punk cuisine 6, 411–22
pupsi 319–20
Puritan tradition 20
purity 49–50

Rabelais, F. 72
race see ethnicity/race
Raj Cookbook, The 302
Ramos, A. 445
Ranga Rao, S. 301, 302
rationalization 403
raw foods: culinary triangle 36–7, 38, 40, 42, 412;
   raw as a critique of cooked 416–18
Reagan, R. 575
receipts 84; see also recipes
recipes: collecting and cultural food colonialism
   334–7; written 78, 82–90
Red Sorghum Restaurant 504, 505, 508
Reejhsinghani, A. 293, 299–300
regional cuisines 290; implicit rather than written
   recipes 86–7; India 290–1, 296–7, 299–302
Rehm, R. 449
relaxation 34
religious significance of food: medieval women 4,
   121–40; see also food taboos, Jewish dietary laws
repressive state apparatus 222
reproductive labor, converting to productive
   work 361–2

re-shaping 487
resistance 315–16
resource, Other as 333
responsibility 494–5
restaurants: Bangkok 524, 529–30, 531–2, 533–4, 535–6; Beijing 506–7, 512–13, 516, 518; Black Cat Café 6, 411, 413–14, 416, 418, 419, 420, 421; classification of 501; Coon Chicken Inn 347; India 293–4; menus 81–2; *see also* fast foods
retrospective lists 79
Reyes, H. 448–9
rice, boxed (*hefan*) 502
Richie, D. 224
Richter, C. 21
Rifondazione Communista 391, 392
Rios, A. 444
risk 384; GMOs and food aid 547–8; 'risky behavior' frame and obesity 490
Rita of Cascia 130–1
rituals 229
Ritzer, G. 403, 452, 464
roasting 37–43
Roberson, J.E. 203, 204
Roberts, J. 346
Rock, C. 342, 348–51
rock badger 60–1
Rodale 187
Roden, C. 329, 333, 334, 335–7
Rohlen, T.P. 230
Roker, A. 492
Rolle, R. 128
Rolls of Parliament 84
Rome, McDonald's in 382, 393, 406
Ronghua Fried Chicken 504
rotted food: culinary triangle 36–7, 38, 40, 42, 412; dumpster diving 419
Round Robin 79, 80
routines 228–9
royal rat 320
Rudd Institute 486, 488
rum 94, 96
ruminants 46, 56–7, 60–3
running 170–1
rural repatriation drives 557–8
Russia 508
Ruybal, H. 354, 356–7, 358–63
Ruysbroeck, J. van 128
Rwezaura, B. 559–60

S-curve 178–9
Sacks, M. 172
sacred cow 60, 66
Saint Quentin's *grand mal* 74

saintliness 154–5
Sakai, M. 208
salad bar 426–8, 436
salads, children's 6, 423–37
salaryman 203
Salone del Gusto 391
Salvation Army 577
San Francisco 281, 283
San Luis Valley, Colorado 5, 354–68
Sanchez, D. 445
Sandoval, C. 356
sandwiches 195
Santa Anita Packers 452, 455–6, 459–60, 460–1, 463
schedules 269–72
schizophrenia 108–9, 115
schools: children's salads 6, 423–37; lunch-boxes for nursery school children 5, 221–39; state, subjectivity and school 225–6
Schumpeter, E.B. 97
science: men's fitness magazines 192–3; and risk 547–8
Scouting for Food drive 572
Scudder, V. 154
sea creatures 63
seasonings 43
Second Harvest 573, 577
second-order myths 224–5
Second World War 37; problem of changing food habits 3, 17–27
secret eating 145, 152
See Faa 531–2
Ségur, Comtesse de 73
self-definition 343–4
self-inscribed lists 79–81
self-insurance 447, 449
Selvini, M.P. 104–5, 114, 115
Sen, A. 555–6, 564–5
sensory perception 72
Seremetakis, N. 384
Seri Indians 5–6, 369–71, 374, 378–9
service, relations of 243–7
sex 44–5, 51; in men's fitness magazines 196–8; street children and sexual exchanges 561–2
sexual development 148
sexuality: control of 130, 178; hunger and 169; medicalization of 165
shaphan 60–1, 62–3
sharing of household work 241–3
Shaw, S. 173
sheep 57, 66
Shelton, A. 501
Sheridan, R. 97
Sheth, A. 304

Shiva, V. 494
shopping: habits 21; lesbigay families 265–9
shopping centers 531
shopping lists 83
Shroff, V. 300
sick role 146
signification 29–32
Simenon, G. 72
simulacra 532–3, 535
Singapore 468, 527
single-child policy 517
single-parent families 250
situations, significance of food and 32, 33–4
slaughterhouses 6–7, 441–51
slavery 94–5, 345–6
slenderness, tyranny of 173–4, 483
Slow Food Movement: awards for biodiversity
    405–6; and *lardo di Colonnata* 6, 381–99;
    manifesto 393–4; and Mexico's peasant
    cuisine 6, 400–10
slow-release foods 372, 373–5, 378–9
small foodshops (Bangkok) 524, 528–31, 536
SMAPXSMAP 205
Smith, D. 171–2
Smith, W. Robertson 91
smoking, as form of cooking 40–3
snack bars 33–4
'snack taxes' 489–90
social capital 486
social functions 489–94
social interaction: new middle class in India
    291–5; sociality in Beijing restaurants 501,
    517–18
social movements 2, 385; *see also* Slow Food
    Movement
social order 230, 235
social space 500–22; appropriation of 513–14;
    and mass consumption 515–18;
    multidimensional 510–13, 515, 518
socialist canteens 506–7, 518
Socialist Party 391
sociality 501, 517–18
solidarity 575–6
Somboon Pattakarn 531–2
South Africa 545
South Indian cuisine 300
Southeast Asia: status of women 526–8
Southern African Development Community
    (SADC) 545
soy 193
space, social 500–22; spatial context of fast-food
    consumption 505–14
specialization 298–302
spices 96–7

'spirit' of food 31
spirituality: medieval women 4, 121–40;
    Victorian femininity 153–5
Spivey, D. 348
split-hoof, cud-chewing formula 46, 56–7, 60–3
St John's College, Cambridge 79–81
StarLink corn 542
starvation 376; externally enforced 107
state: ideology *see* ideological state apparatus;
    organized sociality 517–18; school, subjectivity
    and 225–6
status: Japanese masculinity and 210–11; obesity
    and 491–2; street credit 554–71; Victorian
    anorectics 155–6; women in Southeast Asia
    526–8
steaming 43
Stein, C. von 151
Stephenson, P. 510
stereotypes: black Americans 342, 343, 344–51;
    obesity 490, 491
Sternbach, N.S. 355–6
stigmatization 489–90
Stoddard, E., *The Morgesons* 155, 158
Stoyer, J. 349
street children 7, 554–71
street credit 554–71
Suan Parichat housing development 532–3
subcultures 18–19
subjectivity 225–6
subordination 256–8
subsidies: corn in Mexico 402–3; GMOs and
    549–50
subsistence intake 74
Sudan 542, 545
suffering 125–6, 132
suffragettes 106
sugar 3, 91–103; consumption in USA 28
Sumerians 65; early physician's pharmacopoeia
    84
supplies, monitoring 269–72
surplus, agricultural 572–3; disposal and food
    aid 540, 541, 549
surplus commodity distribution program 573,
    576
surveillance 228–9
Suso, H. 121, 128
sustainability 477–8
Suzuki, N. 203, 204
swarming creatures 46
Swaziland 543
sweet treats 198, 199
sweetness 3, 91–103
Swinburn, B. 378
symbolism of fast food 507–8

Syndrome X 371, 376
Syria 95
system, food as a 30

table, fitness for the 45–7, 50
table manners 152
Taco Bell 6, 401, 403–5
Tambiah, S.J. 44, 51
Tanzania 7, 554–71
tapeworms 56
Task Force on Biotechnology 546
taste 29; and hierarchy in Belize 316–18, 321–2;
    income level and 30–1
Tauler, J. 122, 128
tea 92, 93–4, 97, 99, 100, 383
technology 294
teenagers 511–12, 516–17
television, Japanese 5, 202–20
Teller, E. 574
temple sacrifice 48–9
testimonios 355–6
Texas 446
Texas Workers Compensation Reform Act 1989
    447–9
Thailand 44–5; Bangkok foodscape 7, 523–38
Thanksgiving dinner 327
theft: African Americans 345–6; punk cuisine
    419; recipe collecting and 334–7; street
    children 562
themes, significations of food and 32–3
theory 8
Third World Action Group 469
Thomae, H. 104–5, 115
Thomas, N. 576
Thompson, E.P. 100, 383
Thompson, T. 493
Thorburn, A. 374
thrifty gene 375–6
Tia Tana Chocolate 405–6, 407
tiffin networks 528
tobacco 94, 100
tokenism 576–7
Toledo, F. 406
Tomasita Project 453–5
tomato food chain 7, 452–66
Tonnerus' Minasan no Okage Deshita 205
Torode, A. 99
tortilla factories 402–3
tortillas 401, 403–5, 408
torture 165
trade: Caribbean and Europe 94–5, 97–8; imports
    to Belize 316–18, 319; unclear rules for trade in
    GMOs 545–7
trading, petty 344–5

transformational analysis 30
transgendered masculinities 214–15
transnational corporations 461–2, 549–50
travel 321
trichinosis 55–6
trickster 346, 349
Trollope, A. 152, 155, 156
Tum Nuk Thai 532, 534
Twain, M. 152
tyranny of slenderness 173–4, 483

Uganda 542
UNICEF 476; WHO/UNICEF Code for the
    Marketing of Breastmilk Substitutes 472–5
unionization 456–7
United States (USA) 313; African Americans see
    African Americans; Black Cat Café, Seattle 6,
    411, 413–14, 416, 418, 419, 420, 421; charity and
    food insecurity 7, 494, 572–81; children's
    salads in elementary schools 6, 423–37;
    consumption of sugar 28; cultural symbolism
    of fast food 508; Department of Agriculture
    (USDA) 549; food aid and GMOs 539–53;
    food policy and changing food habits 3, 17–27;
    Hispanic women in southern Colorado 5,
    354–68; and Marketing Code for Breastmilk
    Substitutes 473; obesity and American society
    493–4; obesity epidemic 7, 482–99; obesity and
    political process 494–5
urban food habits: Bangkok 7, 523–38; fast food
    in Beijing 7, 500–22
USAID 542, 544, 549–50
utensils 37, 39, 41
Uvin, P. 541

variety, seductiveness of 294–5, 303–4
Vauchez, A. 122, 123
Veblen, T. 155
veganism, punk 415–16
vegetables 43; peeling 73
vegetarianism 415
Venanzio Restaurant, Colonnata 382, 386–7
Verne, J. 72
Verney, E. 97
Victorian anorectics 4, 141–61
virtuous globalization 394
visual research 6, 423–37
vitamins 74
voice, appetite as 4, 141–61
vomiting, self-induced 110
voraciousness 177–8
vowel triangle 36

wages: meatpackers 443; obesity and 486

waivers 447–8
Walker, M. 533, 534
Wallet, M. 142
wallowing in mud 55, 58
Warner, L. 149
Warren, C. 281
waste: dumpster diving 419; food waste and
    abundant production 572–3
water: animals of water and Jewish dietary laws
    46, 52, 63; culinary triangle 37–43
Watson, J. 501
Watts, A. 167
Weber, C. 371, 372
weight loss: men's fitness magazines 193–4; and
    status mobility 492
Weinroth, L.A. 115
welfare cuts 575
West, C. 274
West, E. 163, 168, 169, 175, 177
White, M. 376
Whites Only Cuisine 348
Whyte, R.O. 59
Williams, C. 468
Williams, Eric 95
Williams, Eva 155
wine 197
Winfrey, O. 492
witchcraft 178
wives 243–7
women: everyday culinary practices 4, 67–77;
    fast-food outlets in Beijing 510, 512–13, 516,
    517; gender relations in heterosexual families 5,
    240–58; in Japanese TV cooking shows 208,
    210–11; medieval 4, 121–40; race, class and
    American consciousness 350–1; status in
    Southeast Asia 526–8; Victorian anorectics 4,

141–61; workers in tomato food chain 7,
    452–66
Women's Franchise League 176
Wood-Allen, M. 148
Woods, J.H. 450
work see employment/work
workers' compensation 446–51
World Alliance for Breastfeeding Action (WABA)
    476
World Anti-Slavery Conference 175
World Breastfeeding Week (WBW) 476
World Food Program (WFP) 540–1, 542, 543–4,
    545, 546–7
World Health Assembly (WHA) 474–5
World Health Organization (WHO) 483;
    WHO/UNICEF Code for the Marketing of
    Breastmilk Substitutes
    472–5
World Trade Organization (WTO) 544, 546, 550
written lists 4, 78–90
Wu Fang Zhai 503
Wyandot Indians 38

Xiangfei Chicken 504

yávos 31
Yellow Wallpaper, The (Gilman) 176
young people 511–12, 516–17

Zambia 408, 539, 543–4, 548, 550
Zapatista rebellion 405
Zimbabwe 543, 545
Zimmerman, D. 274
Zimmerman, T. 448
zines 415, 416
Zola, E. 72